Use of property rights in fisheries management

FAO
FISHERIES
TECHNICAL
PAPER

404/2

Edited by
Ross Shotton
Fishery Resources Officer
Marine Resources Service
Fishery Resources Division
FAO Fisheries Department

Proceedings of the FishRights99 Conference
Fremantle, Western Australia
11-19 November 1999
Workshop presentations

FISHERIES
WESTERN AUSTRALIA

Food
and
Agriculture
Organization
of
the
United
Nations

Rome, 2000

The designations employed and the presentation of the material in this information product do not imply the expression of any opinion whatsoever on the part of the Food and Agriculture Organization of the United Nations concerning the legal status of any country, territory, city or area or of its authorities, or concerning the delimitation of its frontiers or boundaries.

ISBN 92-5-104530-5

All rights reserved. Reproduction and dissemination of material in this information product for educational or other non-commercial purposes are authorized without any prior written permission from the copyright holders provided the source is fully acknowledged. Reproduction of material in this information product for resale or other commercial purposes is prohibited without written permission of the copyright holders. Applications for such permission should be addressed to the Chief, Publishing and Multimedia Service, Information Division, FAO, Viale delle Terme di Caracalla, 00100 Rome, Italy or by e-mail to copyright@fao.org

© **FAO 2000**

PREFACE

The FishRights99 Conference, Use of Property Rights in Fisheries Management, was held from 11 to 19 November 1999 in Fremantle, Western Australia in cooperation with the Food and Agriculture Organization of the United Nations (FAO). Thanks to the efforts of the 352 participants from 49 countries, the conference was a marvellous success. I believe that we all learned more about the spectrum of rights-based management strategies and how these strategies may be used, and I am convinced that this knowledge will help us to better meet our obligations as stewards of the fish resources, part of the common heritage of mankind.

I believe the conference provided the perfect opportunity to address a challenge facing us all – the sharing or allocating of our finite fisheries resources through means that are equitable, socially acceptable, and efficient. As the executive director of Fisheries Western Australia (FWA), one of Australia's larger fisheries management agencies, I am constantly aware of the importance of developing management mechanisms to ensure that the exploitation of our marine resources is ecologically sustainable and accommodates the increasing resource demands from increasing diverse stakeholders. Issues of security, durability, exclusivity, and transferability are at the heart of our daily fisheries management activities, regardless of whether we are managing few or many fishermen, regardless of whether their harvest is of a few or many species and regardless of whether this occurs in low or high-valued fisheries.

The conference benefited from financial support of many organizations, including: The Government of Western Australia, Primary Industries and Resources, The Fisheries Research & Development Corporation, Pearl Producers Association, NSW Fisheries, Agriculture Fisheries Forestry, Australian Fisheries Management Authority, M G Kailis Group, Western Australia Fishing Industry Council Inc., Queensland Department of Primary Industries, Austral Fisheries Pty, Lobster Australia (Kailis and France), Queensland Fisheries Management Authority, Nor-West Seafoods Pty Ltd, The New Zealand Seafood Industry Council and Sealanes Food Services. A number of national governments also contributed to the success of the conference by sponsoring speakers. These included: Fisheries and Oceans Canada, the Ministry of Fisheries, Iceland, the Ministry of Agriculture, Nature Management and Fisheries, Netherlands, the Ministry of Fisheries, New Zealand, Sea Fisheries, Environmental Affairs and Tourism, South Africa. Other sponsoring agencies were The World Bank and the International Centre for Living Aquatic Resource Management.

I would like to take this opportunity to thank all have contributed to the success of the conference. Special mention goes to those who supported and drove the content and quality of the conference through their roles on the Organizing Committee: Mr Peter Millington (FWA), Chair; Mr Ulf Wijkström (FAO); Dr Gary Morgan (PISA); Dr Jim Penn (FWA); Mr Guy Leyland (Western Australian Fishing Industry Council); Mr George Kailis (M G Kailis Pty Ltd); and the Program Co-Chairs, Drs Rebecca Metzner (FWA) and Ross Shotton (FAO). Furthermore, it is only with the support of the FAO and the dedication of Dr Ross Shotton that we have these proceedings volumes in addition to the papers found on the FWA-maintained FishRights99 web site (http://www.FishRights.com.au). Finally, I must thank the FWA staff for their generous contributions of time and energy, which helped to keep the conference running in a timely and smooth manner.

As we look back at FishRights99, Use of Property Rights in Fisheries Management, I hope that we are standing on a more durable and secure platform from which to base our fisheries management. It is also my hope that we will continue to build on the information exchanged at the conference so that, half a decade later, when we revisit the subject, we have pushed the boundaries of how we use property rights to manage our fisheries in ways that are ecologically sustainable and that we are closer than ever to ensuring that we have Fish for the Future.

Peter Rogers
Executive Director
Fisheries Western Australia

FishRights99 Sponsors

 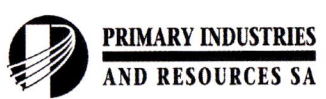 

   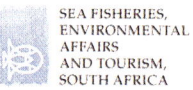 

FOREWORD[1]

From the beginning, the Programme Committee recognized that a third and essential part of the Conference would consist of sessions at which participants could present their experiences and views, negative or positive, on the implementation of property rights in fisheries management. It was also evident early on, that in the time that was available, there would be no alternative other than to run a number of parallel sessions. The challenge for the committee then became how best to anticipate the range of topics that papers offered for presentation would cover, so as to minimize the frustrations among participants when two, or more, interesting papers were presented at the same time. I take it as a complement to the quality of the papers (rather than a reflection of the lack of skills of the Programme Committee) that I frequently found myself in this situation during the workshop sessions.

Originally, the Committee had identified twenty-two major areas of rights-bases management, clearly a number that was too large to be comfortably handled once we had agreed that there would not be more than three parallel sessions. Further, once the abstracts of a significant number of papers had been received, it became apparent that the range of issues that would be discussed exceeded even that which the Programme Committee had anticipated, and a more general grouping was considered preferable. Thus, the session topics selected in the end by my colleague Rebecca Metzner were:

An introduction to rights-based management
What are property rights?
Evolution of rights-based management
Co-management and rights-based management
Multiple communities and rights-based management
Applying rights-based management
Globalization and rights-based management
Responsibilities and rights-based management
Denominating rights
Looking forward: challenges and opportunities

Not all of the papers offered for the Conference were accepted. And, all but a few, have been included in the proceedings. As the reader can attest, the papers were, in the main, of excellent calibre. The criterion that I personally used in coming to this conclusion was the number of papers that offered me new insights and information concerning situations about which I thought I was personally well informed. Many presentations described the same situation, but from different perspectives. Of great interest in this situation was the apparent convergence in views as to the attributes and weaknesses of the various applications under study despite the differing authors' backgrounds - industry, government, etc. As editor of the proceedings, I have no doubt that those who are not intimidated by the large number of papers, and do read them all, will be well rewarded and not regret the effort.

It has been my uniform experience that all those present for the workshop sessions in Fremantle found the period intellectually exciting, not least the opportunity to meet and exchange views face-to-face with those responsible for this form of management in some many countries. There was no doubt that a special atmosphere of common interests and intellectual exchange of experiences and views. And I believe that a number of new and constructive working arrangements came to pass.

Acknowledgements are due to a wide range of people. First to my colleague Rebecca Metzner, Fisheries Western Australia, who played the major part in the organization of the workshops. To each of the workshop chairmen (see the List of Contents for names and sessions), a grateful thanks is also owed for the management of the various sessions. The number of staff at Fisheries Western Australia that strove to make this part of the conference the success it was prevents a personal mention of all. However, deserving of particular mention are Carli Gettingby, and other support staff provided by Fisheries Western Australia. To my secretary, Marie-Thérèse Magnan and script editor, Mike Mann, special thanks are given for the major effort involved in bringing this volume to publication.

Ross Shotton
Editor, Proceedings, Fishrights99
Marine Resources Service
FAO, Rome

---

[1] See also the footnote to Volume I of these proceedings.

Shotton, R. (ed.)
Use of property rights in fisheries management. Proceedings of the FishRights99 Conference. Fremantle, Western Australia, 11-19 November 1999.
FAO Fisheries Technical Paper. No. 404/2. Rome, FAO. 2000. 462p.

ABSTRACT

Part 2 of the proceedings contains papers of presentations made during the Workshop Sessions of the Conference, which were held during the last two days of FishRights99. Seventy-six papers were presented in three parallel sessions. Workshop sessions addressed the themes of:

Introduction to Rights-based Management
What are Property Rights?
Evolution of Rights-based Management
Co-Management & Rights-based Management
What are Property Rights?
Multiple Communities and Rights-based Management
Applying Rights-based Management
Applying Rights-based Management to Developing Countries
Responsibilities and Rights-based Management
Denominating Rights
Looking forward: Challenges and Opportunities.

Thus, the workshop papers addressed national experiences in the design, implementation and modification of rights-based systems of fisheries management. The presentations included those made from the perspective of the fishing industry, government policy makers and administrators, legal implications as a consequence of national systems of law. Those concerned with the social and economic implications of this form of management reviewed the implications for communities affected by such changes in fisheries management approach.

Many papers described specific national implementation experiences, both positive and negative, and national programme successes and 'less-than-successes'. Other papers dealt with the social, economic and legal theory appertaining to this form of management.

Of the 76 papers presented during the Workshop part of the Conference, two were withdrawn after presentation and three were given only as oral presentations or in outline form.

Keywords: Fisheries Management, Property Rights, ITQs, Individual Transferable Quotas, Fisheries Policy, Fishery Access Rights

Distribution:

Conference participants
FAO Regional Fishery Officers
Conference Sponsors
FAO Fisheries Departments
FIRM Fisheries Mailing List
Fisheries Western Australia

TABLE OF CONTENTS

AN INTRODUCTION TO RIGHTS-BASED MANAGEMENT
Chairman: Peter Millington, Fisheries Western Australia, Perth

The Politics of Enclosures with Special Reference to the Icelandic ITQ System
H.H. Gissurarson 1
Introducing Property Rights into Fisheries Management: Governments cannot Cope with Implementation Alone
T. Craig 17
The Common Fisheries Policy of the European Union and Fishing Rights
C. Nordmann 23
The Global Environment Facility: a Partner in the Sustainable Management of Transboundary Fisheries
A. Merla 26

WHAT ARE PROPERTY RIGHTS?
Chairman: Mark Pendlebury, Clayton Utz, Perth

Are ITQs Property Rights? Definition, Discipline and Discourse
R. Connor 29
The Legal Nature of Australian Fishing Licences – Are they Property Rights?
B. McFarlane 39
Rights Based Systems: Sovereignty and Property
C. Jensen 47
Property Rights in Relation to Fishing Licences in Australia from a Legal Perspective
D. Fitzpatrick 53
Balancing Security and Flexibility in Granting a Right to Catch Fish
B. Wylynko and L. McIntosh 57
ITQs – New Zealand and United States: Allocation Formula and Legal Challenges
W.J. Nielander and M.S. Sullivan 59
Fishing Rights: a Multidimensional Perspective
N. Taylor-Moore 72
Rights-based Fisheries Management in New South Wales, Australia
A. Goulstone 78
The Closure of the Port Phillip Bay Scallop Fishery
S. McCormack and R. McLoughlin 84

EVOLUTION OF RIGHTS-BASED MANAGEMENT
Chairman: Lee Anderson, University of Delaware

Enhancing Fisheries Rights through Legislation – Australia's Experience
M. Tsamenyi and A. McIlgorm 88
Development and Implementation of Access Limitation Programmes in Marine Fisheries of the United States
G.H. Darcy and G.C. Matlock 96
Property Rights on the High Seas: Issues for High Seas Fisheries
A. Stokes 107
From Social Thought to Economic Reality: the First 25 Years of the Lake Winnipeg IQ Management Programme
G.S. Gislason 118
The Use of Individual Fishing Quotas in the United States' EEZ
A.C. Wertheimer and D. Swanson 127
Development of Property Rights-based Fisheries Management in the United Kingdom and the Netherlands: a Comparison
G. Valatin 137
Rights-based Fisheries Development in Australia: has it Stalled?
A. McIlgorm and M. Tsamenyi 148

CO-MANAGEMENT AND RIGHTS-BASED MANAGEMENT
Chairman: George Kailis, MG Kailis Pty Ltd, Perth

Co-Management and Rights-based Fisheries
M. Haward and M. Wilson 155
Community Management in Groundfish: a New Approach to Property Rights
F.G. Peacock and J. Hansen 160
The Creation of Property Rights along Lake Kariba through the Formation of a Fishermen's Association
R. Gwazani 170

MULTIPLE COMMUNITIES AND RIGHTS-BASED MANAGEMENT
Chairman: Guy Leyland, Western Australia Fisheries Industry Council, Mt Hawthorn

Bringing the State back in: the Choice of Regulatory System in South Africa's New Fisheries Policy
B. Hersoug and P. Holm 173
Property Rights and Recreational Fishing: Never the Twain Shall Meet?
J. McMurran 184
Negotiating the Establishment and Management of Indigenous Coastal and Marine Resources
D. Campbell 188
Recognition and Provision for Indigenous and Coastal Community Fishing Rights Using Property Rights Instruments
M. Hooper and T. Lynch 199
The Implementation of an Italian Network of Marine Protected Areas: Rights-based Strategies for Coastal Fisheries Management
F. Andaloro and L. Tunesi 206
Estuaries in Western Australia – An Integrated Approach to Management
J. Borg 209
United States' Fishery Cooperatives: Rationalizing Fisheries through Privately Negotiated Contracts
J. Leblanc 215

APPLYING RIGHTS-BASED MANAGEMENT
Chairman: Ross Shotton, Food and Agriculture Organization, Rome

Determining the Impacts of Adopting Property Rights as a Fishery Management Tool in Regulated Open Access Fisheries
J. Ward and W. Keithly 219
The Commercial Geoduck (*Panopea adrupta*) Fishery in British Colombia, Canada – An Operational Perspective of a Limited Entry Fishery with Individual Quotas
S. Heizer 226
Abalone and the Implementation of a Share-based Property Right in New South Wales, Australia
D. Watkins 234
Canadian Scallop Fishery Management: A Case History and Comparison of Property Rights *vs.* Competitive Approaches
F.G. Peacock, J. Nelson, E. Kenchington and G. Stevens 239
The Orange Roughy Management Company Limited – A Positive Example of Fish Rights in Action
G. Clement 254
The Effects of Transferable Property Rights on the Fleet Capacity and Ownership of Harvesting Rights in the Dutch Demersal North Sea Fisheries
W.P. Davidse 258
Trends in Fishing Capacity and Aggregation of Fishing Rights in New Zealand under Individual Transferable Quotas
R. Connor 267
Measurement of Concentration in Canada's Scotia-Fundy Inshore Groundfish Fishery
D.S.K. Liew 279
Indicators of the Effectiveness of Quota Markets: the South East Trawl Fishery of Australia
D. Alden and R. Connor 288

Chairman: Rolf Willmann, Food and Agriculture Organization, Rome

Will Improving Access Rights Lead to Better Management – Quota Management in the Tasmanian Rock Lobster Fishery
W. Ford 289

Individual Transferable Catch Quota: Australian Experience in the Southern Bluefin Tuna Fishery
D. Campbell and T. Battaglene 296

Shark Bay Prawn Fishery – A Synoptic History and the Importance of "Property Rights" in its Ongoing Management
P.P. Rogers and J.P. Penn 297

APPLYING RIGHTS-BASED MANAGEMENT TO DEVELOPING FISHERIES
Chairman: Ulf Wijkström, Food and Agriculture Organization, Rome

Efficient Access Rights Regimes for Exploratory and Developmental Fisheries
A. Cox and A. Kemp 304

Developing New Fisheries in Western Australia
J. Borg and R. Sellers 313

Incorporation of a Fish Property Right: a Hypothetical Example
W. Zacharin and R. Edwards 320

Chairman: Ross Shotton, Food and Agriculture Organization, Rome

Challenges to the Co-Existence of Marine Farming and Capture Fisheries in New Zealand
K. Drummond, P. Kirk and L. Nelson 327

The Role of Property Rights in the Development of New Zealand's Marine Farming Industry
M. Harte and R. Bess 331

The Nature of "Rights" in the Western Australian Pearling Industry
H.G. Brayford and G. Paust... 338

GLOBALIZATION AND RIGHTS-BASED MANAGEMENT
Chairman: John Nicholls, Fisheries Western Australia, Perth

Property Rights as an Alternative to Subsidisation of Fishing and a Key to Eliminating International Seafood Trade Distortions
A. MacFarlane 343

A Proposal for Cost Recovery in the Alaska Individual Fishing Quota (IFQ) Fisheries
P.J. Smith and J.T. Sproul 348

South African Perspectives on Rights in Fishing and Implications for Resource Management
D.J. Bailey 352

RESPONSIBILITIES AND RIGHTS-BASED MANAGEMENT
Chairman: Alain Bonzon, Food and Agriculture Organization, Rome

Making Fishing Rights Worthwhile – Sustainable Fisheries
K. Truelove 355

Evolution of Self-Governance within a Harvesting System Governed by Individual Transferable Quota
M. Arbuckle and K. Drummond 370

Stronger Rights, Higher Fees, Greater Say: Linkages for the Pacific Halibut Fishery in Canada
G.S. Gislason 383

Property Rights and their Role in Sustaining New Zealand Seafood Firms' Competitiveness
R. Bess 390

Chairman: John Nicholls, Fisheries Western Australia, Perth

Why Recover Costs? Cost Recovery and Property Rights in New Zealand
N. Wyatt 402

Fisher Obligations in Co-Managed Fisheries: the Case for Enforcement
J.P. McKinlay and P.J. Millington 405

Enforcement and Compliance of ITQs: New Zealand and the United States of America
W.J. Nielander and M.S. Sullivan 415

DENOMINATING RIGHTS
Chairman: Matt Gleeson, Australian Forestry and Fisheries Administration

A Mechanism to Address Surplus Growth within Quasi-Property Right Systems
C. Annand and F.G. Peacock 428

The Missing T: Path-Dependency within an Individual Vessel Quota System – The Case of Norwegian Cod Fisheries
B. Hersoug, P. Holm and S.A. Rånes 434

The Scalefish Fisheries of Northern Western Australia – The Use of Transferable Effort Allocations in the Management of Multi-Species Scalefish Fisheries
L Cooper and L. Joll 445

LOOKING FORWARD: CHALLENGES AND OPPORTUNITIES
Chairman: Rebecca Metzner, Fisheries Western Australia, Perth

The Fishing-Rights on Marine Resources in China
Z. Wu 454

The Implementation of Fishing-Rights Systems in SouthEast Asia: a Case Study in Thailand
S. Anuchiracheeva 456

AUTHOR INDEX 463

SPECIES INDEX 464

SUBJECT INDEX 465

THE POLITICS OF ENCLOSURES
WITH SPECIAL REFERENCE TO THE ICELANDIC ITQ SYSTEM

H.H. Gissurarson
Oddi House, Faculty of Social Science
University of Iceland, Reykjavik, 101, Iceland
<hannesgi@hi.is>

1. THE EVOLUTION OF THE ICELANDIC[1] ITQ SYSTEM

1.1 Background

While Iceland is a country poor in natural resources, the fishing grounds in Icelandic waters are some of the most fertile in the world. The Icelanders are therefore dependent on the fisheries for their recent affluence, with marine products providing more than 70% of total commodity exports. Demersal fish species, accounting for about 75% of the total value of marine products, include first and foremost cod, but also redfish, haddock, saithe, halibut, plaice and some less important species.

Relatively territorial in nature, cod and other demersal species of fish are found in feeding grounds near the bottom of the shallow continental shelf around Iceland. On the other hand, herring and capelin, which are pelagic species, roam in large schools over wide areas of the sea, usually near the surface. In addition to the demersal and pelagic fisheries, there are the small, but productive, scallop, *Nephrops* (*i.e.* Norwegian lobster) and shrimp fisheries: these species are mostly harvested inshore in clearly identifiable fishing grounds, although some deep-sea shrimp is also found. When it finally began to be understood in the 20th century that fishing grounds were not inexhaustible resources, any attempt to limit the access to those in the Icelandic waters was made difficult by the fact that no single country had clear jurisdiction over them. Indeed, during the period 1952-76 Iceland fought four 'Cod Wars' with the United Kingdom for control over those fishing grounds, unilaterally extending Iceland's Exclusive Economic Zone (EEZ), first to 4 nautical miles, then 12nm, then 50nm, and finally to 200nm. Iceland's two main arguments were that those extensions of the EEZ made the necessary conservation of fish-stocks possible and that the Icelanders, unlike other nations in the North Atlantic Ocean, were totally dependent on fishing.

When the United Kingdom recognised Iceland's 200nm EEZ, and the last British trawler sailed out of Iceland's territorial waters on 1 December 1976, the legal prerequisites for the management of the Icelandic fisheries finally were in place — and not too soon, as subsequent events showed.

1.2 Effort quotas, 1977-83

Because of the difference in nature between the pelagic and demersal fisheries, the two respective fishing fleets also differed in composition. Boats of a similar (medium) size harvested most of the pelagic fish, herring and capelin, whereas the demersal fishing fleet was heterogeneous, comprising large freezer-trawlers, mid-size multi-purpose vessels, small boats, even some undecked rowboats. The relative importance of the two kinds of fisheries also varied by regions. Since the most fertile demersal fishing grounds lay in the northwestern part of Iceland's EEZ, fishing vessels from the Northwest, *i.e.* from the Western Fjords, were better placed to harvest fish there than vessels from other regions. Hence, fishing villages in the Western Fjords relied mostly on harvesting cod and other demersal species of fish. On the other hand, the pelagic fisheries, herring and capelin, being non-territorial, were chased all over the Icelandic waters and even beyond. They were more important to the fishing villages in the East than to those in the Western Fjords.

Another fact undoubtedly had some effect on the evolution of the ITQ system. In the late 1960s, the Icelanders had had a first-hand experience of the dire consequences of over-fishing. After a 'herring boom' of the early 1960s, with annual catches of herring approaching 600 000t, the herring stock collapsed in 1967-8, so that a moratorium was imposed on the herring fishery in 1972 with harvesting resuming on a small scale in 1975. Soon after the extension of the EEZ to 200nm, a special Fisheries Act was passed by Parliament, in 1976 that gave the Minister of Fisheries wide powers to restrict access to the fishing grounds in Icelandic waters, while it was not clearly specified in which ways he should do so.

In 1976 the Icelandic Marine Research Institute (MRI), warned that the cod stock was threatened by overfishing. Fish mortality was alarmingly high and the spawning stock was weak. The MRI recommended a total allowable catch in cod of 230 000t for that year, while the actual total catch turned out to be 350 000t. Vessel owners in the demersal fisheries now were also beginning to realise that the cod stock, the mainstay of the Icelandic economy, accounting for about 35% of the total value of marine products, was in danger of collapse similar to that of the herring stock a decade earlier, still fresh in their memory. Obviously, access to the demersal fishing grounds had to be restricted. There was much discussion whether such restrictions should be in terms of effort or

---

[1] I wish to thank my colleagues Ragnar Arnason, Professor of the Economics of Fisheries and Birgir Thor Runolfsson, Professor of Economics, at the University of Iceland, for their encouragement and assistance in preparing this paper which was originally presented in November 1999 and then revised for publication in March 2000.

of catch. Finally it was decided to restrict effort, *i.e.* allowable fishing time, rather than vessel catch. In 1977, effort quotas in the demersal fisheries were introduced. While entry remained more or less free, and there were no restrictions on the catch of each fishing vessel, allowable fishing days were to be reduced until the desired result in terms of total allowable demersal catch had been reached.

The Minister of Fisheries in 1974-8 came from the Western Fjords, where support for effort quotas was strongest. Because fishing villages in the Western Fjords were closest to the most fertile cod grounds, vessel owners there thought that they would always be at an advantage in competition in terms of unlimited harvesting during a limited period of time. However, it soon became clear that effort quotas were wasteful. This system induced owners of fishing vessels to start a 'Derby', *i.e.* a competitive rush to harvest as much fish as possible during allowable fishing days regardless of cost. Since entry remained almost free, this meant not only that existing fishing capacity was not utilised economically, but also

Association of Fishing Vessel Owners whose leader, Kristjan Ragnarsson, was becoming convinced, with many of his members, that effort quotas did not work. In late 1983, the MRI found that the cod stock was still weakening. The spawning stock was at an all-time low, estimated at only 200 000t; and fish mortality was very high. Even if the total actual catch of cod had gone down from 461 000t in 1981 to 294 000t in 1983, it exceeded that recommended by the MRI by 100 000t. It was also becoming ever clearer that there was massive over-investment in the fisheries. This is shown in Figure 1: in 1945-83, fishing capital increased by well over 1200%, while real catch values only increased by 300%. Thus, the growth of fishing capital exceeded the increase in catch values by a factor of more than four. At the same time as vessel owners in the demersal fisheries could observe massive over-investment there, a sharp reduction in the number of allowable fishing days, and a clear decline in the cod stock, they witnessed the relative success of vessel catch quotas in the pelagic fisheries.

Figure 1
Fishing capital and catch values 1945-1997
(index 1960=100). Source: National Economic Institute

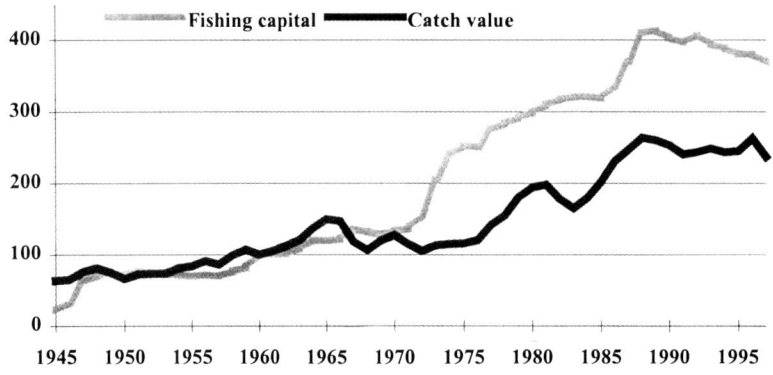

that there was an incentive to add to it. The already too-large fishing fleet became still larger, while the number of allowable fishing days had to be reduced almost every year. For deep-sea trawlers the number of fishing days declined from 323 in 1977 to 215 in 1981. Moreover, total annual actual catches consistently, and by far, exceeded the total annual allowable catches recommended by the MRI.

1.3 The introduction of vessel catch quotas, 1983-4

In Iceland, 1978-83 were years of weak governments, political upheavals and uncertainties. But in the summer of 1983 a strong coalition government of the Independence Party (Iceland's conservative party, with 35-40% of the votes) and the Progressive Party (with rural roots and about 20% of the votes) was formed. The new Minister of Fisheries, Halldor Asgrimsson, who came from the Eastern region, was to remain in office for the next eight years. He worked closely with the powerful

After the herring moratorium of 1972-5, it had been decided to set an annual total allowable catch (TAC), of herring over each year's fishing season, and to divide this TAC equally between the herring boats in operation. This was a simple and non-controversial rule of initial allocation since the herring boats were all of roughly equal size and with a similar catch history. In 1979, those vessel catch quotas had been, at the initiative of the herring-boat owners, made transferable: they had become ITQs. Arguably, this was one of the first ITQ systems in world fisheries. Similarly, in the capelin fishery, vessel catch quotas had been introduced in 1980, at the initiative of the capelin boat owners, to be made transferable in 1986. In both of those pelagic fisheries, such vessel catch quotas had had the effect to reduce boats at the same time as catch increased.

The most vocal support for the introduction of vessel catch quotas in the demersal fisheries came from the

East, whereas vessel owners in the Western Fjords continued to favour effort quotas. In 1983 the supporters of vessel catch quotas finally gained the upper hand in the Association of Fishing Vessel Owners, and at the annual meeting of the Icelandic Fisheries Association—a broad collection of interest groups in the fisheries—in December 1983, a resolution was passed calling on the Minister of Fisheries to experiment with vessel catch quotas in the demersal fisheries, especially in the all-important cod fishery. The Minister of Fisheries promptly proposed an amendment to the original *Fisheries Act 1976*, giving him discretionary power to issue individual quotas for each vessel employed in the demersal fisheries for the year 1984. After much heated discussion, the Icelandic Parliament passed the amendment at the end of December 1983, in the Upper House with a majority of one vote. Consequently, the Minister of Fisheries set a TAC for each demersal species of fish for the year 1984 and issued shares in those TACs to each and every fishing vessel.

The catch vessel quotas were allocated on the basis of catch history over the preceding three years, from 1 November 1981 to 1 October 1983, with exceptions to correct for certain situations, *e.g.* if a vessel had entered the demersal fisheries during those three years or if it had been under repair for part of this period. New vessels could choose between the new kinds of quotas and the old effort quotas (restrictions in terms of allowable fishing days). The new vessel catch quotas were partly transferable. Transfers of quotas between vessels under the same ownership or vessels from the same port were allowed, but transfers between vessels from different ports were only allowed if they were exchanges (e.g. a quota in redfish for a quota in cod), otherwise such transfers had to be approved by the Minister of Fisheries. Small boats, under 10 gross registered tonnes (GRT), were exempt from the quota system; they could harvest fish at will until they reached a total quota set for this type of vessel.

1.4 A mixed system, 1985-90

It is easy to see why vessel catch quotas were initially allocated differently in the demersal and pelagic fisheries. While the herring and capelin boats were of roughly the same size, making an equal initial allocation of vessel catch quotas between them fairly straightforward[2], there were vast differences between individual vessels in the demersal fisheries, so the more complicated rule of catch history over the preceding three years had to be adopted, with small boats exempted altogether from the system. At the end of 1984, when the experience of the previous twelve months under a system of vessel catch quota was reviewed, it was generally accepted that this kind of quota had been much more effective in halting over-fishing than effort quotas. It was therefore decided to extend the amendment to the *Fisheries Act 1976*

for one more year, allowing the Minister of Fisheries to issue vessel catch quotas for 1985. The opposition to vessel catch quotas from the Western Fjords remained strong, however, so, as a compromise, vessel owners were allowed to choose between vessel catch quotas and effort quotas. This meant that a typical vessel owner could either retain the share of the TAC they had received at the end of 1983 and harvest fish up to the limit set by that share; or he could give up his vessel catch quota and try instead to harvest as much as he could in the allowable fishing days, whose number was set by the Minister of Fisheries on the basis of predictions about their contribution to the TAC. This mixed system of vessel catch quotas and effort quotas was in effect for the next six years, until the end of 1990.

At the end of 1985, when the experience of the previous two years was reviewed, it was decided to write the system into a special law, the *Fisheries Management Act*, instead of passing an amendment to the Fisheries Act of 1976, as had been done in 1983 and 1984. It was also decided to issue the vessel catch quotas for two years, 1986 and 1987, instead of for one year. Earlier restrictions on access to certain fishing areas (*e.g.* spawning grounds) and on allowable fishing gear (*e.g.* mesh size) also continued to apply; and in addition to catch quotas, owners of fishing vessels had to hold special fishing permits which were in effect restricted to those who had operated vessels in the first years after the introduction of quotas.

When the Fisheries Management Act came up for review in the Icelandic Parliament at the end of 1987, difficult negotiations began, extending into the first weeks of 1988. The Icelandic Social Democrats (with about 15% of the votes), in a rather weak coalition government with the Independence Party and the Progressive Party since 1987, now insisted on inserting a declaration into the Act to the effect that the fish stocks were 'the common property' of the Icelandic nation. It was also decided in 1988 to extend the duration of the vessel catch quotas from two to three years, from 1988 to the end of 1990, and to make an extensive review of the system in 1990. Another important change in the *Fisheries Management Act 1988* was that it now applied not only to the demersal fisheries. The vessel catch quotas developed in the herring and capelin fisheries from 1975 have already been briefly described. But the *Nephrops*, shrimp and scallop fisheries were quite unlike the demersal and pelagic fisheries. They were confined to certain well-defined inshore fishing grounds and from their beginning in the 1960s and 1970s they were subject to local entry restrictions. In 1973, a TAC for *Nephrops* was first set and catch quotas issued to vessels. A year later, two of the seven inshore shrimp grounds were already subject to vessel catch quotas. In 1975, vessel catch quotas were issued in the inshore shrimp and scallop fisheries. Since boats operating in the *Nephrops*, shrimp and scallop fisheries were all of roughly similar size, vessel catch quotas were initially allocated equally.

---

[2] In the capelin fishery, for complicated historical reasons, two-thirds of the vessel catch quotas initially were allocated equally, and one-third on the basis of the vessel's hold capacity.

Another important change in the *1988 Fisheries Management Act* was that it was made difficult, or even impossible, for vessels which had chosen to operate on effort quotas to increase their share in the TAC. A further problem addressed in the *1988 Fisheries Management Act* was that of the great increase in the number of small boats, under 10 GRT, which had taken place since 1983-84 in response to their exemption from limits on entry (most of the new boats being just under 10 GRT in volume). It was now decided to subject boats between six and ten GRT to fishing permits and to issue no new permits to new boats of this size, unless they replaced old ones.

1.5 A comprehensive system of ITQs, 1990

When the *Fisheries Management Act* was revised in the spring of 1990 it was the first time this was done without the threat of an immediate collapse of any fish stock. The discussion therefore centred on the main objectives of fisheries management. Most of those concerned recognised that vessel catch quotas had turned out to be superior to effort quotas. A vessel owner who received a given share in the TAC, in the form of ITQ, could concentrate on harvesting this share in the most efficient way over each season; if he was successful in doing this, he would have an incentive to buy additional quota from other less successful vessel owners. In a book which I published on this issue in the Spring of 1990, while the Icelandic Parliament was discussing the revision of the *Fisheries Management Act*, I argued that the system of ITQs was reasonably efficient and that it should be developed as far as possible into a system of private property rights (Gissurarson 1990)[3]. The two Icelandic specialists in fisheries economics, Professor Ragnar Arnason of the University of Iceland, and Professor Rognvaldur Hannesson of the Norwegian Business School in Bergen, also argued, in reports to the Parliament, that the ITQs should be maintained, but that limits on their transferability and duration should be abolished. Perhaps most importantly, the Association of Fishing Vessel Owners, under the forceful leadership of Kristjan Ragnarsson, also supported ITQs and argued for their increased transferability.

The opposition to ITQs was strong, however. First, vessel owners in the Western Fjords still preferred effort quotas. Second, there were those who wanted small boats to remain exempt from any quotas, often for romantic reasons. In the third group which had been slowly forming over the preceding few years, there were those who opposed what they perceived to be trends towards the development of private property rights in the fisheries. Some members of this third group wanted to impose a special tax on the fisheries aimed at expropriating the economic rent which holders of quotas would derive from the exclusive access to, and utilisation of, a scarce resource (Jonsson 1975); others called on government to take the ITQs from vessel owners and to rent them back to them, in special auctions (Gylfason 1990).

In 1990, the Icelandic Parliament passed a new *Fisheries Management Act*. It took effect in the beginning of January 1991 at which time the fishing season was redefined from 1 September each year to 31 August of the next year[4]. The three important changes in the system were that effort quotas in the demersal fisheries were abolished, their holders receiving vessel catch quotas instead, that the quotas were issued for an indefinite period of time and that they became fully transferable. In essence, a comprehensive system of ITQs now replaced a mixed system of vessel catch quotas and effort quotas. By the *Fisheries Management Act 1990* fishing vessels between six and ten GRT were also integrated into the ITQ system, receiving share quotas in place of the effort restrictions under which they had previously operated.

Opposition to the ITQ system remained strong, however, and in the *Fisheries Management Act 1990* two concessions were made. First, boats under six GRT remained exempt from the system and subject, for a limited adjustment time, to effort restrictions (a given number of fishing days). Second, at the insistence of the Social Democrats, a paragraph was inserted into the *Fisheries Management Act* to the effect that no assignment of ITQs by this law could constitute any permanent property rights to such quotas or become the ground for compensation if the quotas were taken from their holders. While neither of these concessions seemed important at the time, they both turned out to be unfortunate. The exemption of small boats from the ITQ system created a loophole in the 'fence' erected around the Icelandic fishing grounds and the paragraph in the *1990 Fisheries Management Act* about the impossibility of permanent property rights in ITQs left the legal status of quotas unclear.

1.6 Further developments in the ITQ system, 1990-2000

When the new and comprehensive Fisheries Management Act was passed in 1990, it was stipulated that it should be revised after three years. In 1991 a new and strong coalition government of the Independence Party and the Social Democrats was formed with former Prime Minister Thorsteinn Palsson replacing Halldor

---

[3] In September 1980, I had first argued for the development of private property rights in the fisheries, at a conference on 'Iceland in the Year 2000', organised by Iceland's Management Society. In April 1983, almost a year before individual quotas were first introduced in the demersal fisheries, I argued for recognising the traditional and existing fishing rights as property rights and making them marketable (Gissurarson 1983).

[4] This was done in order to direct harvesting of fish away from the summer months, when quality suffers more quickly and regular factory workers are on vacation. There are a few exceptions, for instance, in 1999-2000 the fishing season for Icelandic herring is set from 1 September 1999 to 1 May 2000 and for inshore shrimp it is 1 October 1999 to 1 May 2000. In the capelin fishery, the TAC applies from 20 June 1999 to 1 May 2000. Harvesting of herring from the Atlanto-Scandian stock, of oceanic redfish in the Irminger Sea and of deep-sea shrimp on the Flemish Cap is also subject to special regulations set by international agreements.

Asgrimsson as Minister of Fisheries. Palsson was to remain Minister of Fisheries for the next eight years, contributing, like his predecessor, much to the development of the ITQ system. In 1993 the two government parties, supported by the Social Democrats, worked out a compromise over the vocal demands for some form of special taxation of quotas. This compromise was that a small 'service fee' was imposed on quota holders. This revenue was used to facilitate the reduction of the fishing fleet. In the same year a public commission on fisheries management came to the conclusion that the ITQ system worked quite well but that some minor changes would make it more efficient.

The commission recommended the integration of small boats under six GRT into the system and the making of ITQs transferable, not only between vessels, but also to fish processing plants. It also recommended that certain privileges of boats using longlines in winter should be abolished and that holders of ITQs should not be allowed to depreciate quotas that they had bought, since fish stocks were renewable natural resources. The Association of Fishing Vessel Owners opposed the idea that quotas should be transferable to others than vessel owners, and this recommendation was not accepted by the Icelandic Parliament.

The commission's other recommendations, after much deliberation, were mostly accepted. In 1996 the privileges of boats using longline in winter were abolished; but those who had enjoyed those privileges received additional ITQs in compensation. Since 1998, holders of ITQs have not been allowed to depreciate quotas that they have bought. The most difficult political change has been the integration of boats under six GRT into the system. The owners of small boats, mainly live in fishing villages in the countryside and with disproportionate representation in the Parliament they form a strong interest group in Iceland. They managed to extend their adjustment period from 1994 to 1996 when they were allowed to choose between receiving vessel quotas, thus entering the ITQ system, or to remain subject to effort quotas (which became less and less attractive, as the number of allowable fishing days was reduced year-by-year). Another compromise was reached by government and owners of small boats in 1997 further facilitating their integration into the ITQ system. However, some small boats (about one-third of the total fleet of about 1100 small boats) still remain outside the ITQ system.

Some further minor additions and amendments have been made to the *Fisheries Management Act 1990*. In 1997 two fish stocks harvested by international agreements outside Iceland's EEZ were integrated into the ITQ system: oceanic redfish in the Irminger Sea, southwest of Iceland's territorial waters, and deep-sea shrimp and Flemish Cap east of Canada. Since 1998 two new rules have been applied to discourage speculation in quotas. One rule is that while a vessel may transfer some of her quota between fishing seasons, she will forfeit all her quota if she catches less than 50% of her total quota in two subsequent years. The other new rule is that within each year, the net transfer of quota (*i.e.* the annual catch entitlement, not the permanent share of the TAC) from any vessel must not exceed 50%.

Another rule has been adopted to try to counter the possible concentration of quotas: no fishing firm may control more than a 10% of the ITQs in cod and haddock and more than 20% of the ITQs in saithe, redfish, Greenland halibut, herring, deep-sea shrimp and capelin. In 1998, after bitter complaints from fishermen's unions that the crew of fishing vessels were forced to participate in quota purchases (*i.e.* to have the cost of renting quota deducted from the total net revenue shared at the end of the fishing season by the vessel owner, captain and crew), it was decided to establish a special Quota Exchange. It is an institution for recording all quota transactions to ensure that they are transparent and public. All quota transfers have to take place through the Quota Exchange except transfers from one vessel to another owned by the same fishing firm, or exchanges of quotas of the same value (but in different species of fish) or transfers that are deemed by the Minister of Fisheries not to have a market value.

1.7 Legal decisions on ITQs

The ITQ system has further evolved in a series of decisions by the Icelandic courts and other authorities on the legal status of ITQs. One problem arises from the fact that holders of ITQs can either sell their right to harvest a given share in the TAC (their TAC-shares), or they can rent it over a season (their annual catch entitlement, the multiple of the TAC and the TAC-share). How should the incomes and outlays generated by such transfers be taxed? In 1993, the Supreme Court decided that the transfer of a permanent TAC-share should be taxed as transfer of property, but that the transfer of the right to harvest a given amount over one season (the annual catch entitlement) should be taxed as income for the seller and cost for the buyer.

Another problem was caused by the fact that the Icelandic Parliament has not been ready to recognise the use of quotas as collateral, despite proposals to that effect from the Minister of Fisheries. Predictably, banks and other lending institutions have circumvented this problem by writing into contracts with vessel owners that quotas issued to vessels used as collateral cannot be transferred from those vessels without the lenders' consent. In 1996, a district judge decided that ITQs could not be used as such indirect collateral since the fish stocks were the declared common property of the Icelandic nation. The Supreme Court, in two decisions in 1999, did however recognise ITQs as indirect collateral of the fishing vessels to which they were issued. It has also been decided, although not in court cases, that inheritance tax has to be paid on the (market value) of ITQs and that they should also be treated as property in the case of divorce.

These cases were all about clarifying the legal status of the ITQs for purposes of taxation and financial transactions. But opponents of the ITQ system have referred two matters of principle to the courts. In late 1998 the Supreme Court decided that requiring people who wanted to harvest fish in the Icelandic waters to hold not only ITQs but also special fishing permits (which were non-transferable and in effect confined to [owners of] fishing vessels operating in the first years of the ITQ system, in 1984-8, or to their replacements) was indeed unconstitutional. According to the Court, to restrict entry into the fisheries in this way to a mostly closed group of people who happened to operate fishing vessels over a given period of time violated the two constitutional principles of economic freedom and equal treatment under the Law. While the special fishing permits were not an integral part of the ITQ system (and only imposed as a short-term measure to try to control the enlargement of the fishing fleet), its opponents rejoiced at this decision. The government promptly changed the law, so now fishing permits are not confined to (owners of) vessels in operation in 1984-8.

The other case was much more important because it was about the ITQs themselves. In early 2000, a district judge in the Western Fjords decided that the initial allocation of ITQs in the demersal fisheries, on the basis of catch history in 1981-3, had violated the constitutional principles of economic freedom and equal treatment before the Law. According to the judge, this method of allocation unfairly discriminated between the group of quota recipients and other Icelanders. In the spring of 2000 the Supreme Court reversed this decision. It decided that the initial allocation of ITQs, on the basis of catch history, had not included any arbitrary or unconstitutional discrimination against those who did not receive such ITQs.

In the initial allocation, it was, the Supreme Court stated, quite fair and relevant to treat differently those who had a vested interest in continuing to harvest fish in the Icelandic waters, and all the others who had no such clear interest. Moreover, unlike the fishing permits, ITQs were transferable so they were not confined to any narrow group of people in the same way as the fishing permits had been. In the same decision, the Supreme Court stated that the general restriction of access to the Icelandic waters to holders of ITQs did not seem to violate the constitutional principle of economic freedom since this restriction had clearly been necessary in the face of collapsing fish stocks and unprofitable fishing firms.

1.8 Who cares whether the commons is privatised?

The evolution of the Icelandic ITQ system was a process of gradual discovery and difficult bargaining. Initially, politicians, marine biologists and vessel owners were mainly concerned about the conservation of fish stocks. It was only later that they came to realise the economic problem of unlimited access to a limited resource, the 'tragedy of the commons' (Hardin 1968). From an economic point of view over-fishing is similar to pollution: where access to a fishing ground is free, the cost of adding one more vessel (or another unit of fishing capital) to the fishing fleet on the ground is not borne solely by the vessel owner. Its activity has harmful effects on others. The consequences are over-capitalisation and excessive fishing effort. The fishing fleet is much larger than would be most efficient. As an illustration, sixteen boats may be harvesting a lesser catch than that which eight boats could easily harvest.

There is one big difference, however, between pollution and overfishing. Pollution is visible, whereas the economic costs that owners of fishing capital impose on one another are invisible. Those costs can be, and have been, demonstrated by economists (Gordon 1954, Scott 1955), but vessel owners usually come to realise the problem when it is too late—when fishing is exceeding not only the level of highest return on outlays, but also the maximum sustainable yield. Memories of the collapse of herring in the late 1960s may however have facilitated the acceptance by Icelandic vessel owners of what was in effect the enclosure of fishing grounds. Desperation lessens transaction costs (Libecap 1989). Another factor lessening transaction costs is homogeneity.

Because Iceland's pelagic fisheries were relatively homogeneous, with similar vessels, the introduction of vessel catch quotas and later ITQs was relatively easy. The bargaining process was much more difficult in the heterogeneous demersal fisheries. Owners of small boats, some of them working part-time, did not think, for example, that they had much in common with owners of large freezer trawlers. Indeed, as I have noted, some small boats are still outside the ITQ system. And vessel owners in villages close to the most fertile fishing grounds also thought that they had different interests from other vessel owners, and their strong opposition delayed the introduction of a comprehensive ITQ system for many years.

The main lesson to be learned from this process is that the introduction of ITQs in a fishery, however necessary it may seem to politicians, marine biologists and economists, is by no means a simple task. There are all kinds of interests that may oppose it. A commons like the fish stocks in Icelandic waters will only be enclosed if the private interests of those utilising it coincide with the public interest. It was probably crucial for the evolution of the Icelandic ITQ system that the Association of Fishing Vessel Owners repeatedly took the initiative in the process, and that government worked closely with it (Jonsson 1990), although it inevitably led critics to say that government was in the thrall of the Association of Fishing Vessel Owners. The important question is: 'Who Cares Whether the Commons is Privatised?' (Buchanan 1997).

It is difficult to see, for example, how vessel owners in the Icelandic demersal fisheries would have agreed to any other initial allocation of quotas in late 1983 than that which was based on catch history. This was the only way for them to continue using the fish stocks without much disruption. In this way they could maintain the value of their investments and human capital whereas it would have become almost worthless if government had auctioned off individual quotas to the highest bidders as some economists proposed. In essence, the problem in the Icelandic fisheries was the same as in all fisheries using modern technology and operating under free access to fishing grounds: it was, to return to our illustration, that sixteen boats were harvesting even less than eight boats could easily harvest. The task therefore was to reduce the number of boats from sixteen to eight. In theory, this could be accomplished by outbidding the owners of eight excessive boats, by taxation or in an auction of quotas. But in practice, this would have been difficult, if not impossible. In the Icelandic case, what was done was to assign transferable quotas sufficient for the profitable operation of eight boats, to the owners of sixteen boats. Over time, the eight boat owners who wanted to continue harvesting fish would have a great incentive to buy quotas from their eight colleagues who wanted to leave the fishery. Thus, people were not outbid; they were bought out.

2. THE NATURE AND PERFORMANCE OF THE ICELANDIC ITQ SYSTEM

2.1 Enclosure of the Commons

Economists analysing the 'tragedy of the commons'—the over-utilisation of non-exclusive natural resources—generally agree that the tragedy is caused by the absence of private property rights to those resources. In the costly race to extract value from such resources, whether they are plots of land, oilfields, mines, or fish stocks, the rent which could be derived from them is dissipated. 'The business of everybody is the business of nobody.' It was only with the enclosure of land, for example, that the problem of overgrazing was solved, and cultivation replaced simple extraction.

The EEZs which fishing nations have established in the 20th century may be regarded as important steps towards the enclosure of marine resources. At first sight, however, private property rights in areas of the sea or in individual specimens of fish do not seem technologically feasible, at least not in deep-sea fisheries; such rights would require techniques of fencing or branding, either non-existent or difficult to develop. ITQs may however go far to solve the fisheries problem (Arnason 1990) precisely because they have some characteristics of private property rights. They are exclusive, which means that only those who hold them may harvest fish; they are individual so that the responsibility for their utilisation is clearly defined and lies with individuals; they are divisible which enables fishing firms to freely decide how much of them to hold at any given time; they are transferable which means that market forces are allowed to select the most efficient fishing firms; and they are permanent, making long-term planning possible.

ITQs are not too difficult to administer or enforce, either, although the political problem of their introduction and initial allocation should not be minimised. Therefore, it is not surprising that ITQs are increasingly being used in world fisheries. Between 5-10% of world total catches are presently harvested under some kinds of vessel catch quotas. Nevertheless, Iceland and New Zealand are the only two countries to have developed a comprehensive ITQ system although ITQs are also widely used in the Netherlands, Australia and some other countries. Despite some weaknesses, the Icelandic ITQ system does not seem too different from the system described by economists that goes far to solve the fisheries problem.

2.2 Total allowable catches

The two pillars of the Icelandic system are total allowable catches (TACs), and individual transferable quotas. TACs are set annually by the Minister of Fisheries for each of the commercially valuable species of fish in the Icelandic waters, on the basis of recommendations from the Marine Research Institute (MRI). Economic considerations—receiving the maximum return on fishing capital—do not seem to play an important role in the setting of TACs although that may change in the future. In the first few years after the introduction of ITQs in the demersal fisheries, the Minister of Fisheries tended to set somewhat larger TACs than recommended by the MRI, mainly because as a politician he was concerned about adverse effects on the economy by sharp reductions in TACs, especially in the fishing villages scattered around Iceland's coastline. This has gradually changed, especially after 1991. In 1995, government even adopted a special rule about the annual TAC in cod: it is to be set at 25% of the fishable biomass, estimated by the MRI. Thus, the TAC is determined by an annual stock assessment. By applying this rule, marine biologists estimate that the chances of stock collapse are less than 1%. Table 1 reproduces the recommendations by the MRI in 1984-99 for the cod TAC, the decision by the Minister of Fisheries and the actual total catch.

The sharp reductions in TACs for cod in 1994-6 are noteworthy. If the members of the Association of Fishing Vessel Owners had not by then begun to think of themselves as stakeholders in the cod fishery, it is doubtful that such sharp reductions could have been accomplished relatively peacefully in a country as heavily dependent on fishing as Iceland.

2.3 Individual transferable quotas

ITQs constitute the other pillar of the Icelandic fisheries system. ITQs are shares in the TAC of a fish stock. They are issued to each vessel for an indefinite period of time, in the demersal fisheries initially on the basis of catch history in 1981-83. The only vessels partly exempt from the system are boats under six GRT whose owners chose to operate under effort restrictions (a given number of allowable fishing days). However, they harvest only a small proportion of the total demersal catch.

The ITQs are transferable both annually and permanently. A legal distinction is therefore made between two kinds of transferable quotas issued to a vessel: her TAC-share, given in percentages, and her Annual Catch Entitlement (ACE), given in tonnes. The ACE is a multiple of the TAC for the fishery and the vessel's TAC-share. For example, if a deep-sea trawler initially received a

0.1% share of the TAC in cod, and if the TAC in the fishing season 1999-2000 is 250 000t, then the vessel owner may harvest 250t of cod in the given year and expect to harvest 0.1% of future TACs. His TAC-share is 0.1%, and his ACE in 1999-2000 is 250t. He can do one of three things with his quota: (a) he can harvest 250t over the 1999-2000 season; (b) while keeping his TAC-share, he can sell his ACE, or a part of it, to the owner of another vessel, *i.e.* the right to harvest 250t, or a part of it, over the 1999-2000 season; (c) he can sell his TAC-share, *i.e.* the right to harvest 0.1% share in the TACs set now and in the future.

Table 1
Recommended and set TACs in cod and total actual catches, 1984-2000 (tonnes)

Year	Recommended TAC (MRI)	Allocated TAC (Ministry of Fisheries)	Actual total catch
1984	200 000	242 000	281 000
1985	200 000	263 000	323 000
1986	300 000	300 000	365 000
1987	300 000	330 000	390 000
1988	300 000	350 000	376 000
1989	300 000	325 000	354 000
1990	250 000	300 000	333 000
1991	240 000	245 000	245 000
1991-2	250 000	265 000	273 000
1992-3	190 000	205 000	240 000
1993-4	150 000	165 000	196 000
1994-5	130 000	155 000	164 000
1995-6	155 000	155 000	169 000
1996-7	186 000	186 000	201 000
1997-8	218 000	218 000	227 000
1998-9	250 000	250 000	N. A.
1999-2000	247 000	250 000	N. A.

Source: Marine Research Institute.

Both the TAC-shares and the ACEs are perfectly divisible. The TAC-shares are also perfectly transferable. There are some restrictions on transfers of ACEs, however, with the objective of stabilising local employment. While ACEs can be freely transferred between vessels under the same ownership or within the same region, their transfer between vessels in different regions has to be approved by the Minister of Fisheries after a review by the regional fishermen's union and local authorities. Since few transfers are blocked, in practice the ACEs can be regarded as freely transferable. Over time most of the ITQs have indeed changed hands: In February 2000 only 19% of the quotas initially allocated in the demersal fisheries were still held by those who originally received them (Morgunbladid 2000).

Since the Icelandic fisheries are mixed fisheries, vessels are bound to capture different species of fish on the same fishing trips. The TAC-shares in different fish stocks therefore have to be interchangeable. But species of fish differ in value: one tonne of cod is *e.g.* worth much more than one tonne of capelin. Cod is used as the common denominator of the system. The term 'cod equivalent' denotes the relative market value of different species of fish, set by regulation every year. The total quota for each vessel having a quota for several species may be calculated in cod equivalents. Quota transfers between vessels are also often measured in cod equivalents. In the fishing season from 1 September 1998 to 31 August 1999, the cod equivalent values were, for example: cod 1.00, haddock 1.05, saithe 0.65, redfish 0.70, plaice 1.20, Greenland halibut 2.15, ocean catfish 0.85, witch 1.20, dab 0.65, long rough dab 0.60, capelin 0.08, herring 0.14, nephrops 8.55, shrimp 1.20 and scallops 0.40.

While the ITQs are perfectly divisible, and easily transferable, their use and transfer are restricted as noted earlier: All transfers of TAC-shares (permanent quotas, in percentages) have to be registered with the Fisheries Directorate. Most transfers of ACEs (quotas over a season, in tonnes) have to go through the Quota Exchange. The owner of a vessel will lose his quota, measured in cod equivalence, if his vessel harvests less than 50% of the vessel's total quota in two subsequent years. The net transfer of quota from the vessel in any given year must not exceed 50% of her quota. Moreover, no fishing firm may hold more than a given fraction of quota in each species of fish.

2.4 Harvesting outside Iceland's EEZ

The ITQ system applies, as far as is possible, in those fisheries which either straddle Iceland's EEZ or are outside it. The general rule is that Iceland negotiates with the other countries concerned, a TAC in each such stock, and then Iceland's share of this TAC is allocated as vessel catch quotas. Capelin and herring are migratory stocks that move in large schools over the Northeast Atlantic Ocean. Iceland has negotiated a TAC in capelin with Norway and Greenland, by which Iceland receives the bulk of the TAC (since most of the capelin is found and harvested in the Icelandic EEZ). Iceland's share is allocated to individual vessels, on the basis of catch history.

After its collapse of the late 1960s, the Atlanto-Scandian herring, suddenly reappeared in the Northeast Atlantic in 1994, and since then Iceland has negotiated a TAC in this stock with other members of the Northeast Atlantic Fisheries Commission (NEAFC). As there was no catch history on which to base an initial allocation of quotas, Iceland's share in this TAC (which has usually been about 15%) in 1994-7 was not subject to individual quotas but to effort restrictions: entry was free until Iceland's share in the TAC had been reached. On the basis of this catch history, and vessel-hold capacity, vessel catch quotas or ITQs were then allocated for the period 1998-2000.[5] Iceland has also negotiated within NEAFC a TAC

---

[5] This was obviously an uneconomical way of allocating the ITQs, since it created an incentive for fishing firms to engage in a 'Derby' for a few years, *i.e.* to undertake 'strategic harvesting'

in oceanic redfish which is harvested in the Irminger Sea in international waters southwest of Iceland's EEZ. Since 1997 Iceland's share in the TAC has been allocated as vessel quotas on the basis of catch history (the three best years of the six years in which this fishery had been in operation, with 5% of the total set aside for those who had started the harvesting, a so-called pioneers' quota).

There have been two kinds of disputes between Iceland and other fishing nations in the North Atlantic Ocean. In the deep-sea shrimp fishery which started in 1993 on the Flemish Cap in international waters east of Canada, Iceland has refused to participate in an agreement reached by the North Atlantic Fishing Organisation (NAFO) because they try to manage this fishery by restrictions on effort, *i.e.* allowable fishing days. Iceland is opposed to this for reasons already explained. Instead, since 1997 Iceland has unilaterally set a TAC for its own fishing vessels on the Flemish Cap; this has then been allocated as ITQs to fishing vessels on the basis of their catch history. The other NAFO countries have accepted this unilateral action, while not endorsing it.

In the Barents Sea fishing grounds of the so-called Loophole between Norwegian and Russian territorial waters, Iceland has had a dispute with Norway and the Russian Federation from 1993, when Icelandic vessels began to fish cod there, until May 1999 when the three countries settled their differences. Iceland agreed to stop fishing in the Loophole in return for small quotas in Norwegian and Russian territorial waters, an option to buy quotas from Russian vessels and the issuing of small quotas to Norway and the Russian Federation in Icelandic waters. During the dispute, Iceland did not try to control the activities of Icelandic trawlers in the Loophole. However, in 1997-8 catches there collapsed, at the same time as the TAC in cod in Icelandic waters was increased. Icelandic vessels have therefore largely ceased harvesting fish in the Barents Sea although they made quite a difference in the difficult 1994-5 period.

2.5 Administration and enforcement

Two government agencies, under the direction of the Minister of Fisheries, are mainly concerned with administering and enforcing the ITQ system. The Marine Research Institute (MRI), investigates the state of fish stocks and makes recommendations about annual TACs of different fish species to the Ministry of Fisheries. The MRI operates research vessels and collects additional information on fishing from vessel skippers. It also undertakes basic research in marine biology. The MRI has a staff of about 170. Approximately one-third of its costs are covered by its own revenues. The Fisheries Directorate (FD) oversees the day-to-day administration of the ITQ system, especially the collection of data on catch and effort. It has a regular staff of about 60 and about half of its budget is covered by its own revenues. In addition, the FD employs observers on vessels fishing in distant waters outside Iceland's EEZ.

The ITQ system is in effect enforced by controlling landings. All marine catch is required by law to be weighed on officially approved scales at the point of landing. Municipal authorities operate the weighing stations and they collect weighing fees from the vessels to cover their costs. The officials of the weighing stations record the landings and verify species compositions. There are 67 such controlled landings ports in Iceland and some major foreign export ports are controlled as well. A sophisticated computer system links the port data systems to the FD enabling the transmission of daily catch data to the FD's computer department. All catch data are transmitted to the FD twice a day and the information is disseminated through the FD's Web pages, through monthly publications and by telephone to skippers and vessel owners who wish to check their catch status. Status reports are sent to vessel owners regularly and upon request. The FD's Web pages of fisheries data show in detail the catch status of individual vessels, quota transfers between different vessels or in different species, quota shares and landings.

A third government agency, The Icelandic Coast Guard, is under the direction of the Minister of Justice and has a staff of about 130. It monitors fishing vessels at sea and enforces regional closures. As noted, extensive nursery grounds are permanently closed to fishing vessels and the spawning grounds of cod are closed for a few weeks in late winter during the spawning period. Moreover, the Minister of Fisheries, on the advice of the MRI, has the right to declare the immediate temporary closure of areas with excessive juvenile fish. There is also a 12nm limit for large trawlers in most areas.

In addition to the surveillance provided by the FD and the Coast Guard, the Ministry of Fisheries itself employs marine observers, some of whom take trips on fishing vessels and some of whom travel between fishing ports. Those observers try to ensure compliance with regulations on mesh size, bycatch, *etc.* Net mesh size must be at least 135mm and in the shrimp fishery a sorting grid is mandatory to avoid catch of juvenile fish. In the demersal fisheries devices for excluding juveniles are also mandatory in certain areas. The Ministry of Fisheries itself has an office staff of about 20. The Ministry charges holders of ITQs a low fee for the costs of administering and enforcing the ITQ system, with an upper limit of 0.4% of the estimated catch value. The revenue from the fee is about $8-9 million/yr. In addition there is revenue from fishing permit fees of about $2 million/yr. The net costs of enforcing and administering the ITQ-system are less than $30 million/yr, which includes the costs of marine biological research and guarding territorial waters.

---

to establish a catch history. The reason the quotas were not auctioned off was probably that there were already loud demands from some opponents of the ITQ system for auctioning off the existing quotas. The Minister of Fisheries may have felt that by such an auction he would only encourage those people. It is ironic if the only impact the of government auctions of quotas had on policy-making was to hinder an auction where it may have been justifiable.

This does not seem huge in comparison to the value of the catch value in the Icelandic fisheries which, in the late 1990s, was on average about $800 million/yr. Violations of the *Fisheries Management Act* and the corresponding regulations of the Ministry of Fisheries carry heavy fines, expropriation of catch and gear, and cancellation of fishing permits. While the Ministry of Fisheries has wide discretionary powers in assessing such penalties and a proven willingness to use them, alleged violators have recourse to the courts if the Ministry's decisions are unacceptable.

2.6 Are the Icelandic ITQs property rights?

On land, fencing techniques such as barbed wire have enabled individuals to establish property rights (*i.e.* to exclude others from the utilisation of land and other immovable objects, whereas branding has enabled them to establish property rights (*i.e.* to exclude others from their utilisation) over animals and other movable objects. However, fences can hardly be erected around different areas of the deep sea (although some kinds of fencing may be possible in inshore fisheries). It is also difficult to see how individual fish in the sea can be branded (at least the cod, herring and other species of fish that Icelanders harvest). It may be argued therefore that ITQs are substitutes for property rights based on fencing or branding. They are not exclusive rights to the utilisation of particular areas of the sea, or of particular fish, but rather exclusive rights to harvest a given share of a given total catch of a species of fish. They are rights of extraction rather than property, comparable to rights to extract a certain quantity of timber from a given forest, or to harvest a certain number of deer from a given colony (Hannesson 1994).

While such rights provide incentives to cut the timber and to catch the deer in the most efficient way, they may not be sufficiently strong to provide optimal husbandry of the forest or the deer colony. Nevertheless, ITQs as described in fisheries economics literature have many of the efficient features of individual property rights. They are exclusive, individual, divisible, transferable and permanent. Holders of such rights have a clear interest in the long-term profitability of the resource. There would be a crucial difference in the behaviour of two groups of quota holders, where the members of one group each have a permanent quota expressed in a given quantity of fish, *e.g.* 250t of cod/yr. Members of the other groups would each have a permanent quota expressed in a given share of the total catch, *e.g.* 0.1% of the TAC for cod. The latter group would be concerned not only with minimising harvesting costs, but also with setting the TAC in such a way that the long-term profitability of the fish stock in question would be maximised.

Arguably, ITQs, as described in fisheries economics literature, come as near to being private property rights as is feasible in deep-sea fisheries. But what about the Icelandic ITQs? Those ITQs are certainly individual and divisible. They are also exclusive although their exclusivity is somewhat reduced by the continuing existence of exemptions from the system for boats, under six GRT. But it is a minor exemption and sooner or later all small boats will probably be integrated into the ITQ system. The Icelandic ITQs are also mostly transferable: the restrictions on quota transfers are not very important. Nevertheless, they are restrictions.

For the system to be more efficient most economists would argue that ITQs should not be issued to fishing vessels, but to individuals and firms and they should be freely transferable. No restrictions should be imposed either on the relative or absolute amount each individual firm could hold, as is now the case. The ITQs should also be fully recognised by the Law as possible collateral that they are not at present. There should not be conditions on their use, such as the rules to discourage speculation in ITQs. More speculation would facilitate transfers in the ITQ market, hasten the reduction of the fishing fleet and enable quota holders to be more flexible in their operations.

The main problem in the Icelandic fisheries is that the ITQs, even if issued to individual vessels for an indefinite period of time since 1990, are not really permanent and secure. As described earlier, a paragraph was inserted in the *Fisheries Management Act 1990* to the effect that no assignment of ITQs by this law could constitute any permanent property rights to such quotas or become the ground for compensation if the quotas were taken from their holders. While it is unlikely that the ITQ system would be abolished, or the quotas taken from their present holders especially since in early 2000 only 19% of the quotas are still in the hands of those to whom they were initially assigned. The unwillingness of the Icelandic Parliament to take any steps legally to recognise the ITQs as property rights, even if they are taxed and for all purposes treated as such, has added to the uncertainty facing their holders.

2.7 Non-territorial rights in response to harmful effects

The emergence of ITQs in the Icelandic fisheries has interesting similarities to the emergence of property rights amongst Indians in Labrador, as analysed by Harold Demsetz (1967). For centuries, before the arrival of Europeans, the Indians had hunted beaver primarily for food and the few furs they needed. Since the beaver stock was a non-exclusive resource, the Indians did not have a vested interest in increasing or maintaining it. However, as their needs were small and the technology primitive the negative effects of beaver hunting were insignificant. When European traders arrived, hunting technology improved, and demand for furs greatly increased. The scale of hunting greatly increased so the harmful effects which each hunter had on others by his hunting became significant. Consequently, the Indians divided themselves into several bands in order to hunt more efficiently. Each band appropriated pieces of land, roughly similar in quality, for it to hunt exclusively. By the middle of the 18th century, the privately-allotted territories were relatively stabilised. Thus, the fur trade had encouraged the husbanding of beaver and the prevention of poaching which such husbanding requires. Demsetz tells this tale to illustrate his main point about property rights: they emerge when harmful or beneficial effects of economic activity emerge, enabling individuals to take them into account.

Consider pollution: if I pollute a river in which you swim, or fish salmon, or from where you get your drinking water, with the consequence that you cannot continue your use of the river, it is typically because neither you nor anyone else owns the river, and is able to hold me responsible for my activities. While the pollution I cause harms you, it costs me nothing. The solution would seem to be to define property rights to the river, just as the Labrador Indians established property rights in different pieces of land. Sometimes, however, the definition of property rights is not feasible as the costs of establishing them are higher than the gains. Demsetz points out that the Indians of the Southwest plains who came into contact with the European market at the same time as the Labrador Indians, did not establish new property rights in response to increased demand for the animals they hunted and improved hunting technology. The reason was that the animals of the plains, such as the buffalo, were primarily grazing animals wandering over wide areas. The cost of husbanding those animals (fencing or branding) was therefore much higher (at least until the introduction of barbed wire) than the cost of husbanding beavers in Labrador which were confined to relatively small areas.

The pelagic species of fish in Icelandic waters, herring and capelin, are rather similar to the animals of the Southwest plains described by Demsetz: clearly, any territorial rights to those two fish stocks would have been unfeasible. Neither fencing nor branding would have been possible. On the other hand, cod and other demersal fish are similar to beaver in the Labrador forests in that they are relatively territorial. The fishing grounds where those species are found are known and rather well-defined. Unlike branding, fencing would in theory have been possible in the demersal fisheries (and even more in the inshore shrimp and nephrops fisheries, confined to small and clearly demarcated areas).

The interesting question is then why territorial rights were not established in those stocks. Several answers may be suggested. First, there were hardly any legal precedents or possibilities available to fishing-vessel owners or legislators. While non-territorial fishing rights in the form of ITQs had already been tried in the pelagic fisheries, and seen to work, ideas about property rights in areas of the sea would have been dismissed as pure fantasy. Secondly, demersal fishing grounds are large in scale, creating possible economic inefficiencies of their own as independent units of operations, while vessel catch quotas are perfectly divisible. Third, fencing each fishing ground would have been costly. Instead, under the ITQ system only the Icelandic EEZ is really fenced off. Moreover, the Icelandic fishing fleet includes many multi-purpose vessels so it was economical to have a comprehensive quota system within which a vessel might switch from harvesting one species to another without many problems.

It is also convenient that the quotas are expressed in terms of cod equivalents so fishing vessels can easily solve the problem of bycatch. On the whole, the evolution of the Icelandic ITQ system can be interpreted as the practical response to the problem of vessel owners imposing economic costs on one another by excessive fishing effort and over-capitalisation—costs which should not have been blamed on them, but rather on the lack of property rights and thus the lack of information about those costs (Coase 1960). It amounts to the enclosure of the fish stocks in Icelandic waters—an enclosure not yet completed.

2.8 The performance of the ITQ system

When access to a resource, such as the fish stocks in Icelandic waters, suddenly becomes exclusive, the behaviour of those utilising the resource should be expected to change greatly. When an ITQ system is introduced in deep-sea fisheries the fish stocks in question are taken into custody, so to speak, by the quota holders. Certainly there has been a marked change in the behaviour of Icelandic vessel owners since the introduction of the ITQ system. Even if their rights of extraction from the fish stocks are by no means as clear or certain as they could be, quota holders within the powerful Association of Fishing Vessel Owners have begun to look upon themselves as custodians of the fish stocks, taking a long-term view of their utilisation and supporting a cautious approach to the setting of TACs. Note how Table 1 shows that the TACs set by the Minister of Fisheries have gradually approached the TACs recommended by the MRI. This is not least because of the increased sense of responsibility within the ranks of vessel owners.

Since the introduction of ITQs most stocks in Icelandic waters have slowly increased, in particular the valuable cod stock (at the same time this stock has collapsed in other parts of the world). Harvesting has also become much more efficient, especially in the pelagic fisheries, as can be seen in Figure 2. In the herring fishery, catch per unit of effort is now roughly 10 times higher than it was when ITQs were first issued. In the capelin fishery, the number of vessels has gone down, and fishing effort has been reduced; at the same time there has been no downward trend in their catches. The evidence also suggests that harvesting in the small *Nephrops*, shrimp and scallop fisheries has become more efficient.

In the demersal fisheries for the first few years after the introduction of ITQs fishing effort and fishing capital indeed increased, but this can be explained by factors such as the unfortunate re-introduction of the effort-quota option in 1985, the partial exemption of small boats from the system, and a structural change in the fisheries, namely the increase in the number of freezer trawlers, in effect moving fish processing from land to sea. Nevertheless, since 1991 when the ITQ system became comprehensive and most exemptions from it were

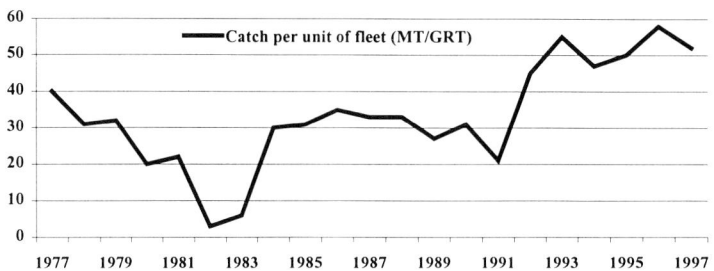

Figure 2
CPUF for the Purse Seine Fleet
in the Pelagic Fisheries 1977-97
Source: National Bureau of Statistics

removed, both fishing capital and fishing effort have been significantly reduced, as can be seen in Figure 3. The reason why fishing capital has not gone down as dramatically as fishing effort may be that many vessel owners want to retain their hold capacity in the hope that with stronger fish stocks future TACs will increase.

Since the introduction of the ITQ system, there has been considerable readjustment in the Icelandic fisheries. Unprofitable firms have gone out of business while other firms have merged and rationalised their operations. The impact of the ITQ system on the structure of the fishing sector has been interesting. In the beginning, it was feared that the system would lead to increased concentration. In a sense, this is what has happened. While the ten largest firms in the demersal fisheries held 24.6% of the quotas in 1991-92, they held 37.6% in 1998-89. But, in the meantime, almost all those firms have become public corporations. Companies previously owned by small families, or sometimes by municipalities, are now owned by 10-20 000 shareholders. So, the ITQs are in the hands of fewer fishing firms, but those fishing firms are in the hands of many more people than before.

It was also feared in the first years of the ITQ system that there would be a net transfer of quota from the small fishing villages scattered around the coastline, to the urbanised Southwest of Iceland where the capital city of Reykjavik is located. This has not happened, on the contrary, there has been a net transfer of quota from the Southwest, and especially to the Northeast. In 1984, firms in the Southwest controlled 29.7% of quota, and firms in the northeast 14.9%, but in 1998-99, firms in the Southwest controlled 25.7% of quota, and firms in the Northeast 21.2%. The impact of the ITQ system has indeed been to strengthen the local economy in the small fishing villages.

The prevailing regional distribution of quota has interesting political consequences. If a special tax would be imposed on quota holders to extract the rent from the fisheries, as has been proposed, then this tax would probably mean a transfer of resources to the Southwest from the rest of the country. While about 75% of the quotas are held outside the Southwest, about 75% of the population resides in the Southwest. This may become a powerful factor in a possible political conflict over rent expropriation in the fisheries though on the whole the ITQ system can be said to have performed quite well (Runolfsson 1999).

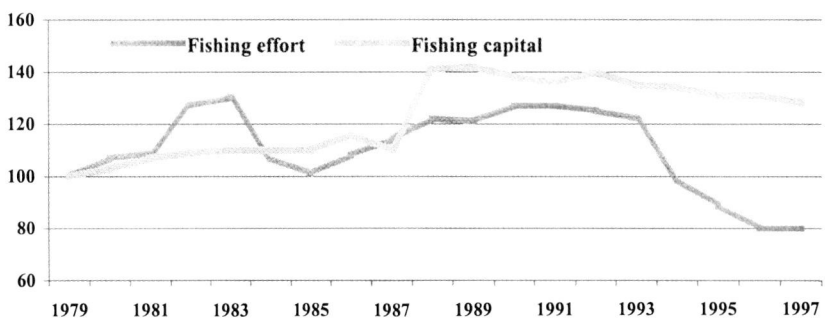

Figure 3
Demersal Fishing Effort and Capital 1979-1997
(index 1979=100). Source: National Bureau of Statistics

2.9 Remaining problems

While the ITQ system in the Icelandic fisheries has performed as well as could be expected and without any serious social consequences some problems remain. Some of them are institutional and can be corrected, but probably at a political cost: the partial exemption of small boats from the system; some remaining restrictions on transfers; and the uncertain legal status of the quotas.

A further problem lies in the fact that all quotas have to be expressed in tonnes over the fishing season whereas the values of two tonnes of catch are not always equal, either because they come from different species of fish or because specimens of one species differ in value. Discarding may therefore occur. However, discarding, the throwing away of non-targeted species, is not much of a problem in the Icelandic ITQ system because a quota in one species is easily transferred to a quota in another through their common denominator, cod. Highgrading, the throwing away of specimens of the targeted species, because they are too small to be of much value, is a greater problem, even if its extent is exaggerated by critics of the ITQ system. In a 1993 government commission report it was estimated that highgrading of demersal fish ranged from 1-6% of the total catch volume depending on the type of gear and vessel used (Arnason 1994). Moreover, according to the report there had been no detectable increase in highgrading since the introduction the ITQs. One reason for the relative insignificance of highgrading is undoubtedly the strict surveillance of fishing vessels. It should also be noted that some highgrading occurs because it is difficult to differentiate between specimens of different value during operations at sea. In the future, improvements in fishing gear will reduce this problem.

3. CURRENT CONTROVERSIES ABOUT THE ICELANDIC ITQ SYSTEM

3.1 Transparency of extraction rights

While the Icelandic fisheries present a strikingly different picture from that in many neighbouring countries whose fisheries are unprofitable, heavily dependent on government subsidies or deplete their fish stocks, the ITQ system is still controversial in Iceland. Its most unpopular aspect is the transferability of quotas. Regularly, there is a public outcry when a holder of a quota sells it, even if this can be seen as a positive step, since it means that the seller leaves the over-capitalised fisheries: this is readjustment by trade, not by force. But public debate raises more general philosophical objections to the ITQ system. One common objection is ITQs mean at least partial 'enclosure' of fish stocks. They imply the development of exclusive extraction rights to fish stocks that share important features of private property rights.

It is argued that the initial allocation of quotas at the end of 1983 in the demersal fisheries was unjust because it constituted a gift to their recipients—owners of fishing vessels operating in the three preceding years—excluding all others. The critics of the ITQ system say that the fish stocks in Icelandic waters are the declared common property of the Icelandic nation, and that it is unjust that individual fishing firms should reap the profits from exploiting them. They propose that the quotas should be taken from their present holders and auctioned by government, or that a special tax should be imposed on their holders to capture the rent that can be derived from the resource. This said, opposition to the ITQ system in Iceland has been no stronger than would be expected in a country so dependent on fishing. In Iceland, almost everyone lives close to the fisheries and all events are well reported in the media. In most other countries, fishing is marginal to the economy and is usually given scant public attention. Therefore, less opposition should be expected from the general public in most other countries to the introduction of ITQ systems in fisheries.

3.2 Is the ITQ system unjust?

The arguments against the initial allocation of quotas are directed solely at the allocation of quotas in the demersal fisheries at the end of 1983. But, if that was unjust, so must have been the initial allocation of quotas in the pelagic fisheries in 1975-80 and in the small *Nephrops*, shrimp and scallops fisheries in the 1970s. It is difficult to see how the demersal quotas could have been allocated in any other way than on the basis of catch history. To return to the illustration in Section 1.8, the task was to reduce the fishing fleet from sixteen to eight boats. In theory, government could do this by taxing, or pricing, eight boats out. But in practice it would be better done by assigning transferable quota sufficient for the successful operation of eight boats to the existing sixteen boats so that the more efficient could, over time, buy out those who wanted to leave the fisheries. In this way the necessary adjustments could occur peacefully. After all, vessel owners had invested in their vessels, gear and practical knowledge (human capital) in the belief that the fishing grounds in Icelandic waters would remain open to them.

When it was necessary to restrict access, it seemed natural to restrict it to those who had made such investments because they were the only ones to lose from the restriction, not those who had made no such investments. It was easier and less costly not to enter the fisheries than to leave them. Put differently, this was the only posssible Pareto-efficient change. A change in institutions is Pareto-efficient if all benefit from some benefit or no one loses (Buchanan 1959). If government had auctioned the quota, it would itself have benefited. Those eight boat owners who would have been able to purchase quotas would have neither benefited nor lost. But those eight who would have been outbid at the auction would have lost because their capital, being specific to the fisheries, would have become worth little. On the other hand, when quota was assigned to the existing owners of fishing capital, and made transferable, as was done, no one lost. Those who remained in the fishery would, over time, have bought quota from the other eight boat owners who then would have gained from the sale of their quota. Even the government would have benefited from the increased productivity in the fisheries and higher tax revenues.

The crew of the eight boats that would have had to leave the fishery would, under an initial assignment of quota to vessel owners, have longer adjustment period than under a government auction. This would have made them redundant overnight. As their skills were not specific to fisheries they could then seek employment elsewhere without losing much of their bargaining power.

The initial allocation of catch quotas on the basis of catch history harmed no one. On the contrary, a system of transaction rules was developed by a group of people who had been imposing economic costs on one another by over-utilising the fish stocks, to end this. Unlike pollution, the harm was invisible: it was benefit forgone, the potential rent from a fertile resource dissipated in over-capitalisation and excessive fishing effort. The ITQ system internalized an externality. Its introduction consisted in assigning responsibility for the fish stocks to individuals and thus enabling them to eliminate the harmful effects that they had previously had on one another.

It is, therefore, misleading to speak about a 'special gift' to the owners of vessels in the demersal fisheries when they received vessel catch quotas at the end of 1983. What government did for them was what it had previously neglected to do, and what is usually regarded as its duty: to define and uphold a system of rules under which people could settle their differences peacefully and to mutual advantage. This system of rules certainly enabled Icelandic vessel owners to create wealth and this is what property is supposed to do.

At this point, opponents of the Icelandic ITQ system may point to the declaration, in the *1990 Fisheries Management Act* about the fish stocks being the common property of Iceland. It is an interesting question what this declaration, inserted at a late stage in the evolution of the ITQ system to facilitate a vote in the legislature, precisely means. Legal experts in Iceland answer that the concept of 'common property' is vague (Lindal 1998). They say that this declaration should not be interpreted as if the fish stocks belonged to government as some buildings and cars do. Rather, it should be regarded as a declaration to the effect that Iceland has full jurisdiction over the fish stocks in Icelandic waters and that their utilisation has to serve the long-term interests of the nation. Certainly, they say, legislators did not mean to nationalise the fish stocks by inserting this declaration into the Fisheries Management Act.

3.3 The demand for a resource rent tax

Some Icelandic economists have argued for a special resource-rent tax in the fisheries, on the grounds that such a tax would not have any distortional effects, unlike most other taxes, that the owners of fishing vessels do not deserve the rent from the fish stocks, and that such a tax might make ITQs more acceptable to the general public (Gylfason 1990, Moller 1996). Unlike pollution fees, for example, such a tax would not be corrective (*i.e.* serve to internalise an externality). The ITQ system has already accomplished the necessary correction by enabling vessel owners to reduce fishing capital and fishing effort in their transactions to the most profitable level. The proposed resource rent tax would therefore be redistributive. While a resource rent tax might seem plausible if it could replace other more distortional taxes, it is quite optimistic to think that it would do so. It is more likely that in the long run it would simply broaden the basis for taxation in Iceland, adding yet another source of income to government. Moreover, there are reasons to believe that such a tax would have distortional effects on operations in the fisheries (Johnson 1995, 1999).

Consider the possibility that government would gradually take the quotas away from their present holders over a period of 10 years or so and rent the quotas again to them, perhaps for 2-3 years. This would mean that the incentives and therefore the behaviour of vessel owners would change. They would no longer think of themselves as having an interest in the long-term productivity of the resource. The responsibility for the resource would lie with government. Therefore, the vessel owners might support higher TACs than would be optimal. Monitoring would also become more difficult. One of the great advantages of a ITQ system is that as the quota holders have a well-defined share in the resource they have an incentive to co-operate and to monitor harvesting. In short, the difference between the behaviour of vessel owners under an ITQ system and under a system of resource rent taxes is that which exists between owners and tenants.

It may be argued that owners of fishing vessels do not deserve the rent that they will, under the ITQ system, be able to derive from the fish stocks in Icelandic waters. Rent from a natural resource is by definition created not by the firms utilising the resource, but by the limited supply of the resource. In a sense, the generation of vessel owners receiving the initial quotas are indeed enjoying a windfall profit. But it is also the consent and active co-operation of this generation which is crucial to the success of the change in institutions. It is difficult to see any others who deserve the rent, either. It may also be quite difficult to isolate the full rent derivable from a resource in such a way that it will not decrease in the process of isolating it, as we have seen. Moreover, if the rent derived from the exclusive access to the fish stocks in Icelandic waters is to be captured by a special resource-rent tax, then it would seem only fair that the rent derived from other resources in limited supply, including land, hot springs, and human talent, should also be taxed. This would however be difficult, both for technical and political reasons. It is by no means certain, either, that a special resource rent tax on the fisheries would make the ITQ system more acceptable to the general public.

The most unpopular aspect of the system is that holders of quota can sell it and leave the fisheries with a large sum of money. As the adjustment process goes on, this is likely to happen less and less frequently. More and more people have also become shareholders in fishing firms. The holders of quotas, although much less vocal than the opponents of the ITQ system, may, in the end, be a much stronger interest group. They have a special interest in the system that is clear and concentrated, whereas the interest of each taxpayer in a small share in the revenue from a possible resource rent is rather weak.

When the evolution of the ITQ system is studied, it becomes clear that it would never have been introduced if it had not been in the interest of owners of fishing capital to accept it. The ITQ system was politically possible, unlike a special resource-rent tax or a government auction of quotas, because it did not work against the private interests of vessel owners. It is no worse for that. Economists since Adam Smith have told us that there is nothing wrong with private interest, if and when it coincides with the common good. It is the great advantage of the ITQ system in the fisheries that it directs the private interest of each vessel owner towards the public interest in profitable fisheries and conservation of fish stocks.

3.4 Possible future developments

If a resource-rent tax were imposed on the Icelandic fisheries it would be a double irony. First, the fisheries problem was that of harmful effects of economic activity. The over-capitalisation and excessive fishing effort, leading to dissipation of the resource rent arose because vessel owners did not operate under an efficient set of rules. The ITQ system enabled them to escape from this 'tragedy of the commons' and to capture the rent previously dissipated. If government would then remove the rent by a tax it would have replaced one set of harmful effects for vessel owners, namely rent dissipation in the form of over-capitalisation and excessive fishing effort, with another, namely the tax. What is the point of legislation if not to remove harmful effects of economic activities possible for those who were, in the first place, affected by those harmful effects? Second, much of the revenue from such a tax would be dissipated through the efforts of interest groups to secure part of it for themselves. Rent dissipation offshore through the costly process of over-investment in the fisheries would be replaced with rent dissipation onshore through the costly process of political redistribution.

Be that as it may, the Icelandic government, in response to the public dissatisfaction with the ITQ system appointed two commissions to make suggestions on improvements of the ITQ system and on other aspects of resource management. The work of these two commissions was delayed by these court cases, first on the constitutionality of the fishing permits, and then of the ITQs. But, after the decision by the Supreme Court in the spring of 2000 that the ITQ system was constitutional, the two commissions started again and should deliver their final reports in late 2000 or early 2001. It is difficult to predict what recommendations the two commissions will make, and also which, if any, of their recommendations Parliament will accept. The evolution of the ITQ system in New Zealand since its introduction in 1986, which in many ways parallels that in Iceland, may offer some guidance.

Initially, the New Zealand system differed from that of the Icelandic in two respects. First, vessel catch quotas were issued in terms of tonnes, not fractions of the TAC for each fish species, the idea being that government would buy or sell quotas to make up for changes in the annual TACs. Second, government imposed a resource rent tax on quota holders. Both those measures were later abandoned apparently because the government felt that closer co-operation with fishing firms was necessary. The quotas became TAC-shares as in Iceland; and a cost recovery charge replaced the resource rent tax (Major 1999). The rule now in New Zealand is that fishing firms bear the full costs of administering and enforcing the ITQ system. This is also a possible, and likely, outcome of the process of reconciling the public in Iceland with the ITQ system.

If a cost recovery charge would be imposed on Icelandic quota holders, presumably they would also get a larger say in the administration and enforcement of the system, which would enhance their sense of responsibility for the resource. It would be an important step towards the self-management of the fisheries and probably also serve to strengthen the rights of quota holders. At present, their rights are imperfect, not only because of the uncertain long-term status of the quotas, but also because those rights are narrow in scope, being by definition rights of extraction rather than property.

In the near future, the two most important tasks in ITQ fisheries systems will be to find ways of setting TACs in different fish stocks to enhance conservation - not at the maximum sustainable yield, but at the somewhat lower level of maximum profitability - and to create incentives to increase the value of the fish stocks. These two tasks can only be undertaken by real stakeholders in the fisheries. One of the main arguments for private property rights is that owners have strong incentives to experiment and innovate in the utilisation of their resources. New techniques in fencing and branding, and in fertilising fishing grounds or genetically improving individual fish, might make fish stocks much more valuable than they are now (De Alessi 1998). Instead of being hunters and gatherers, fishermen might become cultivators. A process of such experiment and innovation in the fisheries is not likely, however, to take place unless ITQs are strengthened into some forms of legally recognised private property rights.

4. LITERATURE CITED

Arnason, R. 1990. Minimum Information Management in Fisheries. Canadian Journal of Economics. **23**:630-53.

Arnason, R. 1994. On Catch Discarding in Fisheries. Marine Resource Economics. **9**:189-208.

Buchanan, J.M. 1959. Positive Economics, Welfare Economics, and Political Economy. Journal of Law and Economics. **2**:124-38.

Buchanan, J.M. 1997. Who Cares Whether the Commons Are Privatized? Post-Socialist Political Economy. Selected Essays, Cheltenham: Edward Elgar. pp.160-7.

Coase, R.H. 1960. The Problem of Social Cost. Journal of Law and Economics. **3**:1-44.

De Alessi, M. 1998. Fishing for Solutions. IEA Studies on the Environment No. 11. London: IEA Environment Unit.

Demsetz, H. 1967. Toward a Theory of Property Rights. American Economic Review, Papers and Proceedings. **57**:347-59.

Gissurarson, H.H. 1983. The Fish War: A Lesson from Iceland. The Journal of Economic Affairs. **3**:220-3.

Gissurarson, H.H. 1990. Fiskistofnarnir vid Island: Thjodareign eda rikiseign? Reykjavik: Stofnun Jons Thorlakssonar.

Gordon, H.S. 1954. The Economic Theory of a Common Property Resource: The Fishery. Journal of Political Economy. **62**:124-42.

Gylfason, T. 1990. Stjorn fiskveida er ekki einkamal utgerdarmanna. Th. Helgason and O. Jonsson (ed). Hagsaeld i hufi, Reykjavik: Haskolautgafan og Sjavarutvegsstofnun Haskolans. pp.120-5.

Hannesson, R. 1994. Trends in Fishery Management. Managing Fishery Resources, World Bank Discussion Paper No. 217. E.A. Loyayza (ed).

Hardin, G. 1968. The Tragedy of the Commons. Sci. **62**:1243-8.

Johnson, R.N. 1995. Implications of Taxing Quota Value in an Individual Transferable Quota Fishery. Marine Resource Economics. **10**:327-40.

Johnson, R.N. 1999. Rents and Taxes in an ITQ Fishery. In R. Arnason and H.H. Gissurarson (eds.), Individual Transferable Quotas in Theory and Practice. pp. 205-13. Reykjavik: University of Iceland Press.

Jonsson, B.B. 1975. Audlindaskattur, idnthroun og efnahagsleg framtid Islands. *Fjarmalatidindi*,.**22**:103-22.

Jonsson, H. 1990. Akvardanataka i sjavarutvegi og stjornun fiskveida. *Samfelagstidindi*. **10**:99-141.

Libecap, G.D. 1989. Contracting for Property Rights. Cambridge University Press.

Lindal, S. 1998. Nytjastofnar a Islandsmidum — sameign thjodarinnar. In H.H. Gissurarson *et. al.* (eds.). *Afmaelisrit David Oddsson fimmtugur*. Reykjavik: Bokafelagid. pp.781-808.

Major, P. 1999. The Evolution of ITQs in the New Zealand Fisheries. In R. Arnason and H.H. Gissurarson (eds.). Individual Transferable Quotas in Theory and Practice. pp. 81-102. Reykjavik: University of Iceland Press.

Moller, M. 1996. Fyrirkomulag veidileyfagjalds. Visbending. 29 February.

Morgunbladid 2000. 80% kvotans hafa skipt um hendur. 18 March.

Runolfsson, B. T. (1999): 'ITQs in Iceland: Their Nature and Performance'. In R. Arnason and H.H. Gissurarson (eds.), *Individual Transferable Quotas in Theory and Practice*, pp. 103-140. Reykjavik: University of Iceland Press.

Scott, A. (1955): 'The Fishery: The Objectives of Sole Ownership', *Journal of Political Economy*, Vol.63, pp.116-24.

INTRODUCING PROPERTY RIGHTS INTO FISHERIES MANAGEMENT: GOVERNMENTS CANNOT COPE WITH IMPLEMENTATION ALONE

T. Craig
New Zealand Seafood Industry Council
Private Bag 24-901, Wellington, New Zealand

1. INTRODUCTION

New Zealand's commercial fisheries have been managed under a property rights system - known as the Quota Management System (QMS) - since 1986. The introduction of the QMS was a radical change, from the previous free-for-all competitive fishing regime with some regulatory control, to a property rights system where each fisher has a right to take a defined share of the total catch.

Since its introduction, the QMS has improved certainty and security for all participants in the fishing industry and there has been substantial growth in the seafood processing and marketing sectors. There is a general recognition that the QMS has played a significant role in improving the biological status of the fisheries resource and commercial return to fishers (Annala 1996). Nevertheless, it is questionable whether New Zealand's fisheries management system allows the ecological and economic potential inherent in a rights-based fisheries management system to be fully realised. This paper argues that in itself, a property rights system - although an essential prerequisite of an biologically sustainable and economically viable fisheries management regime - does not provide all the tools and mechanisms necessary to achieve successful fisheries management. And, that by itself, the government cannot develop and implement reforms to fisheries property rights systems - the direct involvement of the rights holders is essential.

Countries implementing rights-based regimes to address problems such as depleted fisheries resources, excessive fishing capacity, low incomes for fishers, heavy dependence on government support and regulation, and conflict among fishing groups, have watched the introduction, evolution and use of the QMS in New Zealand with interest. The New Zealand seafood industry has identified a number of lessons that can be learnt from the implementation of property rights fisheries management systems. The three main lessons are the need to:

i. define, at the outset, clear, appropriate and enforceable rights and responsibilities for all users of fishery resources
ii. clearly define and separate the roles of the government as sustainability manager from the roles of rights holders as fisheries managers, and
iii. understand that successful property rights management can only be achieved through "bottom up" or local initiatives.

Some of these issues are starting to be addressed through recent reforms to New Zealand's fisheries legislation, but there is still a considerable way to go. So long as the implementation of the property rights regime remains largely in the hands of the government, it will continue to be subject to both political interference and bureaucratic resistance to change and will fail to capture the full benefits that come with durable, flexible and exclusive rights and responsibilities.

The remainder of this paper explores in more detail these three lessons in the context of the implementation of New Zealand's fisheries management regime and draws some conclusions about the respective roles of governments and fisheries rights holders in the implementation of property rights fisheries management regimes. First however, a brief background on New Zealand's QMS is provided.

2. NEW ZEALAND'S QUOTA MANAGEMENT SYSTEM

New Zealand's *1996 Fisheries Act* gives fisheries management the dual purpose of providing for the utilisation of fisheries resources while ensuring sustainability. Under the Act "Ensuring sustainability" means -(a) maintaining the potential of fisheries resources to meet the reasonably foreseeable needs of future generations; and (b) avoiding, remedying, or mitigating any adverse effects of fishing on the marine environment. "Utilisation" means conserving, using, enhancing and developing fisheries resources to enable people to provide for their social, economic and cultural well-being. Most of New Zealand's commercial fishing (85% of the total known fish catch in the Exclusive Economic Zone) occurs under the QMS.

The founding aims of the QMS reflect the dual purpose of the legislation. They include:

i. rebuilding inshore fisheries where required and ensuring that catches are limited to levels that could be sustained over the long term, and
ii. ensuring that catches are harvested efficiently with maximum benefit to the industry and to New Zealand (Luxton 1997).

For each fishstock covered by the system a Total Allowable Catch (TAC) covering commercial, recreational and customary Maori fishing activity is set and reviewed annually. From this, a Total Allowable Commercial Catch (TACC) is established. Commercial fishers acquire rights to harvest fish by being allocated (on the basis of catch history), purchasing or leasing an Individual Transferable Quota (ITQ). ITQs are allocated in perpetuity and can be bought, sold and leased. Quotas are expressed as a proportion of the TACC for each fishery and therefore change as the TACC increases or decreases in response to the assessed health of the fishery.

The QMS currently covers 33 species (over 180 fishstocks) and it is the government's intention to bring more fish into the system over the next few years. To assist in the management of the fisheries, a number of Quota Management Areas (QMAs) have been set up for each species in the QMS. Each QMA corresponds with a particular fishstock - for example, snapper is managed as 6 separate fishstocks in 6 QMAs. A fisher with access to quota for a particular fishstock may harvest the fish anywhere within the relevant QMA unless there are areas, such as marine reserves or areas closed to protect juvenile fish, from which commercial fishing has been excluded. Further discussion of New Zealand's QMS can be found in Clark *et al.* (1988), Dewes (1989), Memon and Cullen (1992), Sissenwine and Mace (1992).

Many aspects of the *1996 Fisheries Act* - designed to refine aspects of the QMS - have yet to be implemented. A recent independent review of the legislation (Hartevelt 1998) found that if this Act were to be implemented in its current form, it would be highly likely that:

i. the purpose of the Act - to provide for the use of fisheries resources while ensuring sustainability - would be undermined
ii. significant compliance costs associated with administering the Act would be imposed on commercial fishers, and yet this cost cannot be justified by any additional cultural, social or environmental gains from the Act, and
iii. the fisheries management regime: (a) would be highly centralised and inflexible, (b) would be contrary to Government's decision to devolve the delivery of non-core government fisheries services to the fishing industry; and (c) would remove incentives for all stakeholders to take a constructive role in the management of the national fisheries resources (Hartevelt 1998).

Following on from this review, the government initiated a number of changes to the 1996 Act, but these changes go only part of the way towards addressing some of the concerns expressed by the reviewer (and shared by the seafood industry). For example, under the recent reforms the provision of fisheries registry services has been devolved to the industry, industry groups are able to purchase directly from providers some required fisheries services (*e.g.* research), and area-based fisheries plans are able to be prepared. Fisheries management in New Zealand could therefore be said to be in a state of limbo. On the one hand, the government is still equipped with a full range of fisheries regulatory mechanisms but is increasingly reluctant to employ them without consensual support from rights holders. On the other hand, fishers - especially ITQ owners - face increasingly strong incentives to manage their own affairs but, on the whole, lack mechanisms to make rules, collect funds and purchase most management services, except on a totally voluntary basis.

3. THREE LESSONS LEARNT
3.1 Definition of rights

So, what lessons can the international fishing community learn from the evolution and implementation of a property rights based fisheries management system in New Zealand?

Lesson 1: Define clear, appropriate and enforceable rights and responsibilities for all users of fisheries resources. It is essential that, from the outset of the implementation of the property rights regime, the rights and responsibilities of all fishers are clear, appropriate (*i.e.* reflect the interests of the fishers, contribute to sustainable use) and enforceable.

In New Zealand's commercial fisheries, although ITQs are relatively well specified compared to the rights of other fisheries stakeholders, the ability of commercial fishers to fully exercise their rights remains restricted. As Copes (1986) notes, rights to the fish stock bestowed by the individual quota - even in the form of ITQ - are still far from fully specified property rights.

Four problems can be identified. First, in spite of the ability of the Total Allowable Catch mechanism to achieve a sustainable level of harvest, many regulations (currently over 4000, many from before the advent of the QMS) governing how and when fishing activity takes place, still exist and restrict the ability of fishers to determine how they will harvest their share of the catch. Second, with a few exceptions, the retention of management rights by the government has prevented quota owners from taking greater responsibility for fisheries decision making, implementation and enforcement. Third, the quota forfeiture provisions of the fisheries legislation mean that ITQ are not the bankable asset that fishers originally anticipated they would be. This has reduced the incentive for fishers to undertake long-term investment in the well being of the fishery. Fourth, ITQ rights are non-exclusive - *i.e.* other groups of rights holders (*e.g.* recreational and customary) can harvest fish from fisheries covered by ITQs - and this can create access conflicts between different user groups.

Marine farming rights relate to the occupation of areas of seabed granted under Resource Management legislation in combination with a relatively poorly defined right to harvest farmed fish from the marine farm structures. These marine farming harvest rights are not well integrated with harvest rights for wild fisheries and confusion over the boundaries between Resource Management and Fisheries legislation means that access issues can sometimes arise between marine farmers and commercial fishers.

Customary Maori marine fishers have territorial use rights which are held by the *iwi* (tribe) occupying the adjoining land. Harvest rules can be developed and observed by the holders of customary rights. However, in most cases these rights are not exclusive and so they are affected by, and in turn affect, extractions from the same stock by other sectors.

Recreational marine fisheries operate as open access fisheries, subject to lightly enforced regulations. These rights are by far the most poorly defined of New Zealand's various fishing rights. Recreational fishers have no certainty about their right to a share of a fishery or the right to access that share. Recreational fishing rights are therefore easily eroded by government and by commercial and customary fishers and other coastal users. And, they are often inconsistent with commercial and customary stewardship regimes, and are poorly enforced. The uncertain nature of the right also means that there is little incentive for recreational fishers to act co-operatively or to invest in the sustainability of the fishery.

Apart from specifying clear, appropriate and enforceable rights where different groups of fishers are active in a fishery (commercial, recreational, customary, marine farming) their rights must also be well integrated. Integration is especially critical in many inshore fisheries where there is competition between fishers for the resources. Integration does not necessarily mean that the rights should be specified identically - for instance, transferability of rights between individuals may not be an issue for recreational and customary Maori fishers. Existing fisheries legislation mechanisms for dealing with integrating different types of rights (*e.g.* areas closed to commercial fishing), are blunt instruments for dealing with conflict and fail to provide incentives for co-operation between fishers. More often than not they result in an intensification of antagonistic relationships between government and fishers, and between fishers.

In summary then, failure to define clear, appropriate and enforceable rights and responsibilities for all fishers can increase the risk of:

i. Conflicts between different groups of rights holders over their share of a fishery and their ability to access their share. This risk is exacerbated by the absence of incentives and mechanisms to encourage different groups of rights holders to work out durable solutions where conflicts arise.
ii. Gradual erosion of fishers' ability to access fisheries in order to exercise harvest rights through competition for space with both other fisheries activities and non-fisheries activities (*e.g.*, marine reserves and other exclusive uses of coastal space). This risk is related to the non-exclusive nature (in terms of spatial allocation) of most types of fishing rights.
iii. Failure of fishers to develop an ethic of "responsible management" of the fishery resource because they lack the opportunity to collectively manage the fisheries in which they have harvesting rights. Without exclusive rights, fishers have little incentive to curb their actions as they know that their conservation efforts will be ineffective and serve only to swell the catch of other fishers. This problem is exacerbated by the lack of integration of commercial, recreational and customary rights, which means there is limited scope beyond direct government intervention and regulation for mechanisms to bind existing and new fishers to agreements reached between the various groups of fishers.

3.2 Roles of government and industry

Lesson 2: Clearly define and separate the roles of the government as sustainability-manager from those of rights holders as fisheries-managers.

New Zealand's fisheries management regime currently entails the government intervening extensively in detailed areas of operational management and enforcement. The result is a highly centralised management system. In part this reflects the old industry structure that emphasises the role of the government as referee, monitoring and enforcing the activities of small fishers. The focus on detailed management means that governmental fisheries management is currently largely input driven and as a result has become distant from the objective of fisheries management - to provide for the utilisation of fisheries resources while ensuring sustainability.

The recent independent review of the fisheries legislation found that these centralised management regimes were becoming increasingly complex, inflexible, unworkable and costly to administer. The reviewer concluded that the regulatory and management principles of the fishing sector need to be aligned with other areas of the economy, based on efficient allocative mechanisms involving both self-management and a less interventionist approach by the government (Hartevelt 1998, p.27).

This view is reinforced by developments within the fishing industry. For instance, the industry's management resources and skills are evolving and the industry is making advances in managing fishing stocks on a sustainable basis, including taking responsibility for avoiding, remedying or mitigating any adverse effects of fishing. Even in the absence of appropriate legislation and devolution of management responsibilities, quota owners have begun to organise themselves into quota management associations and similar fisheries management organisations. For instance, the NZ Seafood Industry Council now recognises 16 commercial fisheries stakeholder organisations and by March 2000 we expect that a further 4 stakeholder groups will establish themselves. Together these groups will represent the interests of 95% of all quota owners. Some of these quota owner associations are already operating successfully on a collective basis and taking on significant management responsibilities. The example of the Challenger group of companies is discussed in more detail below. Responsibility for the management of commercial fisheries registry services (an essential component of the operation of the QMS) is in the process of being devolved from the government to the industry.

Another characteristic of the government's active role in fisheries management has been that it tends to divide stakeholder groups, resulting in conflict between the groups and forcing the government into the role of arbitrator of disputes. This means that fisheries management continues to be subject to intense political input from various interest groups. Both the bureaucratic, inflexible nature of government fisheries management and its susceptibility to political interference increase the risk of unsustainable fisheries management outcomes.

Hartevelt recommended that the government focus on fisheries management outcomes rather than inputs, and have responsibility for:

i. establishing the sustainability and overall management framework for the utilisation of fisheries resources (including continuing to meet Treaty of Waitangi and international obligations)
ii. allocating and ensuring the integrity of fisheries harvesting rights
iii. facilitating and encouraging rights holder-based management
iv. establishing information requirements for the sustainability, framework, co-ordinating the collection and dissemination of relevant information to fisheries management stakeholders, and
v. monitoring and ensuring the integrity of sustainability and fisheries management frameworks and supporting systems and services.

Under this model - referred to as a "co-management" regime - commercial harvesters would be responsible for managing commercial fishing and aquaculture harvesting activities within the sustainability and management frameworks established by the government. Recreational and customary fishers would have similar responsibilities for their sectors of the fisheries.

All interests in fisheries will need to change their current attitudes to fisheries management if co-management is to be implemented successfully (Hartevelt 1998, p. 32). The government will have to be prepared to move away from its current micro-management role and focus on an outcome-oriented, framework setting and monitoring role. This has significant implications for the size and structure of the Ministry of Fisheries. Fishers will have to become more pro-active in the development of longer term plans for fishing activity and take increased responsibility for the stewardship of the resource to which they have harvesting rights. They will have to come to terms with a new operating environment and recognise the opportunities and responsibilities associated with ownership of harvest rights and management of harvesting activity. Finally, all stakeholders and interest groups will have to give greater recognition to the rights and responsibilities of customary, recreational and commercial fishers and to the wider interests of society in the sustainable utilisation of the fisheries resource.

The recommended realignment of the roles of the government and fisheries rights holders has been only partially implemented by the recent reforms to the Fisheries Act. Although mechanisms such as fisheries plans, which can be prepared by rights holder groups to facilitate local area-based management of fisheries, are a step forward the government still retains many of the management functions that could be carried out more effectively and efficiently by rights holders. From the industry's perspective, further realignment of roles must continue, including devolution of day-to-day management functions to the rights holders within a wider sustainability framework established by the government. Failure to do this will mean that:

i. fishers will be required to continue to meet the costs of an overly complex and bureaucratic management system
ii. fisheries management will continue to be constrained by the "single-model" approach required by centralised management, and there will be few incentives for flexible, fishery-specific or area-specific management systems to be developed, and
iii. the dual objectives of the fisheries legislation - efficient use and sustainability - may not be able to be met in the most effective manner.

3.3 Need for stakeholder input

Lesson 3: Successful property rights management can only be achieved through "bottom up", localised initiatives.

Because of the centralised nature of fisheries management in New Zealand, a "one model fits all" approach to fisheries management tends to be applied regardless of the nature of the particular fishery in question. Flexibility of management options and incentives to adopt innovative approaches specific to local areas or particular fishstocks are limited.

The management mechanisms of the QMS tend to operate at a macro level - *i.e.* on the scale of Quota Management Areas. This scale of management is not suitable for addressing local issues which commonly occur in inshore, mixed species fisheries, such as access arrangements between various user groups. The QMS also generally treats fishstocks on an individual basis - an approach that is not always sufficient for multi-species fisheries. Further, government attitudes to management have not always acknowledged, or built upon, the fact that fisheries management is heavily dependent on the positive and willing involvement of commercial fishers and requires the co-operation of industry both as a source of information about the effectiveness of management initiatives and for compliance with the management regime.

The main lesson to be learnt here is that the establishment of a property rights regime is not the be-all and end-all of successful fisheries management - it needs to be overlaid with, and complemented by rights holder-driven, locally-based management initiatives. In relation to ITQ, Copes (1986) comments that the advocates of individual property probably have made too much of the property rights aspects of the scheme and goes on to say that experience so far suggests that we should be non-dogmatic in our choice of management technique and that we should select from the array of available fisheries management devices the combination that is most beneficial and least deficient in any particular set of circumstances.

It is therefore misleading and limiting to think of the QMS as a single management approach; different fisheries management regimes will suit different fisheries and local conditions - biological, economic, social and cultural. Within the basic framework of the QMS, a flexible, non-centralised approach to the management of particular fisheries is required.

In general, the government is not in the best position to select the most appropriate fisheries management techniques for particular fisheries. This choice is best made, within agreed sustainability specifications set by the government, collectively by the rights holders themselves. Far from being incompatible with an ITQ based management system, collective action is reliant on the allocation of ITQs to help provide a framework within which devolution and decentralisation can occur. As Scott (1993) notes, in many fisheries the ITQ will be less a new instrument of regulation, less a kind of individual property right, than a membership card in a self-governing fishery group.

There is therefore considerable scope for area-based fisheries management planning to supplement the basic QMS regime at a local level, and indeed this approach has been facilitated in the recent reforms to the Fisheries Act through provision for fisheries plans. Fisheries plans provide a mechanism for the various rights holder groups to collectively develop a multi-year approach to managing fishstocks, thereby improving sustainability outcomes and reducing conflicts between stakeholder groups. Ultimately however, until the actual management responsibilities set out in such plans are devolved from the government to the rights holders themselves, fisheries plans will fail to be as effective as they could be in facilitating effective fisheries management.

4. TWO OTHER FACTORS INFLUENCING THE IMPLEMENTATION OF PROPERTY RIGHTS REFORMS

Progress is slowly being made towards improving the definition of property rights and fine-tuning the operation of the QMS. There is also considerable momentum-building for the transition to a "co-management" regime where government and rights holder roles are clearly and appropriately defined. However, two factors are still undermining the implementation and future development of the property rights regime:

i. the susceptibility of government-driven fisheries management regimes to political interference, and
ii. bureaucratic resistance to change.

There is no denying that fisheries management is a political issue - it has to have regard to and give effect to various conflicting interests, values and world views. As noted above, government-run fisheries management regimes tend to divide stakeholder groups resulting in conflict both among stakeholder groups and with the government. A more efficient and effective means of resolving disputes is for stakeholders to talk directly with each other. For this to occur, all stakeholder groups must have well defined rights and shares in the fishery as well as properly mandated groups to represent their interests. In such a regime the government would no longer have to act as referee or arbitrator. Fisheries management decisions would become less dependent on lobbying and position taking and more dependent on a co-operative, negotiated approach to developing robust management policies based on sound science.

Another barrier to reforming and implementing fisheries property rights regimes is bureaucratic resistance to change. In particular, the seafood industry needs to be able to convince the government and its officials of the desire and ability of fishers to manage their own interests in the fishery. To do this, the industry needs to demonstrate a high level of organisational and management ability. It also needs to demonstrate that it is capable of assuming the stewardship responsibilities that go alongside its harvest rights.

In the context of resistance to change, it is interesting to note that in spite of the lack of formally devolved management responsibilities, there are already examples in New Zealand of successful, localised, industry- driven collective approaches to the management of inshore fisheries. The most well known of these is the Challenger group of companies in the Nelson-Marlborough area which have responsibility on behalf of the quota owners for the management of all commercial inshore finfish and shellfish quota stocks (with the exception of rock lobster and paua) in Quota Management Areas 7 and 8.

The Southern Scallop Fishery managed by the Challenger Group has been lawfully exempted from the normal sustainability criteria which apply to other similar fisheries. The scallop fishery is now managed under rotational fishing and enhancement programme whereas other similar fisheries are managed by a Total Allowable Catch set at an estimate of the Maximum Sustainable Yield for the fishery. Even in its development phase, when it was faced with high reporting obligations and information requirements, the Challenger group was still able to operate a more cost effective and better targeted management framework than would be achieved if the government had undertaken a centralised and direct management role in the fishery (Harte *et al.*, 1998).

While critics of industry self-management sometimes hold that the Challenger companies are an aberration rather than a model that could be applied more widely to other fisheries, the fishery in QMAs 7 and 8 is in fact more complex than most. It is an inshore multi-species fishery with large numbers of quota owners and significant non-commercial (recreational and customary) interests. It also operates in an area with high natural values and considerable marine farming development. As such, it could be argued that a rights holder-based approach to management could be implemented just as effectively, if not more so, in some of New Zealand's other fisheries where there are fewer quota owners with larger holdings and no, or few, recreational and customary Maori interests (*e.g.* some deepwater fisheries).

5. CONCLUSION - THE ROLE OF GOVERNMENT

It should be clear from the above discussion that the government cannot successfully implement property rights fisheries management systems on its own - the industry must play a central role, particularly in the development and implementation of day-to-day fisheries management regimes within the broader framework of the QMS and sustainability objectives. It is appropriate for the government (in consultation with all interested parties) to:

i. set the framework for property rights reforms
ii. make a range of appropriate tools available, and
iii. then step back and allow a bottom-up approach, initiated by rights holders, to develop within the established framework.

Those rights holder groups who are appropriately motivated, funded and skilled will then take on management responsibilities at their own initiative. The government's motto might be: "initiate nothing, motivate no-one". Its focus should be on the fisheries management outcomes sought by society as a whole (security of property rights, sustainability of fisheries resources, Treaty obligations etc). Within this framework rights holder groups will be able to develop and implement a range of appropriate management mechanisms to achieve the agreed objectives - and these management mechanisms will be based on, but not necessarily be limited to, ITQ systems.

6. LITERATURE CITED

Annala, J. 1996. New Zealand's ITQ system: have the first eight years been a success or failure? Reviews in Fish Biology and Fisheries, 6, pp. 43-62.

Clark, I., P. Major and N. Mollet 1988. Development and implementation of New Zealand's ITQ Management System, Marine Resource Economics, 5, pp. 325-349.

Copes, P. 1986. A Critical Review of the Individual Quota as a Device in Fisheries Management, Land Economics, 62(3), pp. 278-291.

Dewes, C. 1989. Assessment of the Implementation of Individual Transferable Quotas in New Zealand's Inshore Fisheries, North American Journal of Fisheries Management. 9(2), pp. 131-139.

Harte, M., M. Arbuckle and T. McClurg (in press). Property rights and the evolution of fisheries management in New Zealand. Forthcoming, in Private Rights and Public Benefits: Proceedings of the Environment and Property Rights Conference, Lincoln University, Canterbury New Zealand, November 1998.

Hartevelt, T. 1998. Fishing for the Future: Review of the Fisheries Act 1996. Report of the Independent Reviewer of the Fisheries Act 1996 to the Minister of Food, Fibre, Biosecurity and Border Control.

Luxton, J. 1997. Stakeholder management of recreational fisheries. Address to the Recreational Fishing Council Annual General Meeting, Bay of Islands, July.

Memon, A.P. and R. Cullen 1992. Fisheries policies and their impact on the New Zealand Maori. Marine Resource Economics, 7, pp. 153-167.

Scott, A. 1993. Obstacles to Fishery Self-Government. Marine Resource Economics. 8, pp.187-199.

Sissenwine, M.P. and P.M. Mace 1992. ITQs in New Zealand: The era of fixed quota in perpetuity. Fisheries Bulletin, 90, pp. 147-160

THE COMMON FISHERIES POLICY OF THE EUROPEAN UNION AND FISHING RIGHTS

C. Nordmann
European Commission
200, rue de la Loi, Brussels - B1049 Belgium
<christoph.nordmann@cec.eu.int>

1. EUROPEAN POLICY

The Fisheries Policy of the European Union is one of the few real "common" policies, meaning that competence in this field has been completely transferred to the Union and Member States therefore conserve competence only so far as the Union does not legislate or when it delegates part of its competence expressly to Member States. In addition to that, Member States normally manage the day to day implementation of common rules, as the Union does not have local or regional administrations.

2. DEVELOPMENT OF MANAGEMENT POLICY

Management and conservation of fish resources has been expressly indicated as one of the tasks of the Common Fisheries Policy since the accession to the European Community of the United Kingdom, Denmark and Ireland in 1973. Nevertheless, it took nearly ten years before a complete system could be established at the Community level in 1983. This long period of time shows how difficult it was to find a compromise between the existing national schemes and the diverging interests of the different Member States' fishing industries. The two most crucial problems to be solved (which periodically reappear in political discussions) were the rights of access to waters and the question of allocating catching possibilities among the fleets of the Member States.

3. ACCESS RULES

The Community, with its original Members, France, Germany, Italy, Belgium, the Netherlands and Luxembourg, did not provide for access limitations for vessels flying the flag of one of the Member States. Access was, in principle, free up to Member States' beaches before 1973. In the Accession Treaty of the United Kingdom, Denmark and Ireland, a provision opened the possibility for Member States to restrict access up to 6 miles, to the benefit of the national fleet only, for a transitional period of ten years. In the 1983 Community management scheme, this zone was extended to 12 miles for an additional 10 year period, while guaranteeing the continuation of the historical fishing rights of vessels from other Member States. This regime was maintained under the revised scheme in application since 1992.

4. ALLOCATION OF FISHING POSSIBILITIES

The distribution of catching possibilities has been governed since 1983 by the so-called "relative stability", which consists of a permanent allocation formula applied to stocks under TAC and quota arrangements. This key was determined according to historical catching data. Fishing was free (outside the 12 miles zone) for species or stocks for which no TAC and quotas were established. For some stocks, only TACs were fixed. In this case fishing again was free as long as the TACs were not exhausted at which point the fishery was closed.

The situation in the Mediterranean Sea is quite unique because the European Union's Member States do not claim an EEZ in this area (recently Spain only declared a fisheries protection zone of 50 miles). In addition, the continental shelf is very narrow and the main fisheries resources are inside the zone to which access is restricted to the benefit of the national fleets. A TAC (and as a consequence, quotas) has only recently been established for tuna fishing in accordance with decisions by ICCAT. The main instruments for regulating fisheries in this area are therefore effort-control and technical measures.

5. THE 1992 BASIC REGULATION

The last thorough review of the Common Fisheries Policy took place in 1992, ten years after the adoption of the first comprehensive management system. The next review has been scheduled for 2002, again after a ten year period, as foreseen by the basic fisheries Regulation.

The 1992 reform did not modify the basic elements (access, TACs and quotas, relative stability), but tried to modernise the system, taking into account world-wide developments in fisheries management and to achieve a more coherent and flexible regime.

6. MANAGEMENT OBJECTIVES

The 1992 regulation describes the objectives as follows:

"As concerns exploitation activities the general objectives of the common fisheries policy shall be to protect and conserve available and accessible living marine aquatic resources, and to provide for rational and responsible exploitation on a sustainable basis, in appropriate economic and social conditions for the sector, taking account of its implications for the marine eco-system, and in particular taking account of the needs of both producers and consumers."

With regard to the implementation it is stipulated that:

i. "Management objectives" (may be established) "on a multiannual basis, for each fishery or group of fisheries in relation to the specific nature of the resources concerned. Where appropriate these shall be established on a multi-species basis. Priority objectives shall be specified including, as appropriate, the level of resources, forms of production, activities and yields;"

ii. "for each fishery or group of fisheries where management objectives have been set, management strategies (shall be established), where appropriate on a multiannual basis, to achieve the management objectives including the specific conditions under which exploitation activities shall be pursued."

7. MANAGEMENT INSTRUMENTS

More detailed specifications state that:

"In order to ensure the rational and responsible exploitation of resources on a sustainable basis,...Community measures laying down the conditions of access to waters and resources and of the pursuit of exploitation activities." (shall be established) "These measures shall be drawn up in the light of the available biological, socio-economic and technical analysis"...

"These provisions may, in particular, include measures for each fishery or group of fisheries to:

i. establish zones in which fishing activities are prohibited or restricted
ii. limit exploitation rates
iii. set quantitative limits on catches
iv. limit time spent at sea taking account, where appropriate, of the remoteness of the fishing waters
v. fix the number and type of fishing vessels authorised to fish
vi. lay down technical measures regarding fishing gear and its method of use
vii. set a minimum size or weight of individuals that may be caught, and
viii. establish incentives, including those of an economic nature, to promote more selective fishing."

Furthermore, the regulation provides for the obligation, for each Member State, to operate a national system of fishing licences for which minimum requirements are established at the Community level.

8. TACS AND THEIR ALLOCATION

The 1992 regulations also specify concrete measures to be adopted:

i. "shall determine for each fishery or group of fisheries, on a case-by-case basis, the total allowable catch and/or total allowable fishing effort, where appropriate on a multi-annual basis. These shall be based on the management objectives and strategies where they have been established in accordance with paragraph 3;

ii. shall distribute the fishing opportunities between Member States in such a way as to assure each Member State relative stability of fishing activities for each of the stocks concerned; however, following a request from the Member States directly concerned, account may be taken of the development of mini-quotas and regular quota swaps since 1983, with due regard to the overall balance of shares;

iii. shall, where the Community establishes new fishing opportunities in a fishery or group of fisheries not previously prosecuted under the common fisheries policy, decide on the method of allocation taking into account the interests of all Member States;

iv. may also, on a case-by-case basis, determine the conditions for adjusting fishing availabilities from one year to the next;

v. may, based on scientific advice, make any necessary interim adjustments to the management objectives and strategies."

9. MULTI-ANNUAL GUIDANCE PROGRAMS

The other instruments in use to regulate fishing effort are the multi-annual guidance programs for the fishing fleets of the Member States which are decided by the European Commission on the basis of objectives set for "re-structuring the Community fisheries sector with a view to achieving a balance on a sustainable basis between resources and their exploitation". These multi-annual guidance programs fix maximum levels for the fishing capacity and/or fishing effort, specific segments of the national fleets have to reach at the end of the program's period, and prescribe intermediate levels of reduction to be respected.

10. IMPLEMENTATION BY MEMBER STATES

The rules Member States apply for the domestic allocation of national fishing possibilities, decided at the Community level, remain the basic responsibility and competence of Member States, but they have to be in conformity with Community law and the Common Fisheries Policy rules. Member States have to inform the European Commission, each year, of their allocation criteria and the detailed rules for the use of fishing possibilities.

In fact, these criteria and rules differ greatly from one Member State to another not only because of the variety of fishing traditions and patterns but also because of the different political and socio-economic options which are not subject to common rules.

The only Member State applying a straightforward ITQ system is the Netherlands. In some other Member States, systems are applied which are quite close to individual transferable quotas in practical and economic terms, as licences are sold with the attached quota allocation, even if there are no legally recognised property rights and no guarantee that the future allocations will follow the existing pattern. The majority of Member States still keep closer to the traditional view of fisheries as a common resource. Others are looking for intermediate solutions.

11. THE 2002 REVIEW OF THE COMMON FISHERIES POLICY

11.1 Present status

In general, the discussion of the possible introduction of individual transferable quotas has only started in the Community and probably will take some time before clear choices are made. The issue is part of the wide consultation process the European Commission has conducted over the last two years in preparation for the 2002 review of the Common Fisheries Policy for which proposals are expected at the beginning of 2001.

These proposals will be based on a report on the fisheries situation in the Community and, in particular, on the economic and social situation of coastal regions, on the state of the resources and their expected development, and on the implementation of the scheme adopted in 1992.

11.2 Consultation process

The consultation started with the issuing of a questionnaire, which was sent to 350 representative organisations and associations with an interest in fisheries in all the Member States of the European Union. The questionnaire contained 33 questions related to the different aspects of the Common Fisheries Policy ranging from access to waters and resources, to resource management and conservation, through to international co-operation, market policy and structural measures. The Commission received 175 replies to its questionnaire which were often critical of the Common Fisheries Policy and highlighted the main concerns of the fisheries sector and of the other interested groups with respect to the future of fisheries in the European Union.

The second phase of the consultation process on the Common Fisheries Policy after 2002 involved the organisation of 30 regional meetings in Member States with an agenda based upon the issues raised in the questionnaire.

11.3 The question of ITQs

The relevant question in the questionnaire was formulated as follows:

"At present quotas are assigned by the Council to the Member States, which make allocations from them to fishermen or their associations. It sometimes been suggested that quotas should be assigned directly to fishermen's organisations or to the fishermen themselves. They would then be able to trade them among themselves (ITQ (individual transferable quota) system).

Would ITQs have advantages over the present way of doing things? What would be the main difficulties in setting up the system? If ITQs are introduced who should administer quota transfer and utilisation?"

Most of those who replied or commented during the consultation exercise were **against** ITQs, their main arguments being:

i. concentration of fishing rights in a handful of enterprises
ii. monitoring problems
iii. difficulty in finding a reliable allocation system not dependent on the authorities of each Member State
iv. impracticability for mixed fisheries
v. incompatibility with Community principles such as "equal access", "shared resources" and "relative stability" and
vi. danger of overfishing.

Most Spanish, Dutch and Danish organisations were in favour of ITQs. Favourable views were also noted from Swedish, Finnish and Italian organisations. Arguments in favour of ITQs included:

i. more responsibility put on fishermen
ii. better matching of supply with demand and,
iii. advantages resulting if ITQs were tied to a co-management system.

Some organisations thought that, for some fisheries, quota transfers between enterprises in different Member States could be permitted under the supervision of the authorities of those Member States. A public authority supervisory role in quota transfers was also favoured by some organisations which supported ITQs. Some organisations expressed support for their present national quota allocation systems.

The general conclusion to be drawn from the consultation process is that it seems unlikely that a majority of Member States will opt for the introduction of individual transferable quotas into the Common Fisheries Policy. Unless major changes occur between now and 2001, the most probable outcome will be that the internal allocation of fishing rights will remain a matter of national choice for the Member States.

THE GLOBAL ENVIRONMENT FACILITY: A PARTNER IN THE SUSTAINABLE MANAGEMENT OF TRANSBOUNDARY FISHERIES

A. Merla
Global Environment Facility
1818 H Street N.W., Washington, D.C. 20433, USA
<amerla@worldbank.org>

1. INTRODUCTION

The Global Environment Facility was established in 1990 as a pilot financial mechanism to support global environmental protection in four "Focal Areas": Climate Change, Biodiversity, International Waters, and Depletion of the Ozone Layer. In March 1994 it became a permanent mechanism to forge international cooperation and fund projects addressing global environmental problems. It draws its strength from the commitment of donor member countries, which have so far allocated over 4 billion dollars for GEF financed projects. It builds on the different skills, experience and organizational structures of its Implementing Agencies, the World Bank, UNDP and UNEP, and of a wide range of Executing Agencies: Regional Development Banks, UN Agencies, NGOs, the private sector, national Governments. Recipient countries participate on equal terms to its governing body, the GEF Council, and their commitment to, and ownership of, the GEF is one of the elements of GEF's success.

One hundred and sixty four countries currently participate in the GEF, and the Organization has financed over 500 projects in 120 nations. Leveraged co-financing, from other donors and national Governments, more than doubles GEF allocations. Project eligibility for GEF financing is regulated by its Operational Strategy, approved by Council in 1995, which defines overall objectives, fields of intervention and operational guidance (Operational Programmes).

The issue of fisheries depletion is of growing global concern and is well within the mandate of the GEF. Reversing unsustainable trends in the exploitation of aquatic living resources including fisheries, both artisanal and high seas commercial fishing, is in fact central to GEF Strategy in the International Waters and the Biodiversity Focal Areas. While biodiversity concerns relate to the protection of the diversity of species, the focus of the three International Waters Programmes is on sustainable management of transboundary fisheries, enforcement of international agreements, and removal of barriers to the introduction of environmentally benign technologies and policies. GEF grant financing is currently the only significant financial mechanism available to support developing countries and countries with economies in transition in their efforts to reverse the long term declining trends in fish and fishing.

2. SELECTED GEF FISHERIES PROJECTS

2.1 A strategic programme of actions to address transboundary environmental problems of the Pacific Small Island Developing States ($12.2 million)

The Pacific Small Island Developing States (SIDS) are part of the 200 high relief islands and 2500 low relief islands and atolls spread throughout 38.5 million km^2 of the South Pacific. Most are entirely coastal in nature, with limited freshwater resources but abundant access to coral reefs, mangrove forests, seagrass beds, and lagoons. These coastal habitats support enormous amounts of biodiversity and are the basis for significant fisheries, both subsistence and commercial. Fishing is an integral part of the economy of these islands, providing the major source of protein for many of their 6.5 million people. Oceanic fishing contributes comparably little to local diets with only 1% of the fish landed entering local economies. A large percentage of the overall tuna catch, 50 to 60%, comes from the exclusive economic zones of the Pacific SIDS, yet only 4% of the dollar value of the catch goes to the local countries. Despite the important role that marine resources play in the ecology and economy of the region, degradation is occurring through the overexploitation of resources, introduction of pollutants, and modification and destruction of critical habitat. A strategic programme of actions has been developed for the Pacific SIDS with the ultimate objective to restore, conserve and manage in a sustainable manner the coastal and oceanic resources of the Pacific SIDS. This ambitious objective will be pursued through demonstration projects, enhancement of transboundary management strategies, assessments of methods to increase domestic benefits from the tuna fishery and its associated bycatch, with the intention of reducing pressure on overexploited near shore resources. National and regional capacities for fisheries management and assessment will be strengthened and lessons learned will be widely disseminated through an active education and information exchange network.

2.2 Biodiversity protection in Lake Malawi, Malawi, Mozambique, Tanzania ($10 million)

Lake Malawi, located at the southern end of the African Rift valley, is one of the largest freshwater lakes in the world. It plays a substantial nutritional and economic role in the riparian communities of the three littoral countries. The diversity in native fish in the lake is unparalleled, with between 500 and 1000 distinct species occurring within its bounds. Fish, which account for the majority of protein consumed by local people, are primarily caught by artisanal means. Because of pressure placed on the ecosystem by over-fishing, sediment and nutrients pollution, species extinction is a pressing concern particularly for the cichlid species. The GEF project complements existing fisheries projects in the area by providing much needed information on the distribution, abundance, and ecology of the rare endemic species. This information will be used in the establishment of protected areas which will potentially serve as reservoirs for the lake's biodiversity. Training provided to local enforcement and technical staff will strengthen the riparian countries' ability to achieve long term sustainable management of the lake. Finally, a review and evaluation of existing environmental legislation in the three littoral countries will provide recommendations for strengthening enforcement abilities, allowing for compatibility of approaches among the three nations.

2.3 Ghana: Coastal wetlands management project ($7million)

In Ghana, a growing concern for environmental issues led the Government to produce a National Environmental Policy Statement and a National Environmental Policy. A broad range of individuals and institutions collaborated in the production of these documents and it was this process that led to widespread awareness on the issues involved and the consensus on a need for action. The major environmental problems identified through this process included soil degradation and erosion, deforestation, and degradation of habitats within the coastal zone. The objective of the GEF project is to mainain the ecological integrity of critical wetland areas through the integral involvement in management of people who earn their livings from these ecosystems. GEF funding provides for the management of five coastal wetland areas that are registered under the Ramsar Convention. An increase of management capacity of both the Government and the local people is achieved through the strengthening of institutions involved with environmental resources management and providing skills to workers in sectorial and local government agencies. Further, the project provides for monitoring of wetland areas and the fostering of public awareness of environmental issues. These actions are coordinated with efforts to provide for the sustainable use of natural resources through improved management practices and community involvement in the minimization of land degradation, all of which contribute to the maintenance of critical nursery habitat for fisheries species.

2.4 Lake Victoria environment project ($35million)

Crossed by the Equator and bordered by Kenya, Tanzania, and Uganda, lake Victoria is the second largest lake in the world and the largest in the developing world. As with the neighboring large lakes to the South, Lake Victoria has been host to a tremendous burst of speciation among fish, particularly cichlids. Lake Victoria is unique, however, in that this change happened more recently and rapidly than in the other instances, and with fewer opportunities for genetic isolation. The lake catchment area provides for the livelihood of one third of the bordering nation's population, who have a high reliance on subsistence fishing and agriculture. The basin provides sources of food, energy, water, transportation, and as a sink for waste of many forms. The diverse users within the system have come into increasing conflict as the population continues to grow at one of the fastest rates on Earth. These multiple pressures have left the lake's ecosystem unstable and have directly contributed to significant systemic changes. Biodiversity and artisanal fishermen are both threatened by overfishing and hypoxic conditions in the lake's deeper regions. More than 200 indigenous species are near to extinction. Human activites are at the root of these problems, through actions such as heavy nutrient inputs and the disastrous introduction of the Nile perch. The GEF Project is designed to contribute to the rehabilitation of the lake's ecosystem, providing both economic and ecologic benefits. In the first stage of work, fourteen pilot zones have been selected in which to restore the local hydrology, decrease nutrient inputs and fecal coliform levels, determine contamination levels in food fish, stabilize the catch of the Nile perch, and reduce water hyacinth densities to more manageable levels. Lake-wide efforts will improve fisheries research, environmental monitoring and enforcement, as well as address pollution issues from industrial and municipal waste. The objective is to use practical, self-sustaining remedies to solve existing problems, while building the capacity of the bordering nations to conduct ecosystem management on a lake-wide scale.

2.5 Argentina: Coastal contamination prevention and sustainable fisheries management ($8.7million)

The Atlantic coast of Argentina is an area in which the demands and impacts of its multiple users are reaching a point of critical conflict. Patagonia is home to coastal ecosystems and habitats that are unique, but the biodiversity and productivity of these areas are at great risk due to anthropogenic influences. Contamination is being introduced into coastal ecosystems from a variety of sources, Most major cities along the Patagonian coast do not have sufficient waste-water treatment facilities. Increasing nutrient levels has caused eutrophication, which has been linked to the disruption of migration patterns of marine species in the region. Inadequate household and industrial solid waste facilities leach materials that cause significant mortality among marine organisms as the contaminants move through the watersheds into the sea. Pollutants are commonly dumped directly into the ocean. Heavy use of coastal areas in the production and transportation of oil has led to repeated oil spills and discharges of oily ballast and bilge waters. Chronic oil contamination has had strong adverse impacts on marine life, including the death of tens of thousands of seabirds. At the same time, Patagonia supports one of the fastest growing fishing industries in the world, the total catch of which has reached 1 million tonnes per year. There is evidence to suggest that the current levels of capture are well beyond the point needed for a sustainable fishery. Overfishing has depleted marine stocks and may be having an impact on population of seabirds and large marine mammals, as well as ecosystem structure and health as a whole. These massive problems will be approached through a series of projects. Baseline efforts will first address coastal pollution through the strengthening of Argentinian capabilities to diagnose and prioritize pollution problems and select optimal solutions both at national and local scale. In a series of subsequent projects, the diagnosis of pollution hot spots will be undertaken and an environmental atlas, including oil spills trajectory modeling will be produced. Monitoring, control, and surveillance functions will be augmented and a network for communication among monitoring networks will be established. Over-fishing will be addressed through the strengthening of authority to preserve threatened stocks, establishing marine fisheries reserves and an improvement of the data collection networks within coastal areas. Finally, electronic navigation systems will be introduced, which will enhance safety, environmental protection, enforcement and monitoring.

2.6 Other activities

A number of other projects including components related to fisheries are under way or in preparation. They address entire ecosystems or water-bodies, such as the Mediterranean Sea, the Black Sea Basin, the Baltic Sea, the Bay of Bengal, the Benguela and Canary Currents, or specific issues, such as reducing the impact of tropical shrimp trawling.

All these GEF efforts are based on an ecosystem approach and share a goal of increasing the recipient country's ability to reverse environmental degradation and effectively manage their own resources in a manner that is sustainable in the long term. Further, the projects specifically utilize and assist the local scientific communities, making use of their talent for monitoring and their regional knowledge in addressing globally significant problems. This approach will allow the Organization to work in concert with the new directions taken by fisheries professionals.

3. CONCLUSION

In recent decades, fisheries management has undergone a significant paradigm shift. Through the integration of ecological principles into planning, fisheries resources are increasingly viewed as part of finite, multi-species, interconnected ecosystems through which cascading effects may flow. Further, there is an ongoing change in long-held perceptions that allows for the viewing of humans as a significant part of an ecosystem, rather then a force somehow above natural processes. These changes are evident in new management approaches such as rights-based fishing which consider long-term effects and are being applied on varying scales to diverse fisheries resources around the world. Local managers and scientists must be supported in the accurate assessments of populations dynamics and for the setting of appropriate total allowable catches. The fostering of linkages between nations sharing transboundary fisheries stocks can facilitate the exchange of information and assist in the development of complementary management approaches in adjoining countries. Projects targeted at specific problems within a region can result in solutions or approaches that can be used in similar systems worldwide.

GEF funds fisheries-related projects in regions as diverse as the Baltic Sea and the Patagonian Shelf, in climate regimes from the sub-Arctic Bering Sea to tropical Pacific Small Island States. Despite this, we realize that we have not yet scratched the surface of global problems such as depletion of fisheries stocks, loss of critical habitat and land based sources of pollution. More needs to be done.

The GEF is attending the Conference in Fremantle as part of an awareness building initiative to inform recipient countries and the fishing community at large that fisheries related projects are eligible for funding by our Organization under the Focal Areas of International Waters and Biodiversity, consistent with our Operational Strategy. We are committed to substantially expand our involvement in the fisheries sector, along with the complementary ongoing and growing efforts to reduce land-based transboundary pollution, primarily from nutrients and sediments, and to facilitate the phase out of persistent toxic substances.

GEF funding is presently available for fisheries projects addressing several key sectors, including:

i. The assessment and testing of new management systems and environmentally benign fishing technologies
ii. The strengthening of the capacity and structures of recipient countries
iii. The facilitation of decommissioning and of access to alternative livelihoods
iv. The introduction of precision electronic navigation and monitoring systems
v. The removal of barriers to the introduction of ecosystem-based sustainable fisheries management and,
vi. The establishment of protected areas and corridors and the enhancement of their long term sustainability.

The GEF recognizes the immediate need to work as a catalyst to coordinate action with countries as well as specialized agencies, non-governmental organizations, the private sector and the scientific community. We also need to mobilize the full potential of the GEF Implementing Agencies and to establish strategic alliances with new partners and executing agencies, including the private sector, and the donor community. We are convinced that the Global Environment Facility can play an important and unique role in fostering a collaborative response to this new global challenge. The GEF stands ready to work with all towards the common objective of the sustainable use of the living resources of our freshwaters, our seas and our oceans.

ARE ITQs PROPERTY RIGHTS? DEFINITION, DISCIPLINE AND DISCOURSE

R. Connor
Centre for Resource and Environmental Studies
The Australian National University, Canberra, ACT 0200, Australia
<rconnor@cres.anu.edu.au>

1. INTRODUCTION

Individual transferable quota (ITQ) has existed for two decades now as a mature management concept. In the last ten years concern has been focused on the issue of whether ITQs (and variations known as IFQs, IVQs, etc.) should be considered property rights or whether this possibility should be specifically legislated against. The tone of the language relating to fishing quota instruments varies among implementing jurisdictions as do the legal implications of the various regimes for fishers and regulators.

In the US, where federal fisheries legislation now reflects the outcome of a major policy debate on this issue, the legal characterisation of quotas is as revocable privileges. In Australia, Statutory Fishing Rights have been created and in New Zealand, explicit property rights language is used for ITQs.

For those considering the introduction of quota-based management, the situation can appear confusing. Arguments from either side often seem to come from completely different bases with little cross-acknowledgment, except in opposition. Different fisheries can require specific management system design, but this accounts for little of the variation observed in the characterisation of the instrument at the generic, national level. There is a correlation in the three jurisdictions mentioned between the definition of ITQs and provisions in their respective constitutions regarding compensation for appropriation of property rights, but much remains different.

This paper examines the normative assumptions behind the contesting views of ITQs, and how extreme positions are created by the contestors. I use the device of a simple discourse analysis to characterise a few influential positions in the debate. I then attempt a conceptual mapping of the policy terrain, using the discourses as directional forces, and consider management alternatives and potential social, economic and ecological outcomes.

The first part of this paper recounts the legal and economic views of what constitutes property. Second, a characterisation is made of five positions on the spectrum of opinions on what ITQs are, or should be with respect to property rights. Three distinctive discourses that shape these positions and policy outcomes in fisheries management are outlined. The fourth section deploys the discourses as forces influencing policy instrument choice and management outcomes. This device is used to explore how competitive tensions between the discourses can create dichotomies and polarise debate on fisheries policy. An alternative conceptual model is suggested that may allow for more inclusive debate of sustainable fisheries management policy. Finally, conclusions are drawn on the struggle to define rights in fishing and on the requirements for a more unified policy discourse.

2. PROPERTY RIGHTS

Two key informing disciplines in the property rights debate are economics and law. Barzel (1997:3) distinguishes legal and economic property rights. "Economic rights are the end (that is, what people seek), whereas legal rights are a means to achieve the end. ... Legal rights play a primarily supporting role – a very prominent one, however, for they are easier to observe than economic rights."

This is an economist's view. It asserts that property rights are about defining and protecting economic interests; it seems a reasonable position. Although property rights can be used to protect what might be argued to be non-economic values, a majority of economists do not admit that non-economic values exist. For resources, rights are used to allocate their beneficial use to individuals or groups, excluding others. Property can be vested in the state, defined groups, or individuals. But what are the attributes of property?

From a legal view:

i. Property is a right not a thing. It is not something waiting to be discovered, but is a socially constructed convention supported by institutions of social choice such as the courts and legislative statute. However, rights can be submerged or ignored. This can happen due to the lack of examination of established doctrine – as in the case of indigenous land rights. In a sense, such rights exist, waiting to be discovered, are implicit in historical use, and in the precedent, logic and principles of the legal system, await recognition.

ii. Property is a bundle of rights or interests in an asset to use it and manage it. Where these are not held by the same individual, they can take a range of forms. These property interests have a long history in common law in relation to land, as freehold, leasehold, easements, usufruct, *profit à prendre*, rights of fishery in land (rivers and lakes), water appropriation, etc. However, there is little common law development with respect to sea fisheries in the western tradition due to principles established both by the Romans and later in the Magna Carta that prescribed open access (Scott 1988).

iii. Rights can be established and supported locally without official sanction, only to be fully recognised when tested by argument and precedent in the courts. The courts will use a range of criteria to make a judgement on whether property rights

iv. pertain, such as the ability to benefit by ownership, and the ability to capture changes in asset value, particularly by sale.
v. Rights can also be established or extinguished by statute. Statute and common law may come into conflict and the Courts will decide which will be dominant. The way Courts ultimately decide can be influenced by the make-up of the bench and the moral or philosophical inclination of judges. Hence the interpretation of the law is both normative and empirical as is the construction of legislation.

Due to the complexity of the law, what might be recognised as a property right by the ultimate judicial authority of a jurisdiction is often uncertain. The recent scallop fishery closure in Victoria provides an example of opposite interpretations by different judges, and of the power of the legislature to extinguish common law rights[1]. Yanner also provides an example of conflicting interpretations, the complexity of interactions of different aspects of the law relating to property rights, and the way the common law changes to (eventually) reflect social conceptions of what is required for resource management or the restoration of submerged rights[2].

The economic view of property rights is more precise. This is because it is about what people want the world to be like, and does not take account of all the issues recognised at law[3]. From an economic perspective:

i. Property rights are analytical and prescriptive. The economist analyses the impacts of the dimensions of property rights and how they are defined in terms of economic incentives and outcomes. These dimensions include exclusivity, transferability, divisibility, duration, flexibility and quality of title (Scott 1988). Where such dimensions are restricted, rights are said to be attenuated. Each dimension has its effect, but limited or uncertain duration is a key feature of concern in the definition of ITQs.
ii. The concept of the economic externality is critical and the existence of externalities[4] is taken to be evidence of incomplete property rights. By incomplete, what is meant is that not everything is included in the total set of property interests. If they were, the owners would defend their property against the imposition of costs by others, forcing them to internalise those costs. Unfortunately, in fisheries and their associated ecosystems, this is impossible (because of the difficulty in specification) except in the limit where a single owner controls all. In this case, defending the whole system means excluding others, or policing their behaviour to minimise costs imposed, which is in itself very costly. This is the situation the coastal state faces in being sovereign over resources.
iii. Property rights theorists propose that rights systems emerge and evolve when resources become scarce (Demsetz 1967) and economic scarcity is indicated by a positive value. However, the transaction costs of establishing systems of rights can be considerable, particularly where the resource users are dispersed and heterogeneous. When the benefits of establishing rights exceed the transaction costs of doing so, things should start to happen, as long as the costs can be shared appropriately. The activity undertaken by potential beneficiaries to have rights established has been called contracting (Libecap 1989). Critical aspects of contracting for rights are carried out in the law courts, in seeking judgements on how disputes over resources should be settled between claimants and whether particular interests comprise proprietary rights. Where litigation over resources increases, the development of property rights can be predicted.
iv. Economists recognise that property comprises of a bundle of rights. For example, in common pool resources such as fisheries, there may exist rights to beneficial use; a right to determine who has access; a right to decide management rules; and a right to alienate the whole bundle (Schlager and Ostrom 1992). These can be, and often are, split up and vested in different individuals or groups. It is asserted that for the full beneficial effect of property rights in defining economic interests to be accrued, all the sticks in the bundle must be held by a single owner (Edwards 1994). In most, if not all, cases of individual rights in fisheries resources, some of these rights are retained by the state or other social governance institution and the right to alienate the bundle is socially proscribed.
v. Most, if not all, economic property rights are attenuated to some degree, either by law (eg planning or licence rules), or by uncertainty about parameters, that is the rights are too costly to measure, police, enforce or exchange (Barzel 1997).

The economic analysis of property rights presents an agenda for legislative reform, based on an ideal theoretical model. There is also ample historical evidence that property rights do provide part of the necessary conditions for the development of all but the most basic economic activity, for (and this is essential to the arguments) without some surety that they will be able to capture the benefits, people will be reluctant to invest in creating the activity (North 1990). Secure property rights encourage longer time planning horizons and thus, it is argued, more responsible stewardship of resources such as fisheries.

However, questions may be asked as to whether social goals are always served by striving for unattainable

---

[1] Stockdale & Anor v Alesios & Ors [1999] VSCA 128 (25 August 1999).
[2] Yanner v Eaton [1999] HCA 53 (7 October 1999).
[3] Professor Anthony Scott has for many years, and again at this conference, demonstrated that economists do study the legal view of property rights, but it is the specificity of legal consideration of circumstances in case law to which I refer here.
[4] Externalities are costs or benefits to others that are not bourne by the firm or individual generating them and thus not counted by them. Economic efficiency (optimal allocation of scarce resources) cannot be achieved when these costs or benefits are not taken into account in the decision to carry out the economic activity in question. Negative externalities (costs imposed on others) will encourage more of the economic activity to occur than is socially optimal, and thus more of the externality will be generated.

perfectly defined and complete property rights. The normative basis of much advocacy of the economic model is that economic growth is good. Property rights allow increases in economic activity and surpluses, and this drives growth. In the age of full exploitation and sustainability, the growth model is again being questioned as the only alternative, and the purported need to squeeze the last dollar from each activity is ever more under scrutiny for uncounted costs of change, and trade-offs with other (non-market) values. This can lead to social resistance to the introduction of ITQs, and a contest to "frame" the discussion or constrain the implementation of ITQs through definition.

3. CONCEPTIONS OF ITQs

3.1 Fully privatised fisheries model

ITQs are a first step towards fully privatised fisheries, whereby certain stocks of fish, or large parts of ocean habitat, become private property including their full management rights. Some government role may remain to ensure residual public interests are protected.

3.2 Property rights in fishing model

ITQs should be defined as full and unattenuated (as far as possible) property rights – permanent, transferable, divisible, subject to registration of third party interests, etc. They should be subject to compensation when appropriate but not for changes in TACs or other sustainability management actions. ITQs are rights in fishing, rather than in the fish themselves and ownership of fish stocks is vested in the state (*e.g.* the current model in New Zealand)[5].

3.3 Quantified licence model

ITQs are a fishing licence endorsed so the holder can take a specified amount, or share, of available fish. The licence and the quota may transferred, sometimes independently and the quota may be divided. But the licence with quota is subject to revocation for breaches of conditions without compensation (as in the Australian Commonwealth[6]).

3.4 Revocable privileges model

ITQs are revocable privileges granted by the responsible agent of the state to use a resource that rightly belongs to the general public. Such resources are held only in trust by the government and can not be alienated[7] (*e.g.* USA federal system).

3.5 To hell with the model

ITQs should not exist at all. They are an instrument of neo-liberal progressive ideology that endangers the fabric of both community and ecology. They commodify both labour and natural resources and recast human relations in socially negative ways. They entrench patriarchal dominance in fishing by vesting resource capital in existing male participants, thus locking out women. They are opposed by the "independent fisherman" to whom open access is a precious tradition.

4. EXPLAINING THE DIVERSITY OF VIEWS

Is this diversity of definitions of ITQs a problem? For those considering the application of ITQs, there is plenty of scope for confusion about how a framework should be structured and enabled. Second, the duration and security of rights is an essential issue for stakeholders and their incentives to invest in sustainable use. And last, this diversity may be symptomatic of something more basic going on beneath the surface. So how is the diversity to be explained? A first reaction might be that each fishery has its own problems and different answers are needed. But these policies are mainly set at the national level. For example, Canada uses a diverse range of quasi-property rights instruments in its fisheries, but there is political proscription of property rights language in the policy system. This explanation does not offer much light.

A second possible explanation lies in the legal implications and constraints on governments. Two legal issues are "constitutional takings" clauses and the "public trust doctrine". Many tensions exist within the above conceptions of ITQs and their practical implementations, particularly in the sections 3 and 4 models. Much effort has been taken to avoid calling ITQs property rights in these models and the policy action is to attempt to guard against property rights emerging in the common law where they might become recognised. This could amount to derogation or even extinguishment, of common law rights. This statutory action is ostensibly to avoid the issue created by the takings clause of the 5th Amendment to the US Constitution, and an adapted version of that clause used in section 51(xxxi) of the Australian Constitution that invokes compensation for the appropriation of property. However, the circumstances that might lead to

---

[5] There is room for a practical implementation between items as in the definitions given in Sections 3.1 and 3.2. This is the current goal for the New Zealand quota management system: the sole ownership corporation model, where quota holders become shareholders in management companies based on individual species, stocks, or quota management areas.

[6] The South East Trawl Fishery (SETF) the only fishery under Commonwealth jurisdiction to which ITQs had been introduced in the first seven years empowering legislation in 1991 using this model initially. The legislation provides for more secure Statutory Fishing Rights (SFRs), but these are conditional on a management plan being agreed under specific requirements. This occurred for the SETF during 1998 and the process of allocating SFRs to replace endorsed permits commenced. SFRs are potentially more secure, with language in the Act referring to absolute ownership. However, the word "property" has been studiously avoided in policy and legislation and provision for discretionary revocation exists in the legislation. The statutory rights model would sit between those described in Sections 2 and 3.

[7] These conceptions of ITQs that explicitly guard against the development of property rights tend not to consider the "bundle" analysis (Schlager and Ostrom 1992). They attempt to maintain a position that any form of private interest created in fishing comprises alienation of a trust resource. The explicit nature of the statutory construction of ITQs as non-property, seems to indicate a belief that the characteristics of property are in fact exhibited by ITQs, so that without such language they would be in danger of being declared to be property by the courts.

this are rare particularly with proportional quotas[8]. In the Australian case, even with careful wording and definition, and before the mandated statutory rights were effected, the Commonwealth Government was compensating fishers for changes in the initial ITQ allocation formula for the South East Trawl Fishery. Further payments were made, 5 years after the event, for adversely affecting the value of licences by moving from tradable vessel capacity units to ITQs (Trebeck *et al.* 1996).

The common law public trust doctrine has been extensively developed in the USA and has appeared recently in case law in Australia. This doctrine applies to a range of public assets for which it is argued, there is such universal public interest that they should never be alienated by the state. They are thus held in trust by the government for the citizenry. This doctrine has been applied repeatedly in case law in the US against claims to property in fish (Slade *et al.* 1997) and this may have an inhibiting effect on the development of property rights in fishing. However, the development of common law rights in fishing has only recently begun in earnest, as resource scarcity has become evident. As different arrangements are tried in different jurisdictions to achieve economic gains and resolve conflicts over use, the character of property rights in fish and fishing under the law will no doubt change. In the interim, the public trust doctrine has been used as an argument against property rights in fish, but this is not wholly convincing as an argument against recognition of ITQs as property.

5. SUPPORTING DISCOURSES
5.1 Nature of discourses

If these legal arguments are not satisfying, how else to explain the diversity of views in relation to ITQs? In this section, I look at accounts of cause and effect that contribute to fisheries policy. This "discourse analysis" is too brief to provide firm explanatory links, but it offers a way to think about the policy dynamic that can provide insight into how such diversity in positions can arise, and whether this is a problem.

"A discourse is a shared way of apprehending the world. Embedded in language, it enables those who subscribe to it to interpret bits of information and put them together into coherent stories or accounts. Each discourse rests on assumptions, judgements, and contentions that provide the basic terms for analysis, debates, agreements, and disagreements... The way a discourse views the world is not easily comprehended by those who subscribe to other discourses. However, ... interchange across discourse boundaries can occur, however difficult it may seem." (Dryzek 1997)

There are many discourses in fisheries management, three are (a) economic rationalism; (b) administrative rationalism; and (c) social justice.

5.2 Economic rationalism

While the law is the arbiter of truth in what constitutes property rights, it is the insights of economics from which the ITQ debate arises (Gordon 1954, Moloney and Pearse 1979, Scott 1955, Scott 1979, Scott 1988, Scott 1989) and it is one set of economic ideas that drives the agenda toward the top of the discourses rated in Section 5.1. The apex to which economic rationalist arguments on the environment converge is to privatize everything - water, air, fish, etc. - and allow markets to allocate resources to their highest valued use. At this extreme, these ideals are clearly unachievable. However, economic rationalists advocate the development of institutional frameworks that bring one as close as possible to the ideal, namely, property rights and markets.

The logic of rationalist arguments can be compelling. One of the key concepts on which it rests is the idea of incentive compatibility, whereby institutions (particularly property rights and markets) should be constructed so as to provide incentives to economic actors to behave in ways that, in aggregate, fulfil the policy goals of society. This is Adam Smith's "invisible hand" principle, and is surely worth applying where it can be made to work. But, it does require society to have agreed goals first. Economic rationalism defines the single social goal as a bigger economic pie, which makes things easy, and the more the economy grows and the greater the efficiency the more there will be for everyone. Distributional issues are excluded as non-economic problems[9]. The natural relationship recognised by economic rationalism is competition (Dryzek 1997).

In common with many other aspects of modernism, economic rationalism tends not to acknowledge the issues of pervasive and irreducible uncertainty. The assumption seems to be that, given enough forward momentum, all information problems are solvable, and all system behaviour is determinate and therefore ultimately predictable. This is simply not the case, and undermines the economic rationalist position.

---

[8] I query the circumstances under which the state's wish to use the power of uncompensated revocation could be justified. Under proportional quota, TACs can be adjusted as required to protect fish stocks. In Australia's South East Trawl Fishery, the TAC for gemfish was dropped from 3000t in 1988 to zero in 1993 following a stock collapse. Quota owners retain their share of the fishery as before and their tonnage entitlement will be restored as the fishery recovers. In the Port Phillip Bay scallop fishery, where ITQs are not applied, the cancellation of all licences and closure of the fishery was a result not of problems with the stock, but of competing values for use of the Bay's waters for navigation and recreation. Not only were licence holders compensated for the full market value of their licences, but if the state had not introduced specific legislation limiting their liability, findings from subsequent legal challenges strongly suggest that claims for compensation for total economic losses would have been upheld (*supra* note 2).

[9] Different original distributions of resources result in differing efficient configurations of economic activity. Welfare economics explores the issues of which particular distribution might maximise social welfare, and recognises the fact that efficiency gains are not necessarily gains in social welfare. This issue is conveniently neglected by economic rationalists (Perman *et al.* 1996).

5.3 Administrative rationalism

Those who determine what gets decided and implemented as ITQ policy are the administrative and policy bureaucrats and government fisheries managers. Many have disciplinary backgrounds in economics or natural sciences, many others are career bureaucrats whose expertise is in machinery of government, program administration, etc. This is the natural home of the discourse of administrative rationality, which argues that complex social problem solving can only be achieved through problem reduction and disaggregation and the application of expertise-based hierarchy. The motivation of this discourse is of public interest. It philosophically opposes dispersed decision making such as market mechanisms. It believes in the legitimacy of administrative power and that the strengthened hand of government is what is required to deal with environmental problems. Based on the use of specialised expertise-based planning and decision making, administrative rationality also opposes discursive democratic mechanisms such as those based on the co-management. Bureaucratic hierarchies are not conducive to the free flow of information, and misunderstandings and duplication are common. The natural relationship recognised is that people should be subordinate to the state (ibid.).

Administrative rationalists are in a a bind in the tug of war between social stability and economic progress. The tendency is to defend the administrative state by attempting to retain control. As ITQs threaten to disperse some of that control to stakeholders, administrative rationalists would initially oppose them. Where ITQs become inevitable, the discourse would support conditioning of the instrument so as to retain a measure of administrative power. Administrative rationalism reveals its interests in retaining control in comments such as: "..the replacement, alteration or revocation of licences is integral to flexible administration. If licences are declared to be proprietary by the courts in Australia, the power to replace, alter or revoke them will be greatly impaired, as will the flexibility and long-term competitiveness of Australia's fishing industry" (McCamish 1995). This latter conclusion is precisely the opposite to that of the economic rationalist who advocates increased legal security of rights for the same reasons.

The administrative rationalist's position aligns to some extent with the interests of small-scale producers, but an inherent part of the small independent producers outlook is to hold the administrative state in contempt. Conflict with a range of stakeholders is inevitable for subscribers to this discourse. Where it holds sway in management agencies, a siege mentality may be discernible.

5.4 Social justice

The forces opposing privatisation are poised at the other end of the spectrum. On one hand, the anthropological and sociological critique of ITQs by academic authors, for example Davis (1996), Davis and Bailey (1996), Jentoft *et al.* (1998), McCay (1995a), Munk-Madsen (1998) Palsson (1998) Palsson and Helgason (1995), speak on behalf of concerns for social justice and equity, gender relations, and express concern over the extension of corporate control at the expense of small-scale local interests. This is a socially conservative position valuing stability of social relations over increased consumption and growth. This position reflects concern that, under ITQs, fisheries access will become controlled by the captains of industry, not the captains of fishing boats.

Allied with these concerns are small-scale producers who, often have a firm belief in the natural order and social justice of open-access regimes. In the USA this is supported by elements of Jeffersonian philosophy of a republic of independent producers, and the common law public trust doctrine (Macinko 1993, McCay 1995b). In Iceland it is supported by the constitution, which protects a right to work for all citizens. With fishing such a common source of employment in that country as to be almost synonymous with work, the denial of a citizen's right to fish by exclusion through a property rights regime is contentious and legally problematic. The motivation of the social justice discourse is the threat to local stability and social relations posed by the globalising economy and the rationale is to defend these values. The mechanisms used are community organisation for collective action and various forms of political action. The natural relationships recognised are community and self reliance.

It is important to note that the discourse that one subscribes to is not determined by the nature of one's employment, although there may be correlations. Not every bureaucrat is an administrative rationalist, and there are quite a few economic rationalists in government bureaucracies. In many sections of government there are also subscribers to a discourse of social justice. The relative strength of these discourses and others depends on the situation both within the policy system and in society at large. This has in turn been shaped by their histories. Further, the discourses described here are only a sample of the many active in fisheries policy debates.

6. MAPPING THE POLICY LANDSCAPE
6.1 Policy terrain

Figure 1 which can be constructed as a kind of conceptual experiment offers an aid to understanding how this sampling of discourses might contribute to policies on ITQs.

Assume a policy object P exists, with a mass proportional to the number of administrative rationalists in the policy system. This gives the system some inertia. Now a force (blue) can be applied to the policy object toward the right, by the discourse of economic rationalism. To the left a countervailing force, red, for social justice can be applied. To complete the map co-ordinate system, there is a green vertical axis, for "ecological rationality", a concept which the reader may define.

There are four quadrants towards which the policy object may move, depending on the definition of rights resulting from the contest between the forces. Social justice and economic rationality forces determine the horizontal position. But what are the requirements to move, *e.g.* into the upper right-hand quadrant. These include: property rights, good information, low transaction costs, and a low ratio of the discount rate to the growth rate of the stock. These factors contribute to good efficiency and ecological beneficial outcomes. But good social justice outcomes should be possible in combination with private rights too, as has been demonstrated, *e.g.* with Maori fishing rights in New Zealand. Perhaps another dimension to the diagram is needed?

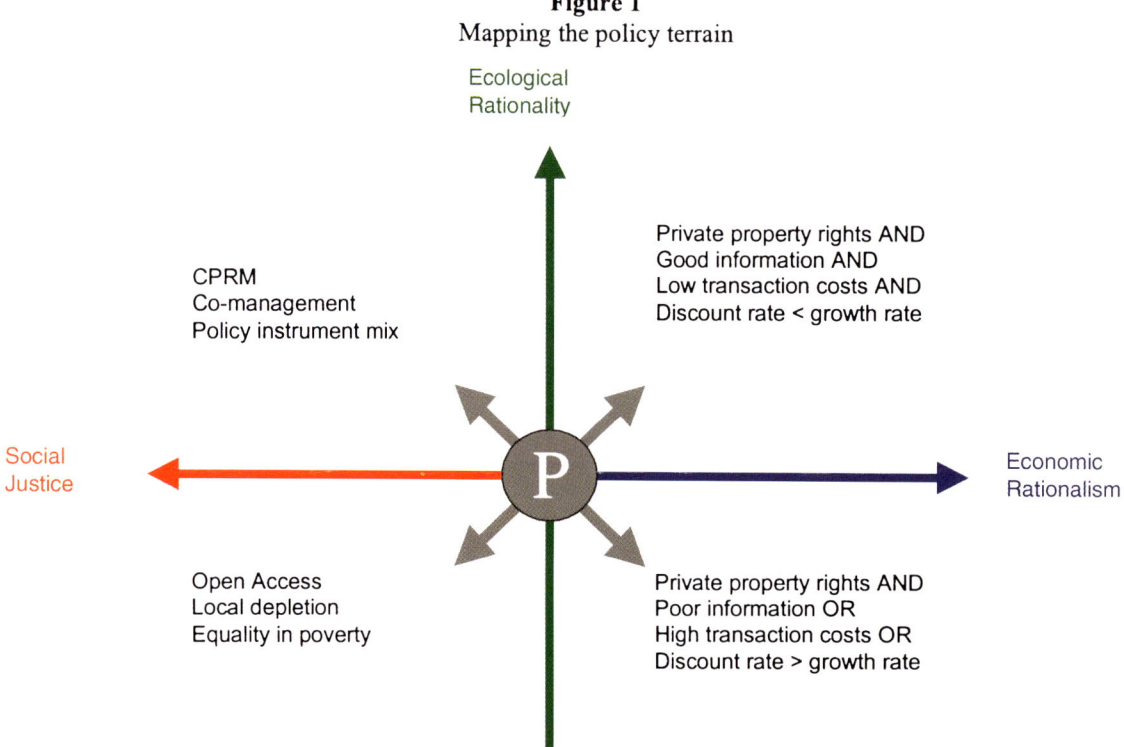

Figure 1
Mapping the policy terrain

If some of the necessary conditions were to fail badly, the end result might be in the lower right hand of the quadrant. This could happen with a species like orange roughy while other species under the same management regime remain in the top right quadrant. Where there is significant uncertainty over stock status or biology, or the degree of compliance of other fishers, or where high discount rates and low biomass growth rates imply economic losses from stock conservation, overfishing may occur for economically rational reasons. This would not necessarily be worse for ecological rationality than the lower left quadrant, which represents the open access result. Here, fishers are free to exploit at will, unencumbered by government interference. This cannot be called an economically rational outcome, despite individual actors' decisions being rational. However, a sense of social equity can be maintained as whatever benefits there are in fishing are available to all. However, ITQs are not the only threat to the stability of this situation and declining incomes relative to the rest of the economy are a feature of this quadrant. There is also a continuing threat to ecological sustainability from open access, accentuated by technological developments that lower fishing costs.

The upper left hand quadrant is the domain that values social justice and stability, but can offer enhancement of both economic and ecologically desirable outcomes in comparison to open access through common property systems and various co-management arrangements. Co-management is a concept with broad application and does not necessarily involve privilege of social stability values over economic efficiency. However, it does offer a basis on which to develop solutions that embed local group values where they might be threatened by completely individualised rights. In general, action in this quadrant of the map is a trade-off of potential efficiency gains to be more certain about the future of both social and ecological stability. This in no way excludes the use of property rights, individual or group held, well defined or otherwise. Management is likely to involve a mix of instruments, with fewer critical assumptions than apply in the upper right quadrant.

6.2 Property rights and social goals

Although the model suggests that social justice and economic efficiency are diametrically opposed, in reality individual rights can be combined with other group-based institutional arrangements to achieve policy goals. For example, within the near comprehensive ITQ management system in New Zealand, the remote Chatham Islands community maintains group ownership of some fishing quota through a trust. The quota is leased to community members and trust income is used for community infrastructure projects. Some dispute over what trust income should be spent on has arisen in this case, which is inevitable. Of course, it helps to have a set of social goals for which a consensus can be found and structure and process are required to build and maintain institutions at the community level.

The force exerted by particular discourses on policy direction will depend on circumstances and may be amplified or attenuated by such factors as the history of policy, agency rivalry, or policy fashion. In New Zealand, ITQs were introduced in a policy atmosphere of economic liberalisation in all economic sectors unleashed by the change of government in 1984. Regardless of the origins and foundation of ITQs, or the nature and history of the problems they were aimed at addressing, in New Zealand in the mid-1980s, if a policy proposal involved property rights, markets, and promises of reduced government regulation, it would have taken a great deal to stop. The nature of the New Zealand political system, having a unicameral parliament and (at that time) a two party system with enforced voting on party lines, gave strong and direct power to executive government to implement policy. In this environment, the social justice discourse was virtually drowned out by calls to leave it to the market. The small boat sector had certainly had some influence on policies under quota, but this has diminished with time. Few small communities were heavily dependent on fishing and little study has been made of resource dependency such as it is (Fairgray 1985 is the exception).

In the case of the Chathams, the government wanted desperately to stop the long standing direct subsidisation of the Islands' infrastructure. Endowment of the community with quota assets was seen as a means to reduce this ongoing and, in the prevailing ethos, rather distasteful liability. Hence the broader interests of economic rationalism assisted the local community to establish a means of protecting and allocating access of their own members to the fishery within the ITQ framework. The settlement of Maori fisheries claims in New Zealand is another example of the potential compatibility of individual rights and group interests, and one where the property rights nature of ITQs became the means to the final settlement of 150 year old social justice grievances of the indigenous people. Maori quota assets are currently managed for the benefit of all Maori people and will in the near future be allocated to 78 tribal (*iwi*) authorities to manage locally in their own interest.

6.3 Stability versus progress

As Seth Macinko (1993) has pointed out, the ideological struggle between small-scale producers and progressive capitalism is not new. He traces it through the development of the public trust doctrine in fisheries cases in the US in the 1820s. Neither is it confined to fishing. The industrial revolution made modern farming, forestry and fishing productive and efficient. In fishing this has allowed expansion to global ecological boundaries. But progress waits for no one and global competitiveness is now overtaking older industrial structures and production methods. The same spectre of decline now faces small-scale enterprises in these three primary industries around the world particularly in the industrialised western democracies. In the globalising economy, inefficiency is punished without mercy. Other producers with lower costs will seize markets, and industries producing substitutes for fish, such as chicken, pork and beef, have adjusted to reduce costs and increase productivity. Without the development of more effective institutions for managing fisheries, things will only get worse.

But traditions, employment and community are highly valued. The individualisation of fishing access rights without accompanying development of institutions for local collective action make small producers and their communities more vulnerable. Economy of scale and market power of large corporates will enable them to pick off individuals one by one as they hit a tough patch and sell their quota to get by. Internal conflict is promoted within communities as certain individuals become the legal holders of what have been seen as community assets. Resistance to enclosure does not solve the fisheries problem, but the concerns are real and understandable. The upper left hand quadrant of the policy map is where solutions for fisheries dependent communities need to be worked out. This does not preclude the use of ITQs, or other fishing rights, but it mandates the development of mechanisms that are able to provide local control of the trade-offs between stability and the advantages of change. Local co-management systems surely can work within wider system of individual rights, given support to develop the necessary institutions for collective action.

6.4 Interaction and interdependence

The world is interconnected and this is nowhere more apparent than in attempting to share fisheries across disparate value systems. Future social outcomes for small communities are less predictable under progressive management, but certainty over access for individual firms is increased under a property rights prescription. It is more difficult to predict reductions in biological uncertainty. ITQ systems create a stronger demand for information for the setting of TACs because of the value of rights, and incentives for stakeholders to pay for more research information are created by their ability to capture increases in the value of rights. However, this does not protect against the situation where, even with perfect information, the economic optimum would be achieved by mining the resource (Clark 1973). In this case appropriate decisions are vulnerable to the uncertainty surrounding fish stock assessments. Dryzek (1992) is not the first political scientist to comment to the effect that the "capitalist market imprisons both liberal democracy and the administrative state by ruling out any significant actions that would hinder business profitability". In this context it means one better be sure of one's ground before crossing the industry by reducing TACs. This may seem extreme, but an element of truth must be acknowledged, even if it is just to defuse the claim that ITQs magically create power for industry over governments – it exists in all economic sectors.

Could science and economics project us universally into the upper right-hand quadrant of the policy map even if social justice issues were ignored? There are serious doubts about such an assumption. The collapse of the northern cod stocks of Newfoundland and Labrador, one of the most productive, long standing, continuously harvested commercial fisheries on the planet has become a classic case of the failure of conventional science-based fisheries management (Finlayson and McCay 1998).

Given that Canada and the US share the highest average living standards in the world and are arguably the best placed countries to bring modern scientific management and institutional change to bear on fisheries, this situation does not bode well.

However, it is institutional development, not science, that must bring about change, and the better definition of rights undoubtedly has an important role to play, whatever the social outcome desired. Hence the question is, if the conceptual model used here an adequate representation of the issues? The diagram is not far wrong in terms of the way people think. We have been raised in

Figure 2 offers the potential to imagine win-win-win outcomes; it offers a new space for solutions that accommodate economic and ecological and social issues. Some imagination, or at least a suspension of disbelief, may be required to accept that this is possible. Many will cling to the cherished flat earth models on which their beliefs and careers in fishing, advocacy, management, discipline or politics have been built. We need to accept that the development of institutions to sustainably manage fisheries lags behind the current rate of their exploitation. This requires some humility, the acknowledgment of both the immense complexities of fishery systems and pervasive

Figure 2
Another dimension to fisheries management

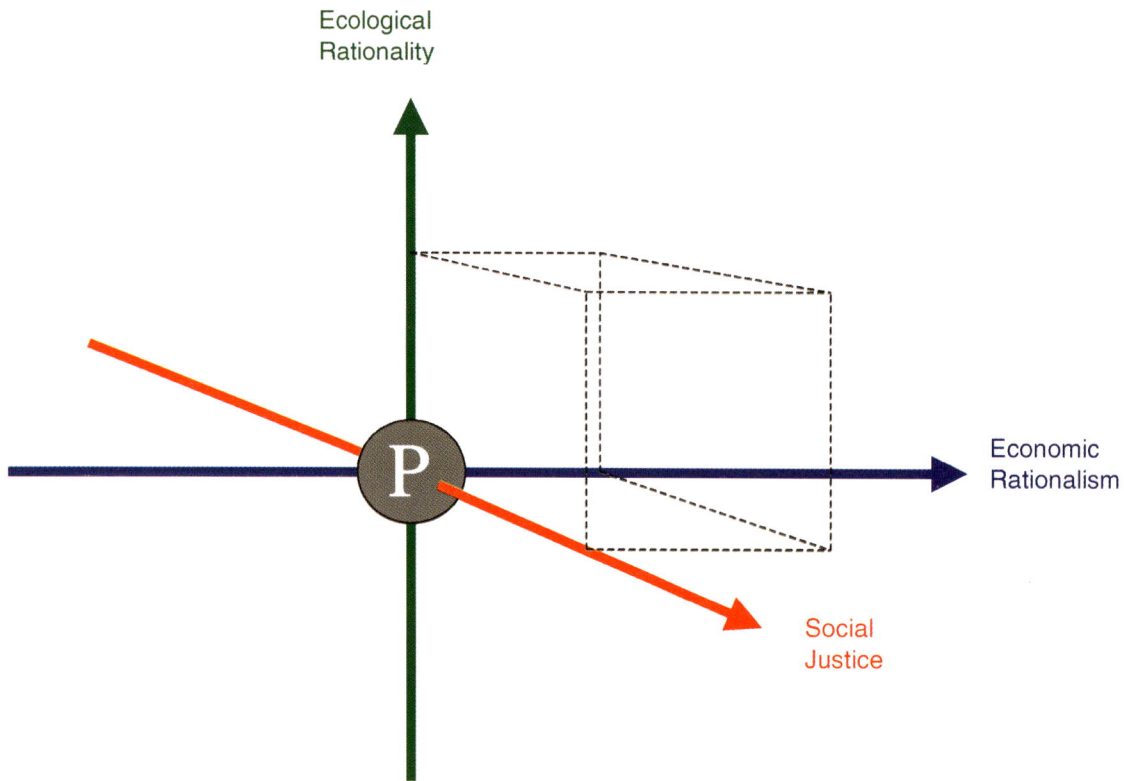

an era of the red and blue forces slogging it out for supremacy. But I propose that the model, although useful in thinking about the issues, is really the flat earth model. It places social equity and economic efficiency, justice and progress, in opposition. This may be a common way of thinking about the issues, or it may be a way of ignoring them. Few deny the need to include the efficiency criterion in fisheries policy, and if this means that social equity cannot be considered, then it will be ignored. On the other hand, where social issues dominate the political landscape, as for example in Atlantic Canada, it is difficult for property rights and efficiency issues to be admitted to the debate. Another dimension is needed in the framework when thinking about sustainable fisheries management. That might enable the opportunity for win-win-win outcomes instead of the best being a win-win-lose.

uncertainty in science, economics and politics. Pretending to have a universal answer will only set back the process of searching for real solutions to unique sets of local conditions.

7. CONCLUSIONS

Discourse analysis has provided some purchase on the forces underlying the fact that different jurisdictions define ITQs differently. In some contexts this amounts to a take-no-prisoners ideological contest. The potential win-lose outcomes are unlikely to be sustainable in the long-term and governments trying to hedge their positions by undermining security of rights in fishing as property may well be making things worse. The answer to management problems in particular fisheries is more likely to be found at the local level, if strongly protected property rights are able to be used when required as part of a mix of management instruments, neither hamstrung by, nor universally imposed from, the national level.

So, are ITQs property rights? There is no doubt that pressure is mounting for rights systems for fisheries management to evolve. This threatens certain established interests and values, which are represented in discourses with the policy debate. These discourses compete to frame the issues and define the terms, with ITQs being one of the bones to fight over. Given the lack of institutional development at the community level for the defence of group values in an environment of better-defined individualised exclusive access rights, resistance to enclosure is understandable. In many cases however the pressure for change will intensify and new property interests specific to fisheries are likely to evolve and gain recognition in law. Statutory definitions may inhibit and constrain specific conditions of rights, but this may be analogous to attempting to constrain fishing effort by prohibiting or regulating one technology. "Rights creep" may come to seriously challenge attempts by administrative rationalists and social justice advocates to stem the tide of property rights contracting.

The fisheries policy community must work diligently to create a more general discourse, a discussion, language and a shared way of apprehending the world that is inclusive of all voices. Leaving it to the market, the experts or the people alone can not cope with the complexity of the issues of modern fisheries management. All the human resources available are needed if a sustainable future is to be secured. The problem for fisheries policy makers is how to specify, or encourage, the development of institutions in a way that creates incentives to both protect resources from overexploitation and to use them more efficiently, but does not undermine other values that have co-evolved with fisheries systems. To move in this win-win-win space, policy makers must integrate economic theory and empirical evidence, specific social, economic and biological conditions, what is possible in statute and likely to be supported by the common law, and the cultural conditions and political realities of their constituencies. This complexity places high demands on policy systems and it is unlikely that traditional, centralised, administrative bureaucracies will cope with problem solving under these conditions. Decentralisation of decision making is required and this is the logic behind both market-based allocation systems and co-management models. Expertise and high level co-ordination are still required. Output controls will not stop being information demanding, but pervasive and irreducible uncertainty need to be more widely acknowledged, and flexibility, resilience and precaution built into management systems.

Expertise is also required in new areas. Institutional design, in structure and process even more than in the rules themselves, requires broadly informed expertise that takes in its purview more than just one disciplinary outlook. Serious attempts at long term social learning with such strategies as adaptive management need to be more widely applied. Skills in facilitation and communication are required to enable stakeholders to work through and resolve issues in ways that provide incentive and are supported by existing institutionals. The fear of a future in which rights in fishing are defined needs to be dissipated through identification and reduction of threats to existing values. Some values will inevitably be given up, but this should be the choice of the holders of those values.

Enlarging the product of our ecologically bounded economy is a broadly embraced imperative, but distributional issues loom ever larger in modern society, both within and between nations. Each deserves our attention, and the challenge is to develop the win-win-win institutional solutions that will take fisheries management into new positive policy spaces in the 21st century.

8. LITERATURE CITED

Barzel, Y. 1997. Economic Analysis of Property Rights. Political Economy of Institutions and Decisions. Cambridge.

Clark, C.W. 1973. The economics of over-exploitation. Science, 181: 630-634.

Davis, A. 1996. Barbed wire and bandwagons: A comment on ITQ fisheries management. Reviews in Fish Biology and Fisheries, 6(1): 97-107.

Davis, A. and C. Bailey 1996. Common in Custom Uncommon in Advantage: Common Property, Local Elites, and Alternative Approaches to Fisheries Management. Society and Natural Resources, 9: 251-265.

Demsetz, H. 1967. Toward a Theory of Property Rights. American Economic Review, 57(2): 347-359.

Dryzek, J.S. 1992. Ecology and Discursive Democracy: Beyond Liberal Capitalism and the Administrative State. Capitalism, Nature, Socialism, 3(2): 18-42.

Dryzek, J.S. 1997. The Politics of the Earth: Environmental Discourses. Oxford University Press.

Edwards, S. 1994. Ownership of Renewable Ocean Reasources. Marine Resource Economics, 9: 253-273.

Fairgray, J.D.M. 1985. ITQ implications study: first report, Northland fishing communities 1984. FMP Series No. 13, New Zealand Fisheries Management Division.

Finlayson, A.C. and B.J. McCay 1998. Crossing the Threshold of Ecosystem Resilience: The Commercial Extinction of Northern Cod. In: C. Folke and F. Berkes (Editors), Linking Social and Ecological Systems: Institutional Learning for Resilience. Cambridge University Press.

Gordon, H.S. 1954. The Economic Theory of a Common-Property Resource: The Fishery. Journal of Political Economy, 62(2): 124-142.

Jentoft, S., B.J. McCay and D.C. Wilson 1998. Social theory and fisheries co-management. Marine Policy, 22(4-5): 423-436.

Libecap, G.D. 1989. Contracting for property rights. Cambridge University Press, Cambridge.

Macinko, S. 1993. Public or private?: United States commercial fisheries management and the public trust doctrine, reciprocal challenges. Natural Resources Journal, 33(Fall 1993): 919-955.

McCamish, C. 1995. Fisheries Management Act 1991: Are ITQs Property? The Federal Law Review, 22.

McCay, B.J. 1995a. Social and Ecological Implications of ITQs: An Overview. Ocean and Coastal Management, 28(1-3): 3-22.

McCay, B.J. 1995b. "That's Not Right": Resistance to Enclosure in Newfoundland Fisheries, Meetings of the International Association for the Study of Common Property, May 24-28 1995, Bodo, Norway.

Moloney, D.G. and P.H. Pearse 1979. Quantitative rights as an instrument for regulating commercial fisheries. Journal of the Fisheries Research Board of Canada, 36: 859-866.

Munk-Madsen, E. 1998. The Norwegian Fishing Quota System: Another Patriarchal Construction? Society & Natural Resources, 11: 229-240.

North, D. 1990. Institutions, Institutional Change, and Economic Performance. CUP, Cambridge.

Palsson, G. 1998. The vitual aquarium: Commodity fiction and cod fishing. Ecological Economics, 24(2-3): 275-288.

Palsson, G. and Helgason, A. 1995. Figuring fish and measuring men: The individual transferable quota system in the Icelandic cod fishery. Ocean and Coastal Management, 28(1-3): 117-146.

Perman, R., Y. Ma and J. McGilvray 1996. Natural Resource & Environmental Economics. Addison Wesley Longman, London and New York, 396 pp.

Schlager, E. and E. Ostrom 1992. Property-Rights Regimes and Natural Resources: A Conceptual Analysis. Land Economics, 68(3): 249-62.

Scott, A. 1955. The fishery: the objectives of sole ownership. Journal of Political Economy, 63: 116-124.

Scott, A. 1979. Development of Economic Theory on Fisheries Regulation. Journal of the Fisheries Research Board of Canada, 36: 725-741.

Scott, A. 1988. Development of Property in the Fishery. Marine Resource Economics, 5: 289-331.

Scott, A. 1989. Evolution of Individual Transferable Quotas as a Distinct Class of Property Right. In: Campbell and Waugh (Editors), The Economics of Fisheries Management in the Pacific Island Region. ACIAR.

Slade, D.C., R.K. Kehoe and J.K. Stahl 1997. Putting the public trust doctrine to work. Coastal States Organisation.

Trebeck, D., T. Battaglene, M. Exel, O. Harasymiw, and G. Hewitt 1996. Report of the South East Fishery Adjustment Working Group, To the Minister for Resources and Energy, Canberra.

THE LEGAL NATURE OF AUSTRALIAN FISHING LICENCES – ARE THEY PROPERTY RIGHTS?

B. McFarlane
Piper Alderman Lawyers
GPO Box 65, Adelaide, SA 5001 Australia
<bmcfarlane@piper-alderman.com.au>

1. INTRODUCTION

In order to determine the legal nature of the rights that fishers currently enjoy, it is necessary to consider what rights the State itself has in relation to fish stocks and fisheries. The term "State" here includes the Commonwealth and the Northern Territory. The State's rights in this area will vary depending on the geographic location of the fishery and the source of their rights. The focus of this paper is licensing in sea fisheries.

The paper examines the nature of an Australian fishing licence. It attempts to determine whether the licence is a property right, a right with some of the legal characteristics of property, or whether it is some different, and perhaps lesser, form of interest. Case law seems to cloud, rather than clarify the issue.

In the majority of instances, fishing licences are creatures of statute and are therefore susceptible to change. However, there are some rights to fish which are recognised at common law. For example, the right to take fish in inland waters can be subject to a *profit à prendre*. However, does such a concept extend to wild sea fisheries? An attempt is made to reconcile the common law and statutory rights to take fish and to determine if, in fact, Australian fishing licences are a property right.

2. THE RIGHT TO FISH STOCKS IN AUSTRALIAN STATES[1]

The Australian States, which commenced life as British colonies, derived their law from the common law system of England. They share that heritage with a number of other so called "common law countries" such as Canada, New Zealand, India, South Africa and Malaysia. The theory was that the first settlers in a new colony brought with them "*all the English laws then in being... which became immediately in force*"[2]. However, that general proposition was severely limited and it is probably more correct to say that they brought with them "*... only so much of the English Law, as is applicable to their own situation and condition as an infant colony*"[3]. Indigenous peoples and their traditional laws were ignored by the adoption of the legal fiction of *terra nullius*, empty land.

Although Australia was colonised from 1788 onwards, and sovereignty claimed, each of the colonies derived their power from, and were beholden to, Great Britain. To the extent hypothesised by Blackstone, the common law of the parent (Great Britain) became the common law of the child (Australian Colonies). Obviously, common law inheritance has been severely modified by two centuries of judicial consideration, both in Australia and other parts of the world.

Given the comments of Blackstone, it can be argued that the English common law position with respect to fish and fisheries would, at least initially, have been in force in each of the Australian colonies. Was there an English common law right to take wild fish or to own wild fish?

Prior to the Magna Carta, there seems to be no doubt that at common law the public had a right to fish in the tidal reaches of all rivers and estuaries, and the sea and arms of the sea within the territorial waters of the kingdom[4]. The exceptions to this principle were where the Crown, or some subject, had acquired a proprietary interest exclusive of the public right, or where Parliament had restricted the common law rights of the public. Following the Magna Carta, the public right could only be excluded or modified by an act of the legislature[5].

Prior to colonisation of Australia, it was also settled law in England that "*there is no absolute property in living fish, other than oysters, mussels, cockles and clams on certain land, for in their natural state they are wild animals, and are not goods and chattels; there may, however, be a qualified property in them as in other wild animals*"[6]. There is nothing to suggest that, subject to modification by statute, this was not also initially the common law position in Australia. Therefore, it is probably not unreasonable to conclude that unless individual States have claimed "*absolute property*" in wild fish by legislative fiat, none exists. As will be discussed later, such a claim has its inherent difficulties.

Each of the States eventually achieved self-government[7] and subsequently received their own

---

[1] Unless otherwise specified, the term State(s) includes the Northern Territory.
[2] Blackstone, *Commentaries* (18th ed), Bk. 1 pp. 111-112.
[3] Ibid.
[4] See, for example, a discussion by the Privy Council of the origin of the right in *A-G for British Columbia v A-G for Canada* [1914] AC 153 at 169. For a discussion of what constitutes the territorial waters of the kingdom, see n10 and following.
[5] Ibid at 170. See also *Harper v Minister for Sea Fisheries* (1989) 168 CLR 314 at 330 which confirms that because it is a public and not a proprietary right, it is amenable to abrogation or regulation by a competent legislature.
[6] Halsbury's Laws of England (4th ed), Vol 18 para 652.
[7] New South Wales and Victoria 1855, Tasmania and South Australia 1856, Queensland 1859 and Western Australia 1890. This process was achieved by means of the various letters patent for each State granted by Great Britain. At that stage the Northern Territory was annexed to South Australia. It did not become

constitutions under which to operate[8]. Although none of these constitutions specifically dealt with the power to legislate with respect to fishing, or the right to fish or ownership of fish stocks, that power was derived from the general power to legislate in a manner that promoted peace, order and good government. However, that general power was limited in that any laws which were repugnant to the laws of England would be struck down. Evidence of each of the colonies exercising that general power can be found by reference to a range of early fisheries legislation, none of which purported to claim ownership of the fish stocks.

The independence of the States was limited by choice in 1901. The *Commonwealth of Australia Constitution Act 1900* although the result of many conventions and draftings convened by the Australian States in the 1890's, was passed by the United Kingdom Parliament in July 1900. The Constitution contained within the Act did not come into force until 1 January 1901. On the same day the Commonwealth of Australia was established, which created a Federation of the six original States and a central government. The States were left with their own existing constitutions, as modified by the Commonwealth Constitution. That document set out the special powers conferred on the central government and its Parliament, its Executive and its Courts by the States, as well as declaring certain guarantees and prohibitions. The rest of the general powers remained with the States.

One of the specific powers transferred to the Commonwealth dealt with fishing. The Constitution vested power in the Commonwealth to make laws for:-

"... peace, order and good Government of the Commonwealth with respect to:

(x) Fisheries in Australian waters beyond Territorial Limits"[9].

The reference to "beyond Territorial Limits" is interesting. It implies that the States wished to retain their right to legislate inside that limit.

3. THE STATES' TERRITORIAL LIMITS

As a matter of Customary International Law, a State retains sovereign rights in relation to waters of the sea that are waters of, or within any, bay, gulf, estuary, river, creek, inlet, port or harbour within the limits of the State. This customary position was confirmed by the United Nations Convention on the Law of the Sea "UNCLOS", which refers to these maritime areas as internal waters[10]. Australia is a signatory to that convention and has ratified its position, so that the convention's provisions apply to all Australian States and Territories[11].

But what was meant, then and now, by the Territorial Limits referred to in the Constitution? At the time of colonisation of Australia, Britain asserted sovereignty over both the land and the sea[12]. Following Federation, although the issue of sovereignty was clear, there were continuing tensions between the State and Commonwealth Governments as to who could exercise that sovereignty over the sea adjoining the Australian coastline. The matter came to a head in 1973 when the Commonwealth Parliament enacted the *Seas and Submerged Lands Act* 1973 (Cth). The Act purported to settle the sovereignty issue by including a specific provision which declared that:

"It is by this Act declared and enacted that the sovereignty in respect of the territorial sea, and in respect of the air space over it and in respect of its bed and sub-soil, is vested in and exercised by the Crown in right of the Commonwealth"[13].

That claim for sovereignty also extended to the contiguous zone and the continental shelf of Australia[14]. The effect was that the territorial seas adjacent to the States, which the States had previously claimed sovereignty over, were now vested in the Commonwealth. In the case of fishing, this seemed to shift the boundary referred to in Section 51(x) of the Constitution back to the States' coasts and internal waters boundaries. That is, those areas which could be considered part of the State. Or perhaps the term *Territorial Limits* always had this meaning.

Although all States challenged the power of the Commonwealth to legislate in this manner, the High Court of Australia upheld the validity of the Act[15]. In particular, the majority of the Court found, inter alia, that:

i. the low watermark constituted the seaward boundary of the States

---

a State in its own right and after Federation its legislative capacity was controlled by the Commonwealth as one of its Territories.

[8] The following Acts or their predecessors: *The Constitution Act 1902(NSW), The Constitution Act 1975(Vic.), The Constitution Act 1934(Tas.),* together with the Australian Constitutions Act (No.2) 1850 (Imp), *The Constitution Act 1934(SA), The Constitution Act 1867(Qld.),* and *The Constitution Acts 1899 (WA).*

[9] Section 51(x). There is some evidence in the convention papers that some States identified specific fisheries over which they wished to retain legislative competence.

[10] United Nations Convention on The Law of the Sea UN Doc. A/CONF.62/122, 21 ILM 1261(1982) Part 11 Article 7 which states *"waters on the landward side of the baseline of the territorial sea form part of the internal waters of the state".*

[11] Although the Convention was entered into in 1982, it didn't come into force until 16 November 1994.

[12] But did that claim of sovereignty amount to a claim of ownership or property in the sea? In Mabo No2 (*Mabo and Others v The State of Queensland (1992) 175 CLR 1*) the High Court found that it did not.

[13] *Seas and Submerged Lands Act* 1973 (Cth) at Section 6. The territorial sea extends 12 nautical miles to sea from the low water line along the coast, except where it follows baselines deliniating the internal waters of a State or the internal waters of the Commonwealth.

[14] Ibid - Sections 10 and 11 respectively.

[15] *New South Wales -v- The Commonwealth* (1976) 135 CLR 337.

ii. the jurisdiction of the Commonwealth extended over fisheries both in the territorial sea and on the continental shelf, and
iii. the Commonwealth derived the power to legislate over off-shore areas from Section 51 (XXIX) of the Constitution[16].

In the following years, there were a number of cases dealing with the ability of State legislation to have extra territorial effect (operate outside the States' boundaries), so far as off-shore regions were concerned[17]. Those cases consistently upheld the position that whenever there was an inconsistency between State and Commonwealth legislation caused by the State legislation entering an area (legislative) covered by the Commonwealth, the Commonwealth legislation would prevail. There were also a number of fishing cases that helped to define the boundaries of the States internal waters[18].

In 1979, the Commonwealth and the States reached agreement on the settlement of off-shore constitutional issues. The settlement relied on a whole raft of complimentary legislation being enacted by both the Commonwealth and the States[19]. As part of the overall settlement the off-shore regions were divided up between the States and the Commonwealth.

4. CONCEPT OF SEA ZONES

By reference to particular articles in the UNCLOS, specific zones were established:

i. **the adjacent territorial sea** (or State territorial sea or Coastal waters), which extends 3 nautical miles from the baseline established in accordance with the provisions of UNCLOS[20]. The normal baseline used for establishing the breadths of the territorial sea is the low water line along the coast[21]. This coincides with the definition of the States' boundaries used by the Commonwealth in the *Seas and Submerged Lands Act* 1973 (Cth). The States were given concurrent power to legislate with respect to this area[22]. They were also given limited title to the sea-bed in these coastal waters[23].

ii. **the territorial sea**, which is now 12 nautical miles seaward from that baseline, having been extended from the original 3 nautical miles. The States' powers remain limited to the coastal waters

iii. **the contiguous zone** is the area between 12 nautical miles and 24 nautical miles seaward from the baseline from which the breadth of the adjacent territorial sea is measured[24]

iv. **the exclusive economic zone (EEZ)** extends for a distance of 200 nautical miles from the baselines which establish the territorial sea[25]. It vests sovereign rights to the coastal State for the purposes of exploring and exploiting natural resources in the waters super adjacent to the sea-bed, on the sea-bed and below the sea-bed[26]

v. **the continental shelf** is the area between 12 nautical miles and 200 nautical miles seaward from the territorial sea baseline and any areas of physical continental shelf beyond 200 nautical miles. Australia has the right to explore and exploit the living and non-living resources of the shelf.

Much reference is made to the territorial sea baseline and where the boundaries of the various States and the Northern Territory begin and end. This is particularly important with respect to the internal waters[27]. Where the baseline is the low water line, it is easily defined. Where it crosses bays and gulfs, it is more problematical. The importance of this concept is that those areas on the landward side of this baseline form part of the State in accordance with customary international law. This principle is recognised by UNCLOS and has been confirmed by Commonwealth Legislation[28].

5. THE CONCEPT OF SOVEREIGNTY

Sovereignty has been referred to as "*a legal, categorical and absolute condition. A territory either has*

---

[16] The external affairs power.
[17] For example *Pearce -v- Florenca* (1976) 135 CLR 507 at pp 513-521 and *Robinson -v- The Western Australian Museum* (1977) 130 CLR 283.
[18] For example *A Raptis & Son v The State of South Australia* [1976-1977] 138 CLR 346.
[19] This was collectively known as the "Offshore Constitutional settlement" (the "OCS").
[20] Supra n 10 at Part II, Territorial Sea and Contiguous Zones, Articles 5 and 7.
[21] Infra n22 s1 and s4. It specifically excludes any area resulting from an increase in the width of the Territorial Sea (s4 (2)). The **Coastal Waters of a State** is defined to mean:
"That part of the territorial sea 3 nautical miles seaward from the baseline AND any sea that is on the landward side of any part of the territorial sea of Australia and is within the adjacent area in respect of the State but is not within the limits of the State".
[22] *Coastal Waters (State Powers) Act* 1980 (Cth) No. 75 of 1980. The State is also given power to legislate for areas outside coastal waters where there is an arrangement with the Commonwealth in place. Those powers include power to make laws with respect to fisheries, as if the waters were within the limits of the State (section 5(a)). A recent example of a State exercising a power to legislate in this in-shore area is the declaration by the Government of South Australia of the Head of the Great Australian Bight as a conservation zone dedicated to the preservation of the Southern Right Whale.
[23] *Coastal Waters (State Titles) Act* 1980 (Cth) No. 77 of 1980. The Act gave to the States the same right and title to the property in the sea-bed beneath the coastal waters of the State and the same rights in respect of the space (including the space occupied by water above that sea-bed), as would belong to the State if that sea-bed were the sea-bed beneath waters of the sea within the limits of the State. However, the grant did not extend to complete sovereignty.
[24] Its relevance is that it allows Australia to exercise the control necessary to take enforcement measures for breaches of such things as customs, immigration and sanitary laws.
[25] Supra n 10 at Article 51.
[26] In addition, UNCLOS specifically deals with a whole range of other rights and responsibilities in this area.
[27] Supra n10.
[28] Seas and Submerged Lands Act 1973 Section 14.

sovereignty, or it does not have sovereignty. There is no halfway point for the sovereign conditions. A demonstrated capacity for self-government remains central for sovereign Statehood. Sovereignty is therefore a property of States. It reflects effective control over territory and independence from other States" [29].

Sovereignty has also been defined as:

"By 'exercising de facto administrative control' or 'exercising effective administrative control', I understand exercising all the functions of a sovereign government, in maintaining law and order, instituting and maintaining Courts of justice, adopting or imposing laws regulating the relations of the inhabitants of the territory to one and another and to the government. It necessarily implies the ownership and control of property whether for military or civil purposes, including vessels, whether lawships or merchantships. In those circumstances it seems to me that the recognition of the government as possessing all those attributes in a territory or not subordinate to any other government in that territory is to recognise it as sovereign, and for the purpose of international law as a foreign sovereign State" [30].

The term, and perhaps its distinction from acquisition of property, was clarified in the Mabo (No.2) case by BrennanJ, who notes that:

'The acquisition of territory is chiefly the province of international law; the acquisition of property is chiefly the province of the common law. The distinction between the Crown's title to territory and the Crown's ownership of land within a territory is made as well by the common law as by international law'" [31].

Sovereignty is merely a right to control and not ownership. The High Court in *Harper v Minister for Sea Fisheries* [1989] 168 CLR 314 at 330 confirmed that the competence of the State Legislature to make laws regulating a right of fishing in such waters is not dependent upon the State's possession of a proprietary right in the bed of the seas or rivers over which such waters flow. The Court repeated with approval Lord Herschell's comments in *Attorney-General (Canada) v Attorney-General's (Ontario, Quebec and Nova Scotia)* [1898] AC 700 at 709 where he states:

"There is a broad distinction between proprietary rights and legislative jurisdiction."

6. THE CONCEPT OF NATURAL RESOURCES

What rights do the States claim over the natural resources? All States have traditionally asserted the right of ownership of their minerals and some other terrestial resources. Recently, every State in Australia has confirmed ownership of their natural resources in their respective Native Titles Acts[32]. But what is included within the term natural resources? Butterworths Australian Legal Dictionary gives one definition of natural resources:

"The stock of naturally occurring, as opposed to manmade, tangible and intangible substances which are capable of exploitation for commercial purposes. Examples are timber, land, oil, gas, minerals and mineral ores, coal, lakes and submerged lands. It includes native features of benefit for health, welfare and wellbeing such as parks and heritage items. The United Nations General Assembly has established a regulatory regime for the use of natural resources and recognise that States and peoples have permanent sovereignty over natural resources: GA RES 1803 (XVII). Resolution is accepted as customary international law: for example Texaco Overseas Petroleum Co & California Asiatic Oil Co v Libya (1977) 53 ILR 389." [33]

The concept of permanent sovereignty over natural resources is further defined by Butterworth as:

"The principle that, under international law, people and nations have the right to own and control their natural wealth and resources; GA RES 1803 (XVII). Permanent sovereignty over natural resources is a basic constituent of the right to self-determination. The utilisation, development and naturalisation of natural resources must be pursuant to the national development and well being of the people and international economic co-operation must be based on respect for the sovereign right to natural resources; GA RES 1803 (XVII).

Article 56[1][a] of the United Nations Convention on the Law of the Sea confirms that in the exclusive economic zone (200 nautical miles seaward) the coastal state has sovereign rights to explore and exploit, conserve and manage the natural resources, whether living or non-living, of the waters superjacent to the seabed".

There seems to be an acceptance of ownership of natural resources in the wide General Assembly Resolution. However, that appears to be limited by the more

---

[29] *International Environmental Law and World Order Guru Swammy*, Palmer & Weston 1994 West Publishing Co St Paul Minneapolis at p396.
[30] *The Arantzazu Mendi* [1939] AC 256 at 263-265, per Lord Atkin.
[31] Supra n 12 at 44.
[32] *Native Title Act 1994 (ACT)* – Section 11, *Native Title Act 1993 (Cth)* – Section 212, *Native Title (NSW) Act 1994* – Section 17, *Native Title (Qld) Act 1993* – Section 17, *Native Title (SA) Act 1994* – Section 39, *Native Title (Tas) Act 1994* – Section 13, *Land Titles Validation Act 1994 (Vic)* – Section 14 and *Titles Validation Act 1995* – Section 13
[33] *Butterworths Australian Legal Dictionary.*

specific UNCLOS. It certainly acknowledges all the elements of legislative, administrative and extractive control but seems to fall short of ownership.

A search of Australian Legislation relevant to offshore areas suggests that the only definition of natural resources contained in any legislation is in Part I Section 5 of the *Quarantine Act 1998*. That provision defines natural resources as *"the mineral and other non-living resources of the seabed and its subsoil"*. This seems to suggest that, at least at some levels, the term is limited. The only judicial comment on the subject appears to support that position.

7. THE STATE FISHERIES LEGISLATION

But does it include fish? Except for Tasmania and Victoria, none of the other States, the Northern Territory or Commonwealth Fisheries legislation have asserted a claim to ownership of the living marine resources. Tasmania made this claim in 1995 in their living marine resource legislation[34]. Under Section 9 of that Act it asserts:

> *"(1) All living marine resources present in waters referred to in Section 5(1)(a), (b) and (c) are owned by the State."*[35]

Living marine resources are defined in Section 3 of the Act as *"fish and their environment"*.

The waters referred to in Section 5 are as follows:

> *"(1) State waters are:*
>
> *(a) Any waters of the territorial sea of Australia that are:*
>
> *(i) within 3 nautical miles of the baseline by reference to which the territorial limits of Australia are defined for the purposes of international law; and*
>
> *(ii) adjacent to the State; and*
>
> *(b) Any marine of tidal waters that are on the landward side of that baseline and are adjacent to the State, except inland waters; and*
>
> *(c) Any land which is swept by those waters to the highest landward extent."*

It is interesting that Tasmania and Victoria are the only States to assert such ownership. However, from where do they derive the power or right to do so? Although the *Coastal Waters (State Titles) Act 1980* of the Commonwealth vested title in the seabed beneath the coastal waters adjacent to the State, and that vesting gave the States the same rights as would have belonged to them if that seabed were the seabed beneath waters of the sea within the limits of the State, the grant did not extend to complete sovereignty. Nor did it pass title to the living marine resources within its area, principally it is submitted, because the Commonwealth did not itself have title to these resources. Similarly, although the *Coastal Waters (State Powers Act) 1980 (Cwth)* gave the States power to legislate within the coastal waters of the State, it fell well short of conferring any ownership of the fish stock. A query then is whether the claim to ownership would be sustainable if it was subjected to legal challenge? Perhaps so, if the terms *"own"* and *"ownership"* were confined by the courts to the sovereign rights referred to above rather than absolute property in them.

8. THE CONCEPT OF PROPERTY

What then of the term *"property"*? It appears from the above that neither the States (apart from Tasmania and Victoria) or the Commonwealth claimed property in the wild fish stocks as they did with such things as minerals and forests. Rather they merely asserted a right to legislate with respect to the resource.

Earlier in this paper I identified that at common law there were only limited rights to ownership of wild animals and that the common law treated fish as analogous to wild animals. It was also recognised that the Crown did not assert ownership of the fish while they were in the wild. In effect they belonged to no one until they were caught and reduced into the possession of someone. At that point they became the property of that person.

A recent example of the difficulty with the Crown asserting *"property"* in wild things is the *Queensland Fauna Act*, the terms of which were considered by the High Court of Australia in *Yanner v Eaton*[36]. That case involved the prosecution of an Aboriginal person for taking a protected species (crocodile) under the legislation. The defendant claimed that he was exercising traditional native title rights and accordingly, was not subject to the legislation.

In the judgement, the High Court considered the concept of property. It noted that the word *"property"* is often used to refer to something that belongs to another. But in the *Fauna Act,* as elsewhere in the law, property *"does not refer to a thing; it is a description of the legal relationship with a thing"*. It refers to a degree of power that is recognised in law as power permissibly exercised over the thing.

The concept of *'property'* is elusive and I do not propose anything more here than a brief overview for the purposes of clarification. Usually Property is treated as a "bundle of rights". But even this may have its limits as an analytical tool or accurate description, and it may be, as Professor Gray has said, that *"the ultimate fact about property is that it does not really exist; it is merely illusion"*[37]. So too, identifying the apparent circularity of reasoning, from the availability of specific performance and protection of property rights in a chattel to the

---

[34] *Living Marine Resources Management Act 1995 Tasmania.* Victoria asserts ownership of all wild fish, fauna and flora found in Victorian waters (s10(1) of Fisheries Act 1995 (Vic). Consistent with the common law position, it then passes that ownership to any person who lawfully catches any wild fish (s10(2)).
[35] *Ibid* Section 9.
[36] *Yanner v Eaton* [1999] 166 ALR 258.
[37] Ibid at 9.

conclusion that the rights protected are proprietary, may illustrate some of the limits to the use of "*property*" as an analytical tool. No doubt the examples could be multiplied.

Nevertheless, as Professor Gray also says, *"An extensive frame of reference is created by the notion that property 'consists primarily in control over access'. Much of our false thinking about property stems from the residual perception that 'property' is itself a thing or resource rather than a legally endorsed concentration of power over things and resources"*[38].

It is clear that *"Property"* is a comprehensive term that can be used to describe all, or any, of many different kinds of relationships between a person and a subject matter. The High Court in *Yanner* decided that there were several reasons to conclude that the *"property"* conferred on the Crown in that case is not accurately described as *"full beneficial, or absolute, ownership"*. They did so on a number of bases. First, there is the difficulty of identifying what fauna is owned by the Crown. Second, assuming that the subject matter of the asserted ownership could be identified, or some suitable criterion of identification could be determined, what exactly is meant by saying that the Crown has full beneficial, or absolute, ownership of a wild bird or an animal? They confirmed that at common law, wild animals were the subject of only the most limited property rights. There could be no *"'absolute property' but only 'qualified property' in fire, light, air and water and wild animals"*[39].

In the same judgment the High Court quotes Roscoe Pound[40] and his hypothesis of why wild animals and other things not the subject of private ownership are spoken of as being publicly owned. Pound states:

"We are also tending to limit the idea of discovery and occupation by making res nullius (e.g.: wild game) into res publicae and to justify a more stringent relationship of individual use of res communes (eg: of the use of running water for irrigation or for power) by declaring that they are the property of State or are 'owned by the State in trust for the people'. It should be said, however, that while in form our Courts and legislature seem thus to have reduced everything but the air and the high seas to ownership, in fact the so-called state of ownership of res communes and res nullius is only a sort of guardianship for social purposes. It is imperium, not dominium. The State as a corporation does not own a river as it owns the furniture in the State house. It does not own wild game as it owns the cash in the vaults of the Treasury. What is meant is that conservation of important social resources requires regulation of the use of res communes to eliminate friction and prevent waste, and requires limitation of the times when, places where, and persons by who res nullius may be acquired in order to prevent their extermination. Our modern way of putting it is only an incident of the 19th Century dogma that everything must be owned[41].

This approach is entirely consistent with that adopted by Mason CJ, Dean and Gaudron JJ in *Harper v Minister for Sea Fisheries* when they considered the nature of a fishing licence. They noted:

"Under the licensing system, the general public is deprived of the right of unfettered exploitation of the Tasmanian abalone fisheries. What was formally in the public domain is converted into the exclusive but controlled preserve of those who hold licences. The right of commercial exploitation of a public resource for personal profit has become a privilege confined to those who hold commercial licences. This privilege can be compared to a profit à prendre. In truth, however, it is an entitlement of a new kind created as part of a system for preserving of limited public natural resource in a society which is coming to recognise that, insofar as such resources are concerned, to fail to protect may destroy and to preserve the right of everyone to take what he or she will may eventually deprive that right of all content.

In that context, the commercial licence fee is properly to be seen as the price exacted by the public, through its laws, for the appropriation of a limited public natural resource to the commercial exploitation of those who, by their own choice, require or attain commercial licences"[42].

Later in the same judgment and in the same vein, Brennan J observed:

"If the right to fish for abalone were created in diminution of proprietary rights of the owner of the seabed and without the owner's consent, some question as to the validity of the law might have arisen, that a legislature of a State may not be competent to create proprietary rights over property beyond the boundaries of the State and to which the State has no title. That problem does not arise in this case, however, for the management of the fishery in accordance with Tasmania law, is arranged between the Commonwealth and Tasmania. **If title be needed to support the fishing rights conferred on the abalone licenceholders,** *the arrangement made under the Act and the Commonwealth Act testifies to the consent of the Crown in right of the*

---

[38] Ibid.
[39] Ibid see Gleeson CJ, Gaudron, Kirby and Hayne JJ at pp8-13.
[40] Pound *Introductions of the Philosophy of Law (rev ed)* (1954).
[41] Ibid at 111.
[42] At 325.
[43] Ibid at 335.

Commonwealth and of Tasmania to the creation of those right."[43]

In making the statement highlighted in bold the High Court appears to acknowledge that the Crowns right to grant a licence is not necessarily dependant on title or ownership of the resource itself.

9. FISHERIES "PROPERTY" CASES

Although there have been a number of cases to which reference is often made as a basis of saying that the courts have accepted that fishing licences are property, it is submitted that those cases really do not take the matter to any conclusion. Rather they tend to use an analysis which is purposive. That is, the courts have analysed it within the confines of the particular case. The cases have therefore tended to confuse rather than assist with a clear understanding of the issue. Perhaps the High Court in *Harper* was close to the mark when it noted:

> "*This privilege can be compared to a profit à prendre. In truth, however, it is an entitlement of a new kind created as part of a system for preserving of limited public natural resource in a society which is coming to recognise that, insofar as such resources are concerned, to fail to protect may destroy and to preserve the right of everyone to take what he or she will may eventually deprive that right of all content*"[44].

It is also worth noting that an analysis of other High Court "*property*" decisions seem to suggest that while the word "*property*" is to be understood broadly, it generally (or, on one view, necessarily) includes rights recognised by, or founded on, the general law as opposed to rights which find their sole source in statute[45]. There have been cases when purely statutory rights have been held to be property, but principally when they constitute choices in action.

Perhaps one way of viewing a fishing licence is to consider it as a dispensation from a general prohibition rather than a right. For completeness I have included some of the other fishing cases which have discussed various proprietary interests said to give rise to property rights as a basis for fishing licences[46]. Some support that proposition while others find no such interest. For example, Olsson J. in the case of *Edwards*[47] has defined licences, registrations, authorities and permits as property, the beneficial ownership of which can be form the subject matter of legal relationships such as trusts[48]. In doing so he confirmed the earlier decision of *Pennington*[49].

However neither case defines what that "*property*" is or its nature. In reality, it is probably in the nature of some sort of a proprietary interest, or perhaps as the High Court suggests in *Harper*, it is a new right, which has not yet reached legal maturity by way of a clear definition. Certainly there are a string of cases dealing with the diminution of, or removal of, Commonwealth Fishing entitlements in which the High Court has held that they constitute property for the purposes of the Constitutional guarantee in s51(xxxi) dealing with "*taking property on just terms*". However, as noted above, it is analysed in this manner for the purposes of the case at bar and it is submitted that it is unhelpful to extrapolate these determinations to cover situations where there is no common law or statutory support.

10. WHAT THEN DO FISHERS HAVE?

As a fundamental principle of property law, you cannot give more than you have got. That means that if you have a limited form of title, you cannot grant to any other person any greater form of title than that which you currently have. For example, in pure property law terms, a person who holds a lease for a fixed term of years cannot grant to another person a lease of a greater number of years. The reason that issue is raised is because the State, in terms of fisheries, can only grant licenses which are consistent with their own title.

It is also clear that there are no common law rights which underpin fishing licences. Rather they derive all their authority from the statute that grants them. The nearest property concept that describes the right to take fish resources by way of a licence is that of a *profit à prendre*, although the concept is one which relates to land or something related to land. For example, the right to take wood or soil from the land. It is acknowledged that fish in inland waters are also capable of being the subject of a *profit*. However, the concept requires that the thing to be taken be capable of being "owned" at the time of the taking, and that it be in some way related to the land. Therefore, for these and other reasons, a *profit* is unlikely to assist in the analysis.

At present a licence holder appears to have the right to:

i. take fish from a defined region using defined equipment
ii. sell the fish taken
iii. have the licence renewed (arguably), and
iv. sell or transfer the licence.

The other rights, such as excluding others, managing the resource and protecting it appear to reside with the Crown. Added to that is the fact that the licence is solely a creature of statute and liable to derrogation. The list of rights above may give rise to remedies against the

---

[44] Supra n 43.
[45] *Minister of State for the Army v Dalziel* (1944) 68 CLR
[46] *Pennington –V- McGovern* - SA Abalone, *Fitti –v – Minister for Primary Industries, Davey –v- Minister for Primary Industries, Bienke –v- Minister for Primary Industries* – Cth Prawn, *Kelly –v- Kelly* – SA Abalone, *Austell –v- Commissioner of State Taxation* – WA Rock Lobster, *Pike –v- Duncan* – Vic Scallops and *Harper –v- Minister for Sea Fisheries* – Tas Abalone.
[47] *Edwards and Deep Sea Arc Pty Ltd -v- Olsen & Ors* (1996) 67 SASR 266e.

[48] Provided there is no statutory provisions to the contrary.
[49] *Pennington -v- McGovern* (1987) 45 SASR 27e.

granting body, but only limited rights against third parties. It is submitted that the lack of rights against third parties highlights the limitations of analysing fishing licences in terms of property rights. If a fisher is disadvantaged in some way, the issue of compensation becomes one of social justice or equity, rather than a matter based on the licence being a property right.

It may be that a fishing licence is better described as a right in the nature of a *profit à prendre*, the defence of which relies on principles of equity rather than property law.

Whatever the ultimate characterisation of the rights, it is submitted that the focus on them being some form of property right, is generated from the protection sought rather than the source of the right in Australia. As Justice Gummow said in Yanner:

"*Although appropriate to describe it as having a proprietary character, that is not because property is the basis upon which protection is given; rather this is because of the effect of that protection*"[50].

11. CONCLUSION

While there is no doubt that the use of rights-based notions has assisted in developing effective fishery management regimes in Australia, continued secure access to the resource remains a critical concern. It is submitted that what is needed is a more clearly defined legal framework (both legislative and judicial) to support that development. It may be that the issue of native title and claims to the sea and its resources will provide the stimulus for the Courts to clarify and possibly redefine the legal nature of Australian Commercial fishing licence.

---

[50] Supra n 37 at 36; albeit describing intellectual property rights.

RIGHTS BASED SYSTEMS: SOVEREIGNTY AND PROPERTY

C. Jensen
167 East Park Avenue
Wasilla, Alaska 99654 USA
<cjensen@alaska.com>

1. INTRODUCTION

A successful rights-based system must draw a distinct line between sovereignty and property. The distinct goals of each principle must be defined and, because they compete with each other, these goals must be prioritized to avoid conflicts. Next, the system must develop management rules to achieve these goals: one set of rules for sovereignty, another for property and the third for interaction between the two. This structure stimulates a healthy tug-of-war between the competing principles. It produces a dynamic balance that makes the current management system reliable and, looking forward, reasonably predictable in the face of normal changes and therefore durable. If this lively interaction is lost, the game is over for the rights-holders.

2. CURRENT THEORY
2.1 Economic model

Figure 1 is a tool Dr. Scott has provided to help gauge the quality of property rights created through individual transferable quotas (ITQs) and other transferable licences in a rights-based system[1]. We can also use this tool to compare the quality of rights created by different systems.

Figure 1
Economic model (Scott 1988) - Characteristics of property rights

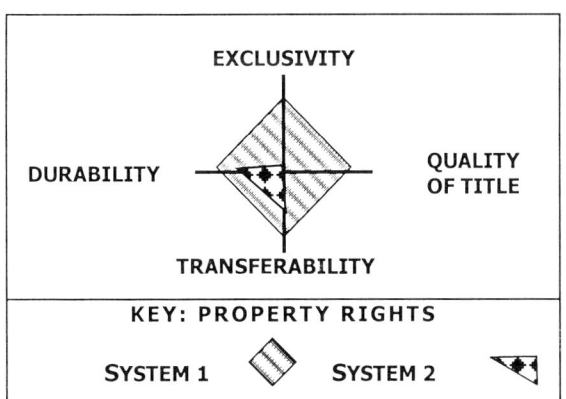

An important finding, documented by using this model to study existing fisheries management programmes, is that systems that create the most property rights for their fishers have healthier fisheries and greater economic efficiency.

2.2 Working hypothesis

Many opinions have been offered to explain this finding. Most believe there is a relationship between private ownership, through rights-based allocations, and healthier fisheries. That view suggests fishers take better care of the resource, independently or through co-management with government, because they have an ownership stake in it.

In my view, better science and increasing environmental regulation are the more likely explanation for healthier fisheries. To some degree, limiting participation helps attain environmental goals. But, the common purpose of every rights-based system is economic: to limit access to increase the likelihood the remaining participants will make a living at the licensed activity. In turn, these licensees must comply with environmental regulations to maintain their licence rights. Compliance achieves sustainable stocks and other environmental goals.

As environmental goals increasingly encroach on economic rights, the licencees have a greater incentive to influence what goals are selected by government and the process by which these goals are to be achieved. To preserve the greatest economic rights, licensees develop an increasing capability to participate in the government management process. But, industry participation is the typical response to increasing government regulation of any activity. Thus, environmental stewardship and co-management are normal industry responses, whether or not the activity is controlled through rights-based allocations.

In this respect, the relationship between ownership and the environment seems tenuous. Increasing environmental regulation, as a cause, and a healthier resource, as an effect, seems a more likely explanation. Whether one participates as an owner or not is largely irrelevant. In any event, the ownership that a rights-based system provides to its licensees depends completely on the success of the system, as this paper demonstrates.

We can only progress if we examine rights-based systems in a valid framework. Today, the private sector has a huge investment that directly depends on the success of systems that use the rights-based model. Once the investment is made, the private sector has a limited ability to adapt to extreme changes in government management. For this reason, we have all our eggs in one basket.

We need to identify flaws in the management structure and fix them before they become serious problems and the solution is taken out of our hands. At the end of the day, without the licence, or the rights we

---
[1] More recently, Dr. Scott (1999) has added flexibility and divisibility to this model.

thought we had bought, there is no business, only a pile of sticks and bricks, vessels and gear and miscellaneous equipment that has no equally productive use.

2.3 Question: how many property rights are enough?

Our task is to chart the future for rights-based systems. A loud cry for more (or, its synonym, better) private rights has resounded throughout this conference. It is based on an assumption: if some private rights rebuild resources and promote economic efficiency, more property rights will provide an even better result. To test this assumption, we must ask how many property rights does a system need to function optimally? Once we find the answer, we will have a viable rights-based model.

2.4 Perfect property eliminated

An important point can be quickly made using an extreme example. Here, we will use the concept of perfect property. Figure 2 below shows us what perfect property looks like in the economic model[2].

Figure 2
Perfect property

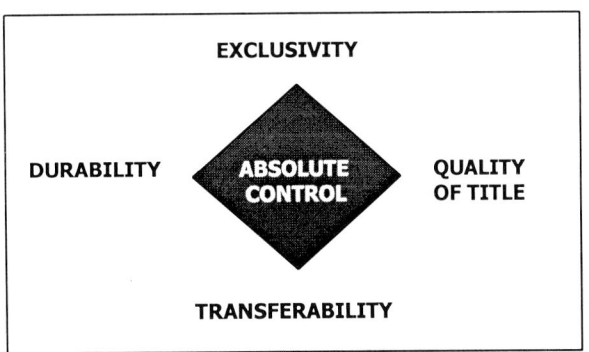

Participants from Namibia, Zimbabwe, South Africa, Estonia, plus the Aborigines immediately recognized the meaning of the big black rectangle. Perfect property gives absolute control to the owners. It enables these owners to manage the activity to achieve their goals. Everyone else is excluded.

Absolute control is not a characteristic of property in the real world. Rights at the extreme ends of the model are not property rights. Thus, the optimal property rights will fall short of this extreme in the economic model.

2.5 Framework: the relationship between sovereignty and property

There has been a great deal of analysis on the subject of property in rights-based systems. There has been less analysis on how property works outside a rights-based management system. Perhaps that inquiry can point us in a fruitful direction.

If your government fell today, what would your property be worth? The answer is nothing. A government creates property and, as it goes, so goes your property. Hence, owners have a vested interest in the stability of the government that creates their property.

If your government allocates economic rights to only a few, what will happen? It will fail. Why? The huge group of excluded have no reason to support the government. So, a government must manage for the good of its community, not just its property owners. Sometimes, there are conflicts between the sovereign community and the property owners. Then, the government must make choices between the two. A government will thrive if it creates enough rights and, when deciding conflicts, its choices preserve a balance between the community and the property owners.

These examples illustrate the dynamic relationship between sovereignty and property. The same rules apply to a rights-based system, which functions according to the same principles. It might help to think of a rights-based system as a mini-sovereign system.

A successful system must incorporate the sovereignty-property dynamic if it is to endure. We will consider that general relationship here. But, this analysis does not evaluate particular systems, engage in comparisons of different systems, or recommend any specific course of action. Instead, it leaves that difficult task to those with a direct stake in the outcome, an approach which, in my experience, consistently yields the best result.

3. PROPERTY

3.1 Properties of property

Many speakers have examined property rights in specific systems. We have also heard about the historical development of property, including the rights-based fisheries model. This section takes the generic approach. It implies that property is the same in every nation of the world. The distinctions are simply a matter of degree. As such, property is a concept that is easily understood by everyone.

3.2 What is property?

Property is a man-made concept. Its purpose is to organize private relationships. Sovereigns create property. In the broad sense, a sovereign is a community that bands together for the common good of its members. In a technical sense, a sovereign exercises dominion over a specific group. This word is typically associated with government and I use this term here in that context.

A government usually creates property by making laws that create property rights. Standing alone, a property right is insignificant unless the owner has a way to enforce it. For this reason, the government must also make laws that give individuals a way to protect their property rights.

3.3 Property is a right, not a thing

Property is the legal right, not the thing to which the right refers. Specifically, property is a relation between an owner and others in reference to a thing (Cohen 1927). An example helps us understand the difference:

i. **My neighbour.** When I am leaning over my fence, talking with my neighbour, I point to the house behind me and ask, "How do you like my house?" In my relationship with my neighbour, I have all the rights to the house; to use, possess, encumber and

---
[2] Scott (1999) did not design the model for this purpose.

convey it. That is why I refer to it as "my house" when I talk to him.

ii. **My spouse.** When I sit down with my spouse, we talk about "our house" because, in this relationship, my spouse and I share the legal rights to it.

iii. **My banker.** Later, I drop off my mortgage payment to my banker. In this relationship, my banker has property rights in the house. I gave him those rights when I signed my loan contract. My banker can enforce his property rights if I do not pay my debt. However, I keep my property rights to the house as long as I make my payments on time. So, my banker and I each have property rights to the house that are enforceable, by one or the other of us, but only if certain events occur.

These examples shows us that different people have different property rights to the same thing at the same time. We also see that a property right depends on the specific relationship: here, the owner, a neighbor, a spouse and a banker. The word "property" is just a shorthand term that means "property rights". Usually, a person who refers to a thing as "my property" has the current right to possess or use it. But, the accurate term is property rights, not property.

A property right gives the holder the power to exclude others. If I have the legal right to possess a thing, I have the power to exclude others from possessing it. The same is true with the right to use, convey and encumber. This power is the way private relationships are organized. But, this power is never absolute. It competes with other property rights and, as we will see later, this private power is subordinate to public law.

3.4 Fisheries management systems
3.4.1 Property rights are created by licensing rules

New property rights are created by the licensing rules. All fisheries management systems create new property rights, not just systems that use transferable licences or allocations. A few examples demonstrate this point:

i. Today many systems manage fisheries through transferable licences, such as ITQs. If a licence rule says an ITQ is transferable, that rule creates certain property rights. For example, the rule creates the private right to receive payment when the ITQ is transferred.

ii. Although less obvious, a licence rule can create a property right even though a licence is not transferable. For example, there is a licence buy-back programme established under United States' fisheries law. Under it, the buy-back fund will pay money to retire access rights. So, this licence rule creates a property right to receive payment, even though the licence, itself, is not transferable.

iii. When a new rights-based system is designed, licence rules set out the requirements a fisher must meet to qualify for an access permit. Typically, one requirement is catch history. This licence rule creates a property right if it allows catch history to be transferred. Specifically, that rule authorizes a fisher to sell his catch history to another person before any licences are issued. Later, the buyer can apply for a licence in his own name.

These examples show us that licence rules create property rights. However, the licence is not property. It is only the thing to which these property rights refer.

3.4.2 Rights-based systems do not recognize property rights or provide rules to enforce those rights

Today, no rights-based system expressly states that licence rules create property rights. Instead:

i. Most systems define their licences as privileges, even though all recognize they are bought and sold in the marketplace everyday. And, a few of these systems record liens for the private sector on their licences. A lien cannot attach to a privilege.

ii. The regulations in a few systems assert that their licences are property, which is wrong. As between the sovereign and the licensee, the licence is always a privilege. The reason is explained in the following section on sovereignty. Even in a private relationship, the licence is only the thing that refers to the property rights, not the property.

Further, no rights-based rule specifically provides that individuals enforce their property rights through property law. Property rules serve a specific purpose - to order private relationships. For that reason, they are peculiar to their purpose. Property rules are not designed to manage sovereign activities, like fisheries. Finally, fisheries management rules are not designed to organize private relationships, like spouses, debtor-creditor and business partnerships.

These two deficiencies threaten both the management system and private investments made in reliance on those systems for obvious reasons. They leave key parameters to be decided outside the licensing system; namely, by the courts or in the political arena. These deficiencies can be corrected without disrupting the management system or the property rights it creates.

A successful system must expressly state that property rights are created by its licence rules. It must, also, provide that property law controls the enforcement of those rights with two exceptions:

i. Only the manager has the authority to transfer a licence. Private voluntary or involuntary transfers are prohibited. This rule makes sure the manager can control participation in the way that is required by the management-system rules. It also solves serious property problems that otherwise arise, including when a private transfer is made to an ineligible person. And, it prevents a secondary market solely for passive investment, speculation and manipulation.

ii. The manager will transfer a licence if the licence is eligible for transfer and the licensee is eligible to hold the licence under the management-system rules. This rule tells the private sector that the manager will not impermissibly intrude into private relationships or impose burdens that unnecessarily restrict property. It will work only if the system designers can accurately identify a limitation on property rights that is required to manage the fishery compared with a limitation that is not required to manage the fishery.

These tasks are not as simple as they sound. The type of rights created should be generically defined, not specifically stated, because, over time, specific rights change. On the property rules, managers and industry are always tempted to write a special set of property rules that only apply to licences for that particular system. Fisheries managers and industry are not competent to craft property rules. An attempt to craft property rules in a licensing system will restrict the market and upset settled rules that organize private relationships outside the context of business activities without providing any corresponding benefit to the management system. That approach just creates more headaches for everyone.

4. SOVEREIGNTY
4.1 The purpose of sovereignty

Sovereignty is the same in all nations. The distinctions are simply a matter of degree. As such, sovereignty is a concept that should be easily understood by all. The purpose of sovereignty is to impose limitations on individuals for the common good. These limitations create an orderly society. They are made through public law.

Public law is unlike property law, which orders private relationships. For example, a private law may allow you to buy a car, but public law can prevent you from driving it without a driver's licence. Or, private law may allow you to buy a factory, but public law can prevent you from operating the factory if it pollutes the air. In our context, private law may allow you to buy a vessel, but public law prevents you from going fishing unless you first get a government licence. So, the distinction between public and private law generally depends on the object to be achieved. Is the law trying to protect the community or to protect a particular individual?

4.2 The real reason licences are not property

In the big picture, the purpose of a sovereign is simply to govern. It has no other central purpose, such as to own property or be profitable. The power to govern is inherent and, as logic confirms, a sovereign cannot give that authority away. For example, a government can make a contract with a company to provide police services. But, a government cannot contract with a company to make the public safety laws. This basic principle of sovereignty is an absolute barrier to full privatization of commercial fisheries. Industry should recognize this barrier and adjust investment accordingly.

4.3 I paid for it, it's mine

A law is unenforceable if it gives away the government's power to rule. For example, a government can choose, but not be required, to pay fishers when the government decides to limit or end access-rights. Recently, the government of New Zealand went into the market place and bought back its licences to satisfy a court order related to Maori rights. Luckily, a good price was obtained. I will use this event to consider the big picture.

Certainly, a government can choose to buy-back its own licences. The economics of that decision are not entirely clear since, in most nations, the government initially issued transferable licences to fishers for free or a nominal charge. And, there seems to be no benefit to, later, buy those licences back at the taxpayers' expense. We can also see why a government might buy-back resources in a different situation. If a government owned natural resources in its jurisdiction and sold them, it might pay the new owner to buy the resources back at a later time. In this kind of transaction, the government is acting in a private capacity and the normal property rules apply.

However, no one owns the free-swimming fish. The futility of trying to exert ownership over something that swims away clearly makes this point. Instead, a government's authority over its fisheries is based on its power to manage affairs within its jurisdiction. A government is acting solely in its sovereign capacity in this arena. So, the important question is whether a government can be forced to pay licensees before it can manage its fisheries in a different way.

A fishing licence or an ITQ, is simply a tool to manage fisheries in the changing national interest. These licences regulate private participation to achieve national goals, such as sustained yield, economic efficiency and social benefits. The process of governing is dynamic. National goals change over time and, as they change, the sovereign adjusts its rules accordingly. For this reason, the sovereign retains the sole power to create, expand, limit or end private licence rights and does not have to pay the owner if their rights are diminished or terminated as a result.

What does this mean for rights holders? It means their property rights are only as good as the current licensing rules. It also means that fishing licences, or their allocation, is a privilege given by their government; a sovereign permission to engage in conduct that otherwise would be illegal. In a contest with the government, property law does not apply. Instead, one's remedies are limited to those expressly provided in the fisheries management system and, then, only if they do not require the government to buy back its right to govern.

4.4 The benefits of sovereign limitation on property rights
4.4.1 Sovereign community

To round out the picture, we must recognize the benefits we obtain from the sovereign principle of limitation. The ability of a government to change the way a country is run from time to time allows progress. It acts as a pressure valve when public needs shift. We know that governments that lack effective pressure valves fail.

More to the point, public rules create social order. Social order creates property value. Therefore, property rights are only as reliable and durable as the government system on which those private rights depend. So, what we lose in certainty on the private rights side, we gain through stability on the public side. These simple principles illustrate the mutually dependent relationship between sovereignty and property.

The sovereign community consists of those on whom a government depends for support. In our case, it includes commercial fishers, for-hire sport fishing businesses, recreational fishers, subsistence fishers, other marine users, environmental advocates, consumers, the public, other nations and so forth.

Each sector of the community has needs. Particular needs change over time. So, the government must remain able to adjust its management methods to respond to changing needs in its community. A rights-based system must have effective mechanisms to allow dynamic sovereignty to work with the least disruption to established property rights.

Rights holders should assist government in designing these mechanisms and work to put them in place before problems polarize the sovereign community. An essential requirement may be a strong connection to the local communities in which specific fisheries are conducted. And, licencees should recognize that changes implemented to achieve new sovereign goals may affect the licence conditions under which they can conduct their business operations.

4.4.2 Overcapitalization and aggregation of rights

"Overcapitalization" is the popular word for too much effort, measured by units of gear or in some other way. It was one scourge that limited access programmes and, later, rights-based systems were designed to remedy. Did they achieve this goal? The number of participants clearly diminished. So, the goal was achieved in that respect. But, limitation on the number of participants created something new. The value of a fishing operation shifted from the vessel to the newly created licence and, as rights aggregate into relatively fewer hands, the licence value increases proportionately. Was one form of overcapitalization traded for another?

Now, we will add another dimension. In open-access, assume one begins with a fishery that is not sustainable and suffers various problems, which is not always the case. With limited-access, the fishery begins to rebuild. With exclusive transferable allocations through the rights-based model, the fishery reaches sustainability. As the allocations continue to aggregate, the fishery may even become bountiful. At this point, there may be only a few fishers that control a vibrantly productive fishery; perhaps, one of the most productive fisheries in the world for the particular species. We have some fisheries today that fit these specifications.

One can all see how the picture has changed from open-access to aggregation of rights. The end result could easily be perceived as a government-created and controlled monopoly for a favored few. Of course, the fishers devoted their lives to building the fishery and their livelihoods depend on the level of allocations they acquired. But, the excluded think they should share or, at least, start paying rent. There are producers in South Africa today who have gone through this process and lost property rights. Until the entire sovereign community is included, the system is not complete. Whatever the reason, and in every case, the government gets to make the call.

5. SUMMARY

Today, there are six billion people on Earth. Twelve years ago there were one billion less. The United Nations projects that twelve years from now there will be one billion more. By any measure the world is experiencing staggering growth and the political response of nations, on a local and a global scale, may range from ominous to fruitful. Either way, it is going to be expensive.

Ways must be found to accommodate the population increases starting with the basic necessities of food, housing, health-care and energy. Today, many nations with the greatest projected growth have trouble feeding their current population. The role other nations play is important. And, issues of genetically modified foods, sustainable agriculture, forests and fisheries, clean air, water, land and energy, including non-fossil or more efficient fuels, biodiversity, global warming and the like, are real problems of today and not intellectual exercises.

On a more practical note, we must recognize that, with population growth, competition for everything (not just profit-making activities) increases and, in response, governments will be required to allocate increasingly scarce opportunities to a relative few. To this requirement, we must factor in the tendency for businesses to grow, which is a particular concern in many nations. The result, as we can already been seen, is that competition becomes increasingly ferocious among sectors.

The necessary policies will be set by governments and the solutions will be developed by the private sector or free market. In my view, our burgeoning population and growing businesses are the catalysts to develop new management techniques simply because what we use today will not work tomorrow.

At this juncture, rights-based systems offer the management tool of choice. Today, many nations control a large number of activities using the rights-based model. Nuclear power plants, hydro-projects, telecommunications, broadcasting, air carriers and trucking, to name a few. Use of this model is a global trend that is strong and expanding with no end in sight. As opportunities diminish, there will be a premium for systems that divide opportunities and allow reallocations in a fair way. Of all the systems I have studied, I believe fisheries systems are the most likely to develop a rights-based model that will best meet the challenges of this millennium, along with producing a bountiful harvest for all to share.

6. LITERATURE CITED

Cohen, M. 1927. Property and Sovereignty, 13 Cornel L.Q. 8.

Scott, A.D. 1988. Conceptual Origins of Rights Based Fishing. *In* Neher *et al.*(eds.) *Rights Based Fishing*. Kluwer Academic Publishers, Dordrecht.

Scott, A.D. 2000. Introducing Property in Fisheries Management. *In* Use of Property Rights in Fisheries Management. Proceedings of the FishRights99 Conference. Fremantle, Western Australia, 11-19 November 1999. FAO Fisheries Technical Paper No.404/1, FAO, Rome.

PROPERTY RIGHTS IN RELATION TO FISHING LICENCES IN AUSTRALIA FROM A LEGAL PERSPECTIVE

D. Fitzpatrick
Fitzpatrick Teale, 380 Bourke Street, Melbourne
Victoria, Australia 3000
<higteale@ozemail.com.au>

1. INTRODUCTION

From a legal perspective there are many forms of property. Property[1] may be real[2] or personal, tangible or intangible, legal[3] or equitable[4]. Examples of property include land, a chattel (eg. motor car), a debt, shares in companies, patents, trademarks, copyright, and depending on the circumstances, fishing licences.

This paper discusses a number of legal cases over the last 10 years in Australia and examines when a fishing licence will be property and when it will not. The most unusual case of a state government in Australia taking away a fishers licence by legislation and the issue of compensation is then discussed. Finally the paper outlines what should be addressed in an Act of Parliament regulating fishing so that a licence can be described as property.

2. CASE EXAMPLES

Over the years the Victorian Scallop Fishers have found themselves in Court on a number of occasions in an effort to protect and preserve their licences and livelihoods. The first case was **John Manias & Ors V. Crabb & Ors (No. 7374 of 1991 Unreported decision of Marks J. of the Victorian Supreme Court)** where the issue was whether the Victorian Minister for Fisheries had power under the Fisheries Act to prohibit licence holders from dredging for scallops in Port Phillip Bay. The Minister relied upon a regulation in the Scallop Regulations to prohibit fishing in Port Phillip Bay during 1991. A declaration was sought from the Court by the Port Phillip Bay licence holders that this regulation was beyond the power of the governor in council to make such a regulation. The regulation was held by the Court to be void. Justice Marks of the Victorian Supreme Court said:

"Dredging for scallop requires a vessel conformed and fitted out for the purpose. An essential assumption of the Act is that a licence and payment of its fees provides some security and safeguard for investment by its holder in boats and equipment required to operate it.

It would frustrate entirely any purpose of the Act if the law were to uphold the validity of delegated legislation which utterly frustrated and effectively confiscated the rights and property for which the Act provides.

The evidence shows that the licence itself is valuable property for which high sums may be paid."

The second case to be discussed is **Springall v. Kirner & Ors (1988) VR 159** where the Victorian Fisheries Minister attempted to prevent abalone divers from taking abalone in the waters adjoining Wilson's Promontory Marine Reserve. The Minister asserted that she was empowered by provisions in the Victorian National Parks Act to stop divers taking abalone. There was a conflict between provisions in the Fisheries Act, which allowed the abalone diver to fish in the Marine Park and provisions in the National Parks Act, which stopped him from fishing. The Court found that a diver's right to take abalone was one of considerable value. The Court referred to a principle of statutory construction that, unless it is unavoidable, an enactment should not be construed in a manner that would lead to the loss of a person's valuable rights without payment of compensation. On this basis, the Court granted an injunction against the Minister preventing her from enforcing the provisions of the National Parks Act against the abalone diver.

In **Harper v. Minister for Sea Fisheries and Ors (1989) 168 CLR 314** a Tasmanian abalone diver challenged a regulation made under the Tasmanian Sea Fisheries Act exacting a substantial licence fee payable each year for a commercial abalone licence and contended that it amounted to an excise[5] and therefore was contrary to section 90 of The Australian Constitution. In the course of dealing with this contention, the High Court examined the provisions of the Tasmanian Fisheries Act. It held

---

[1] Property is a word which can be used to describe every type of right (*i.e.* a claim recognised by law), interest, or thing, which is legally capable of ownership and which has value (*Butterworths Concise Australian Dictionary*).

[2] "Real property" means land and "personal property" means chattels *e.g.* a motor car, a book etc. The Dictionary of Modern Legal Usage, 2nd Edition, Brian Gardner, sets out the difference as follows:

"The distinction between real property and personal property is as old as Roman Law, but the curious terminology is much more recent. From the early 17th Century on, land was commonly called real property and chattels were called personal property merely because land could be recovered specifically in a real action but chattels could be made the subject only of a damages action."

[3] Legal property is the full and complete title as regards ownership or possession.

[4] Equitable property is an interest in property enforced and created by the Courts in the situation where it would have been unconscionable for the legal owner of the property to retain the benefit of the property for him or herself. An example of legal and equitable interests in property is where land is in the name of A who holds the land on behalf of B who is the beneficial or true owner of the land. In this situation, A has the legal interest in the land and B has the equitable interest in the land. Both interests are a form of property.

---

[5] To impose an excise or tax upon.

that a statutory right to exploit a limited natural resource (such as taking abalone or scallops) from the sea was a right akin to property. Brennan J. of High Court of Australia said:

> "When a natural resource is limited so that it is liable to damage, exhaustion or destruction by uncontrolled exploitation by the public, a statute which prohibits the public from exercising a common law right to exploit the resource and confers statutory rights on licensees to exploit the resource to a limited extent confers on those licensees a privilege analogous to a profit a prende in or over the property of another."

A "profit à prendre" is a proprietary right to take the produce or part of the soil from the land of another person (*e.g.* trees, minerals, clay or soil).

In the case **Pennington v. McGovern (1987) 45 SASR 27**, the Supreme Court of South Australia had to determine whether an abalone licence issued under the South Australian Fisheries Act constituted property and, therefore, was capable of being the subject of a trust. In arriving at its conclusion the Court held that the licence was proprietary in nature. Mr Justice King said:

> "It is clear from the provisions of the Act and regulations to which I have referred that the licence in question is no mere personal, inalienable right. It is a transferable right, which is contemplated as having value. The limit of six licences renders it likely, as a matter of commonsense, that a licence will possess value".

The Judge confirmed:

> ".... the provisions or the regulations to which I have referred as to the contemplated value and transferability of the licence and as to the right to hold it notwithstanding that its exercise is subject to the direction and instructions of another, are all, to my mind, indicia of rights or property and I have no difficulty in reaching the conclusion that the rights conferred by the licence are proprietary in character."

Kelly v. Kelly (1990) 64 ALJR 234, is a case which involved a question for determination by the High Court as to whether an abalone licence issued under the provisions of the South Australian Fisheries Act was capable of being partnership property under the South Australian Partnership Act. The High Court found that after analysing the South Australian Fisheries Act, the abalone licence could constitute partnership property. The High Court said:

> "Whatever the position with the abalone permit, there can be no doubt that the abalone authority gave rise to valuable rights which were capable of being held for the partnership in such a way as to constitute partnership property: see Amber v. Bolton (1872) LR 14 Eq 427; O'Brien v. Komosaroff (1982) 150 CLR 310. Despite the fact that it could only be done *indirectly and with the consent of the Director of Fisheries, it was plainly possible to make what was effectively the transfer of an authority for consideration, thus enabling a value to be placed upon it. This was so, notwithstanding that there were certain requirements in respect of an abalone authority which were personal to the holder, such as the requirement that he be medically fit to dive."

A further case in this regard is that of **Austell Pty Ltd v. Commissioner of State Taxation (1991) 4 WAR 235**. This case involved the Supreme Court of Western Australia deciding whether a transfer of a rock lobster licence should be subject to stamp duty. It was argued that the fishing licence being a limited entry fishing licence was not property within the meaning of provisions of the Stamps Act. The Court rejected this argument and said:

> "I must say I would have thought that if a person bought this particular licence, he had an interest that could be called property in the ordinary meaning of the word as used by a layman: see Pollock B in The Smelting Company of Australia Ltd v. The Commissioner of Inland Revenue (1896) 2 QB 179 at 184."

There are series of legal cases in Australia involving the Northern Prawn Fishery Management Plan. These are:

i. Minister for Primary Industries & Energy & Ors v. **Davey** & Or. (1993) 47 FCR 151
ii. **Fritti** v. The Minister for Primary Industries & Energy & Anor (1993) 47 FCR 151
iii. **Bienke** & Ors v. The Minister for Primary Industries & Energy & Anors 135 ALR 128

These are decisions of the Federal Court of Australia. The issue in these cases arose from amendments to the Northern Prawn Fishery Management Plan. Under the Plan units of fishing capacity could be issued to individual licensed boats. A certain number of units were required to fish in the Northern Prawn Fishery (NPF) then amendments were made to the NPF which cancelled units thereby reducing the total number of units of fishing capacity for the NPF. As a result of the cancellation of units, one of the Applicant's boat had insufficient units to allow it to fish in the NPF.

The Court held that a fishing boat licence granted under the Fisheries Act 1952 (the old Commonwealth Fisheries Act) does not create an interest based on antecedent proprietary rights recognised by the general law. The licence represents a new species of statutory entitlement, the nature and extent of which depends entirely on the terms of the legislation. The Federal Court held that the units of fishing capacity allocated under the Northern Prawn Fishery Management Plan conferred only a defeasible[9] interest, subject to valid amendments to the Northern Prawn Fishery Plan under which they are issued. The making of such amendments is not to deal with property; it is the exercise of powers inherent at the time of its creation and integral to the property itself.

Another case is that of **Gordon Laidler & Associates v. Hocking (Supreme Court of New South Wales), Young J. unreported 6 March 1995)**. This case involved a dispute between two joint venture parties. The issue was whether a fishing boat licence issued under the New South Wales Fisheries & Oyster Farms Act 1935 was property. The Judge held it was. This case contains a useful examination of the cases in Australia. The Judge highlighted the older view of a licence which was a permission to do something which would otherwise be illegal. If the licence was purely personal to the person to whom it is issued and could not be transferred, it was hard to categorise it as a proprietary right. The Court in referring to these cases stated that the classifications as to whether a licence was property tends to depend on whether the licence is transferable. Thus in R V Toohey; Ex Parte Meneling Station Pty Ltd (1982) 158 CLR 327 the High Court had to consider whether a grazing licence issued under Northern Territory Legislation constituted a right of property. The licence was not assignable, nor was there any applicable market for the transfer of the licence. The Court held that such a licence was <u>not</u> property.

The last case of relevance is the recent High Court was that of **The Commmonwealth v. WMC Resources Limited (1998) HCA 8**. The fact in this case was that WMC Resources was the holder of an interest in an exploration permit issued under Federal legislation to permit and encourage exploration for petroleum in defined areas of the Australian continental shelf. Subsequently, the Commonwealth agreed with the Republic of Indonesia to establish a zone of co-operation in an area of the disputed sea bed boundaries between the Island of Timor and Australia known as the Timor Gap. Some of the areas of exploitation provided by the permit fell within the zone and by subsequent Federal law were extinguished in order that new permits might be granted within the zone by a joint authority constituted by Australia and Indonesia. In relation to the permit, in the end it was acknowledged by the Commonwealth that it was proprietary in nature. It was noted that the rights of the permit was susceptible of exercise during the currency of the permit, the permit could be transferred and the interest in the permit may be created or assigned subject to approval. The High Court acknowledged that these qualities of the permit and WMC's interest in it are indicative of the proprietary character of the rights possessed by the WMC.

3. DISCUSSION

From the cases discussed above, the following is indicative of a statutory licence that is proprietary in nature:

i. Whether the licence is saleable or transferable (whether subject to approval or not).
ii. Whether an interest in the licence may be created or assigned.

Indicia of a statutory licence which is not proprietary include:

i. Where the licence can be terminated upon notice by the Minister.
ii. Whether the licence is personal in nature.

The mere fact that a licence is statutory, and the statute can be amended does not make it not proprietary (otherwise no statutory licence could be proprietary in nature). It is clear from discussing legal cases in Australia that one needs to examine the bundle of rights conferred by the statute to ascertain whether such a licence is property or not.

The last case discussed here is the recent Scallop case in Victoria. **Alesios v. The Honourable Stockdale & Ors (Supreme Court of Victoria, Unreported decision of Cummins J. 15 April 1988)** and **The Honourable Alan Robert Stockdale & Ors v. Alesios & Ors (1989) VSCA 128**. In late 1996 the Victorian Parliament enacted legislation which had the following effect:

i. It cancelled scallop licences to take scallops in Port Phillip Bay in Victoria.
ii. It provided that a licence holder is entitled to be paid a sum of money which sum is to be determined by the Treasurer and the Minister for Fisheries.

There were no guidelines legislated as to how the Minister and the Treasurer would determine the sum to be paid. The Minister and Treasurer when carrying out their function determined that each scallop fisher should be paid the sum of $120 000 being in their view the value of the Port Phillip Bay scallop licence.

A group of scallop fishers issued legal proceedings against the Ministers claiming that the Ministers must determine and pay full and proper compensation to each individual licence holder for the cancellation of their Port Phillip Bay scallop licences. The case was heard in March 1998 before Mr. Justice Cummins of the Victorian Supreme Court. His Honour found that the Ministers were wrong by not considering each licence holder individually and comprehending the consequential loss to each licence holder as a result of the cancellation of his or her Port Phillip Bay scallop licence.

It was argued on behalf of the scallop licence holders that these licences were property and therefore attracted the common law principle that a statute will not be construed to take away property without compensation unless the statute says so unequivocally. The Court held that the purpose of the common law principle of compensation is to protect the rights of subjects and the principle is to be scrupulously defended by the Courts and with vigilance. Such principle, however, will not avail licence holders unless the licence is property in nature. His Honour, after an analysis of the scallop licence, held that the licence was property in nature and accordingly attracted the common law principle of full compensation upon cancellation.

The Court further held that Parliament intended that:

i. the payment to be made to each licence holder reflects his or her loss of the benefit of the preannouncement market value of the licence, scallop boat, scallop equipment and commonwealth permit and the post announcement market value of same
ii. there be compensation for loss of a licence holder's business of dredging for, or taking and selling, scallops pursuant to the scallop licence.

The matter went to the Court of Appeal where two of the Judges said that the common law rule did not apply in this case because on a proper reading of the statute it excluded its application. These Judges stated that the Common Law Rule might be called in aid if the Treasurer and the Minister had made determinations of altogether arbitrary amounts such as $A50 or even $A5000. The third Judge held that the question was not whether the statute expressly, or by implication, excluded a particular type of compensation, but rather what is the nature of the compensation which the statute contemplates.

This case is currently on appeal to the High Court. The case is a most unusual one where the Courts had to deal with the property of a citizen being taken away by a state government without payment of full and proper compensation. In the event that property were taken by the Commonwealth Government a fisher may be able to rely on Section 51(xxxi) of the Australian Constitution which allows for the acquisition of property on just terms.

There may well a number of ways of implementing property rights-based fisheries. In my view in order to have strong property rights in a traditional limited-entry licensed fishery at least the following rights, entitlements and matters should be present in legislation:

i. The entitlement of the licence must be clearly defined.
ii. The licence should be automatically renewed when it expires.
iii. The licence should be freely transferable to another person (subject only to eligibility criteria) set out in legislation or a management plan.
iv. The licence should be able to be used as security for financial accommodation. Further, financial institutions should be able to register an interest over the licence and be protected under the provisions of an Act of Parliament.
v. The licence should become an asset of the holder's estate upon his death.
vi. In the case of a fishery involving a total allowable catch, the Minister should determine it after receiving advice from the management advisory committee in relation to that fishery.
vii. The management tools for the fishery should be set out in a management plan which can be enforced, where necessary by regulation, or as licence conditions.
viii. Full and proper compensation should be payable and a proper mechanism for payment of compensation should be set up where a licence holder can establish that there has been a reduction or diminution in the value of the licence (other than on biological grounds). An example of compensation could include any policy decision by a government to reduce, or diminish, fishing grounds or rights.
ix. These matters should be set out in an act of parliament and not in delegated legislation (*i.e.* regulation or management plan).

BALANCING SECURITY AND FLEXIBILITY IN GRANTING A RIGHT TO CATCH FISH

B. Wylynko and L. McIntosh
Mallesons Stephen Jaques, Solicitors
Level 10 Central Park, 152 St George's Terrace
Perth WA 6000 Western Australia
<brad.wylynko@msj.com.au>

1. INTRODUCTION

Of the 26 major commercial fisheries in Western Australia 23 have been rated "fully" or "over" exploited (FWA 1998)[1]. As governments grapple with this type of problem, they are turning to alternative methods of regulation including the provision of private property rights to fish. This paper addresses the availability of using private property rights in the commercial fishing context.

2. CURRENT REGIME

The right to fish has traditionally been considered a form of property known as a *profit à prendre*[2]. Classifying the right to fish as a profit meant that the owner of the land under the water containing the fish could grant fishing rights to anyone he chose. However, the notion that the right to fish at the seashore was a purely private right was qualified by another ancient notion - the common law right to fish[3].

The courts have recognized that every citizen has a common law right to fish in tidal waters. This right has been accepted in Australia[4] However, the right is not unrestricted - it may be abrogated by competent legislation[5]. This means that a *profit à prendre*, granted by the government, may still be possible in tidal waters. If possible, would it be desirable?

3. *PROFIT À PRENDRE*

A *profit à prendre* is a property right which encompasses the right of a person to enter onto another's land and take part of the produce of that land[6]. The owner of a *profit à prendre* has a right of action against any person who disturbs that right. In addition, a profit may be registered and the holder may be able to mortgage against the profit. From a fisher's point of view; therefore, a *profit à prendre* would be very desirable. But, is a fishing right capable of being classified as a *profit à prendre* today?

4. THE RIGHT TO FISH

For a fish right to be eligible to be a *profit à prendre*, it must satisfy two criteria:

i. the fish must be capable of being classified as part of the land under the water and
ii. the fish must be shown to have been "taken from" the land.

Generally the right to take wild animals is incapable of being a profit. However, the common law has traditionally treated fish as capable of forming the subject matter of a *profit à prendre*, despite the fact that are migratory and wandering[7].

At the same time, while it is clear that fishers take fish where they find them and fish do move around, nevertheless, the fishers are located within waters overlaying Australian land. Therefore, the fish are "taken" from the land. On these two grounds, the right to fish seems capable of being considered a *profit à prendre*[8].

5. REGULATION OF THE FISH RIGHT

However, a private property fish right in the form of a *profit à prendre* would probably be limited. In order to ensure sustainability, governments would likely regulate the amount or type of fish which could be caught.

For example, under the *Endangered Species Protection Act 1992* (Cth) seven species of fish are listed as endangered, and six as vulnerable. It is an offence to take any of these fish from a Commonwealth area[9]. When the *Environment Protection and Biodiversity Conservation Act 1999* (Cth) comes into force, it will be an offence to take an action that will have a significant impact upon a listed species regardless of whether the species was in a Commonwealth area or not[10]. If any fish species are added to this list in the future, and a fisher has a right to fish a quantity of that fish, that right will be effectively expropriated. Any of these regulations may have the effect of effectively expropriating or seriously affecting a fish right.

---

[1] Fisheries Western Australia, *State of the Fisheries Report 1997/1998*. Literature cited.
[2] *Wickham v Hawker* (1840) 7 M & W 63.
[3] *Neill v Duke of Devonshire* 8 AC 135 at 177.
[4] *NSW v Cth* (Seas and Submerged Lands Case) (1975) 135 CLR 337 at 419-420.
[5] *Harper v Minister for Sea Fisheries* (1989) 168 CLR 314 at 330.
[6] *Australian Softwood Forests Pty Ltd v Attorney-General (NSW)* (1981) 148 CLR 121

[7] *Willimas v Hawker* (1840) 7 M & W 63.
[8] Also see *Harper v Minister for Sea Fisheries* (1989) 168 CLR 314 at 335.
[9] "Commonwealth area" means all waters more than 3 miles off the coast, to the edge of the Continental Shelf s. 87.
[10] s. 18.

6. CONCLUSION

The right to fish was initially a *profit à prendre* granted by the owner of the land under the waters containing the fish. However, the courts recognised that in the tidal zones and offshore, the public had exercised a common law right to fish since time immemorial and this limited the Crown from granting an exclusive right to anyone. Nevertheless, the public common law right to fish could be overridden by specific legislation in effect bringing back the notion of a *profit à prendre* in fish.

The aim of a fish right is to ensure security for fishers and to ensure a sustainable fishery. The *profit à prendre* notion may provide a basis from which to do this, but regulators will have to read a fine line to achieve both these objectives.

ITQs – NEW ZEALAND AND UNITED STATES: ALLOCATION FORMULA AND LEGAL CHALLENGES

W.J. Nielander and M.S. Sullivan
Fisheries Management Consultancy International Ltd
116 Interlake Blvd, Lake Placid, Fla. 33852, USA
<wnielander@htn.net> and <mikesu@oceanlaw.co.nz>

1. INTRODUCTION

The United States has taken a cautious approach to implementing Individual Transferable Quotas (ITQs) in their fisheries by only implementing three ITQ programmes. New Zealand, on the other hand, decided that their fisheries management was going to be based on ITQs and created the legislation and management framework accordingly. As a result, there are only three examples in the United States to analyze and little case law. In New Zealand, most of their fisheries are managed by ITQs through the Quota Management Systems (QMS) and the regulatory and legislative framework was created by the same governmental entity, unlike the United States where the management plans are created by separate fishery management councils in conjunction with States. As a result, New Zealand's QMS legislation manages most of the fisheries under similar legislative and regulatory framework.

2. BACKGROUND - UNITED STATES
2.1 General fisheries management

In the United States, fisheries management in federal waters is managed primarily by Total Allowable Catches (TACs). Once the TAC for a particular fishery is reached for the fishing season, the fishery is closed for the remainder of the season. This form of fisheries management works well in some fisheries but not in others. In those fisheries where the season would be extremely short, ITQs or some other form of allocation of fisheries resources are more efficient.

In the United States, States regulate their fisheries within three nautical miles of shore and the federal government regulates fisheries from this limit out to the 200 mile, i.e. for the EEZ[1]. New Zealand has no states or provinces so the government manages the fisheries throughout their range. This situation in the United States causes more complexity since there may be different authorities and laws pertaining to the same fish that move through federal and state waters. If an ITQ programme is implemented in the United States in federal waters and not in state waters, enforcement to determine where the fish are caught can be expensive and difficult.

Each fishery in federal waters is managed by fishery management councils[2]. The North Pacific Fishery Management Council in Alaska has implemented regulations for their fisheries far different than those of the South Atlantic Fishery Management Council for their fisheries off Florida. Therefore, the ITQ allocation methodology and applicable regulations in each ITQ fishery are very different. New Zealand's ITQ programmes, on the other hand, were based on the same allocation methods and are similar.

In the United States, the predominant federal marine fisheries that are managed under ITQs are wreckfish, surf clams, Pacific halibut and sablefish. Of these fisheries, the surf clam fishery management plan, which initiated ITQs, has been in existence the longest, since 1990. The wreckfish ITQ plan has been in existence since 1992. The Pacific halibut and sablefish ITQ fishery management plans have been in existence since 1993. As a result, there are few court cases regarding federal ITQ legislation and regulations. However, this paper will analyze the allocation formula utilized in each fishery, as well as the few cases challenging ITQ allocation and administration.

2.2 Atlantic surf clam and ocean quahog fishery

The regulations pertaining to Atlantic surf clam are in Volume 50 CFR (Code of Federal Regulations) 648.70. The original final rule implementing the surf clam ITQ fisheries set forth the application requirements to determine whether an owner would receive an ITQ quota[3]. Those conditions require that applicants must have had reported landings of surf clams or ocean quahogs between 1 January 1979 and 31 December 1988[4]. Initial allocations were made by species in the form of an allocation permit issued to the vessel owner, specifying the total number of bushels he or she was entitled to harvest based

---

[1] Off the west coast of Florida and the states along the Gulf of Mexico, state waters extend to nine nautical miles.
[2] *The Magnuson Fisheries Management Act* established eight regional fisheries management councils to manage fisheries in federal waters. 16 U.S.C. 1852. The purpose set forth in the Act is: "to establish Regional Fishery Management Councils to exercise sound judgment in the stewardship of fishery resources through the preparation, monitoring, and revision of such plans under circumstances (a) which will enable the States, the fishing industry, consumer and environmental organizations, and other interested persons to participate in, and advise on, the establishment and administration of such plans, and (b) which take into account the social and economic needs of the States." 16 U.S.C. 1801.
[3] 55 Fed. Reg. 24184 (June 14, 1990). The Comments and Responses summarized in the first six pages of the federal register cited herein sets forth the rationale for most aspects of the ITQ program for surf clam and ocean quahog. See pages 24184 through 24189, supra.
[4] The original regulations were set forth at 50 CFR 652.20 (1990).

on the allocation percentage calculated pursuant to the regulations. There were several components involved in the initial allocation. These components were based on the historical catch (80%) and vessel size (20%).

The historical performance component was based on the log book reports for the years 1979, 1980, 1981, 1982, 1983, 1984, 1985 (counted twice), 1986 (counted twice), 1987 (counted twice), and 1988 (counted twice), resulting in a total history of fourteen data years for each vessel. The two years with the vessel's lowest landings were deleted from each vessel's history, leaving a total of twelve data years in each vessel's catch history. The historical performance of each vessel relative to the entire fleet was calculated by dividing the individual history totals by the total for the fleet, resulting in a 'historical ratio'.

The vessel size component was determined by a cubic factor for each vessel, calculated from the vessel's length, width and draught. This factor was summed to give a fleet total. The relative size of each vessel to the total for the whole fleet was calculated by dividing each vessel's value by the fleet total. The vessel's historical performance contributed 80% of the allocation index and the vessel's physical size contributed 20%.

The result of the implementation of the ITQ system was that the number of vessels and the surf clam off-shore fleet shrank by 41%, to 75 vessels in 1991, an historical low since 1980. However, average productivity of the offshore surf clam fleet under the ITQ system increased to a record level, as the fleet reduced its excess capital in the fishing capacity. In addition, three of the largest owners of surf clam ITQ increased their ownership from 48.5% to 50.6% of the total fishery[5].

In summary, the surf clam and ocean quahog ITQ programme has been deemed a success by NMFS in that it reduced the overcapitalization of the fishery and helped increase the value of the fishery[6].

2.3 Wreckfish fishery

Wreckfish are a deep water, grouper-like fish taken by a directed fishery that began with two vessels landing fewer than 30 000lb in 1987. By 1989, over 2 000 000lb of wreckfish were caught by roughly 25 vessels. This increased to 4 000 000lb for the calendar 1990, landed by more than 40 vessels. This extreme increase from 30 000lb and two vessels in 1987 to 4 000 000lb and 40 vessels in 1990 gave rise to the Fishery Management Council determining some form of limited-entry programme was necessary (Gauvin, Ward and Burgess 1994). By 1991, there were approximately 90 vessels permitted for wreckfish, and a 2 000 000lb TAC was put into effect. Wreckfish appeared to be a ripe fishery for ITQs, since it was a small fishery with a small number of fishing vessels and few ports.

The final rule implementing the wreckfish ITQ programme became effective in April 1992. When the programme was implemented, shares of the fishery were allocated to historical participants based primarily on catch history. The eligibility criteria was that catches of at least 5000lb of wreckfish in either 1989 or 1990 had to be documented. Applicants were responsible for providing fish-house receipts and affidavits from fish-houses for their catches. Official landing records were used to verify submitted records. The initial allocation formula divided 50 of the 100 available percentage shares in direct proportion to the applicants' documented catch from 1987 to 1990. The remaining 50 shares were divided equally among eligible applicants; a total share for an applicant was the sum of the two sub-allocations.

In the initial allocation, a single business entity could not receive more than ten of the 100 percentage shares. The rationale for placing limits on the share sizes at the outset was to prevent an entity from receiving an initial share which might create an unfair advantage in terms of purchasing other shares entities[7].

The ITQs are calculated by the Regional Director of National Marine Fisheries Service (NMFS) each year. Each ITQ is the product of the wreckfish TAC, for the ensuing fishing year and each wreckfish shareholder's percentage share which reflects share transaction reports on forms received by the Regional Director during the previous year. The Regional Director then provides each wreckfish shareholder with ITQ coupons. The Regional Director assigns a percentage of shares, which is the same each year, calculated from the TAC.

Fishermen's catches are tallied through the catch coupon system. Fishermen are issued coupons each year in 500-lb and 100-lb denominations, equaling the weight of wreckfish corresponding to the shareholder's percentage share of the annual TAC. Annual catch coupons are transferable among wreckfish shareholders only.

As a result of the initial allocation, 49 individuals received shares in 1992. Consolidation of the share began immediately. The number of shareholders declined to 37 by August 15, 1992, to 31 by June 1993, and to 26 by May 1994. Currently, there are 25 shareholders in the fishery, of which only three actively landed wreckfish in the 1998-99 season.[8] Information from the industry suggests that some of the wreckfish vessels are involved in other, more profitable fisheries. There is no information to suggest that the fishery has declined. In fact, although only 210 800lb were landed in 1998, the TAC remains set at 2 000 000lb[9].

The overall analysis of the wreckfish ITQ programme is that it decapitalized the fishery and allowed the fishermen to increase the ex-vessel price of the fish

---

[5] Id. At 23.
[6] Id.

[7] Id.
[8] Snapper Grouper Assessment Group Wreckfish Report, South Atlantic Fishery Management Council, February 2, 1999.
[9] *Id*. at 3.

while controlling the total allowable catch each year. However, any further conclusions are difficult to analyze since few fish have been landed in the past few years[10].

2.4 Pacific Halibut and Sablefish

The most complex ITQ programme in the United States' federal waters is the fixed-gear Halibut and Sablefish Fishery off Alaska which went into effect in December of 1993 and January of 1994. It was called Individual Fishing Quotas (IFQ). Quota share (QS) is the percentage of share for the area. In order to qualify for quota share, a person had to prove, by means of registration, documentation, or bill of sale, that they owned the vessel and that the vessel had documented landings in 1988, 1989 or 1990.

The quota shares were issued in several vessel categories. These categories were freezer vessels of any length, catcher vessels greater than 60ft, catcher vessels lesser than or equal to 60ft, and catcher vessels lesser than or equal to 35ft. There were actually different IFQ regulatory areas in different-sized vessel quotas. Each qualified person's QS was assigned to a vessel category, based on the length of the vessel.

The annual allocation of IFQ to any person in the IFQ regulatory area is equal to the product of the total allowable catch of halibut of sablefish by fixed gear for that area, and that person's QS divided by the QS pool for that area. Overages[11] are subtracted from a person's IFQ. Expressed algebraically, the annual IFQ allocation formula is:

$$IFQ_{pa} = [(\text{fixed gear TAC a} - CDQ\ RESERVE_a) \times (QS_{pa}/QS\ POOL_a)] - \text{Overage of } IFQ_{pa}$$

where CDQ = Community Development Quota.

Although the Alaska IFQ programme was complex when implemented, there were only two federal court challenges.

3. NEW ZEALAND ITQ FISHERIES
3.1 The Fisheries Act 1983

Unlike the US, ITQs have been introduced for most of New Zealand's major commercial fisheries. A total of 26 species were initially included in the quota management system in 1986, within a total of up to 10 separate quota management (fish stock) areas for each species. The 10 quota management areas incorporated all New Zealand fishery waters out to the limit of the EEZ. These 26 species accounted for approximately 83% by weight of all finfish taken in the commercial fishery in 1985 (Bogel and Dewees 1992). Since that time the number of ITQ species has increased to 42. The ITQ system is now the dominant management system, and it is stated Government policy to bring all future commercial fisheries into the QMS system.

The New Zealand ITQ system was introduced by the provisions of the *Fisheries Amendment Act 1986* which substantially altered the *Fisheries Act 1983*. The New Zealand ITQ system involved the allocation to fishermen and fishing companies of individual transferable quotas, which were that person's or entity's share of the overall total allowable catch in a particular fishery[12].

The background to the introduction of the ITQ system and the allocative process under the amended *Fisheries Act 1983*, has been extensively described (Bogel and Dewees 1992; Clark, Major and Mollet 1985; Christy 1979; Moloney and Pearse 1979; Clark and Duncan 1986; Clark, Major and Mollett 1988). The ITQ system as it was initially introduced can be conveniently summarised as follows: the Minister of Fisheries declared by notice published in the Gazette that the taking of a species of fish in an area was subject to the management system[13]. The Minister also specified the total allowable catch in respect of each species of fish for the area[14], and the periods in respect of which fishing returns were to be used for determining provisional maximum individual transferable quota (PMITQ) which formed the first and fundamental step in the allocative process[15]. The *Fisheries Act 1983* then required the Director-General of Agriculture and Fisheries to make a determination as to amount of PMITQ to be allocated to persons holding fishing permits primarily on the basis of their catch returns for the period specified by the Minister[16]. The

---

[10] The landings were 4 161 965lb in 1989; 1 970 299lb in 1990; 1 926 088lb in 1991; 1 270 557lb in 1992; 1 144 726lb in 1993; 1 203 265lb in 1994; 644 887lb in 1995; 396 868lb in 1996; 248 084lb in 1997; and, 219 800lb in 1998. There were 308 trips in 1991 and only 36 reported trips in 1998. *Id*.

[11] An overage occurs when an IFQ holder catches more than is entitled.

[12] ITQs were allocated upon entry to the ITQ system. For most finfish this was 18 September 1986, for paua, squid and jack mackerel it was 1 October 1987, for rock lobster 1 April 1990, and for southern scallop 1 October 1992.

[13] s28B(1) Fisheries Act 1983.

[14] s28C(1) Fisheries Act 1983.

[15] s28C(3) Fisheries Act 1983.

[16] s28E(1) Fisheries Act 1983. The procedure that was in fact adopted to work out provisional maximum individual transferable quotas was somewhat different to that contemplated by the legislation. Work began before the Amendment Act came into force on 1 August 1986. Regional objection committees were established, as was a national committee. The regional committees heard representations from fishermen. Primarily, the approach of the committees was to determine the historical catches of individual fishermen. Some adjustments were made to the figures of actual catches for reasons such as a change in the vessel used or the laying-up of the vessel for a certain period of the year. It is doubtful whether the committees regarded themselves as being competent to consider the commitment and dependence mentioned in s28E(3)(a), or whether they in fact did so. It seems that the attitude taken by the committees was such as to discourage the fishers from in fact placing any reliance on matters of that kind. When it came to the Director-General making his final decision on the provisional maximum individual transferable quota, the Director-General largely adopted the recommendation of the committee, confirmed in the meantime by the national committee, and may not himself have gone through the two-stage process contemplated by ss(1) and ss(3)

periods initially specified by the Minister in the introduction of species into the ITQ system were any two of the fishing years commencing 1 October 1981, 1 October 1982 or 1 October 1983 as chosen by the commercial fisherman[17].

The key component of the allocative mechanism under the *Fisheries Act 1983*, and the one that would ultimately prove the source of most litigation, was Section 28E which set out the criteria for granting PMITQs. The basis for determining each individual's or company's allocation was relatively straightforward, being:

> *"the proportion that the commercial catch of the person in that quota management area of that species or class of fish as shown in the fishing returns of that person bears to the total commercial catch in that quota management area of that species or class of fish in previous years"*[18].

Allocations could only be made only to:

> "(a) persons who held fishing permits issued under the 1983 Act at the date of the declaration under s28B of this Act; and
> (b) persons who held such permits within the previous 12 months or such longer period as the Director-General considers appropriate for special reasons relating to any particular case"[19].

Under the 1983 Act, the Minister was then authorised to enter into arrangements with fishermen to "buy back" all or part of an allocated PMITQ, the purpose being to equate the final PMITQ totals with the total allowable catch[20]. If this is not achieved then there would be a proportionate reduction of each PMITQ, but not below an earlier determined guaranteed minimum individual transferable quota (GMITQ)[21]. The end result of this process was the allocation of an individual transferable quota, which was a perpetual and transferable right to take the species to which it refered from a particular Quota Management Area (QMA).

Following the initial introduction of the ITQ system in 1986, problems arose concerning the Crown's continuing obligations to Maori under the Treaty of Waitangi and the unresolved status of customary fishing rights under S88(2) of the *Fisheries Act 1983*[22]. Following the

interim settlement reached on Maori fishing rights in the *Maori Fisheries Act 1989*, specific provisions were added to the *Fisheries Act 1983*, providing for the introduction of rock lobster into the ITQ system, which substantially mirrored the general allocative process set out above[23].

The other principal species introduced into the ITQ system under specific statutory amendments to the 1983 Act was the southern scallop. This fishery, based at the top of the South Island, was managed under a rotational enhancement scheme. Due to the unique nature of the fishery, it was initially introduced under a separate quota management scheme designed for the enhancement nature of the fishery[24]. The allocative mechanism was fundamentally different to those that went before, as each fisher received a statutorily imposed, and predetermined, allocation without any associated right of appeal or review[25].

3.2 The Fisheries Act 1996

In August 1991 the Minister of Fisheries began a comprehensive review of the ITQ system and fisheries management by appointing an independent task force to make recommendations on the future development of fisheries legislation and associated structures in New Zealand. Although the task force delivered its report in

---

of s28E. See *Wardle v Attorney – General* [1987] 1 NZLR 296, at 4.
[17] Fisheries (Quota Management Areas, Total Allowable Catches, and Catch Histories) Notice 1986.
[18] s28E(1) Fisheries Act 1983.
[19] s28E(2) Fisheries Act 1983.
[20] s28E(5) and s28L Fisheries Act 1983.
[21] s28F Fisheries Act 1983.
[22] The QMS was introduced at a time of growing recognition of Maori culture and the interests preserved by the Treaty of Waitangi. Following the landmark case *Te Weehi v Regional Fisheries Officer* [1986] 1 NZLR 680; (1986) 6 NZAR 114 (which held that s88(2) Fisheries Act 1983 had effectively preserved Maori fishing rights), and the decision in *NZ Maori Council v A-G* 8/10/87, Greig J, HC Wellington, CP 553/87, the

High Court considered the effect of the promulgation and operation of the QMS on possible Maori fishing rights protected by s88(2) (see *Ngai Tahu Maori Trust Board v A-G* 2/11/87, Greig J, HC Wellington, CP 559/87; CP 610/87; CP 614/87). The Court concluded that it was arguable that s88(2) made the Treaty of Waitangi directly enforceable in an active rather than passive sense. With the wider provisions of Article 2 of the Treaty of Waitangi directly conflicting with the proprietary nature of the rights being conferred under s28C Fisheries Act 1983, the High Court issued a series of injunctions preventing the Minister of Fisheries bringing any further species under the QMS. As a result of these developments, and in order to bring rock lobster into the QMS, the Government passed the Maori Fisheries Act 1989 as part of an interim settlement, which required the Government to purchase 10% of quota under the QMS. The partial settlement cleared the way for the rock lobster fishery to be introduced into the QMS via the transitional use of Term Transferable Quota (TTQ). TTQs were subsequently converted into full ITQ. While the QMS initially ignited indigenous claims to large areas of fisheries, it also proved an effective means of resolving those claims. Following the temporary solution set out in the Maori Fisheries Act 1989 it was agreed that, after protracted negotiation by the Crown and Maori negotiators, Maori interests in commercial fisheries would be increased to 20% in a final settlement of all claims (colloquially referred to as "the Sealord deal").
[23] Refer sections 28BA, 28CA, 28DA, 28EA, 28FA, 28GA, 28HA, 28JA, 28KA, 28NA and 28OA Fisheries Act 1983.
[24] Part IIB Fisheries Act 1983 as inserted by the Fisheries Amendment Act (No 2) 1992. Part IIB has been largely repealed, primarily by the Fisheries Amendment Act 1995 (1995 No 51) and apart from specific provisions relating to the enhancement aspects of the fishery, is now administered under the same provisions as other ITQ fisheries.
[25] Schedule 1D Part III Fisheries Act 1983.

April 1992[26], legislative and policy initiatives did not begin until late in 1994. On 6 December 1994 a new Fisheries Bill was introduced to Parliament and referred to the Primary Production Committee[27]. It was not until 13 August 1996, however, that the new *Fisheries Act 1996* was subsequently passed.

Although the *Fisheries Act 1996* has been passed, the *Fisheries Act 1983* remains partly in force and continues to be the primary Act in a number of respects. The 1996 Act is coming into force in incremental stages as supporting systems, procedures, forms and regulations are developed to support it.

Part IV of the 1996 Act, which governs allocation of ITQs, reflects the Government's policy of bringing all commercially harvested species into the QMS. Those species, which previously came under the QMS introduced by Part IIA of the *Fisheries Act 1983*, will continue to be subject to the provision of the ITQ system under the 1996 Act. Species that are brought under the ITQ by the 1996 Act also continue to be allocated on the basis of catch history[28].

On 1 October 1997 certain aspects of Part IV of the 1996 Act came into force introducing new procedures governing the allocation of Provisional Catch History[29], allocation of quota for new quota management species[30] and attendant Appeals process[31]. Currently, the management system concerning ITQs consists of a legislative hybrid, with the day-to-day operational aspects of the system still governed by the *Fisheries Act 1983* Act while the allocation procedures are those set out in the *Fisheries Act 1996*.

The basic structure, however, of the *Fisheries Act 1996* as it relates to introduction of new species and allocation of ITQs, is described in Sections 3.3-3.7.

3.3 Declaration of stocks as subject to the quota management system[32]

Like the 1983 Act, fish stocks are declared subject to the QMS by notice. The Minister is empowered to specify, for a stock introduced to the QMS:

i. the QMA to which the notice relates
ii. the fishing year in respect of the stock (i.e. its start and end)
iii. whether the total allowable commercial catch (TACC) and the annual catch entitlement (ACE)[33] are to be expressed in meatweight or whole weight and
iv. such other matters as may be contemplated by the Act.

3.4 Quota allocation[34]

Part V is far more prescriptive than its predecessor, providing little in the way of discretionary powers or options. The criteria for allocation of ITQ was significantly changed, with the removal of the equivalent of the "commitment to and dependence on" provisions of s28E(3) of the *Fisheries Act 1983*. In addition, in respect of new fish stocks, PMITQ has been replaced with provisional catch history (PCH) and is now the sole basis for allocation. Twenty percent of all quota continues to be allocated to the Treaty of Waitangi Fisheries Commission and any residual quota is allocated to the Crown.

In respect of fisheries where there are no persons eligible to receive provisional catch history (by virtue of the undeveloped nature of that fishery or where there are remaining quota shares in a stock that have not been allocated), the remaining unallocated shares in the stock are allocated to the Crown[35].

Provisional catch history is based on either:

i. an individual catch entitlement (ICE) being a catch limit before the introduction of the stock to the QMS allocated by permit, licence, regulation or notice is an annual amount of any stock to be taken exclusively by that fisher as at the date of the declaration bringing the species into the QMS or
ii. an eligible catch history over the first consecutive 12-month period commencing 30 September 1992 or such other period as set out in s33 of the Act.

Eligibility for provisional catch history is again premised on the holding of a controlled fishery licence (where appropriate) and fishing permit at statutorily specified times[36].

3.5 Calculation of provisional catch history

For those fishstocks not controlled by ICE at the time of introduction, PCH is calculated from the total

---

[26] Sustainable Fisheries, Report of the Fisheries Task Force, April 1992.
[27] This Fisheries Bill was the subject of considerable and detailed criticism. In December 1995 the committee made an interim report to the House, recommending substantial changes to the Bill.
[28] This is subject to the allocation of 20% of the quota of each species to the Treaty of Waitangi Fisheries Commission (he Commission) in accordance with the Government's commitment to Maori under the Deed of Settlement and the provisions of the Treaty of Waitangi (Fisheries Claims) Settlement Act 1992.
[29] Sections 30-41 Fisheries Act 1996.
[30] Sections 44-49 Fisheries Act 1996.
[31] Sections 51-55 Fisheries Act 1996.
[32] Sections 18 and 19 Fisheries Act 1996.
[33] Sections 65 to 74 Fisheries Act 1996 establish one of the fundamental changes to the QMS since its introduction in 1986. ACE - an annual catching right expressed in kilogrammes, is distinct from the underlying ITQ property right from which it is derived. All fishing of QMS species will be done under the authority of an ACE rather than an ITQ. In general terms, an ACE is generated at the commencement of each fishing year and is equivalent to the total ITQ held. The ACE held by a fisher then represents the proportion of the relevant TACC which the fisher may catch in that fishing year. The ITQ simply becomes a tradable perpetual harvesting right in the particular fishery which generates an annual right to an ACE.
[34] Sections 29 to 41 Fisheries Act 1996.
[35] s49 Fisheries Act 1996, subject to the allocation of 20% to the Treaty of Waitangi Fisheries Commission.
[36] s 32 Fisheries Act 1996.

weight of eligible catch[37] reported in the person's lawfully completed and furnished catch landing returns or catch effort landing returns in respect of the qualifying year or years as the case may be[38]. For the majority of cases, the qualifying years are set by the Act and are those commencing 1 October 1990 or 1 October 1991[39].

3.6 Individual catch entitlements

In respect of stocks previously managed by ICEs, a commercial fisher is only eligible to receive provisional catch history if that fisher held ICEs for that stock at the date the stock is declared by notice under s18 to be subject to the QMS. The commercial fisher's provisional catch history is the equivalent of the commercial fisher's ICE for that stock for the fishing year in which the notice under s18 of the 1996 Act is published in the Gazette[40].

3.7 Fourth schedule species (fully developed)

In respect of certain species listed in the Fourth Schedule to the Act, the requirement to allocate to the Treaty of Waitangi Fisheries Commission 20% of quota in each new stock introduced into the ITQ system[41] gave rise to fishing industry opposition, on the basis that these fisheries were fully developed and that such an allocation would adversely affect existing commercial fishers. As a result, the 1996 Act provides that where, in respect of those stocks listed in the Fourth Schedule to the Act, the Chief Executive of the Ministry of Fisheries considers that the total amount of ICEs for that stock held by eligible commercial fishers will, or is likely to, exceed the equivalent of 80% of the shares in that stock, no further steps may be taken under the Act to allocate quota for the stock concerned[42]. A further Act of Parliament will be required to introduce these species into the QMS.

4. APPEAL PROCESS FOR ITQ INITIAL ALLOCATIONS

4.1 United States ITQ Appeal Process

4.1.1 Regional characteristics

In the United States, each regional fishery management council developed an appeal process for the initial allocation of ITQs. These process were developed in the fishery management plans creating ITQs as a management tool in a particular fishery. A fishery management plan (FMP) is developed first, which may include ITQs or other management regimes, and the regulations are developed thereafter to implement the provisions of the FMP. Therefore, each ITQ programme has its own appeal process. This section will primarily address the appeal processes for Alaskan Pacific halibut and sablefish since wreckfish and surf clam programmes had, by comparison, limited appeal processes.

4.1.2 Alaska IFQ Programme Appeal Process

The appeal process in Alaska's halibut and sablefish ITQ fishery is specific compared to the appeal processes in the other ITQ fisheries in the United States[43]. Initially, any person who is "directly and adversely affected by an initial administrative determination" may file a written appeal[44]. The appeal must be submitted to the Regional Administrator of the particular region within sixty days after the date the determination was made[45]. Before an appeal will be considered, the applicant must submit a concise statement of the reasons the initial determination[46] has a direct and adverse effect on the applicant and should be reversed or modified[47]. If the applicant requests a hearing on any issue presented in the appeal, the request for a hearing must be accompanied by a concise written statement raising genuine and substantial issues of fact for resolution and a list of available and specifically identified evidence[48].

In the Alaska IFQ programme, the appellate officer was appointed by the Regional Administrator. The appellate officer has complete discretion to deny the appeal, issue a decision based on the merits of the appeal or allow the appeal but deny the oral hearing. Then the applicant may dispute the appellate officer's decision to the regional administrator[49]. The regional administrator may then affirm, reverse, or modify an appellate officer's decision[50]. The regional administrator's decision is considered to be the "final agency action", which can then be appealed to the United States' District Courts.

There were several appeals to the initial allocation of IFQs in the Alaska halibut and sablefish fishery. The appeals range from contentions that the fishermen did not receive adequate shares to contentions that the wrong owners received the shares.[51] One case, *Foss v. National Marine Fisheries Service*, was appealed to the United States District Court[52].

---

[37] For the purposes of s34 and Part XV of the Act, the term "eligible catch" is defined as the total weight of all the catch of the relevant stock lawfully taken and lawfully reported as landed, or otherwise lawfully disposed of by a person eligible to receive provisional catch history under s32 during the applicable qualifying years. This includes fish, aquatic life, or seaweed of the stock in question that was reported as bait.
[38] s34 Fisheries Act 1996.
[39] s33 Fisheries Act 1996.
[40] s40 Fisheries Act 1996.
[41] s44 Fisheries Act 1996.
[42] a39 Fisheries Act 1996.

[43] See 50 CFR 679.43
[44] 50 CFR 679.43(b)
[45] 50 CFR 679.43(d)
[46] There are many reasons why an applicant may appeal the initial determination. However, most appeals are because the applicant did not receive any ITQ quota shares or because they did not receive the quantity of quota share anticipated.
[47] 50 CFR 679.43(f)
[48] Id.
[49] 50 CFR 679.43(o)
[50] 50 CFR 679.43.(o)(5)
[51] The Alaska Regional Office of the National Marine Fisheries Service has a web site that lists the appeals decisions relating to the IFQ program. It is a very thorough site that lists the issues involved in the appeals.
See: www.fakr.noaa.gov/appeals/default.htm
[52] 161 F.3d 584 (9th Cir. 1998).

The *Foss* case involved a fisherman, Foss, who fished in the halibut and sablefish fishery during the time period that qualified him for an IFQ permit share. However, Mr. Foss was fishing in the South Pacific for several years on and off, and did not receive notice from the National Fisheries Service that he was eligible for IFQ, or a quota share, of the halibut and sablefish fishery. Foss further contended that the IFQ permit was property for purposes of his constitutional procedure of due process rights and that such constitutional rights were violated.

The Fifth Amendment to the United States Constitution provides: *"...nor shall any person be deprived of life, liberty or property, without due process of law; nor shall private property be taken for public use, without just compensation."* In order to be deprived of procedural due process, there would have to be (a) a protectable liberty or property interest in obtaining the IFQ permit; and (b) a denial of adequate procedural protections[53].

The court, in *Foss*, analyzed whether Foss had a constitutionally protectable property interest in acquiring an IFQ permit. The court found that there could be no doubt that an IFQ permit is "property." It is subject to sale, transfer, lease, inheritance, and division as marital property in dissolution. The court went on to state that property interests are created and their dimensions are defined by existing rules or understandings that stem from their independent source, such as state law rules. Accordingly, the court held that, for procedural due process purposes, Foss had a protectable property interest in receiving the IFQ permit. The Court did not address the issue of whether an IFQ permit is "property" for purposes of a "taking" just compensation under the Constitution[54].

The court analyzed the government's notice process. The government made several attempts in government publications and official publications to put thousands of fishermen on notice that the IFQ programme would be restricting access to the fishery. In addition, the government sent close to 10 000 letters to potential IFQ holders. In short, the court found that the government was not arbitrary and that the government's notice procedures were more than ample to satisfy due process concerns.

4.1.3 Atlantic surf clam and ocean quahog ITQ appeal process

The appeal process in the Surf Clam fishery was simple compared with the Halibut and Sablefish ITQ situation. In the surf clam fishery, the only grounds set forth in the regulations for appeal of the initial ITQ allocations is that data used by the Regional Administrator or the calculation is incorrect[55]. All appeals made for initial allocations are not made to an appellate officer. Instead, they are made to the Regional Administrator who is the same person who made the initial determination. This system, obviously does not provide a great deal of confidence to the applicants. The applicant may request an appeal. The Regional Administrator's decision of the appeal is final agency action, which can then be appealed to the United States' District Court.

4.2 New Zealand ITQ Appeal Process
4.2.1 Fisheries Act 1983

The *Fisheries Act 1983* provided for the establishment of a Quota Appeal Authority (QAA)[56] to which persons could appeal against allocation or failure to allocate PMITQ[57]. The grounds of Appeal were broadly set out as being[58]:

i. the amount of the PMITQ allocated to that person and
ii. the failure or refusal of the Director-General to allocate any PMITQ to that person.

An appeal to the QAA was conducted by way of a *de novo* hearing[59] and, as soon as practicable after hearing each appeal, to determine whether to grant or not and inform the parties to each appeal of its decision and the reasons for the decision. Every decision of the QAA was final unless challenged by an application for review under Part I of the *Judicature Amendment Act 1972*[60]. Initially there was no time limit specified for challenging the decision of the QAA in the High Court but, as the result of a number of cases where reviews were commenced some years after the original QAA decision[61], the 1983 Act was amended to introduce a time limitation of 3 months[62].

While the 1983 Act stopped short of giving a direct right of appeal to the Courts, it had expressly allowed for judicial review. The purpose of any such review was to ensure the Authority acted:

i. in accordance with law and principle

---

[53] *Id.* at 588. See Board of Regents v. Roth, 408 U.S. 564, 569-71 (1972).
[54] It is not the intention of this paper to address the issue of whether ITQs are "property" for purposes of "taking without just compensation." This issue could be the topic of a paper by itself.
[55] 55 Fed. Reg. 24194 (June 14, 1990).
[56] The Quota Appeal Authority consists of a Chairman (being a person who has held a practising certificate as a barrister or solicitor for at least 7 years), one member appointed after consultation with the Fishing Industry Board and one member who could not be an officer or employee of the Ministry. The members of the Quota Appeal Authority were appointed by the Minister for a term of 3 years or less (s28A *Fisheries Act 1983*).
[57] s28H initially provided that an Appeal had to be made within 28 days of notification of the disputed allocation or such longer period as the Quota Appeal Authority allowed. This was subsequently narrowed by the *Fisheries Amendment Act 1991* to 3 months.
[58] s28H. *Fisheries Act 1983*, both for affected persons and the Director-General.
[59] *Jenssen v Director-General of Agriculture and Fisheries* 16/9/92, CA313/91 (per Cooke P, at p2). See also *Wardle v A-G* [1987] 1 NZLR 296, 300.
[60] s28I *Fisheries Act 1983*.
[61] See for example *Gunn v Quota Appeal Authority* [1993] NZAR 102.
[62] s28I(4) as inserted by s2 Fisheries Amendment Act (No 3) 1992.

ii. in accordance with natural justice so far as procedure is concerned and
iii. on an assessment of the facts that was open to a reasonable Authority (the Courts could not, however, simply substitute its own view of the facts)[63].

The relief that might be granted under the *Judicature Amendment Act 1972* is also discretionary. The applicant was, in the absence of hardship or injustice, required to exercise his or her statutory appeal rights before judicial review proceedings could be invoked[64].

4.2.2 Fisheries Act 1996

The appeal procedures under the 1996 Act are significantly more limited in scope or nature when compared with the procedures under the *Fisheries Act 1983*.

The QAA ceased to exist and is replaced by a Catch History Review Committee (CHRC) established under Part XV of the Act[65]. The Act restricts the grounds on which any person, including the Chief Executive, may appeal to the CHRC or apply to the High Court for a review. The grounds on which a person can appeal are defined as[66]:

"(a) in the case of any stock for which PCH was allocated:

(i) a decision of the Chief Executive to the effect that the person is or is not a commercial fisher who has an ICE entitling the person to an allocation of PCH; or

(ii) an allocation of PCH that is different from the amount to which the person is entitled under the Act:

(b) in any other case:

(i) a decision of the Chief Executive to the effect that the person is or is not eligible to receive PCH either because that person held or did not hold a fishing permit or controlled fishery licence at the appropriate qualifying time; or

(ii) a decision of the Chief Executive to the effect that the person has, or does not have, eligible catch in the qualifying year or years entitling the person to be allocated PCH; or

(iii) a decision of the Chief Executive as to the quantum of eligible catch reported in any eligible returns made by any person eligible to receive PCH, on the ground that -

(a) the information on the relevant returns held by the Chief Executive has been incorrectly recorded by the Chief Executive; or

(b) the Chief Executive has excluded fish, aquatic life, or seaweed that was lawfully taken and lawfully reported in eligible returns from the person's eligible catch; or

(iv) an allocation of PCH that is different from the amount to which the person is entitled under s34 of this Act".

There is also a narrow right to apply directly to the High Court for a declaration as to whether that person is, or is not, an overseas person and therefore ineligible for allocation. The determinations of the Catch History Committee remain subject to judicial review under the *Judicature Amendment Act 1972*, but the limited grounds under the Act clearly provide little fertile ground compared to the *Fisheries Act 1983*.

5. JUDICIAL CHALLENGES TO UNITED STATES' ITQ LEGISLATION

Unfortunately for legal analysts, there have been few challenges to ITQ programmes in federal waters in the United States. There are only two federal cases analyzing allocation of federal ITQ fisheries management measures. The first case to determine a federal ITQ allocation programme is *Sea Watch International, et al. v. Mosbacher*[67]. The *Sea Watch* case was initiated by fishermen and seafood processors alleging serious economic harm from the new ITQ management plan in regulations. The plaintiffs argued that the decision to limit access to the quahog and surf clam fisheries was not supported by the evidence in the administrative record, and therefore was arbitrary and capricious and not in compliance with the United States *Magnuson Fishery Conservation and Management Act*[68].

One of plaintiffs' arguments in the *Sea Watch* case was that the implementation of an ITQ system amounts to privatization of the surf clam and quahog resource, and that such a transfer of private ownership interest in a fishery was unauthorized by the *Magnuson Fishery Conservation and Management Act*. The court found this argument to be unpersuasive, and held that the Magnuson Act did allow for the possibility of dividing total allowable catches into shares and quotas when Congress was debating the Act in its legislative history[69]. The court went on to discuss briefly the proprietary nature of ITQs and stated:

> "The new quotas do not become permanent possessions of those who hold them, any more than landing rights at slot-constrained airports become the property of airlines, or radio frequencies become the property of broadcasters. These interests remain subject to the control of the federal government which, in the exercise of its regulatory authority, can alter and revise such schemes, just as the Council and the Secretary have done in this instance[70]."

---

[63] *Jenssen v Director-General of Agriculture and Fisheries* 16/9/92, CA 313/91 per Cooke P, at pp 2, 3.
[64] *Wardle v A-G* [1987] 1 NZLR 296, 300 (CA).
[65] Sections 283 to 293 Fisheries Act 1996.
[66] s51 Fisheries Act 1996.

[67] 762 F.Supp. 370(D.D.C. 1991).
[68] 16 U.S.C. 1801 et seq (1998).
[69] Supra, note 70 at 375.
[70] Supra, note 70 at 376.

As a result, the court dismissed the contention that the ITQs were a transfer of private ownership interest in the fishery.

The plaintiffs in *Sea Watch* further contended that the implementation of the ITQ programme was in violation of the National Standard of the *Magnuson Fishery Conservation and Management Act*. National Standard 4 provides that conservation and management measures shall not discriminate between residents of different states[71]. If it becomes necessary to allocate, or assign fishing privileges among various United States fishermen, such allocation shall be fair and reasonable, and reasonably fair and equitable to all such fishermen, reasonably calculated to promote conservation and carried out in such a manner that no particular individual, corporation, or other entity requires an excessive share of such privileges[72]. In addressing the Plaintiffs' contention that the ITQ system was in violation of National Standard 4, the court found that such National Standard does not require that allocations of quotas to fishermen be made by calculating the exact historical catch of each fisherman on an individual basis. However, the court further found that the plaintiffs/fishermen failed to demonstrate that the use of past histories was irrational or in violation of the *Magnuson Fishery Conservation and Management Act* or other applicable law.

The plaintiffs further contended that the ITQ system was intended to drive a particular group of individuals out of the fishery and it caused small fishermen, who lacked the capital to purchase ITQs, to operate their vessel to full capacity and ultimately would drive them out of business. The court found that it was quite possible that economies of scale and transferability of ITQs would produce some consolidation. However, the court found that there was nothing intentionally unfair in the plan adopted by the government.

The final argument by the plaintiffs that the ITQ plan violates the National Standards of the *Magnuson Fishery Conservation and Management Act* in that the new regulations resulted in consolidation, contrary to National Standard 4 (that prohibits excessive shares). Plaintiffs alleged that two fishermen now hold ITQs totaling 40% of the annual catch quota for ocean quahogs, and that fragmentation of the remaining shares would necessarily result in further consolidation. The court found that: "This figure does give pause, although the broad number may not be economically significant[73]." The court went on to discuss that there is no definition of "excessive shares," and that the Secretary of Commerce's judgment of what is excessive is given significant weight by the court[74]. Therefore, the court found that the administrative record reflected that the government considered the problem and addressed it by providing an annual review of the industry concentration of quotas. The court in the *Sea Watch* found that the government had provided an adequate administrative record of the decision-making process and, therefore the ITQ programme was held not to be arbitrarily capricious.

The second case addressing ITQs in the United States is *Alliance Against IFQs v. Brown*, 84 F.3d 343 (9th Cir. 1996). This case involved the sablefish and Pacific halibut fisheries. In *Alliance Against IFQs*, the court reviewed the sablefish and halibut fishery management plan and regulations that provided for individual fish quotas (IFQs). The regulatory scheme for halibut found at 50 C.F.R. 679.40 is complicated compared to other ITQ schemes.

The court analyzed the allocation scheme whereby the qualifying fishers had to have had landings of halibut or sablefish during 1988, 1989, or 1990. The quota share was based on the person's highest total landings of halibut during 1984 to 1990. The court, in this case immediately commented that it appeared to be unfair against the fishermen. The court stated in the first few pages of the case that:

> "The regulatory scheme has a practical effect of transferring economic power over the fishery from those who fished to those who owned or leased fishing boats. For these reasons, among others, the case is troubling and difficult[75]."

The court once again reviewed the statutory framework which provides that any fishery management plan which is prepared by a Fishery Management Council or the Secretary of Commerce with respect to any fishery may establish a system for limiting access to the fishery if, in developing such system, the Council and Secretary take into account the present participation in the fishery, historical fishing practices in, and dependence on the fishery, the economics of the fishery, the capability of the fishing vessels used in the fishery to engage in other fisheries, the cultural and social framework relative to the fishery, and any other relevant considerations[76]. Unfortunately, there was a substantial delay between the formulation of the fishery management plan and the promulgation of the Rules. The plaintiffs argued that the cutoff of fishing history was 1990, but the plan did not go into effect until the end of 1993. However, the court reviewed the administrative record and determined that there was substantial reasoning in support of using previous years for the allocation cutoff.

The plaintiffs, in *Alliance Against IFQs* argued that use of data for 1990 could not be considered "present participation" under the *Magnuson Fishery Conservation and Management Act*, therefore it was in violation of the Act. Yet, the court found that, while the length of time

---

[71] 18 U.S.C. 1851
[72] 16 U.S.C. § 1851(a)(4).
[73] Sea Watch, supra, at 377.
[74] Id. at 377.

[75] Alliance Against IFQs v. Brown, 84 F.3d 343, 345 (9th Cir. 1996).
[76] 16 U.S.C. § 1853(b)(6)(A).

between the end of the participation period considered and the promulgation of the Rules "pushed the limits of reasonableness," the court was unable to characterize the use of 1988 through 1990 as so far from "present participation" as to be arbitrary and capricious, which is the standard of review[77].

Plaintiff also argued that the ITQ system was unfair and inequitable because allocation of ITQ was only made to vessel owners and leasees of vessels, and not to crew of vessels. This provision, they argued, was unfair and contrary to the Magnuson Act, which states that any allocation of quota shall be "fair and equitable" to all such fishermen[78]. Yet, once again, the court found that the government had reviewed the issue and believed that equity to people who invest in boats and the greater ease of ascertaining how much fish boats, as opposed to individual fishermen, had taken, favored allocating quota shares according to the owner and leassees of boats. The rationale for the Fishery Management Council was that, vessel owners and leaseholders are the participants who supply the means to harvest the fish, and suffer the most financial and liability risk to do so, and direct the fishing operations. The Council did consider allocating quota share to crew members, but decided against it because of the practical difficulties in documenting crew shares. As a result, the court found that the Secretary was not arbitrary and capricious in its determination to exclude crew members from obtaining quota shares[79].

The final argument by the plaintiffs was that it was inappropriate for the Secretary of Commerce to add Bellingham, Washington as a port in which clearances would be made by the NMFS. The court reviewed the general scheme for landings of fish, which involved vessel clearances with the NMFS's inspection of individual fishing quota permits. The Rules also only allowed sixteen primary ports in Alaska and Washington, for unloading halibut and sablefish. The plaintiffs were arguing that, since most of the fish were caught in Alaska, a State of Washington port should not have been utilized. The court dismissed this argument.

The Court obviously had difficulty with holding in favor of the government because it was apparent that fishermen would be significantly affected by the IFQ programme. The Court stated:

> "This is a troubling case. Perfectly innocent people going about their legitimate business in a productive industry have suffered great economic harm because the federal regulatory scheme changed. Alternate schemes can easily been imagined. The old way could have been left in place, but whoever caught the fish first, kept them, and seasons were shortened to allow enough fish to escape and reproduce. Allocation of quota shares could have been on a more current basis, so that the fishermen in 1996 would not have their income based upon the fish they caught before 1991. Quota shares could have been allocated to all fishermen, instead of to vessel owners and lessees, so that the non-owning fisherman would have something valuable to sell to their vessel owners. But we are not the regulators of the North Pacific halibut and sablefish industry. The Secretary of Commerce is. We cannot overturn the Secretary's decision on the grounds that some parties' interests are injured[80]."

In the *Alliance Against IFQs'* case, as well as the *Sea Watch* case, the courts appear to review ITQ programmes cautiously since the initial allocation process may cause enormous economic harm to individual fishermen. However, the courts' role in the United States in such cases is only to review the administrative record developed by the Fishery Management Council and the government in developing the Fishery Management Plan and applicable regulations. The test in the United States is that courts may only set aside the Secretary of Commerce's decision if the decisions are found to be: arbitrary, capricious, an abuse of discretion or otherwise not in accordance with law; or, contrary to constitutional right, power, privilege, or immunity; or, in excess of statutory jurisdiction, authority, or limitations, or short of statutory right; or without observance of procedure required by law.

6. JUDICIAL CHALLENGES TO NEW ZEALAND ITQ LEGISLATION

6.1 Fisheries Act 1983

As was to be expected, a number of technical legal challenges initially arose regarding the eligibility of persons or companies for allocation. One of the first successful challenges concerned the "ownership" of catch histories and therefore the entitlement to allocation of PMITQs. In *Montgomery v A-G*[81], the High Court held that the Director-General could not refuse to make an allocation to a person who came (or should have been recognised as coming) within the terms of s28E(2) only because the relevant returns were furnished in the name of another person. The Court helpfully noted that how returns furnished in the name of one person were to be treated when there are two or more competing claimants was a matter for the Director-General to consider and decide (and, one might be tempted to add, inevitably be reviewed on).

The most serious technical challenge arose in the case *Gunn v Quota Appeal Authority*[82]. Prior to the introduction of the ITQ system, many part-time fishers (many of whom were Maori) were refused fishing permits on the

---

[77] Alliance Against IFQs v. Brown, 84 F.3d 343, 347 (9th Cir. 1996).
[78] 16 U.S.C. § 1851(a)(4).
[79] 84 F.3d, 343, 349.

[80] Alliance Against IFQs v. Brown, 84 F.3d 343, 349 (9th Cir. 1996)
[81] 28/3/88, Henry J, HC Auckland CP 1445/86, p15.
[82] [1993] NZAR 102.

basis of the Director-General's determination as to the meaning of "commercial fisherman" (Ackroyd, Hide and Sharp 1990)[83]. The Director-General purported to exclude any person who was intending to engage in fishing for sale, either throughout the year or during a specified part of the year, but who did not otherwise satisfy a requirement of substantiality. The High Court ruled that the Director-General's determination was repugnant to the definition of "commercial fisherman" under S2 (as it was at the time of the declaration under S28B). This decision, at least potentially, opened a floodgate of technical challenges to past determinations as to the eligibility of numerous affected persons.

Notwithstanding the above appeals, it was the clause "commitment to, and dependence on" contained in S28E that would provide the real challenge to the allocation system under the 1983 Act. Section 28(3) provided that:

"(3) In determining any provisional maximum individual transferable quota the Director-General may, where the Director-General is satisfied in a particular case that the provisional maximum individual transferable quota determined under subsection (1) of this section would be unfair having regard to —

(a) the commitment to, and dependence on, the taking of fish of that species or class in that quota management area by the person at that date of the declaration under section 28B of this Act; and

(b) the other provisional maximum individual transferable quota (if any) allocated to that person, —

allocate a different provisional maximum individual transferable quota to the person".

It did not take a number of fishers long to identify that s28E(3) offered a significant avenue for overcoming the lack of a substantial catch history in the qualifying years and the corresponding PMITQ allocations made by the Director-General. Appeals to the Quota Appeal Authority quickly followed. Of the 1800 individuals notified of their PMITQ allocations, 1400 lodged initial objections, and in excess of 1100 Appeals to the QAA were subsequently lodged by 1988 (Clark, Major and Mollet 1989). Ultimately, some of those who failed in their appeal to the QAA turned to the Courts for remedy.

The Appellate Courts focused on the meaning to be attributed to the phrase "commitment to, and dependence on".

Initially, the Courts took a relatively cautious approach to reviewing the decisions of the QAA. In *Jenssen v Director-General of Agriculture and Fisheries*[84], the High Court considered an appeal against the QAA's refusal to grant the appellant an ITQ for orange roughy. The appellant had no qualifying commercial catch history and was reliant on the provisions of s28E(3). However, the appellant had spent in excess of $NZ100 000 outfitting his fishing boat in preparation for long-term fishing of orange roughy. The Authority declined to issue a quota under s28E(3) as the appellant had not exhibited sufficient commitment to justify allocation of a quota. The Court, after reviewing the history of the legislation, held that the Authority's conclusion was justifiable and one that was open to it on the facts.

In *Esperance Fishing Co Ltd v Quota Appeal Authority*[85], the Courts began to herald a view that the QAA had approached its task incorrectly when it looked at what was actually allocated and then asked whether that was unfair finding instead that it should have considered whether the allocation based on the actual catches was unfair. Subsequently, in *Wylie v Director-General of Agriculture and Fisheries*[86], it was held that under ss(3)(a) of s28E the principal inquiry was essentially that of unfairness to a person who has a commitment to, and dependence on the taking of fish of the particular species or class in question in that quota management area at the time the allocation is made. The Court noted that words of such an imprecise and unmeasured nature could be only guidelines for the essential inquiry of unfairness. The High Court held that QAA had taken a restrictive interpretation of the words in ss(3)(a) and concluded that they placed only a modest fetter on the general discretion open to the Authority to cure injustice by allocating different quota. The words "commitment to" and "dependence on" were to be viewed as words of general purport only and could be translated loosely as covering those persons seriously engaged in making a living from the taking of that species at the relevant time. In the Court's view, the QAA's restrictive interpretation requiring some form of financial hardship threshold to be crossed was not justified when one considered the scheme of the Act and the emphasis placed on the importance of establishing a fishing history.

Shortly after the decision in *Wylie v Director-General of Agriculture and Fisheries*, the New Zealand Court of Appeal heard the appeal from the more favourable decision to the traditional approach of the QAA in *Jenssen* (above). The Court of Appeal in *Jenssen v Director-General of Agriculture and Fisheries*[87] held that the words chosen by the Legislature were deliberately wide and the correct interpretation was that "commitment" extended to a firm intention to fish for a species, evidenced by the taking of significant practical steps to that end; and "dependence" refers to the economic sig-

---

[83] Notwithstanding that part-time fishermen were not seen as the major cause of the decline of fish stocks, or as landing a significant amount of fish, the group was perceived as having the greatest potential to increase effort and that it was appropriate to remove this possibility before the fisheries were rebuilt. In the event, some 2260 licence holders were to be excluded. (see Ackroyd, Hide and Sharp 1990).

[84] 14/10/91, McGechan J, HC Wellington, CP 1035/90.
[85] 10/3/92, Barker J, HC Auckland M714/90.
[86] 18/3/92, Heron J, HC Wellington CP 892/90.
[87] 16/9/92, CA 313/91.

nificance of the species in the person's fishing history or plans. In addition, in respect of species other than orange roughy for which the Authority had granted the appellant quotas, the Court of Appeal was of the view that the arithmetical approach taken in respect of the appellant's actual catches in the 1985 and 1986 fishing years, ignoring dumped, or confiscated fish, and was too narrow under ss(3). The case was then remitted to the Quota Appeal Authority for further consideration.

The successive judgments in *Montgomery v A-G, Gunn v Quota Appeal Authority* and *Jenssen v Director-General of Agriculture* cumulatively established that entitlement to a grant of PMITQ under s28E of the Act depended on a lower threshold than the Director-General or the Quota Appeal Authority had employed and that entitlement could be declared retrospectively. Following these judgments, the Government responded and a Supplementary Order Paper was introduced on 19 November 1992 proposing amendments to *the Finance Bill (No 2)* which was then before the House. These became, after further amendments, s28I(4) and s28ZGA and effectively overruled the combination of judgments in *Montgomery, Gunn* and *Jenssen* and prevented them from having further effect[88]. The provisions imposed a condition precedent to a fisher's receipt of PMITQ, requiring a fisher to be, or to have been at the time of subjection of a species to the ITQ system, the holder of a fishing permit for that species issued under the 1983 Act and introduced a time bar on taking review proceedings.

Subsequently, in *Cooper v A-G*[89], a direct challenge was launched against the constitutionality of the restrictions imposed under s28ZGA The plaintiffs contended that:

"(a) the amendment effected by the Fisheries Amendment Act (No 3) 1992 had purported to deprive them of access to the Court to secure a declaration or decision concerning claimed substantive rights, infringing a fundamental constitutional principle;

(b) they retained such rights under the original ITQ system introduced by the Fisheries Amendment Act 1986 which, as interpreted by the Court of Appeal in Jenssen, entitled them to allocation of quota; and

(c) Parliament lacked power to deprive the Court of its authority to hear a citizen's claim to have a legal right enforced".

The High Court held that the effect of s28ZGA was to reverse the effect of *Jenssen* and to overrule the decisions in *Montgomery v A-G* and *Gunn v Quota Appeal Authority* and that the intention of Parliament was to exclude further entrants into the QMS who were not permit holders, and the necessary intention of the Act was to remove the rights which, in terms of the *Jenssen, Montgomery* and the *Gunn* decisions, they had previously enjoyed.

The High Court found it unnecessary to respond in detail to an alternative submission that Parliament had no power to remove the plaintiff's substantive rights, as no authority was advanced in support of that proposition and there being no protection of property rights equivalent to the Fifth Amendment to the US Constitution against uncompensated "takings" of personal property, s28ZGA was held to meet the relevant constitutional safeguard for property rights in New Zealand. The safeguard in question was from ch 29 of the Magna Carta (by virtue of s3(1) and the First Schedule to the Imperial Laws Application Act 1988):

"No freeman shall be ... disseised of his freehold or liberties, or free customs ... but ... by the law of the land".

By any normal test, the High Court concluded that s28ZGA was "the law of the land". Having no effective answer to s28ZGA, the plaintiff's proceedings were struck out.

On the issue of the retrospective effect of the legislation, the High Court adopted the decision of Mason CJ in *Polyukhovich v Commonwealth of Australia*[90] which held that, in the absence of a constitutional prohibition, it is not beyond the powers of Parliament to enact retrospective laws.

6.2 Fisheries Act 1996

In October 1997, 11 new species were brought into the ITQ system under the new allocative provisions of the *Fisheries Act 1996*[91]. While various appeals have been filed against PCH allocations made by the Chief Executive and appeals are being currently heard and determined by the CHRC, no substantive appeals to the High Court have yet occurred.

Given the prescriptive nature of the legislation and allocative process and the narrow grounds of appeal, which are essentially technical or procedural in nature, it is unlikely that the *Fisheries Act 1996* will provide the same scope for legal challenges as its predecessor.

Even the past scope for technical challenges as to whether persons have been wrongly excluded from eligibility for PCH has been severely curtailed by the provisions of the Act retrospectively validating past permitting decisions[92]. There may well be some challenges launched by persons who filed reviews of past permitting decisions within time to avoid the retrospective validating provisions of the Act, but these by their very nature will

---

[88] This section was inserted, as from 18 December 1992, by s3 Fisheries Amendment Act (No 3) 1992 (1992 No 137). Sections 3(2) and (3) of that Act provided that nothing in s28ZGA(a) or s28ZGA(b) affected any application for review or other civil proceedings made or commenced before 16 September 1992 and nothing in s28ZGA(d) or s28ZGA(e) affected any civil proceedings filed before 5 October 1992.

[89] [1996] 3 NZLR 480.

[90] (1991) 172 CLR 501, 534.

[91] Fisheries (Declaration of New Stocks Subject to Quota Management System) Notice 1997.

[92] s329 Fisheries Act 1996.

ultimately be exausted and will have little on-going precedent value.

7. CONCLUSION

Legislation in the United States prohibited approval and implementation of ITQ programmes in federal fisheries until 1 October 2000[93]. However, the study required by the *Sustainable Fisheries Act* to be performed by the Academy of Sciences has been completed and recommends a lifting of the ban. Therefore, the United States may have more ITQ programmes in federal waters within a few years. This will provide more diversity to study the affects and perhaps further litigation.

New Zealand has developed a response to litigation challenging the exercise of discretion and factual determinations under the earlier ITQ legislation by introducing new legislation which statutorily pre-determines the allocation formulas and the application of those formulas to any given situation. The New Zealand approach, however, may eventually result in unfair and unjust allocations since the allocation formula is not flexible enough for future years. Also the 1996 Act does not recognize the issues of fairness in the allocation process that were accommodated under the commitment and dependence provisions of the earlier legislation. This will inevitably push the focus of legal challenges in New Zealand to allocation issues in non-ITQ fisheries before they become subject to the quota management system.

There is no doubt that every effort should be made during the legislative and regulatory drafting stages to minimize possible litigation exposure. Unfortunately, ITQs are, at times, so valuable that individuals risk litigation costs for the possibility of obtaining initial or additional quota. Therefore, regardless of the allocation formula adopted, the following principles should be followed to minimize the scope for successful legal challenges:

i. A thorough record leading to the allocative decisions should be meticulously documented
ii. the allocation formula and resulting process should allow the exercise of some discretion to address gross examples of unfairness and
iii. consideration should be given to implement more stringent time frames within which legal challenges must be timely filed and concluded.

8. LITERATURE CITED

Ackroyd, P., R. Hide and B. Sharp 1990. New Zealand's ITQ System: Proposal for the Evolution of Sole Ownership Corporations: A Report to MAFish, August 1990, p. 21.

Bogel, R. and C. Dewees 19.. Putting Theory Into Practice: Individual Transferable Quotas in New Zealand Fisheries, Society and Natural Resources, Vol 5, pp. 179-198.

Christy F.T. Jr. 1979. Fisherman Quotas: A Tentative Suggestion for Domestic Management, Occasional Paper No 19, Rhode Island: Law of the Sea Institute.

Clark, I., P. Major and N. Mollet 1985. The Development of New Zealand's ITQ Management System, Rights Based Fishing, NATO ASI Series, Kliwer Academic Publishers, 1989, pp. 117-145.

Clark, I., P. Major and N.Mollett 1988. Development and Implementation of New Zealand's ITQ Management System, 5 Mar Res Econ 325.

Clark, I. and A. Duncan 1986. New Zealand's Fisheries Management Policies – Past, Present and Future: The Implementation of an ITQ-Based Management System. Proceedings of the Workshop on Management Options for North Pacific Longline Fisheries 118.

Gauvin, J.R., J.M. Ward and E.E. Burgess 1994. Description and preliminary evaluation of the wreckfish (*Polyprion americanus*) fishery under individual transferable quotas. Marine Resource Economics 9(2) 99-118.

Ministry of Agriculture and Fisheries (New Zealand) 1985. ITQs A Chance for Change, Fisheries Bulletin, MAF Fisheries Management, Vol 1, No 3, December 1985.

Moloney, D.G. and P.H. Pearse 1979. Quantitative Rights as an Instrument for Regulating Commercial Fisheries 36 J. Fish Res Board Can 859.

Wang, S.D. and V.H. Tank 1993. The Performance of US Atlantic Surf Clam and Ocean Quahog Fisheries under Limited Entry and Individual Transferable Quota Systems. Fishery Analysis Division. National Marine Fisheries Service. May 1993.

---

[93] 16 USC 1853(d)(1)(A) 1996.

FISHING RIGHTS: A MULTIDIMENSIONAL PERSPECTIVE

N. Taylor-Moore
Fisheries Group, Department of Primary Industries
P.O. Box 3129, Brisbane, Queensland 4001, Australia
<taylorn@dpi.qld.gov.au>

1. INTRODUCTION

Society now demands ecologically sustainable use of aquatic resources. However, a growing number of different user groups are seeking greater access to these resources. As this access expands, overcrowding, ecosystem degradation and community dissatisfaction occurs. The fishing sector cannot be managed in isolation from these other user groups. Use of aquatic resources by any one sector has repercussions on all other users and managing the effects of these intra-sector and inter-sector demands requires a holistic strategic approach. An approach with a multidimensional perspective is the widely accepted ecological sustainable development (ESD) paradigm, which provides a framework for understanding the complex nature of fishing rights.

For the purpose of this paper, 'sector' means a group of users seeking similar objectives from aquatic resources, such as commercial fishers, recreational fishers, scuba divers, reef walkers, fish farmers and marina developers. Intra-sector means within specific fisheries, for example, commercial and recreational fishers targeting the same species or trawl and crab fishers targeting the same species. Inter-sector means different user groups, such as fish farmers, irrigators, tourists, amenity users and habitat modifiers using these aquatic resources to meet different wants and needs, which are not always compatible.

Therefore, each user group has a desired set of outcomes arising from the use of aquatic resources. Each sector has a set of 'rights' associated with these desired outcomes. These sets of 'rights' are a continuum of natural justice rights, customary rights, statutory rights, specific sector rights and property rights. Each sector demands their 'rights' but needs to understand that rights are by nature dynamic, relative to circumstance and time, not absolute, and can be rescinded. Each sector's set of rights may differ to those of other sectors. For example, rights related to a fishing ethic may differ from those rights based on the conservation philosophy of the users of the Great Barrier Reef. Rights are pervasive, complex and multidimensional in nature and thus not easily defined.

Generally, the significance of aquatic sector rights in the context of fisheries management depends on the degree of exclusivity inherent in the right granted to one sector that may determine the type of impact affecting the 'rights' of another. This degree of exclusivity is the key to understanding fishing rights.

2. WHAT ARE FISHING RIGHTS?
2.1 Operational aspects of rights

Fishing rights are nested within aquatic resource rights. These aquatic resource rights can be categorised by the degree of exclusion attached to them (after van de Elst 1997 and Symes 1998) as:

i. open access rights – no exclusion rights attached to any user group
ii. limited access rights – specific user groups have limited exclusion rights
iii. private property rights – private sector has exclusive rights
iv. communal rights – specific communities have exclusion rights
v. nation state rights – exclusion rights held on behalf of its citizens and
vi. global rights – exclusion by agreement of nation states.

Fishing rights are found within all categories of this continuum of aquatic sector rights and therefore have varying degrees of exclusivity attached to them. The demand for 'fishing rights' and rights-based fisheries management has stemmed from the acceptance of ESD as a cornerstone of aquatic resource management. This relationship has spilled over to the fishing sector through fishing rights being generated, or demanded, through fisheries legislation, statutory fishery management plans and the inclusion of stakeholders in the decision making process. The principles of ESD noted below require this relationship.

Four generic aquatic resource rights are identified to illustrate the complex nature of fishing rights: harvest rights, use rights, conservation rights and management rights. Aquatic resource user groups, and specifically fishers, would expect these rights to exist under a modern rights-based management regime. These rights are briefly explained below with examples of the exclusive actions, which are statutory explicit, implied or socially acknowledged.

Harvest rights are mainly the right to take fish (recreational fishing, commercial fishing and charter fishing licences), for ownership of the resource (ITQs), to sell fish (commercial fishing licences) and to own transferable licences (commercial fishing licences). Others rights include access to specific species (recreational fishing licences) and access to specific locations (indigenous fishing areas and recreational only fishing areas). Harvest rights also include the right to water supply (off-stream irrigation permits), the right to collect aquaculture broodstock (fish farm permits), the right to enter new fisheries (exploratory and developmental fishing licences) and the right to compensation for loss of harvest rights.

Use rights are made up of: the right to modify aquatic habitat (mangrove permits); right to visit specific sites (tourist permits); right for passive recreation (dive permits); right to destroy aquatic resources (land planning approvals); right to modify stream flows (water planning approvals); and rights to obtain amenity (assumed rights).

Conservation rights consist of the right to maintain ecological flows in riverine systems; the right to provide fish-ways; the right to set aside conservation areas in aquatic ecosystems; the right to have multiple-use areas; the right to protect endangered species; and the right to manage threatened species.

Management rights include: rights-based management; the right for stakeholders to be recognised in the management planning decision making process, the right of formal conflict resolution processes, the right of fishers to negotiate, the right for compensation for impacts of exclusion; and the right for all aquatic sectors to be involved.

2.2 Fishing rights and the ecological sustainable development paradigm

The challenge facing Australian fisheries management is the adoption of a holistic and sustainable approach to the management of aquatic resources through a fisheries ecosystem management framework. This framework is the ESD paradigm: *conserving our ecosystems in the pursuit of current and future welfare of Australian citizens.* Underpinning this framework are seven guiding policy principles, accepted by Australia. These are:

i. decision making processes should effectively integrate both long and short-term economic, environmental, social and equity considerations
ii. where there are threats of serious or irreversible environmental damage, lack of full scientific certainty should not be used as a reason for postponing measures to prevent environmental degradation
iii. the global dimension of environmental impacts of actions and policies should be recognised and considered
iv. the need to develop a strong, growing and diversified economy, which can enhance the capacity for environmental protection, should be recognised
v. the need to maintain and enhance international competitiveness in an environmentally sound manner should be recognised
vi. cost effective and flexible policy instruments should be adopted such as improved valuation, pricing and incentive mechanisms and
vii. decisions and actions should provide for broad community involvement on issues which affect them" (Environment Australia 1992).

A balanced approach to aquatic resource management is required that takes into account these principles, the multidimensional nature of ESD and the associated objectives of fisheries management and which leads to the attainment of the desired outcomes and goals related to the ESD paradigm. The application of the ESD paradigm provides the framework for understanding the complex nature of aquatic resource rights. The, values and examples of associated objectives of a rights-based fisheries management regime associated with the major dimensions of ESD should include at least the following:

i. *ecological dimension* (sustainable ecosystems, by-catch reduction, threatening processes modified, endangered and threatened flora and fauna species protected)
ii. *biological dimension* (sustainable fisheries, fisheries habitat protection)
iii. *industrial dimension* (diversification of fishing operations, changes in fleet structure and industry infrastructure)
iv. *economic dimension* (viable fishers and seafood industry, seafood supply, compensation for adjustments, resource security and planning security)
v. *social dimension* (increased recreational fishing opportunities, fair access by all user groups)
vi. *governance dimension* (system of property rights, formal fisheries management planning processes)
vii. *political dimension* (reduced conflicts, acceptance of decisions)
viii. *cultural dimension* (maintaining communities and lifestyles, indigenous community co-management) and the
ix. *psychological dimension* (aesthetic and amenity values maintained, maintaining sense of community).

Depending on the circumstances other dimensions and objectives would need to be incorporated into the policy analysis.

2.3 Rights-based management paradigm

Rights-based management objectives such as the above can be achieved through the implementation of a series of aquatic resource management strategies. Examples from the range of these strategies include:

i. *access controls* (areas and fishing platforms such as beaches, zoning through marine and national parks)
ii. *tenure controls* (land-use planning approvals, licences, fish shares, permits, approvals to modify habitat)
iii. *input controls* (gear, fishers and vessels)
iv. *output controls* (on quantity to be taken such as bag limits and quotas, periods and transferability)
v. *temporal and spatial controls* (seasonal and area closures, fisheries reserves, environmental reserves, refugia)
vi. *species controls* (species protection, spawning areas, translocation strategies, totem species and size limits) and
vii. *finance controls* (such as access fees, quotas levies, cost recovery, economic rent, peak body support, and licence fees).

How these interrelationships, i.e. linkages of resource management objectives based on ESD dimensions and the range of potential management measures, are managed is both strategic and tactical in nature. Outcomes of these relationships can create, or lead to, a redistribution of fishing rights. For example, access taken away from fishers through species controls used to increase the "ecological rights" for fish stocks to achieve a stated goal of biodiversity protection.

Such outcomes can be planned and unplanned. The management planning process needs a set of guiding principles to achieve these desired outcomes. Guiding principles should highlight the uncertainties and risks of applying inappropriate measures and of setting unachievable or unmeasurable objectives. This strategic approach is proposed as a rights-based management paradigm. The linkages are illustrated in Figure 1.

The ten following guiding principles should underpin a rights-based management paradigm for aquatic resources and therefore for fisheries management. This paradigm is not found as an entity in any Australian fisheries legislation.

i. *Formal planning processes* with specific management objectives, measurable outcomes and formalised Management Plans, including a monitoring, compliance and surveillance program, as the basis of aquatic management
ii. *risk management* with decision rules based on multiple performance indicators and reference points and best available information
iii. *clarification of resource security* for all users of the marine environment
iv. *conflict resolution processes* available for all parties
v. *equity and social justice* for all affected parties
vi. *a formal and systematic inclusive consultation process*, which will lead to better and 'owned' decisions through equal opportunity for greater stakeholder and civic participation in decision making using formal negotiating frameworks
vii. *community empowerment* through extension/education/awareness action
viii. *flexible decision making organisational arrangements* incorporating options such as self governance, market-driven models, expert policy groups, co-management and integrated planning
ix. *formal institutional learning* for fisheries agencies and stakeholders by case studies and
x. *multidimensional evaluation frameworks* for management planning and for evaluation" (Taylor-Moore 1995, 1998).

An effective management system based on these guiding principles should lead to a better understanding and effects of fishing rights such as harvest rights, use rights, conservation rights and management rights mentioned above.

3. A MULTIDIMENSIONAL PERSPECTIVE OF FISHING RIGHTS

3.1 Introduction

Changes in policy based on the above rights-based management paradigm may lead to a redistribution of

Figure 1
A rights-based aquatic resource management Matrix of ESD based outcomes

Ecological Sustainable Development (ESD)	Aquatic Resource Management Measures (*e.g.* licence fees - *Finance controls*)
Dimensions and Objectives of ESD (*e.g.* Financial viability of fishing fleet (*Economic dimension*))	**Type I and Type II Outcomes** (creation or redistribution of fishing rights)

fishing rights resulting in the long-term ecological sustainable development of aquatic resources. Changes of fishing rights as an outcome of these regimes or policies are inherent in many aquatic resource management and policy regimes. For example, a policy change such as the introduction of a reef tax on tourist operators in exchange for long term access rights to specific reefs, does not affect on the rights of commercial fishers: *i.e.* a *Type I outcome*. However, a policy to give mooring rights to tourist operators on a specific reef will extinguish or reduce the rights of reef fishers to access that same reef; *i.e.* an inter-sector redistribution of fishing rights or T*ype IIa outcome*. On the other hand, the creation of recreational only fishing areas will transfer commercial fishing effort to other areas, an intra-sector redistribution of fishing rights or a *Type IIb outcome*. This paper does not consider Type I outcomes as fishing rights are neither created nor redistributed. The emphasis lies on Type II outcomes, the creation or the redistribution of fishing rights.

The implementation of this paradigm can be illustrated through a holistic framework. This enables managers and stakeholders to identify and evaluate the relationships between the measures available for managing aquatic resources, the multidimensional context within which these management decisions are made and the Type I and Type II outcomes for the fisheries sectors (Figure 1). Changing a sector-specific aquatic resource policy, which creates Type II outcomes, will increase the demands by all affected parties for the development of rights-based fisheries management.

Consider the ESD approach given in Figure 1. Policy changes and their impact on fishing rights can be explained through two general cases. Case 1: using the matrix rows to show changes in fishing rights within a specific ESD dimension/objective using a range of management measures. Case 2: using the matrix columns to show changes in the impacts of a specific management measure on fishing rights given a range of ESD dimensions and objectives. Table 1 provides a broad range of these potential types of changes, specific cases and fishing rights inherent in the outcomes arising from aquatic resource management.

3.2 General case 1

Each row of Figure 1 and Table 1 provides a range of measures available to achieve specific ESD-based objectives for the management of aquatic resources. For example, in the *ecological dimension* an objective such as maintaining aquatic biodiversity within aquatic ecosystems could be achieved through a range of aquatic resource management strategies, all of which create a reduction in the right of commercial fishers to access fish stocks - Type IIa outcomes.

Some strategies include using *access controls* such as conservation zones in marine parks to protect threatened habitats, which leads to exclusion of fishers; or applying *tenure controls* through limited entry licences to exclude classes of fishers to fish stocks. Other strategies include *input controls*, through gear modification requirements, such as turtle excluder devices (TEDs) and bycatch reduction devices (BRDs) that reduce access by fishers to potentially profitable fish biomass. Or the use of *output controls*, such as total allowable catches (TAC) and individual transferable quota (ITQs), which reduce the take of specific species. The application of *temporal controls* reduces conflicting use of specific waters resulting in reduced take during fish aggregation periods. Other means are *financial controls* that impose permit fees on certain classes of fishers and species controls that place certain species on restricted export schedules, both of which transfer effort to other species.

3.3 General case 2

Each column of Figure 1 and Table 1 specifically highlights the ESD dimension that may be affected by the application of a specific control measure and the potential changes in fishing rights. For example the application of *access controls*, through the creation of recreational-only fishing areas changing, or removing, the rights of commercial fishers to those fish stocks.

Access controls on the main natural resource example of the ESD matrix are affected by many factors:

i. *ecological dimension* where the protection of endangered species rights creates a reduction of commercial fisher rights to bycatch species
ii. *biological dimension* where the lack of access to national parks or beaches could lead to a reduction of commercial fisher rights to target specific species
iii. *industrial dimension* – loss of access to marine parks causing a structural adjustment of fleet infrastructure and thus loss of commercial fisher rights to harvest specific species
iv. *economic dimension* – the need for compensation for the loss of fishing rights through creating refugia
v. *social dimension* – where fisher family rights are reduced as fishers have to move to other fishing because of the closure of their current location
vi. *psychological dimension* – the right of fishers to obtain satisfaction from certain fishing styles is diminished when fishers are forced to change their fishing location and
vii. *cultural dimension* – where the rights of indigenous people are lost through the granting of fishing access to sacred sites.

3.4 Specific cases

Table 1 is proposed as an example of the combination of strategic options available to policy makers and the potential outcomes of these interactions. It is a useful approach to analysing policy options and changes leading to a redistribution of fishing rights. These include:

i. *rights to modify aquatic habitats* through local authorities and other coastal management planning processes that increase the number of marinas, canal estates, golf courses, etc.
ii. *rights to use fisheries habitat* shifting from the fishing sectors to other users to achieve tourism benefits
iii. *rights of fishers to security of access to marine park* resources in exchange for a reduction of fishing effort to enhance biodiversity
iv. *rights of coastal communities to economic* development as population shifts create demands for broader social objectives where fishing rights can be reduced as local authorities seek to close rivers to commercial fishers or where they are increased through demands for more seafood outlets and fisheries infrastructure such as boat ramps
v. *rights of indigenous communities for decision making empowerment* as local fish resources are sold to commercial fisheries under quota
vi. *rights of fishing sector power and decision making* shifting to include other aquatic resource users
vii. *right access to fishing areas* reduced as the commercial sector adjusts or restructures through rationalisation and buy-back schemes to accommodate the demands of conservationists to protect endangered species and habitat

Table 1
Impacts of aquatic resource management policy on fishing rights

ESD dimensions	Aquatic resource management measures						
	Access controls	Tenure controls	Input controls	Output controls	Temporal, spatial controls	Species controls	Financial controls
Biological	preditor/prey relationships maintained sustainability of target species	sunset clause leading to long term target biomass levels	latent effort realised pressure on target species	sustainable limits; stock size uncertainty	protection of spawning stocks; single user area	commercial species protected; health and disease reduceds	research levies financing monitoring programs
Ecological	ponded pastures habitat loss by development national park access	tenure of access building stewardship of ecosystem	bycatch reduction devices reducing key threatening processes	bycatch of protected species reduced	key habitats protected by marine park zoning	inappropriate translocations reduced	internalising of externalities
Economic	loss of income and regional growth from area closures needing compensation and resource security	windfall gains from buy back schemes	reduced financial viability of fishers through input control inefficiencies	concentration of ownership seafood supply reduced	closures affecting tourism and seasonal incomes	fishing gear adapted to commercial only species	cost recovery or economic rent providing adjustment incentives
Governance	marine park management planning	property rights to protect long term seafood supply	licences to use types of fishing gear	quota management regime	surveillance programme	compliance and enforcement	payment of fishery rents
Social	family dislocation to meet increased amenity demands	dislocation and continued conflict with recreational sector if property rights not in place	competitive fishing gear and different sector rules	bag limits and angling licences changing style of operations	sectoral expectations based on areas set aside for recreational fishers	creed of greed	consultation costs
Political	community support based on fisher and community expectations	self adjustment reducing conflict and changing community perceptions	shark meshing to protect beaches changed due to capture of dolphins	recreational TAC to match commercial TAC	local Government area closures causing resource partitioning	'cuddly' species protected	peak organisation support for input into planning processes
Psychological	community acceptance of less access to sensitive sites	husbandry ethic expected as trade off for long term licences	less fishers that can be seen in sensitive areas	fish stocks are better off	closures lead to better visual effects	feel-good effects from endangered species protection	willingness to pay for controls
Cultural	indigenous sacred sites access limited seen as loss of historical activity	community commercial fisheries	traditional hunting gear redefined to meet modern activities	indigenous TAC as part of commercial TAC	indigenous fishing areas	totem species	support for indigenous input into planning

viii. *rights to offstream water users* increased to meet the social and economic objectives of water needs for irrigation and human consumption through water infrastructure projects and

ix. *rights of the community to a cleaner environment increased* as psychological objectives are meet through environmental impact studies required from aquatic resource based infrastructure such as aquaculture farms and cage culture in coastal waters.

4. CONCLUSION

The nature of fishing rights is complex and far more significant than the narrow concept of property rights in fisheries management. An holistic approach to understanding their significance in aquatic resource management can be achieved by using ecological sustainable development (ESD) as the policy making framework. The paper provides a general introduction to this concept and a few examples of how this process could work.

5. LITERATURE CITED

Environment Australia 1992. National Strategy for Ecological Sustainable Development. Australian Government Publishing Service.

Symes, D. 1998. Property Rights and Regulatory Systems in Fisheries. Fishing News Books, 1-16.

Taylor-Moore, N. 1995. Fisheries Management Planning in the 1990s. Proceedings of the Second National Fisheries Managers Workshop, Bribe Island Queensland. Queensland Fisheries Management Authority, 1-14.

Taylor-Moore, N. 1997. The Allocation of Inshore Marine and Estuarine Fish resources in Australia: the need for a Precautionary Decision-making Paradigm? In Hancock, D.A. Developing and Sustaining World Fisheries Resources - The State of Science and Management. Proceedings Second World Fisheries congress, Brisbane. Australian Society for Fish Biology, 352-357.

Taylor-Moore, N. 1998. Adjustment of Queensland fisheries: a draft policy for restructuring the Queensland commercial fishing fleet. Queensland Department of Primary Industries. pp. 37.

Van Elst, R. 1997. How Can Fisheries Resources Be Allocated? In Hancock, D.A. Developing and Sustaining World Fisheries Resources - The State of Science and Management. Proceedings Second World Fisheries congress, Brisbane. Australian Society for Fish Biology, 426-427.

RIGHTS-BASED FISHERIES MANAGEMENT IN NEW SOUTH WALES, AUSTRALIA

A. Goulstone
Commercial Finfish, NSW Fisheries
Fisheries Management Division
PO Box 21, Cronulla 2230, New South Wales, Australia.
<goulstoa@fisheries.nsw.gov.au>

1. INTRODUCTION

New South Wales (NSW) is located on the east coast of Australia and has a coastline approximately 1350km long. The prevailing East Australian Current which runs from Queensland in the north to Victoria in the south mixes warm tropical waters with cooler temperate waters. The continental shelf adjacent to NSW is narrow compared to other States in Australia and indeed other countries. These environmental conditions provide for the existence of a range of commercial fisheries that is relatively small by volume but extremely diverse in terms of the number of species taken (over 130) and the gear types used.

Such diversity in species, fishing methods and environmental conditions makes designing a sound and equitable rights-based management regime a difficult task. The current regime in NSW involves about 30 different types of commercial fishing endorsements, each of which authorises a different type of fishing within nine defined commercial fisheries. There are approximately 1800 licensed commercial fishers.

Despite the introduction of management rules for the first time through the enactment of the Fisheries Act 1865, the development of a rights-based scheme is a very new development in NSW. Limited licensing was only introduced in 1982 and most fisheries have been largely open access up until very recently.

Significant changes were made to the management regime in 1994 with the introduction of a new *Fisheries Management Act*. This Act provided a radical new framework for managing commercial fisheries by issuing a perpetual right to licence holders in the form of shares – the new framework is termed 'share management fisheries'. Not surprisingly, the debate over the proposal to introduce a full property rights system for commercial fisheries was a lively one.

As a comprehensive analysis of the share management fishery framework has already been undertaken (Young 1996), I will provide more of a comparative review of the two management frameworks currently available. I will also outline the novel approach that I believe was taken in issuing fishing rights in NSW through the allocation of validated catch history. But first, a brief history of management in NSW to put it all into context.

2. HISTORY OF FISHERIES MANAGEMENT IN NEW SOUTH WALES

Commercial fishing commenced in NSW estuaries in the mid 1800s. The introduction of the *Fisheries Act 1865* was a response to concerns of overfishing. This saw the commencement of seasonal and area closures on commercial fishing. A Royal Commission on Fisheries in 1879 resulted in the introduction of the *Fisheries Act 1881*, which provided for *inter alia* the regulation of fishing gear including controls on the mesh sizes of nets, and, importantly for rights-based management, for the licensing of fishers and fishing boats (NSW Fisheries 1999; adapted from Wilkinson 1997).

Fishing licences were relatively easy to obtain throughout most of the 20th Century and limits on individual fisheries only commenced in real terms from 1980 onwards. While the licensing of fishers and their boats had been underway for over 50 years, it wasn't until 1982 that a freeze on the issue of new boat licences was introduced. This freeze was consolidated in 1987 when the Government decided not to issue any further personal fishing licences in order to prevent speculation by investors.

Throughout this period, licences were renewed annually and were subject to cancellation if a fisher failed to spend the major portion of their time, or earned the major portion of their income, from fishing. This provided little security for the industry as fishers were sometimes required to prove their financial details and licences could potentially be cancelled at each annual renewal. The requirement forced fishers to work harder than they may have wanted or apply more fishing pressure on the resource than was desirable.

The abalone dive fishery was the first fishery in NSW to which entry by new fishers was limited; this occurred in 1982. Later, in 1989, a quota system was introduced for abalone and divers were issued an equal amount of individual quota. Prawn trawling in the five major estuary systems was limited in the mid-1980s, and at the same time steps were taken to restrict prawn trawling in offshore oceanic waters (beyond 3 nautical miles). The next of the State's fisheries to be put under restricted access was the rock lobster fishery (a trap fishery) which saw limits on the number of participants introduced in 1993 and a individual transferable quota scheme introduced in 1994.

A common theme in all of the above fishery restrictions was that future access was defined by examining the historical involvement of fishers in each fishery through historic catch records.

3. FISHING RIGHTS UNDER THE NEW LEGAL REGIME

After a long public consultation period during which the issue of property rights for commercial fisheries received a great deal of attention, the *Fisheries Management Act 1994* was passed by the Parliament. The Act was controversial because it contained a new "share management fishery" scheme that enabled the issue of a full property right (or 'shares') to eligible fishers, and was the first scheme of this type developed in Australia. The Act also retained provisions for an alternative limited access management framework termed "restricted fisheries" which provides a less secure fishing right.

The old scheme where licences were subject to administrative assessment and possible refusal of renewal on an annual basis was abolished and relaced by an automatic licence renewal process. Despite the inclusion of the new share management provisions, this change to the licensing regime was a significant enhancement to the rights and security of the industry. No longer are fishers subjected to annual licence renewal assessments and the Minister now has limited grounds upon which to refuse a renewal application, primarily related to breaches of the Act.

The primary difference between the share management fishery scheme and the restricted fishery scheme is that shares are issued in perpetuity and shareholders have a statutory right to compensation if the Government decides to close the fishery and cancel the shares. No such compensation provisions apply in restricted fisheries. A summary comparison between the restricted fishery and share management fishery frameworks is provided in Table 1.

While the share management scheme provides a greater property right, this comes at a cost to shareholders. The increased cost is a result of the Government's current policy to collect the full attributable costs of management in share management fisheries and a legal requirement for shareholders to pay a "community contribution" (synonymous with the payment of a resource rent). The community contribution

Table 1
Comparison of the restricted fishery and share management fishery frameworks

Variable	Restricted fishery	Share management fishery
Property right	Validated catch history which gives rise to an "Entitlement"	Shares
Access right	Endorsement	Endorsement
Property right tradeable?	Yes, subject to transfer policy	Yes, subject to the management plan
Statutory compensation payable?	No	Yes, if shares are cancelled
Statutory management plan required?	No	Yes, 5 year plans (the plan can only be reviewed if criteria for review set within the plan are met)
Appeal mechanism	Statutory review panel	Statutory review panel
Management cost recovery	Partial, moratorium on additional cost recovery for term of 1st plan	Full cost recovery
Community contribution payable?	No	Yes

is a payment for the privileged and secure access of shareholders to a community-owned resource and the revenue is deposited directly into the Government's general consolidated fund.

In a share management fishery, it is the ownership of shares that determines whether a person is eligible to hold a commercial fishing licence and an endorsement to operate in a particular fishery. The management plan for a share management fishery can:

i. set a minimum holding of shares required before fishing is permitted to take place;
ii. set a higher minimum shareholding for new entrants than for existing fishers and/or
iii. periodically increase the minimum holding of shares required by all fishers.

In a fishery managed by quota, the quota issued to individuals (or companies) must be allocated proportionally on the basis of shareholdings. In an input control fishery, shares may simply determine who is eligible for access to the fishery or, in a more complex manner, they can equate to the amount of gear able to be used. For example, the number of shares held might determine how many traps could be used, the length of net permitted or even the number of days that could be fished.

In the way that a share fishery management plan can use shareholdings to determine relative access levels, 'catch history' can be used to determine the level of access in a restricted fishery. For example, the level of catch history for a species could determine the quantity of quota allocated to a person or the amount of gear able to be used. In other words, both regimes are forms of fractional licensing schemes whereby the property right is kept separate from the right to access the resource (see Figure 1).

Catch history has been the primary mechanism for allocating property rights and access rights in NSW fisheries. It can form the basis for the issuing of shares in a share management fishery, although the formula used to issue shares can also take into account any other entitlements existing in a fishery before it becomes a share management fishery.

To address this complexity, a policy was first developed in consultation with industry representatives. Given the diversity in fishing operations, the policy evolved substantially over time to cater for unforeseen circumstances. Particular difficulties were faced when determinations had to be made with respect to whether catch history had transferred with boat licences during the 1986 to 1993 validated catch history period. The policy included (NSW Fisheries 1994, 1996):

i. Identifying the scope of each person's fishing business[2]. This entailed an assessment of each fishers' operation to determine if they held one or more fishing businesses and to identify the boat licences that were associated with each business. Two or more businesses were normally only awarded if the fisher had clearly separate ocean and estuary operations, or if they owned multiple large vessels that had been operated separately in the past;

Figure 1
Diagrammatic representation of the three tiered rights-based system in NSW

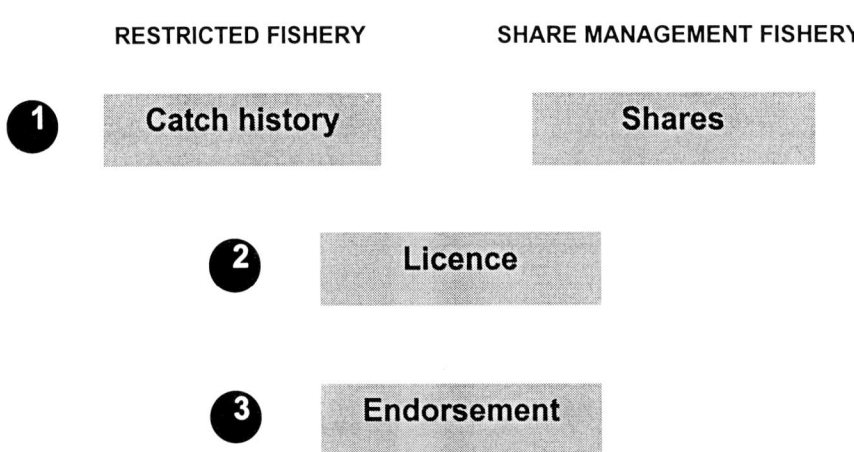

4. ALLOCATION OF CATCH HISTORY

The process of allocating catch history to fishers involved a close examination of the historic catch records that had been submitted to NSW Fisheries[1] and the development of quite detailed and complex policies to address the diversity in fishing operations that exist in the State. The traditional concept of using exclusively either the catch history of a boat or the catch history of a fisher for allocating rights would have resulted in an inequitable allocation. For instance, while the owner of a large, high investment ocean trawler would expect to receive the underlying history of that vessel upon purchase, a typical estuary fisher who uses multiple small punts (of relatively little value) would not expect to pass any catch history on with the sale of each boat.

ii. Examining the catch history attributable to each business, which fell into two categories:
- personal history – generally, personal catch history applied in circumstances where the fisher was the primary unit of effort. This included estuary and beach fishing, and boats involved in mixed estuary/ocean activities. A fishers personal history comprised all of his/her catches taken from *general purpose vessels* and
- boat history – boat history applied where the boat was the primary unit of effort. This included estuary prawn trawlers and larger ocean going vessels. These boats are referred to as *boat history vessels*.

---

[1] Catch returns were required to be submitted on a monthly basis by each licensed fisher, including information about the boats they had fished from during the month.

[2] From the 1900 or so licensed fishers and some 3500 licensed fishing boats.

iii. Making a determination in relation to each licensed boat as to whether it was a *boat history vessel* or a *general purpose vessel*;
iv. Allocating all the catches taken from a boat history vessel to the current owner of the vessel (including catches recorded by employee skippers). Catches taken from boats categorised as general purpose vessels are considered part of the personal history of the fisher who took the catch, irrespective of the boat owner.

The catch validation process was instigated by the introduction of a new Licensing Policy (NSW Fisheries, 1994). This policy abolished the freeze on the issue of new fishing licences which had been in place since 1987 (albeit considerably weakened through exemptions and variations) and established new requirements for people wishing to enter the commercial fishing industry.

Under the 1994 Licensing Policy, a person had to purchase an existing fishing business that contained a minimum level of validated catch history to obtain a new commercial fishing licence. The policy aimed at consolidating fishing businesses and ensuring that new participants had to replace actual fishing effort, rather than purchasing a licence that had been unused, or little used, since 1986. The policy proved successful in forcing the amalgamation of many smaller and inactive fishing businesses and is still in place today. The policy promoted the trading of validated catch history and established catch history as a vital foundation for the development of property rights in NSW.

5. THE IMPLEMENTATION OF SHARE MANAGEMENT FISHERIES AND RESTRICTED FISHERIES

Upon the introduction of the new Act, the Liberal Government of the day decided to apply the share management fishery scheme to all major NSW commercial fisheries. The first step in the process was to define specific boundaries for each fishery (by species, method, area, etc), then establish advisory committees to develop criteria for issuing shares. The defined fisheries that were established are listed in Table 2.

In March 1995, a new Labor Government was elected into power. They were more cautious about issuing long term property rights in fisheries and quickly established a review into the implementation of share management fisheries.

The review was conducted from June to August 1995 and it found that, *inter alia*, the abalone and lobster fisheries should proceed directly to a share management scheme as the rights were already well defined and the management direction well established. The review noted the substantial amount of administration and time required to establish a full share management fishery and recommended that the priority for the other fisheries should be to limit access in the first instance (using the restricted fishery framework) to prevent opportunistic increases in fishing effort. The review also recommended that a scoping process should be undertaken for these fisheries during which the long term management direction should be determined before perpetual property rights are issued. Needless to say, the review recommendations were accepted. Table 3 summarises the results of the share management fishery review.

The advantage of implementing a restricted fishery regime prior to becoming a share management fishery is that the management structure or direction can be altered if needed without having to redefine the fishery (and expose the Government to large-scale compensation) or undertaking a major review of the plan as would be required under the share management scheme. This is important in newly managed fisheries as fishing activities can change, often unpredictably, in response to new management rules and an adaptable regime is obviously preferable.

6. CONTINUING DEVELOPMENT OF RIGHTS

Since the review into the implementation of share management fisheries, the abalone and lobster fisheries have proceeded through the provisional stages of share management and the development of a management plan is nearing completion. At present, shares are issued on a provisional basis in those fisheries, however, they will be replaced with final shares when the management plans commence.

Table 2
Defined NSW commercial fisheries

Abalone	Ocean trap and line
Lobster	Ocean hauling
Ocean prawn trawl	Estuary prawn trawl
Ocean fish trawl	Estuary general

Some fisheries, such as abalone, lobster and estuary prawn trawl had been previously defined and subjected to limited access so no changes were required to their definition. New definitions were needed for some of the other large and more complex fisheries such as estuary and ocean trap fishing because access to these methods was still available to all licensed commercial fishers.

A process was undertaken to declare the remaining fisheries as restricted fisheries, ending in March 1997. The process included establishing steering committees for each fishery to establish eligibility criteria and an application and assessment process. In most cases validated catch history was used as the basis for

Table 3
Action following the review into the implementation of share management fisheries

Abalone Lobster	Share management fishery
Estuary prawn trawl Ocean hauling – beach hauling sector Ocean prawn trawl – offshore sector	Remain as a restricted fishery
- inshore sector - deepwater sector Ocean hauling – purse seine sector Ocean trap and line Ocean fish trawl Estuary general	New restricted fishery

determining access to (part of) each new restricted fishery.

The restricted fishery entry criteria were set at a relatively low level in order to be inclusive rather than exclusive. This generated concern that too many endorsements would be issued and that licences previously operated at a low level would be transferred to fishers who could operate at much greater levels of effort. It was this concern that led to the introduction of an "interim transfer policy" which requires fishing businesses to be transferred as an entity (i.e. no licence splitting) and limits each business to only one transfer (NSW Fisheries 1999b).

The purpose of the interim transfer policy is to allow the limited transferability of fishing businesses while NSW Fisheries, in consultation with each Management Advisory Committee, develops longer term criteria for the transfer of entitlements. Discussions are now taking place in some fisheries about the development of more stringent catch history criteria to be applied upon the transfer of a fishing business to prevent the activation of latent (or unused) fishing effort and restructure fisheries over time.

7. DISCUSSION

Despite the differences between the share management fishery and restricted fishery frameworks, both schemes embody a form of property right. There is no doubt that the concept of share management provides a stronger and more secure property right for commercial fishers, but this comes at a financial cost to fishers and results in a system that arguably is less adaptive to change.

Shares are clearly the 'property' in a share management fishery, just as validated catch history can be considered the property in the majority of restricted fisheries. Catch history is limited, tradeable, and more often than not, determines whether access to a fishery is granted. Indeed, despite not benefiting from statutory compensation provisions, validated catch history has assumed a significant monetary value ever since it was used as a criteria for new entrants to gain a commercial fishing licence.

The use of catch history as a basis of determining access to a fishery or for the allocation of shares in a share management fishery is favoured because:

i. It is equitable, as fishers who correctly recorded their catches throughout the years gain advantages over fishers who black-marketed product and failed to properly declare their catch. It also recognises the rights of fishers who have historically relied on a fishery compared to those who only fish in a fishery periodically.
ii. It can be used effectively to define scaled access to resources, by allocating more access to fishers with a greater catch history.
iii. With the development of appropriate policies, it can be used to prevent the activation of latent fishing effort or to facilitate an industry funded restructure program for a fishery.

Using the concept of a combination of boat history and personal catch history, rather than the traditional catch history allocation method, has been an effective way of addressing the allocation of rights in a diversified fishery system and has proved an equitable method of allocation. However, if such a system is to be adopted, it is recommended that firm and clear policy guidelines are prepared from the outset, and developed in close consultation with the industry.

One of the detractions of using catch history as a rights allocation tool is that future data are likely to be biased by fishers who over-report catches in the hope that they will be considered in future allocations. Despite using an historical validated catch history period in NSW (i.e. 1986 to 1993) over-reporting is suspected to be occurring. The validation of catch data with market records, observer data or fishery independent data will most probably become important over the coming years for monitoring fish stocks.

8. LITERATURE CITED

Fisheries Management Act 1994. *http://www.austlii.edu.au/au/legis/nsw/consol_act/fma1994193/*

NSW Fisheries 1994. Licensing Policy – June 1994.

NSW Fisheries 1996. Licensing Policy – Version 2 – November 1996.

NSW Fisheries 1999a. Fishery profile: Estuary General Restricted Fishery.

NSW Fisheries 1999b. Discussion paper – Options for new transfer rules in the Ocean Trap and Line Fishery – August 1999.

Wilkinson, J. 1997. Commercial fishing in NSW: Origins and development to the early 1990s. NSW Parliamentary Library Research Service.

Young, M.D. 1996. The design of fishing-right systems – the NSW experience. *Ocean and Coastal Management.* **28**(1-3): 45-61.

THE CLOSURE OF THE PORT PHILLIP BAY SCALLOP FISHERY

S. McCormack and R. McLoughlin
Fisheries Victoria
GPO Box 500 East Melbourne
Melbourne, Victoria 3002, Australia
<Stephen.McCormack@nre.vic.gov.au> and <richard.mcloughlin@nre.vic.gov.au>

1. INTRODUCTION

Port Phillip Bay is a large shallow embayment in southeast Australia adjacent to the cities of Melbourne and Geelong. A combined population in excess of four million people reside around the Bay and it is a major recreational area supporting a large recreational fishery and many other aquatic sports. A variety of species of fish and shellfish have also been commercially harvested for over a century and in 1963 a winter/spring dredge fishery for scallops was established. Initially scallop-dredging was carried out without attracting adverse comment as most of the boats engaged in the fishery were relatively small and not highly visible.

A rapid increase in the use of Port Phillip Bay for boating and recreational fishing occurred during the 1960s and 1970s. A large benthic-feeding snapper (*Pagrus auratus*) was one of the main recreational target species. While there is no specific point in time when it began, it was clear that by the mid 1980s there was a high level of concern in the community about the effect that scallop-dredging was having on the seabed of Port Phillip Bay. Much of this concern related to the (perceived) indirect negative impacts of scallop-dredging on recreational fishing for species like snapper although, increasingly, there was sound evidence that scallop-dredging causing severe effects on benthic communities in the Bay (*e.g.* Currie and Parry 1994).

In response to these concerns an attempt was made by the Victorian Government to close Port Phillip Bay to scallop-dredging in 1991. This action was successfully challenged in the Supreme Court by the scallop industry and fishing resumed in the following season.

In 1996 it was announced that, as part of an initiative to promote Port Phillip Bay as a recreational and tourist area, scallop-dredging was to cease in the following year. Licences would be cancelled and they would be bought back from operators in the fishery. This time however, the Victorian Government legislated for the closure and a process by which compensation would be paid. This action was again successfully challenged by the industry in Supreme Court when, while not arguing against the ability of the Government to close the fishery, they argued on the basis that the amount of compensation was inadequate and had not taken into account the differing interests of individuals. A subsequent appeal by the Victorian government overturned this decision.

This paper does not attempt to fully document either the complete history of the Port Phillip Bay Scallop Fishery or the arguments about the property-right component of the commercial fishing licences that provided for this activity. However of significance is that the Port Phillip Bay scallop fishery is probably the first, and certainly the most valuable commercial fishery in Australia, to be closed by specific legislation. Thus the events that led to this closure combined with arguments that followed in the Supreme Court are of widespread interest to both participants in, and managers of, commercial fisheries.

2. HISTORY OF THE FISHERY

Despite there being some knowledge of the existence of scallops beds in Port Phillip Bay and there being lucrative domestic and overseas markets for scallops it was not until September 1963 that a commercial dredge-fishery was established. In that year the then Fisheries and Wildlife Department published information on the extent of the beds and described the methods used in Tasmania for the capture and processing of scallops. A number of boats left Tasmania where the scallop fishery was in decline to try their luck in Port Phillip Bay. The results could be described as spectacular and many boats - possibly in excess of 200 - joined the fishery, often from other sectors of the fishing industry.

For the period 1963/4 the total Australian landings for scallops totalled 2953t live shell weight with the bulk of the catch coming from Tasmanian and Queensland waters. The Port Phillip Bay scallop-dredge fishery, which started in September 1963, increased the level of national production to 6988t the following year. At its peak in 1966-67 landings of 13 289t shell weight were recorded from Port Phillip Bay.

In 1968 a new *Fisheries Act* was introduced which provided for a specific Scallop Fishing Licence. Boats that were employed in the fishery immediately prior to the 18 December 1967 were entitled to be issued with a scallop licence for dredging in Port Phillip Bay. Some 160 licences were issued. Many of these licences were subsequently not renewed, presumably because the holders had concerns about the long-term prospects for the fishery.

With the future obviously uncertain many fishermen and their boats returned to their former fisheries or left the industry altogether. At this time the majority of boats in the fishery did not exceed 15m in length with a variety of hull design and wheelhouse configurations. There was considerable investment by the processing sector in plant and equipment as demand for scallops remained high in local and overseas markets.

The decline in the fishery in 1970 was the catalyst for boats to look elsewhere in Victoria for scallops and eventually extensive new beds were found off Lakes Entrance. While many of the boats that worked in Port Phillip Bay were unsuitable for the demands of commercial fishing in Bass Strait, approximately 70 made the journey to Lakes Entrance and commenced dredging for scallops from that port. Licences were issued to about 30 local boats to fish for scallops in Bass Strait waters adjacent to Lakes Entrance. The boats that stayed in Port Phillip Bay were eventually denied the opportunity to work Bass Strait as that component of the Victorian Scallop Fishery was also closed to additional entrants.

Eventually Port Phillip Bay again produced scallops although the peak catches of 1966 and 1967 were never repeated. The combination of landings from Port Phillip Bay and Lakes Entrance increasingly became necessary to sustain the catching and processing sectors of the industry. Many of the smaller boats that were adequate for working in the relatively sheltered waters of Port Phillip Bay found the going a lot tougher at Lakes Entrance. Fishermen, realising that to be viable in the fishery which now had separate but interlinked components, started to upgrade their boats to be able to work more distant and rougher waters. Thus the larger, more sea-worthy boats capable of working scallop beds that were being discovered some distance from Lakes Entrance also started entering the Bay fishery by replacing the smaller Port Phillip Bay boats.

This change in the fleet configuration occurred gradually during the 1970s but received a major impetus in the early 1980s when a speculative investment company purchased a large number of the existing licences and scrapped the boats they were attached to. New steel boats capable of fishing for scallops anywhere in Bass Strait now entered the fishery. Most of the licences that were purchased also had the right to dredge in Port Phillip Bay as part of their entitlement and they fished there when it was more profitable than operating in Bass Strait.

The investment company eventually folded but the boats that were built and the licences attached stayed, leaving the fishery with a fishing capacity well in excess of the available stock (McLoughlin 1994).

3. FLEET STRUCTURE

The Victorian scallop fishery had three distinct licensed sectors :

i. Some 30 generally small boats, less than 15m in length, and mainly of wooden construction that stayed in Port Phillip Bay and chose not to make the journey to Lakes Entrance. When that component of the fishery was closed in 1971 their licences were limited to fishing in Port Phillip Bay.
ii. Approximately 64 boats that, by virtue of the fact that they originally fished in Port Phillip Bay and then relocated to Lakes Entrance in 1970, were issued with a licence to fish both components of the fishery. They held what became known as All Victoria Waters Scallop Licence.
iii. A third group of approximately 30 boats were licensed to fish only in waters adjacent to Lakes Entrance. These were "local" boats that were issued licences in 1970/71 to fish the newly found scallop beds adjacent to their home port.

Concerns about the impact that scallop-dredges were having on the bed of Port Phillip Bay started to appear in the 1970s. The substantial upgrading of the fleet raised the profile of the fishery and made its operations more obvious. The tendency of the fleet to work patches of scallops as a group drew attention to the fishery and concerns were expressed at the effect that 30 or 40 boats dredging a confined part of Port Phillip Bay for scallops was having.

Bigger boats towing larger dredges, often the same dredge that was used in the ocean, were in contrast to the original operators in the fishery that worked smaller boats with small dredges. Many of the dredges were fitted with a tooth-bar which attracted specific criticism. No doubt there were a range of issues that led to growing concern about the dredge fishery and by the mid 1980s the fishery was regularly subject to criticism about the adverse effect that dredging had the bed of Port Phillip Bay. Scientific work in the early 1990s (McLoughlin *et al.* 1991) showed that dredges also had a deleterious effect on scallops as well as non-target species and added fuel to the growing debate about the impact of the dredges in the Bay.

4. COMMUNITY DEBATE

Irrespective of the changes that happened to the configuration of the scallop fleet, the growing community awareness of environmental issues combined with the popularity of Port Phillip Bay as a major recreational amenity for the state, focused attention on the dredge fishery. Recreational fishers in particular blamed scallop-dredging for what was perceived to be a decline in their catches of fish of which snapper was the main species.

As an adjunct to the scallop-dredge fishery in Port Phillip Bay some boats also dredged for the common blue mussel and in some years annual catches exceeded 2000t (shell weight). This fishery was closed in the mid 1980s because of the concern over the effect that dredging was having on the hard, or reef substrate, where mussels were found. Also aquaculture of mussels was developing as a new industry and it was expected that in time it would be able to replace the dredge fishery. Those boats that had fished for mussels received no compensation and were required to fish exclusively for scallops if they wished to remain in the industry.

After a good season in 1987 which followed a long period of profitability, the fishery went into a slump in 1988 and was closed entirely for 1989 and 1990. The then Fisheries and Wildlife Department had been periodically surveying scallop beds in Port Phillip Bay since the early 1970s and was able to predict fairly accuracy what the prospects would be annually for the industry. With general support from industry it was agreed that the fishery would not open until it was clear that the beds had recovered and could support a resumption of dredging.

As with the catching sector of the fishery, management of the fishery had evolved over the years and had, by necessity, become quite dynamic. In any one year there may have been 6 separate openings and subsequent closures of sectors of the fishery, all of which required changes to the Fisheries Regulations which controlled the operations of the fishermen. Changing Regulations quickly to respond to the constantly changing state of the fishery became an expensive, onerous and time-consuming task. Given the rapidity of necessary changes to the Regulations, fishery managers often required specific exemptions from normal regulatory processes, including the need for a Regulatory Impact Statement which, in Victoria, is required for most subordinate legislation.

In the late 1980s the *Fisheries Act* was amended to provide for the issue of Fisheries Notices, which enabled the responsible Minister, after consultation with the industry, to issue Notices to open and close the fishery, thereby greatly reducing the workload in managing the fishery.

In 1991 surveys showed that there were commercially harvestable beds of scallops in Port Phillip Bay. Scallop fishers were expecting that the then Minister responsible for fisheries would duly issue a Fisheries Notice "opening" the season and importantly set the daily catch-limit. The industry had been consulted as required by the relevant legislation, was aware of the abundance of scallops, and prepared itself to commence fishing.

In May of 1991 the Minister announced that he would not open the fishery. The basis for this decision was a response to community concerns about the impact of scallop-dredging on the bed of Port Phillip Bay. The scallop fishermen found themselves with a licence issued under the *Fisheries Act 1968* that enabled them to employ a boat to take scallops for sale by dredging in Port Phillip Bay. However without there being a Fisheries Notice on issue there was no opening date set for the commencement of fishing and no daily catch limits in place.

This decision not to allow scallop fishing in Port Phillip Bay was successfully challenged by the industry in the Supreme Court. While a number of factors were taken into account by the Court, it found that the licences were issued for the fishery (*i.e.* to specifically dredge for scallops in Port Phillip Bay) and that licences were issued to provide for the commercial harvesting of fish (subject to the provisions of that Act and Regulations, Fisheries Notices, *etc.*) and that it was reasonable for fishermen to expect to be able to go fishing. That is, that the Minister could not issue licences and then not allow licence holders to fish when there was ostensibly no stock management issue.

Scallop fishing continued in Port Phillip Bay on a seasonal basis until 1996 when on 11 March a joint announcement was made by the Premier of Victoria and the Minister for Conservation and Environment relating to a number of issues concerning Port Phillip Bay. It was announced for the scallop fishery that - "*the Government has pledged to end scallop dredging in the Bay by repurchasing all existing licences by the end of next year, at an estimated cost of $10 million*". The Minister for Conservation and the Environment stated in the announcement that "*Because of heightened community concern about its impact, no new licences to dredge scallops would be issued and all existing scallop licences repurchased*".

In 1996 legislation was passed to :

i. Close the fishery
ii. Cancel all relevant licences and,
iii. Pay an amount to licensees to be determined by the Minister and the Treasurer.

The amended legislation came into effect on the 31 March 1997, while an exhaustive process led to a compensation amount determined by way of payment of $A120 000 per licence. This amount considered only the price of the licence and assumed that the value of each licence was identical. Again the industry headed to court with the validity of this determination challenged in the Supreme Court on the basis that the payment did not take into account consequential losses on boats, gear, *etc*. On 15 April 1998 the Supreme Court found in favour of the scallop fishermen.

In an article in the Professional Fishermen's Magazine which was published in June 1998, industry solicitor David Fitzpatrick summarised the judgement as follows – "*The Court found that the Ministers made a mistake by not considering each licence-holder individually and they should have comprehended the consequential loss of each licence-holder.*

The Court also found that the most fundamental determinative flaw in the conduct of the Ministers is that it is predicated on the erroneous view of the nature and incidents of the licence which was cancelled by the legislation. The Ministers argued that the scallop licences were not propriety in nature. The Court found that a scallop licence issued under the provisions of the Fisheries Act 1968 for dredging for or taking scallops for sale is propriety in nature and thus attracts the common law principle of full compensation upon cancellation."

In a footnote to the above article it was stated that licence holders would probably receive between $A300 000 and $A400 000 over and above the $A120 000 already awarded.

This time the Government appealed the decision and the judgement was overturned in the Court of Appeal on 25 August 1999. No attempt is made here to summarise the judgement which runs to 23 pages but the following key issues are presented below:

i. The Court of Appeal considered that the Common Law could not be applied to the licence cancellation because it was specifically authorised by statute, which overrode common law entitlements in this case.
ii. There could have been limited access to common law entitlements had the Ministers made an arbitrary, or unreasonable, decision on the amount to be paid in respect of each licence. In this case the Minister had established a Working Party of senior officers of the Department which heard submissions from all licensees who wished to make representations and who also commissioned independent valuations of the licences. The Court considered that acting on advice from the Working Party, the Ministers had made a reasonable determination and that the requirements for procedural fairness had been met.
iii. An important issue was that the case relied upon the proper interpretation of the statute and that the Government's interpretation of the statute was correct. In reaching this conclusion the Court considered the wording of the statute itself and also the Parliamentary debates, in particular the Minister's Second Reading Speech, a statement of intent as to the purposes of the amended legislation.

5. DISCUSSION

In hindsight, it is clear that the 25-year history of this fishery is littered with numerous examples of reactive and aggressive stakeholders, litigious industry participants, a focus on industry management rather than fish stock management and ultimately a decisive role for community interests in Government decision- making. Notwithstanding this however, it is clear that once community interest in sustainability and resource- sharing issues was reflected in government policy-making, the fate of the fishery was clear. This is reflected in the bulk of the recent legal debate, with the debate shifting from argument about the 'rights' afforded by the licence to argument about the appropriate level of compensation for the loss of the licence. In general terms the courts have decided in this case that a form of property-right exists in a licence until specific legislation removes that property-right.

That the bulk of recent argument has been about the financial aspects of the decision to close the fishery is consistent with experience elsewhere (*e.g.* National Research Council - USA 1999). Capture fisheries are difficult to manage in economic terms, mainly because of poorly-defined property-rights. This has two main consequences. Firstly, there is little incentive to fish conservatively to maintain stocks at some optimum level, and instead the pressure is on fishers to catch their maximum share of the available resource. Secondly, fleet and processing over-capacity becomes the norm as fishers look to maximise their catch or efficiency of catching. Moreover, fisheries management becomes increasingly difficult in the face of declining margins and profitability of the participants in the fishery who become overly reactive to changes that may lead to reduced incomes. It is clear that both these consequences occurred in the Port Phillip Bay scallop fishery and that many lessons need to be heeded.

6. LITERATURE CITED

Currie, D.R. & G.D. Parry 1994. The impact of scallop dredging on a soft sediment community using multivariate techniques. Memoirs of the Queensland Museum 36(2): 315-326.

McLoughlin, R.J., P.C. Young, R.B. Martin & J. Parslow 1991. The Australian scallop dredge: estimates of catching efficiency and associated indirect fishing mortality. *Fisheries Research* 11: 1-24.

McLoughlin, R.J. 1994. Management of the Bass Strait scallop fishery. Memoirs of the Queensland Museum 36(2): 307-314.

National Research Council - USA. 1999 'Sustaining Marine Fisheries'. Committee on Ecosystem Management for Sustainable Marine Fisheries, Ocean Studies Board. 164pp.

ENHANCING FISHERIES RIGHTS THROUGH LEGISLATION – AUSTRALIA'S EXPERIENCE

M. Tsamenyi
Centre for Maritime Policy, University of Wollongong, NSW 2522, Australia
<martin_tsamenyi@uow.edu.au>
and
A. McIlgorm
Dominion Consulting PTY Ltd, suite 7&8, 822-824 Old Princes Highway, NSW 2232, Australia <mcilgorm@tradesrv.com.au>

1. INTRODUCTION

This paper has two main objectives. First, it reviews judicial approaches to the concept of property and the recognition of fisheries entitlements as property. It will be shown that Australian Courts have recognised various types of fisheries entitlements as property. Second, the paper reviews Australian Commonwealth, states and territory fisheries legislation to assess the extent to which fisheries property rights are recognised. The review will be conducted under the following headings: (i) Types of Access Rights; (ii) Duration of entitlements; (iii) Transferability; (iv) Recognition of dealings in entitlements; (v) Payment of compensation. Following the summary, a general assessment will be made regarding the extent to which the legislative framework as a whole enhances fisheries property rights.

2. WHAT IS PROPERTY IN LAW?

The term 'property' is commonly used to refer to a 'thing' or to denote ownership of a 'thing'. Legally, however, this approach has been rejected. Property is not a 'thing' but 'a description of a legal relationship with a thing (see: *Yanner v Eaton* [1999] HCA 53 Per Gleeson C.J., Gaudron, Kirby and Hayne J.J. at para. 17 citing Bentham, 'An Introduction to the Principles of Morals and Legislation' in W. Harrison (ed.) (1948) sy 337, note 1; K. Gray and S. F. Gray, 'The Idea of Property in Land', in Bright and Dewar (eds.), *Land Law: Themes and Perspectives*, (1998) 15 at 15 and 27-30. See also *Yanner v Eaton* [1999] HCA 53 at 86 per Gummow J., citing the observations of Finkelstein J., in *Wily v St. George Partnership Banking Ltd.* (1999) 84 FCR 423 at 431, Hohfeld 'Some Fundamental Legal Conceptions as Applied in Judicial Reasoning' (1913) 23 *Yale Law Journal* 16 at 21-22). Neither can property be equated with ownership, as it is a far more complex relationship.

This legal relationship has been described by Australian courts as a 'bundle of rights' (first use by H. Maine; *Minister of State for the Army v Dalziel* [1944] 68 CLR 261 at 285 per Rich J. Also *Yanner v Eaton* [1999] HCA 53 at para 27 per Gleeson CJ, Gaudron, Kirby and Hayne JJ although they recognise that 'this may have its limits as an analytical tool or accurate description'). Such an approach is consistent with that of Honoré who has suggested 11 indicia of property. Australian courts have, at various stages, attempted to identify these rights, or incidents, of property. Indicia identified have included identifiability, transferability, a degree of stability, a right to exclude and a right to use and enjoy. For example in *R v Toohey; ex parte Meneling Station Pty. Ltd.* Mason J noted that 'before a right or interests can be admitted into the category of property, or of a right affecting property, it must be definable, identifiable by third parties, capable in its nature of assumption by third parties, and have some degree of permanence or stability' (*National Provincial Bank v Ainsworth* [1965] AC 1175 at 1248 per Lord Wilberforce; adopted by Mason J in *R v Toohey; ex parte Meneling Station Pty. Ltd.* (1982) 158 CLR 327 at 342). The Courts have held that these indicia are not all necessary to establish 'property' but neither are they individually sufficient. By way of example, in *Yanner v Eaton*, Gummow J noted that transferability is not itself a necessary incident of property (*Yanner v Eaton* [1999] HCA 53 at para 85 per Gummow J).

Australian courts have taken a broad view of property, recognising possession, managerial control, common law rights and privileges, (*Georgiadis v Australian & Overseas Telecommunications Commission* (1994) 179 CLR 297 where it was held that property exists in a chose in action), and statutory rights and privileges as property. For example, a statutory right to payment was considered property in *Health Insurance Commission v Peverill* (1994) 179 CLR 226. It has also been said that property extends 'to every species of valuable right and interest including real and personal property, incorporeal hereditaments ... rights of way, rights of profit or use in land of another, and choses in action' *(Minister for the State of the Army v Dalziel* (1944) 68 CLR 261 per Starke J at 290). With respect to statutory rights which receive the status of 'property', the limits of the property are defined by the statutory instrument creating them. Accordingly, the content of the term 'property' becomes a question of statutory interpretation[1].

3. JUDICIAL APPROACH TO FISHERIES ENTITLEMENTS AS PROPERTY RIGHTS

Australian courts have generally acknowledged that various fisheries entitlements in the form of licences are capable of being considered 'property', although in a sometimes restricted sense. In *Harper,* the High Court considered the statutory right to fish to be analogous to a profit-à-prendre[2]. In the Northern Prawn Fishery cases a single judge of the Federal Court held that a fishing licence can be considered as property: it brings with it a privilege and a right that is proprietary in nature, subject

---

[1] *Yanner v Eaton* [1999] HCA 53 at para 85 per Gummow J.
[2] *Harper v Minister for Sea Fisheries and Others* (1989) 168 CLR 314.

only to constraints in the legislation[3]. However, on appeal to the Full Court it was said that the right to fish is based upon Commonwealth sovereignty rather than a private law proprietary right. The right to fish was held to be a public right, although amenable to change by a competent legislature. Each judge was, however, prepared to assume that the units were property. In *Bienke*, the full court of the Federal Court held that a fishing boat licence does not create an interest based on antecedent property rights. Rather, the licence is a new species of statutory entitlement dependant on the terms of the statute[4]. In *Gasparinatos*, the Tasmanian Supreme Court held that fishing rights were 'capable' of being valuable property rights[5]. Finally, in *Pennigton* the South Australia Supreme Court held that a fishing licence did confer proprietary interest given that it had the indicia of property under the relevant statute[6].

From these cases it can be observed that Australian courts have been willing to view fisheries licences as property.

4. FISHERIES RIGHTS IN LEGISLATION
4.1 Status

Fisheries entitlements are creatures of statute conferring a statutory right to fish. The scope of these rights as property must therefore be considered with reference to the relevant statute. This Section reviews Commonwealth, state and territory legislation to determine the extent to which the legislation accords property rights to fisheries entitlements.

4.2 The Commonwealth
4.2.1 Types of access rights

The governing legislation is the *Fisheries Management Act (Cth.) 1991*. Rights are categorised in two ways. The first is a Statutory Fishing Right (SFR) (Div. 5) (s21); and the second is a Fishing Permit (Div. 5) (s32).

SFR apply to both the right to take certain fish and the right to use certain equipment for fishing. These rights include the right to take a particular quantity, type and proportion of fish, and the right to use a particular type and size of boat and equipment (s21(1)). Where a management plan terminates, the holder of a SFR has the option of exercising his or her entitlements in another fisheries subject to a management plan; in which case the SFR becomes a SFR Option. A Fishing Permit, on the other hand, provides for the use of an Australian boat for fishing within a managed area (s32(1)).

Statutory fishing rights are issued by the Australian Fisheries Management Authority (AFMA) (s22) under management plans imposed by AFMA which the holder of the fishing right must comply with (s22(3)(a)). The rights may be granted by auction, tender, ballot or other procedure prescribed by AFMA (s25(b)). Fishing Permits are also granted by AFMA subject to specified conditions being met (s32(5)(a)-(e)).

SFR are renewable if the management plan remains in force until the date specified on the SFR. This is conditional on the holder having committed no offences in contravention of the SFR. The permit ceases to have effect if the holder surrenders the permit by written notice to AFMA (s32(9)).

4.2.2 Duration of entitlement

The duration of the SFR may be specified (s22(4)(b)). However if not specified, it remains in force until cancelled or surrendered or otherwise ceases to have effect under the Act (s22(4)(c)). The duration of a fishing permit is specified, but must not be greater than five years (s32(6)(c)).

4.2.3 Transferability

A SFR may be transferred provided conditions specified by AFMA in the SFR are met (s22(4)(a)). In addition, the holder of a SFR option also has the option to deal with the option, subject to giving a good discharge for any such dealing (s31J(1)). Except where a fishing permit is stated to be non-transferable, it can be transferred subject to the approval of AFMA (s32(10)).

4.2.4 Recognition of dealings in entitlements

Interests can be created in SFR options (s31F). Those interests, having the effect of creating, assigning, transmitting or extinguishing an interest must be registered with AFMA (s31F(3)). If the dealing creates a charge over the assets of those registered as having an interest, they must be notified of proposed changes, transfers etc. (s31F(9)). AFMA can only refuse to register an interest when it would be contrary to the proposed management plan, which the option relates to (s31F(7)). The act is silent with respect to dealings relating to fishing permits.

4.2.5 Payment of compensation

The act provides that no compensation is payable because a SFR is cancelled, ceases to have effect or ceases to apply (s22(3)(e)). Likewise with respect to fishing permits (s32(5)). However if the operation of the act results in acquisition of property otherwise than on just terms, the Commonwealth must pay reasonable compensation to that person (s167a).

4.3 Queensland
4.3.1 Types of access rights

The regulating Act, the *Fisheries Act 1994* classifies entitlements as either a Statutory Fishing Right (Div. 5) (s21); or a Fishing Permit (Div. 5) (s32). An Authority can be a licence, permit, quota or other authority in force under this Act. A licence may relate to fishing, crew, boat, storage or buyers licence. An Authority is renewable if the agency is satisfied the application is in the best interests of the management, use and development of the

---

[3] *Fitti and Others v Minister for Primary Industries and Energy and Another;* and *Davey and Others v Minister for Primary Industries and Energy and Another* (1993) 40 FCR 286 at 292. O'Loughlin J; *Davey and Others v Minister for Primary Industries and Energy and Another;* and *Minister of Primary Industries and Energy and Another v Davey and Another;* and *Minister for Primary Industries and Energy and Another v Fitti* (1993) 47 FCR 151.
[4] *Bienke and Others v Minister for Primary Industry and Energy and Others* (1996) 135 ALR 128 citing as authority *Harper v Minister for Sea Fisheries and Others* (1989) 168 CLR 314.
[5] *Gasparinatos v State of Tasmania* (1995) 5 Tas. R. 301 citing *Harper v Minister for Sea Fisheries* (1989) 168 CLR 314.
[6] *Pennington v McGovern* (1987) 45 SASR 24.

protection of fisheries resources (s59(1)). Conditions can be imposed on renewal (s61). A permit provides for a variety of activities, including possessing regulated fish, permit the use of boats, or permit the removal or destruction of marine plants (s51(1)).

4.3.2 Duration of entitlements

The tenure for an Authority is for a term as specified on the Authority (s53(c)).

4.3.3 Transferability

An Authority (other than a permit) is transferable unless otherwise specified under a Regulation (s65(1)). Entitlements stated as not transferable are a fisher licence and crew licence, and a licence bearing fish or shell symbols (Cl. 59 *Fisheries Regulations 1995*). A permit is not renewable (s57(1)).

4.3.4 Recognition of dealings in entitlements

The holder of an Authority can apply to have a third party interest noted on the Register of authorities (s73(6)).

4.3.5 Payment of compensation

No compensation is payable if a fisheries agency cancels or suspends an Authority (s68(7)), if the agency refuses to issue or renew an Authority (s59(2)), or if the agency amends the Authority (s63(7)). Likewise no compensation is due amends or repeals a management plan (s4040(1)). However the Act does provide that compensation may be payable if specified under a particular management plan or regulation (s68(8)). Compensation can be awarded for fishery resources or property destroyed in an emergency if the Chief Executive so decides (s103).

4.4 New South Wales
4.4.1 Categorisation

The governing legislation for fisheries in New South Wales, *Fisheries Management Act 1994*, does not provide an explicit statement with respect to fishing licences as property. Fishing rights are granted through a Commercial Fishing Licence or Shares. There are five classes of commercial fishing licences (Cl. 141). These include:

Licences can also be endorsed for fishing in restricted areas such as lobster fishery (Div. 2). All fishers must have a commercial fishing licence. Shares are issued with respect to any fishery declared as a share-Managed Fishery (s42) under a management plan.

Without a commercial fishing licence no person can take fish for sale from waters to which the Act applies (s102). In addition, a fisher cannot use a boat for commercial fishing unless it is licensed (s107). A commercial fishing licence can be endorsed to fish in a share-Managed Fishery if the licensee holds enough shares (s68). Licences are issued by the Minister to an eligible person upon application and meeting the criteria prescribed by the Regulations(s104). Shares are also issued by the Minister when a fishery becomes a share-Managed Fishery (s46(1)).

A commercial fishing licence is renewable, upon application in writing (Cl. 139(1)) at the discretion of the Minister if all the conditions prescribed in the Act are met (Cl. 139(3)). Shares are also renewable after their initial term if there has been no new management plan issued (s73(2)(3)).

4.4.2 Duration

A commercial fishing licence remains in force for a period of one year or such other time as prescribed on the licence (Cl. 139(4)). Shares are issued for an initial period of ten years, calculated from the commencement of the management plan (s73(1)). Shares are renewable for a further ten years after this initial period (s73(2)(3)).

4.4.3 Transferability

Licences are not transferable, as an applicant for a licence must prove he is an eligible person to the Minister (Cl. 135). A shareholder may transfer his share holding in a fishery to any other shareholder in the fishery subject to any restrictions imposed by the management plan (s79(1)).

4.4.4 Recognition of dealings

All dealings with respect to shares must be registered in the Share Management Fisheries Register (s89). Shares may be held by persons other then the holders of a commercial fishing licence (s49(1)). A share can also be mortgaged and assigned (s71(1)).

4.4.5 Payment of compensation

Compensation is not provided for in the Act with respect to licences. Compensation is not payable if a fishery ceases to be restricted (s115). Compensation is only provided for if the Minister cancels shares in a share-Managed Fishery (s44(3)).

4.5 Northern Territory
4.5.1 Types of access rights

The relevant legislation is the *Fisheries Act* (as amended by the *Fisheries Amendment Act 1997*). The Act provides for the following categories of entitlements: (a) licences (s11), (b) permits (s16), and (c) special permits (s17). A licence enables the holder to take any fish to sell, process for sale, for the purposes of aquaculture, or exhibiting any of them for profit, and to use certain equipment (s10(7)).

A permit does not entitle the holder to do things that affect the marine environment; *e.g.* among other things, to release fish, or pollute waters or use an electric fishing device (s15(1)). A permit may be subject to conditions as the Director considers appropriate or as may otherwise be prescribed (s16(3)).

A special permit allows fish to be taken and certain fishing gear to be used for the purposes of education, research, sport or recreation in the case of a disabled person who, in the opinion of the Director, would otherwise be unable, by reason of the person's disability, to fish by the methods permitted by this Act; or any other purpose approved by the Minister. All fish or aquatic life taken must be disposed of as the Director directs or as specified in the permit (s17(2)).

A licence is renewable (s12(1)) subject to a charge and conditions as outlined in s 12. The Director must be satisfied that the applicant has a commitment to the fishery in respect of which the applicant is licensed and to the fishing industry generally; that the conditions of the licence have been complied with (s12(c)); and that nothing in an instrument of a legislative or administrative character made under the Act prevents it (s12(d)).

If the applicant is a corporation, an application for renewal of a licence must contain a statement indicating the current nominal and beneficial ownership of the shares in the corporation and each sale or transfer of that ownership since the grant or transfer of the licence to the corporation, or the last renewal of the licence, whichever is the later (s12(3)). The corporation must also provide statements as to the share structure of the corporation (s12(4)(b)). If these criteria are satisfied, the Director must renew the licence on payment of the prescribed fee (if any) (s12(4)(d)).

Every renewal of a licence is granted on the same terms and conditions as apply to the original licence, unless the terms and conditions have been or are amended pursuant to Section 11 (s12(5)). There act is silent on the issue of renewing permits and special permits.

4.5.2 Duration of entitlements

A licence may be granted for a period of not more than 5 financial years on payment of the prescribed fee (if any) for each financial year of the licence (s10(7)). A permit may be issued for such period as the Director thinks fit (s16(3)). A special permit may have a time specified in the permit (s17(1)(a)(iii)) but can be revoked at any time by the Director serving notice in writing on the holder (s17(3)). A licence or registration can be suspended or cancelled by the court (s20(1)), or a licence can be suspended by Director if the holder is found guilty of a related offence (s20(2)).

4.5.3 Transferability

The Act provides for temporary licence transfer (s12A) or a permanent licence transfer (s12B). A licensee may only permit another person to use the licence with the approval of the Director in writing (s2A(1),(8)), and on payment of the prescribed fee (s12A(3)). This is subject to the regulations, a fishery management plan or condition of the licence (s12A(1)). This decision is 'in the Director's absolute discretion' (subject to Section 12C) (s12A(4)). This temporary transfer expires at such date, if any, specified in the agreement or at the end of the financial year in which it was entered into, whichever is the sooner (s12A(6)). This approval can be revoked at any time on application by either party (s12A(10)).

A licensee may permanently transfer the licence, on approval of the Director and payment of the prescribed fee (s12B(2)). The transfer is subject to Section 12B, the Regulations or a fishery management plan made in respect of the fishery for which a licence has been granted or a condition of the licence (s12B(1)). This decision is in the Director's absolute discretion (subject to 12C) (s12B(3)).

A licence or permit can only be transferred to Australian residents or Australian corporations (s12C). The act is silent on the transfer of permits except for the above statement in Section 12C.

4.5.4 Recognition of dealings in entitlements

Section 9 provides for the maintenance of a register to record, among other things, any interests held in a licence, permit or vessel (s9(1)). Section 9(2) outlines the conditions on which persons may have access to the register.

4.5.5 Payment of compensation

The act is silent on the issue of compensation.

4.6 South Australia

4.6.1 Types of access rights

The relevant legislation considered includes *Fisheries Act (SA)1982; Fisheries (Gulf St. Vincent Prawn Fishery Rationalization) Act (SA) 1987;* and *Fisheries (Southern Zone Rock Lobster Fishery Rationalization) Act (SA) 1987* The act uses the term 'authority' to refer to a licence, permit, registration or lease provided for by the Act (s5(1)).

Specifically of interest is the licence (s34(1)) and registration (s34(2)). Licence means a fishery licence and registration means registration of a boat by endorsement of a fishery licence or registration of the master of a boat by endorsement of a fishery licence (s33). A licence is required to engage in a fishing activity of a type that constitutes a fishery for the purpose of trade or business (s34(1)). The Director can impose, vary or revoke conditions for the licence, as set out in s37. Registration is required for a boat to be used to engage in a fishing activity for the purpose of trade or business (s34(2)).

4.6.2 Duration of entitlements

Licences and Registrations remain in force until the expiry of the term prescribed for licences in respect of the fishery (s39(1)). The licence and registration 'run together' so that if a licence is cancelled or surrendered the registration is automatically cancelled or surrendered (s39(2)(a)). If a licence is suspended, the registration is also suspended for that period (s39(2)(b)). An authority may be cancelled or suspended by the Minister if it is shown that the authority was obtained improperly, or, the holder has been convicted of a particular offence (s57(1)). An authority may be cancelled or suspended by the Court if the holder has been convicted of an offence against the Act (s56). The holder of an authority may at any time surrender it to the Director (s61). The *Fisheries (Gulf St. Vincent Prawn Fishery Rationalization) Act 1987* allows the Minister to cancel licences until there are no more than 10 licences in force in that fishery (s5).

The *Fisheries Act* does not contain a direct provision regarding the renewal of authorities. However, the act does refer to renewal indirectly when it states that any Authority that has been suspended by the Minister in accordance with s57 may be renewed, but remains subject to suspension until the expiration of the period of suspension (s57(3)). And again, any Authority that has been suspended by the court in accordance with s56 may also be renewed but remains subject to suspension until the expiration of the period of suspension (s56(8)).

4.6.3 Transferability

Generally, a licence is not transferable. A licence can, however, be transferred if the scheme of management for a fishery allows a licence to be transferred with the consent of the Director (s38(2)). In the event that a licence is transferred, the registration of the boat may also be transferred (s38(4)). If a licence is transferable and the holder dies the licence vests in the personal representative of the deceased and forms part of the deceased's estate. However, it cannot be transferred in the course of administration of the estate except with the consent of the Director (s38(5)).

If the deceased was also registered as master of the boat then, while the licence is vested in the personal representative, the boat can continue to be used for fishing if it is in the charge of a person acting with the consent of the Director (s38(6)). If the licence is not transferred within two years (or a further period approved by the Minister) of the death of the licence holder, the licence is suspended, pending such transfer (s38(7)).

The *Fisheries (Gulf St. Vincent Prawn Fishery Rationalization) Act 1987* permits the transfer of a licence with the consent of the Director. This consent must be given if the criteria prescribed by the regulations are satisfied and the licensee's accrued liabilities by way of surcharge under the Act are paid to the Director (s4).

4.6.4 Recognition of dealings in entitlements

The Director is to keep a register of all authorities granted under the Act (s65(1)). The Director can make notations on the register that a specified person nominated by the holder of the licence has an interest in the licence (s65(3)). Where the register includes such a notation the interested parties must consent to the surrender (s61(2)) or transfer of the licence (s38(2)(b)).

4.6.5 Payment of compensation

The *Fisheries Act 1982* is silent on the issue of compensation. However, the *Fisheries (Gulf St. Vincent Prawn Fishery Rationalization) Act 1987* and *Fisheries (Southern Zone Rock Lobster Fishery Rationalization) Act 1987* both establish legislative schemes for the provision of compensation for the cancelled and surrendered licences with respect to the rationalisation of those fisheries.

The *Fisheries (Gulf St. Vincent Prawn Fishery Rationalization) Act 1987* provides for compensation in s6 in the event that a licence is cancelled by the Minister in accordance with the rationalisation scheme. The amount of compensation is to be determined as outlined in s 6. Under the *Fisheries (Southern Zone Rock Lobster Fishery Rationalization) Act 1987* the amount is to be determined under s10.

4.7 Tasmania

4.7.1 Types of access rights

The governing legislation is the *Living Marine Resources Management Act (Tas) 1995*. The *Living Marine Resources Management Act (Tas) 1995* explicitly says that the State owns all living marine resources present in Tasmanian waters (waters defined by s5(1)(a)-(c)) (s9(a)). Further, any fish specifically provided for under a marine farming licence are not owned by the State but are the property of the holder of that licence (s9(b)).

The act establishes a system of 'licences' and 'permits'. The three categories of licence are the Fishing licence (Pt. 4 Div. 1, s60), the Marine farming licence (Pt. 4 Div. 2, s64) and the Fish processing licence (Pt. 4 Div. 3, s67). There can be many different classes of fishing licence (s34).

A Fishing Licence authorises the holder to carry out fishing in accordance with the licence (s61). The licence is subject both to the rules of a management plan applicable to that licence and any condition specified in the licence (s62(a)-(b)). Generally, a fishing licence allows a person to participate in fishing in state waters and to take fish, or use apparatus, for the purpose of fishing or take any other action permitted under the licence (s60(1)(a)-(d)).

A Marine Farming Licence authorises the holder to carry out marine farming in accordance with the licence (s65). The licence is subject to marine farming development plan and conditions specified in the licence (s66(a)-(b)). Generally, it enables a person to carry out marine farming in State waters or to take live fish for that purpose, or operate a fish hatchery or breed, culture or farm fish in inland waters or on land where fish would have ended up in state waters (s64(1)(a)-(c)).

A Permit allows the holder to take action which would otherwise contravene the Act for the purpose of scientific research, promotion of fishing or fish products, development of fishing technology, educational and community awareness programs, fish stock depletion or enhancement, collection, keeping, breeding, hatching or cultivating rare or endangered fish or sport and recreation purposes due to the holder's disability, that require methods otherwise illegal under the Act (s12(1)(a)-(h)).

A licence holder can apply to the Minister for the licence to be renewed (s81(1)). The Minister must renew the licence if the applicant has complied with the conditions of the licence in the previous 5 years, has not been convicted of an offence under the Act which the Minister considers relevant to holding the licence, has not been disqualified from holding the licence, is a fit and proper person, where there are no environmental or resource constraint on doing so and the Minister thinks it appropriate (s81(2)). The licence can only then be renewed on the payment of the prescribed fee (s81(2)).

4.7.2 Duration of entitlements

A Licence remains in force for a period not exceeding 10 years as specified in the licence (s80(1)). A Permit, on the other hand, remains in force for a period not exceeding 12 months, specified in the permit unless the Minister sooner revokes it (s16). The Act is silent on the issue of renewing a Permit.

4.7.3 Transferability

The holder of a licence can apply to the Minister to transfer either the licence to another person (s82(1)(a)) or a quota or under the licence to another licensee (s82(1)(b)). The Minister may grant the application for transfer on payment of the prescribed fee (s82(2)). With respect to entitlements under a licence, rules may be made in relation to the authorisation of the temporary transfer of any entitlement (s36(g)). A fishing licence can be leased, sub-leased, or lent under with the Minister's approval (s87).

4.7.4 Recognition of dealings in entitlements

The Secretary is to keep a register of details relating to grant, renewal, variation, transfer etc. of authorisations (s298(1)). A person may apply for registration of an interest in a deed of agreement (s101).

4.7.5 Payment of compensation

Compensation is not payable to the holder of an authorisation if a management plan is amended or revoked, or limitations are prescribed for fishing, or there is a reduction in total allowable catch, if the Minister takes any reasonable action under the act or as a result of any requirement complied with under an order made under Section 272 (s300 (1)). However, if the Minister takes action, which is not consistent with the purpose of the Act, s300(1) does not apply (s300(2)).

4.8 Victoria
4.8.1 Types of access rights

The governing legislation is the *Fisheries Act (Vic) 1995* which explicitly states that the Crown in the right of Victoria **owns** all wild fish and other fauna and flora found in Victorian waters (s10(1)). The Act provides that property passes: to the holder of licence or permit when taken in accordance with the licence or permit (s10(2)(a)). to any other person when lawfully taken or where no licence or permit is required under the Act for that purpose (s10(2)(b)).

The Act provides for two categories of access rights: 'Licences' and 'Permits'. Licences can be either fishing licences or any category of licence created by the regulations under Clause 3.2 of Schedule 3 (s4).

The main fishing licence is the Access Licence (s38). The regulations may create different classes of access licences and may specify that the holder of an access licence of a particular class can do certain things. Other licences include the Aquaculture Licence (s43), Recreational Fishing Licence (s45 (individual), s46 (group)) and Fish Receivers Licence (s41). The main permit is the General Permit (s49). Other permits include the Protected Aquatic Biota Permit (s72) and Noxious Aquatic Species Permit (s81).

An Access Licence may enable the holder to take specified fish and, or, fishing bait for sale, or use a boat and certain equipment for fishing (s38(1)(a)-(g)). The Secretary must give a quota notice setting out details of individual quota allocated to the licence (s65 (1)).

A General Permit may authorise the holder to take fish for research, education, fish management, aquaculture, compliance or scientific purposes; to take fish from a developing fishery; to carry out research, exploitation, work or operation for the purpose of developing any fishery or aquaculture; to investigate any species of fish or any fishery or any device; to sell or dispose of any fish obtained under the permit; or to use certain equipment (s49(2)).

An Access Licence is renewable in accordance with s57 (s38(6)). The holder must continue to satisfy the eligibility criteria, be a fit and proper person and be actively, substantially and regularly engaged in the activities authorised by the licence. In addition, the holder must show sufficient cause for renewing the licence. If the Secretary considers these criteria satisfied and the holder has a record of compliance with the Act, the Secretary must renew the licence (s57(2)-(6)).

A General Permit is not renewable. However, the Secretary can issue another general permit to a person whose permit is about to expire or a person who has previously held a permit (s49(6)).

4.8.2 Duration of entitlements

An Access Licence continues in force for up to one year, as specified in the licence, unless it is cancelled or suspended in accordance with the Act (s38(4)-(5)). A General Permit continues in force for up to three years, as specified in the permit, although it may be cancelled at any time without notice (s38(3),(4),(8)).

4.8.3 Transferability

Fishery licences (of which an Access Licence is a specific type) of a particular category or class are not transferable unless the regulations allow (s50B). If the regulations allow a particular category or class of fishery licence to be transferable, the transfer is dealt with by s56.

If the regulations allow an Access Licence to be transferable, when the holder dies the benefit of the licence is deemed to be an asset of the estate of the deceased (s38(7)(a)). The personal representative of the deceased is deemed to be the holder of the licence until it is transferred from the personal representative to an eligible person in accordance with the Act (s38(7)(b)). An Access Licence permits the transfer of quota subject to the approval of the Secretary (s65(3)). A General Permit is not transferable (s49(7)).

4.8.4 Recognition of dealings in entitlements

A person who is not the holder of an access licence, but who has a financial interest in that licence can register details of that financial interests with the secretary (s59(1)). The Secretary must notify the holder of a registered financial interest within 21 days of receiving an application to transfer the licence (s59(3)). The holder can then give his or her approval or disapproval of the proposed transfer (s59(4)). If the holder disapproves, the Secretary must not transfer the licence (s59(5)). If a transferable licence is cancelled by the court the secretary must notify each holder of a registered financial interest in the licence of the cancellation (s60 (1), (2)).

4.8.5 Payment of compensation

The Minister may issue directions for licence reduction arrangements and requiring the Secretary to cancel licences (s61(1)(b),(c)). If licences are cancelled in this way, compensation is payable to the person who held the licence, and any person who held a registered financial interest in the licence at the time it was cancelled (s63(2)).

Compensation is for the financial loss suffered as a natural, direct and reasonable consequence of the cancellation of the licence, in proportion to the extent of their respective interests (s63(2)). The amount of compensation is to be determined by the Secretary in accordance with the regulations (s64(4)) and Parts 10 and 11 and Section 37 of the *Land Acquisition and Compensation Act 1986*, with necessary modifications also apply as if the claim were a claim under Section 37 of that Act (s64(6)). Apart from this, no compensation is payable by the Crown to any person for any loss or damage as a result of the enactment of the Act and the repeal of the *Fisheries Act 1968*.

4.9 Western Australia
4.9.1 Types of access rights

The governing legislation are the *Fish Resources Management Act 1994* and the *Fishing and Related Industries Compensations (Marine Reserves) Act 1997*. All Access Rights are generally referred to as an 'authorization' which can either be a licence or permit (s4(1)). Authorization' is used to refer to two particular types of licence and permit in relation to managed fisheries and interim managed fisheries in particular: the Managed Fishery Licence (s53(a)) and the Interim Managed Fishery Permit (s53(b)). The Act also allows for the issue of an Exclusive Licence (s251).

Generally, a Managed Fishery Licence and an Interim Managed Fishery Permit authorise a person to engage in fishing or any fishing activity in a Managed Fishery or interim Managed Fishery (s66(2)). A commercial fishing licence, or any other licence, does not authorise a person to engage in fishing in a Managed Fishery or interim Managed Fishery (s73). A Managed Fishery Licence and an Interim Managed Fishery Permit are subject to conditions in the relevant management plan and any conditions specified by the Executive Director in accordance with s69 (s69(1)).

An Exclusive Licence allows a person to take fish from a specified area of coastal waters and the foreshore above high-water mark (s251(1)). It can be granted subject to such terms and conditions as the Minister things fit (s251(3)). These conditions can include the period/s which fish can be taken, the type and quantity of fish and the method and fees, etc. (s251 (4)).

Both Managed Fishery Licences and an Interim Managed Fishery Permits are issued by the Executive Director (s66). An Exclusive Licence is issued by the Minister (s251 (1)). An application can be made to the Executive Director for the renewal of authorizations generally (s135). It must be accompanied by the fee prescribed (if at all) or specified in the relevant management plan (s135(1)(b)). An authorization can be renewed if an application is made within 60 days after the day on which it expired (139(1)). In this event, the authorisation is of no effect between the date of expiry and the date of renewal (s139(2)(b)). The regulations may prescribe, or a management plan may specify, an additional fee payable by way of penalty if renewed after the date of expiry (s139(3)).

A Managed Fishery Licence and an Interim Managed Fishery Permit are renewable. If a person applies for a renewal the Executive Director must renew them (subject to s 143 which relates to times when the Executive Director does not have to renew an authorisation generally) (s68). However, the Managed Fishery Licence and Interim Managed Fishery Permit can be renewed subject to such conditions as the Executive Director thinks fit and specifies in the authorisation (s69(2)). An Exclusive Licence may be renewed by the Minister from time to time for any further period or periods not exceeding 7 years in each case (s251(2)(b)). It may be renewed subject to such terms and conditions as the Minister thinks fit (s251(3)).

4.9.2 Duration of entitlements

Managed Fisheries Licences and Interim Managed Fishery Permits remain in force for 12 months, or such other period as is specified in the relevant management plan, from the day on which it is granted or renewed (s68). However, if the management plan is revoked, or expires, the authorisation also ceases to have any effect (s70).

A Management plan can specify a period for which a Managed Fishery Licence and an Interim Managed Fishery Permit remain in force after it has been granted or renewed under Section 58(2)(i). If a subsequent management plan is determined for that fishery, the fact that a person had a previous Managed Fishery Licence or Interim Managed Fishery Permit does not confer any right to the grant of another (s72(1)). However, the Executive Director must take that fact into account when determining whether or not to grant the person another Managed Fishery Licence or Interim Managed Fishery Permit (s72(2)). An Exclusive Licence may be granted for an initial term not exceeding 14 years (s251(2)(a)). The Minister may vary or revoke an Exclusive Licence in the manner provided for in the licence (s251(5)).

4.9.3 Transferability

An application can be made to the Executive Director for the transfer of any Authorization or part of an entitlement under an Authorization (s135). It must be accompanied by the fee prescribed (if any), or specified, in the relevant management plan (s135(1)(b)). The Executive Director must transfer the Authorization or part of the entitlement if the Director is satisfied that the requirements under s 140(2) and (3) are satisfied (s140(1)).

4.9.4 Recognition of dealings in entitlements

The *Fish Resources Management Act* recognises 'security interests' in authorizations. A security interest, in relation to an authorisation, is defined as an interest in the authorization (however arising) that secures payment of a debt, or other pecuniary obligation, or the performance of any other obligation (s4).

The Act establishes a register of authorisations and exemptions (s126). A holder of an authorization may apply to the Registrar to have noted on the register that a specified person has a security interest in the authorization (s127). Upon application and payment of the prescribed fee (if any), the Registrar must make a notation on the register (s128) with the details required in s128(2).

The effect of the registration is that a person with a security interest in an authorization must be notified if the holder is convicted of a prescribed offence, if there is an application for transfer or a partial transfer of the authorization, if the Executive Director proposes to cancel or suspend, or not renew the authorization or, if a fisheries adjustment scheme is established in respect of authorizations of that class (s130).

4.9.5 Payment of compensation

The *Fish Resources Management Act 1994* is silent on the issue of compensation except to say that no compensation is payable in respect of anything done or omitted to be done in good faith relating to the register (s133).

The *Fishing and Related Industries Compensation (Marine Reserves) Act 1997* establishes a scheme for the provision of compensation to holders of leases, licences and permits under the *Fish Resources Management Act 1994* on account of the effect of marine nature reserves and marine parks constituted under the *Conservation and Land Management Act 1984* (CALM). The *Compensation Act* specifies the particular events which cause an entitlement to compensation, including the CALM Act coming into operation (s4). The Act states that a person who holds an authorization is entitled to fair compensation for any loss suffered as a result of a relevant event (s5(1)). Generally, a person is considered to suffer 'loss' if and only if the market value of the authorisation is reduced because an authorization will not be able to be renewed (s5(2)(a)), or, that the authorisation relates to an area and will only be able to be renewed in respect of part of that area or another area (s5(2)(b),(c); or because an area will not be available for commercial fishing after the renewal of the authorization (s5(2)(e)). However, in the latter case, it is noteworthy that this only applies to a person who obtains a certificate from the Executive Director stating that, in the Executive Director's opinion, the history of the authorization shows that the area has been fished under the authorization on a long term and consistent basis (s5(5)). Section 5 outlines the method for determining the amount of compensation.

5. CONCLUSIONS

In respect of property rights, the following conclusions can be drawn from the review of cases and fisheries legislation:

i. Generally, Australian courts have generally acknowledged that various fisheries entitlements in the form of licences are capable of being considered 'property.'

ii. Fisheries licences are 'capable' of being property, but, as statutory rights, this is entirely dependant on the terms and interpretation of the relevant statute.

iii. The issue for industry is not so much whether fisheries entitlements constitute property rights in the legal sense, but the extent to which the legislative framework enhances such rights. In this respect, then generally it can be said that current Australian fisheries legislation provides for weaker property rights. The factors contributing to the lack of stronger rights include:
   a) the discretionary powers to intervene granted to fisheries administrators
   b) the limitations on transferability of entitlements (generally transferability is subject to the consent of the fisheries administrator)
   c) the various Fisheries Acts provide for the suspension, or cancellation, of entitlements for the commission of specified or unspecified offences
   d) inadequate provisions for the payment of compensation for loss of entitlements and
   e) the limited duration of most entitlements (one year in many cases).

6. THANKS

We thank the Fisheries Research and Development Corporation (FRDC) Project 99-161 "Sustainable fisheries management through enhanced access rights and resource security" Part I. We thank the South Australian Fishing Industry Council (SAFIC) "Fishing Rights Benchmarking Project" project, Dr A. Mcilgorm, principal investigator and Prof. Martin Tsamenyi, University of Wollongong, co-investigator. We thank the Fish Rights'99 Committee for the invitation to the Conference.

DEVELOPMENT AND IMPLEMENTATION OF ACCESS LIMITATION PROGRAMMES IN MARINE FISHERIES OF THE UNITED STATES

G.H. Darcy and G.C. Matlock
Office of Sustainable Fisheries
National Marine Fisheries Service, 1315 East-West Highway, Silver Spring, Maryland, 21042, USA
<George.Darcy@noaa.gov> and <Gary.Matlock@noaa.gov>

1. INTRODUCTION

The management of marine fisheries of the United States has changed dramatically in the last 25 years, much of it brought about by important legislation that has reflected the increasing public interest in the protection and conservation of American fisheries and the resources that sustain them. The fairly straightforward biological and economic aspects of commercial fisheries that were the initial concern of most managers and scientists have evolved and expanded to include such issues as: biological integrity and sustainability of living marine resources; biodiversity, ecosystem, and habitat protection; economic and social implications of management decisions and resource allocations. Not only have the issues changed, but the participants in the fishery management process have become more diverse, the public's knowledge of, and interest in, marine issues has increased and numerous technological changes have significantly increased the efficiency and effectiveness of fishing effort. The result is an extremely complex landscape of interests, constituencies and conflicts that must be taken into account in making management decisions.

While many tools are available to fishery managers, expanding fishing effort in U.S. fisheries and other fisheries worldwide in the last several decades has focused attention on use of effort and access limitation measures to deal with problems such as overcapacity, safety at sea, gear conflicts, overfishing, bycatch, and habitat damage. This paper reviews the development and implementation of access limitation measures in the federally-managed U.S. marine fisheries, examines the current status of those measures and provides an assessment of the effectiveness of some of the access-limitation schemes that have been implemented.

2. HISTORY OF MANAGEMENT

Prior to the mid-1970s, there was little Federal regulation of U.S. marine fisheries; activities were instead focused primarily on exploration, development, monitoring and assessment of living marine resources. Within the territorial sea (generally from shore to 3 nautical miles seaward), the individual coastal states had management authority. With the establishment of the Fishery Conservation Zone (now known as the Exclusive Economic Zone (EEZ)) and the passage in 1976 of the *Fishery Conservation and Management Act,* the United States took a major step in "Americanizing" its marine fisheries. In so doing, the U.S. Congress implemented a unique system of management of public resources, based on eight Regional Fishery Management Councils (Councils), to provide a mechanism for bringing diverse interests together to develop and recommend fishery management plans (FMPs) and management measures for approval and implementation by the National Marine Fisheries Service (NMFS) on behalf of the Secretary of Commerce (Secretary).

Early fishery management actions in the EEZ were relatively simple and were intended to address the most obvious problems in some of the major commercial fisheries, such as gear conflicts and reductions in certain stocks. Both the Councils and NMFS faced a necessary, but substantial, learning curve in dealing with their new responsibilities. A management process was required that involved the public, protected the resources, sustained the fisheries in the long term and complied with numerous Federal statutes. Further, substantial new efforts were needed to gather information necessary for the management of these fisheries, many of which had never been managed before. The first FMP, for the Atlantic surf clam and ocean quahog fishery, was developed by the Mid-Atlantic Council and was approved and implemented by NMFS on behalf of the Secretary, in 1977.

In the late 1970s and 1980s, an increasing number of fisheries came under management through the Councils' development of new FMPs. The number of approved FMPs grew from nine in 1980 to 32 by 1990; over the same time period more than 100 FMP amendments also were implemented. But as more and more management measures were developed it also became clear that many of those measures were not adequate to accomplish the FMPs' objectives. Although some of the problems that had been identified in the FMPs had been successfully addressed, many continued or worsened despite the good intentions of scientists, fishermen and managers. Most of the early FMPs regulated the fisheries primarily through standard management measures such as gear restrictions (*e.g.* mesh-size restrictions in New England groundfish fisheries), or on the basis of outputs of the fishery (*e.g.* quotas). However, because there were few, if any, barriers to entry in most U.S. fisheries, as foreign effort exited domestic effort entered and expanded. Quota management and gear restrictions often proved inadequate on their own to address the biological problems caused by this increasing effort. And quota management, though useful in capping total catch, did nothing to address the problem of increasing fishing pressure and competition for the quotas. The result was continued overfishing and proliferation of fishing effort and capital investment in many fisheries.

The failure to restrict effort adequately and the increased participation in many fisheries led to other problems as well. Biological productivity, even under ideal management, is finite and open-access fisheries create an incentive to fish faster and harder to get as large a share of the total allowable catch as possible before the quota is filled and the fishery closed. Technology "helped" by offering fishermen increased efficiency in locating, identifying and catching fish. The result was often an abnormally compressed commercial fishing season, market gluts (with subsequent low prices for the fishermen and poor product quality and/or reduced availability for the consumer), unsafe fishing conditions, poor match of capital investment to the available resource and a variety of other problems. There were repercussions in the processing sector (*e.g.* the need for far greater capital equipment investment and difficulty in matching labour to large pulses in landings), as well as the retail markets. Pressure from competing groups vying for quota share, including commercial and recreational sectors of many fisheries, continued to increase, as did pressure from those outside the fisheries (*e.g.* environmental groups and non-consumptive users) who became increasingly vocal about their concerns regarding ecological implications of expanded effort, such as habitat damage, large bycatches and decreased biodiversity. In short, the well known "tragedy of the unmanaged commons" associated with open access resulted.

Because of the problems associated with open-access fisheries in the 1980s and 1990s, NMFS began encouraging the use of limited access (NMFS 1991). NMFS could not unilaterally develop a limited access programme for a fishery subject to Council jurisdiction because Section 304(c)(3) of the *Magnuson-Stevens Fishery Conservation and Management Act (Magnuson-Stevens Act)* requires a Council recommendation (NMFS 1996). However, with encouragement, the Councils amended many of their FMPs to supplement the existing measures with at least some controls on commercial fishing effort. By October 1999, the number of approved FMPs had risen to 41 and more than 300 FMP amendments had been implemented. Several new FMPs were also under development by the Councils. Hundreds of regulatory amendments, emergency rules and other management actions had also been undertaken. Although biological problems, such as overfishing continued to be addressed through many of these actions, most dealt with allocational issues including measures to limit access to all or a portion of commercial fisheries.

Because management actions that restrict or allocate access to fishery resources typically are controversial and can have a broad range of economic and social impacts, many of which are difficult to assess thoroughly, they are often more difficult and time consuming to design and implement than the more traditional management measures. The fishery management process created by Congress in 1976 is complex and often deliberate, but it was designed to involve the broad range of interested parties such that management and conservation decisions could be made in full view of the public. Both the Councils and NMFS are critical players in this process – the Councils to provide a public forum for decision-making and NMFS to provide scientific support and review of the Councils' recommended FMPs and management measures. How, and in what time frame, these activities are conducted is determined by available resources and numerous statutory requirements and constraints. Several of the access limitation programmes took many years for the Councils to develop and, even when approved, required years more for NMFS to implement.

The single most important statute that relates to management of fisheries in the EEZ is the *Magnuson-Stevens Act*, which was substantially amended by the *Sustainable Fisheries Act (SFA)* in 1996. Among other things, the SFA placed increased emphasis on ending and preventing overfishing of all stocks, reduction of waste and bycatch in the fisheries, identification and protection of essential fish habitat and consideration of effects of fishery management decisions on fishing communities. In passing the SFA, Congress sent a clear message that there were unacceptable problems in many U.S. fisheries, that changes needed to be made and a longer-term view taken in their management.

The SFA added three new national standards to the seven existing standards in the *Magnuson-Stevens Act* (Appendix A), to focus attention on specific areas of concern. The national standards are statutory criteria with which all FMPs and amendments prepared by the Councils and the Secretary must comply. Existing standards required, among other things, that overfishing be prevented, that best scientific information be used and that efficiency be considered in selecting management measures. However, Congress, in creating the new standards, reflected public concern about three other issues: impacts of management actions on fishing communities, bycatch and safety at sea.

The three new national standards are relevant to the discussion of access limitation in that such measures, if carefully chosen and creatively applied, can address many of the problems the new standards were created to solve. National Standard 8 requires that conservation and management measures shall, consistent with the conservation requirements of the *Magnuson-Stevens Act* (including the prevention of overfishing and rebuilding of overfished stocks), take into account the importance of fishery resources to fishing communities in order to provide for the sustained participation of those communities and, to the extent practicable, minimize adverse economic impacts on the communities. Under appropriate conditions and with thoughtful development and incorporation of suitable constraints, access limitation measures can play a role in maintaining viable fisheries that support local communities and that take into account the economic and social needs and heritage of those communities. National Standard 9 states that conservation and management measures shall, to the extent practicable, minimize bycatch and, to the extent bycatch cannot be avoided, minimize the mortality of that bycatch. Certainly there are technological and fishing behavior modifications that can be (and have been) applied to address bycatch problems. However, effort controls and measures that achieve a more rational and efficient application of effort should also be explored, developed, and adopted, as appropriate and effective effort, applied more efficiently to target stocks, would reduce bycatch in the fishery as a whole. Finally, National Standard 10 requires that conservation and management measures shall, to the extent practicable, promote the safety of human life at sea. Measures that encourage fishermen to fish as fast and hard as possible have been shown to create risks to safety at sea; measures that allow fishermen to decide when to go to sea, with less pressure to catch fish before someone else does, reduce such risks.

3. ACCESS LIMITATION IN U.S. FISHERIES
3.1 Status of implementation

While there remain numerous challenges in fishery management today, among the most difficult is dealing with overcapitalization and excessive effort. Solving the biological problems of overfishing and bycatch are an imperative and the recent changes to the *Magnuson-Stevens Act* through the SFA reflect that. However, a root cause of those problems and a major impediment to solving them is the enormous effort, both active and latent, that exists in the U.S. fisheries. The more capital that is amassed in the fisheries, particularly capital that is out of balance with the available natural resources, the greater the pressure to overfish, to take undue risks in setting quotas, to prematurely reopen or release restrictions on effort in rebuilding fisheries and to engage in allocation battles (Ginter and Rettig 1978).

In recognition of the need to address problems associated with overcapitalization and excessive effort, the Councils have recommended, and NMFS has implemented, access limitation measures in every major commercial fishery in the United States except the penaeid shrimp fisheries in the Gulf of Mexico and South Atlantic region. As of October 1999, of the 41 approved FMPs, 27 contain measures that restrict access to at least some extent and three others allow no harvest at all in the EEZ (Appendix B). Approved measures range from licence and/or vessel moratoria to freely transferrable individual transferrable quotas (ITQs). Licence/vessel reductions, restrictions of effort through Days at Sea (an allocation of effort to individual vessels in the fishery) or other constraints, effort buyout programmes and non-transferrable IFQs have also been implemented. In addition, the Councils have informed the public through announcement of "control dates" that access may be limited in the future in many other U.S. commercial fisheries.

The United States has also begun to limit access of its citizens participating in international commercial fisheries. A treaty (South Pacific Tuna Treaty) was negotiated with the Forum Fisheries Agency in 1987 that included authorization for a limited number of permits (50) to be issued to U.S. tuna purse seine vessels to fish in the domestic waters of 16 Pacific island nations; longliners also may apply for these permits. About 35 U.S.-flagged tuna purse-seiners have actively fished in the last few years in the western tropical Pacific. The Inter-American Tropical Tuna Commission has recently begun discussions that would limit the number of tuna purse-seine vessels (including U.S. vessels) fishing in the eastern tropical Pacific.

3.2 Control dates

Control dates have been used by the Councils and NMFS to inform the public that a limited access programme is being considered for a particular fishery and that anyone entering the fishery after the control date is not assured that he/she will be given access to the fishery once the limited access programme is adopted. The intent of control dates is to reduce speculative entry into fisheries. The public is notified of the control date and its significance through publication of an Advance Notice of Proposed Rulemaking in the *Federal Register*. As of October 1999, control dates had been established for 26 federally managed fisheries and others were under consideration. Legal considerations require that control dates be relatively recent in order to be valid and, when a control date is established, there should be a reasonable expectation that an access limitation programme will be developed and implemented in the near future. Control dates that are several years old generally must be rescinded or replaced with more recent dates, if a Council still intends to develop an access limitation programme in that fishery.

3.3 Licence/vessel moratoria and limitations

Licence and vessel moratoria and limitations are designed to cap or reduce the number of participants and, or, vessels in a fishery by establishing criteria for their continued inclusion, such as historical participation at some threshold level. As of October 1999 there were 15 moratoria and 11 licence/vessel limitation programmes in place in federally managed U.S. marine fisheries. The extent of use of these programmes varies considerably among regions of the country. In the northeastern United States, seven of the 11 approved FMPs contain licence or vessel moratorium provisions and another (Atlantic salmon) allows no harvest in the EEZ. But in the Southeast, only three of the 16 approved FMPs contain such provisions; two others (South Atlantic and Gulf of Mexico Red Drum) allow no harvest in the EEZ; and one (Atlantic Billfish) allows no commercial harvest. Five out of seven FMPs on the Pacific Coast and in the Western Pacific region include licence or vessel moratoria (and a sixth, for the salmon fisheries, has limited entry programmes run by the states); four out of five Alaska FMPs include these provisions. This distribution reflects, in part, the choices of the regional Councils, since they are given considerable flexibility under the *Magnuson-Stevens Act* to recommend management measures to address problems and issues in fisheries in their areas of jurisdiction. The North Pacific Council, for example, has chosen to implement moratorium measures even though the majority of stocks in that area (the EEZ off Alaska) are not overfished. The New England and Mid-Atlantic Councils have taken similar actions, but in response to problems resulting from severely overfished and overcapitalized fisheries. The extent to which such programmes actually reduce effort varies, as is discussed below.

3.4 IFQs/ITQs

IFQs and ITQs (a subset of IFQs, distinguished by their transferability) are among the management tools that can be used to constrain effort and to achieve rational application of capital to the available fish resources. Their primary purpose has generally been to achieve a better match of capacity to resource productivity and to address economic inefficiencies by eliminating derby fisheries and allowing for consolidation and distribution of fishing effort. IFQs are unique in their reliance on market-based forces to distribute effort and benefits. Traditional fishery management tools have focused on input controls (*e.g.* restrictions on gear, days at sea, seasons) and output controls (*e.g.* quotas, bag and, or, trip limits). These traditional tools, particularly input controls, seek to limit the productivity of fishing vessels with resultant effects on the efficiency and profitability of both fishing vessels and processors. The input/output control approach also places a significant administrative burden on fishery managers as they are often required to issue "command and control" measures for the fishery to replace decisions that might better be made by the individual entrepreneur (*e.g.* When and how should I fish?) or by the market. Fishery management based on transferable individual shares of the harvest meets the same goal of output controls, but with the added efficiency of private ownership and market transferability of the access to that output.

The United States currently has five fisheries under IFQ management: Surf clams and ocean quahogs (Mid-Atlantic), halibut (Alaska), sablefish (Alaska), wreckfish (South Atlantic) and the bluefin tuna purse seine fishery (North Atlantic). IFQs are currently subject to a moratorium in the United States, which was imposed by Congress through the SFA in 1996. The Gulf of Mexico Council had developed, and NMFS had approved but not yet implemented, an ITQ programme for a sixth fishery – the commercial red snapper in the Gulf of Mexico – when the SFA moratorium was imposed. That programme has, therefore, not been implemented. The major impetus behind the moratorium appears to have been concern over the design and implementation of the North Pacific halibut/sablefish ITQ programme though there are a number of social and economic issues associated with IFQs (*e.g.* afffects on fishing communities and on vessel crew members) that have fueled vigorous public debate in the United States. The *Magnuson-Stevens Act* also includes a prohibition on the sale of initial ITQs, which is particularly troublesome to both economists and environmentalists – albeit for different reasons. Many economists feel that the most expeditious and efficient method for allocation of quota shares is by open sale to the highest bidder(s). Some environmentalists, and others, are opposed to what they view as a "giveaway" of the fish, which are a public resource. A more detailed discussion of these IFQs is provided by the National Research Council (1999) and Wertheimer and Swanson (2000).

4. FISHERY COOPERATIVES

Fishery cooperatives are relatively new in U.S. fisheries. Although developed as alternatives to IFQs, cooperatives have thus far not faced the same level of public and political resistance and, at least under certain circumstances, may not be subject to the SFA moratorium on IFQs. Instead of allocations of fish being given to individual vessels, participants in cooperatives agree among themselves on how to share an allocation made to the cooperative or to the industry sector operating as a cooperative. Vessel owners or operators then coordinate their fishing activities to achieve economic efficiency or to meet other mutual objectives.

There are currently two cooperatives operating in the United States – one for whiting vessels in the Pacific Northwest and one for Alaska pollock vessels fishing in the Bering Sea. Four companies holding limited entry permits in the catcher-processor sector of the offshore whiting fleet off Washington and Oregon voluntarily formed a cooperative in 1997 to allocate among themselves the quota available to that sector of the fishery; there was no government involvement. Their main objective was to eliminate derby fishing and to reduce bycatch of other species. The U.S. Department of Justice was consulted regarding potential antitrust violations; the Department concluded that the cooperative did not appear to have an anti-competitive effect (NRC 1999). The Bering Sea cooperative resulted from the *American Fisheries Act of 1998*, which included statutory authority for catcher-processors, shore-based processors and motherships in the Bering Sea pollock fishery to form cooperatives. Cooperatives are also being considered for possible use in the Gulf of Alaska pollock fishery.

5. BUYBACK PROGRAMMES

These programmes have been used to reduce capacity in U.S. marine fisheries by purchasing permits, vessels, and/or fishing gear from fishermen. Latent or active effort, or both, can be reduced through buybacks, depending on the criteria of the programmes. Since 1976, the U.S. Government has authorized 10 buyback programmes under various statutory authorities, including the *Interjurisdictional Fisheries Act 1986*, the *Sustainable Fisheries Act 1996*, and the *American Fisheries Act 1998*. Most programmes have focused on buying back commercial fishing permits, though some have bought vessels or placed restrictions on their future use. Some have also provided economic assistance to fishermen exiting the fishery. Congress has recognized the potential value of buyback programmes; the *Magnuson-Stevens Act* explicitly provides for fishery capacity reduction programmes, including industry-funded programmes under certain conditions (Section 312). Five U.S. marine fisheries have thus far been the subject of buybacks: Pacific Northwest salmon, New England groundfish, Texas shrimp, Bering Sea groundfish, and Alaska Dungeness crab. The summaries below are based on a U.S. General Accounting Office report prepared for Congress in 1999 (GAO 1999).

The Pacific Northwest salmon fishery has recently been plagued by excessive harvesting capacity, declining stocks, and increased numbers of salmon stocks listed as threatened or endangered under the *Endangered Species Act*. There have been five separate buyback programmes; none are currently active. All of the programmes were used to buy back state-issued commercial limited access permits, primarily in the State of Washington. The Washington State buybacks were largely in response to shifts in allocation to tribal fishermen or were a form of disaster relief. One of the programmes purchased vessels; another paid vessel owners a portion of their vessels' value in exchange for a 10-year abstinence from commercial salmon fishing. Most of the purchased vessels were resold, but with restrictions on their future use in the fishery.

Many of the major stocks in the New England groundfish fishery (*e.g.* cod, haddock, yellowtail flounder) have been severely overfished and are in the process of recovery. To accomplish this, drastic cutbacks in effort (through an annual Days at Sea effort allocation) have been required. At the same time, the number of vessels landing groundfish in New England has doubled from the 1980 level (Kitts *et al.* 1998). Two programmes have been used to buy back permits and vessels since 1995, neither is currently active. Vessels were purchased through a reverse auction (lower bids, weighted by groundfish catches, were accepted first in order to maximize effort purchased per dollar) and were scrapped, sunk or converted to uses other than commercial fishing. All Federal permits associated with purchased vessels, whether for the multispecies fishery or other fisheries, were also surrendered.

The Gulf of Mexico shrimp fishery is heavily overcapitalized and participants have had to face issues such as incidental takes of sea turtles and large bycatches of juvenile red snapper and other commercially and recreationally important finfish. Imports and aquaculture of shrimp have also created economic problems in the U.S. industry. A buyback of Texas state bay and bait shrimping permits began in 1996 to reduce the shrimp trawl effort in the western Gulf of Mexico.

Stocks in the Bering Sea groundfish fishery off Alaska are not overfished, but fishing capacity is large and there are significant allocational and other socio-economic issues (*e.g.* large non-Alaskan vessels competing with smaller Alaskan vessels and with fishery-dependent communities in western Alaska). A programme to buy back nine large pollock vessels and their associated permits was undertaken in 1998. Eight of the vessels were scrapped and the ninth is prohibited from fishing in U.S. waters.

The Dungeness crab fishery in Glacier Bay, Alaska, takes place largely within the boundaries of a national park. Because the National Park Service is interested in reducing commercial fishing activities in the park, a buyback programme, administered by the Park Service, was begun in 1999. The programme will purchase state permits and possibly vessels and gear.

The total cost of these buyback programmes was approximately $140 million. More than 3000 permits have been bought back and about 600 vessels have been purchased or had their use in the fisheries restricted. Additional buyback programmes have been proposed in 1999 by the commercial fishing industry and the State of Washington, which would potentially affect Bering Sea/Aleutian Islands crab, Northwest salmon, Atlantic swordfish, Atlantic sea scallop, and Atlantic shark fisheries. A programme proposed by West Coast groundfish fishermen is also being considered though recent reductions in quotas due to declining stocks may make it difficult for the industry to afford to fund the cost of such a programme in that fishery.

6. EFFECTIVENESS OF ACCESS LIMITATION MEASURES

Limited access programmes are now an integral component of the U.S. fishery management system. However, their success in addressing the biological, economic and social problems of our fisheries has varied among the regions, fisheries and type of approach selected.

Control dates, in themselves, are the weakest tool employed to control access in the U.S. marine fisheries in that they have no direct regulatory effect and impose no obligation on a Council or the Secretary to use those dates as cutoffs for entry into the fishery. Thus, declaration of a control date does not in itself cap effort. Whether or not control dates have a significant effect in reducing the number of fishermen that might have entered a fishery, or in inhibiting increases in vessel numbers, size and/or fishing power that might otherwise have occurred, is difficult to assess. There may be some discouragement of speculative entry or investment by putting the public on notice that, after a certain date, a person's participation in a fishery may not result in his or her inclusion, should a limited entry system be implemented. However, there is also some evidence that, at least in some fisheries, there has actually been an increase in effort by fishermen hoping to establish a history that may result in their being "grandfathered" into the fishery should the Council or NMFS ultimately choose not to apply the published control date.

Those fishermen who pay attention to Council actions appear to take control dates seriously; control dates are discussed on VHF radio at sea and even on the internet. Anecdotal information suggests that when a control date is anticipated many fishermen make sure that their landings of the species in question are increased and, or, recorded and that they hold the appropriate permits. In the West Coast groundfish fishery for example, there is evidence that the control date and Pacific Council discussions of the qualifying period for inclusion in the fishery under a limited access programme influenced fishermen's actions. One factory trawler entered the whiting fishery 15 days before the end of the qualifying period in an attempt to qualify for a permit. At least two vessels were purchased and several more were built with the intent of qualifying for the fishery in a larger size class and several fishermen sold vessels without their "fishing history", *i.e.* they reserved the qualifying history for use on another vessel.

In the Alaska halibut fishery, the North Pacific Council established the first "cut-off" date (equivalent to a control date) for limiting entry as December 31, 1978. However, while the date was being discussed and a limited entry programme developed, effort was increasing in the fishery and the length of the commercial fishing season in area 3A decreased from about 150 days in 1970 to less than 100 days in 1979. In 1982 the Council voted to establish a moratorium on new entrants to the halibut fishery, and established a new cut-off date of December 31, 1981. The proposed moratorium was ultimately disapproved, but effort in the fishery had further increased and the season shrank to only about 20 days even though quotas were actually increasing. By the time the Council adopted a halibut ITQ programme in December 1991 the season in areas 3A and 2C had been reduced to only a few days. Although the cut-off date for the ITQ programme was 31 December 1990, it was not implemented until March 15, 1995. A study by the State of Alaska's Commercial Fisheries Entry Commission in 1996 (P. Smith, NMFS Alaska Regional Office, Juneau, Alaska, pers. com.) found that between 1991 and 1995, 836 new operations entered the area 2C halibut fishery, 1023 entered the area 3A fishery and 366 entered the area 3B fishery in spite of the well publicized cut-off date. Thus, the public process of development of an access limitation system was paralleled by spiraling effort, much of which ultimately was included in the fishery.

In the Atlantic highly-migratory species (HMS) fishery, establishing a control date encouraged some permit holders, even those who did not fish their permits, to apply for permit renewals year after year to increase the likelihood that they would be included in any access limitation programme that might be developed. In effect, latent effort was inadvertently maintained. In June 1999, NMFS implemented a limited access programme in the HMS fisheries for swordfish, sharks and the pelagic longline fishery for Atlantic tunas. As of 31 August 1999 the number of permit holders in the fleet was 498 for swordfish, compared to 1000 in 1997; 906 for sharks, compared to 2257 in 1996; and 505 for tunas, which was unchanged as a result of the limited access programme. However, while there was a significant reduction in the number of permits issued in the swordfish and shark sectors of the fishery there may have been little reduction in the number of active permits in the fishery and, thus, little immediate reduction in effort in the fishery.

The fact that control dates are often established well after the public has become aware of a Council's intention of declaring a date and long after the first indications that the Council may ultimately limit access in a fishery, likely results in many fishermen taking action to establish some history in that fishery, either through fishing or acquisition of a permit. By the time the Council is ready to adopt final measures for a limited access programme, years may have elapsed and there may be considerable pressure to include those who entered the fishery after the control date. Under these circumstances the control date has very limited effectiveness. Control dates that are set without years of prior public discussion, that are announced widely, and that are followed quickly by access limitation programmes are probably more effective. There may also be some reduction in the willingness of lenders or prospective buyers to purchase vessels or equipment in fisheries for which control dates have been declared, especially if the borrower cannot demonstrate a solid history in the fishery.

The mere imposition of limitations on the number of permits has generally succeeded in eliminating the expansion of fishing effort, at least in terms of the number of vessels in the fishery. However, it usually has not solved the problem of overfishing. Often, the limitation followed a moratorium that resulted in little, if any, reduction in the authorized number of participants (*e.g.* the Hawaii longline fishery, the Gulf of Mexico red snapper fishery, and the Alaska groundfish and scallop fisheries). In many (probably most) cases, the Councils have been inclusive when it came time to approve a limited access programme and set criteria for continued participation in the fishery. The New England Council for example, has in some fisheries (Northeast multispecies, summer flounder, scup, black sea bass, American lobster) accepted proof of a single landing of any quantity of a species in that fishery as proof of historical participation (H. Goodale, NMFS Northeast Regional Office, Gloucester, Massachusetts, p.c.). There are also cases however, including the Atlantic swordfish and shark fisheries, in which qualification criteria were set such that significant reductions in permits resulted. In addition, Councils have sometimes established long qualification periods that have allowed the inclusion of a large number of fishermen and vessels (*e.g. Illex* and *Loligo* squid, and butterfish fisheries of the Mid-Atlantic). The inclusion of large numbers of historical participants can mean that a large amount of latent effort may be built into the limited access programme. Under those circumstances, when the fishery rebuilds, dormant effort could be reactivated and the potential benefits that might otherwise have accrued to those fishermen who remained in the fishery through the hard times could be dissipated.

In establishing the moratorium on the development and implementation of new IFQ programmes in the United States through the SFA, the Congress also required that the National Academy of Sciences (NAS) review the existing IFQ programmes in the United States and to make recommendations on their future use. The results of that study (NRC 1999) are given by Wertheimer and Swanson (2000). In general, the study concluded that U.S. IFQ programmes are meeting most of their objectives and are successful, though not without problems. Clearly, IFQs are not a universal solution to all the problems of fishery management. They require a considerable initial investment in design and implementation and, in most cases, are subject to significant implementation and monitoring costs. IFQ management requires sound science for determination of quotas and considerable enforcement to ensure adherence to regulations. When appropriate however, IFQs can be a powerful tool in ensuring effective and efficient fisheries management.

IFQs are consistent with the precautionary approach to fisheries management to the extent that they allow permit holders a vested interest in the fishery through ownership of the right to harvest. One of the greatest concerns regarding the use of IFQs is their impact on the historical and cultural aspects of the fishery. For some fishery participants, and even environmental groups, privatization of fishing rights smacks of corporate ownership, concentration of wealth by powerful vessels and, or, worse yet, foreign ownership. Yet the Alaska halibut and sablefish ITQ programmes are examples of closely regulated IFQ systems that were designed specifically to address the safe-guarding of the social and cultural fabric of this fishery.

Cooperatives have shown promise in filling the gap between: permit or vessel-limitation programmes that are so inclusive that derby-style fisheries continue, and true IFQ programmes. The first few years of experience of the West Coast whiting cooperatives indicated that economic efficiency, including product recovery rates, improved, and bycatch of other species that were of biological and regulatory concern decreased. Success of the whiting cooperative has been attributed at least in part to the relatively small number of homogeneous participants. There have been negative consequences as well: some excess capacity from the whiting fishery moved into other groundfish fisheries (*e.g.* the yellowfin sole and rock sole fisheries in the Bering Sea) that were already overcapitalized (NRC 1999).

While there are no data yet available on the effectiveness of the majority of buyback programmes, results of the capacity reduction programme for the Northeast multispecies fishery have been examined (Kitts *et al.* 1998; NMFS 1998). As of October 1995 there were 5128 vessels with associated permits for the Northeast multispecies fishery; of these, 2451 were known to have sold at least some marine products (not necessarily groundfish) (Kitts and Thunberg, unpublished). The vessel buyback programme removed 79 vessels from the fishery between 1995 and 1997, in two phases, which accounted for 10.1% of baseline physical capital, 4.9% of all allocated days-at-sea (DAS), and 16.8% of all DAS that were actually used in the 1996 fishing year. The vessels bought back were, on average, larger and more active than other vessels holding permits in the fishery. Kitts and Thunberg (unpublished) examined the economic considerations of the programme and concluded that it was unlikely that the buyback programme would significantly reduce incentives for input substitution or "capital stuffing" by those remaining in the fishery, nor would it be likely to reduce existing fishing power enough to constrain, or ameliorate, the derby effects of a quota management system (though the New England Council has yet to adopt a "hard" quota for the fishery). However, they also concluded that the vessel buyback could slow the pace of a derby and capital-stuffing.

While conservation benefits of the Northeast multispecies buyback programme were difficult to distinguish from management measures implemented by the New England Fishery Management Council during the same period (*i.e.* Amendment 7 to the FMP), the programme likely did result in reduced fishing mortality, at least in the short term (NMFS 1998). Further, the programme achieved its goal of providing a means for distressed fishermen to exit the fishery (Kitts *et al.* 1998). Substantial latent effort remained in the fishery, even after accounting for the capacity removed through the vessel buybacks and there is continuing concern that a substantial amount of this latent effort could be reactivated as the stocks rebuild (*e.g.* only 21.1% of the groundfish fleet's allocated DAS were actually used in the 1996 fishing year (Kitts *et al.* 1998)). Owners whose vessels were bought-back were not required to surrender the right to reenter the fishery provided they purchase a vessel with the necessary permits. An additional benefit of the programme was that, since all Federal fishery permits associated with each purchased vessel were also retired, a total of 463 permits (*e.g.* for American lobster and Atlantic sea scallops) were also removed from service through the purchase of only 79 vessels (Kitts *et al.* 1998), thus reducing effort in those fisheries.

7. EXPANSION OF LIMITED ACCESS MEASURES

Most of the U.S. marine fisheries limited access management experience to date has been with commercial fisheries. However, that is changing. In 1999 NMFS published at the request of the Gulf of Mexico and South Atlantic Councils control dates for the for-hire (charterboat and headboat) sectors of the Gulf of Mexico reef fish fishery and the Gulf of Mexico and South Atlantic king and Spanish mackerel fisheries. For-hire vessels are transitional between purely commercial and purely recreational endeavours, and may be the next sector to which limited access measures are applied in the United States. The extent to which limited access programme will actually be implemented in for-hire fisheries and what the programmes will look like is yet to be determined.

8. CONCLUSIONS

Fishery management in the United States has evolved from relatively simple responses to biological problems in the 1970s to the much more complex problems of allocation and overcapitalization in the 1990s. Access limitation schemes have been developed in response to these problems and span a range from control dates, which may have little real impact in controlling effort, to full-blown IFQ programmes, which allocate rights to the resource to a finite number of participants on an individual basis. The Councils have developed, and NMFS has implemented, numerous FMPs and FMP amendments with the intent of limiting effort and, or, reducing capitalization in the commercial fisheries. However, the Councils have generally adopted inclusive programmes, such that actual effort has seldom been significantly reduced upon implementation. Limited access permit programmes have provided for further evolution of the concept of rights-based fishery management in the United States and experimentation with several forms of its application. Examples of the range of management schemes that have resulted include: co-management, as reflected in the Alaska Community Development Quotas; industry-based programmes, with minimal government intervention, such as the West Coast whiting cooperatives; IFQs in fisheries on both coasts of the United States; and government and, or, industry-funded buybacks of fishing effort. There is evidence that limited access may next be applied to the for-hire sectors of many fisheries. Given that the US population continues to increase and that the productivity of living marine resources is finite, it is very conceivable that some form of access limitation may ultimately be applied in the marine recreational fisheries. The heat of public debate can be expected to directly correlate with the effectiveness of limited access measures in controlling effort and capitalization in the fisheries.

9. ACKNOWLEDGMENTS

The authors thank the following for their comments regarding the effectiveness of control dates in U.S. fisheries: Rod Dalton, NMFS Southeast Regional Office, St. Petersburg, Florida; Peter Fricke, NMFS Headquarters Office, Silver Spring, Maryland; Jay Ginter, NMFS Alaska Regional Office, Juneau, Alaska; Hannah Goodale, NMFS Northeast Regional Office, Gloucester, Massachusetts; Rod McInnis, NMFS Southwest Regional Office, Long Beach, California; Richard Raulerson, NMFS Southeast Regional Office, St. Petersburg, Florida; Phil Smith, NMFS Alaska Regional Office, Juneau, Alaska; John Ward, NMFS Headquarters Office, Silver Spring, Maryland.

10. LITERATURE CITED

GAO 1999. Commercial fisheries buyback programmes, United States General Accounting Office, GAO/RCED-00-SR, Washington, D.C.

Ginter, J.J.C. and R.B. Rettig 1978. Limited entry revisited, in R.B. Rettig and J.J.C. Ginter, eds., *Limited entry as a fishery management tool*, Proceedings of a national conference to consider limited entry as a tool in fishery management, Denver, Colorado, 157-174.

Kitts, A. and E. Thunberg. Economic considerations in the design of northeast U.S. fishing vessel buyout programmes, National Marine Fisheries Service, Northeast Fisheries Science Center, Woods Hole, Mass., 43 p., *unpublished*.

Kitts, A., E. Thunberg and G. Sheppard 1998. The Northeast groundfish fishery buyout programme, *NOAA Technical Memorandum*, NMFS-NE-115, 39-45.

National Marine Fisheries Service. NMFS 1991. Strategic Plan of the National Marine Fisheries Service, National Marine Fisheries Service, Silver Spring, Maryland, 21 p.

National Marine Fisheries Service. NMFS 1996. Magnuson-Stevens Fishery Conservation and Management Act as amended through October 11, 1996. *NOAA Technical Memorandum*, NMFS-F/SPO-23, 121 p.

National Marine Fisheries Service. NMFS 1998. Report to Congress on Northeast multispecies harvest capacity and impact of Northeast fishing capacity reduction, 19 p.

National Research Council. NRC 1999. Sharing the fish: toward a national policy on individual fishing quotas, National Academy Press, Washington, D.C., 422 p.

Shotton, R. 2000. (Ed.) Current Property Rights Systems in Fisheries Management. *In* Use of Property Rights in Fisheries Management. Proceedings of the FishRights99 Conference. Fremantle, Western Australia, 11-19 November 1999. FAO Fisheries Technical Paper No. 404/1, FAO, Rome.

Wertheimer, A.C., and D. Swanson 2000. The use of individual fishing quotas in the U.S. EEZ, [this symposium].

Appendix 1
National Standard Guidelines of the *Magnuson-Stevens Act*

1. Conservation and management measures shall prevent overfishing while achieving, on a continuing basis, the optimum yield from each fishery for the United States fishing industry.

2. Conservation and management measures shall be based on the best scientific information available.

3. To the extent practicable, an individual stock of fish shall be managed as a unit throughout its range, and interrelated stocks of fish shall be managed as a unit or in close coordination.

4. Conservation and management measures shall not discriminate between residents of different states. If it becomes necessary to allocate or assign fishing privileges among various United States fishermen, such allocation shall be (a) fair and equitable to all such fishermen; (b) reasonably calculated to promote conservation; and (c) carried out in such manner that no particular individual, corporation, or other entity acquires an excessive share of such privileges.

5. Conservation and management measures shall, where practicable, consider efficiency in the utilization of fishery resources; except that no such measure shall have economic allocation as its sole purpose.

6. Conservation and management measures shall take into account and allow for variations among, and contingencies in, fisheries, fishery resources, and catches.

7. Conservation and management measures shall, where practicable, minimize costs and avoid unnecessary duplication.

8. Conservation measures shall, consistent with the conservation requirements of this Act (including the prevention of overfishing and rebuilding of overfished stocks), take into account the importance of fishery resources to fishing communities in order to (a) provide for the sustained participation of such communities, and (b) to the extent practicable, minimize adverse economic impacts on such communities.

9. Conservation and management measures shall, to the extent practicable (a) minimize bycatch and (b) to the extent bycatch cannot be avoided, minimize the mortality of such bycatch.

10. Conservation and management measures shall, to the extent practicable, promote the safety of human life at sea.

Appendix 2
Limited Access Management Measures in U.S. Fisheries

Fishery Management Plan	Licence or vessel moratorium	Licence or vessel limitation	ITQ or IFQ	Control date	Buyback programme	Comments
Atlantic Sea Scallops	Yes	No	No	Yes	No	Days at Sea (DAS) effort controls.
American Lobster	Yes	No	No	Yes	No	
Northeast Multispecies	Yes	No	No	Yes	Yes	DAS effort controls.
Atlantic Salmon	No	No	No	No	No	No fishery allowed in the EEZ.
Atlantic Surf Clam & Ocean Quahog	Yes	No	ITQ	Yes	No	First ITQ program in the U.S.
Atlantic Mackerel, Squid & Butterfish	Yes	No	No	Yes	No	Control dates for squid and butterfish; different control date for Atl. mackerel. License moratoria for squid and butterfish; no license moratorium for Atl. mackerel.
Summer Flounder, Scup & Black Sea Bass	Yes	No	No	Yes	No	
Atlantic Bluefish	No	No	No	No	No	
Spiny Dogfish	No	No	No	No	No	
Atlantic Herring	No	No	No	Yes	No	
Monkfish	Yes	No	No	Yes	No	
South Atlantic Snapper/Grouper	No	Yes	Yes	Yes	No	Wreckfish fishery is ITQ; implemented April 1992.
Atlantic Coast Red Drum	No	No	No	No	No	No harvest allowed in EEZ.
South Atlantic Shrimp	No	No	No	Yes	No	Control date for rock shrimp only.
South Atlantic Corals	No	No	No	No	No	No harvest in the EEZ for most species.
South Atlantic Golden Crab	No	Yes	No	Yes	No	Controlled access regime; specific criteria for permits for 3 fishing zones.
Gulf/S. Atlantic Spiny Lobster	No	Yes	No	No	No	FMP has adopted Florida's trap reduction program for the EEZ.
Coastal Migratory Pelagics	Yes	No	No	Yes	No	Control date for commercial king and Spanish mackerel (1995). Commercial license moratorium for king mackerel. Control date for commercial gillnet fishery for Atl. Group kings N. of Pt. Lookout, NC (1999). Control date for dolphin/wahoo commercial fisheries in S. Atl. (1999). Control date for recreational for-hire fisheries in Gulf of Mexico (1999).
Gulf of Mexico Corals	No	No	No	Yes	No	Control date for live rock (1994). No harvest in the EEZ for most species.
Gulf of Mexico Red Drum	No	No	No	No	No	No harvest allowed in the EEZ.
Gulf of Mexico Stone Crab	Yes	No	No	Yes	No	Temporary moratorium on Federal permits for vessels, ends no later than 06/30/02.
Gulf of Mexico Shrimp	No	No	No	No	No	
Gulf of Mexico Reef Fish	Yes	Yes	No	Yes	No	Red snapper ITQ program approved in 1995, but not implemented due to Congressional action to freeze ITQ programs. Reef fish fishery permit moratorium through the end of the year 2000. License limitation for red snapper only. Control date for recreational for-hire fisheries for reef fish (1999).

Evolution of Rights-based Management

Fishery Management Plan	Licence or vessel moratorium	Licence or vessel limitation	ITQ or IFQ	Control date	Buyback programme	Comments
Caribbean Spiny Lobster	No	No	No	No	No	
Caribbean Shallow-water Reef Fish	No	No	No	No	No	
Caribbean Corals/Invertebrates	No	No	No	No	No	
Caribbean Queen Conch	No	No	No	No	No	
Washington, Oregon, California Salmon	No	Indirect, through state programs	No	No	Yes	No Federal permits, but states have limited access programs that serve to control effort in these fisheries through restrictions on commercial landings.
Coastal Pelagic Species	No	Yes	No	Yes	No	Limited entry system in the commercial fishery (except for squid), south of 39° N. lat.
Pacific Coast Groundfish	No	Yes	No	Yes	No	Limited entry and open access fisheries. Cooperative in the offshore whiting fishery.
Western Pacific Crustaceans	No	Yes	No	No	No	Limited access permit for Permit Area 1; harvest guideline for Permit Area 1.
Western Pacific Precious Corals	No	No	No	No	No	No harvest allowed except by 1 permittee.
Western Pacific Bottomfish	No	Yes	No	Yes	No	Limited access permit for Ho'omalu Zone and Mau Zone.
Western Pacific Pelagics	Yes	Yes	No	Yes	No	Limited access permit for longline only; former control date; 1991-94 moratorium on new entry.
Gulf of Alaska Groundfish	Yes	No	Yes	Yes	No	ITQ is for Pacific halibut and sablefish fixed gear only. Vessel moratorium.
Alaska High Seas Salmon	No	Yes	No	No	No	License limitation is by the State of Alaska.
Bering Sea/Aleutian Is. Groundfish	Yes	No	Yes	Yes	Yes	ITQ is for Pacific halibut and sablefish fixed gear only. Vessel moratorium. Cooperative in pollock fishery.
Bering Sea King and Tanner Crab	Yes	No	No	No	No	Vessel moratorium.
Alaska Scallops	Yes	No	No	Yes	No	Vessel moratorium.
Atlantic Highly Migratory Species (Sharks)	No	Yes	No	Yes	No	Directed and incidental limited access permits.
(Bluefin Tuna)	No	Yes (purse seine, longline)	Yes (purse seine only)	Yes	No	Individual vessel transferable quotas for purse seiners. Longline permits are limited access; must have limited access swordfish/shark permit to longline for tunas
(Yellowfin, other tunas)	No	Yes (purse seine, longline)	No	Yes	No	Longline permits are limited access; must have limited access swordfish/shark permit to longline for tunas.
(Swordfish)	No	Yes	No	Yes	No	Directed and incidental limited access permits
Atlantic Billfish	No	No	No	No	No	No commercial sale is allowed.

PROPERTY RIGHTS ON THE HIGH SEAS: ISSUES FOR HIGH SEAS FISHERIES

A. Stokes*
Department of Agriculture, Fisheries and Forestry
GPO Box 858, Canberra ACT, Australia 2601
<adam.stokes@affa.gov.au>

1. INTRODUCTION

The open access characteristics of the world's fisheries that existed up until the last few decades of the 20th century contributed to the overexploitation of many of the world's major fish stocks as demand and catching capacity outgrew biological productive capacity. This forced a rethink of the appropriateness of preserving the concept of 'freedom of the seas' which was the pillar of the open access argument. Some concerned coastal states took unilateral action by making claims to adjacent territorial seas and in effect expanded the areas and resources over which they claimed sovereign jurisdiction. Other states entered into agreements for various high seas fisheries to promote cooperation.

The culmination of these events was the development of the 1982 United Nations Convention on the Law of the Sea 1982 (1982 Convention), which seeks to balance the preferences of coastal and distant water fishing nations and provide a better foundation for the conservation and effective use of fish stocks. The Convention provides for the establishment of a 200nm Exclusive Economic Zone (EEZ) in waters adjacent to coastal states, granting the state sovereign rights of access, exploitation and management within the zone. In addition the Convention reinforces the rights of parties to fish on the high seas subject to the duty to cooperate (either directly or through sub-regional, regional or international organisations) in the conservation and management of high seas living resources.

While the 1982 Convention provides a more stable and certain environment for around 90% of the world's marine fish resources, events with respect to transboundry and discrete high seas stocks[1] made it clear that even this Convention was inadequate to prevent overfishing of high seas stocks and conflict between fishing nations. The fleets of distant water fishing nations, now banished from fishing grounds within the EEZ of coastal states, intensified their fishing of high seas stocks. Improved harvesting and processing technology assisted this intensification. The ambiguous nature of the rights of nations under the 1982 Convention with regard to high seas stocks perpetuated this situation by reducing the incentive to cooperate and in effect creating an open access environment. Further international reform was required.

Agenda 21 of the Rio Declaration 1992 called for the convening of a United Nations conference on straddling and highly migratory stocks to assist the implementation of the 1982 Convention on the high seas. The United Nations Conference on Straddling Fish Stocks and Highly Migratory Fish Stocks commenced in April 1993 and concluded in December 1995. The resulting agreement (known here as the UN Fish Stocks Agreement, or UNFSA), which is yet to come into force, attempts to improve the application of the 1982 Convention provisions relating to high seas fisheries by:

i. promoting the adoption of compatible management strategies between EEZs and the adjacent high seas, most notably through the establishment of Regional Fisheries Management Organisations (RFMO)
ii. establishing special access rights to those states participating in or abiding by the RFMO and
iii. establishing cooperative enforcement rules.

While UNFSA is as yet untested, there is growing interest in its proposed effectiveness (e.g. Munro 1996, 1998). Most of this interest centres on the tools that UNFSA will employ to achieve better conservation and utilisation. The right to fish qualified by the obligation to cooperate parallels a system of property rights. However, how strong these rights are is questionable, and there are doubts surrounding the capacity to maintain cooperation, with the principle thesis being that the *value* of the right must be maintained if there is to be any hope of success. This issue is the topic of this paper.

The paper first briefly reviews the history of high seas management up to the development of a uniform framework under the 1982 Convention. The purpose of this is to highlight the historical forces that moulded the debate between coastal and distant water fishing nations over access to resources. This provides insight into the formulation of the framework of the 1982 Convention. In doing so, I examine the importance of effective property rights for the management of fisheries. Second, the paper provides an overview of both of the 1982 Convention and UNFSA to more precisely outline the issues. Third, I examine whether either agreement establishes an effective system of property rights over high seas fish stocks. The paper will argue that the arrangements mandated by the agreements do not in themselves sufficiently fulfil the criteria for an efficient property rights structure, and the structure and policies of the individual RFMO's will largely determine the success of the agreements. The underlying argument is that without an effective right there is unlikely to be cooperation between members and

---

*This paper was written with the assistance of Matthew Gleeson and Jennifer Doust both of the Department of Agriculture, Fisheries and Forestry, Australia. However, all errors are the responsibility of the author alone. The views expressed in this paper are those of the author alone and do not necessarily represent the views of the Department.

[1] The stocks which are of issue include straddling stocks (found in both EEZ and adjacent high seas waters), highly migratory stocks (primarily tunas, with the capacity to travel across many EEZ's and high seas areas), and discrete high seas stocks (are exclusively a high seas stock).

discord will undermine property rights further. Fourth, the paper uses a number of economic concepts (such as strategic behaviour theory) to determine the factors that impinge upon cooperation, and suggest ways in which the activities of the RFMOs could be conducted to improve the cooperative environment. Last, the issues are considered in an Australian context by examining the management arrangements of two of Australia's main high seas fisheries, the southern bluefin tuna fishery and the Tasman Rise orange roughy fishery.

2. BACKGROUND
2.1 A brief history of high seas management

The principle of freedom of the seas has been a powerful catch-cry, resisting attempts by various nations throughout the centuries to monopolise access to the sea and its increasingly important fish stocks. By the late 1890s a number of factors arose to challenge the uncompromising notion of the right to uninhibited access to marine resources. Most notably improved technology had resulted in a massive increase in fishing effort that exceeded beyond the productive capacity of a number of important fisheries. It was the overfishing of these shared stocks and the inability of any one nation to enforce control that was the catalyst for integrated action, resulting in a number of multilateral and bilateral management agreements for various Atlantic and Pacific fish stocks. During this period a number of coastal states had made claims to territorial seas to provide solitary access to, and control over, stocks. The debate over property rights cumulated in the Hague Conference for the Codification of International Law in 1930. This lead to the recognition of the rights of coastal states to claim territorial seas, although the extent of such a claim was not resolved, while maintaining freedom of the high seas (OECD 1997). The problem of unregulated open access in high seas fisheries was therefore unresolved. However the notional principle of demarcated rights was firmly enough entrenched within the international arena to set the direction for future debate.

The period after the Second World War saw further deterioration of fish stocks due to excessive effort and overexploitation. This was the result of a number of factors, including (a) insufficient scientific data resulting in excessive yield expectations, (b) technical progress in harvesting and processing, (c) expansion of distant water fleet, particularly due to depleted local stocks, (d) increased market demand for seafood, (e) domestic political pressures (notably pertaining to unemployment and the push for industrial development) which resulted in the use of such instruments as subsidies, (f) the lack of success of traditional management measures, and (g) the lack of well defined property rights (OECD 1997). While catches grew rapidly, rents were dissipated by competition and excess capital as the lack of property rights discouraged incentives to invest in the natural capital of the fishery to improve future returns.

Throughout this period there was increasing recognition that the concept of freedom of the seas was not conducive to maintaining either the biological or economic integrity of fisheries. The obsolescence of this concept lead many coastal states to implement extended fishery jurisdictions to promote effective management of coastal stocks. Importantly, during the 1970s there was a growing consensus among nations that a 200nm jurisdictional zone was required for effective management given that around 90% of the global fish harvest were taken within this area (Munro 1998). Nations also exhibited an increasing reliance on international and regional regulatory fishing bodies and conventions in an endeavour to provide consistency and cooperation in fisheries management. However, while coastal waters were afforded more effective management frameworks, fisheries beyond the territorial sea still remained relatively unregulated. To the frustration of many, the international regimes established by the Geneva Convention on Fishing and Conservation of Living Resources of the High Seas (1958) and the United Nations Convention on the Law of the Sea (1960) failed to rectify the problem of open access in high seas fisheries (OECD 1997).

The first real attempt to consolidate the piecemeal reforms implemented to overcome the common property problems created by open access in fisheries, particularly on the high seas, came with the United Nations Third Conference on the Law of the Sea 1973-1982 and the subsequent ratification of the 1982 United Nations Convention on the Law of the Sea in 1994. This Convention formalised the trend of expanded fishery jurisdictions by establishing a 200nm Exclusive Economic Zone for the sovereign rights of coastal states. While this provided exclusivity of management for the majority fish stocks, the issue of transboundary and discrete high seas stocks remained.

2.2 The 1982 United Nations Convention on the Law of the Sea

The 1982 Convention entered into force on the 16 November 1994. Parts V and VII deal with the issue of rights and obligations in fisheries. Part V establishes the EEZ under Article 56, providing the coastal state with essentially what are full property rights to the fishery resources within the zone. Articles 61 and 62 identify coastal state responsibilities with respect to the conservation and utilisation of living resources within the zone. Article 61 requires the coastal state to:

i. determine the allowable catch
ii. take into account the best available scientific evidence to be utilised to determine measures to avoid over-exploitation, including cooperating with the appropriate international organisations
iii. collect and share scientific information and
iv. install measures which will maintain and restore fish populations at a level allowing the maximum sustainable yield to be produced, subject to economic and environmental factors (including species interdependence).

Article 62 requires the coastal state to:

i. promote optimum utilisation of living resources and

ii. determine the harvesting capacity of its fleet and make agreements to permit other states to access the surplus.

The focus of the 1982 Convention is the conservation of resources. Economic considerations relate to the minimisation of adverse impacts on fishing communities and the full utilisation of fishing capacity. There is no mechanism mandated to provide for the economic efficient utilisation of resources within the zone, and in particular, the articles do not allude to the provision of property rights at the domestic level through quota or similar arrangements.

Articles 63 and 64 have provided for a continuing area of confusion in relation to high seas management: they attempt to seek a compromise to the conflicting wishes of coastal and distant water states with regards to the rule of law over highly migratory species, most notably the high valued tunas (Kaitala and Munro 1993). Article 63(1) identifies the need for states to coordinate measures, either directly or through subregional or regional organisations, to ensure cooperation and development of *shared* stocks (stocks which are found within the EEZs of two or more coastal states). Article 63(2) calls for similar measures to be applied for *straddling* stocks (stocks that are found within the EEZ and in the high seas adjacent to the zone). Unusually however the conservation of stocks is not raised and the focus of cooperative efforts is in the area adjacent to the EEZ and not over the whole stock (inside and outside of the EEZ).

Article 64 provides a somewhat dubious separation between the Article 63 species and *highly migratory* species (listed in Annex 1 of the Convention, which may be defined as moving across multiple EEZs and high seas areas). Paragraph 1 calls for cooperation in the conservation and optimum utilisation of these species either directly between nations or through appropriate international organisations. The much stronger focus on international management forums and utilisation than for the Article 63 species is in line with the wishes of distant water fishing nations not to allow coastal states to dominate the management regimes for these species (Munro 1998), which seems to be the case in relation to straddling stocks.

The direct guidance for high seas fisheries management is provided by Part VII. Article 87, subparagraph 1(e), institutes the freedom to fish on the high seas subject to exercising due regard for the interests of other states and abiding by the rights and obligations laid down under Articles 116 to 120 (Bernaerts 1988: 48-9). Article 116 qualifies high seas fishing freedoms by the requirement to take into account the interests of relevant coastal states and other treaty obligations. Articles 117 and 118 stress the requirement to cooperate and negotiate management measures to ensure the conservation of living resources. Article 119 details the process for determining allowable catches to achieve conservation. Article 120 applies to the conservation of sea mammals.

The role of Articles 116-120 in clarifying the rights of states on the high seas has been the subject of much debate, particularly the question of whether coastal states or distant water fishing nations interests dominate. Kailala and Munro (1993: 316) summarise this debate with the view that the Articles are "a model of vagueness and imprecision". Property rights for straddling and highly migratory fish stocks on the high seas are not clarified by the 1982 Convention. Instead heavy reliance is placed upon the ability of states to voluntarily cooperate, either directly or through regional organisations, to obtain a working compromise in the face of ever increasing conflict on the high seas. As alluded to above, Article 63(2) adds to only further confusion and internal contradiction.

Finally, the strain began to show. The decade after the Law of the Sea Conference concluded is littered with examples of conflict over fish resources. Disgruntled coastal states aired their discontent with threats to expand the EEZ beyond the 200nm limit to prevent what they saw as continual overfishing by high seas fleets (OECD 1997), which negatively affected the conservation efforts by the coastal states for these same stocks. High seas fish stocks came under increasing pressure as cooperation gave way to competition and the 1982 Convention became under threat of becoming an obsolete international instrument. Agenda 21 of the Rio Declaration (1992) called for a United Nations Conference to be held to resolve the issue of overfishing of straddling and highly migratory stocks on the high seas and thereby assist with the implementation of the 1982 Convention. Program C of Agenda 21 identified the problems associated with high seas fishing as:

i. the lack of regulation
ii. overcapitalisation and excessive fleet size
iii. vessel re-flagging to escape controls
iv. insufficiently selective gear
v. unreliable databases and
vi. a lack of effective cooperation between states.

By identifying these problems, and with the general moral authority which was provided by the Rio Declaration, Agenda 21 set the focus for the Conference.

2.3 The United Nations Conference on Straddling Fish Stocks and Highly Migratory Fish Stocks

The role of the Conference was to provide an agreement that would establish the implementation regime of the 1982 Convention in relation to straddling and highly migratory fish stocks. The failures of the 1982 Convention as identified under Agenda 21 (inadequate scientific data, lack of enforcability, re-flagging problems and the problems relating to the ambiguous interpretation of its provisions) established the agenda for the development of the new agreement (Barston 1995).

From the onset of the Conference there was division between the coastal states and the distant water fishing nations (Barston 1995, Munro 1998). Coastal states aimed at securing a binding agreement that would protect their rights under the 1982 Convention in relation to high seas fisheries (as defined by Article 116(b)). Distant water fishing nations, in contrast, wanted a non-binding agreement that would provide broad guidelines in relation to high sea management. They also sought to have any proposed management arrangements conducted by regional or sub-regional organisations in an attempt to mitigate the influence of coastal states in adjacent waters. An important aspect of this was the push for the coherence of measures across the whole of the migratory range of straddling stocks. Two opposing views had developed: those who wanted the management regime in the adjacent area to be consistent with that of the coastal State (the coastal State regime therefore being dominant), and those who wanted to avoid 'creeping jurisdictionalism' by having the cooperative management regime administered through an international organisation (Kaitala and Munro 1993). The aims of the distant water fishing nations were not surprising given that at that stage they accounted for around 65% of all high seas fishing (Munro 1998), making them major stakeholders.

Despite this division, positive steps were promptly taken to set a credible agenda for negotiations. It was agreed that:

i. there would be consistency between the management regimes of the EEZ and adjacent high seas area
ii. stocks would be managed on a sub-regional basis and
iii. management would be conducted by RFMO's (Munro 1998).

The Conference sat for six sessions. During the initial negotiations spanning the first three sessions a number of issues emerged as focal points, including:

i. the level of coherence and compatibility of conservation and management measures inside and outside areas of national jurisdiction
ii. principles of fisheries cooperation and management
iii. international cooperation through regional and subregional organisations
iv. flag and port state duties
v. action with regard to non-parties
vi. the form of compulsory dispute settlement and review procedures and
vii. special requirements for developing countries (Barston 1995).

Of the issues, three emerged as significant sources of dispute: (a) the compatibility of management measures, (b) enforcement and (c) flag state responsibilities. These divisions were driven by the reluctance of high seas fishing states to grant any concessions which would undermine the traditional regime[2] of flag states; ambiguity over the form of any agreement and a reluctance to reach consensus; and a lack of support for developing a precise definition for adjacent area. This last point skewed the focus of the Conference towards the 'open ocean' and subsequently resulted in the failure to deal with rights within areas of dispute (Barston 1995) and therefore the fish stocks. This was to manifest itself in ambiguity within the subsequent agreement over where the powers of resource management reside within the RFMO (Munro 1998).

---

[2] Flying flags of convenience masks the identity and origin of vessels and provides a means for states to by-pass enforcement responsibilities.

A consensus position over the form of the agreement was finally attained. The agreement would be binding (reflecting the will of the coastal states) and would apply to stocks within and without of the EEZ (emolliating the concerns of distant water fishing nations over coastal State dominance of management arrangements (Munro 1998).

2.4 The United Nation Fish Stocks Agreement

The Agreement for the Implementation of the United Nations Convention on the Law of the Sea of 10 December 1982 Relating to the Conservation and Management of Straddling Fish Stocks and Highly Migratory Fish Stocks (known here as UNFSA or the Agreement) will come into force thirty days after it has been ratified by thirty countries. To date, 23 countries have ratified the Agreement and a number of others are close to finalising legislation to ratify.

The UNFSA is intended to provide rights, obligations and fisheries management principles for the long term conservation and sustainable use of straddling and highly migratory fish stocks. The terms of the Agreement require nations to join and cooperate with each other to sustainably manage these stocks.

Part II contains the general principles for conservation and management. Parties to the Agreement are obliged to adopt measures to ensure:

i. the long term sustainability and optimum utilisation of stocks (Article 5(a))
ii. the best scientific evidence is used to retain or restore stocks to sustainable levels (Article 5(b))
iii. a precautionary approach is applied to the management of fish stocks (Articles 5(c) and 6(1)) and
iv. the application of comparable conservation and management measures with respect to highly migratory and straddling fish stocks through direct cooperation or through a RFMO (Article 7).

Part III contains the mechanisms for international cooperation. Most important is the dominant role given to RFMOs under Article 8. Relevant fishing states have the duty to join existing RFMOs (Article 8(3)) or establish such organisations (Article 8(5)), particularly where stocks are threatened by overexploitation. Importantly, Article 8 seems to define *access* (if not *property*) rights and obligations at the regional or fishery level. Paragraph 3 emphasises that states "having a *real interest* in the fisheries concerned" may become members, although the terms of participation shall not preclude or discriminate against any states with a "real interest". Paragraph 4 goes further by granting those member states, or states agreeing to abide by the measures adopted by the RFMO, exclusive access to the fisheries resources over which the measures apply.

The reason why it may be more appropriate to describe the UNFSA as providing *access* rights, as opposed to *property* rights, is because the Agreement provides no formula to distribute the total allowable catch among RFMO members. This is left to the discretion of each individual RFMO, it may be subject to strategic manoeuvring by individual members and is unlikely to be based upon economic efficiency criteria.

In contrast, the participatory right of new members or participants, which includes the allocation of allowable catch, is subject to a number of considerations. Article 11 grants new members or participants in the fishery qualified participatory rights (which may, but not necessarily, be interpreted as a loose and convoluted set of factors that define real interest) subject to such factors as:

i. stock status and existing effort
ii. the interests, fishing practices and patterns of these states
iii. their contribution to management goals, including level of research
iv. the needs of coastal fishing communities and states whose economies are highly dependent upon the exploitation of the stock and,
v. the special needs and interests of developing states.

Within Part III there exist a potentially damaging contradiction that may limit the ability to establish and maintain an effective property rights structure. Relevant participatory parties are granted exclusive access to the resources governed by the RFMO. However the 'exclusiveness' of the access right is compromised by the inability to discriminate and exclude those with a "real interest" who are prepared to join or abide by the regional agreement. The concept of "real interest' is not defined, but is likely to depend upon economic need as much as catch history. It shall be shown later that such an arrangement adversely affects the incentive to create or join an RFMO, particularly in fisheries where stocks are being rebuilt.

The UNFSA mandates a number of obligations to ensure compliance with regional management arrangements. Articles 12, 13 and 14 seek to make decision-making processes transparent, oblige parties to collect and exchange scientific and other information and cooperate to improve the general effectiveness of RFMO's. Enforcement measures (Part VI) identify flag State responsibility and international cooperation procedures in relation to enforcement. An extension of enforcement measures to the high seas allows any RFMO member, not just the flag State, to take action against vessels suspected of breaching regulations.

Part VII gives special consideration to developing states and their capacity to implement conservation and management measures[3]. Cooperation that is required with such states includes financial assistance, technology transfer and consultancy services. The special consideration granted to developing states has obvious implications for the ability to exclude such states from RFMO membership for non-compliance. Part VIII outlines dispute settlement procedures, provides the framework for grievances to be negotiated and conciliated, and where need be, compulsorily arbitrated by various international or UNCLOS-constituted tribunals and courts.

In summary, while it is clear that the UNFSA provides a more appropriate instrument for management of stocks on the high seas, it may not establish a strong property rights structure.

3. PROPERTY RIGHTS IN HIGH SEAS FISHERIES

3.1 The theory of property rights

For economists, property rights refer to a bundle of entitlements defining the owner's rights, privileges and limitations for use of the resource (Tietenberg 1988:39). Property rights affect people's incentives to use a resource by providing certainty as to the impact of resource use on the users own welfare. When rights are well defined, people who own the rights can be certain how their actions will affect their current and future welfare and therefore they will have an incentive to use resources efficiently.

For renewable resources such as fish, the certainty which property rights provide enhance both conservation and utilisation outcomes[4]. The ability for long term planning, which is nonexistent under an open access regime, will promote more sustainable use. Fishers will reap the benefits of their actions to invest in the natural capital stock without the risk of having the benefits appropriated by others. Therefore, where there are property rights in fisheries there is likely to be better scientific research, market research and development, investment in management systems, better and more sustainable gear and lower catch rates in the short term[5].

This example shows an important distinction between a right and the *value* of that right. It is the value of the right, not the right in itself, which determines fishers behaviour. The right of access to a stock is meaningless to the extent that it merely provides a fisher with the opportunity to compete for a scarce resource. The benefits from fishing will depend solely on how much they can catch before the total allowable catch or economic and, or, biological extinction is reached. However, a right that allocates a certain percentage of the stock has an estimated value against which planning can be undertaken. This value will be affected by the actions of the individual. Where a group of individuals share rights, there are incentives to reach consensus on how the stock should be managed for the wider common good[6]. This common goal will provide the catalyst for cooperation, which will be assisted by the existence of a stable group of participants, providing for more efficient communication and administration channels.

---

[3] One concern is that this may provide the incentive for other nations to invest in the fleets of developing countries to by-pass more stringent access conditions and reduce fishing costs.

[4] This is the rule rather than the exception. In the rare cases where the discount rate is greater than the biological growth rate of stock, a purely rational individual may have the incentive to fish the stock to extinction (assuming price does not fall below a threshold level). However, if the ecosystem affects are considered, this is an unlikely scenario and regulation to prevent extinction will be the norm.

[5] In the long term, catch rates will grow with the expansion of the stock.

[6] Even if rights are well-defined, shared rights may be subject to externalities such as crowding.

3.1.1 Efficient property rights structures

The effectiveness of property rights in governing the efficient use of resources is a function of four characteristics (Tietenberg 1988:39):

i. Universality: all resources are privately owned and all entitlements completely specified
ii. Exclusivity: all benefits and costs from owning and using the resource accrue to the owner either directly or by sale
iii. Transferability: all property rights are transferable between owners in a voluntary exchange
iv. Enforceability: rights should be secure from involuntary seizure or encroachment.

It is when these criteria are met that true property rights are established. Criteria i, ii and iv promote certainty that only the owners of the property right will suffer the consequences of their actions. Criterion ii provides the incentive to internalise costs. Criterion iii ensures that rights and the resources they oversee will be allocated to the most efficient (highest valued) end use.

3.2 The structure of property rights under the 1982 Convention and UNFSA

Both the Convention and UNFSA attempt to manipulate party's rights and conditions of access to entice better conservation and management of stocks. The Convention successfully establishes rights at the national level by creating EEZs, bringing the majority of the world's fish resources under single jurisdiction management. However, the rights and obligations pertaining to stocks on the high seas (straddling, highly migratory and those uniquely high seas in location) are less well defined. Consequently, a number of the world's high seas fisheries have seen problems closely resembling common property outcomes.

The UNFSA seeks to deal specifically with the problems that have affected these resources. The principle instrument for resolving these problems is the establishment of RFMOs to conduct management arrangements in a manner consistent with the principles of the 1982 Convention and UNFSA. These management bodies have a high level of discretion over the entire range of the stock. Significantly, the UNFSA specifies that:

i. only members of the RFMO, or those abiding by its rules, shall have access to the stock and
ii. membership is qualified by a "real interest" in the fishery and obligations to cooperate, share information and operate transparently are explicit to membership.

However, it is unlikely that the arrangements under RFMOs qualify as full property rights and are more likely to resemble access rights, which do not in themselves have the capacity to resolve common property problems[7]. This is supported by several factors:

i. The proportion of total allowable catch allocated to individual members is a function of the total number of members. Because membership of or participation in an RFMO is flexible and indiscriminate, membership can readily vary. This situation increases the degree of uncertainty with respect to the allocated catch that each party will acquire and prevents long term planning of investment in production and processing capacity, skill enhancement and management at the national level. In a fishery where the number of participants is stable the allowable catch is determined by the state of the fish stock, which is a function *inter alia* of the level and selectivity of fishing. Fishers understand that their current fishing effort will affect future catch rates and economic returns through changes in the total allowable catch, and they will have an incentive to better manage the stock (including as a cooperative). Fishers will also be in a position to plan more effectively because they know in advance the likely catch quota. When the number of participants is unstable, fishers' catch allocation is related less to their own catch history and more on the level of membership, which is uncertain. This increases the risk of individual members losing their investment in the fishery and therefore they do not have the same incentives to use the stock efficiently. The value of the right is therefore likely to be highly unstable.
ii. Catch allocations are only seen as one of a range of possible management tools and are not explicitly mandated (paragragh 10(b)).
iii. The total allowable catch and its allocation lack an explicit economic dimension other than to maintain an industrial base. No effort is made to direct restructuring in the long term, the likely outcome being pressure to maintain the *status quo*.

It is useful to consider the capacity of the Agreements to determine whether they can implement and uphold property rights to meet each of the criteria that yield an efficient property rights structure. Table 1 makes this comparison for a range of different stocks. Five stocks are highlighted. Discrete zone stocks are those occurring explicitly within the EEZ of a State. Shared stocks travel between two or more EEZs but do not venture onto the high seas (covered by Paragraph 63(1) of the 1982 Convention). Straddling stocks occur within a State's EEZ and on the high seas and are the same stock body (covered by Paragraph 63(2) of the Convention). Highly migratory stocks are defined under Annex I of the Convention (Article 64). Discrete high seas stocks are unique to the high seas.

The universality criterion is divided into two components: (a) whether the resource is owned and (b) whether the entitlements are specified. The two are not mutually inclusive. For example, a shared stock is owned by the fact that it can be found within adjacent EEZs. However, the entitlements to the stock (allocation) are an outcome of the bargaining process between the owners.

The Table attempts to indicate if the Agreements' conditions are consistent with the criteria or at least establish a framework that is potentially consistent. To this end, there is a high degree of subjectiveness with regard

---

[7] Access rights equate to property rights when access is restricted to one party.

Table 1
Efficient property right structures and the United Nations agreements

Criteria	1982 Convention			UNFSA		
		Owned	Entitled		Owned	Entitled
Universality All resources are privately owned and all entitlements completely specified	Discrete zone stocks	✓	✓	Territorial stocks	✓	✓
	Shared stocks	✓	?	Shared stocks	✓	?
	Straddling stocks	✗	✗	Straddling stocks	?	?
	Highly migratory	✗	✗	Highly migratory	?	?
	Discrete high seas	✗	✗	Discrete high seas	?	?
Exclusivity All benefits and costs from owning and using the resource accrue to the owner either directly or by sale	Discrete zone stocks	✓		Discrete zone stocks	✓	
	Shared stocks	?(✓)		Shared stocks	?(✓)	
	Straddling stocks	✗		Straddling stocks	?(✗)	
	Highly migratory	✗		Highly migratory	?(✗)	
	Discrete high seas	✗		Discrete high seas	?(✗)	
Transferability All property rights are transferable between owners in a voluntary exchange	Discrete zone stocks	?(✓)		Discrete zone stocks	?(✓)	
	Shared stocks	?		Shared stocks	?	
	Straddling stocks	?(✗)		Straddling stocks	?	
	Highly migratory	?(✗)		Highly migratory	?	
	Discrete high seas	?(✗)		Discrete high seas	?	
Enforceability Rights should be secure from involuntary seizure or encroachment	Discrete zone stocks	✓		Discrete zone stocks	✓	
	Shared stocks	?		Shared stocks	?	
	Straddling stocks	✗		Straddling stocks	?	
	Highly migratory	✗		Highly migratory	?	
	Discrete high seas	✗		Discrete high seas	?	

to the interpretation of the Agreements and their respective elements. One factor that muddies the interpretation of the Agreements in relation to these criteria is Paragraph 63(1) and 64(2) of the Convention, which refer to shared and highly migratory species respectively. Both specify that the actions taken to ensure conservation of these stocks apply "without prejudice" and "in addition to the other provisions of this part" (Part V). This may be interpreted as meaning that the management arrangements detailed for discrete zone stocks under Articles 61 and 62 also apply to highly migratory species. If this is the case, then the management arrangements of transboundary and high seas stocks may be seen in a more positive light against the criteria. However, this does not take away from the thrust of the argument, which is that property rights are not effectively defined within the UN Agreements and will be determined by the structure, capacity and willingness of the RFMOs to implement property rights that maintain the value of the right.

The first point to note is that discreet zone stocks and shared stocks meet most of the conditions for property rights to evolve. This is not surprising given the special treatment these stocks receive. The 200nm zone established under the 1982 Convention gives coastal states exclusive rights over resources within the zone. The management practices within the zone are generally specified in terms of sustainable catch rates and the consideration of interdependent species in addition to the level of research and general competency of management. Sub-paragraph 62(4b) specifies the use of a quota system. However, as the table indicates, the transferability of quota is not dictated but is highly likely to occur.

Shared stocks also tend to have recognisable ownership (or at least access) rights. However, it is uncertain whether the arrangements for these stocks meet all the criteria, hence the question marks in the Table. This depends greatly upon whether Articles 61 and 62 of the Convention can be interpreted as applying to shared stocks. Overall, however, there are no outright negatives for shared stocks in relation to the criteria because the outcomes depend on the level of interplay and cooperation between the coastal states which share the stocks.

Turning to the criteria, we see that the Convention generally fails to provide for universality in relation to straddling, highly migratory and discrete high seas stocks. In contrast, UNFSA does have the capacity for universality. Resource access (if not ownership) under UNFSA is conditional upon membership of, or abiding by, an RFMO, which are also responsible for setting and allocation of the total allowable catch. However, there is a distinct ability to discriminate in relation to membership and consequently catch entitlements are likely to be uncertain. The issue that is likely to evolve in relation to universality is to identify the "real interest" as this may provide a mechanism for maintaining stability and cooperation within an RFMO.

Exclusivity is not guaranteed under the 1982 Convention for straddling, highly migratory and high seas stocks. There is the capacity to free-ride because cooperation and quota are neither mandated nor enforceable. Arrangements under UNFSA are potentially more conducive of exclusivity, however this is not a foregone conclusion. Unless all externalities can be identified, apportioned and internalised, exclusivity does not hold in a strict sense. As noted above, the capacity for the number of parties to an agreement to change will amplify the difficulty in dealing with these issues. Arrangements to promote internal stability and cooperation within the RFMO will assist in establishing exclusive rights.

Although a quota system is envisaged, transferability is not mandated in relation to any stock. The issue of transferability of fishing rights is at the discretion of the RFMOs and will depend upon the preferences of the individual members. However, conditions may not be favourable to entice participation in an international transferable quota market. At the national level, quota is traded between domestic fleets (ignoring joint-access arrangements in which quota may be leased[8]). These fleets tend to use domestic ports, processors, net and boat makers, provide fish to the domestic wholesale, retail and restaurant trade, and even support a domestic tourist industry. These inter-industry linkages produce multiplier effects that give the fisheries resource a value far in excess of the quota value[9]. This value is not recognised at the international level as domestic linkage effects are not considered and therefore it is unlikely that a country will willingly sell its quota[10].

Enforceability of the 1982 Convention fishing within discrete zones is determined by the domestic management arrangements. Article 73 of the Convention provides for strict enforcement arrangements pertaining to foreign vessels within the EEZ. Outside of the EEZ, the Convention attempts to build a transparent operating environment through the provisions for information sharing, research and dispute settlement procedures. UNFSA simplifies enforcement by defining a strict (yet flexible) division of access rights so that parties are readily identifiable. While this may assist in dealing with illegal fishing by non-members, the ability to enforce the value of recognised fishing rights is disputable. Because it is relatively easy to acquire access rights (and a proportion of the allowable catch) under the UNFSA there is likely to be encroachment on parties rights that negatively affect fishermen's behaviour because any redistribution of the allowable catch equates to a reduction in the value of the right to fish. This is likely to erode the willingness to cooperate and make enforceability difficult.

From this simple analysis we can draw two conclusions:

i. While the 1982 Convention fails or is unlikely to meet the criteria for establishing an effective property rights structure, the UNFSA provides a more sound foundation for success.

ii. Even though the chance of meeting the criteria is higher it depends on whether the value of the right is maintained, which is closely associated with the degree of cooperation within the RFMO which depends upon whether RFMO's can implement programs which induce or enforce cooperation, which in turn depends upon: (a) how closely the goals of the parties coincide, and (b) the benefits of cooperation, as determined by the value of the right to fish, relative to the benefits of non-cooperation.

3.3 Strategic behaviour and property rights

The application of strategic behaviour theory to high seas fisheries is a relatively recent academic pursuit (see Munro 1996, 1998; Hannesson 1995). It is an interesting area given the importance of maintaining cooperation and thereby the value of fishing rights. The standard analysis is based around dynamic game theory, which investigates situations where a party may have the capacity to undertake actions that improve their own welfare to the detriment of other parties. Game theory can be applied to the two situations that are likely to arise within the institutional framework of the UN Fisheries Agreements: competitive and cooperative. In competitive (noncooperative) games, it is assumed that parties will act in self-interest to maximise their own welfare, assuming that the actions which the other parties adopt are also be undertaken out of self-interest. In this situation aggregate welfare is minimised while the losses to the individual are minimised. Generally the outcome of this game is for all rents to be dissipated as competition drives the fishery to the common property equilibrium (Kaitala and Munro 1993, 1997).

In cooperative games, players act jointly to maximise the aggregate welfare available for distribution between the parties. The games core is based around the decision to cooperate or otherwise as determined by the relative benefits of cooperation and noncooperation. The game is complicated by many factors, including:

---

[8] Quota leasing does not provide the same level of certainty as when quota is purchased outright.

[9] Interrelated or co-dependant utility functions at the community level are likely to reinforce the high value of rights as the economic and social benefits of fishing have a reciprocal affect. Under these circumstances, communities are likely to be *welfare maximisers*, whereby Pareto improvements for the community as a whole may arise even if an individual is made materially worse off (eg. giving away some quota). The marginal utility of income of the individuals in the community will of course be important. In contrast, international fishing fleets will be *profit maximisers*, owned by multi-nationals, or a large body of socially independent shareholders whose utility is derived from the profits reaped from fishing.

[10] This depends upon whether the quota is handled domestically by an administrative body or is given outright to fishers. Fishers's willingness to sell quota will be determined by their fishing costs and the price of fish relative to the value of quota.

i. the homogeneity of goals across the players... capacity to adjust production
ii. the relative bargaining power of parties and degree of collusion
iii. the number and stability of membership (exit/entry costs)
iv. the stability or enforceability of cooperative arrangements and
v. the capacity for transfer payments or side payments as a means of bribe (Hannesson 1995; Munro 1996, 1998).

3.3.1 Homogeneity of goals

The goal of players in fisheries tends to be reflected in the desired harvesting rate (level of effort) as determined by the relative costs (labour, capital, fuel, etc.) and benefits (price) of fishing and the discount rate. All else equal, the lower (higher) are the fishing costs, the higher (lower) are the benefits and the higher (lower) is the discount rate, the higher (lower) will the harvesting preference be. Given that these factors are likely to differ between countries due to differences in *inter alia* social preferences, labour costs, subsidies, exchange rates and trade barriers, there will be a multitude of harvesting preferences that will encourage noncooperation and impede negotiation and settlement of the distribution of the allowable catch between parties.

In addition, there are likely to be other factors that influence the level of harvesting. The political persuasion of the various harvesting, processing and wholesale/retail sectors is important. The emphasis on employment policies also tends to impede the willingness of countries to negotiate and restructure fleets.

3.3.2 Bargaining power

Where one country has more bargaining power, there is likely to be a negotiated settlement that can be easily enforced. Bargaining power comes from a number of sources, including market access or dominance (which determine net benefits) in fisheries and other markets, access to infrastructure (such as ports) and fishing costs. Collusion also raises the bargaining power of the colluding parties. In general, the more a party has to lose from any settlement, the less bargaining power they have (Kaitala and Munro 1997). For this reason, the issue of new entrants is particularly important as they have little to lose from entry into an agreement.

There is an ambiguous allocation of bargaining power between coastal states and distant water fishing nations institutionalised within the UN Agreements. Article 116 of the 1982 Convention requires management arrangements for high seas fisheries to account for the interests of relevant coastal states, although as Munro (1998) points out such interests are ill-defined. This is not clarified by the UNFSA, which contains two contrary clauses under Articles 7(1) and 7(2) suggesting the power lies within coastal states and distant water fishing nations respectively.

This confusion of relative interests may be more inhibiting to a final solution than the imbalance of power itself. This is because, as Coase's Theorem specifies, the outcomes are independent of the assignment of property rights, suggesting that an efficient outcome will be achieved if either coastal states or distant water fishing nations have the dominant bargaining power. Therefore, this implies that explicitly instilling power within one party may be an appropriate basis for reaching an agreement without compromising efficient outcomes. Kaitala and Munro (1993:324) suggest that it may be optimal for the management preferences of the partner placing the highest value on the resource to be dominant[11].

3.3.3 Number and stability of membership

A large number of parties will make agreement difficult, not only because the aims of the parties are likely to be different but also because of the administrative burden involved in negotiation and enforcement. When membership is variable, as is the case under the UNFSA, the renegotiation of allocation and responsibilities under any agreement is likely to be arduous, cause conflict and ultimately place stress upon parties continuing to abide by any agreement. This is particularly the case given the problems relating to the changing value of parties' rights with changing membership. As the numbers of new members increase and the value of members allocation subsequently declines, parties will weight up the benefits of continued cooperation with those of noncooperation (Kaitala and Munro 1997). The fact that new entrants have a certain level of power given their bargaining position (with nothing to lose), there may be an incentive to enter into an agreement after it has been negotiated, especially since new entrants can free-ride on the work already undertaken by existing members. Kaitala and Munro (1997) suggest that the use of side-payments as a bribe to new or potential entrants is one possible mechanism to promote stability within the organisation.

One possible avenue open to existing members to ensure stability of membership is through Article 11 of the UNFSA, which places conditions on access rights. Through article 11(a) it may be possible for members to declare a fishery fully utilised, in which case additional membership is likely to be inconsistent with the principle of optimum utilisation (article 5(a)).

3.3.4 Side payments and transfer payments

It may be possible for parties to offer payments as a means of limiting the number of members or limit the catch. The level of payment that would be made would be equivalent to the marginal benefits received from the payment. A problematic issue arises when there is more than one existing member: a transfer payment by one member will benefit all existing members (free-riders), therefore there is likely to be less incentive to provide a

---

[11] How the true value of the resource is determined is a difficult question. In a market system, preferences are reflected in the level of effective demand as represented by the amount of money a party is willing to forgo. A country with an advanced market economy will be in a better position to indicate preferences than a country with less wealth, even though the preferences of the later may be greater than the former. This is particularly the case for subsistence fishers, who are unlikely to actively participate global management organisations.

transfer payment unless the benefits can be appropriated by the payer.

4. THE AUSTRALIAN CONTEXT
4.1 Introduction

Australia participates in a number of fisheries in which the 1982 Convention and UNFSA are important regulatory instruments: the southern bluefin tuna fishery and the Tasman Rise orange roughy fishery are two examples.

4.2 Southern Bluefin Tuna[12]

The Southern bluefin tuna (SBT) is a slow growing, highly migratory species that forms a vast single stock distributed throughout the southern ocean between 30°S and 50°S. Individuals live for up to 40 years and grow to around 200kg. They reach maturity at around eight years of age and aggregate to spawn in an area between Java and north Western Australia between September and March. Juveniles move south along the Australian coast where they disperse into deeper waters after reaching maturity (Cox, Stubbs and Davies 1999).

SBT is a high valued fish. The market for SBT is dominated by Japan, which consumes around 95% of the world catch, primarily in the sashimi market. Small niche markets have recently developed in the Republic of Korea, and in Taiwan - Province of China (Cox, Stubbs and Davies 1999).

Fishing for SBT has traditionally been dominated by Japan and Australia, and to a lesser extent New Zealand, because of the geographic location of both the stocks and the markets. In 1997, these countries accounted for around 36%, 34% and 2% of the global catch respectively (Cox, Stubbs and Davies 1999). However recent decades have seen an increase in the catches by other countries (Republic of Korea, Taiwan - Province of China, Indonesia).

Since May 1994 the activities of the three major SBT fishing nations have been managed under the Convention for the Conservation of Southern Bluefin Tuna (CCSBT). This agreement was prompted by the decline in stocks since the 1960s, and followed an informal trilateral agreement developed in 1984 between these nations to limit and allocate the total allowable catch. The objective of the CCSBT is 'to ensure, through appropriate management, the conservation and optimum utilisation of southern bluefin tuna' (Article 3). The aims of the CCSBT are pursued and managed by a Commission of member countries which aims at promoting cooperation and coordination in annually setting and allocating the allowable catch between parties. Parties are obliged to provide and exchange scientific and fishing information (for national and non-party nations) as the basis for decision-making.

One of the most important pressures on the CCSBT has been the presence of non-member fleets, whose catches in 1997 reached 28% of the global catch (Cox, Stubbs and Davies 1999). These non-members are acquiring the benefits of conservation efforts under the CCSBT, and reducing the value of the rights under it. It has also been suggested that the information-gap on catches that results from the inability to assess non-party impacts could make it difficult to detect cheating by members (Cox, Stubbs and Davies 1999). The mechanisms for dealing with non-member states are different to those for member states under the UNFSA. The enforcement rights of RFMO nations against the activities of non-members are fairly weak in relation to non-signatories of the UNFSA. If countries are signatories they have an obligation to join, or abide, by any RFMO, while RFMO members have the legal capacity to board and inspect boats of signatory nations to enforce RFMO arrangements. The gap in the law in relation to non-signatory states may close with time since the UNFSA is adopted into customary international law, thus providing extended coverage for its provisions.

While there have been moves to entice non-members to join the CCSBT, with some countries expressing an interest to do so, there are still obstacles to overcome. The size of the allocations required to persuade countries to join may become a sticking point to existing members, who will have to relinquish part of their share. Cox, Stubbs and Davies (1999: 44) suggest that non-parties may be increasing their catches to ensure a greater allocation when they join, assuming that such an allocation is based upon catch history. This suggests that the definition of 'real interest' in a fishery will prove a difficult concept to define and apply.

4.3 Orange roughy

Orange roughy is a slow growing, long-lived species that is widely distributed in temperate deep waters areas of both the southern and northern latitudes. Australia fishes for orange roughy in domestic and international waters. The international fishery in which Australia actively participates, along with New Zealand, is the South Tasman Rise fishery which extends from the Australian Fishing Zone into the adjacent high seas south of Tasmania. The area is significant in that it attracts spawning aggregations that are easily targeted. However, the slow growth rate and low fecundity of the species makes it susceptible to overfishing.

The fishery has been managed under a memorandum of understanding (MOU) between Australia and New Zealand since 1988. The MOU was a reaction to the increased fishing pressure exerted by both nations. The MOU is a non-binding cooperative management arrangement that establishes and distributes a total allowable catch between the two countries. The MOU lapsed in February 1999 as a result of a dispute over the share of the total allowable catch, which is based upon catch history. In negotiating new arrangements, both sides have agreed to respect the established total allowable catch (some 2400t), which is deemed sustainable, although New Zealand has refused to adhere to the previously agreed shares (75:25: Australia 1800t: New Zealand 600t). Subsequently, while the Australian Fisheries Management Authority closed the fishery to Australian vessels in April 1999 just short of the allowable national share of catch,

---

[12] This section is based primarily upon work undertaken by Cox, Stubbs and Davies (1999).

New Zealand continued to fish beyond their agreed allocation and eventually caught more than 1900t. Concern about this overfishing has lead both parties to agree to a ban for fishing for orange roughy in the South Tasman Rise until March 2000 while new arrangements are being made.

In late June and early July four foreign vessels conducted unregulated fishing in the area while the self-imposed Australian-New Zealand precautionary ban was in place. The vessels were identified as being South African and Belizean flagged vessels. Surveillance and diplomatic action disrupted the boats fishing efforts (although it did not prevent them from fishing) and the boats eventually departed. The impact that this fishing activity had on the stock is unknown.

Both Australia and New Zealand have a duty and right under the 1982 Convention to sustainably manage the stock in the South Tasman Rise. The MOU is also likely to meet provisions under UNFSA, given that the parties have developed a cooperative research program aimed at providing information on stock structure, status and productivity. In relation to the issue of allocation between Australia and New Zealand both parties have a number of channels for resolution, although a conciliated outcome is preferable. Resolution of this issue is likely to rest upon the definition of 'real interest'. Consideration of 'real interest' under UNFSA is convoluted and no real formula exists for defining access rights, let alone the distribution of allowable catch. One factor that may help to clarify the relative interest of parties is the issue of whether the stock straddles the area between the Australian EEZ and the high seas fishing grounds. If so, then Australian interests may take precedence.

In terms of the unregulated fishing, UNFSA will provide a greater basis for Australia and New Zealand to defend their rights in the region. Enforcement measures will be firmer and parties wishing to participate in the fishery will be obliged to join or abide by the MOU arrangements. If the UNFSA had been in force during the incident of unregulated fishing, Australia would have been able to board, inspect, and potentially take enforcement action against the vessels. Since the incident of unregulated fishing South Africa has expressed its interest in joining the Australian - New Zealand arrangement. However it would be uncertain whether South Africa has a 'real interest' in the fishery. How this would equate into excluding such a party from fishing given the provisions under Article 8 of the UNFSA is unclear.

5. LITERATURE CITED

Barston, R. 1995. United Nations Conference on straddling and highly migratory fish stocks, *Marine Policy*, 19(2), 159-166.

Bernaerts, A. 1988. *Bernaerts' Guide to the 1982 United Nations Convention on the Law of the Sea*. Fairplay Publications, Coulesdon, UK.

Cox, A., Stubbs, M. and L. Davies 1999. *Southern Bluefin Tuna and CITES: An Economic Perspective*, Report for the Fisheries Resource Research Fund and Environment Australia, ABARE Research Report 99.2, Canberra.

Hannesson, R. 1995. Fishing on the high seas: Cooperation or competition? *Marine Policy*, 19(5), 371-377.

Kaitala, V. and G.R. Munro 1993. The Management of High Sea Fisheries, *Marine Resource Economics*, 8(4), 313-329.

Kaitala, V. and G.R. Munro 1997. The conservation and management of high seas fishery resources under the new Law of the Sea, *Natural Resource Modelling*, 10(2), 87-108.

Munro, G.R. 1996. Approaches to the economics of the management of high seas fishery resources: a summary, *Canadian Journal of Economics*, 29,S157-S163, Special Issue April 1996.

Munro, G.R. 1998. The Management of High Seas Fisheries and the United Nations Conference on Straddling Fish Stocks and Highly Migratory Fish Stocks: A Review, (unpublished mimeo).

OECD 1997. *Toward Sustainable Fisheries: Economic Aspects of the Management of Living Marine Resources*, Paris, France.

Tietenburg, T. 1988. *Environmental and Natural Resources Economics* (2nd edition). Scott Foresman. Glenview 111 and London. pp. 559.

FROM SOCIAL THOUGHT TO ECONOMIC REALITY
THE FIRST 25 YEARS OF THE LAKE WINNIPEG IQ MANAGEMENT PROGRAMME

G.S. Gislason
GSGislason & Associates Ltd.
PO Box 10321, Pacific Centre, 880 - 609 Granville Street
Vancouver, BC, Canada V7Y 1G5
<gsg@gsg.bc.ca>

1. INTRODUCTION

Individual quota (IQ) fisheries management was introduced on Lake Winnipeg in 1972 and became the first IQ fisheries programme in Canada. Lake Winnipeg is a large, 15 000km^2 freshwater lake located in central Canada in the province of Manitoba. It has supported a commercial gillnet fishery for over one hundred years. The annual catch of approximately 5000t is taken by over 700 licence holders. Initially, the IQ Programme had a strong social development and income distribution focus. However, at the instigation of fishermen, adjustments have been made to the initial design to enhance the long term economic viability and sustainability of the fishery.

This paper outlines the unique political, management and resource components in place at IQ programme inception and charts the evolution of IQ management over the past 25 years. The Lake Winnipeg IQ programme, being one of the first in the world, can provide important lessons for other fishery management programmes.

2. BACKGROUND
2.1 History to 1970

Judson (1961) and Gislason *et al.* (1982) outlined the development of the commercial fishery on Lake Winnipeg. The fishery started on a large scale in the 1880s after the construction of the Canadian Pacific Railway and the arrival of Icelandic immigrants on the southwest shore of Lake Winnipeg. From the beginning, most of the catch has been exported to the United States, particularly the northcentral and northeast regions. In the early years, fishing took place during the summer season only and was restricted to regions with a ready transportation network to move the product from lakeside to the US market. Sailboats were used as fishing platforms and gillnets were the predominant gear used.

Around 1900, a winter fishery developed with the invention of the jigger, which made setting nets under winter ice easier. In winter, fish were hauled by horse teams to southern lakeside communities from which the railway moved the fish to southern markets. In the 1920s the gasoline engine was introduced and universally accepted by the larger whitefish boats and smaller skiffs alike. Tractor trains gradually replaced horse teams in transporting winter fish. Nylon nets were estimated to be more than three times as efficient as the cotton nets that they replaced in the 1950s.

Initially, the primary commercial species was whitefish, a species with high fat content that could easily be smoked or salted. Preservation was critically important. With the advent of speedier transportation and the introduction of refrigeration, other species became commercially important.

The technological changes and the increasing penchant of the consumer for convenience items such as fillets have resulted in a gradual shift in the focus of the fishery from whitefish to walleye and sauger. All three species are important economically today and comprise over 90% of the catch in weight and over 95% of the landed value. Other species harvested include pike, perch, mullet and goldeye.

Several new fisheries have emerged during the past 100 years in response to the increasing demand for walleye and sauger, *e.g.* fall fisheries. During the Second World War, in addition to meeting a high demand for fish, several "pocket" summer skiff fisheries were opened out of concern for the economic welfare of aboriginal people living in Lake Winnipeg lakeside communities.

Catches[1] declined during the 1950s and 1960s and the economics of the fishery were poor by end of the period (see Figure 1). The aggregate catches of walleye, sauger and whitefish had fallen to close to 2000t and the landed value[2] had fallen by two-thirds since 1950. Some fishermen were resorting to using small mesh nets or setting excessive amounts of gear thereby violating fishing regulations. Historical catches by individual species groups are given in Figure 2 with prices by species given in Figure 3.

In April 1970, Lake Winnipeg was closed to commercial fishing because of mercury contamination in fish arising from pulp mill discharge into a major tributary to the lake.

2.2 The Freshwater Fish Marketing Corporation

Initially, entrepreneurs and fish companies purchased fish from fishermen and then handled the processing and distribution of fish to markets. Several complaints emerged, in part because of the monopolistic buying power of the few brokers and processors, namely: (a) fishermen received only a minimal price and were not paid on a timely basis, (b) prices fluctuated significantly during the season inhibiting business planning, (c) banks would not finance fishing operations so many fishermen became indebted to fish companies (to pay back equipment and cash grubstake advances) and, (d) the system did not provide high quality fish to the market. In turn, the

---

[1] In this paper, all catches are reported in round weight.
[2] All landed values are reported in real, inflation-adjusted 1998 Canadian dollars.

Figure 1
Lake Winnipeg Commercial catch and landed value, 1950 to 1998

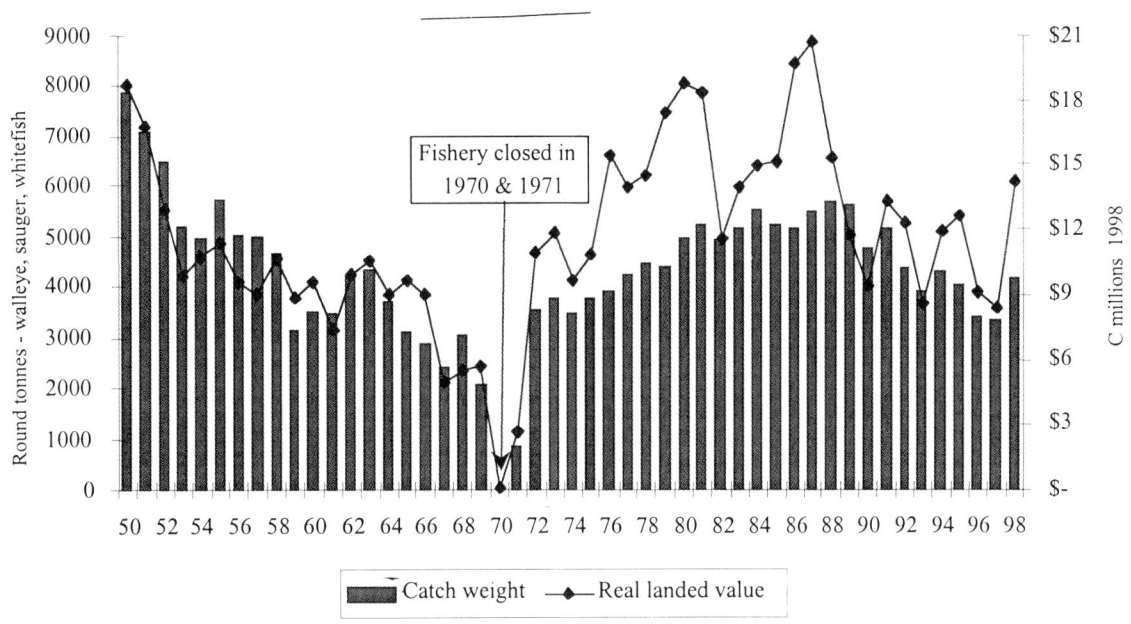

Figure 2
Lake Winnipeg Commercial Catch by Species, 1950 to 1998

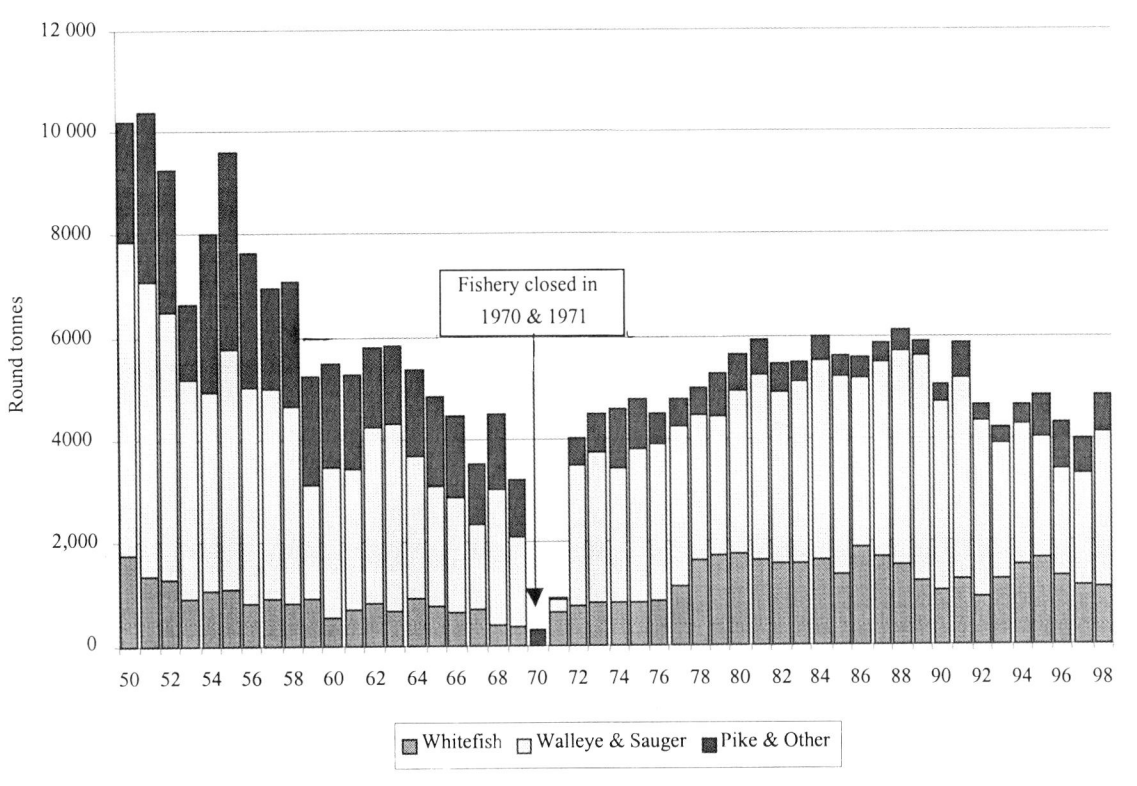

Source: Manitoba Conservation and Freshwater Fish Marketing Corporation

Figure 3
Real, inflation-adjusted landed fish prices for Lake Winnipeg, 1950 to 1998

Whitefish

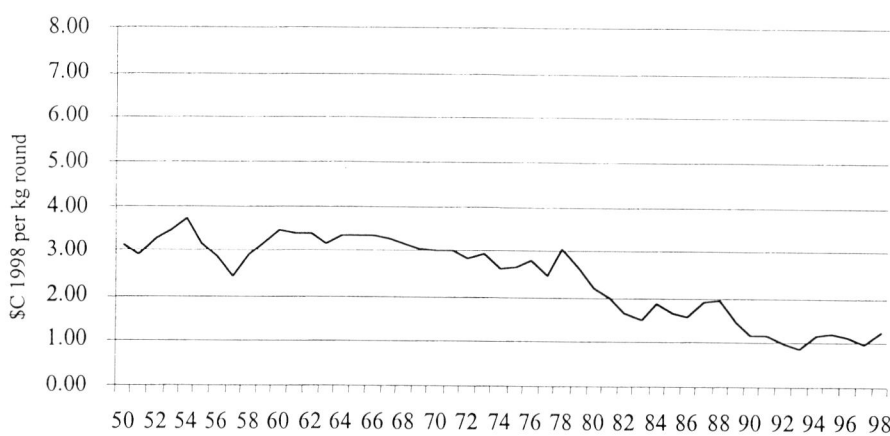

Walleye

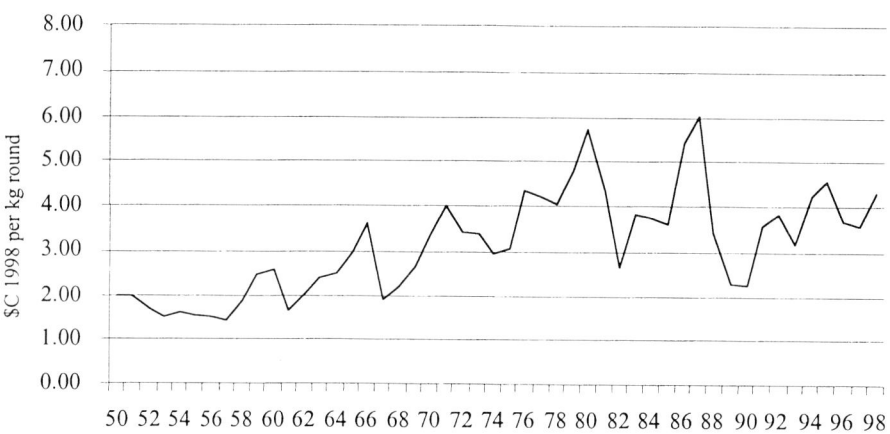

Sauger

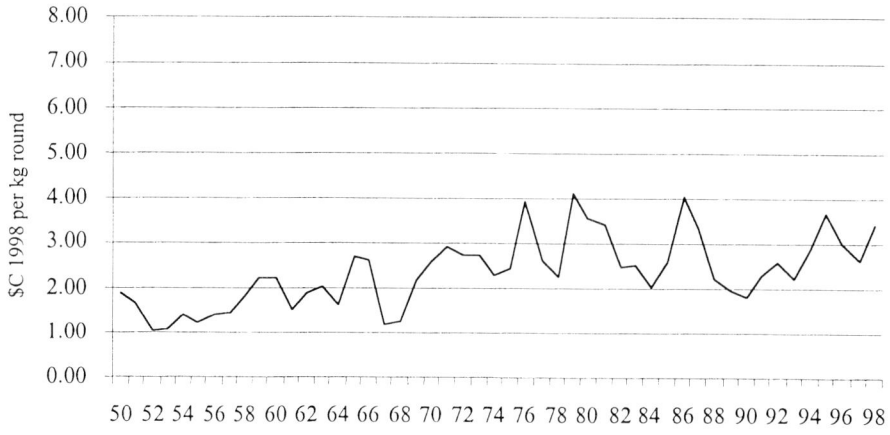

Source: Manitoba Conservation and Freshwater Fish Marketing Corporation

companies argued that the fish business was a high risk and unstable industry and they were paying people what they could afford.

The generally poor economics of the fishing industry resulted in a persistent call for government intervention. At least half a dozen government Commissions of Inquiry or investigations were held between the late 1880s and the 1960s. The federal McIvor Commission Report of 1966 called for the formation of a government-run single desk selling agency for freshwater fish. This recommendation was enacted into law.

In May 1969, the Freshwater Fish Marketing Corporation (FFMC) was created as a federal Crown Corporation and given exclusive jurisdiction over the interprovincial and export trade in freshwater fish for Western Canada. Fishermen could still sell direct to the final consumer in province. The FFMC has three goals: (a) to increase returns to fishermen, (b) to promote orderly marketing and, (c) to increase interprovincial and export trade in freshwater fish.

2.3 Fishing methods

By law, gillnets are the only allowable fishing gear on Lake Winnipeg. Open-water fishing occurs in the summer and fall seasons and ice fishing during the winter season. Three broad classes of operations exist today, two types of "open-water" fisheries and a "winter" fishery.

i. The summer whitefish fishery occurs in the North Basin of the lake using 12m to 18m steel or aluminum boats equipped with diesel engines, hydraulic net lifters and essential living quarters (crews usually consist of 2 to 4 hired workers in addition to the owner-operator).
ii. The summer and fall skiff fishery occurs throughout designated areas in the lake using 5m to 8m aluminum open boats equipped with an outboard motor of 70 to 150 horsepower; the boats have no living quarters and nets are set and lifted by hand (usually the owner-operator works alone but some have a hired worker).
iii. The winter fishery occurs throughout the lake with normally a large 6m long two-tracked enclosed van, (referred to as a "bombardier") or less frequently, a recreational snowmobile towing a small caboose for transportation over lake ice. Both types of operations use an auger and jigger to set nets by hand under lake ice (the owner-operator may have one or two hired workers).

Open-water skiff and winter fishermen return to their home base each day after lifting and setting nets. In contrast, the operators in the much larger whitefish boats may be away from home a month or more (they deliver their fish to a lakeside station several times a week). Summer whitefish fishermen catch mainly whitefish and open-water skiff and winter fishermen catch mainly walleye and sauger.

The fishery is, by design, labour intensive. Gislason (1977) and others have characterized fishermen as of advanced age, low formal education, and few transferable skills. In remote areas of the Lake, particularly aboriginal areas, the fishery is seen as a generator of employment and regional development. Today, about two-thirds of Lake Winnipeg fishermen are aboriginal.

2.4 Regulation of fisheries

In Canada, two levels of government exercise control over non-tidal fisheries such as the Lake Winnipeg fishery. Prior to 1930 the federal government had both proprietary rights and legislative jurisdiction. Under the *Resources Transfer Act of 1930*, the Province of Manitoba acquired proprietary rights to natural resources of the province including the fishery and the responsibility for administration of federal fisheries legislation was delegated to the provincial government. Nevertheless, the federal government maintains a constitutional responsibility to provide for the regulation, protection and preservation of all fisheries in Canada. That is, the management of the fisheries of Manitoba, including Lake Winnipeg, is undertaken as a co-operative endeavour between federal and provincial governments. The provincial government recommends changes to fisheries regulation. The federal government then reviews and approves such changes, and enacts them into legislation.

Commercial fisheries are regulated on the basis of seasons, areas, allowable gear (gillnets), amount of gear, minimum fish size, etc. The Manitoba Department of Conservation is responsible for fisheries management in the province and can determine who can fish and what fees, if any, will be paid. Each person seeking to fish commercially must be licensed (the individual rather than the vessel is licensed). Prior to the closure of the lake in 1970, entry to the commercial fishery generally was not restricted and area quotas, rather than individual quotas, applied. Any resident paying the nominal licence fee was allowed to fish commercially. Table 1 shows the evolution of fisheries management on Lake Winnipeg.

3. THE 1972 INDIVIDUAL QUOTA SYSTEM – A FRESH START

3.1 The setting

In the early 1970s a set of unique ingredients came together to allow the launch of the individual quota (IQ) system on Lake Winnipeg. These included:

i. the continuing low profitability of the commercial fishery – the combination of low catches, low prices, and poor returns led to widespread discussion and debate as to a fresh approach to management
ii. the closure of the fishery for 2 years – this gave time for reflection as to desirable fisheries management changes, and a "window of opportunity" for consultation and study
iii. the FFMC Daily Catch Record (DCR) monitoring system – this made possible the tracking of individual catches and provided the compliance mechanism necessary for an individual quota system and
iv. the incubation of new ideas and energy, from the newly-elected provincial government and from senior fisheries management created an atmosphere of innovation.

These conditions, in combination, provided the will and the means to effect profound change in Lake

Table 1
The evolution of property rights in the Lake Winnipeg commercial fishery

Era	Year	Fishery management regime
Pre-IQ development	1867	Canada becomes country under the British North America Act
	1870	Province of Manitoba joins Dominion of Canada
	1880+	Management of open access gillnet fishery mainly through season, area, amount of gear, and mesh size restrictions (three seasons – summer, fall, winter)
	1930	Provincial government can exercise property rights to fisheries under Resources Transfer Act
	1969	Launch of single desk fish marketing agency (the Freshwater Fish Marketing Corporation or FFMC)
Closure	1970 to 1971	Lake Winnipeg commercial fishery closed due to mercury contamination. Consultations with licensed fishermen indicate support in most but not all quarters for individual quota system
Non-transferable IQs	1972	Non-transferable individual quota (IQ) system introduced with 12 area/season combinations 　grandfathering of licence holders from late 1960s 　all licence holders in same area/season get equal quota 　licences restricted to "owner-operator" individuals 　licences from fishermen who die or retire revert to the Crown 　"points system" based on experience, dependency, and training used to reallocate available licences to applicants from same area
	1976	Licence transfers allowed between fishermen and son or daughter
	1976	"Retirement Licences" with modest quota introduced for those 55+ years of age who relinquish regular licence to son/daughter or Crown (can hold retirement licence for one season and fish regular licence in another)
	1981	Additional licences issued to some aboriginal communities to spur native economic development (now 17 area/season combinations)
	1985	Referendum of licenced fishermen indicate support for transferable quota system
Transferable IQs	1986	Licences & quotas made fully transferable, *i.e.*, "Quota Entitlement" system 　freely transferable subject to (area) residency and experience requirements 　can acquire a second quota for a season (max. of 2 per season) 　can redesignate season of purchased quota 　can rollover uncaught non-winter quota to next season 　maximum number of quotas per person limited to 4 to 6 (depending on area)
	1993	Property rights under individual quota system entrenched in provincial legislation
	1995	Possible for community organizations, as opposed to individuals, to hold quota
	1997	Restrict retirement licence to one per person per year (cannot hold regular licence)
	1998	Government provides funds for native community to purchase quota

Winnipeg fisheries management when there was no industry consensus about moving to an IQ system. In particular, the Lake closure was a "big event" that provided the impetus for change. Several meetings and discussions were held with fishermen in communities around the Lake. When Lake Winnipeg was fully re-opened to commercial fishery in 1972, a non-transferable individual quota system was adopted.

3.2 Initial allocation

Individuals received licences and quotas in 1972 through the "grandfathering" of participants in the fishery from the late 1960s. To obtain a licence and quota a person must have held a licence to fish the season in 1968 or 1969 or have held a licence for the season in six of the last seven years prior to 1968. Under this initial allocation, a total of 690 people received 1222 licences/quotas with one quota per licence (439 summer, 551 fall, and 232 winter).

The level of individual quota for each season was established in steps. Firstly, biological data collected during the lake closure period were encouraging so area quotas were pegged at a level 50% higher than the average catch levels over the late 1960s. Then, the area quota was allocated to summer, fall or winter seasons based on the historical contribution of seasonal catch to annual catch. Finally, the level of individual quota for each season/area combination was established by dividing the total quota by the number of qualifying licence holders.

The quota level varied by season and, or, area, but, apart from the summer whitefish fishery, was relatively small at 3200kg on average (the summer whitefish IQ was 12 000kg). All quotas were for the aggregate of whitefish, walleye and sauger.

3.3 Transferability and new entry

There was no provision for the transfer or consolidation of quota rights. It was illegal to sell or to lease a quota and the licence holder had to be on the lake during all times the quota was being fished, *i.e.* there was an "owner-operator" clause. Licences were automatically renewed for the same season in the subsequent year unless they were temporarily revoked for fishing infractions or low levels of production. In essence, licences and quota belonged to the Province of Manitoba, which issued them to fishermen on a temporary basis subject to renewal for satisfactory performance.

A "points system" was devised to allocate licences vacated through death, retirement or other reasons. Points were allocated on a declining scale for fishery participation in each of the last 10 years, and for taking courses at a government-sponsored Fisheries Training Centre. Applicants also had to satisfy certain criteria such as residency, dependency on fishing (one could not hold a full-time job) and access to equipment. Those with the most points in each of twelve Community Licencing Areas were allocated the vacated licences.

3.4 Programme adjustments

The provincial government made two adjustments to the IQ programme in 1976. The first allowed the transfer of a licence and quota between a fisherman and son or daughter. The second change was more substantial. The government created a new category of licence, the "Retirement Licence". Fishermen aged 55 years of age or older could relinquish their regular licence to a son or daughter, or the Crown, but could continue to fish under a retirement licence and modest quota of 650kg. Fishermen holding a Retirement Licence for one season could still hold a regular licence in another season.

In 1981, the provincial government created several new fall fishing areas and allocated new licences to fishermen (the Grand Rapids and Poplar River areas). The government also created new winter licences. The fishermen lived in aboriginal communities and the intent of the licence expansion programme was to spur aboriginal economic development.

The total lake quota of whitefish, walleye and sauger in aggregate increased from 4360t in the early 1970s to 5950t by the early 1980s (in response to these licence additions and certain IQ increases, *e.g.* the summer whitefish quota increased from 12 000kg to 15 880kg per licence holder).

3.5 Programme issues

According to Gislason *et al.* (1982) and Scaife (1991), several problems or issues emerged with the non-transferable IQ programme. These included:

i. A very low turnover in licences – only ten or so licences out of more than 1000 licences changed hands each year under the "points system" for issuing new licences and as a result, the fishermen population was aging rapidly (see Table 2). It was very difficult for new people to enter the industry, *e.g.* in some areas it would take 8 to 10 years of building "points" to get a licence.

ii. The inability to expand and tap economies of scale – a good fisherman could catch his or her seasonal quota in well under half a season but there was no mechanism to increase in-season quota holdings. The low turnover in licences meant that it was difficult for individuals to acquire a licence for another season.

iii. A heavy administrative burden – this was borne by the provincial government who recorded and verified "points" by individuals, wrote letters of rejection to the applicants not successful in acquiring a licence, etc. In addition, fishermen were constantly lobbying the government to increase quota levels as this was the only legal way to expand fishing opportunities.

iv. The lack of equity or resale value in fishery businesses – this was a constant source of irritation to fishermen. A set of fishing equipment without a licence or quota as part of the asset bundle was worth little. Fishermen worked their whole life in the industry and had nothing to retire on.

v. The strategic and illegal behaviour of fishermen – through fishing other people's quotas, "black market" sales, etc. became increasingly common as individuals strove to increase their production base.

These problems became apparent soon after the non-transferable IQ programme was implemented in 1972, but it took the better part of 15 years to address them.

4. THE 1986 QUOTA ENTITLEMENT SYSTEM

4.1 The setting

Most of the problems with the existing IQ management system stemmed from the non-transferability provision. There was support amongst government fishery managers and fishermen for making quotas transferable. In fact, the Lake Winnipeg Fishermen's Advisory Board, which consisted of 12 fishermen from around the lake, had endorsed the concept and had drawn up implementation guidelines for a transferable quota system. However, the concept of transferable quotas met considerable resistance at the political level and appeared to be shelved indefinitely.

What broke the impasse was a strong-willed Minister of Natural Resources (now called the Minister of Conservation) who believed passionately in transferable IQs, who was not going to seek re-election and who pushed the concept through caucus. Caucus insisted that: (a) a consultant hold a series of meetings around the lake on the topic and table a recommendation and (b), if the recommendation was favourable, that a referendum of licence holders on the topic be held. The consultant's report and the results of the following referendum both endorsed the concept of transferable quotas. In 1985 the government announced that licences and quotas would become transferable under the new "Quota Entitlement" programme.

Table 2
Socio-economic characteristics of Lake Winnipeg licence holders, selected years

	Year			
	1961	1969	1973	1999[a]
Age group[a]				
< 25 years	11%	8%	4%	4%
25 – 34 years	31	22	15	18
35 – 44 years	20	30	32	33
45 – 54 years	21	15	20	20
55 – 64 years	13	17	18	15
65+ years	4	8	11	10
All	100%	100%	100%	100%
Average age	40.3	42.9	46.6	45.0

	1973	1999		
		Regular licences[a]	Retirement licences[b]	All
Gender				
Male	>99%	91%	96%	92%
Female	<1%	9%	4%	8%
All	100%	100%	100%	100%

Source: Gislason et al. (1982) and Manitoba Conservation
[a] 751 Regular Licence holders (excludes licence holders fishing the 17 summer whitefish licences at Norway House). [b] 116 Retirement Licence holders.

The year 1986 was essentially the first year that the new programme took effect.

4.2 Initial allocation

Everyone who had licences or quotas received the same quotas under the Quota Entitlement (QE) programme. The government also decided to allocate a quota to the top point holder in each season for each community area under the now obsolete "points system" (as some fishermen had built up significant points but had been unable to obtain quotas because of the extremely low vacancy rate).

4.3 Transferability and new entry

The Quota Entitlement (QE), or transferable individual quota system, unbundled the licence and the quota. A licence holder could sell a QE to anyone in the same Community Licensing Area who meets experience and residency criteria, provided the buyer had no more than two QEs per season and four overall (except in certain areas where the local fishermen's association voted to have a limit of six).

In addition, fishermen not filling their open water quota may roll over the unspent portion for which they hold a licence (but those not filling a winter quota cannot roll it over into the next open water season). Also, individuals can redesignate the season of a purchased quota (except for the winter season licences in the northern part of the Lake), e.g. it is common for a person to purchase a fall quota and redesignate it as a summer quota.

The QE system has greatly increased the annual turnover of licences. The QE system also has taken government out of the business of reissuing vacated licences and decreased their administrative load significantly. Table 3 presents a snapshot of the Lake Winnipeg fishery, before and after the move to transferability.

4.4 Programme adjustments

The fishermen of Lake Winnipeg were generally satisfied with the QE system but constantly feared that the government would revise or cancel the programme without due notice and fishermen, through the Lake Winnipeg Fisheries Management Advisory Board, pushed the government to entrench their property rights in legislation. This, the government did in 1993. The Government of Manitoba (1993) recognized in legislation that "the allocation of an individual quota entitlement to a fisherman...constitutes a property interest of the fisherman in a right to fish the specified quota". In addition, the government under stated policy could not cancel the QE programme without giving five years notice. In 1995 the provincial government, through a policy change, made it possible for a community organization in addition to individuals to hold quota. Community organizations needed the support of 75% or more of fishermen in the area.

In 1998 the Norway House Fishermen's Co-operative became the first community organization to hold quota by acquiring the QEs associated with 17 summer whitefish licences (the government funded the purchase). The Co-operative chose not to fish the 270 aggregate tonne quota using 17 large summer whitefish boats but rather to allow more than 50 individuals to fish the quota with smaller skiffs under a "catch as catch can" non-IQ system.

The government has also allowed other areas or communities, where many quotas had not been fully used, to divide the local season into two. The initial phase operates under the normal QE system. The second phase, if the local fishermen support the concept, is run as a "catch as catch can" open season for the remaining aggregate quota, i.e. licensed fishermen can catch as much as they want subject to the integrity of the overall area quota.

This system has been implemented in several aboriginal communities for fall and winter fisheries in the northern part of the lake.

A major change to the Retirement Licence provision occurred in 1997. Many individuals greater than 55 years of age were selling or transferring their quotas and taking out retirement licences (many individuals held two or three retirement licences). By the mid-1990s, these retirement quotas represented an appreciable amount of quota in total. Starting in 1997, the individual was restricted to one retirement licence and no new retirement licences were to be issued.

The restriction on retirement licences was recommended by Symbion (1996). Other aspects of the Symbion report were more controversial. Fishermen in certain communities held diametrically opposite views to those of fishermen in other communities. One consequence was the disbanding of the Lake Winnipeg Fisheries Management Advisory Board.

4.5 Programme issues

The move to a transferable IQ system in the mid-1980s addressed most of the pressing issues of the day. For example, the number of licences changing hands increased from 10 a year pre-1985 to 200 or more a year in the late 1980s. Young people now can enter the fishery through buying an existing holder's licence rather than applying to government. However, a number of new issues have emerged in the 1990s. These include:

i. The long term sustainability of the resource – the approximate 40% increase in the aggregate lake quota since the early 1970s has raised the concern that the lake quota may be too high (catches declined during the 1990s, but the commercial catch has been very good in 1998 and in 1999 to date). Less than 80% of the aggregate quota has been taken in recent years.

ii. The wisdom of having one aggregate quota for whitefish, walleye and sauger – the biology of the three species is different and the price of walleye and sauger is much higher than for whitefish (see Table 3). High grading has become a problem. But moving to separate species quotas, however biologically sound, dilutes the property rights of certain segments of the fishermen population and therefore is controversial. At present, the summer whitefish fishery is the only one with a walleye-sauger tolerance. Up to 4545kg of the 15 880 QE can comprise walleye and sauger.

iii. The maximum allowable number of quotas held by an individual, of four to six depending on the community area, may be too low – advances in technology over the past 25 years such as larger skiffs, more powerful motors, monofilament nets, Global Positioning Systems (GPS), etc. mean that the catching power of the industry is much greater today. Fishermen have circumvented the restriction by having family members (*e.g.* their wives) acquire licences but this strategy, in many cases, is dependent on the government not rigidly enforcing the licensing provision that the quota holder must be on the lake during all times the quota is fished.

iv. The concentration of harvest in open water seasons – as a result of the rollover clause and the season re-designation clause of purchased quotas, many fishermen will try to catch all their quota in open water seasons (to minimize the risk their winter quota will not be caught, to not require a second set of winter equipment, etc). This practice inhibits the marketing of fresh fish by the FFMC throughout the year and can create processing bottlenecks, *e.g.* the FFMC did not accept deliveries of fish for several days this fall due to an overtime labour dispute at a time of peak deliveries.

v. The demise of the Lake Winnipeg Fisheries Management Advisory Board and its consultative process created a void for implementing change – the Board was critical to ushering in the QE system of the mid-1980s. Without the Board, or some effective substitute, it is not clear how needed change to QE management can occur.

vi. The low licence fee may impinge on maintaining rights in the future – the commercial fishermen of Lake Winnipeg collectively pay less than C$25 000 in licence fees annually, or less than 0.2% of the revenue base of $C14 million (currently the fees are $C52.50 per summer whitefish licence, $C17.50 per other open water licence, and $C22.50 per winter licence with $C2.50 of each representing an enhancement surcharge). The basic licence fee in nominal dollar terms has not increased in over 50 years. The low licence fee may create problems in maintaining property rights to the fishery in the face of encroachment of other fish and, or, water users that pay more, *e.g.* licence fees paid by recreational anglers, royalties or water rentals on hydro-electric development. Gislason (1999) has asserted that those who pay more have greater say and, by implication, have stronger rights.

5. CONCLUSIONS

The Lake Winnipeg individual quota system has been successful in large measure due to its ability to evolve and adapt since its inception in 1972. Each fishery around the world has its unique characteristics; what works in one fishery is not necessarily advisable for another fishery. However, the Lake Winnipeg situation does offer several "lessons learned".

Lesson #1: An individual quota (IQ) management system must be flexible and must continue to evolve as issues emerge, technology changes, and markets fluctuate.

Lesson #2: To launch an IQ programme and to make substantial changes over time takes the collective will and energy of three interests – the fishermen, the fisheries managers, and the elected politicians. The fishermen drive the process, the managers facilitate the process and the politicians execute the process. Leadership is required from all three groups.

Lesson #3: A formal process of dialogue for collating fishermen's input is essential to sound fisheries manage-

Table 3
Overview of the Lake Winnipeg Commercial Fishery, selected years

	Non-transferable IQs		Transferable IQs	
	1973	1984	1986	1999
No. of quotas held				
Summer[a]	439	468	535	891
Fall	551	642	626	375
Winter	232	309	318	213
All	1222	1419	1479	1479
Total quota tonnes[b]				
Summer[a]	1 530	1 970	2 320	3 650
Fall	2 100	2 450	2 280	1 390
Winter	730	1 530	1 570	1 130
All	4360	5950	6170	6170
Activity measures				
Catch (tonnes)[b]	3780	5540	5170	4150[d]
Landed value ($ million)[c]	11.9	14.9	19.7	14.2[d]
No. of fishermen	690	720	750	768

Source: Manitoba Conservation (formerly Manitoba Natural Resources) and Gislason et al.(1982)
[a] Includes 44 summer whitefish quotas at 12 000kg each in 1973 and 15 880kg each in other years, but excludes quotas associated with Retirement Licences and the Mossy Bay fishery. [b] Refer to whitefish, walleye and sauger only. [c] $C millions of real, inflation-adjusted 1998 dollars. [d] 1998 figures.

ment initiatives, including the development of IQ programmes. This process generally takes at least two forms: (a) a regular, meaningful advisory process and (b), an industry-wide vote or referendum on major issues, *e.g.* the move to a transferable IQ system.

Lesson #4: The impetus for change to existing IQ management programmes will be driven largely by business planning and economic issues of licence holders. The "agent of change" will be the economic circumstances of fishermen and not the broad social goals of government.

These lessons are broad and should be applicable to fisheries around the world.

6. ACKNOWLEDGEMENT

The author has benefited from discussions with and information provided by several people, namely: Ken Campbell, Sherman Fraser, Worth Hayden, Stephen Kendall, David Olson, Karen Olson, Richard Peters, Barbara Scaife, Ellen Smith, and Gordon Wakeling. Edna Lam provided valuable comments on drafts of the paper. Notwithstanding this assistance, the author has final responsibility for the analysis and conclusions of the study.

7. LITERATURE CITED

England, R.E. and J.R. Peters 1971. *Fisheries Adjustment Study*, Manitoba Department of Mines, Resources and Environmental Management, Winnipeg.

Gislason, G.S. 1977. Socio-economic characteristics of selected Manitoba commercial fishermen 1974-75, Natural Resource Institute, University of Manitoba.

Gislason, G.S., J.A. MacMillan and J.W. Craven 1982. *The Manitoba Commercial Freshwater Fishery: An Economic Analysis*, University of Manitoba Press, pp. 311.

Gislason, G.S. 2000. Stronger rights, higher fees, greater say: linkages for the Pacific halibut fishery in Canada. *In* Use of Property Rights in Fisheries Management. Proceedings of the FishRights99 Conference. Fremantle, Western Australia, 11-19 November 1999. FAO Fisheries Technical Paper No. 404/2. pp.383-389, FAO, Rome.

Government of Manitoba 1993. *The Fisheries Act (Manitoba)*, Part V.

Judson, T.A. 1961. The commercial fishing industry of Western Canada, Doctoral Dissertation, University of Toronto.

Manitoba Natural Resources 1994. *Five Year Report to the Legislature: Year Ending March 31 1994*.

McIvor, G.H. 1966. Commissioner, *Report of Commission of Inquiry into Freshwater Fish Marketing*, Government of Canada.

Scaife, B. 1991. Evaluation of the Lake Winnipeg commercial fishery quota entitlement system, Manitoba Department of Natural Resources, pp. 12.

Symbion Consultants 1991. An investigation into the status of issues affecting the Manitoba commercial fishery, Report prepared for the *Ad Hoc* Committee on the status of Manitoba commercial fisheries, pp. 111.

Symbion Consultants 1996. Third party review of Lake Winnipeg commercial fishery management issues, report prepared for Manitoba Department of Natural Resources, pp. 39.

Thompson, P.C. 1974. Institutional constraints in fisheries management, *Journal of Fisheries Research Board of Canada*, 31 (12). 1965-1981.

Thompson, P.C., L.E. Anderson and D.E. Topolniski 1994. Lake Winnipeg: quota entitlement programme, in Experience with Individual Quota and Enterprise Allocation (IQ/EA) Management in Canadian Fisheries 1972-1994, Canada Department of Fisheries and Oceans, November 1994.

Wysocki, W. 1981. Property rights and the Lake Winnipeg commercial fishery: a case study, Practicum submitted to the Natural Resource Institute, University of Manitoba, pp. 134.

THE USE OF INDIVIDUAL FISHING QUOTAS IN THE UNITED STATES' EEZ

A.C. Wertheimer
Auke Bay Laboratory, National Marine Fisheries Service
11305 Glacier Highway, Auke Bay, AK 99801 USA
<Alex.Wertheimer@noaa.gov>
and
D. Swanson
International Fisheries Division, National Marine Fisheries Service
1315 East-West Highway, Silver Spring, Maryland, 20910 USA
<Dean.Swanson@noaa.gov>

1. INTRODUCTION

Concern over declining marine fishery catches and clear signs of overexploitation of many important fish stocks worldwide have led governments and international organizations to seek new management policies and approaches to rebuild and sustain marine fisheries (Fujita *et al.* 1998). Traditional management tools have not always been effective at conserving fish populations and preventing exploitation (*e.g.* NMFS 1999a), or at avoiding overcapitalization and maintaining employment and fishing communities (Munro *et al.* 1998). One of the new management approaches is the granting of exclusive privileges to harvest portions of an overall quota of marine fish or shellfish. Such quota privileges have been given a variety of names depending on the characteristics of the management application, including individual quota, individual transferable quota, and individual vessel quota. In this paper we will use the term individual fishing quota (IFQ), consistent with its use and definition in the *U.S. Sustainable Fisheries Act (SFA) 1996*.

The SFA included amendments to the *Magnuson-Stevens Fishery and Conservation and Management Act*, the law that establishes the regional-council management system for regulating the fisheries in the United States. EEZ. This law, originally passed in 1976 as the *Fishery Conservation and Management Act*, has been amended several times as the emphasis and course of management policy has evolved (Darcy and Matlock this symposium). Initially, the principal policy goals were to assert U.S. authority and "Americanize" the fishery in the U.S. 200-mile EEZ. These policy objectives were achieved, but the rapid growth of the domestic fisheries led to many problems associated with open access and the race for the fish, including overcapacity; economic inefficiency; short, dangerous fishing seasons; ghost-fishing by lost gear and excessive bycatch; and continuous pressure to maintain high exploitation rates when faced with scientific uncertainty. To redress, or avoid such problems, management councils began developing and implementing limited access programmes, including IFQs.

The rationale for IFQs is to provide incentives to avoid the negative consequences of open access fisheries and limited entry systems and overcapitalization by individual fishers. Resource economists have long held that for a common-property natural resource, the incentive for the individual is to harvest as much as possible as quickly as possible and to compete for the greatest possible share (Gordon 1954, Scott 1955). Under open access fisheries with no overriding community constraints, the result can be the well-known "tragedy of the commons" (Hardin 1968), gross overcapitalization that is economically inefficient and threatens the biological viability of the resource. Even under limited entry systems, the 'race for fish' drives overcapitalization as the participants compete for their shares with improved technology and increased capacity (Grafton 1996). To address these problems, IFQs are intended to (a) improve economic efficiency by providing incentives to reduce excess harvesting and processing capacity; (b) improve conservation by creating incentives for stewardship activities, such as reduced bycatch and lost gear; and (c), improve safety by reducing incentives to fish in dangerous conditions. It must be recognized that IFQs do not remove all incentives for individual fishers to circumvent regulations for their own short-term benefits; highgrading of catch to increase its value and illegally exceeding individual allocations ("quota-busting") are concerns in enforcement of these programmes (Grafton 1996, Fujita *et al.* 1998).

The use of IFQs is controversial. Concerns have been raised about the basic equity of gifting a public trust resource, the fairness of initial allocations, increased costs for new fishermen to gain entry, and decreased employment and the disruption of fishing communities (McCay 1995, McCay *et al.* 1995). In the United States, these concerns have resulted in legislative intervention. The U.S. Congress, as part of the SFA, placed a moratorium on the implementation of new IFQ programmes in the United States' EEZ. In the SFA, Congress also directed the National Academy of Sciences (NAS) to undertake a review of existing IFQ programmes and to make recommendations on their use.

The SFA did much more than establish the moratorium on IFQs in U.S. fisheries; it enacted changes to the *Magnuson-Stevens Act* that reemphasized its goal of conservation and sustainability of living marine resources (Darcy and Matlock this symposium). The SFA established an explicit mandate to end overfishing and to rebuild overfished stocks, and to protect essential fish habitat and reduce bycatch. Conservation and management measures must consider efficiency in utilization, but

Figure 1
Landings of western Atlantic bluefin tuna by U. S. purse seiners, 1964-1998.
The IFQ programme was implemented in 1982. Source: NMFS (1999c).

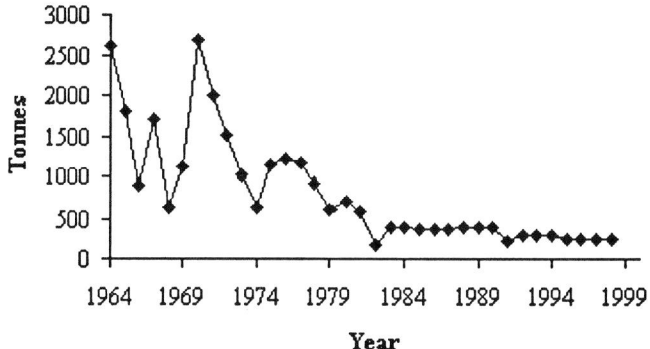

they must also consider the impact on fishing communities and the equity and fairness in the distribution of the benthey must also consider the impact on fishing communities and the equity and fairness in the distribution of the benefits from fishing.

This paper reviews the recent application of Individual Fishing Quotas (IFQs) to fisheries in the United States' EEZ and reviews the findings and recommendations from the NAS study evaluating the role of IFQs in achieving the mandates of U.S. fisheries law.

2. REVIEW OF EXISTING PROGRAMMES
2.1 Fisheries under IFQ management

Five U.S. fisheries are under IFQ management: the purse seine fishery for bluefin tuna off the northeast Atlantic coast of the U.S.; the surf clam/ocean quahog fishery off the mid-Atlantic Coast; the wreckfish fishery in the South Atlantic region; and the halibut fishery and sablefish fishery in the Pacific Northwest and Alaska regions. Prior to implementation of IFQs, most of these fisheries were characterized by overcapitalization and problems associated with the race for available fish. The objectives of the IFQ programmes included biological conservation (effective implementation of the TAC, reduction in ghost fishing and bycatch); economic (reduced overcapitalization, increased availability and value); social goals (increased safety, preserving traditional fishing patterns); and administrative improvements (better enforcement, more cost-efficient administration).

2.2 Bluefin tuna purse seine fishery

The first fishery to come under a form of IFQ management in the United States' EEZ was purse seining for western Atlantic bluefin tuna. Vessels using purse seines have landed bluefin tuna regularly since the 1950s, with some landings as early as the 1930s (Sakagawa 1975). Up to 21 purse seine vessels participated in the fishery, in some years landing in excess of 2500t (NMFS 1999b). The fishery primarily targeted small, schooling fish. Catches generally declined during the 1970s (Figure 1) and by 1982, five vessels were participating in the fishery. In response to concerns over declining catches and stock assessments, the International Commission for the Conservation of Atlantic Tunas (ICCAT) imposed quota restrictions on the western North Atlantic stock in 1982. While the total catch of all U.S. fisheries for bluefin tuna was limited by the quota, only the purse seine fishery was placed under limited access. Individual vessel quotas were assigned to the five vessels then participating in the fishery with each vessel receiving equal portions of the purse seine allocation. The purpose of the specific allocation to the purse seine fleet was to maintain the economic viability of the traditional fishery while meeting the mandated harvest constraints. Under current quota allocations, catches in recent years have been held to around 250t (Figure 1).

Since the implementation of the ITQ system, the purse seine fishery has been able to enhance its economic return by switching from targeting low-value, schooling bluefin to high-value, giant bluefin. Average size of fish in the landings has increased from less than 10kg in the late 1970s, to over 170kg in the late 1990s (Figure 2). Fishermen are now able to target the larger fish at times when their market value is maximized. Prices for purse seine catches have increased from less than $1/kg in the 1970s to as high as $19/kg in recent years (Figure 2).

The IFQ programme has met the objective of maintaining the traditional fishery while greatly reducing catch. The individual vessel quotas have been transferable in whole since 1983, and in whole or in part since 1996, but there has been no permanent consolidation of quota shares. The major controversy over this IFQ fishery has been the fairness of allocating such a large portion of a highly valued resource to a small number of permit holders (NMFS 1999b). While the purse seine fishery has been limited to five vessels, the number of vessels permitted in other gear categories (commercial and recreational) has increased to over 20 000.

2.3 Surf clam/ocean quahog fishery

These two closely related fisheries are conducted by vessels using hydraulic clam dredges (Serchuk and Murawski 1997). Surf clam fishing began in the 1940s and ocean quahog fishing began in the 1970s. Prior to IFQs, the fisheries were managed with a combination of size

Figure 2
Average size and price for western Atlantic bluefin tuna landed by U. S. Purse seiners, 1976-1978 and 1996-1998. Source: NMFS (1999b, 1999c).

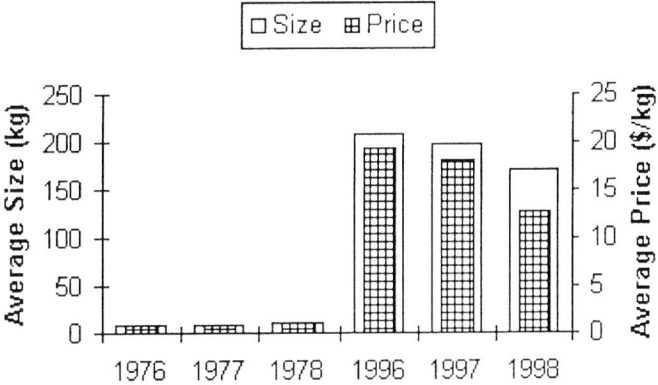

limits, quotas, and time restrictions to spread out the fishing season. There is no discernable stock and recruit relationship and the TAC is set conservatively to allow sustained harvest of the occasional large year classes. A moratorium on new entrants was imposed in 1977 as part of an effort to rebuild surf clam populations depleted due to overharvesting and poor environmental conditions. Under the moratorium the number of permitted vessels remained constant at approximately 140, but the harvesting capacity continued to increased due to vessel upgrades and gear improvements (Wang and Tang 1996). The moratorium was successful in reducing overharvest of surf clams, but regulation was cumbersome and enforcement with short, derby-style openings spaced throughout the year was difficult and costly. Overcapacity, enforcement, and safety issues relating to fishermen' feeling they had to fish under poor weather conditions during derby-style openings were major factors in considering IFQs as a management alternative.

The IFQ programme was implemented in 1990. Its objectives were to conserve and rebuild the resources by stabilizing annual harvest rates; to simplify regulatory requirements and minimize the cost of administration and compliance; and to provide for economic efficiency and reduction of overcapacity. Initial allocations of quota shares were made to owners of permitted vessels based on catch history. Shares are transferable and there is no maximum limit to accumulation except as determined by U.S. antitrust law. Allocation permit fees are collected to help defray administrative costs.

The programme appears to be meeting most of its stated objectives (Wang and Tang 1996, Serchuk and Murawski 1997). TAC overruns have been reduced (Heaton and Hoff 1999) and the number of discards are down, which has been attributed to IFQs providing incentives to target on relatively pure concentrations of large clams. Overcapacity has been reduced as the number of vessels has declined by 74% in the surf clam fishery, and by 40% in the quahog fishery (Figure 3a). The remaining vessels operate more efficiently, make more trips per year and harvest more clams per trip (Figure 3b). Regulations have been simplified, derby openings have been eliminated and harvesters have more flexibility in fishing operations.

Major controversies with the programme include the concentration of quota share, which is uncontrolled except by antitrust law; the equity of the initial allocation, which did not recognize crew participation; and the ability to track and enforce quotas (McCay et al. 1995). Commensurate with the decline in vessel participation has been a decline in employment for crew and a concentration of shoreside processors. Effects among coastal communities have been mixed, depending on where the harvest and processing power was consolidated.

2.4 Wreckfish fishery

Wreckfish, a member of the temperate bass family, are caught with specialized hook-and-line gear in a relatively small area in the U. S. South Atlantic region, in deep waters approximately 100 miles offshore (Sedberry et al. 1999). The fishery began in 1987 when concentrations of these fish were first located in the region. Catch and participation increased rapidly. Catch jumped from 13t in 1987 to 1887t in 1989 (Figure 4) and the number of vessels increased from two in 1987 to 90 in 1991 (Gauvin et al. 1994). A TAC of 907t was imposed in 1990, a level that is considered conservative and the biological characteristics of the landed fish have remained relatively constant. The rapid expansion of the fishery and the uncertainty about the population dynamics of this long-lived, but poorly understood fish were the most compelling factors for developing an IFQ programme. By 1990, capacity had increased to the point of requiring reduction of fishing season length, thus resulting in a race for fish. The South Atlantic Fishery Management Council identified this fishery as an opportunity to "rationalize" a fishery at its early stages (Gauvin et al. 1994).

The IFQ programme for wreckfish was implemented in 1992. Its objectives were to create incentives for conservation and regulatory compliance, avoid over-

Figure 3

Number of vessels (top) and number of bushels of clams harvested per trip (bottom) for the surfclam and ocean quahog fisheries off the northeast Atlantic coast of the U. S., 1979-1998. The IFQ program was implemented in 1990. Source: Heaton and Hoff (1999).

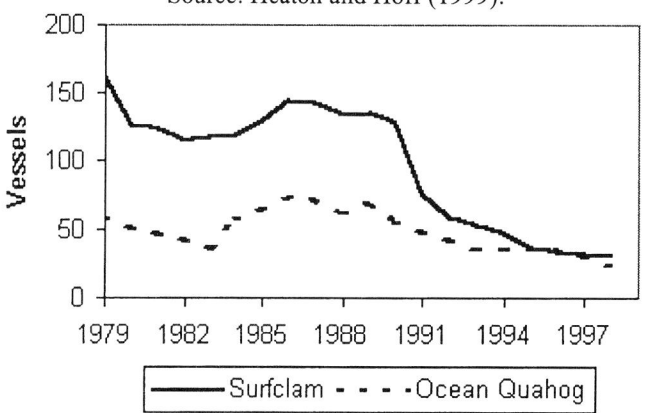

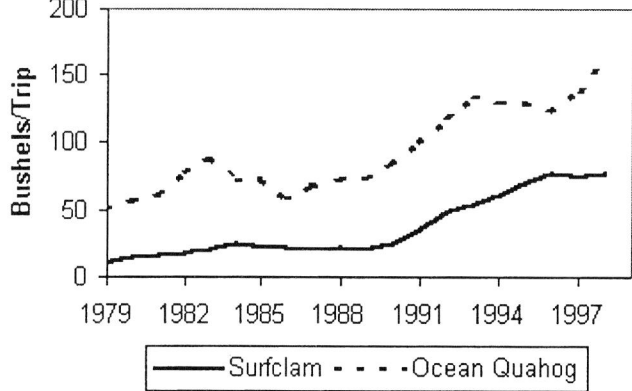

capitalization and encourage economic efficiency, minimize gear conflicts, provide a cost-effective management regime, and maintain product quality and increase total producer and consumer benefits from the fishery (Gauvin *et al.* 1994). Initial allocations of shares were issued to permit holders who had landed more than 2273kg of wreckfish in 1989 or 1990. No single business entity could receive more than 10% of the initial shares. The shares are fully transferable and there is no limit on the accumulation of shares.

Most of the objectives of the IFQ programme are being realized to some degree (Gauvin *et al.* 1994). The programme has reduced capitalization in the fishery. The number of participants in the fishery dropped from 90 to 49 with implementation of the programme. Further consolidation has occurred, with the number of shareholders declining to 25, of which only eight landed wreckfish in the 1996-1997 season. The small number of IFQ holders has made administration, monitoring, and enforcement much easier and more cost-effective than prior to IFQs. Catch has been constrained since implementation of IFQs. Full utilization of the TAC occurred in 1990 and 1991, prior to IFQs, but landings then declined continuously; in 1998, only 10% of the TAC was landed (Figure 4).

The consolidation of quota shares and the decline in landings are controversial aspects of this IFQ programme. Consolidation of shares can be viewed as market efficiency in operation, but it is unclear why effort and landings have continued to decline. Under IFQ management, improved availability to better match market prices was reflected by increased ex-vessel prices, from around $2.90/kg to $4.07/kg (Gauvin *et al.* 1994); prices have continued to increase and now are around $5.00/kg. Some shareholders may have decided to pursue other more profitable fisheries, and may be either holding their shares as a reserve, or they may be willing sellers but there may be no market demand for quota. For whatever reason, underutilization fails to meet the stated objective of maximizing total producer and consumer benefits from the resource.

2.5 Alaska halibut and sablefish fisheries

Commercial fisheries for Pacific halibut and sablefish occur off the coast of the U.S. Pacific Northwest, British Columbia and Alaska. These fish range from the Sea of Japan, through the Bering Sea and Gulf of Alaska and along the Pacific coast of North America to central California. The distribution of sablefish extends even farther south to Baja California. Each species is considered to be a single stock throughout its range. The

Figure 4
Landings of wreckfish, prior to and post-implementation of individual fishing quotas (IFQ), and the total allowable catch (TAC) for the fishery. Source: Sedberry et al. (1999).

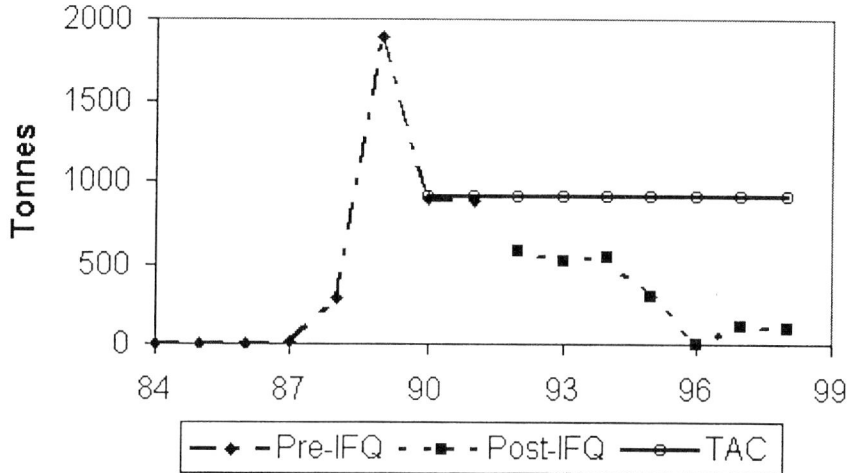

directed U.S. fishery for halibut uses longline gear, while the directed U.S. fishery for sablefish includes longline, pot, and trawl gear. Although these fisheries are managed under separate IFQ allocations they share many of the same characteristics and the programmes for the fisheries off Alaska were developed concurrently by the North Pacific Fishery Management Council (Pautzke and Oliver 1997).

Catches of halibut and sablefish have been historically controlled with a combination of area, season, gear and TAC limitations. There are highly refined stock assessment programmes for both species that include fishery-independent surveys and advanced population dynamics models (Sullivan et al. 1999, Sigler et al. 1999). The halibut fishery has long been recognized as a successful example of international cooperation and of scientific management maintaining sustainable catches from an exploited fish population. The United States and Canada negotiated the Halibut Treaty of 1923 and subsequently established the International Pacific Halibut Commission (IPHC). The IPHC sets the TAC for a number of management subareas. With the extension of U.S. and Canada EEZs, fishermen from each country were excluded from the waters of the others; allocation decisions for a specific subarea are made by the country with jurisdiction over that sub-area. For sablefish, allocations of TAC have also been made between the gear types.

The circumstances that led to development of IFQs in these fisheries are a classic litany of the problems associated with the race for fish. Perhaps the most striking example is the reduction in season length in the fisheries; the halibut season declined from about 50 days in the 1970s to an average of 2-3 days in most areas from 1980-1994 (Figure 5). The North Pacific Fishery Management Council noted that while traditional management measures could keep catch within biologically acceptable limits, substantial waste, economic inefficiency and unsafe fishing conditions would continue under such conditions (Pautzke and Oliver 1997). The Council identified ten problems that the IFQ programme was intended to address: allocation conflicts; gear conflicts; deadloss due to lost gear; bycatch loss; discard mortality; excess fishing capacity; product quality; safety; economic stability in the fishery and communities; and rural coastal community development of a small-boat fishery.

The IFQ programmes were implemented in 1995. Quota allocation was specific to management area, gear, and vessel size categories. Shares less than the equivalent of 9090kg have been "blocked" so they cannot be subdivided on transfer. Quota shares were allocated to vessel owners and leaseholders who had verifiable commercial landings during 1988, 1989 and 1990. More than 5000 fishers were allocated halibut quota shares and more than 1000 fishers were allocated sablefish quota shares. These numbers exceed the maximum number of vessels participating in any pre-IFQ fishing year. There are limits on transferability between vessel size categories; vessels in larger size categories cannot fish quota initially allocated to a smaller vessel size category. There are also limits on total accumulation of quota shares. For halibut, the limit is 0.5-1.0%, depending on the area; for sablefish the limit is 1% within specified management regions.

The implementation of IFQs in these fisheries has changed both the biological and socio-economic characteristics of the fisheries. In general, the programme has met most of its stated objectives. Overfishing the TAC, common before IFQs, has been eliminated (Figure 6). The IPHC estimates that halibut fishing mortality from lost and abandoned gear has decreased by over 75% and that discard of halibut bycatch has decreased by over 80%. There has been no evidence of high-grading and statistical analysis of size distributions indicate it has not been a problem. Season length has now increased from less than 5 days to 245 days per year for both species

Figure 5
Season length for the longline fisheries in the Gulf of Alaska for halibut (area 3A) and sablefish (West Yakatat). The IFQ program was implemented in 1995. Sources: IPHC (1998); NMFS (1999).

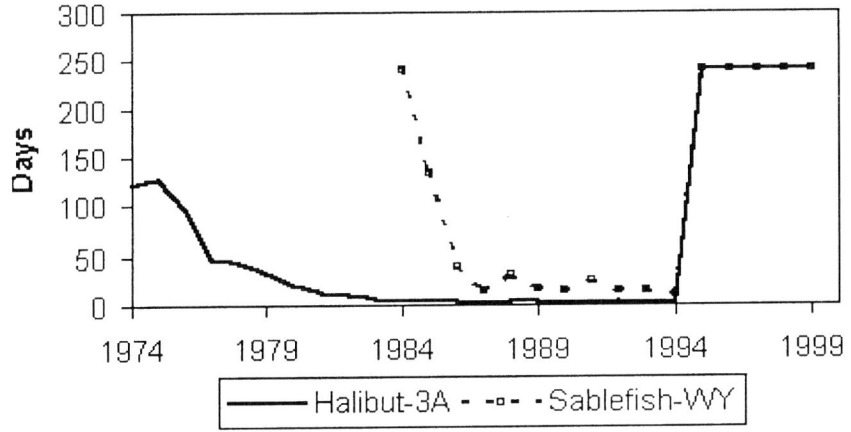

Figure 6
Percent deviation from total allowable catch (TAC) for the longline fisheries for halibut (Alaska region) and sablefish (West Yakatat region). The IFQ programme was implemented in 1995. Sources: IPHC (1998), NMFS (1999).

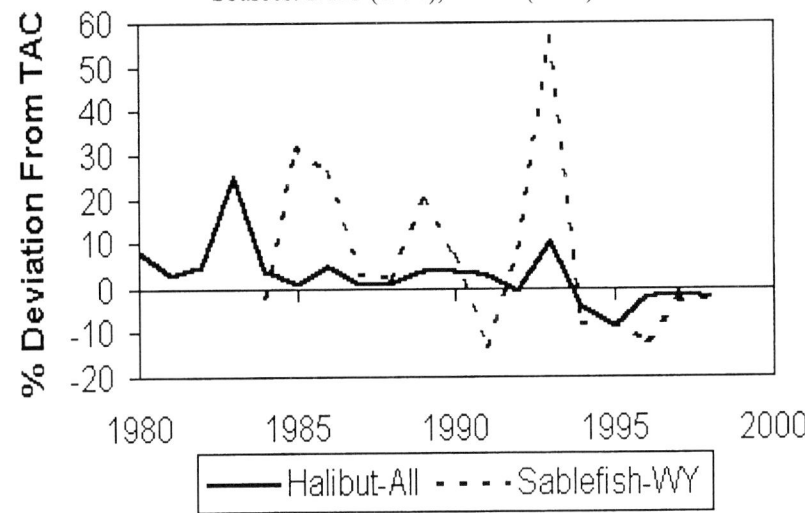

(Figure 5), decreasing gear conflicts and improving product availability, quality and value. In the sablefish fishery, age at catch has increased as reduced gear concentrations on the grounds have allowed fishers to target larger, older fish (Sigler et al. 1999). The numbers of vessels in the fisheries have declined by 40-50% (Figure 7). Some consolidation of shares has occurred, with a 24% decline in halibut quota holders and an 18% decline in sablefish quota holders (Smith 1999). The elimination of the race for fish has improved safety by reducing the pressure to fish under dangerous conditions. The number of Coast Guard rescues in the 3 years since implementation has been 31, less than half the 83 rescues in the 3 years prior to implementation (Figure 8). The complicated restrictions on accumulation and transfer of quota shares has maintained the diversity of small and large vessels and has provided opportunities for entrance into the fisheries.

The fairness of the initial allocation process remains an issue of contention. Crew members and processors are discontent that the process rewarded only vessel owners. With consolidation and reduced vessel participation, crew numbers have been reduced with vessels using smaller crews and some shareholders crewing for each other (Pautzke and Oliver 1997). The complexity of the system and the long season has resulted in increased costs for administration and enforcement. At present, these costs are not recovered from the fisheries, but a plan is being developed to assess up to 3% of the ex-vessel value of landings to cover these costs (Smith this symposium).

Community impacts have been mixed. While there has been a net increase in quota shares in Alaska fishing communities, there has been a decline in the proportion of shares held in smaller, village communities (Smith 1999). Changes in fishing patterns in response to IFQs have

Figure 7
Number of vessels participating in halibut and sablefish directed longline fisheries in Alaska prior to (1992-1994) and after (1995-1997) implementation of the IFQ programme. Source: Smith (1999).

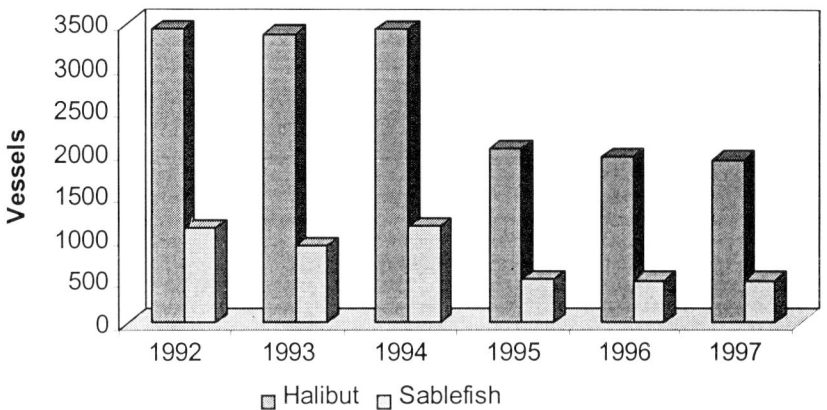

Figure 8
Number of search and rescue cases and associated deaths in the Alaskan IFQ fisheries prior to (1992-1994) and after (1995-1997) implementation of the IFQ programme. Source: NRC (1999).

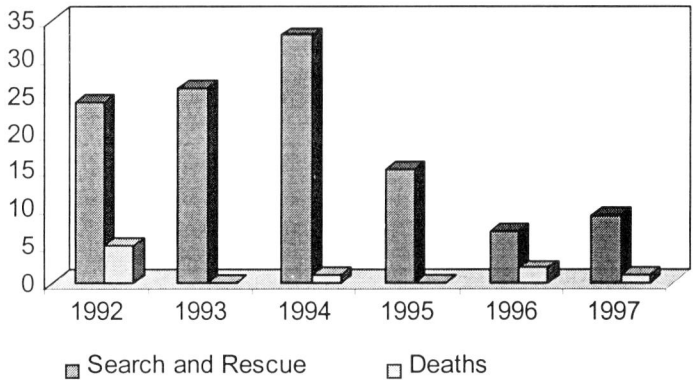

occurred (Gilroy 1996) and there is concern about local depletion and loss of recreational and personal use opportunities around some communities to the point where specific management actions are being considered to address such concern (DOC 1999).

3. NATIONAL ACADEMY OF SCIENCE STUDY
3.1 Findings and recommendations

The National Research Council of the U.S. National Academy of Sciences (NAS) has completed its report on IFQs (NRC 1999a). The report, "Sharing the Fish: Toward a National Policy on Individual Fishing Quotas", is comprehensive, detailed and presents a balanced perspective on the outcomes of IFQ programmes in both the United States and other nations. We highlight only some of its principal findings and recommendations for the purpose of this report.

The NAS analysis of U.S. programmes, as well as the experience of other nations, found that IFQs can effectively address some fishery management problems, especially issues associated with the race for fish - overcapacity, economic efficiency, product quality and safety. IFQs are not primarily a conservation measure; application of a scientifically determined TAC, combined with protection of essential fish habitat, size limits, gear restrictions, and other such management measures remain the main conservation tools in many fisheries, including those with IFQ programmes. There are numerous examples of meeting conservation goals for heavily exploited fish stocks without the application of IFQs. However, the NAS found that IFQs can contribute towards conservation by developing incentives for stewardship of the resource, decreasing TAC overruns and reducing wasteful fishing mortality sources such as ghost fishing and bycatch.

Because of the effectiveness of IFQ management systems at addressing long-standing problems in modern fisheries management, the NAS report recommended that the U.S. Congress lift the moratorium on the development and implementation of IFQs in U.S. fisheries as they considered IFQs an important tool that fishery managers should be allowed to consider and use.

3.2 Stakeholder participation

Successful application of an IFQ programme requires specific objectives and broad stakeholder support. The NAS found that IFQs have had different effects in different fisheries, partially due to differing objectives and partially due to unanticipated outcomes. They concluded that biological, social, and economic objectives should be clearly defined in a process that invites public and stakeholder participation. In the development of an IFQ programme, consensus among all those affected is unlikely, given that limited access programmes, including IFQs, involve a restructuring of the social and economic characteristics of a fishery. The NAS noted that every effort should be made to address issues of equity and fairness, and that the social and economic impacts on individuals and communities be considered, to ensure public support and industry compliance. The NAS also noted that it is equally important to consider these impacts in the context of alternative, or no, action as IFQs are typically proposed to either help avoid overfishing and its negative biological and economic impacts, or as an attempt to reverse the effects of overfishing and fleet overcapacity.

The NAS recommended that regional fishery management councils be allowed the flexibility to adjust existing IFQ programmes, consider IFQs as a management option for other fisheries and develop new IFQ programmes where they are deemed appropriate. The *Magnuson-Stevens Act* established the regional councils as a mechanism for the development of fishery management plans by those knowledgable in the fisheries of each region of the U. S. (Darcy and Matlock this symposium). Because there is no "one size fits all" IFQ programme, each fishery must be approached on a case-by-case basis. As noted above, broad stakeholder participation and support are needed in the development and implementation of the programmes. In complex, dynamic systems, it is best to address problems close to their source. The national responsibility for stewardship is not ignored in this process, however and the Secretary of Commerce retains oversight and must approve the management plans.

Although IFQ programmes can sometimes reduce management costs *(e.g.* as in the wreckfish fishery), implementation of these programmes has also resulted in large increases in administration and enforcement costs. Tracking and monitoring individual quotas can be complex and expensive and enforcing compliance can be difficult. In the Alaska halibut and sablefish fisheries, one of the major benefits has been extension of the season from a few, to over 200 days, improving product quality, availability and value. However, this has resulted in a substantial increase in resources needed for monitoring and enforcement throughout the longer season (Smith *this symposium*). Also, while IFQs provide incentives to the individual for conservation, considerable public and stakeholder concern exists about the potential for quota-busting, highgrading and poaching. Adequate enforcement is required to address these concerns and in many cases, on-board observers may be a necessary part of monitoring and enforcement. The NAS recommends that these management costs should be borne by those who are granted the exclusive privilege to harvest the resource, as is allowed under U.S. fisheries law.

3.3 Allocations of rights

The NAS study found that the initial allocation is the most contentious and controversial aspect of the implementation of IFQ programmes. Eligibility for quota shares has typically been limited to the catch history of vessel owners, which has been perceived as inequitable to captains, mates, deckhands and processors, all of which have large vested interests in the fishery. To meet the equity goal of the *Magnuson-Stevens Act*, the NAS recommended that a broad range of allocation criteria should be considered in structuring the allocation, such as extent of participation and dependence on the fishery.

Unlimited accumulation of transferable quota shares may result in consolidation of shares and regional changes in access to a fishery and community structure. The NAS found that limits on the transferability and accumulation of shares may be necessary in the context of the objectives of the programmes, especially where there is intent to promote an owner-operated fishery or conserve the geographic and community structure of a fishery. Such limits may include overall limits, restrictions by area or vessel category and restriction of ownership to *bona-fide* fishermen. The Alaska halibut and sablefish fisheries are examples of incorporating specific restrictions on transfers to maintain fleet diversity.

The NAS recommended that quota allocation programmes also consider alternatives to IFQs for accomplishing conservation and socio-economic objectives. Alternatives to private rights of access exist; communal rights, such as community development quotas (CDQs) may offer advantages for achieving management objectives in some fisheries (Reiser 1997). Allocation of quota share to communities is a possible mechanism for contributing to the sustainability of coastal communities that are heavily dependent on fishing and, or, have little alternative economic opportunity (Ginter 1995). The North Pacific Fisheries Management Council established CDQs in 1992 for certain groundfish species in the Western Alaska-Bering Sea region to coalitions of villages in the area. The programme is considered highly successful in fostering greater involvement in the fishing industry by the communities, with commensurate economic and social benefits including increased employment, infrastructure development and enhanced training and education opportunities for residents. These types of alternatives can be utilized for a fishery in conjunction with, or independent, of an IFQ programme; CDQs for Western Alaska villages have been incorporated into the IFQ management regime for halibut and sablefish in Alaska (Ginter 1995).

4. CONCLUSION

The NAS study and others have noted that IFQs are not a panacea (NRC 1999a, Fujita *et al.* 1998). Not all fisheries will be well suited for IFQ management regimes and alternatives exist (Reiser 1997). But the history of IFQs has shown their potential for addressing some long-standing problems in modern fisheries management. The allocation of permits to harvest a specified portion of a fishery resource can promote economic efficiency, increase benefits to the industry and the consumer and enhance conservation efforts. However, the U.S. experience has also shown that programmes must be developed to meet societal expectation for a public trust resource. As Christy (1996) noted, exclusive use rights, whether limited entry, IFQs, or communal rights such as CDQs, mean that some individuals will gain and others lose, either now or in the future, and questions of equity cannot be avoided. These questions must be considered in the perspective of the dismal history of open access fisheries management. Conservation, fairness and efficiency are all standards mandated by the *Magnuson-Stevens Act*. Policy makers must not only convince themselves and the stakeholders, but also society at large, that application of rights-based systems such as IFQs will meet the national standards and provide optimum utility of a public resource.

5. LITERATURE CITED

Christy, F. C. 1996. The death rattle of open access and the advent of property rights regimes in fisheries. Marine Resource Economics 11: 287-304.

Darcy, G.H. and G.C. Matlock 2000. Development and implementation of access limitation programmes in marine fisheries of the United States. *In* Use of Property Rights in Fisheries Management. Proceedings of the FishRights99 Conference. Fremantle, Western Australia, 11-19 November 1999. FAO Fisheries Technical Paper No.404/2. pp.96-106, FAO, Rome.

DOC 1999. Pacific halibut fisheries; local area management plan for the halibut fishery in Sitka Sound. Department of Commerce, NOAA 50 CFR Part 300. Federal Register 64 (81): 22826-22830.

Fujita, R.M., T. Foran and I. Zevos 1998. Innovative approaches for fostering conservation in marine fisheries. Ecological Applications 8 (1) Supplement: S139-S150.

Gauvin, J.R., J.M. Ward and E.E. Burgess 1994. Description and evaluation of the wreckfish (*Polyprion americanus*) fishery under individual transferable quotas. Marine Resource Economics 9: 99-118.

Gilroy, H., P.J. Sullivan, S. Lowe and J.M. Terry 1996. Preliminary assessment of the halibut and sablefish IFQ programmes in terms of nine potential conservation effects. International Pacific Halibut Commission.

Ginter, J.J.C. 1995. The Alaska community development quota fisheries management programme. Ocean and Coastal Management 28 (1-3): 147-163.

Gordon, J.S. 1954. The economic theory of a common-property resource: the fishery. Journal of Political Economy 62: 124-142.

Grafton, R.Q. 1996. Individual transferable quotas: theory and practice. Reviews in Fish Biology and Fisheries 6: 5-20.

Hardin, G. 1968. The tragedy of the commons. Science 162: 1243-1248.

Heaton, C. E. and T. B. Hoff 1999. Overview of the surfclam and ocean quahog fisheries and quota recommendations for 2000. Mid-Atlantic Fisheries Management Council, Dover, Deleware.

IPHC 1998. Report of assessment and research activities 1998. International Pacific Halibut Commission, www.iphc.washington.edu.

McCay, B.J. 1995. Social and ecological implications of ITQs: an overview. Ocean and Coastal Management 29 (1-3): 3-22.

McCay, B.J., C.F. Creed, A.C. Finlayson, R. Apostle and K. Mikalsen 1995. Individual transferable quotas (ITQs) in Canadian and US Fisheries. Ocean and Coastal Management 29 (1-3): 85-115.

Munro, G., N. Bingham and E. Pikitch. 1998. Individual transferable quotas, community-based fisheries management systems, and "virtual" communities. Fisheries 23(3): 12-15.

NMFS 1999a. Our living oceans. Report on the status of U. S. living marine resources, 1999. NOAA Tech. Memorandum NMFS-F/SPO-

NMFS 1999b. Final fishery management plan for Atlantic tunas, swordfish, and sharks. NOAA NMFS Office of Sustainable Fisheries, Silver Spring, MD.

NMFS 1999c. Fisheries statistics and economics. National Marine Fisheries Service, www.st.nmfs.gov.

NRC 1999a. Sharing the fish: toward a national policy on individual fishing quotas. Ocean Study Board, National Research Council, National Academy of Sciences. National Academy Press, Washington, D.C.

NRC 1999b. The community development quota programme in Alaska. Ocean Study Board, National Research Council, National Academy of Sciences. National Academy Press, Washington, D.C.

Pautzke, C.G. and C.W. Oliver 1997. Development of the individual fishing quota programme for sablefish and halibut longline fisheries off Alaska. North Pacific Fisheries Management Council, www.fakr.noaa.gov/npfmc.

Reiser, A. 1997. Property rights and ecosystem management in U. S. fisheries: contracting for the commons. Ecology Law Quarterlty 24: 813-832.

Sakagawa, G.T. 1975. The purse-seine fishery for bluefin tuna in the Northwestern Atlanticocean. Marine Fisheries Review 37 (3): 1-8.

Scott, A.D. 1955. The fishery: the objectives of sole ownership. Journal of Polit. Economy 63: 116-124.

Sedberry, G., J. Carlin, B. Chapman, B. Eleby, G. Ulrich and G. Wyanski 1999. Fishery biology, life history, and genetic population structure of globally-distributed wreckfish,*Polyprion americanus*. South Carolina Department of Natural Resources, www.dnr.state.sc.us.

Serchuk, F.M. and S.A. Murawski 1997. The offshore molluscan resources of the northeastern coast of the United States: surfclams, ocean quahogs, and sea scallops. Pages 45-62 *in* C. L. MacKenzie, V. G. Burrell, A. Rosenfield, and W. L. Hobart. The history, present condition, and future of the molluscan fisheries of north and central America and Europe: Volume 1, Atlantic and Gulf Coasts. NOAA Technical Report NMFS 127.

Sigler, M.F., J.T. Fujioka and S.A. Lowe 1999. Alaskan Sablefish Stock Assessment for 2000. North Pacific Fishery Management Council.

Smith, P.J. 1999.The Alaska halibut and sablefish individual fishing quota (IFQ) programme: a manager's perspective on its history, implementation, and performance. ICES Journal of Marine Science, vol. 56.

Smith, P. J. *This symposium*. Administration of the IFQ programmes for halibut and sablefish in Alaska.

Sullivan, P.J., A.M. Parma and W.G. Clark 1999. The Pacific Halibut Stock Assessment of 1997. International Pacific Halibut Commission Scientific Report 79.

Wang, S.D., and V.H. Tang 1996. The surf calm ITQ management: an evaluation. Pages 125-128 *in* Our living oceans. The economic status of U.S. fisheries, 1996. NOAA Tech. Memo. NMFS F/SPO-22.

DEVELOPMENT OF PROPERTY RIGHTS-BASED FISHERIES MANAGEMENT IN THE UNITED KINGDOM AND THE NETHERLANDS : A COMPARISON

G. Valatin

Laboratoire d'halieutique

Ecole Nationale Supérieure Agronomique de Rennes (ENSAR)

65, Rue de Saint-Brieuc, CS 84215 - 35042 Rennes Cedex, France

<valatin@roazhon.inra.fr>

1. INTRODUCTION

Two essential characteristics of "property" are that they involve a bundle of both rights and responsibilities and that there is allocation of exclusive rights to a "benefit stream". These rights would have to value and there should be recognition, including willingness to enforce them, by the wider society (Bromley 1991). Because of measures that restrict access, property rights thus represent economic "goods" due to their scarcity and utility (Davidse 1997). Their value is dependent upon characteristics such as duration, flexibility, divisibility, and transferability (Scott 1988), upon regulations that affect these attributes, or the expected flow of benefits accruing from access.

The creation of property rights in fisheries is associated with measures that allocate exclusive access to fishing opportunities, fish stocks or fishing areas. These measures include limited entry, such as restrictive licensing and registration, which limit access to sea areas, as well as measures, including quota allocations, which determine the extent of access to resources. Enforcement and the extent of compliance with regulations governing access is crucial in determining the extent to which exclusive fishing rights are of value in practice.

General vessel licensing and registration requirements, Total Allowable Catches (TACs), area and seasonal closures, taxes and subsidies, and other regulations, may also significantly affect the returns from participation in a fishery and therefore also any market value of associated property rights. However, except where such management instruments confer exclusive access rights, they do not involve the creation of property rights, or lead to rights-based fisheries management.

Fisheries management in the United Kingdom and the Netherlands occurs within the framework of the European Union's Common Fisheries Policy (CFP), which in line with the principle of "subsidiarity", allows for national variations in implementing fisheries management policies. Research for this paper was undertaken within the context of the EU-funded ELSA-fisheries (Ethical, Legal and Social Aspects of Fisheries Management) project, which is reviewing and comparing fisheries management systems in ten European countries and is linked to current reviews on the CFP and potential changes after 2002.

The development of rights-based fisheries management measures in the UK and the Netherlands is described after briefly describing the development of exclusive national fishing rights in the context of international agreements and the evolution of the EU Common Fisheries Policy in particular. The development of property rights in each country is compared focusing upon limited access and quota allocation measures as the main types of management instrument conferring exclusive fishing rights.

2. INTERNATIONAL AGREEMENTS AND THE DEVELOPMENT OF EXCLUSIVE NATIONAL FISHING RIGHTS

The establishment of exclusive fishing zones and curtailing access of foreign fishing fleets was associated with the initial development of national property rights in fisheries, upon which subsequent developments were based. In the United Kingdom, a 3-mile exclusive fishing zone was established in 1878 under *the Territorial Waters Jurisdiction Act*, but although the North Sea Fisheries Convention of 1882 provided for North Sea coastal states to establish such zones, a 3-mile exclusive fishing zone was not established in the Netherlands until 1952 (United States Department of Defense 1997). The European Fisheries Convention of 1964 established the right of coastal states to establish a 12-mile exclusive fishing zone providing that the historic access rights of other countries within the 6-12 mile band continued to be recognised. Accordingly, limits were extended to 12-miles in both the United Kingdom and the Netherlands in 1964 (Davidse 1996) although this did not apply to the Channel Islands (UK).

On creation of the European Economic Community in 1957 by France, Germany, Italy, the Netherlands, Belgium and Luxembourg, the Treaty of Rome provided for the establishment of a common policy on agriculture, including sea and fresh water fisheries, although fisheries were viewed as of minor importance. Only in 1970, with applications to join the Community by the United Kingdom, Ireland, Denmark, and Norway, adjacent to existing Member States' fishing areas, was there sufficient incentive for agreement to be reached on an initial fisheries policy aimed at ensuring that access to traditional fishing grounds was maintained. Much to the dismay of the new applicants, the policy encapsulated in EEC Regulation No. 2141/70 and concluded just prior to their accession, stipulated that equal conditions of access to the fishing grounds under the jurisdiction of each Member State applied to all vessels registered within the Community.

Although providing for the retention of existing exclusive 6-12 mile coastal limits, the policy otherwise effectively prohibited discrimination against vessels from another Member State purely on grounds of nationality, and was to be binding on all Member States for a period of 10 years. While Norway decided not to become a member of the Community, it was reportedly perceived by the other governments at the time to be relatively unimportant compared to the advantages of joining (Davidse 1997, Holden 1996).

In contrast to the principle of equal access agreed within the Community, from the mid 1970s the North East Atlantic Fisheries Commission (NEAFC) began recommending TACs and national quotas for some of the principal shared stocks, further laying the basis for the creation of exclusive fishing rights for member countries, including the United Kingdom and the Netherlands (Davidse 1996). Following the creation in 1976 of a separate Directorate General for Fisheries within the European Commission and the establishment of a 200-mile exclusive fishing zone from 1 January 1977, negotiations began in earnest within the Community to agree to a comprehensive fisheries policy covering the most important non-sedentary fish stocks. National interests campaigned for their government to negotiate the greatest share of access rights possible, and UK ports were blockaded by fishermen in 1975 in support of the establishment of a 100-mile exclusive fishing zone (Holden 1996).

After prolonged negotiations on questions including the retention and size of exclusive fishing zones, and the allocation of TACs into national quotas, the CFP was finally agreed in late January 1983, for a period of a further 20 years. An incentive to reach agreement had been the expiry of existing exclusive 6-12 mile coastal limits under the existing policy at the end of December 1982 (Holden 1996).

The CFP provides for the setting of TACs for all the main stocks in Community waters and their division between Member States on the basis of "relative stability", a formula reportedly established bearing in mind the wider Community principle of allocation based upon need rather than purely contribution to Community resources. The fixed allocation procedure adopted takes into account the historic catches of Member States' fleets, the needs of coastal areas heavily dependent on fisheries (including the northern part of the UK), lost fishing opportunities of certain Member States (including the UK) arising from the declaration of a 200-mile exclusive fishing zone by third countries and national priorities in terms of targe stocks. Where no data were available on which to base scientific advice precautionary TACs were set (Holden 1996).

While investment in the catching sector had previously been encouraged in an attempt to eliminate the Community's deficit in fish supplies, as part of the CFP Structural Policy, the first Multi-Annual Guidance Programme (MAGP) adopted in 1983 aimed to curb growth in the fishing fleets in an attempt to achieve a satisfactory balance between 'fishing capacity' deployed and available stocks. Henceforth, aggregate tonnages and engine powers of national fleets were to be constrained to try and create a balance between national fleets and Member States' shares of the TACs (Holden 1996), to reduce the risks of quota-busting and stock over-exploitation, with subsequent MAGPs further segmenting national fleets and stipulating associated fleet capacity targets. To achieve national MAGP targets required the introduction of limited access through closing national fleet registers to new entrants (except where at least the same amount of "capacity" was simultaneously withdrawn), and the inclusion of the relevant vessel "capacity" characteristics in national fleet registers, resulting in the creation of property rights for existing fishing vessel owners.

3. LIMITED ENTRY IN NATIONAL FISHERIES

Apart from access restrictions within national 3-mile coastal zones, the period prior to the mid 1970s was characterised essentially by open access and a lack of property rights in fishing both within the UK (Hatcher and Cunningham 1994) and the Netherlands. By contrast, the past couple of decades has seen the implementation of a wide variety of limited entry restrictions and the creation of associated property rights. Table 1 provides a summary of some of the principal limited entry measures introduced in UK and Dutch fisheries.

The introduction of limited entry in UK and Dutch fisheries was associated with attempts to protect the fishing opportunities of specific vessels or groups of vessels, and with attempts to ensure a balance between the catching power of national fleets and available resources in line with national quotas and MAGP targets. In most cases access rights were allocated purely on the basis of previous participation in a fishery, thus excluding potential new entrants. For example, in the UK in 1984 "pressure stock licences" were allocated to all vessels which had previously participated in, or held a licence for, the fisheries affected.

Cases of more restrictive allocation of access rights that excluded an important group of vessels which had previously participated in the fishery, unsurprisingly, have tended to be more hotly contested. For example under the *Merchant Shipping Act 1988* an attempt was made to tighten the registration provisions by stipulating that owners of UK fishing vessels had to be British citizens, or if company-owned, 75% of the shares had to be owned by British citizens and 75% of the directors had to be British citizens, excluding vessels operated by nationals from other Member States (Morin 1998). The Spanish and Dutch vessel operators thereby excluded took legal proceedings against the British government. The European Court of Justice ruled in 1991 that the UK government had acted in violation of the right of establishment under Community law (Morin 1998), and the vessels were allowed to re-register; their owners were awarded £30million in compensation (Robinson, Pascoe and

Table 1
Limited entry measures in UK and Dutch fisheries

Year	Country	Limited entry measure
1975	Netherlands	Permits for vessels herring fishing
1975	Netherlands	GK permits for vessels fishing for shrimp in Wadden Sea
1975	Netherlands	GV permits for vessels fishing for shrimp in 12-mile zone
1977	UK	Licences for vessels fishing for herring in northern Irish Sea
1980	UK	Licences for mackerel purse seiners and freezer trawlers
1981	Netherlands	K Permits for vessels mainly targeting cod
1984	UK	Pressure stock licences for vessels fishing for TAC stocks
1986	UK	Limited pressure stock licences for fishing new TAC stocks
1986	Netherlands	List I permits for beam trawling for flatfish within 12-mile zone
1986	Netherlands	List II permits for beam trawling for shrimp within 12-mile zone
1987	Netherlands	R Permits for other vessels mainly fishing for roundfish
1987	Netherlands	S Permits for other trawlers and netters seasonally fishing roundfish
1988	UK	Exclusion from register of vessels operated by non-nationals (temporary)
1990	UK	Miscellaneous species licences mainly for shellfish fleet
1992	UK	North Sea beam trawl pressure stock licences
1999	UK	Genuine economic link criteria for vessel registration

Hatcher 1998). Following subsequent lengthy negotiations (letter from Jacques Santer to Tony Blair dated 17 June 1997) between the British Government and the European Commission, to ensure that vessels maintain a genuine economic link with the country of registration, less restrictive criteria were introduced in 1999 related to landings, crew residence, and operating expenditure[1], which seem less likely to result in existing vessels being excluded from the UK register, or be subject to legal challenge.

Although initial limited entry measures restricted fleet size in terms numbers of vessels in some specific segments it quickly became apparent that this would be far from sufficient to curb fleet catching power, as investment channelled instead into improvements to existing vessels in fisheries subject to restriction, or expanding fleet size in fisheries not subject to limitations. Thus, further limited entry measures were deemed necessary to curb such growth, including restrictions on transfer of access rights to larger vessels. In the UK, this resulted in a multiplication in restrictive licensing schemes, so that by 1992 there were 154 different types of licence; prior to simplification in 1995 licences were categorised into five basic types, with some vessels holding more than one type! (Europeche 1995). Table 2 provides a summary of some of the principal access rights transferability measures introduced in UK and Dutch fisheries.

Fundamental difficulties exist in adequately defining and measuring fishing capacity as the principal determinant of fishing mortality (Valatin 1992, Holden 1996), with its measurement differing between Member States. In the UK a «vessel capacity unit» (VCU) system was adopted in 1990 replacing existing restrictions based upon vessel length. This allowed almost unrestricted transfer of the licences between vessels of different sizes, providing that the total VCUs of the vessel to which one or more licences was transferred was sufficiently below the total of the original vessel(s), with vessel VCUs defined as {length(m) x breadth(m) + 0.45 x engine power (kW)}. In the Netherlands, engine power and tonnage measures were used to measure capacity in line with the units used at EU level for the MAGPs.

Not least due to the definitional difficulties, vessel operators soon found ways to circumvent such measures, so that restrictions on transfer of access rights were soon found to be only partially effective in limiting the growth of fleet catching power. For example, as measurement of engine power took account only of the main engine, this allowed vessel operators to install auxiliary engines. Main engines were de-rated to a fraction of their nominal continuous ratings, and auxiliary engines were installed whose combined power exceeding that of the main engine (De Wilde 1998). Capacity limitation rules resulted in widespread modification of vessel design characteristics to create so-called "rule beaters", increasing catching efficiency but not the level of capacity officially measured.

Once introduced, restrictions on transfer of access rights have not always proved easy to alter. For example, in the Netherlands, a change to measuring engine power using a system of "maximum continuous rating" had to be abandoned due to successful court action by the industry, who claimed the new measure to be unfair as it would have entailed significantly increasing the registered HP of some vessels (LEI 1997).

Decommissioning schemes have also been used in both countries to reduce fleet size in a further attempt to ensure a balance between the catching power of national fleets and available resources. In so far as they reduce the number of vessels allowed to fish, such schemes also

---

[1] Answer from MAFF Minister Nick Brown to Parliamentary Question tabled by Austin Mitchell MP on 30 July 1998 on the introduction of measures to ensure that British registered fishing vessels maintain a real economic link with UK populations dependent on fisheries).

Table 2
Access rights transferability measures introduced in UK and Dutch fisheries

Year	Country	Quota allocation measure
1984	UK	Pressure Stock licence transfer restricted to within vessel length groups
1985	Netherlands	Engine power incorporated into licences, with 10% penalty on transfer
1990	UK	Vessel Capacity Units system introduced with 10% transfer penalty
1992	Netherlands	Prohibition on transfer of >2000HP licences
1992	UK	Capacity penalty increased to 20% on transfer of licence
1994	UK	Capacity penalty increased to 30% if 3 or more licences aggregated
1994	Netherlands	Transfers of roundfish permits allowed without transfer of vessel
1996	UK	Aggregate tonnage and horsepower restricted to previous level
1998	UK	Capacity penalties withdrawn for vessels fishing outside EU waters

represent a limited entry measure, with entry restricted to those existing vessels not decommissioned.

4. QUOTA ALLOCATIONS IN NATIONAL FISHERIES

In addition to limited entry, a wide range of quota allocation measures have been introduced in the past two decades, reserving exclusive rights to part of the national quotas for specific vessels, creating property rights for the boats concerned. Table 3 below provides a summary of some of the principal quota allocation measures introduced in UK and Dutch fisheries:

Both in the UK and in the Netherlands, the implementation of national quota provisions by closing a fishery once the quota was taken, was quickly found to be unsatisfactory where catch rates were high relative to quota levels. Prohibitions on fishing at the end of the year not only caused problems for their shore sector, but, by concentrating landings earlier in the season, reduced the prices obtained by the fishermen.

To overcome the problem of early closures, mechanisms for allocating quotas to individual vessels were introduced. In the UK the first mandatory restrictions on individual vessel landings were introduced in 1977 in the North Sea haddock and Western mackerel fisheries (Hatcher and Cunningham 1994). In the Netherlands, following the early closure of the sole fishery the previous year, an individual quota (IQ) system was introduced for plaice and sole in 1976, with part of the quota retained as a national reserve to allow for any overshooting of quotas. In the Dutch parliament it was announced that allocating quotas to individual vessels would increase operational certainty, allowing vessel operators to plan their fishing activities in advance, discuss their plans with their financiers and maximise their profits (Hoefnagel and Smit 1996).

Besides overcoming problems associated with early closures, introduction of quota allocation measures often further reflected attempts to safeguard the fishing opportunities of individual vessels or specific fleet segments, being closely linked to limited entry measures. For example, in the Netherlands, following the early closure of the cod fishery in September 1979, to protect operators from early closures, from 1981 part of the cod quota (equivalent to 200t per boat) was reserved for the 20 vessels targeting cod that had long track records in the fishery, which were issued with special permits (*K-documenten*). Increased pressure on roundfish quotas, due to larger cod bycatches associated with an expansion of the beam trawl sector in the early 1980s, which led to early closures,

Table 3
Quota allocations measures introduced in UK and Dutch fisheries

Year	Country	Quota allocation measure
1975	Netherlands	Division of herring quotas between freezer trawlers and other vessels
1976	Netherlands	Individual Quota system for sole and plaice
1977	UK	Fixed vessel quota system initiated (North Sea haddock/Western mackerel)
1984	UK	Sectoral quota allocation system initiated (Shetland Fish Producers Organisation)
1985	Netherlands	Individual Transferable Quotas for mackerel freezer trawler fleet
1985	Netherlands	Individual Transferable Quotas for sole and plaice
1986	UK	Individual Quota system initiated (The Fish Producers Organisation)
1988	Netherlands	Division of cod Quota between 3 fleet segments («Kistenregeling»)
1993	Netherlands	Biesheuvel quota management group system inaugurated
1994	Netherlands	Individual Transferable Quotas for cod and whiting
1994	Netherlands	Individual Transferable Quotas for seed mussels
1999	UK	Fixed track record system inaugurated for allocating quotas

resulted in separate quotas being allocated in 1988 to the two other sectors of the roundfish fleet, and the permit system was extended accordingly (Hoefnagel and Smit 1996).

In each case, quotas were initially allocated to owners on the basis of vessel characteristics, such as vessel size, or past catches. In the UK, for example individual vessels' landings of North Sea haddock were restricted on a per trip basis, while landings in the Western mackerel fishery were restricted on a crew member per day basis (Hatcher and Cunningham 1994). Allocations to Producers Organizations (POs) were based upon the historic track records of member vessels, reflecting their past fishing patterns (grounds fished, species targeted, etc.) (Hatcher 1997). In the Netherlands, following the introduction of TACs for herring by NEAFC in 1975, 68% of Dutch herring quotas for areas VIId (Eastern Channel), IV (North Sea) and IIa (Norwegian sea) were allocated to the freezer trawler sector (vessels over 59m length), with the remainder allocated to other vessels and managed by means of weekly landings limits (Davidse 1997, LEI 1996). Dutch quotas allocations for sole and plaice were initially based upon vessel catch records or engine power. Quotas for those vessels fishing prior to 1974 were based upon the highest annual sole and plaice landings during a 3-year reference period, and those for vessels having entered subsequently were based upon average catches in the same engine-power group, or set separately by the Ministry.

Mechanisms for allocating quotas initially adopted have often been contested and subsequently modified. For example, in the Netherlands, the method used in 1976 for initially determining sole and plaice allocations was much criticised as it resulted in large disparities in allocations to similar sized vessels, and was modified in 1977 to take account of both engine power and historic catch records, with bycatch quotas for non beamers over 250 HP capped at the level of 250 HP vessels, creating mini-quotas (Hoefnagel and Smit 1996). In the UK the reference period used to determine sectoral quota allocations was initially the previous five years, but due to pressure from the industry it was progressively reduced to three years in the case of demersal species, and to two years for pelagic and distant-water species (Europeche 1995). In 1999 a fixed track record system based upon 1994 - 1996 as the reference period was adopted to reduce annual fluctuations in allocations and overcome mis-reporting associated with the incentive for vessel operators to over-declare catches ("ghost fish") rather than risk losing future fishing opportunities by failing to fully take a current quota allocation.

To increase industry involvement in fisheries management, responsibilities for managing quota allocations have been largely devolved to industry groups. In the UK a sectoral allocation system devolving quota management to Producer Organisations was initiated in 1984, when the Shetland FPO was given responsibilities for managing its own Area IV and VI haddock quotas. The system gradually extended to other POs and other quotas, so that by 1996 sectoral allocations accounted for 96%, 91% and 84% of demersal quotas in Areas IV, VI, and VII respectively (Hatcher 1997). In the Netherlands, on the advice of the Stuurgroep Biesheuvel, in 1993 eight quota management (*Biesheuvel*) groups were established under the auspices of the *Produktschap Vis*, covering five-sixths of the fleet, to manage the aggregate quota of their members. In order to maximise returns, each prepares an annual fishing plans outlining the envisaged pattern of landings over the year (Ministerie van Landbouw, Natuurbeheer en Visserij 1993, Hoefnagel and Smit 1996).

To further increase the scope for decision-making by industry groups in shaping fisheries management, the UK POs were initially free to choose the quota allocation system applying to their members. Allowing POs to choose whether to manage a sectoral quota, provided an incentive to build up their track record prior to requesting a sectoral allocation so as to obtain a larger share, and except where deemed advantageous to prevent an early closure, to decline taking a sectoral quota in years when their members' track record was worse than that of the non-sector (Slaymaker 1992). This tended to reduce the non-sector allocation. To reduce problems associated with misreporting area of capture (Europeche 1995), the system was made more restrictive from 1995 when POs were obliged to accept sectoral allocations for all demersal species but not pelagics (Hatcher 1997).

Decisions concerning how to allocate quotas between group members after their initial allocation by the government have also been devolved to industry groups, notably in the UK, but also to some extent in the Netherlands. Under the UK sectoral allocation system, most POs tended initially to allocate fixed monthly quotas to member vessels, which in a few cases have varied allocations between different vessel categories (*e.g.* lengths groups) in an analogous manner to the non-sector allocations for vessels not in membership of POs or in membership of POs not having a sectoral quota for the particular stock (Hatcher 1997). However, an individual vessel quota (IQ) system with allocations based upon vessel track records has been used by the Fish Producers' Organisation[2] since 1986, with similar systems increasingly introduced by other POs. IQs are generally allocated annually and sometimes monthly and can usually be combined to give added flexibility where the same owner has several vessels. Within POs, vessel operators are generally free to swap and trade their IQs (Hatcher, Holland and Cunningham 1995), thus resembling an individual transferable quota (ITQ) system for some vessel operators (Europeche 1995). In the Netherlands, mackerel quotas for areas other than the North Sea, which comprise the bulk of those available to the Dutch fleet, were allocated exclusively to freezer trawlers and since 1985 freezer trawler owners

---

[2] The Fish Producers Organization, NFFO Offices, Marsden Road, Fish Docks, Grimsby DN31 3SG, UK.

have divided these quotas amongst themselves without restriction on subsequent transfers (LEI 1994, LEI 1996).

While quota allocation systems have generally been initiated by the government, occasionally there has been a case for them to be inaugurated by the industry. In the early 1990s Dutch industry organizations drafted a plan to limit the impact of the mussel seed fishery on the marine ecosystem by setting a TAC for seed mussels and allocating IQs on the basis of historic fishing performance and culture plot sizes, with the smallest firms allocated an additional amount and a separate allocation made to off-bottom growers (Keus 1994). The initiative was reportedly taken in order to pre-empt more stringent restrictions following the decision in 1991 by the trilateral Ministers Conference on the Protection of the Wadden Sea to close large areas to cockle and mussel fishing to protect seabirds, which at the time were reportedly dying of starvation in large numbers. Despite some initial enforcement difficulties, the system is reported to have lessened the race to fish and led to more efficient use of the mussel seed stock without having reduced the sector's output (Keus 1995).

Apart from providing for the transfer of quota rights on replacing an existing vessel or change of ownership, transferability of quota allocations by individual vessel owners has tended to be restricted initially, and where transfers have subsequently become allowed, this has often been preceded by the development of an unofficial trade in fishing rights. In the Netherlands, for example, sole and plaice IQs could initially only be legally transferred between owners on the transfer of the vessel. However, in practice this restriction proved easy to circumvent (LEI 1997a) as vessel operators soon found that in order to acquire additional quotas, they could simply buy a vessel and its associated IQs and then re-sell it without its quota entitlements (Davidse 1997). Only in 1985 did transfers of quota entitlements by themselves became legal.

Quota transfers between groups have also been subject to restrictions, although generally less stringent than those applying to individual vessel owners. For example, the UK POs were free to swap quotas with other POs, although until 1993 such swaps had to balance in terms of "cod-equivalents" (Hatcher 1997)[3].

Quota allocations to groups create a form of common property rights. In some cases mechanisms have been established to protect group rights. For example, in the UK, since 1994 POs have been allowed to retain and "ring fence" the track records of member vessels who voluntarily surrendering their licence, with the vessel's owners generally being financially compensated by the PO. In this way the PO retains vessel's track record in the event it leaves the PO, and around 20% of the Shetland FPO's cod quota is reportedly ring-fenced in this way (Phililipson 1997). In the Netherlands, members of Biesheuvel Quota Management Groups have to sign an agreement transferring the right to manage their ITQs to the board of the group, committing them to remaining within the group for the year and to submitting an annual fishing plan to the group board showing anticipated quota uptake and days at sea to be utilised during each quarter. This can only be modified with the board's permission (LEI 1996, Langstraat 1997). Members have to report all quota transactions to the group board and offer quota initially to other members of the group before entering into any agreement to rent out part to non-group members although transactions agreed on an exchange basis are exempt (LEI 1995).

To the extent that individual vessel owners retain rights to withdraw from the group taking quota entitlements with them, group common property rights have been of less importance than the private property rights of individual vessel owners. For example, UK vessel owners are always free to resign their membership of a PO if they disagree with the quota allocation method used, and a willingness of vessel owners with relatively high track records to leave POs allocating quotas on the basis of equal shares, may account for the increasing use of allocation schemes based upon individual vessel track records.

5. ENFORCEMENT AND THE VALUE OF PROPERTY RIGHTS

Enforcement within the EU has been described by a former head of the European Commission Fishery Directorate's Conservation Unit, as the "Achilles' heel" of the conservation policy as is neither efficient nor effective (Holden 1996, p.87, p.167). From the outset, the application of control measures within the CFP was devolved to individual Member States, who, by retaining implementation as a national responsibility, reportedly mainly aimed at rendering the policy ineffective (Holden 1996), with lax enforcement combined with widespread non-compliance making the allocations agreed at EU level of little relevance in practice and creating conditions for an international Tragedy of the Commons to develop[3].

Subsequently the ineffectiveness of the Common Fisheries Policy in conserving fish stocks has become increasingly criticised. Calls for more effective enforcement gained increasing support, with a gradual tightening of control measures agreed at EU level. However, Commission fisheries inspectors have no legal powers to enforce legislation, and their role has been described as simply "looking over the shoulder of national inspectors" (Holden 1996, p.163). And, commission proposals to increase the powers of Community inspectors, to grant them real autonomy, or harmonise the sanctions applied,

---

[3] Cod-equivalents are described by Holden (1994:47) as having been created in the course of negotiations with Norway, Sweden and the Faroes, to facilitate quota swaps for different species, by multiplying each quantity by an approximate market value (*i.e.* the "exchange rate") index, taken as 1.0 for cod, haddock and plaice, 0.77 for saithe, and 0.86 for whiting. Thus, one tonne of saithe counted as equivalent to 0.77 tonnes of cod, haddock, or plaice, or 0.90 tonnes of whiting.

have been consistently rejected by the Council of Ministers (Holden 1996 and Fischler 1999). As implementation of control measures remains the responsibility of individual Member States, enforcement continues to depend to a large extent upon national administrations and perceptions of the legitimacy of agreements reached under the CFP with some Member States apparently continuing to regard their national interest as best served by allowing landings by their fleet to continue largely unrestricted.

In part to reflect changes agreed at the EU level and in part a national consideration in the early 1990s. The level of enforcement by the UK and Dutch administrations increased. In the aftermath of the resignation of the Dutch Fisheries Minister over misreported landings, a radical change of emphasis occurrred with the creation of the Biesheuvel Quota Management Groups. This apparently solved the country's "black fish" (quota-busting) problems. This was cited by a European Commission report evaluating implementation of fisheries regulations by Member States in the mid 1980's as the worst case in preventing over-quota landings (European Commission 1986) and the Netherlands is now widely regarded as a model of enforcement within the EU. Although this issue has never resulted in the resignation of a UK Minister, and is perceived to be a far greater problem in other countries, such as Spain, the problem was nonetheless declared a priority by the incoming Labour Government after the last elections, and although clearly not totally solved, recent measures are claimed to have met with considerable success (House of Commons 1998).

In some instances, attempts to improve enforcement involved the introduction of completely different types of regulations. For example, in the Netherlands days-at-sea restrictions (*Zeedagenregeling*) were introduced in 1987 to facilitate enforcement of quotas, with allocations of days to individual vessels dependent upon the fishery, engine power, quota entitlements and type of permit held (LEI 1994). In both countries a system of designated landing ports was introduced to improve enforcement.

Because of widespread non-compliance with regulations, differences in property rights on paper were at first of little significance in practice, and thus of relatively little value. However, as enforcement increased, or expectations of stricter enforcement grew, interest increased in ensuring that landings were made legally, generating a rise in demand for fishing rights and an associated increase in the value of individual vessel property rights. For example, in the Netherlands average prices of sole ITQs are reported to have increased seven-fold between 1986 and 1988 as a consequence of stricter enforcement, from an estimated NLG 1O-15/kg to NLG 7O-80/kg (Davidse 1997).

Fishing rights were initially allocated free of charge and have now acquired extremely high market values - millions of ECU in many cases - notably in Dutch fisheries, but also in some UK fisheries (MacNeill 1998). For example, as Table 4 illustrates, in 1994 the mean ITQ holdings of a vessel in the Dutch cutter fleet of 58t of sole and 173t of plaice was valued at NLG 4.4 million (*i.e.* around ECU 2 million).

As allocation of quotas had been based upon characteristics such as vessel size, or historic catches, the associated market values of individual fishing rights has varied accordingly. For example, as Table 4 shows, the 72 vessels in the Dutch cutter fleet with engine power over 2000 HP had mean ITQ holdings in 1994 of 121t of sole and 345t of plaice, valued at NLG 9.2 million, compared to the value of mean ITQ holdings of the 67 vessels with engine power under 260 HP of NLG 0.4 million. In the UK average values of vessel fishing rights also vary according to size and type of vessel, and past level of landings, with average prices reported of £776, £1222, and £2083 per VCU, for Category A, beam trawler, and purse seiner licences respectively in 1996, and for track record, of £700, £1200, £2500 and around £8000/t for mackerel, cod, hake and sole respectively in the first quarter of 1997 (Nautilus Consultants 1997).

In some cases, notably in the Netherlands, the market value of fishing rights has overtaken that of vessels themselves. For example, the price of a second hand Dutch beam trawler of average size was reportedly around NLG 2.5 million in 1993/4 compared to the value of the average flatfish ITQ of NLG 8.3 million for 103t of sole and 310t of plaice (Davidse 1997).

Amounts spent on acquiring fishing rights have in some cases exceeded investment in boats and equipment. For example, total investment in flatfish ITQs in the Netherlands is estimated to have been NLG 91 million in 1990 and NLG 115 million in 1991, exceeding total investment in boats and equipment, which for the cutter sector as a whole, reportedly amounted to NLG 64 million and NLG 31 million respectively in 1990 and 1991 (Davidse 1997).

Differences in enforcement and other factors, such as opportunity costs and proximity to markets has led to the market values of access rights differing between countries. This has tended to provide incentives for vessel operators to buy fishing opportunities abroad if costs were lower. In the Netherlands, the increasing cost of IQs and heavy enforcement reportedly led Dutch vessel operators to buy fishing rights in the UK, Germany, Belgium and to a lesser extent Norway, where the costs were lower (Davidse 1996). Although the CFP provides for TACs to be divided on a national basis between Member States, the British government's attempts to prevent citizens from other EU countries acquiring quota entitlements in the UK failed as any measures discriminating against other EU nationals are consistently ruled illegal by the European Court of Justice.

Table 4
Mean holdings, estimated prices and values of Dutch flatfish ITQs in 1994

Engine power	Number of vessels	Mean sole ITQ			Mean plaice ITQ			Total flatfish value (NLG million)
		Quantity (tonnes)	Price (NLG /kg)	Value (NLG million)	Quantity (tonnes)	Price (NLG /kg)	Value (NLG million)	
<260 HP	67	4.75	60	0.28	15.13	5.5	0.08	0.37
261-300 HP	99	18.74	60	1.12	62.57	5.5	0.34	1.47
310-1100 HP	25	13.68	60	0.82	53.92	5.5	0.30	1.12
1101-1500 HP	33	65.52	60	3.93	221.52	5.5	1.22	5.15
1501-2000 HP	86	100.65	60	6.04	294.26	5.5	1.62	7.66
>2000 HP	72	120.94	60	7.26	344.60	5.5	1.90	9.15
All cutters	382	57.70	60	3.46	172.73	5.5	0.95	4.41
Flatfish sector	212	93.77	60	5.63	296.13	5.5	1.63	7.26

Source: Davidse, W.P. (ed) (1997). Property rights in fishing; Effects on the industry and effectiveness for fishing management policy (LEI-DLO, The Hague, The Netherlands).

6. SUMMARY AND CONCLUSIONS

With few exceptions, the period prior to the mid 1970s was characterised essentially by open access and a lack of property rights in UK or Dutch fisheries. By contrast, the past couple of decades has seen the implementation of a wide variety of access restrictions, which, often more by accident than design, have created property rights.

Fisheries management in the United Kingdom and the Netherlands occurs within the framework of the EU Common Fisheries Policy, which, from its inception, was largely concerned with the definition and allocation of fishing rights and involved a lengthy process of negotiation of national property rights before agreement was reached in 1983 on a policy to last the following 20 years. In line with the "Subsidiarity Principle", the CFP allows scope for national variations in implementing EU fisheries management policies, which initially was reflected by widespread lack of enforcement that undermined the fishing right allocations agreed at the EU level and allowed fishing pressure to increase largely unconstrained, creating the conditions for an international Tragedy of the Commons to develop.

Development of property rights has occurred in a somewhat haphazard fashion. Fisheries management measures resulting in the creation of property rights in UK and Dutch fisheries had two principal aims. Some were aimed primarily at improving fisheries management by overcoming the race to fish and problems associated with early closures when national quotas were fully taken, and the creation and maintenance of a balance between fleet size and resources. Others were primarily distributional in nature, aimed at the establishment and protection of particular access rights.

The process of regulatory change unwittingly resulted in some outcomes that were generally considered undesirable, such as the escalation in market values of fishing rights and distortionary affects associated with capacity controls and widespread modifications in vessel design to create the so-called "rule beaters". Rather than providing a stable regulatory framework for vessel operators, in many fisheries there have been frequent changes in limited entry and quota allocation measures.

Often governments did not foresee or proved unable to prevent the evolution of property rights. For example, while the introduction of IQs initially reflected an attempt to overcome apparent flaws in the existing pattern of exploitation, notably a race to fish followed by premature closure of the fisheries, the inability to constrain the powerful economic incentives for vessel operators to try to obtain the market value of property rights on transfer and thus the transition to ITQs was not foreseen. Despite official restrictions on transferability, an unofficial ITQ system developed spontaneously following the introduction of IQs in the Dutch fishery for sole and plaice, with regulations only later introduced to put the system on a statutory footing. The system subsequently extended to other fisheries. In the UK quota management was largely devolved to Producer Organisations, with adoption of individual quotas and trade in quotas and track records leading to a system which resembles an ITQ system in some cases. Despite Government statements that there are no plans to introduce ITQs (House of Commons 1999a), on the basis of past trends and the unplanned transition experienced in the Netherlands, it seems most probable that, more by accident than design, UK quota management will increasingly come to resemble an ITQ system.

Due to widespread non-compliance, differences in quota allocations on paper were at first of little significance in practice, but as enforcement tightened, the value of individual vessel fishing rights rose. In the Netherlands a radical change occurred in the early 1990s in the government's approach from being essentially "top-down" to being largely "bottom-up", with quota management responsibilities devolved to groups of vessel operators operating within a PO framework. By empowering groups of vessel operators to decide for themselves on the operational rules and penalties, the raison d'être for regulation and the impact of non-compliance on other group members became apparent, and fishermen's perceptions of the legitimacy of the management system increased. Incentives arose to encourage co-operation, and the associated institutional changes appear to have solved the Netherlands' "black fish" (quota-busting) problems. Quota management changed from a situation where misreporting led to the resignation of the Fisheries Minister, to one currently regarded as a model of enforcement within the EU. In the UK, recent measures, notably the designated ports scheme, are claimed to have met with considerable success in dealing with enforcement problems (House of Commons 1998).

While initially allocated free of charge, fishing rights have acquired high market values - in the Netherlands they are now worth millions of ECU in many cases. Relatively high prices of property rights in the Netherlands, associated partly with increased enforcement, provided an incentive for Dutch vessel operators to purchase fishing rights in other countries. Although the CFP provides for TACs to be divided between Member States, the UK government's attempts to prevent vessels from other Member States from acquiring fishing rights in the UK generally failed, as measures discriminating against other EU nationals were consistently ruled illegal by the European Court of Justice. The principle of common access to resources by nationals from all Member States and unofficial transfers in property rights to vessel operators from other EU states seem likely to increasingly weaken the basis of national quota allocations.

As the EU Common Fisheries Policy is currently under review, a possibility exists that a different management system will be instituted after 2002, altering, or negating the property rights that have developed nationally under the current policy, or changing the distribution of fishing rights between Member States. However, despite its major failings, in view of the difficulties encountered in negotiating the existing CFP, it seems unlikely that fundamental changes will be agreed.

In the UK the "Save Britain's Fish" campaign has gathered strength in recent years arguing that UK waters contain some four-fifths of EU stocks, while under the CFP the UK has been allocated only two-fifths of the TACs, with an associated value of around one sixth of the total, implying that British fishermen failed to receive their rightful share of fishing opportunities and the UK should withdraw from the CFP. Even though the necessity of reasserting national control over fisheries policy apparently now represents the position of the Official Opposition party (the Conservatives) in Parliament, if not the view of all its Members (House of Commons 1999b), it seems highly unlikely that the UK will withdraw from the CFP, thereby increasing its fishing rights. Such action would not be acceptable to the other Member States, probably necessitating complete withdrawal from the EU, a policy inconsistent with other UK interests.

From a situation of virtual open access in the early 1970s, the process of regulatory change in fisheries over the past couple of decades could be characterised as one resulting largely in the privatisation of the marine commons. New entrants have increasingly had to purchase access rights from existing vessel operators in order to enter a fishery.

While quotas have become largely a form of private property, vessel operators continue to have only indirect influence over the level of access these rights provide, as setting TACs remains the prerogative of the Council of Ministers. The TAC and quota system has conspicuously failed to conserve stocks, as TACs have consistently been set higher than justified by the scientific advice due to short-term political considerations and further undermined by non-compliance (Rodgers and Valatin 1997). However it is claimed that the IQ system has prevented more rapid deterioration than would otherwise have occurred. Counter to biological advice, a shift in Dutch policy in 1993 to maintaining the viability of the sector by simply ensuring that stocks remain above a safe biologically levels and otherwise allowing TACs to be determined by the industry (Ministerie van Landbouw, Natuurbeheer en Visserij 1993), was subsequently reflected in the higher TACs set by the EU. This was later reversed when the increased risks of stock collapse were realised (Pers. comm.). An initial public airing of scientific advice on the increased risks of the new policy simply led to the leading Dutch scientist who raised these concerns being suspended and subsequently transferred to another area of research (Köbben and Tromp 1999).

States retain powers to modify existing exclusive fishing rights, but these are tempered by vessel operators' campaigning skills and ability to mount legal challenges. Appreciation in value and the trade in property rights has increased policy inertia mitigating against further fundamental regulatory changes, as owners of such rights can be expected to resist changes which diminish the value of their assets.

Elements of common property have been established. However, these seem likely to diminish in importance as vessel operators continue to attempt to obtain the maximum value from their fishing rights. Creation of property rights, notably ITQs, has altered fishermen's outlook and way of life. Fishermen have been transformed increasingly from being primarily resource hunters to managers of fishing allocations (Davidse 1997).

7. REFERENCES

Bromley, D.W. 1991. Testing for common versus private property: Comment. *Journal of Environmental Economics and Management,* Vol.21, pp.92-96.

Clark, C. 1973. The economics of Exploitation. *Science,* 181, pp.630-634.

Davidse, W.P. (ed) 1997. Property rights in fishing; Effects on the industry and effectiveness for fishing management policy LEI-DLO, The Hague, The Netherlands.

Davidse, W. 1996. Strategic responses of dutch fishermen to limiting measures and property rights. *First European Social Science Fisheries Network Workshop.*

De Wilde, J.W. 1998. Effects of subsidies on the distant water and coastal fisheries of the Netherlands. Proceedings of the first workshop of the Concerted Action on Economics and the Common Fisheries Policy, Portsmouth

European Commission 1986. Rapport de la Commission au Conseil sur l'application de la Politique Commune de la Pêche. COM(86 301), The European Commission, Brussels, Belgium.

Europeche 1995. Inventory of the different systems for managing fishing efforts and quotas in E.U. Member States. DGXIV, European Commission, Brussels.

Fischler, F. 1999. (Commissioner-designate for Agriculture, Fisheries and Rural Affairs) reply to the questionnaire submitted by the European Parliament's Committee on Fisheries. August 1999.

Hardin, G. 1968. The Tragedy of the Commons. *Science*, 162, pp.1243-1247.

Hatcher, A. and S. Cunningham 1994. The development of fishing rights in UK fisheries policy. Research paper 69, Centre for the Economics and Management of Aquatic Resources, University of Portsmouth.

Hatcher A.C. 1997. Producers' Organizations and Devolved Fisheries Management in the United Kingdom : Collective and Individual Quota Systems. *Marine Policy*, 21, (6), 519-533.

Hatcher, A., P. Holland and S. Cunningham 1995. Producers' Organizations in the UK fishing industry. VIIth Annual EAFE Conference, Portsmouth.

Hoefnagel, E. and W. Smit 1996. Co-management experiences in the Netherlands. Report EC 94/60, LEI-DLO (Landbouw-Economisch Institut, Dienst Landbouwkundig Onderzoek), The Hague.

Holden, M. 1996. The Common Fisheries Policy. Fishing News Books, Oxford.

House of Commons 1998. Fisheries debate. Hansard, 15 December, pp. 825-868.

House of Commons 1999a. Written Answers by Fisheries Minister Elliot Morely to Questions from Mr Andrew George. Hansard, No. 1808, 11-14 January.

House of Commons 1999b. Fisheries debate. Hansard, 14 January, pp. 529-546.

Keus, B. 1994. Self-regulation in fisheries: The case of the mussel-seed fishery in the Netherlands VIth EAFE Annual Conference, Crete.

Keus, B. 1995. Shellfish farming and Bird Protection in the Dutch Wadden Sea. Conference on Shellfish farming and the integrated development of Coastal areas, La Rochelle.

Köbben, A. J. F. and H. Tromp 1999. De onwelkome boodschap of Hoe de vrijheid van wetenschap bedreigd wordt. Jan Mets, Amsterdam.

Langstraat, D. 1997. The Dutch co-management system for sea fisheries. Proceedings of workshop on Alternative Management Systems, Brest, 18-20 September.

LEI (Landbouw-Economisch Institut, Dienst Landbouwkundig Onderzoek) 1997. Licensing of Fishing Vessels in five EU Member States. Working Document No.5, EU AIR CT94 -1489, LEI, The Hague.

LEI 1996 (Landbouw-Economisch Institut, Dienst Landbouwkundig Onderzoek). Phase III National Report: Attitudes to PO membership in the Netherlands. EU AIR -2 CT93 -1392, LEI, The Hague.

LEI 1995 (Landbouw-Economisch Institut, Dienst Landbouwkundig Onderzoek). Phase II National Report: Attitudes to PO membership in the Netherlands. EU AIR -2 CT93 -1392, LEI, The Hague.

LEI 1994 (Landbouw-Economisch Institut, Dienst Landbouwkundig Onderzoek). Phase I National Report: Attitudes to PO membership in the Netherlands. EU AIR -2 CT93 -1392, LEI, The Hague.

MacNeill, F. 1998. Dead skipper's family to get £5m for quota. Press & Journal, Aberdeen. 20.10.98

Ministerie van Landbouw, Natuurbeheer en Visserij 1993. Balanced fisheries: A survey of the main points from the Dutch government's Policy Document on Sea and Coastal Fisheries. Den Haag.

Morin, M. 1998. Fisheries Dependent Regions and Fisheries Resources : Reflections on the Principle of Relative Stability. Fifth European Social Science Fisheries Network workshop, Lofoten, Norway.

Nautilus Consultants 1997. The Economic Evaluation of the Fishing Vessels (Decommissioning) Schemes: Report on behalf of The UK Fisheries Departments, Nautilus Consultants, Edinburgh.

Phililipson, J. 1997. The Fish Producers' Organisations in the UK : A Strategic Analysis. Third European Social Science Fisheries Network workshop, Brest, France.

Macneill. F. 1998. Dead skipper's family to get £5m for quota. Press and Journal, Aberdeen, 20 October 1998.

Robinson, C., S. Pascoe and A. Hatcher 1998. Why are the Spanish fishing in our waters? An economic perspective. Research paper 138, Centre for Marine Resource Economics, University of Portsmouth

Rodgers, P. and G. Valatin 1997. The Common Fisheries Poicy after 2002: Alternative options to the TACs and Quota system for conservation and management of fisheries resources. European Parliament Research Directorate, Luxembourg.

Scott, A. 1988. Development of property in the fishery. Marine Resources Economics, Vol.5, pp.289-311.

Slaymaker, J.E. 1992. The role of Producer Organizations in Fisheries Management :The case of South-West England. IVth Annual EAFE Conference, Salerno, Italy.

United States Dept of Defense 1997. Maritime Claims Reference Manual. DoD 2005 1-M, Ocean Policy Affairs, Department of Defense, Washington DC.

Valatin, G. 1992. The Relationship between Fleet Capacity and Fishing Effort. Proceedings of the VIth International Conference of the International Institute of Fisheries Economics and Trade, Paris, France, July.

RIGHTS–BASED FISHERIES DEVELOPMENT IN AUSTRALIA: HAS IT STALLED?

A. McIlgorm
Dominion Consulting Pty Ltd, Suite 7&8, 822-824 Old Princes Highway, NSW 2232, Australia
<mcilgorm@tradesrv.com.au>
and
M. Tsamenyi
Centre for Maritime Policy, University of Wollongong, NSW 2522, Australia
<martin_tsamenyi@uow.edu.au>

1. INTRODUCTION

Australia is a huge continent with a mix of fisheries. The reader should be aware of several characteristics of Australian fisheries:

i. their diversity and geographical extent – from Antarctic, mid-latitudes, to the tropics
ii. their administration, by the Commonwealth Government and 6 States, each with autonomous fisheries departments
iii. the coastline involved - 36 000km and a huge EEZ
iv. the limited continental shelves, few finfish, but many high value crustaceans
v. existence of many small producers in estuarine & inshore fisheries
vi. that the seafood industry is export oriented.

In the 1960s several Australian fisheries moved from open access to become limited entry licensed fisheries. This continued through the 1970s and 1980s by which time most commercial Australian fisheries were managed by limited entry. In 1980 a conference was held on limited entry licensing in Australia. In 1982, Meany reviewed the nature and adequacy of rights in Australian fisheries and since the debate, 20 years ago, we have come further than expected, but have made limited progress in codifying rights for the fishing industry in fisheries legislation.

During the 1970s various regulations were added to limited entry criteria primarily to address the rise in effective effort. In the 1980s the regulated licensing was found to have led to overcapacity and three approaches were taken:

i. *Unitisation:* Units based on different fishing vessel inputs were applied as a measure of vessel capacity and hence fishing effort.
ii. *Buy-back and voluntary adjustment regimes*: These were to address the over-capacity of fishing vessels, which had increased effort in many fisheries.
iii. *Output regimes:* A change of fisheries management to management regimes based on limiting catch, usually by Individual Transferable Quotas (ITQs).

Different rights and management approaches were being developed in different Australian jurisdictions.

The significant influences on the development of different fishing rights in the 1980s were:

i. *Legal experience in the implementation of new fisheries management arrangements:* Several decisions were handed down on fishing licences as "proprietary rights" from non-fisheries legal cases (Pennington v McGovern, inheritance case; and Kelly v Kelly, divorce settlement) in South Australia. Some fishers also took fisheries administrations to court to test their fishing rights (Fitti case, Commonwealth). These test cases became fundamental in subsequent rights development.
ii. *Pressure to go to ITQs:* Fisheries departments came under the influence of "economic rationalism" and the much publicised implementation of ITQs in New Zealand and in the Southern Bluefin Tuna (SBT) fishery in Australia.
iii. *Changing management arrangements:* A statutory authority model was implemented with the formation of the Australian Fisheries Management Authority (AFMA) in 1990.

In the 1990s the following features have been noted:

i. *Legal issues:* Significant legal challenges were made, particularly in association with the implementation of new ITQ arrangements in fisheries previously under management by input controls. The statutory fishing right (SFR) emerged at the Commonwealth level where fishing rights were linked to management plans.
ii. *New rights-based regimes:* Several ITQ schemes were introduced with controversy and legal action over wealth reallocation between the new, and former management, regimes. Most states implemented some ITQ regimes with NSW developing the "share management" system, possibly the most advanced rights fishery management system in Australia.
iii. *New institutional arrangements:* The inclusion of stake-holders in Management Advisory Committees gave them greater participation in the management of fisheries.
iv. *Cost recovery:* Recovering the costs of fisheries management from industry has led to more involvement in the management process by industry.

In spite of these developments the fishing industry has concerns over access security and debates the quality of fishing rights. Are fishing rights secure? Management and academics have been calling for implementation of output based regimes such as ITQs but this may have left fisheries which do not progress to ITQs with little advancement in real rights.

In this paper we will analyse the current state of rights development in Australia and review Australian rights-management regimes. This will show which industry sectors have gone to ITQ and which are under other rights regimes. It will also describe future rights development needs in Australian fisheries.

2. RIGHTS DEVELOPMENT
2.1 Antecedents

According to Scott (1988, 1989) new property rights develop through a process of demand and supply. Demandeurs seek an increase in the characteristics of their rights (*i.e.* duration, transferability, exclusivity, and security [quality of title], also their divisibility and flexibility. Scott notes these historical rights developments have been through:

i. informal processes
ii. violent means and conquests
iii. customary sources and processes
iv. judicial decisions and
v. government, political, bureaucratic means – legislation.

Scott proposes that historically the arena for demandeurs and suppliers to contest fishery rights development was in:

i. villages and manors where custom was law
ii. in conquered and lawless lands
iii. in the courts and
iv. in the legislating organs of government.

In the case of fisheries the villages and manors may be analogous to the culture of the fishing industry and conquering to the "new world" of rights development in the ocean. The courts and government refer to the political and legislative fisheries management process.

2.2 Demandeurs

Demandeurs are parties "that seek relief from the constraints of an existing standard interest" (Scott 1988) wanting change through combining one or more of its characteristics. The timing of change, the extent of change demanded and amount they are willing to pay for it, are crucial issues (Scott 1989). The absence of some characteristic has started to cost them more profit or rent than before (Scott 1989).

2.3 Suppliers

Scott (1989) suggests these are institutions, persons or groups who can add to, or subtract, from the characteristics in existing rights. Why do institutions respond to demands when they do? The extent of the change they provide and the reward they would ask are significant issues in rights development. In the Australian situation, rights are supplied by the policies of government, political opportunities, the courts and the fisheries administration. Government in Australia has jurisdictional divisions between the State and Commonwealth governments, which may affect supply of new rights regimes and these may differ between Commonwealth and State systems.

Management and enforcement costs are also part of the supply equation, as when they are high, the move to make demandeurs pay these under cost recovery may make the supply of increased rights through ITQs less attractive. Administration costs also influence supply with ITQs being desired by government, but contested by fishers who face resource fluctuations and uncertainty in revenues. Governments may also see rights for fisheries in the Exclusive Economic Zone as being desirable to defend national rights. In this paper we wish to analyse the progress towards fuller rights in this demand and supply framework using available empirical data.

The relationship between rent and the need to change rights will be difficult to test empirically as few fisheries have had rent assessments undertaken. Total fishery value may indicate potential rent. In this study average price is taken as potentially the most significant indicator of the existence of rent from the information available. Demandeurs may also be driven by resource scarcity which may be reflected in the market price of fish.

2.4 Analysis – the demand and supply of fuller fishery rights in Australia

The study reviews the data available for 105 fisheries listed in annual reports of all state and Commonwealth Fisheries Departments in Australia for the year 1997-98. These data give the type of fishery, species taken, licences held, value, volume and average price of product. From knowledge of management arrangements it is possible to work out whether a fishery is managed by ITQs or not. The fisheries are grouped by method and industry sector with the main species being indicated.

In this review all fisheries were under limited entry licencing regimes and the move to ITQ is taken as being an increase in rights characteristics (Scott 1989). The fisheries management data are analysed in this framework to appraise the rights status of fisheries and attempt to determine the impediments to rights development.

2.5 The data

The number of licences and gross revenue of product of fisheries at first sale are reported in Table 1 with production statistics. The data for 1997-98 indicate that the managed Australian fisheries sector, not including pearling, aquaculture or production of fish from fisheries not under management, had a Gross Value of Production (GVP) of $A1.47 billion representing 209 600t of product. This had a weighted average price of $A7.03/kg.

In all these fisheries there were a minimum of 14 585 fishing licences (the statistics on multiple endorsements held are complex). These are fishing rights of different qualities (McIlgorm and Tsamenyi 1999, and Tsamenyi and McIlgorm 1999). Some fisheries have gone to output management regimes, usually Individual Transferable Quota. Table 1 also reports ITQs management in Australia as number, value and volume for each production sector.

2.6 Results
2.6.1 ITQs in Australian fisheries

The data in Table 1 show that 28% of the 105 fisheries have gone to ITQ (including some in progress). These constitute 22% of the total value of production and 34% of the total volume of production.

From the sectoral review:

i. The Abalone sector is high value, with a high average price and has 100% management by ITQs;

Table 1
Review of the methods, fisheries licences, production and extent of ITQs in Australian fisheries

Method	Species/sector	Number of licences	Value $A million (1997-98)	Tonnes	Average price $A	ITQ by number	% ITQs by no.	% ITQs by value	% ITQs by weight
Diving	Abalone	294	181	5 249	34.48	7/7	100	100	100
Line/long-line	Snapper/tuna	2 449	131	18 558	7.08	5/9	56	59	61
Purse-Seine	Pilchards, etc.	155	33	41 543	0.79	3/7	43	23	42
Others	Assorted methods	1 460	50	9 150	5.50	5/16	31	63	43
Pots	Rock lobster/crab	2 656	440	19 539	22.52	6/21	29	31	38
Trawling	Finfish	2 024	182	59 125	3.08	2/7	29	36	54
Nets	Scale/fin fish	2 603	64	21 906	2.92	1/21	5	0	2
Prawn-trawl	Prawns	2 944	391	34 547	11.32	0/18	0	0	0
Total		14 585	1 473	209 617	7.03	29/105	28	22	34

x/y denotes that x fisheries out of a total of y are under ITQ management.

ii. The long-line sector targeting tuna and snapper is 56% managed by ITQs
iii. The purse-seine sector landing a low-value product is 43% managed by ITQs
iv. The fish-trawling sector is 29% by number and 54% by weight (South East Fishery) under ITQ management
v. Other fishing methods e.g. higher value scallop species have 31% ITQ management
vi. The valuable Rock lobster and crab potting sectors have 29% by number managed by ITQs
vii. Net fisheries, yielding low average prices (<$A10) have minimal ITQ management (5%)
viii. No prawn fisheries have ITQ management (0%).

The limited number of fisheries under ITQ management in the Prawn and Rock Lobster sectors contributed 56% of the total value of Australian fisheries production.

2.6.2 ITQs and fish prices

From earlier discussions our theoretical expectation would be that fisheries with high market price would be dominant in fisheries managed by ITQs. Figure 1 shows the relationship between the market price of fish and individual fisheries under ITQ or non-ITQ management. All Abalone and some rock lobster fisheries have gone to ITQ, but some rock lobster and scallop fisheries have not, nor have any prawn fisheries. Considering an average price greater than $A20/kg 12 of the 19 fisheries are under ITQ.

In the $A10-20 segment three of twenty of the fisheries are under ITQs management, mainly tuna and scallop fisheries. No prawn, scallop or high value fin fisheries are under ITQs.

In the under $A10/kg range, 14 of 56 fisheries are under ITQ - tuna, crab, pilchard, salmon and mackerel. Those not under ITQ are finfish, scalefish and estuarine prawn fisheries. Figure 2 shows that the frequency of managed fisheries for different average prices and indicates that lower-priced fish have less ITQ management than higher-priced species, but that is not to say all high valued species are managed by ITQ.

The scalefish, finfish and estuarine prawn fisheries with product priced below $A10/kg remain almost untouched by ITQ management, with the exception of the high volume purse-seine fisheries. These "small fisher" sectors represent a large number of licence holders (estimated >30% nationally), but are generally small fisheries having only 20% of national fisheries gross value of production.

From the price analysis it appears we have three distinct groups of rights managed fisheries in Australia:

Group I - ITQ managed fisheries (25% of all licences)
 (a) High-priced/medium-priced -ITQ fisheries, representing 10% of licence numbers
 (b) Low-priced - ITQ fisheries representing 15% of licence numbers

Group II - Individual Transferable Effort Quota (ITE) managed fisheries (14% of all licences)
 High-priced/medium-priced - Other rights managed fisheries (ITE).

Group III - Other rights (61% of licences)
 Low-priced - Other rights - some ITEs

Hence, it can be seen that the majority of fish licence holders in Australian fisheries have not seen any significant improvement in status in the last 20 years of management. Rights-enhancement for these fisheries may be limited by fears of overcapacity due to licence splitting.

Figure 1
Plot of average price and number of fisheries under ITQ or other-rights management in Australia

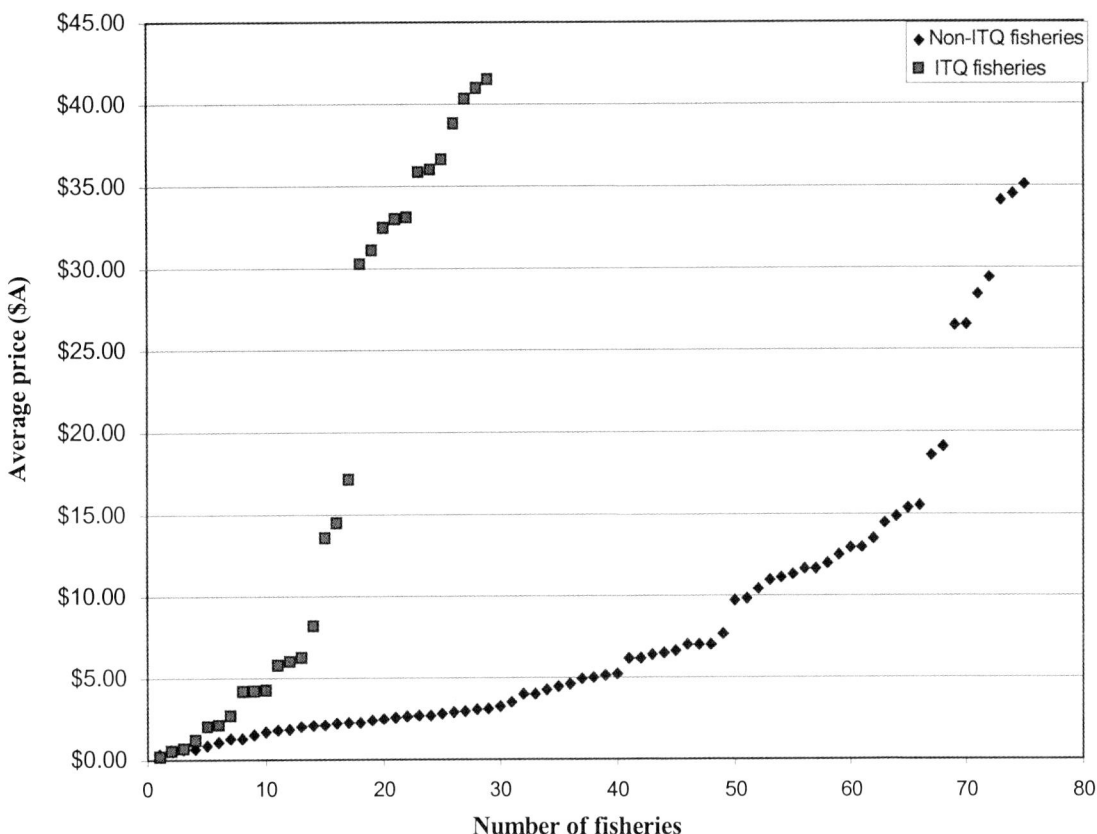

Figure 2
Frequencies of ITQ or other-rights managed fisheries, against average fish-price in Australia

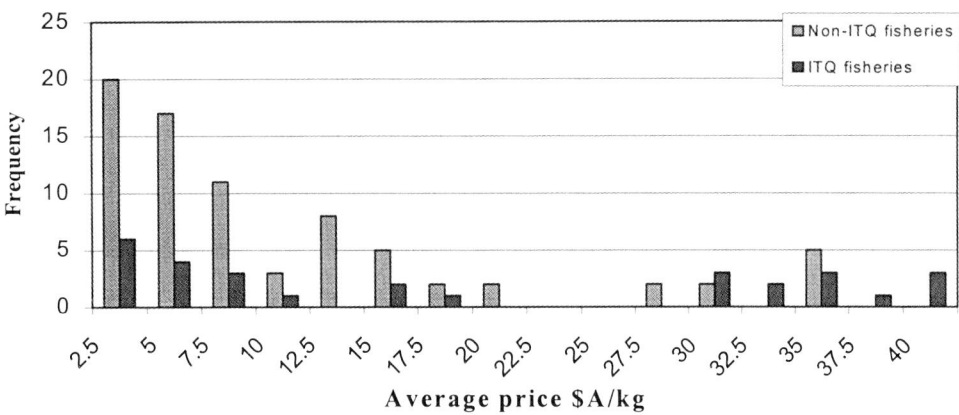

3. DISCUSSION

3.1 Rights development in Australia - has it stalled?

The international impression generated over the last 20 years is that Australia has moved significantly towards fuller fishing-rights regimes. Much attention has been given to ITQ implementation, particularly in the academic literature, yet the extent of ITQs across Australian fisheries is limited (approximately 25% of all licence holders). We can analyse the limitations under Scott's demand and supply framework.

3.2 Demands for increased rights

Scott (1988) predicts that fishers will demand increased rights due to some perceived inadequacy with the current rights or a profit opportunity. There is currently a desire among fishers across Australia for fuller access rights and greater investment security. Some of this position may be driven by potential erosion of their current rights by declaration of marine parks and a potential diminution of the resource through pollution. These groups have been identified in Section 2.6.2 above.

Group I – Moving to ITQs

Industry has been divided by the perceived need to move to ITQ management. Strangely the demand for ITQs has generally come from government. The rate of progress towards management by ITQs may have slowed in the 1990s, primarily due to a lack of demand among fishers. The price of fuller rights through ITQs may arise from restructuring, uncertainty in new allocations, and higher costs of administration, management and enforcement with the potential of paying rent charges to the community. There is a significant impasse because industry sees the move to ITQs being driven by political philosophy as well as administrative and financial expedience of government. Those in ITQ-rights regimes have also queried the rights advances incorporated in an ITQ (rights characteristics of ITQs can be enhanced as illustrated by the NSW share management regime).

Group II – Fisheries managed by ITEs

The prawn and crayfish sector are examples of these where greater rights are demanded, in forms other than ITQs. This is worthy of further investigation, but is probably explained by the uncertainty of their position under a possible new management regime, against the certainty of their current form of management. There may also be a mistrust of government, thought to be implementing rights management as a restructuring and income-reducing tool.

Group III – Licensed fisheries

In these the demand for rights is driven by fear of licence removals or reduction in economic returns due to the declaration of marine parks or pollution incidents. Similarly, the participants wish to maintain their continued access in order to promote investment security and provide more transparency for the banking sector.

3.3 Supply of rights - Impediments and restrictions

Scott suggests that institutions such as government, the courts, social groups and people can add, or to subtract from, fishery rights.

ITQs supplied by philosophy?

The Commonwealth government had a policy in the 1990s to supply fishers with output-based rights regimes under a "One size fits all" policy. At the state government level there has been less explicit promotion of ITQs; recommending them as the "best tool for the job". The meaning of Scott's riddle "When is a right not a right?; when it is a means of administration" has not been understood by many fisheries departments. ITQs may be granted as another form of management, not necessarily as a means of giving improved control and autonomy to industry in order to improve sustainable outcomes.

The push towards ITQ management has left some sectors of industry confused about fishing rights. Why can there not be diversity in fishing rights? There is a need to examine forms such as ITEs, other alternative rights regimes and their relative efficiency. ITEs are considered to be "optimal" on some rational basis by those fishers who endorse them but resist ITQs.

Statutory rights

Today fishing rights come from statute law and can be classed as statutory fishing rights (SFR). The SFR term has been used in Commonwealth fisheries to describe their mode of rights management in connection with management plans. One test of such rights developments is the possibility of compensation should the rights be appropriated. At the Commonwealth level there would have been "acquisition" under the Australian constitution. State jurisdictions vary in their policies to compensation.

Government can supply investment security for industry by making current rights more explicit. A simple increase in duration from a one year period to five or ten years, would assist the fisherman when seeking finance from bankers and financiers using his "rights" as colateral. Until recently Departments have not increased the duration of permits due to a lack of incentive to do so and the possible need for restructuring in the future of the management of the fishery.

3.4 Issues for rights development

Several key questions arise from the previous analysis concerning rights development in Australia.

i. *What rights regimes and options are there?*

There needs to be research on the design of rights systems, not just ITQs. For example prawns fishers see ITEs, contracts, and time and performance based agreements, as preferable to ITQs in their highly variable fishery. This should be investigated and evaluated in the light of Australian industry experience which has not convinced any other prawn fisheries in Australia to move to ITQs.

For the lower-priced fisheries it is recommended that ITQs will always be an expensive form of management. Fishers should investigate socially-based management alternatives using community structure.

ii. *Increasing licence duration*

If licences can be increased to a "5 years +" basis at minimal cost, this should be implemented for industry security.

iii. Statutory fishing rights
SFRs have potential for rights development based around management plans. They are really a permit with more recognised duration. But, are they more than this?

iv. Compensation
Compensation should be part of any rights regime. Adjusting rights systems creates winners and losers and hence the need for compensation to usher in the new regime.

v. Which fisheries can go to corporate governance?
Self-governance should be considered for fisheries which have more defined rights and no restructuring problems. The NSW Share management scheme is illustrative of this next step, as ITQs have been advanced into legally recognised and compensatable shares of resource access with ten years of duration and renewal. Other self governance initiatives may develop. This is another way that rights develop.

3.5 Rights development

The analysis revealed different segments in fishing rights development in Australian fisheries. Is Australia moving towards management by ITQs across all species? It does not look likely. At the "top end", where ITQs predominate, fisheries such as Abalone could consider corporate governance models for management.

Prawn fisheries show major resistance to ITQs, still preferring input regulations and ITEs. The reasons for this may be the variability of the prawn resource, the high cost of quota system management, and possibly wanting to avoid rent under an ITQ regime. Several major rock lobster fisheries have a similar perspective. The experience of the prawn sector is that ITEs are suited to prawn-fishery management.

In our price analysis the cases of low-priced scale fish or finfish present the most difficult management in each State. There is little economic surplus in these fisheries and there are many low income fishers. ITQs would be an expensive form of management. This sector need rights incorporated in a more socially-based management framework.

3.6 Impediments to rights development

Impediments to the development of fuller fishing rights come at different levels and from different sources. Impediments to further ITQ development are the cost of administering these systems and their high social impacts. Many fishers are not demanding ITQs and implementation has been a restructuring ploy rather than a rights development.

The demand for increased fishing rights comes from those seeking security of access and increased duration in their current rights. The granting of fuller rights will come from government and the courts. The administrative annual issue of licences is a major impediment to further rights development. A simple way to change this is to add to the duration of these licences, giving 5, 10 or 15 year licences. The linking of rights from statute (*i.e.* any licence) to a management plan under an empowering Fishery Act leads to a recognised statutory fishing right. These are becoming popular as a mechanism to increase security of industry access.

A fear among government fisheries managers is that fuller rights may augment fishing capacity and hence create an expensive restructuring requirement and sustainability problems. However some sections of industry are questioning the degree to which their rights have been increased. Our analysis indicates that fishing rights have not progressed uniformly for all sectors of industry.

4. CONCLUSION

Rights development has concentrated on ITQs with limited attention to rights development in non-ITQ fisheries. ITEs are an established mode of management in Australia's prawn fisheries where operators are not moving to ITQs.

The linking of licences with management plans have assisted in making statutory fishing rights a way to increase security of industry access.

There needs to be investigation of the benefits and impediments of further fishing rights development. In the high value fisheries under ITQ management, corporate governance experiments could be implemented; e.g. for Abalone. In low-priced fisheries there is a need for greater duration in fishing rights. It is apparent from the study that:

i. Industry needs real improvements in fishing rights and more security in resource access and
ii. More diverse applied research on fishing rights is required.

Has rights development stalled? It will, if we do not continue to go forward across the full range of Australian fisheries.

5. LITERATURE CITED

Meany, F. 1982. The nature and adequacy of property rights in Australian fisheries, Policy and Practice in Fisheries Management, DPI, Canberra, pp 57-76.

McIlgorm, A. 1999. Corporate Governance: an option for the fisheries management. Paper given at the 3rd International Rock Lobster Congress, Adelaide, September 1999.

McIlgorm, A. and M. Tsamenyi 1999. Fishing rights benchmarking project. A report to the South Australian Fishing Industry Council (SAFIC), Adelaide, S.A., October 1999.

Scott, A.D. 1988. Conceptual origins of rights based fishing in Ed. P.A. Neher, R.Arnason and N.Mollett, Rights Based Fishing, NATO ASI Series E: Applied Sciences, 169, pp. 11-38.

Scott, A.D. 1989. The emergence of ITQs as a distinct form of property right, A paper given at an international conference on the economics of fisheries management in the Pacific Islands Region. ACIAR Proceedings, 26.

Scott, A.D. 1993. Obstacles to fishery self-government, Marine Resource Economics, 8, pp. 187-199.

Townsend, R. and S.G. Pooley 1995. Distributed governance in fisheries in S. Hanna and Mohon Munqinghe (Eds), Property Rights and the Environment: Social and Ecological Issues. Beijer International Institute of Ecological Economics and the World Bank. pp 47-58.

Tsamenyi, M. and A. McIlgorm 2000. Enhancing fisheries rights through legislation – Australia's experience. *In* Use of Property Rights in Fisheries Management. Proceedings of the FishRights99 Conference. Fremantle, Western Australia, 11-19 November 1999. FAO Fisheries Technical Paper No.404/2. pp.88-95, FAO, Rome.

6. THANKS

We thank the Fisheries Research and Development Corporation (FRDC) project 99-161 *"Sustainable fisheries management through enhanced access rights and resource security" Part I*. We thank the South Australian Fishing Industry Council (SAFIC) *"Fishing Rights Benchmarking Project"* project, Dr A. McIlgorm, principal investigator and Prof. Martin Tsamenyi, University of Wollongong, co-investigator. We thank the FishRights99 Committee for the invitation to the Conference.

CO-MANAGEMENT AND RIGHTS-BASED FISHERIES

M. Haward
Institute of Antarctic and Southern Ocean Studies, University of Tasmania,
GPO Box 252-77 Hobart Tasmania 7001, Australia
<M.G.Haward@utas.edu.au>
and
M. Wilson
Faculty of Fisheries and the Marine Environment, Australian Maritime College
PO Box 21 Beaconsfield Tasmania 7270, Australia
<M.Wilson@fme.amc.edu.au>

1. INTRODUCTION

Fisheries have traditionally be seen as common property resources rather than as private goods (Gordon 1954). This has led to fishing being considered as a paradigmatic case of the 'tragedy of the commons' (Hardin 1968) where individual self interest, through increases in catch effort, leads inexorably to a stock collapse. Regulation limiting effort, with power exercised by an external authority, was seen as the only way to reduce such tragedies in the fishery (but see Ostrom 1987). Recent literature has, however, noted that the analytic problems deriving from confusing 'common property' with 'open access' have resulted in private property regimes become the only alternative to regulation (Ferguson 1997: 286). Insights from the growing literature on governance over common property resources (McCay and Acheson 1987; McCay 1994, Larmour 1997) have increasingly emphasised co-management as an appropriate response to fishery management dilemmas (Sen and Nielsen 1996). The 'privatarian' argument does, however, remain influential, utilising neo-classical assumptions about individual economic rationality (Ferguson 1997: 286). Ferguson examines self-management as a complementary 'solution' to the introduction of market-based rights approaches, arguing that this solution has the added advantage of increasing the effectiveness of rights-based management.

Australian fisheries are small on a world scale but individual fisheries have developed to the point where they are significant export earners and critical to the economies of specific locations. At the same time a number of Australian fisheries are fully exploited. While current claims of a general crisis in Australian fisheries may be overstated, the present debate on fisheries management has particular salience. Fisheries management involves a complex balance between ecologic sustainability, economic performance and community benefit. This balance, the effective underpinning of co-management arrangements, can be enhanced by increasing the involvement of resource users in management. In Australia the establishment of institutional arrangements facilitating industry involvement in management provide an important adjunct to the use of rights-based approaches to management.

2. CO-MANAGEMENT: THE CONCEPT AND PRACTICE

Co-management can be defined as "an arrangement where responsibility for resource management is shared between government and user groups" (Sen and Neilsen 1996: 46). In Australia this approach was formalised through the introduction of Management Advisory Committees (MACs) and associated consultative groups within Commonwealth fisheries in 1992, and the introduction of similar arrangements within the state and territory fisheries. Rights-based co-management arrangements in Australia have developed at a time when traditional regulatory based fisheries management have been seen to have 'failed' in a number of fisheries around the world. Concern with such failures has re-introduced concepts such 'self governance' as the basis for an alternative approaches (Wilson et al. 1994, but see rejoinders from Parsons and Maguire [1996] and Hilborn and Gunderson [1996]). Cooperative management has been advocated to counter regulatory failure (Wilson et al. 1994), including non-compliance with laws and regulations (Sutinen, Rieser and Gauvin 1991).

Co-management arrangements introduce a further dimension to the debate over rights in fisheries. This debate has centred on the relationship between government and industry in ensuring a balance between short-term economic viability of what are arguably fully exploited fisheries and needs for longer term sustainability of catch. Since most fisheries are classic 'common pool resources', where a limited number of individuals may have access to what is communal property, a central problem is the balance between individual benefits gained by the fisher as opposed to broader community benefits (including conservation of stocks). The processes and outcomes in working towards this balance are commonly known as fisheries management. 'Management' is itself an important element, overlaying the 'conservation' of stocks; 'community' interest; and the 'economic' performance of fishers and the fishery. These latter three elements have been seen as critical 'paradigms' shaping conflict over fisheries management (Charles 1992: 379). Fisheries managers face difficult tasks in balancing demand from different communities of interest[1]. Defining and identifying 'community' becomes a critical element in co-management.

A 'Community' is an 'open textured concept' (Taylor 1982) subject to many definitions and uses, including what has been seen as a symbolic 'spray-on solution' to organisational failure (Bryson and Mowbray 1981). Theoretical bases for co-management derive from Tonnies's concept of *gemeinschaft* (socially constructed order) (Tonnies 1887). More recently Taylor (1982) developed "an idealized community" from "empirical studies of stateless societies to identify the conditions under which people sustained order without centralized, hierarchical forms of government" (Larmour 1997: 386). For Taylor, order was achieved by the community's shared norms and values, the existence of multiple relationships between members of the community and the use of shaming or retaliation against defectors from these norms and values (see also Larmour 1997: 386-87). Effectively co-management creates socially constructed order (including forms of 'rights') from the recognition of the shared interests and values (reciprocity) among the community defined by these interests.

Reciprocity provides the basis of community self-enforcement for mutual obligations, which together with shaming or ostracism, can be powerful instruments. Sturgess, for example, notes that "it is not in the interests of commercial fishers to ignore the damage being done to the catch in one or two years time [through catch of under-size fish]. While nothing might be said on the first occasion, if this behaviour is repeated the offending fisher will be spoken to" (Sturgess 1997).

While this may reinforce community values, the difficulties of community self-enforcement are also significant. As Larmour notes "notions of governance derived from stateless societies or common pool resources do not sit easily with other notions of governance that emphasise human, economic and cultural rights" (1997: 390). Commitments to justice and fairness are seen as fundamental, but, as Healey and Hennessy (1998) note, such commitments can lead to increasingly complex management arrangements to implement such values. It is these institutional frameworks which paradoxically can contribute to management failures in common pool resources (Healey and Hennessy 1998: 109).

Co-management is predicated on individuals and groups (forming a community of interest around a fishery) recognising that such problems arises from their day-to-day activities. In relation to the more practical use of co-management, the attraction of 'community based' management is obvious. Too often the community has not recognised that problems are theirs to own. Responses following recognition and ownership of the problem are likely to have much greater effectiveness that solutions imposed on that community. There are some obvious caveats. It is equally clear that community action cannot resolve all issues. Difficulties arise in dealing with spill-over effects – the actions of fishers who lack kinship or community ties (thus escaping community sanction), or who fish either predator or prey species – affecting the food chain and thus biomass.

One important aspect of community involvement relates to the role of the fishing industry itself. Seeing the industry as a community provides important theoretical and practical insights into fisheries management. In terms of theory the developing literature dealing with common pool resources emphasises community solidarity as a means of providing effective 'self management'. In practice, while fisheries may be fragmented and a unified view of them is difficult to ascertain, the role of industry self-management is an integral component in ensuring compliance with both externally imposed management arrangements and community-based codes of practice. Compliance with management measures is clearly more likely when these measures can be shown to directly benefit the fisher's economic performance.

3. REVISITING THE TRAGEDY OF THE COMMONS

The tragedy-of-the-commons thesis was popularised by an influential article by Garrett Hardin in 1968, although the organisational and administrative issues to which it is concerned have a long history, and were first raised by David Hume in the mid- 18th century[2]. The tragedy of the commons as discussed by Hardin has utility on several levels. It is a metaphor for a catastrophe that can arise from the problems between individuals and groups actions. And, the tragedy of the commons has empirical significance in the case of managing certain types of resources and public goods. Hardin's essay also

---

[1] For example environmental organisations have been critical of an over-emphasis on the interests of commercial fishing and a lack of clarity in the development of ecologically sustainable development in the work of the Australian Fisheries Management Authority. These criticisms were also made by Australian National Audit Office (ANOA) in a performance audit of AFMA in 1995 (ANOA 1995). Following the tabling of the ANOA report the management of Commonwealth fisheries were subject to further inquiry by a House of Representatives Standing Committee on Primary Industries, Resources and Rural and Regional Affairs (HORSCPIRRA) (HORSCPIRRA 1997).

[2] David Hume also considered aspects of this problem. Hume was concerned with the issues which arose from farmers having to organise together to drain a meadow or field, an action which would benefit them all in the long term but would involve each of them in considerable work in the short term. The example has been well described - the problem that arises is a classic free rider problem; it may be rational to stand back and let others do the work rather as the benefits will inevitably come to you. Hume's example reinforces the problems of free riding and the problems between individual and collective benefits. These ideas are more fully raised in a discussion of the tragedy of the commons, see Colebatch & Larmour (1993, 6).

provides the basis for engaging in discussions on the role of individuals and their relations with other members of communities who are also assumed to be 'rational actors'.

Hardin's essay was written at a time (the late 1960s) of considerable concern about global environmental catastrophe and concern at resource depletion and over population. Hardin used an example of the medieval common to illustrate the extent of the problem of self-interest or individual rationality. In seeking individual benefits and acting in a way that is rational (adding extra cows to gain added individual returns) the villagers destroy the common and thus create a catastrophe. Since individuals were responsible for the tragedy Hardin was dismissive of the ability of individuals to organise to resolve it without imposition of externally imposed discipline. This discipline could be imposed either by rules about conduct (changing the nature of human action) or by changing the nature of the common. Hardin emphasised two solutions; (a) externally imposed regulation of conduct or privatisation; (b) the latter utilising market instruments and private rights to ensure that the common remained viable.

4. PRIVATE RIGHTS

Common pool resources have traditionally relied on formal rules – usually through legislation or regulatory instruments – to resolve the 'tragedy of the commons'. More recently market-type instruments have been introduced with the development of rights-based fisheries management. The use of tradeable rights and the creation of quasi-market approaches by such 'trades' in fisheries management has clearly provided an alternative paradigm for both fishers and fisheries managers. In its extreme form this paradigm tackles the 'tragedy of the commons' by creating private property regimes, based on what have been termed 'privatarian' approaches to common pool resources. The development of individual transferable quotas (ITQs) creates quasi property rights, provides an opportunity to utilise market mechanisms and allows the market to determine the value of the quota or its component 'units'. Setting the total allowable catch (TAC) and determining quota and unit shares of the TAC provides a powerful tool for fisheries managers in the control of fishing effort and 'technology creep'.

In the 1990s Australian governments have increased the use of economic instruments, chiefly through the introduction of individual transferable quotas, fishing rights and resource rent recovery, as a means of overcoming regulatory failure associated with an inability to control fishing effort. This in part conforms to the privatisation solution – changing the nature of the common – advocated by Hardin. One effect has been to increase the direct interest and involvement of fishers in the management of their fisheries, but "the Hardin metaphor deflects analytic attention away from the actual socio-organizational arrangements able to overcome resource degradation and make common property regimes viable"
(Ferguson 1997: 295). It is these arrangements which are the basis of co-management.

5. CO-MANAGEMENT AND RIGHTS-BASED APPROACHES

The tragedy-of-the-commons thesis, and the focus on regulatory and/or market regimes as solutions to the tragedy, has had significant influence on fisheries management. Concerns over exploitation of stocks increasingly often lead to fisheries management 'solutions' which cast regulatory arrangements and government control as the only means of protecting fish stocks and controlling fishers. This approach dominated fisheries management arrangements in Australia as elsewhere but, in what may be seen as a classic case of regulatory failure, was less successful in restricting effort and catch levels in fisheries.

As Anthony Charles has noted in relation to Atlantic Canada '[h]istorically fishery management in Canada was based on a polarised view of the worlds, in line with Hardin's 'Tragedy of the Commons': fishers were seen as selfish profit maximisers, versus regulators as protectors of the resources. This perspective, although flawed, actually became self-fulfilling' (Charles 1997, 108). As a result fishers, excluded from decision-making had few incentives to moderate catches. Simply speaking they were seen as rapacious maulers of the commons unable, or unwilling, to reduce short-term profit over long term sustainability of stocks in the fishery.

Increased support for co-management derives first from a reappraisal of view of the inevitability of a tragedy-of-the-commons; and second from the recognition of the limits of government action. The possibility of an alternative solution to those proposed by Hardin has already been suggested. A number of writers have shown that Hardin's pessimistic prognosis for the commons is in fact limited. Berkes, Feeney, McKay and Acheson used a number of brief examples to show "that success [in managing the commons] can be achieved in ways other that privatisation or government control" (1989: 91). Berkes *et al.* point out that "[c]ommunities dependent on common-property resources have adopted various institutional arrangements to manage those resources, with varying degrees of success in achieving sustainable use" (1989: 91). These arrangements can, in fact, be quite complex as shown in studies of community rights-based fishing arrangements in different parts of the world.

6. MANAGING AUSTRALIAN FISHERIES

International debate over fisheries management arrangements is replicated in Australia where the great majority of Commonwealth fisheries are fully, if not over, exploited. Australian reforms in fisheries management provide an important study which has application to other fields and in other countries. Australian fisheries have undergone a period of development in the last five years with export earnings of fisheries products exceeded

$A one billion for the first time in 1992-93. This development occurred at the same time that the majority of Australia's fishery stocks were heavily exploited, emphasising the need for fishery management to balance short-term economic performance with longer-term ecological sustainability. Recreational fishing is also a significant element - surveys indicate that 4.5 million Australians undertake at least one fishing trip a year with over 800 000 people regarded as 'serious' fishers (Industry Commission 1992, 203). Traditional fishing activities by Australia's indigenous peoples raise important management issues such as access to resources and sea claims (Bergin 1991, 1993; Exel 1994), and may provide direct conflicts with commercial fishing interests.

The Australian public is becoming increasingly aware of this nation's responsibility for management of the fishery resources within its fishing zone and exclusive economic zone, the world's fourth largest (7 000 000 km^2 in area, roughly equivalent to Australia's terrestrial land mass). In most cases though, public attention is drawn to these responsibilities by secondary means; the scarcity of fish, high prices in markets or retail outlets, or publicity and conflict surrounding closures or restrictions in different fisheries. These outcomes are the public face of fisheries management. The less public face of fisheries management – institutional arrangements, the impact of legislative and regulatory obligations on government officials (and increasingly on industry), the relationship between science and management and industry government relations - is given little public or academic attention.

The introduction of an institutional basis for industry involvement within Australian Commonwealth fisheries was also part of wide ranging reforms to fisheries in the early 1990s. The legislative reforms in Commonwealth fisheries in 1990-91 created a statutory authority, the Australian Fisheries Management Authority (AFMA) to undertake day-to-day management of Commonwealth fisheries and saw statutory management plans established for all Commonwealth fisheries (Haward 1995). These management plans gave increased roles and responsibilities to the fishing industry at the same time that industry was provided with statutory-based fishing rights and became levied with full cost recovery for management costs. Thus, these reforms radically changed the traditional regulatory-based, input controls, and as a result, the relationship between government and industry (Haward 1995). The relationship between industry's increased responsibilities and the move to output-controls in fisheries management which give greater property rights in the fishery has been critical in enhancing alternative approaches to management.

One cannot understate the influence of these legislative and administrative reforms on the fishing industry. These developments took place during a period in which the Australian fishing industry was undergoing significant structural change, grappling with over-capitalisation and problems in reduced stock levels in major fisheries and an increasing dependence on export markets. Two significant outcomes have been identified. The first was the response of the fishing industry to increasing responsibilities and their recognition of the challenges facing industry[3]. The second relates to the facilitation of the involvement of industry in management, which lead to several parliamentary or government inquiries between 1993 and 1997. The most recent House of Representatives inquiry into AFMA, which reported in June 1997, reiterated the importance of industry involvement in management through the broadening of the role and membership of MACs.

The MAC model is as an example of institutional arrangements that reflect a shift away from traditional government lurnt-dominated management structures and processes. MACs (although established and maintained under a regulatory framework) have introduced a form of cooperative management to these fisheries. The members of a MAC (which usually consist of 6-8 people, comprising 3-4 catch sector representatives; a fisheries manager; a fisheries scientist; an environmental representative; chaired by an independent chairman) are established under relevant legislation to provide 'advice' on the management of a fishery to AFMA. The MACs' focus on consensus-based decision making emphasises the internal dynamics of the MAC as the most critical variable in determining effectiveness in cooperative management.

7. CONCLUSION

The introduction of property rights, cost recovery and user charges have clearly enhanced the 'user pays–user says' relationship. The relationship between such rights and industry involvement in fisheries management is, however, more complex. While current fishery advisory bodies may be viewed as symbolic attempts at co-management where incorporation of industry views, while important, does not reduce the level of government control, in practice fishers have considerable influence on management of their fishery. A positive aspect is the ongoing developing collaborative approaches to management based on recognition of concepts of rights deriving from shared, or community, interests. This is clearly part of the ongoing evolution of such approaches where recognition of communities of interest can be the base for substantive co-management model. This evolution has included representation of non-commercial fishing and environmental interests within Australian MACs and ongoing concerns with balancing ecological sustainability with economic efficiency.

8. LITERATURE CITED

Australian National Audit Office (ANOA) 1995. *Commonwealth Fisheries Management- Australian*

---

[3] One response was the change in name of the peak (*i.e.* representative) industry body from the National Fishing Industry Council to the Australian Seafood Industry Council.

Fisheries Management Authority, Audit Report No 32, 1995-1996 AGPS, Canberra.

Bergin, A. 1991. Aboriginal sea claims in the Northern Territory of Australia, *Ocean and Shoreline Management*, 15 (3), 171-204.

Bergin, A. 1993. *Aboriginal and Torres Strait Islander Interests in the Great Barrier Reef Marine Park*, GBRMPA, Townsville.

Bergin, A. and M. Haward 1995. International environmental conventions and actions–potential implications for the fishing industry, *Outlook 95*, 1, 281-299.

Berkes, F., D. Feeney, B. McKay and J.M. Acheson 1989. The benefits of the commons *Nature*, 340 (13) 91-93.

Bryson, L., and M. Mowbray 1981. Community: the spray-on solution, *Australian Journal of Social Issues*, 16 (4), 255-267.

Charles, A.T. 1992. Fishery conflicts: a unified framework, *Marine Policy*, 16, 379.

Charles, A.T. 1997. Fisheries management in Atlantic Canada, *Ocean and Coastal Management*, 35 (2-3), 101-119.

Colebatch, L. and P. Larmour 1993. *Market, Bureaucracy and Community*, Pluto Press, London.

Exel, M. 1994. Australian fisheries management - resource allocation and traditional rights, *Outlook 94*, 2, 231-237.

Ferguson, J.R. 1997. The expanses of sustainability and the limits of privatarianism" *Canadian Journal of Political Science*, 30 (2), 285-306.

Fisheries Reviewed 1993. Report of the Senate Standing Committee on Industry, Science, Technology, Transport, Communications and Infrastructure Canberra.

Gordon, H.S. 1954. The economic theory of a common property resource: the fishery, *Journal of Political Economy*, 62, 124.

Hardin, G. 1968. The tragedy of the commons, *Science*, 162, 1243-48.

Haward, M. 1995. The commonwealth in Australian fisheries management: 1955-1995, *The Australasian Journal of Natural Resources Law and Policy*, 2 (2), 313-325.

Healey, M.C. and T. Hennessy 1998. The paradox of fairness: the impact of escalating complexity on fishery management, *Marine Policy*, 22 (2), 109-118.

Hilborn, R. and D. Gunderson 1996. Rejoinder - chaos and paradigms for fisheries management, *Marine Policy*, 20 (1), 87-89.

HORSCPIRRA 1997. (House of Representatives Standing Committee on Primary Industries, Resources and Rural and Regional Affairs) *Inquiry into the Management of Commonwealth Fisheries* AGPS, Canberra, 1997.

Industry Commission 1992. *Cost Recovery in Fisheries Management* Report No. 17 Canberra.

Larmour, P. 1997. Models of governance and public administration, *International Review of Administrative Sciences*, 63, 383-394.

McCay, B.M. and J.M. Acheson (eds.) 1987. *The Question of the Commons*, University of Arizona Press, Tucson.

McCay, B. 1994. The ocean commons and community, *Dalhousie Review*, 74 (3), 310-39.

Ostrom, E. 1987. Institutional arrangements for resolving commons dilemmas: some contending approaches, in McCay, B.M., and J.M. Acheson (eds.) *The Question of the Commons*, University of Arizona Press, Tucson.

Parsons, L.S. and J.-J. Maguire 1996. Rejoinder–comments on chaos, complexity and community management of fisheries, *Marine Policy*, 20, (2) 175-176.

Sen, S. and J.R. Nielsen 1996. Fisheries co-management: a comparative analysis, *Marine Policy*, 20 (5), 405-418,

Sturgess, G. 1997. Cooperation comes into its own, *Australian*, January 30, 11.

Sutinen, J.G., A. Rieser and J. R. Gauvin 1991. Measuring and explaining noncompliance in federally managed fisheries, *Ocean Development and International Law*, 21, 335-372.

Taylor, M. 1982. *Community, Anarchy and Liberty* Cambridge University Press, Cambridge.

Tonnies, F. 1957. *Community and Society* translated C. Loomis, Michigan State University Press, East Lancing, [1887].

Wilson, J.A., J.M. Acheson, M. Metcalfe and P. Kleban 1994. Chaos, complexity and community management of fisheries, *Marine Policy*, 18 (3), 291-305.

COMMUNITY MANAGEMENT IN GROUNDFISH: A NEW APPROACH TO PROPERTY RIGHTS

F.G. Peacock and J. Hansen
Department of Fisheries and Oceans
Resource Management Branch, Marine House
Dartmouth, Nova Scotia B2Y 4T3 Canada
<PeacockG@mar.dfo-mpo.gc.ca>

1. INTRODUCTION

Canada's demersal fisheries were based on the cod fisheries of the Atlantic coast, supplemented by haddock, pollock, flatfish, cusk and hake for the Scotia-Fundy sector (the waters off the coast of Nova Scotia). The Atlantic cod collapse had a major impact on all Atlantic Canada and most cod resources remain severely depressed to this day. Up until 1976 these resources were managed by the International Commission for Northwest Atlantic Fisheries (ICNAF). Domestic management of groundfish fisheries on Canada's Atlantic coast began in 1976, with the extension of jurisdiction to 200nm. This newly formed economic zone was to herald bounties from the sea through the replacement of foreign fishing with Canadian effort using management planning to define and direct fishing operations for the betterment of the Canadian public. This approach was expressed in the 1976 Policy for Canada's Commercial Fishery (Canada 1976) which stated the need to move to an economically viable fishery. In the case of groundfish, management planning outlined several objectives designed to accomplish this economic viability policy principle.

The groundfish fishery in Atlantic Canada is arguably the most complex of any fishery in Canada. Groundfish is the generalized term for a series of species of fish, mostly gadoid, which are harvested separately or collectively by many fleets involving thousands of fishermen. In the Scotia-Fundy Region, the community-based approach discussed in this paper depends on nine separate fleets, based on a combination of different gear usage (mobile and fixed-gear) and vessel sizes (vessels range from the very small under-10m to large wetfish trawlers over-30m in length). These fleets harvest eleven groundfish species ranging from cod to redfish in five NAFO divisions (Figure 1). This diversity of the fleets made attainment of the 1976 policy goals exceedingly difficult. The policy approach was to first develop management plans for the large, far ranging offshore-vessel fleet. Management initiatives were gradually introduced to other fleet sectors over the subsequent several years. For the inshore fixed-gear fleets, which use vessels under-19.8m in length and utilize handline (manual and automated), longline and gillnet gear, full management applications, including quotas, were not in place until the 1982 fishing season.

The goal of economic viability has led to a series of crises during the 25 years since 1976, largely owing to a number of factors working at cross purposes within the planning process. While the principle of economic viability remained paramount within the 1976 Policy, the lack of clear definition of what this actually was gave rise to individual interpretations and manifestations. It became clear that management approaches did not meet the policy objectives, and the planned benefits to fishermen and the public of Canada, were thus not forthcoming. This led to a series of studies to define new directions for fleets as crises within these arose. For instance, in 1982, the Kirby Task-Force released a plan to bring stability and viability to the offshore groundfish fleets through the use of Enterprise Allocation programmes. This report again talked about developing plans to allow for economic viability. However, the focus only related to the offshore fleets (vessels over 30m).

The second crisis in the mobile-gear sectors within the Scotia-Fundy region came in 1989. The Scotia-Fundy Groundfish Task-Force (Hache 1989) focused recommendations on capacity/resource imbalances and mechanisms needed to bring stability and economic viability to the inshore components of the groundfish industry. The concept of maintaining an inshore fixed-gear sector outside of the offshore management plan, or through the use of allowances after 1982, permeated thinking of the day. It was generally accepted that inshore, fixed-gear operations, although numerous (3300 vessels), would apply relatively little effort on the large and seemingly fully-resilient groundfish stocks. Management of the larger inshore mobile and fixed-gear and offshore vessels (500 vessels) became the priority. Thus, by 1986 vessels over 19.8m were on company-based enterprise allocation programmes while the mobile gear sector under-19.8m and the fixed-gear sector 13.8-19.8m opted for ITQ based programmes, beginning in 1991. The vessels under-13.8m working fixed-gear (3300 strong) operated under an overall quota that was harvested on a competitive basis. This sector was permitted to continue fishing on a 1500kg/trip limit once the annual fleet quota had been reached.

Following the demise of cod resources in most parts of the Atlantic in 1992-93, widespread closures were implemented in the 1990s. On the Scotian Shelf, groundfish stocks in NAFO divisions 4Vn and 4VsW were closed in August 1993 and remain closed. Cod in NAFO divisions 4X and 5Z continue to be depressed as do haddock, pollock, cusk and hake resources. Many Scotian Shelf stocks have population biomasses at record low levels and it is only because fishers harvest a wide variety of groundfish in the NAFO divisions 4X and 5Z (Southwest Nova Scotia/Georges Bank) that fisheries are allowed to continue. The latest scientific advice suggests that groundfish resources continue to be depressed in Atlantic Canada, the USA and in International waters. At present, fishers are using harvest targets approximately 30% of those in 1990. The devastation of the groundfish resources has

Figure 1
Locations of NAFO Divisions in the Scotia-Fundy Region (5Y, 5Ze, 4X, 4W, 4V) of Atlantic Canada.
NAFO Division 3P is in Newfoundland Region

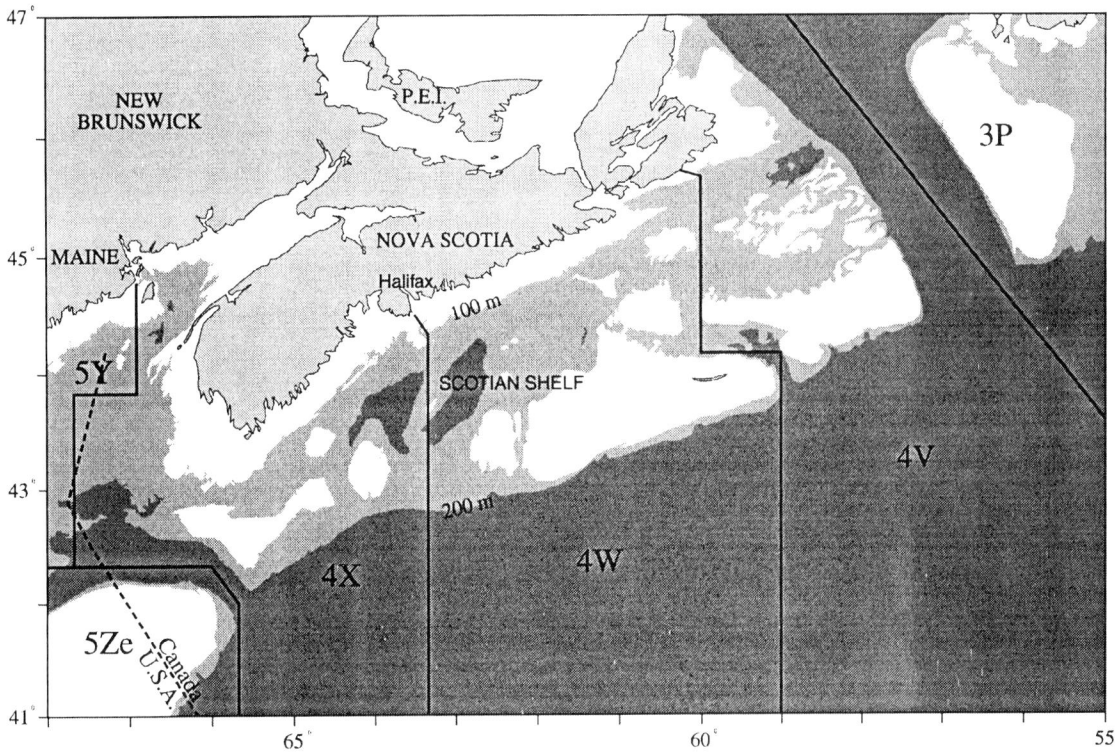

emphasized the need for management mechanisms that would allow fleets to better balance their capacity with the size of resource, for the continued viability of both. By 1995, all fleets (except the fixed-gear, under 19.8m fleet) had developed mechanisms to allow catch adjustments as the resource biomass changed size. There were two prominent events that called for management of the fixed-gear inshore fleet. The first was the dramatic reduction in the resource and the impact that the large active and inactive fixed-gear licences might have on it. The second was the increase in fixed-gear licences as a result of imposing ITQs on the mobile-gear sector.

Attempts began in the mid 1990s to bring management applications to the entire inshore fixed-gear sector. The 1500kg/trip limit provision for the under-13.8m vessels was eliminated in 1994 as one means of bringing the fishery in line with the resource. However, the development of management approaches within this diverse sector proved long and circuitous before the current community-based concept was developed. During the process, many associations representing inshore fixed-gear segments arose, resulting in intense competition within the advisory committees for allocation advantage, and often culminating with demands for the Minister of Fisheries and Oceans to intercede to change sharing arrangements. The priority to ensure continued viability of the inshore fleet resulted in destructive management decisions and resulted in overfishing of the dwindling resources.

Given the many problems with the management of the inshore fixed-gear fleet, a new approach was introduced in 1995 on a trial basis in one area - Halifax West, Sambro, Nova Scotia, and was followed by a three-year test application on a regional basis beginning in 1997. This approach recognized differences within the sector by redefining fleets using either geography or "like-minded" views to better define fleet structures in order to accommodate the demands of the particular fleet-group in question. The term "like-minded" simply refers to the recognition of groups of fishers who have common management objectives. This trial in 1995 became the community-based management approach of today. As will be described below; it received wide acceptance. Industry management boards currently have almost complete control within this process and have moved the management concept well beyond that envisioned in 1995. The community approach may have application to other parts of the world or within other emerging issues such as aboriginal rights. This paper describes the development of community-based management, along with challenges, pitfalls, lessons learned and opportunities for the future.

2. HISTORICAL BACKGROUND AND THE NEED FOR CHANGE IN THE FIXED-GEAR FISHERY

The demise of the groundfish resources underlined the need for a rationalization of fleet fishing-capacity. The fixed-gear 13.8m fleet was the most numerous and although many vessels were marginal in activity, more than 2000 vessels were attempting to make a significant portion of their income from the Scotia Fundy groundfish resources. At the same time groundfish resources were mostly in decline and were being shared by other fleets that had already addressed the problems of viability and over-capacity. However, the size and diversity of the under-13.8m fleet presented unique challenges that prevented a similar management approach to the over-13.8m fleet. Management activities in the over-13.8m fleets had at least started to address the fishing-capacity – resource balance by the mid-1990s. For the under-13.8m fleet in particular, the size of the fleet and the number of landing sites made enforcement very difficult if not impossible, and the large number of participants made it a political issue. Further, the large number of inactive licences raised the issue of the right versus the privilege to fish. There was also the belief that this gear does not harm the resource, *i.e.*, no level of capacity was harmful. This belief was re-enforced by non-government organizations (NGOs) that were supporting an anti-ITQ movement. Further, the bulk of the coastal community infrastructure was invested in this small-boat fixed-gear fleet (Table 1). The large number of landing ports, processing plants and the easy opportunity of direct marketing in the USA also supported the demand to find viable solutions to this problem as they could not be supported by the level of the resource, which was now at 30% of 1980's levels. This section below describes some of the management attempts on this fleet, providing context to the community-based management approach.

Table 1

Characteristics of the Scotia-Fundy groundfish fleets

Mobile-gear fleet (trawl)
Size 12m to 40m
500 vessels (1980)
135 vessels (1995)
2/3 all quotas
ITQ Management (1982-1990)
Fixed-gear fleet (longline, gillnet, handline)
Size less than 12m
330 vessels (1980)
3300 vessels (1995)
1/3 all quotas
Competitive Management
Other
500 landing ports; 300 plants; direct sales to USA fresh market

The under-13.8m fleet was assigned separate fleet-quotas in the 1987 Atlantic Groundfish Management plan for portions of the Scotia-Fundy sector. However, separate fleet-quotas had not been assigned to all groups until 1997. Table 1 provides a summary of the status of the various groundfish fleets operating within the Scotia Fundy sector. While the under-13.8m fixed-gear fleet was not formally defined in all areas until 1997 (it began in 1987 in some areas) the use of limited-entry licensing began in 1976. Therefore, the ability to compare the fixed-gear and mobile-gear fleets through time has been a fairly easy task. However, there is a degree of estimation in these figures as formal handline licensing for non-automated handlines did not occur until 1991. Prior to this, handline activity was either part of any longline operation or as a separate operation designated under regulations. Criteria for this latter form of handline fishing required a fulltime, registered fisher using a vessel licensed under the government system. Therefore one can only assume that the number of participants has not changed over time.

The first major attempt at restructuring the small fixed-gear fleet occurred through the split in quota. This was first attempted in fisheries east of Halifax in the 1980s, and later and more successfully as a trial in 1995 using the community-management approach. From the trial in Halifax West/Sambro(HW/S), Nova Scotia, the development of a complete management approach based on community-allocations for vessels under-13.8m allowed the separate development of these fleets (under-13.8m) from the larger fixed fleet (13.8-19.8m).

At the same time that the HW/S began its community-management project, the remainder of the fixed-gear sector in the under-13.8m group established an allocation system based on the three gear-types of handline, longline and gillnet. In this approach, each licence-holder was required to choose between using one of three possible gear-types. Quotas were assigned based on the participation with these three broad gear categories. However, this approach proved to be unsuccessful with quota depletion occurring rapidly. Within these sectors, there did not develop any system of self-government and consequently the race-for-fish continued, as has been the experience with other competitive fisheries (Hardin 1968).

In contrast, the mobile-gear fleets through the use of quasi property-right programmes of EA's and ITQ's successfully addressed the imbalance between fleet numbers and available resource. The numbers of mobile-gear vessels active today is a reflection of this fleet's ability to adjust to economic and supply situations and ensure their viabilty operations. The inability of the fixed-gear fleet under-13.8m to address over-capacity within its quota share led to severe problems among all fleets as this sector utilized political pressure to change historical sharing arrangements (*i.e.* transfer quota to this sector at the expense of the others). However, rather than making these changes, DFO continually pressed for solutions to be found within the fleet sectors, a practice that in hindsight was ultimately responsible for the development of the community-management approach.

Environmental groups also became prominent in the debate. Along with the continuing support for coastal communities as a "proper way of life", new dialogue on environmentally-friendly harvest methods and ecosystem management became popular. They believed that the fixed-gear applications of gillnet, handline and longline were "more" ecologically friendly and moved the process towards more long-term, sustainable harvest applications and generally enhanced rather than degraded the resource overall. These arguments tended to be used within the sector as justification for the *status quo* as well as justification for the transfer of quota from mobile-gear to this fleet sector. Thus by the mid-1990's rationalization of most groundfish fleets had occurred except for the continuing over-capacity of the under-13.8m fixed-gear sector.

3. THE COMMUNITY BOARD SYSTEM

In 1995, the Halifax West group of fishers was the first to try a new approach – one based on area or community rather than boat size. Approximately 50 active vessels out of a total fleet of almost 100 vessels (Figure 2) requested to be permitted to manage an allocation of vessels in the 13.8-19.8m category to create a separate group which would allow an independent management application to occur. This group was therefore separate and developed an independent harvest plan. For the remaining fleet under-13.8m, the following measures were implemented:

Figure 2
The number of active and total number of groundfish licences by community group from 1996 to 1998

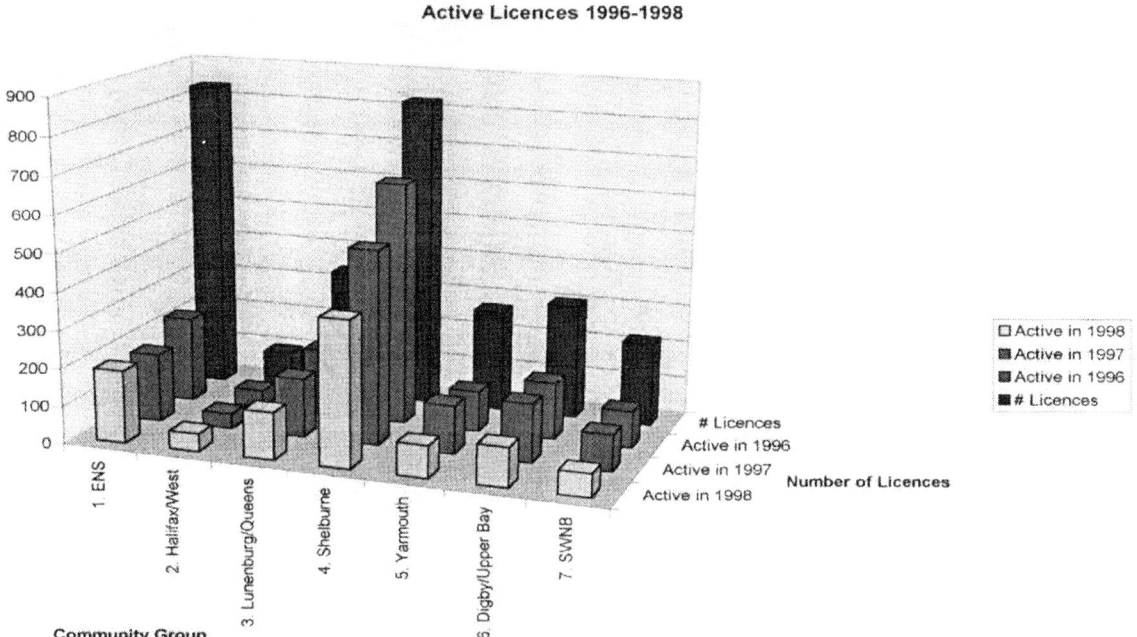

groundfish based on historic shares of the available resource. DFO assigned this group a quota of fish approximately equal to the historic share, applied as a percentage to the annual total quota. This group was given permission to manage this allocation as a "separate fleet" fully independent of the remainder of the fixed-gear fleet. DFO contributed to this exercise by encouraging dialogue on communities of interest for allocation purposes and, for the Halifax West experiment, providing separate allocations, monitoring and undertaking opening and closing season.

This initiative proved interesting to the rest of the fixed-gear fleet (all vessels under-19.8m), which was struggling with the day-to-day problems of an over-capitalized fishery. Through an industry-sponsored workshop and many industry/government meetings, the fixed-gear community decided to establish a programme of community management for the entire fixed-gear under-19.8m sector. As a beginning, vessels in the 13.8-19.8m category were to be excluded. This was done to allow the relatively small number of licence-holders operating

i. A series of seven communities were established based solely on geography (Figure 3). The concept of "like-minded individuals" supported by DFO was rejected in favour of a geography definition as a means of avoiding the concept "corporate concentration/ITQ management" within the inshore, fixed-gear groups. However, the Shelburne community, because of the great differences between highline vessels and smaller inshore operators both in performance and philosophy, subsequently required partitioning into two separate management groups. This necessity provides one of the examples of why the geography decision was, in reflection, somewhat inferior to the like-minded approach.

ii. All fishers were assigned to a community-based on the area of registry of the licence-holder as of 31 December 1996. To ensure choice, an opting-out provision was available which was relatively unattractive but which ensured that a choice would be made by each fisher.

Figure 3
Location of the seven community groups in Nova Scotia and New Brunswick. Each community group has a separate community management board

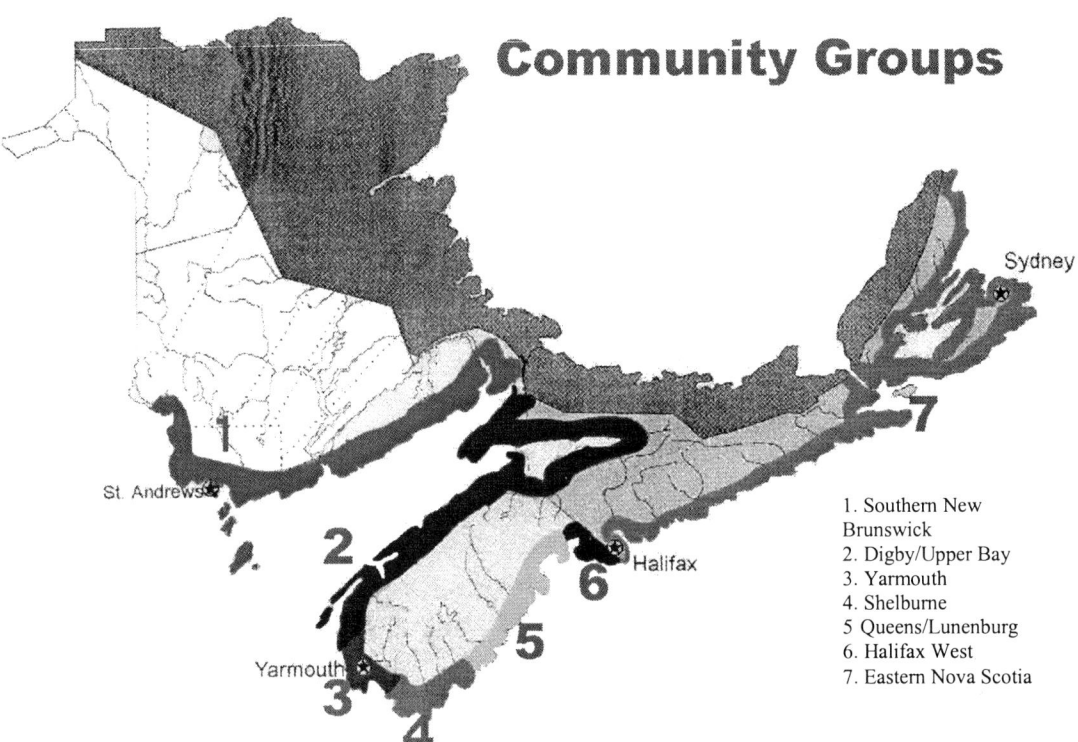

iii. Establishment of Community Management Boards (CMBs), comprised of elected representatives which in most cases are fishers. These are private boards, designed to provide input into in-season management and to develop, implement, control and monitor the activities of the community fleet. Management boards also provide representatives to the public advisory process, including input into the long-term planning process for the gear sector. Terms of reference for the public fixed-gear under-13.8m fleet advisory process stipulate three representatives from each management board. At their inception, the CMBs were thought of as the social/economic driver for the particular community sector and were responsible for all activities associated with these functions, including the development and implementation of co-management approaches. Each CMB developed a Conservation Harvest Plan (CHP), was responsible for controlling fishing activities of members and adopting standardized monitoring and catch controls and created management boards of elected fishers.

iv. Quota was allocated to each of the CMBs on the basis of catch history of each individual using the 1986-1993 period and standardized to 1996 quota levels. These calculations were based on landings that could be attributed to an individual licence-holder plus unidentified (by licence-holder) landings from processors within these various communities. This process utilized numerous input sources, including DFO, for data analysis and a mediator to resolve differences in opinion with respect to community sharing.

v. CMBs could trade quota among communities, trade or exchange members, apply penalties for breach of violations of CHPs and generally conduct a business-like approach to fishing within the conservation umbrella demanded within the precautionary approach.

What the community management approach represented was the development and implementation of a self-governance system that did not reach down to the boat level, as is done in ITQ systems, but nevertheless represents a quasi-property approach, but at the community level.

In order to analyze the community approach and the CMBs, it is helpful to analyze management activities against a model of the planning and implementation process. Figure 4 provides a schematic representation of the type of analysis posed. While it is not intended that this paper go into the details of why this schematic was used (the details can be found in a discussion paper by Burke *et al.* 1996), needless to say the elements interrelate to create a management mosaic.

Figure 4
Schematic representation of the various elements involved in the planning and implementation
of the community management boards

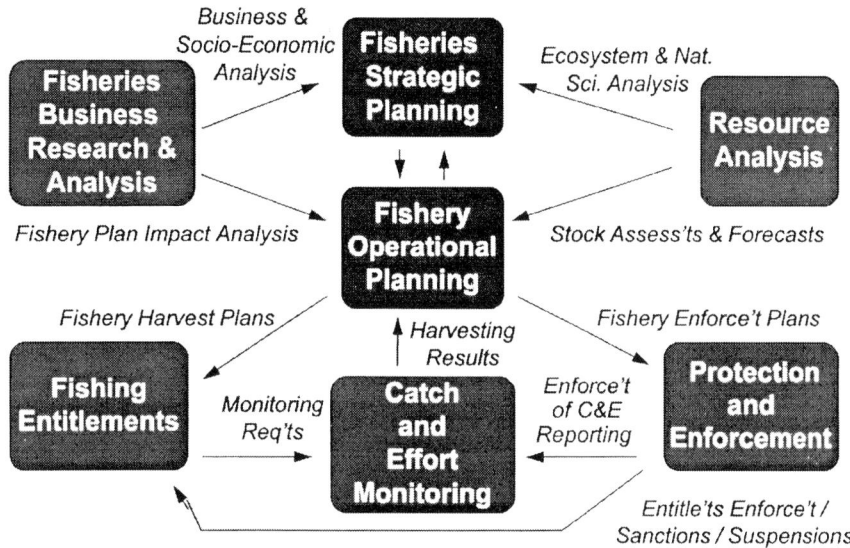

4. STRATEGIC ISSUES

If one compares the community management plans of 1996 and beyond with those previously prepared, one finds several strategic issues, which contributed to the success of this approach. Since the beginning of the process, plans for this sector had largely been developed through a "top down" process using DFO as the lead developer, implementer and controller, with industry relegated to an advisory role. This type of management, which was common in many countries in the late 1970s and 1980s, created an adversarial system in which little positive dialogue occurred. These readily developed situations where no climate for change existed. This was most evident among the smallest of the vessels which, while comprising 30% of the fleet, had over the years been able to effect great protection from the system thus ensuring some degree of viability at the expense of other fixed-gear groups.

The need for a bottom-up approach became evident. However, to move to the current situation several factors were required:

i. The government needed to change its approach to management. In this instance, an internal programme review proposed a change in philosophy to a process where the government facilitated direction and assisted industry uptake.
ii. There was the need for the industry to want to change the process, which in this case was the 1995 experiment.
iii. Government was required to stop pandering to lobby groups, which occurred to a limited extent in this case, but was sufficient to remove the leverage of the small boat owners.
iv. New ideas needed to be developed, which directed the industry to new avenues of approach.
v. The industry needed organizers who could develop plans on behalf of industry and who could work together for the collective good of the fleets in question.

Therefore, while the plan was for something quite different, the end-result was an industry-adopted approach almost fully supported by industry.

5. ANNUAL PLANNING

As described earlier, the process involved the establishment of CMBs made up of elected industry representatives (although in one instance a non-fisher community mayor is a board member). The process of moving to these management-communities could not be described as democratic in its commonly accepted sense. Fishers by nature tend not to be participatory, either in dialogue, or through written expression. While criticism is common among fishers, as is support for election processes, the history in the Scotia Fundy fisheries shows that participation in voting procedures is not strong. In the instance of developing-communities of fishers where quota allocation would rest with assigned licence-holders, it became obvious that all licence-holders would be required to choose one of the community groups, or the default DFO group, before allocation settlement could occur. Therefore DFO forced every licence-holder into a community through a vote, with an abstention indicating the DFO group choice. In this sense, DFO forced the issue.

Once divided into groups the responsibility for developing harvest plans, which reflected social and/or economic interests, fell to the industry. Simply put, without a

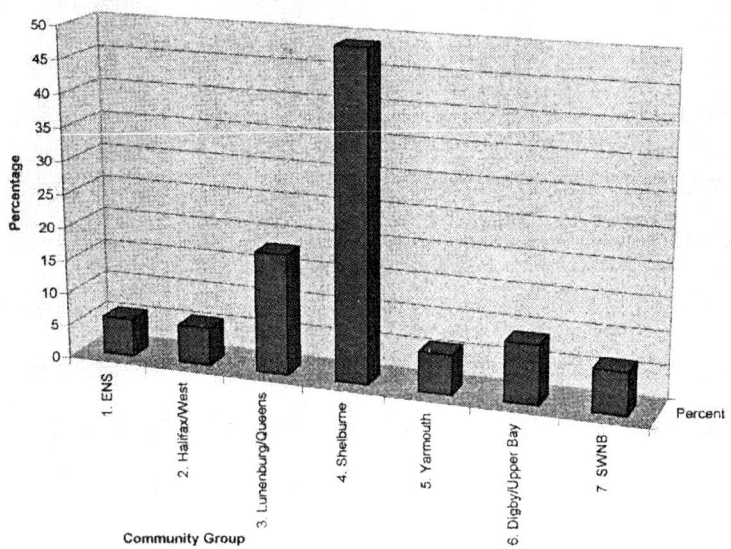

Figure 5
Percent allocation of quota amongst community groups.

plan that respects conservation and delivers, at the same time, industry requirements, no fishery would occur. Industry decided on a community-basis how to deliver the conservation standards as well as the appropriate harvest and performance approach. After 15 years of commonality, suddenly there were eight plans allowing eight individual applications. Industry boards also became responsible and accountable to the membership for the performance of the fleet and the members within the fleet, which in some cases included development and administration of penalty structures.

6. ENTITLEMENTS

Role and responsibility sharing has been a constant theme in most co-management approaches now being introduced into several Atlantic Coast commercial fisheries. In the community-management approach, significant shifts in sharing of responsibilities have occurred, which emphasize the increased role of the fishers in the overall application of the fish management approach.

Within this "new arrangement", government maintains the activities of licensing, registration of vessels, identification and limitation of gear and the description of area to be fished or controlled. Many of these applications occur through DFO-administered licence conditions with delivery through DFO-enforcement.

The CMBs are now vested with the responsibility for defining entitlements on how to harvest the assigned allocation. The seven communities have taken a number of approaches, which range from a competitive fishery (by gear type) within an overall community-quota on a per species basis, to an industry-developed and delivered ITQ initiative. Combinations, or permutations, of these approaches were also used in the other community groups. The approaches can vary and can be independent, or can work in conjunction with each other. The choice is the community's to make.

In order to get to this point, it was necessary to determine initial allocations to each community, which included dividing the initial fixed-gear under-19.8m overall quota into two distinct groups. This partitioning at 13.8m as had occurred many years earlier in NAFO divisions east of Halifax was based solely on catch history. The resultant under-13.8m-quota-allocation was subsequently divided among the eight competing management groups.

Always a contentious issue, the need to define allocations was an essential step in the process of shifting responsibility to the industry for many of the decision-making areas (Figure 5). The process of "initial allocation" utilized mediation as well as a thorough vetting through workshops and meetings. Government, in support of this process, also provided suggestions as to possible approaches, including the development of sharing possibilities. The problem with allocating quota to users is that there are normally winners and losers. In this case, the added concern of unknown catch histories further complicated this issue as all users claimed shares regardless of the source of the historical record. In the end, fishers agreed upon a sharing format using the following criteria:

i. Cumulative catch history on an individual/per species basis by community
ii. Unidentified landings from each community were counted as a percentage share of the community in question
iii. Upon completion of the share calculations, apportioning involved 97% of calculated amounts, the remaining 3% was supplied to individual communities

iii. to address to some degree inequities that resulted in the 1995 fishery
iv. For the Shelburne community groups, elections to determine community of choice for each fisher
v. Provision of transfers of quota among communities, and for instances where fishers move from one community to another, the requirement of both communities' agreement, including whether or not the catch history would move as well
vi. Equal shares for a portion of the overall quota.

Thus industry has moved toward establishing communities that function for the good of the members of that community and participate in the stewardship of the resource along with the government and other private and government agencies.

7. CATCH/EFFORT MONITORING

CMBs control the fleet and individual movements and restrictions through developed and implemented industry harvest-plans. Industry, in developing these plans, follows set standards and principles for conservation while at the same time devising mechanisms that deal with community needs. In all cases, where internal designs require monitoring functions, responsibility for this falls on the fisher and, or, community. To that end, the role of government in the process is to ensure that the overall conservation objectives are met and that the overall agreed community-allocations are respected. This is the government investment into the system, *i.e.* respecting and controlling overall allocations within the conservation rubric. This government audit-function ensures both conservation approaches are adopted and respected and that industry-agreed sharing occurs. The government lists and records seasonal quota-limits developed by the management boards and individual vessel-landings are provided to boards to assist in managing industry-imposed or conservation-dictated limits.

The foundation for this process is the Dockside Monitoring Programme (DMP) that is based on the entitlement. The type of management adopted (ITQ or quota-limited competitive catch), type and size of vessel, contribute directly to the level of recorded detail. DMP is a 3-step process of: (a) hail from sea, (b) verification of unloading amounts at dockside and (c), collection and data entry of catch data on a real-time basis. This service is an independent function of several companies funded by the fisher.

8. ENFORCEMENT

In addition to the monitoring of catch and quota, additional requirements to ensure the conservation of the resource are delivered through a combination of government activities and industry commitment and delivery. From the government perspective, controls involve the use of at-sea boardings, observers and sea/air surveillance to augment the hail/DMP process. Industry supplements at-sea monitoring mostly through the funding of the "at-sea" portion of the costs of observers.

From an industry perspective, in addition to the funding activities, peer-pressure has provided a significant deterrent to illegal fishing activities, including those that compromise conservation or activities that compromise industry harvest-plans. In some communities, industry sanctions have been adopted in an attempt to control illegal activities. Normally, industry boards administrate sanctions but the mechanism can vary. What is common is that all penalties, which are normally reductions in quota and/or time that can be spent at sea, are more draconian than government-penalties using court systems, and secondly all penalties are determined by the fishers. Normally in such processes there is no appeal procedure as in the formal government system, making the effectiveness quite remarkable.

9. RESOURCE ANALYSIS

One can generally assume that in such co-management applications cooperation with science requirements would be high. Fishermen do not fully understand survey techniques used by government scientists and often insist that the information is wrong because it does not reflect what is being experienced on the water. It is also true that in almost all cases fishers believe in some form of conservation. However the definition of what that is, is often difficult to determine. Therefore, in a collaborative setting, industry is often eager to provide assistance to science initiatives, particularly to advance two specific goals: (a) to better understand the resource which might translate into more quota and (b), to ensure that scientific surveys fully cover the management unit.

In the case of community-management, what has occurred to better the science-understanding? First, there has been more dialogue between scientists and industry on the subject of groundfish in the Southwest Nova Scotia area. Partially aided by an industry advisory council (the Fishery Resource Conservation Council), this dialogue has advanced the understanding within the communities of science issues and species interactions. Improved knowledge provides for a better approach to management. Second, the communities have provided additional funds to extend government surveys or have participated in the survey process. In either case, the enhanced knowledge has proved beneficial. This improved knowledge base is manifested in a science-advisory process that is more interactive and more detailed in its analysis. Specifically, community fishers participate in the Regional Advisory Process (RAP), often providing valuable comments on suggested inferences from data sets. An increased industry knowledge-base also contributes to overall knowledge that translates into better community decisions, and by understanding the process in more detail, the delivery of data by fishermen improves.

10. BUSINESS ANALYSIS

While there has been little actual analysis of the community-management approach to date, this is something that will be focused on in the near future. However, initial comments suggest that the cost of fishing has gone up. This is largely due to transaction costs and the additional costs of setting up the systems, complete with controls. The increased cost of management appears to be due to the costs of establishing the community boards, but in the long run, savings are expected to occur as more responsibilities flow to the harvesters.

A second issue surrounds the level of participation. While the community-approach did not reduce licence numbers, the opportunity among communities to address issues in a more business-like manner has resulted in a reduction in total participation. Today, licence numbers are closer to a balance with resource and even in communities where a more socially-oriented approach of sharing has been adopted, the adjustment in participants has occurred as well. The number of active vessels in all communities fishing in the NAFO Division areas of 4X and 5Y has decreased from 1275 in 1996 to 773 in 1999 (Figure 2). Having said this, the large number of licence-holders has dictated that a solution to this resource and licence number imbalance must be found. In the highline fleets, some level of licence-stacking has occurred and the use of informal ITQ arrangements has provided for some balancing. However, in other communities, there are no mechanisms to afford adjustment outside of those identified above (*i.e.* retirement) and therefore, there is a need for a more economic solution if a balance is to be desired. In spite of this deficiency, community-management has afforded opportunities to invoke closure when community-quotas are reached.

11. ISSUES FOR THE FUTURE

In spite of the advances championed by the fishers within these communities, a number of problems persist. First and foremost is the rift among individuals and groups that support a more socially-guided fishery based on a competitive format within an overall quota, and those desiring a more economically-driven fishery using quasi-property-right mechanisms. Those in the social camp argue that community control should not allow evolution in the other direction, which they feel creates a "have *vs.* have-not" syndrome so widely disliked among small boat owners.

The informal quasi-property-right approach adopted by some community-management groups is also criticized by formal ITQ groups who, under the current system, must pay larger access fees to acquire individual quota. Allowing such quasi-property-right arrangements is said to be a mechanism by which the government "subsidizes" the inshore fixed-gear fleet.

The second issue surrounds the resolution of inter-community conflicts. Prior to the introduction of the community-management approach, inter-community conflicts, mainly over allocation issues, were the norm. Today one can see cooperative approaches in projects such as the Bay of Fundy Council, which is a council made up of two CMBs as well as several non-consumptive users and is dedicated to developing an ecosystem management approach in the Bay of Fundy. In addition, one sees cooperation between CMBs in the transfer of quotas and other management related issues, suggesting that the autonomy provided by such a management system provides for ancillary cooperation benefits as well.

The third issue focuses on conservation. Has this approach changed the industry's approach to conservation positively or negatively? In most instances there have been positive responses to conservation approaches. One might say that overall, the fleet is more conservation-oriented under this system than under the previous competitive format. However, problems continue. While there has been a significant reduction in discarding and hi-grading, these practices continue at a level believed to be too high since the introduction of the Community-Management approach. Under the current management framework, this comment could apply to any fleet-group. The declines in groundfish stocks overall and the apparent imbalance in relative quotas contributes to these problems but clearly the industry continues to have some distance to go to be a fully conservation-oriented harvest sector. The husbandry of the resource, while prominent in the minds of most, can be overshadowed by the needs of survival, and in cases where there continues to be imbalance between resource and fleet numbers, the problems of conservation will continue.

12. CONCLUSIONS

The decision-making approach related to the partitioning of the resource among communities, and the flexibility within communities to devise appropriate management applications for each community group, has eliminated virtually all of the criticism and lobbying of previous planning approaches. One of the main developments of the community-approach is self-governance, an essential characteristic of formal ITQ systems.

In providing community autonomy and accountability, this approach allows for community solutions to the problems of fish management, many aspects of enforcement, transfers of quota and catch history, monitoring and conservation of the resource. Over time, remaining issues associated with the imbalance between fleet and resource will be resolved within the context of the community. To accomplish these goals there has been an increased cost-burden on both government and industry. However, as efficiencies continue in the process, we expect that both government and industry will realize significant savings while at the same time benefit from a more enlightened management system that is flexible, being able to change as time and conditions dictate.

13. ACKNOWLEDGEMENTS

Drs. M. Sinclair, E. Kenchington and R. O'Boyle (Science Branch, Department of Fisheries and Oceans, Bedford Institute of Oceanography) provided useful comment on the manuscript.

14. LITERATURE CITED

Burke, D.L., R.N. O'Boyle, P. Partington and M. Sinclair 1996. Report of the Second Workshop on Scotia-Fundy Groundfish Management. Can. Tech. Rep. Fish. Aquat. Sci 2100: vii + 247pp.

Burke, L., C. Annand, R. Barbara, L. Brander, M.A. Etter, D. Liew, R. O'Boyle and G. Peacock 1994. The Scotia-Fundy Inshore Dragger Fleet ITQ Programme Background, Implementation and Results to Date. ICES C. M. 1994/T:35.

Canada 1976. Policy For Canada's Commercial Fisheries, Fisheries and Marine Service, Department of the Environment: 70pp.

Gardner, M. 1988. Enterprise Allocation System in the Offshore Groundfish Sector in Atlantic Canada. Mar. Res. Econ., Vol. 5: 389-454.

Hache, J.E. 1989. Report of the Scotia-Fundy Task-Force, Fisheries and Oceans, Canada: 86pp.

Hardin, G. 1968. The Tragedy of the Commons. Science, 162: 1243-1248.

Kirby, M.J.L. 1982. Navigating Troubled Waters, A New Policy for the Atlantic Fisheries, Department of Fisheries and Oceans, Task-Force on Atlantic Fisheries:152pp.

Halliday, R.G., F.G. Peacock and D.L. Burke 1992. Development of Management Measures for the Groundfish Fishery in Atlantic. Canada Marine Policy, November 1992.

THE CREATION OF PROPERTY RIGHTS ALONG LAKE KARIBA THROUGH THE FORMATION OF A FISHERMEN'S ASSOCIATION

R. Gwazani
Sebakwe Recreational Park
P. Bag 8040, Kwe Kwe, Zimbabwe
<alcom@harare.iafrica.com>

1. BACKGROUND ON LAKE KARIBA

Lake Kariba is a man-made lake on the Zambezi river between the northern border of Zimbabwe and the southern border of Zambia. The lake is 5364km² in area and the third largest man-made lake in Africa after Lake Volta and Lake Nasser. The dam was constructed in the early 1960s for the generation of electricity for both Zambia and Zimbabwe. Due to the large nature of the project, construction of the dam flooded large areas of the catchment area. This lead to the displacement of all the families which were originally settled along the Zambezi river. Prior to construction, the families could settle on either side of the river and freely cross between Zimbabwe and Zambia. This free movement is now restricted because the lake is as wide as 40km in some areas and the two countries are now under different Governments.

The displaced families were promised fishing rights as compensation to land and property lost in the process of movement. Based on this promise the Chiefs and their offspring settled along the lakeshore in fishing villages named after them. The families are growing and their need for financial resources is increasing, thus forcing them to look for alternative sources of income. Some men and women are employed in the different Government and private sectors of the country while others are employed in agricultural production. Since the land along Lake Kariba is non-arable and the rainfall marginal, the people who are farmers always move further inland in search for more arable land. To them fishing is a seasonal activity for the dry season, meanwhile farming is the major activity during the rainy season. In addition, the need for schools and clinics and such social infrastructure also justifies the securing of second homes away from the fishing villages.

Lake Kariba is so large that it is now an attraction for many, providing, employment, income, recreation and residence. The introduction of a pelagic sardine *Limnothrissa miodon* (Kapenta) has provided the basis for a multimillion dollar fishery. The classes of people making a living from Lake Kariba vary because some of the activities require high capital and equipment investments while others do not. For example, the kapenta industry requires high equipment investment. On the contrary, gill-net fishing is conducted with simple dugout wooden canoes and with one or two hand-made nets. The minimal capital investment needed to enter the gill-net fishing is partly the reason for the high influx of people into Lake Kariba, generally from low income groups.

The incoming fishermen have their own negative impacts on the traditional system of operation they find within an area because they tend to ignore and disrespect the traditional leaders' exercise of resource management. Attempts by Central Government to keep the traditional groupings intact through the issuing of annual fishing permits has proven difficult to implement due to lack of manpower and equipment. Fishing permits are issued jointly by the Department of National Parks and Wildlife Management (DNPWM) and the Rural District Council (RDC). The DNPWLM is the responsible authority for all terrestrial and aquatic resources in the country, meanwhile the RDC is the lowest arm of Central Government and operates in rural areas. In this case Lake Kariba is made up of two RDCs: Binga and Nyami Nyami. Several studies show that the numbers of people entering the fishery keep increasing, for instance, the latest study by Songore *et al.* (1998) indicates that:

i. there is overcrowding in some fishing villages, for example, there is a total of 1404 fishermen on Lake Kariba, twice the recommended number of 771
ii. fish catch per unit effort has reduced and
iii. fish size has generally become smaller and thus smaller net mesh sizes are now being used.

The fact that the management problems have been increasing forced the Government to consider creating a sense of ownership within fishermen through creation of some property rights over the fishing grounds (Kwaramba and Nzunga 1994). The DNPWM decided to extend to fisheries the communal area management strategies that are currently being applied to terrestrial animals.

2. ADOPTION OF THE COMMUNAL AREA MANAGEMENT PROGRAMME FOR INDIGENOUS FISHERIES RESOURCES

The Communal Area Management Programme for Indigenous Resources (CAMPFIRE) is a management programme that the DNPWLM has adopted for over 10 years. CAMPFIRE as a management programme involves the various stakeholders, *i.e.* the communities, the DNPWM and the RDC. The communities are required to be aware of the wildlife resources in their area and the value attached to each species by age and sex. The RDC then arranges for training of its communities on game counting, quota setting and resource monitoring. The DNPWM grants Appropriate Authority to the respective RDC to manage its wildlife, through implementing problem animal control, anti-poaching and sustainable utilization. The RDC is allocated a quota in relation to the quota suggested by the communities. The RDC puts to tender the hunting quota and negotiates a lease agreement with the selected professional hunter or private operator. The operator then pays fees for the concession and trophies to the RDC. The revenue collected from the operator is later distributed as benefits to the communities directly suffering the cost of living with wildlife. The CAMPFIRE principle assumes that local communities living with wildlife are the best custodians of that resource as long as they receive benefits accruing from its proper management.

Adoption of the CAMPFIRE principles in fisheries varies slightly in that the fishing permits are issued directly to the fishermen. The local communities are the operators. Such operation by the custodians ensures that they are participating in resource utilization and, through proper management structures and awareness programmes, overexploitation is reduced. The second difference is that all the revenue obtained from fishing goes to the fisherman directly and he has the full responsibility of deciding on how to use it. The third difference is the structure that CAMPFIRE operates under *i.e.* the ward (comprising 6 villages) and village (comprising 100 households) structures of the RDC. In the case of Lake Kariba, fishing villages have been taken as the units of operation. A body that links all wards and encompasses all fishermen is in the process of being formed which will operate as an association and provide fishermen with a forum to communicate with the various relevant authorities and among themselves. In comparison, the Gill-Net Fisherman's Association will be similar to the Kapenta Producers Association currently functioning on Lake Kariba. It is the anticipation of the DNPWM, the RDC and the fishermen that the GNF Association will promote:

i. A team working spirit and provide a sense of belonging
ii. A sense of security in terms of renewal of fishing permits
iii. Infrastructure development of facilities which require joint use and management
iv. A sense of ownership in all fishermen and hence create an environment that facilitates local monitoring of illegal activities and
v. A spirit of joint resource management through putting value to fish recording and data collection.

The objectives of the Association have been set up together with the fishermen. Although this may appear like a top-down approach, it is not because the process has always been with the full participation of fishermen. Effort are being made to avoid the top-down approach since it has proved to be ineffective (Nielsen and Vedsmand 1999). The establishment of functional institutional arrangements and management capacities of fishermen is a process that requires the initial involvement of Government which will later decrease as the fishing communities begin to show increased decision making powers.

3. THE PRODUCERS ASSOCIATION

3.1 Objectives

The objectives of the Association are:

i. To become aware of and exercise fishing rights and rights over land
ii. To increase period of fishing lease and participate in the setting up of lease conditions
iii. To have control on access to fishing grounds
iv. To have powers of inclusion and exclusion on members
v. To set up a constitution and by-laws
vi. To participate in RDC planning meetings and represent fishermen at different forums
vii. To hire services of a lawyer
viii. To establish infrastructure for fish production, processing and marketing
ix. To operate competitively and
x. To have collateral security for borrowing.

3.2 Structure

When established the name of the association will be Lake Kariba Gill-Net Fisherman's Association. So far, the structure formed is at district level and the National level is still to be set up. The structure in place arises from grouping of existing fishing villages into sub-areas. A sub-area is made up of four to five fishing villages as per fisherman's recommendations. In Nyami Nyami RDC there is a total of 16 fishing villages and Binga RDC 20 villages. There are 4 sub-areas in each RDC. The sub-areas are alphabetically named as, A, B, C and D in Nyami Nyami and E, F, G and H in Binga. This structure can be summarized as follows:

	Nyami Nyami RDC	Binga RDC
Village level:	16 villages	20 villages
Sub-area level:	A, B, C and D	E, F, G and H
District level:	All 4 sub-areas	All 4 sub-areas
National level:	Single representation for the 2 RDCs	

3.3 Progress with the formation of the Association

The process involves extensive consultation with fishermen and participatory methods are used before any decisions are reached. The achievements are as follows:

i. Eight management committees have been established in the 8 sub-areas. The management committees are responsible for organizing meetings, forming a link between fishermen and the authorities and ensuring that discussed issues are effected.
ii. Two management committees have been formed for each district to represent sub-areas at district level.
iii. A constitution has been drafted in each district.
iv. Training on management techniques and resource monitoring has been conducted in both districts.
v. Each sub-area has started fund raising.
vi. Fishermen drafted a brochure on what they preferred for the rules and regulations of fishing and compared

these with the rules that were initially set by the DNPWLM.

vii. Fishermen participated in the demarcation of fishing grounds for the lake shore master plan and

viii. The RDCs have agreed to issue all fishing permits through the association and to recognize the association as the body representing all fishermen.

The role of the Association at sub-area and district levels has been discussed and outlined below in terms of fish production, processing, marketing, access rights and resource management. Sub-areas function at village level ensuring that fishermen have easy access to fishing gear, operate at optimal levels, and meet the market demand for fish. The district representation ensures that there are cold rooms and transportation facilities to market fish on a regular basis. There are many responsibilities of the sub-area representation:

i. Production – to sell fishing gear to fishermen and monitor quality of fish caught
ii. Processing – to put up concrete tables and running water for fish processing in fishing villages
iii. Marketing – to buy fish from villages and then sell to district representatives
iv. Access rights – to recommend names of fishermen to the council for renewal of their permits
v. Resource management – to keep the proper catch and effort records for research purposes and
vi. Fund raising – to receive subscriptions and all fees at village level.

The responsibilities of the district representative are:

i. Production – to secure fishing equipment in RDCs for resale in sub-areas
ii. Processing – to put up freezing facilities in sub-areas and in the RDCs
iii. Marketing – to buy fish from sub-areas and then retail them to hotels
iv. Access rights – to spearhead discussions on leases
v. Resource management – to organize training in resource management and
vi. Fund raising – to seek funding through proposal writing.

4. CONCLUSION

The process of forming the Association has proved to be long and dynamic requiring genuine commitment on the part of the facilitating agencies. The fishing communities are large and their level of participation and acceptance of the new management regime varies. This variation demands constant follow-up on all activities to ensure uniformity in understanding. There is also variation in the level of confidence among fishermen, which needs to be addressed through regular encouragement. Fishermen need to be motivated, receive incentives and have the capacity to meet goals and achieve objectives. Relevant and timely training will create the proper environment for a capable team of fishermen.

Before fishermen can be actively involved in assuming the responsibility of managing the resource, the trust lost during the past top-down approach must be restored. The relationship between fishermen and the authorities has always been one of conflict. Fishermen have been on the receiving end of the authoritarian systemand never part of the planning system.

The initial stage of empowering communities requires administrative, technical and financial support from the relevant authorities. Fishermen have always operated as individuals and it will take time before they appreciate the advantages of operating under a single body with set objectives. However, in this case, a joint approach is a prerequisite before any skills can be imparted. Financially, fishermen are not yet in a position to establish the basic infrastructure required for the fish business. They need technical and financial assistance to improve their fishing gear, fish processing facilities and marketing skills. Borrowing from financial institutions will be possible once fishermen can use their joint assets as security.

For fishermen to assume full responsibility for the resource an enabling legislation has to be created. Fishermen should be able to make decisions that can influence policy. Since conditions of the lease agreement are still to be drafted there is an opportunity which should be used to ensure that fisherman's rights over the demarcated fishing zones are outlined and clarified to all parties concerned. The role of Government should be advisory and not instructive. The problem of overcrowding will be solved by fishermen as they gain an increased sense of ownership and as they exercise their power of exclusion to protect and enjoy the resource.

5. LITERATURE CITED

Kwaramba, R. and S. Nzunga 1994. Can the gill-net fisherman's association solve the long dreaded problems on the Lake? CAMPFIRE Fisheries Report N° 1 pp. 1-19.

Nielsen, J.R. and T. Vedsmand 1999. User participation and institutional change in fisheries management: a viable alternative to the failures of top-down driven control. Ocean and coastal management 42 pp. 19-37.

Songore, N., M. Mugwagwa and A. Moyo 1998. Results of the 1998 frame survey Lake Kariba-Zimbabwe shore. Project Report No. 92 pp. 1-38.

BRINGING THE STATE BACK IN: THE CHOICE OF REGULATORY SYSTEM IN SOUTH AFRICA'S NEW FISHERIES POLICY

B. Hersoug and P. Holm
The Norwegian College of Fishery Science, University of Tromsø
<Bjoernh@nfh.uit.no>, <Petterh@nfh.uit.no>

1. INTRODUCTION

Over the last 10-15 years there has been an increasing tendency to characterise modern fisheries management as a failure. According to FAO's leading spokesman on the issue: "The description of the state of the resources and fisheries indicate clearly that, in the developing as well as in developed world, fisheries governance is "sick" (Garcia 1998). The symptoms of "sick governance" is according to Garcia:

i. Lack of political will for difficult adjustments
ii. Persistence of direct and indirect subsidies
iii. Lack of control on fleets by flag states
iv. Ineffective fishery commissions (no power)
v. Lack of control of access and unclear use-rights
vi. Top-down Command & Control management
vii. Disregard for traditional communities
viii. Power of industry lobbies resisting change
ix. Lack of implementation capacity.

As a result of the sickness between 10 and 40% of all the resources available in each FAO fishery region are exploited beyond the point of maximum sustainable yield (MSY) and must be considered overfished and even depleted. On average 75% of all resources are exploited at or beyond MSY level. Even more damning are the characteristics of, *e.g.* the EU's Common Fisheries Policy (CFP). According to Holden, for many years one of the chief administrators of the CFP: "*TACs have been fixed primarily on the basis of socio-economic criteria. ... it has been an almost total practical failure. The possibilities to reduce significantly catches of small fish and the high rates of fishing have been squandered. There are still as many small fish caught and the rates of fishing are still as high now as they were in 1983. .... For political reasons the Community decided not to create an effective system of control and enforcement, which it did successfully, thus contributing to the practical failure of the policy.*" (Holden 1994: 167)

The same story can be told for Canada, the United States, Norway and a number of other important fishing nations. Considering that fisheries management by central government was instituted due to market failure in the first place, government failure should be even more serious. Where economists generally point to inefficient and costly management measures, unable to produce resource-rents, social scientists are usually more concerned with the autocratic nature of management, not taking the fishermens' own input into account and the consequences in terms of illegitimate management structures and measures. The state is being criticized by both groups, sometimes without reservation (Christy 1996, Hannesson 1996, Berkes 1989), at other times with more academic restraint, although the message may still be the same. We have also ourselves contributed to this "wave", in search for better and more just management systems (Hersoug and Raanes 1997). Today, however, it is time for sobering up.

There is no doubt that management by the state is expensive and inefficient in many countries, developed as well as developing. It is, furthermore, beyond doubt that improvements can be made by introducing a clearer specification of rights as in the case of ITQs and/or by introducing various forms of co-management or user group participation. Although the general applicability of these reforms is still under discussion, they are about to be implemented in a large number of countries. Nevertheless, there are situations where only the state can do the job, where market solutions hardly work and where community-based schemes are inappropriate. We refer to the establishment of new fisheries policies where *redistribution* of rights and quotas (favouring previously disadvantaged groups) figures prominently. Although such cases may seem rare, they may become more common in the future, following democratisation of previously autocratic states. In this article we shall use South Africa's new fisheries policy as a case and starting point. Why did South Africa choose to go for a state dominated system in the face of stern opposition from the existing industry and in spite of the general climate in favour of co-management and user-group participation propagated by the large and influential South African sector of non-governmental organisations (NGOs)? In more theoretical terms, the challenge is to delineate the capacity of different management systems to undertake certain key functions, of which redistribution is the central focus here.

In this account of the South African experience we start with a brief description of the three alternative systems and their capabilities for redistribution, followed by a short overview of the South African fishing industry, including the management structure. The section 4 gives a brief account of the policy process leading up to the new fisheries policy and the *Marine Living Resources Act (1998)*. The section 4 deals with the possible options and the one chosen, while the section 6 discusses the refuted market alternative and the neglected community version. The section 7 deals with the extent of actual redistribution, with examples from the economically important hake sector and the equally important (in terms of employment) rock lobster sector. The seventh and last section deals with the practical and theoretical implications of our findings, with particular focus on necessary, but not sufficient, conditions for social development.

2. GOVERNING REDISTRIBUTION

A reform of fisheries policy, like the one presently attempted in South Africa, may be seen as a shift among three broad institutional orders: community, market and state (Streeck and Schmitter 1988, Apostle *et al.* 1998). First, the governance principles suggested by the community metaphor are characterised by close inter-personal ties, egalitarian, and often multiplex, social networks and shared identities. The state, which includes the institutions and structures of policy, law, and governance, is broadly characterised by hierarchical order, bureaucratic structures and authority relations, and professional, uni-dimensional relationships. The third metaphor, the market, suggests competition, economic efficiency and rationality. In contrast to those in the community, relationships in the market are impersonal, task-oriented and without inherent value. In contrast to the state, the market is characterised by decentralised exchange rather than central command and formal authority.

In the fisheries, the three contrasting institutional orders are reflected in each of the three dominant models of fisheries resource management. The state model corresponds to the centralised and bureaucratic form of management that presently forms the basis for fisheries management in most developed fisheries nations. While dominant, however, it is often claimed to be in a crisis (McGoodwin 1990, Crean and Symes 1996). Hence, the models drawing on the two other institutional orders are suggested as solutions. On the one hand, the ITQ model, now implemented in several countries, seeks to redress the inefficiencies of the state-centred model by using the mechanisms of the market for quota allocations. In contrast, the models of local management and co-management seek to (re)embed fisheries management in local community structures, or user-groups, to increase the legitimacy of resource regulations (Jentoft 1989, Dyer and McGoodwin 1994).

These three models have different features, and suggest very different solutions to the problems of the fisheries. The state model emphasises control and fits perfectly situations in which the main problem of the fisheries is over-exploitation and the control of fishing effort has first priority. The market model emphasises efficiency. It fits situations in which the main problems are economic problems due to over-capacity, under-development and inefficient allocation of resources are the main problems. The community model emphasises equitable distribution and legitimacy in the eyes of the fisheries population. It fits situations in which the main problems are lack of equitable and fair access to resources by users located in traditional fishing communities.

In the case of South African fisheries policy, we are interested in the consequences that the choice of governance principles has with regard to redistribution. Each of the three models has a distinct capacity for redistribution, albeit in somewhat different directions. In a market, redistribution will happen in accordance with the principles of efficiency. Those actors, able to exploit the most efficient strategies, will gain control over the resources. The scope and scale of the redistribution that will ensue is difficult to predict. What is important is that if the efficient structure is to be realised, attempts to influence the speed and direction of redistribution should be kept to a minimum. In South Africa, in which access to capital and information is highly skewed in favour of the rich white population, it seems unlikely that the market principle will work in favour of the previously disadvantaged groups.

According to the community principle, redistribution should take place in favour of the small, local and traditional participants. In the absence of formal, hierarchical and bureaucratic controls, it will be difficult for large-scale, rational enterprises to establish the predictability and formal guarantees of access to the resource that they thrive on. In the South African situation, relying on the community model could undermine the big companies in favour of the coastal population, at least in some sub-sectors. It also comes with the cost of not prescribing efficient mechanisms for controlling over-capacity and sustainable resource practices. The governance structures of the state emphasise control, including control over distribution and redistribution. How this potential for redistribution can be used, however, depends on the interests of these who gain control of the state apparatus. In the ideal Weberian[1] situation, the state bureaucracy remains in the hands of the political elite, which in a democracy like South Africa, is controlled by the people through elections. Thus, to the extent the government is committed to redistribution, that is what will be achieved. In modern, complex states, however, the elected politicians do not remain in complete control. Important policy decisions are delegated to the government and government bureaucracy, and may be influenced by lobbying or corporatist negotiation. To the extent the state is captured by organised interests, the capacity for redistribution will decrease.

The type of, and capacity for, redistribution is hence a question of choice of governance structures. In this choice, the goal of redistribution will have to be balanced against other considerations, for instance that of economic efficiency and biological sustainability, without which there will be less to redistribute. Further it is not a choice that can be made from scratch. The process of creating a new fisheries policy happens in a setting in which established user-groups usually know how to make

---

[1] Max Weber (1924), German social scientist, who developed among other things, the ideal type of bureaucracy.

themselves heard. We start, then, by giving a brief sketch of the new fisheries policy process.

3. SOUTH AFRICAN FISHERIES

Even if South Africa is the top fishing nation in Africa, it is usually ranked about 30th on the list of the world's fishing nations, with a total catch of approximately 500 000t/yr. The fishery accounts for only 0,4% of GDP, but its regional and local importance is considerably larger than indicated by the national figures. Some 27 000 workers are employed in the formal sector, while an unknown number of people depend on fishing as a way of subsistence. In addition some 750 000 people are involved in recreational fisheries. Most of the industry is concentrated in Western Cape, in particular to Cape Town. The total (processed) value of the production is 2.1 billion Rands, of which 1 million Rands are generated by export. The industry consists of 19 different fisheries, the most important of which are described here (Table 1).

There are two important rock lobster fisheries in South Africa – one each on the West and South Coasts. The West Coast fishery operates inshore and employs some 3500 people, mainly on a seasonal basis. The fishery is heavily oversubscribed, not only by registered fishermen, but by poachers and recreational fishermen, resulting in reduced quotas every year. The South Coast fishery is in better shape, being concentrated to only twelve companies, who use large vessels harvesting in deeper waters.

The line fisheries target tuna, squid and several demersal species. The fishery is characterised by over-capacity and declining catches, which is partly why this group would like to enter the hake sector. Finally there is a small, but heavily disputed, abalone fishery, formerly concentrated in a few processors and their connected divers. Today the divers have their own quotas and the number has been increased considerably over the last years. The fishery is, nevertheless, completely

Table 1
Nominal catch and estimated value of South African commercial sea fisheries 1997
(Stuttaford 1999)

Industry sector	Volume (t)	Value (R'000)
Offshore trawl	182 321	989 744
Inshore trawl	15 150	68 736
Longlining	4 753	46 373
Pelagic	272 111	439 224
Rock Lobster	2 570	167 021
Squid	3 811	91 464
Line/small nets	17 221	128 239
Abalone	537	79 433
Oysters	708	10 142
Mussel farming	2 145	27 885
Prawns	514	17 044
Seaweed	991	4 971
Grand total	502 832	2070 283

The single most important fishery in South Africa is the demersal (offshore trawl) fishery for hake, dominated by a few vertically integrated companies, employing some 9000 people. Annual catch is around 150 000t, mainly caught by deep-sea vessels. There is a growing interest and demand for greater participation by longliners, which are more easily accessible for new entrants.

The second major fishery in economic terms is the pelagic fishery based on anchovy, pilchard and herring. The fishery employs nearly 5000 people along the West Coast, mainly on a seasonal basis, producing fish for canning as well as the meal and oil industry. Catches are erratic, causing boom and bust cycles that have important economic and social repercussions for the fishing communities.

oversubscribed, partly due to heavy recreational catches in addition to a thriving illegal fishery, stimulated by the high prices that are perceived in the Far East.

Most available fisheries resources are fully exploited and the few potential "new resources" (such as for example orange roughy) need more research before their importance can be accurately assessed. Mariculture, and especially shellfish farming (3000t in 1997) is one alternative, but the number of protected locations are few and the sector will not be able to provide large numbers of jobs. The same applies to increased use of bycatch. There are, consequently, few openings for new entrants, be they fishermen formerly deprived of existing rights or black entrepreneurs who would like to participate in the industrial and semi-industrial fisheries. If new entrants are to be allowed into the established fisheries, old participants must quit. The dilemma is precisely formulated by Cochrane (1995): "The RDP aims of meeting basic needs and building the economy cannot be met by increasing exploitation pressure on these resources and improvement must come from better and broader utilisation". The other alternative is *redistribution* – to the extent this is politically possible. Being a relatively marginal sector of the South African economy the fisheries administration is placed under the Ministry of Environmental Affairs and Tourism as a separate department. All the practical work is performed by Marine and Coastal Management (MCM) (formerly Sea Fisheries), acting as a chief directorate, located in Cape Town. Up to 1998 the Directorate consisted of three units; Administration, Control, monitoring and surveillance and Sea Fisheries Research Institute (SFRI). By the new administrative reform the new chief directorate now consists of four sub-directorates, (a) coastal and inshore resources management, (b) offshore resources management, (c) economics and resource development and (d) support services, with researchers and MCS personnel connected to each of the two management units. Considering that lack of legitimacy and trust were the main problems of the old administration, the change of name and some administrative reshuffling was no doubt a smart move. Whether dividing up of the research institute and the separation of the control responsibilities are similar smart moves remains to be seen.

While the Minister is the ultimate policy maker and is responsible for fixing the TACs and distributing the quotas, he has a Consultative Advisory Forum (CAF) to advise on such matters, including on the use and allocations of the Marine Living Resources Fund. The CAF has 17 members drawn from all sectors of the fishing industry and performs by and large the same duties as the previous Sea Fisheries Advisory Committee (SFAC). The Fisheries Transformation Council (FTC) was also established by the new Act and has six members appointed by the Minister to assist in development and capacity-building of the historically disadvantaged and small and medium-sized enterprises by leasing them fishing rights. To some extent the FTC has taken over some functions of the old Quota Board with the important difference that the FTC is politically responsible to the Minister and ultimately to the Parliament; further, the FTC is considered to be temporary and is to be dissolved when the transformation process is officially finished. At the moment there is considerable confusion over the two procedures for new entrants; they can apply directly to the Minister for new quotas or they can lease quotas from the FTC.[2] While most of the management costs are covered by ordinary state budget allocations, research, monitoring and surveillance is partly covered by the Marine Living Resources Fund, which depends on payments made for permits, licences, levies, fines, etc. Fisheries management is at present in a dramatic financial squeeze because no direct resource fees are paid by users and demands in all sectors by far exceed the funds available.

4. WORKING OUT A NEW FISHERIES POLICY

The process of establishing a new fisheries policy was initiated by the Minister of Environmental Affairs and Tourism at a public launch on 27 October 1994. The immediate background was the unrest among fishermen and fish workers over the then policy, claimed to be corrupt and insensitive to the difficult situation of most coastal communities. By that time some ANC-aligned groups in the fishing industry had already worked out a preliminary programme as part of the electoral manifesto. In December 1994 a new meeting was held in Cape Town to discuss how a new fisheries policy could be developed. It was agreed to set up a Fisheries Policy Development Committee (FPDC), mandated to prepare a Green Paper on the fisheries policy. The Committee consisted of 5 representatives from each of the 13 different sectors of the fishing industry. In addition, one representative was appointed by each of the maritime provinces. With the representative from the Ministry the committee counted 70 members!

The first meeting in the plenary Committee soon found out that a Working Committee was needed to execute the task. This Working Committee originally consisted of 18 members drawn from the participating groups in the Plenary Committee. The Working Committee was headed by Mr. Mandla Gxanyana, General Secretary of the Food and Allied Workers Union, assisted by a small permanent secretariat, originally staffed by five assistants with special qualifications. The Working Committee soon encountered large problems, not only on policy matters, but also on the question of representation. Organised labour claimed to be under-represented, demanding five representatives in the Working Committee (and 20 in the Plenary). After a staged walkout, backed by "big business", organised labour got their demand accepted, in order to get the process going again. By the same time it was agreed that all other sectors should be entitled to ten representatives each, increasing the plenary to 150 members.

FPDC requested all stakeholders to submit their ideas for a first integrated document. This document was discussed at subsequent meetings to identify areas of agreement. On issues where the FPDC was not able to find a common solution, technical teams were set up to provide possible solutions. Six technical or task teams were appointed of which the Technical Team on Access Rights played the most prominent role. Based on its unilateral recommendation of an individual transferable quota system (ITQ) for most South African fisheries, two workshops were held in order to find some common ground. In addition all meetings were opened to interested

---

[2] Due to legal uncertainties the Minister has stopped the practise of allocating FTC quotas for further leasing. Instead FTC members are used to discuss and recommend applications for ordinary fishing rights. By March 2000 the members of FTC have resigned, due to lack of payment and more generally, lack of direction.

parties. Drafts were widely circulated for comments and the participants consulted extensively with their constituencies.

By May 1996 the Working Committee of the FPDC had finalised a draft which was endorsed in principle by the Plenary Committee, although with strong reservations from some sectors during a two-day meeting in Cape Town. By 4 June the final document was delivered to the Minister, who promised to proceed immediately with the drafting of a White Paper. Meanwhile, the transitional government ended and the ANC-controlled government took over. After some hesitation the process of drafting the White Paper started, with a Norwegian consultant hired as an "unbiased expert". The issue of access-rights and transferability continued to be a contentious issue. The FPDC had proposed transferable-rights, granted for perpetuity, but without being very specific about how the nature of access rights and how the ITQ-system was supposed to bring redistribution in favour of disadvantaged groups. To advance the process the Minister nominated a special panel to review the access-right options. With all sorts of allegations floating concerning biased participants, the Minister chose to nominate four "outsiders", *i.e.* persons with no vested interests in the fishing industry, to the Access Rights Panel. Two lawyers, a sociologist and an economist made up the panel. After four months they delivered their report to the Cabinet with clear recommendations as to the nature of the access-rights. They were to be real, long-term, transferable and inheritable property rights. According to the panel, "the stakeholders should be encouraged to behave as farmers and harvesters and not as predators" (Sea Fisheries 1996a). The panel had little understanding for the FPDC principle (introduced mainly by "big business") that there should be no sudden removal of rights and quotas. On the contrary, it was considered in the interest of all parties that the changes to be made were implemented speedily.

With only slight editing, the recommendations of the Access Rights Panel were written into the White Paper, which was presented to Parliament by May 1997 (RSA 1997a). The White Paper was distributed widely to contribute further to the policy debate over the proposed changes. Contrary to the normal procedure, however, the writing of a Bill on the new fisheries policy did not await the responses to the White Paper. Because of the pressure to produce speedy results, the "Marine Living Resources Bill" (RSA 1997b) was prepared in a parallel process with the White Paper. As part of its internal deliberations the Department of Sea Fisheries had appointed a legal task team consisting of local as well as foreign experts to review the existing 1988 Law. The team soon found out that it would be impossible to implement the proposed policy through the existing law and advised the writing of a new bill, a work which was started immediately, with the assistance from the same team. During this process the team writing the White Paper provided the legal task team with steady inputs, "trying to keep one chapter ahead" as one of the participants described the internal process.

By the end of September 1997 the Bill was introduced, adhering strictly to most, if not all of the major recommendations in the White Paper. By then a new political process started, with the initiative squarely placed on the Portfolio Committee of Environmental Affairs and Tourism. The Committee was not happy with all the recommendations of the White Paper and the Bill, in particular with the idea of real, long-term and transferable property rights. After a lengthy process of hearings, with written as well as oral submissions from most stakeholders, the Portfolio Committee was able to reach a compromise on all contentious issues, including that of access-rights (RSA 1998). With this truly remarkable compromise the Bill was able to be tabled by the House of Representatives, ans waited for the final approval in the National Council of Provinces. The Bill was passed with only technical amendments and the Act became effective by June 1998.

In principle, the 1998/99 allocations could then be implemented according to the *Marine Living Resources Act.* In the meantime a large number of new potential entrants had filed applications. In the West Coast rock lobster fishery the allocation of 62 new quotas was challenged on legal and technical grounds. In the Supreme Court the old quota-holders won the case, suspending the utilisation of the new quotas, thereby creating a precedence for most other disgruntled established quota-holders. Even if the administration was able to work out a set of compromises regarding the 1998/99 season, with a certain percentage of the TAC to be set aside for the new entrants, there are signs of a growing administrative chaos. Thousands of quota applications are pouring in, to be handled by a minimal administration with little judicial expertise and even less capacity to control the extensive information required from the new applicants. At the time of writing the MCM administration has been accused of gross financial mismanagement, including corruption, leading to the replacement of the director and the suspension of several other key officials (Cape Argus, 8.12.99)[3]. Nevertheless, allocations according to the new law are now being made and from 1999 onwards one should be able to see the first results of the new fisheries policy.

5. MANY OPTIONS AND HARD CHOICES

The choice of possible management regimes is difficult. The regime is determined by what is regulated

---

[3] As of February 2000 it was clear that the report from the internal audit is very critical as to the procedures followed by MCM regarding tendering and spending from the Marine Living Resources Fund, but none of the accused have been found to have enriched themselves personally. All suspended leaders have been reinstalled in their former positions, but MCM has been reinforced by a special programme manager to oversee that bureaucratic procedures are followed.

(effort or catches), how rights are distributed (giving, selling or leasing), to whom (persons, vessels, communities or firms) and their status regarding transferability and duration. According to Matthiasson (1992) the possible options, which are shown in Table 2, yield some 240 possible regimes. If the impossible combinations (like delegating rights to persons, firms, *etc.* when the resources have already been given to a public enterprise) are excluded, there are still 204 possible outcomes. Some of them were constantly discussed during the policy process.

While there was agreement on regulating the catch in most fisheries (though effort in the case of the squid fishery), disagreement was frequent regarding the distribution principles, where actually all alternatives at some time were discussed. In the White Paper the selling receivers of rights and quotas. The next section sets out to explain why these two options, for simplicity called the market approach, and the community approach were ruled out.

6. WHY NOT A MARKET-SOLUTION OR COMMUNITY-MANAGEMENT?

When the *status quo* was definitely out of the question, there was a gradual movement towards more market-like solutions among the existing rights holders in the fishing industry with the important reservation that most participants were extremely reluctant to pay for their rights. After the special Access Rights Panel had delivered its recommendations the White Paper leaned even more heavily towards a market solution. Nevertheless, this solution lost out in the last instance. Rights were not granted in perpetuity and they were not freely

Table 2
Possible contract regimes

Factor	Nature of allocation	Owning entity	Transferability	Duration
	Public enterprise			
Effort	Handing out rights	**Persons** **Vessels**	**Transferable rights**	**Limited time**
Catch	**Selling rights**	Communities Firms	Non-transferable rights	Unlimited time
	Renting rights			Undefined time

of rights figured prominently (together with a public enterprise to take care of the new entrants), while the Parliament in the end chose to rent out fishing rights (not quotas!) without specifying how much and when the rights-holders should pay. Under the present regime rights can be distributed to persons as well as firms, while the community option was never seriously considered. According to the FPDC document and the White Paper, rights should be freely transferable, while the politicians chose more limited transferability, still to be dependent on ministerial approval. The main battle was over their durability. In the FPDC document the length of the rights-period was just estimated to be "long term", while the White Paper prefered to see rights granted for perpetuity. Even within the Portfolio Committee the alternatives varied between 50 years (proposed by the National Party) and 10 years claimed by ANC. The compromise was 15 years, with even shorter duration for new entrants. In summary, the existing South African system is based on catch controls and the renting of rights (leasing) to persons and firms. The rights are only partially transferable while their duration is limited. Measured against what Matthiasson (1992) considered to be the ideal regime, that is the regime best suited to achieve *efficiency, equity and reversibility* (marked in bold in Table 2) the new South African regime does not fare badly. The contentious issue is the apparent lack of transferability, which is the trade mark of a market solution. At the same time communities are ruled out as transferable. How was it big business stumbled in the run-up?

First, there was in the ANC caucus a general fear of "business as usual". The apartheid legacy, in terms of skewed ownership, lack of participation, *etc.* was so bleak it was felt that something had to be done (Hersoug 1998). The political activists familiar with the fishing industry realised that the possibilities of reallocation would be gone as soon as the initial allocation had been done. With rights granted in perpetuity, new entrants could be bought out relatively quickly and the industry would revert back to square one; an oligopolistic industry mainly dominated by white-owned companies. In addition it was totally unrealistic to expect people who had been discriminated against through all their lives to turn into professional businessmen within five years, which was the stipulated period of transformation. Thus, the playing field was extremely uneven, or as one ANC-activist formulated the problem: "Why should we get rid of the most important instrument for restructuring, having fought for 80 years to acquire state power. Our experience with apartheid was precisely the indiscriminate use of the state to reallocate resources" (pers. comm.).

Secondly, there was not only in the ANC, but in business circles as well, great scepticism regarding the possibility of windfall gains for the new entrants. With a short transformation period, a limited reallocation and then "business as usual", it would be possible to sell the newly acquired quotas for substantial sums of money. "Paper quota holders" were considered a threat to the industry by nearly all participants in the Fisheries Policy Development Committee and they went to great lengths in order to limit the transferability of the rights pertaining to the new entrants (Hersoug 1996).

Finally, there was an acute lack of institutional credit among the small business and potential new entrants, which could impede the selling of rights to previously disadvantaged groups. The South African state has more pressing needs in housing, education and health and it would not be possible to obtain cheap institutional credit of any magnitude. So, existing businesses would be the nearest source of credit, which normally would mean a take-over within a short time, if not formally, then in practical terms. All in all, "business as usual" was not a very tempting option for the party in power, now finally having the possibility of rectifying some of the unjustices wrought by apartheid policy.

To argue why South African authorities did not choose to utilise the community option may for some look like breaking down open doors – it was simply not considered an option. Existing owners of rights and quotas had no intention of handing over even a small part of their assets in "some kind of roulette", as one of our informants claimed in 1996. And they were wholeheartedly supported by the biological establishment in Sea Fisheries (now MCM), which preferred centralised solutions, and by organised labour who strongly resisted any kind of grand-scale social experiments in which their members might lose their jobs. Nevertheless, it is still of interest why this option was excluded right from the beginning, especially in a country where donors as well as NGOs have for years argued in favour of community management and community participation. Seen from the outside there seems to be three factors working against community management.

First, the coastal communities of South Africa and especially of Western Cape, are close knit fishing communities of the type found in a number of other countries. Due to historical circumstances, in particular the forced removals of coloureds and blacks starting back in the 1930s, but more systematically from 1961 onwards, several "communities" are simply a number of coincidentally assembled people with little or no interconnection, with a majority living from means other than fishing and fish processing. Add to this a systematic discrimination against Africans during the "40 lost years", during which nearly all fishing rights and quotas were allocated to whites and the result was a rather distorted fishing community (O'Meara 1996). Even today, a large number of fishermen fishing from some of the typical Cape Town fishing communities are located in entirely different places, and must travel considerable distances to get to sea.

Second, whatever the justice of claims for participation and allocation of quotas, it is nevertheless a fact that capacity and competence to administer and distribute resources of this magnitude is seriously lacking in most coastal communities. As shown by an official inquiry in 1993, the social situation was rather bleak in most coastal communities (De Wet Schutte 1994). Poverty, lack of housing, alcoholism, unemployment and illiteracy were pertinent features of the coastal communities in the Western Cape, and were even worse in the Eastern Cape. With some notable exceptions the level of organisation was also low and consequently, the ability of taking on complicated administrative tasks was absent. Capacity building and training was never a priority task of the previous National Party government, and even in the new dispensation efforts soon stumbled for lack of money and human resources, in addition to new political cleavages, this time between blacks and coloureds.

Last, but not least, most participants in the original FPDC process had some special experiences with community management or, more precisely, with community quotas, allocated for hake in the period 1993-94 (De Wet Schutte [1994] records the history in detail). The dismal state of the coastal communities became problematic by the early 1990s, necessitating a political initiative. Following an inquiry into the socio-economic conditions of the fishing communities of the West Coast, the Minister requested the Quota Board to consider the allocation of hake quotas to specific areas. The idea was not completely new in South Africa since West Coast rock lobster had already been allocated to certain communities. It seems, however, that the impetus came from Alaska, where community quotas for hake were established in 1992 with considerable income generated for community purposes (Townsend 1997). As could be expected, a large number of communities organised as trusts, and although it took time to get them formally registered, it soon became clear that the original 3000t would be inadequate for creating any improvement in living conditions.

Based on the report the Quota Board proposed the set-up of Fishermen's Community Trusts, to provide support for those members of communities who were dependent on fishing but not for the whole community. The original idea was to provide support to those areas where no support was available from the state or other established sources. Even though the Quota Board tried to provide standardised guidelines and conditions, the actual target group (the beneficiaries) were ill-defined and the mechanisms of support ill-conceived.

In the subsequent year the amount of quota set aside was increased, but to little avail. Financial mismanagement, corruption, cash payments to the alleged needy, all contributed to strife, local cleavages and political turmoil – effects that were exactly opposite of what the Quota Board had intended. By the end of 1994 a special committee was established to look into the working of the trusts and reported shortly afterwards that:

"The possible total abrogation of the Community Quota system should be seriously considered." (De Wet Schutte 1994:43). If the system was to continue, the investigation committee recommended a more coherent management framework, no cash payment, and the establishment of an umbrella (mother) trust for all the existing community trusts. Before political action could be taken, the Cape Supreme Court had ruled that this type of trust was not considered a legitimate receiver of quotas according to the *Sea Fisheries Act of 1988* then in effect. Consequently, most trusts were dissolved, leaving both the disappointed fishermen and administrators experience-wise richer. Unfortunately, it also left a negative legacy, not only for community quota schemes, but for community management schemes in general.

7. REDISTRIBUTION – THE PRELIMINARY RECORD

The new *Marine Living Resources Act* was formally signed into law in May 1998. The 1998/99 allocations were then prepared according to the new law and its accompanying regulations. Not all the old quota holders accepted this situation and the case of West Coast rock lobster was brought to the Supreme Court. Due to legal technicalities (applications for the 1998/99 season had been done according to the old Sea Fisheries Law) the rights of the new entrants were suspended and the shares of the old participants restored. In other fisheries, *e.g.* hake and abalone, a negotiated settlement was reached, based on the fact that the Minister could force a stronger cut the next year, if the old rights holders did not accept the "new deal".

By the end of 1999, after only one year with the new Act, is definitely too early to measure the extent of redistribution. It is clearly stated both in the White Paper and in the final MLR Act that the transformation process will take some time, although no specific time frame is indicated. Nevertheless, an assessment at this time can indicate something about the speed, the magnitude and the allocation criteria that are being used, thereby indicating to what extent "a more fair distribution" is taking place. Redistribution could also be seen in a broader perspective. In 1994 a new Quota Board was elected, with prominent members from the previously disadvantaged groups. Partly on its own initiative, and partly pressured by the new forces now coming to power, the new Quota Board started its own restructuring – a process which was highly disputed among the participants of the Fisheries Policy Development Committee. The Quota Board has made a heavy impact on the restructuring process, bringing in a large number of new entrants. For this reason the analysis of restructuring should start in 1994 and, eventually determine the extent to which the new redistribution (according to the 1998 MLR Act) differs from the more "private" initiative of the Quota Board.

We have on an earlier occasion been pessimistic as to the extent of possible redistribution (Hersoug and Holm 1998). Neither the FPDC document, the White Paper nor the MLR Act lay down any specific targets for redistribution. Only in the Access Committee's report is there an indication of 5-10% of the TACs as a reasonable target, a figure which has never been politically confirmed. In contrast, the labour union (FAWU), which organises most of the processing workers, in its submission to the Select Committee claimed 70%!

Considering the number of new entrants there is no doubt that we have been too pessimistic. In the deep-sea hake sector the number of participants has increased from 31 in 1994 to 105 in 1999, mainly due to the new participants in longlining. In the West Coast rock lobster fishery the number of participants has increased from 93 to 192. In squid, abalone, line fishing and the pelagic sector there is also a considerable increase in the numbers of operators, although the reallocated share of the TACs is more modest. There is a clear tendency in the size of the new allocations in the hake fishery: the quotas become smaller and smaller every year. In 1993 the new entrants in the hake sector received on average a quota of 814t, while in 1998 the average quota was down to 215t. The opposite trend applies to West Coast rock lobster fishery, where new quotas in 1998/99 were on average double the size that were allocated in 1993/94. In 1999/00 they are again back to the 1993/94 level. The overall impression is nevertheless that the magnitude of redistribution has been less impressive than the sheer numbers of new participants being brought into the sector. So far 14% of the hake quantity has been reallocated (including all the new allocations done from 1993 to 99), while in West Coast rock lobster the similar figure is 31%. It is well worth noticing that most of the restructuring shown in the following tables took place under the previous Quota Board, while the present restructuring (after the MLR Act) has just started. Table 3 indicates the number of new quota-holders per year in the deep sea hake fisheries; Table 4 gives the similar figures for the West Coast rock lobster industry.

An important question is whether the reallocation produced a transformation of the industry, in terms of broader participation, greater local employment, and ultimately, better living conditions in the coastal communities. Again, it is too early to judge (and much field work needs to be done to answer the question properly). But a few observations are possible. First, few new entrants have been able to acquire equity on their own, that is; for vessels, processing or marketing facilities.

A large number of the new entrants have become "paper quota holders", meaning that they receive valuable quotas which they immediately sell to established fishing companies, possibly as part of a "joint venture". This is hardly surprising considering the small size of the new quotas, nevertheless, this is contrary to the ideals of the FPDC, the White Paper and the MLR Act. An interesting question will arise at the end of their term of tenure. Will they be refused a new lease because they do not fulfil the

Table 3
Hake quota allocation 1993-2000

Year	Number of participants	New entrants	Exit	Average quota new (t)	New % entrants	Total TAC (t)
1993	33	5	0	814	2.77	147 000
1994	31	0	2	0	0	148 000
1995	31	1	1	372	0.25	148 000
1996	42	14	3	367	3.40	151 000
1997	57	16	1	265	2.76	153 702
1998	57	0	0	0	0	151 000
1999	105	52	4	215	7.66	146 120
2000	41*	1*	?*	750*	0.62*	121 210*

Figures based on Stuttaford (1994-99), compiled by D. Baron.
* Only the deep sea hake trawl has been allocated while hake longline has not yet been finalised.

Table 4
Rock lobster quota allocation 1993-2000

Year	Number of participants	New entrants	Exit	Average quota new (t)	% of TAC new entrants	Total TAC (kg)
1993-94	93	46	2	3 031	6.34	2 200 000
1994-95	99	7	0	3 761	1.32	2 000 000
1995-96	104	8	3	3 538	1.89	1 500 000
1996-97	145	43	2	3 720	9.41	1 700 000
1997-98	173	36	8	5 429	10.18	1 920 000
1998-99	192	22	3	7 537	9.32	1 780 000
1999-00	187	14	19	3 571	3.10	1 613 477

Figures based on Stuttaford (1994-98), compiled by D. Baron.

criteria set by the new regulations, or will they, as happened in Namibia, be given an extension, and hence longer time to acquire capital and eventually invest?

Even though the existing quota-owners in the deep-sea hake sector disgruntlingly accepted a modest cut over the last five years and a new negotiated deal of 8.2% of the TAC for new entrants in 1999, the situation is not as simple as this (Hutton et al. 1999). In reality, a considerable part of the redistributed quotas are fished by the old established companies, with the difference that they now must pay the owners of the quotas. And the established companies are more than willing to buy quota, or to enter into joint ventures with the new entrants.

How is used the money paid for such fishing rights? Stories about new 4 wheel-drive cars, satellite TVs and extravagant lifestyles abound, but only careful research can uncover the local investment patterns. So far it seems that little new employment has been created, because the new quotas are being fished by existing vessels and processed by established companies. According to Isaacs and Normann (forthcoming) these new organisational entrepreneurs act as a "filter" between the fishers and the grass-roots level, filtering information from government down and initiatives from the fishers up. Information and support are hence the most important assets, besides for their own organisational capabilities. Among them there are unscrupulous opportunists, who have amassed a number of names to support their "firm" in applications for quotas, to the "true" community worker have organised co-operatives where income (and investments) are shared between participants. Unfortunately the former appears more frequent than the latter.

8. BRINGING THE STATE BACK IN!

Based on the South African case of establishing a new fisheries policy where redistribution of rights and quotas figure prominently, there is little doubt that the state is essential, *i.e.* redistribution takes place through a political/administrative process. For reasons that have been explained above, neither community nor market can perform the task of reallocation in the same manner. A community-based solution was early ruled out because of the lack of community management traditions. Even though the state could have divided the TACs (or part of the TACs) into community quotas, former experience and the highly diversified pattern of most South African coastal communities worked against such a solution. In addition, weak local competence, weak organisations and the general lack of trust, all contributed to the choice between only the two remaining alternatives; state or market. The market solution figured strongly all through the FPDC process and not least in the White Paper. In the special report from the Access Rights Panel it was consistently argued for a once-off selling process, a short transformation period and then "business as usual". This alternative did not survive the Select Committee's treatment of the proposed Act, as it was evident that a market-solution would have accommodated few new entrants and even fewer on a permanent basis. If a market-solution would have been preferred, it would have been necessary to introduce credit facilities, as well as capacity building and regulations pertaining to further sale of fishing rights for a certain period. If not, established companies would probably have quickly bought up most of the new entrants. This also applied to the auctioning of rights - an alternative which was seriously discussed in business circles at one point in the policy process.

Only the state, through its new political force (the ANC), now formally in charge of the administration, could enforce a true reallocation of rights and quotas. For this reason it was important that the political responsibility was invested in the Minister, after having been placed in an independent organisation, similar to the previous Quota Board. It should be remembered, however, that any reallocation takes place as part of a "negotiated revolution", where former stakeholders still figure strongly in the policy process. Reallocations were also met with resistance within the new situation, not least by the workforce of the established companies. The reallocation process was therefore a careful balancing act, where too much change would create havoc in the existing industry and too little change threaten the legitimacy of the new fisheries policy. At the moment, however, the whole process seems to be threatened by lack of administrative capacity to handle all the applications (more than 5000!). Even more important is the lack of oversight to check if the information provided by the new applicants is correct. Finally the administration lacks some kind of watchdog to check how the new entrants are performing on the ground, that is, to what extent their bright business plans are being implemented. Stories of blatant fraud abound (Independent Newspapers 1997).

Although the negotiated revolution in 1994 was unique, South Africa is not a special case in terms of setting up new fisheries policies. A number of former autocratic countries must reorganise their fishing sectors in the future. Depending on the goals and the resource situation, they will have greater or lesser room to manoeuvre and will depend on a strong and committed state if the goal is to increase participation and obtain a more equitable distribution. *Hence, it is much too early to write off the state as the main player in fisheries management.* However, as the development in South Africa clearly shows; a new fisheries policy is not enough: the policy has to be implemented as well, putting mechanisms in place whereby the previously disadvantaged can work their way into the fishing industry. Finally it is a question of how the new entrants behave, in terms of creating new opportunities or just enriching themselves as "paper quota holders". A strong state is clearly *a necessary but not sufficient condition* in order to create a more equitable fishing sector.

9. LITERATURE CITED

Apostle, R., G. Barrett, P. Holm, S. Jentoft, L. Mazany, B. McCay, and K. Mikalsen 1998. *Community, State and Market at the North Atlantic Rim. Challenges to Modernity in the Fisheries.* University of Toronto Press, Toronto.

Berkes, F. (Ed.) 1989. Common Property Resources, Ecology and Community Based Sustainable Development. Belhaven Press, London.

Cape Argus 1999. Moosa acts after audit reveals R24-m scams. December 8.

Christy, F. 1996. The Death Rattle of Open Access and the Advent of Property Rights Regimes in Fisheries. *Marine Resource Economics.* Vol 11.

Cochrane, K.L. 1995. Anticipated Impacts of Recent Political Changes on Fisheries Management in South Africa. *NAGA*, January, ICLARM, Manila.

Cochrane, K.L. and A.I.L. Payne 1998. People, purses and power: developing fisheries for the new South Africa, in Pitcher, T.J., P.B. Hart and D. Pauly (Eds.) 1998. *Reinventing fisheries management,* Kluwer Academic Publishers, Dordrecht, London.

Crean, K. and D. Symes (eds.) 1996. *Fisheries Management in Crisis.* Fishing News Books, Oxford.

De Wet Schutte 1994. Report of the committee of inquiry into fishermen's community trusts. Ministry of Environment Affairs and Tourism, Cape Town.

Dyer, C.L. and J.R. McGoodwin (eds.) 1994. *Folk Management in the World's Fisheries: Lessons for Modern Fisheries Management.* University Press of Colorado, Niwot.

Evans, P.B., D. Rueschmeyer and T. Skocpol. *Bringing the state back in.* Cambridge University Press, Cambridge.

Fisheries Policy Development Committee (FPDC) 1996. *National Marine Fisheries Policy for South Africa. Report to the Minister from the Fisheries Policy Development Committee.* Sea Fisheries, Cape Town.

Garcia, S. 1998. Resource management systems in development countries. Proceedings from the Soria Moria Conference, Research Council of Norway, Oslo.

Hannesson, R. 1996. *Fisheries Mismanagement. The case of North Atlantic Cod.* Fishing News Books, London.

Hersoug, B. 1996. Same procedure as last year? Same procedure as every year! – some reflections on South Africa's new fisheries policy. Paper delivered at the international seminar on National Marine Fisheries Policy for South Africa. UWC, Cape Town.

Hersoug, B. 1998. Fishing in a sea of sharks: Reconstruction and development in the South

African fishing industry. *Transformation,* University of Natal, Durban.

Hersoug, B. and P. Holm 1998. Change without Redistribution: An Intitutional Perspective on South Africa's New Fisheries Policy. Paper delivered to the IIFET Conference inTromsoe, June 1998. Forthcoming in *Marine Policy,* Spring 2000.

Hersoug, B. and S.A. Raanes 1997. What is good for the fishermen, is good for the nation: co-management in the Norwegian fishing industry in the 1990s. *Ocean & Coastal Management.* Vol 35, Numbers 2-3.

Holden, M. 1994. *The Common Fisheries Policy.* Fishing News Books, London.

Hutton, T., J. Raakjaer Nielsen and M. Mayekiso 1999. Government-Industry Co-management Arrangements within the South African Deep-Sea Hake Fishery. Paper delivered on IMF's conference on Co-management in Malaysia.

Independent Newspapers 1997. Crayfish quotas causing major conflict and fraud. http://archive.iol.co.za/Archives/1998/9801/30/tense.htn

Isaacs, M. And A.K. Normann 1999. Experiences from fishing communities in Western and Eastern Cape, South Africa. Is there a basis for co-management? Paper presented at the international workshop on fisheries co-management, August 1999 (forthcoming).

Jentoft, S. 1989. Fisheries Co-Management: Delegating Government Responsibility to Fishermen's Organizations. *Marine Policy* 13: 137-154.

Manning, P.R. 1998. Managing Namibia's marine fisheries: Optimal resource use and national development objectives. PhD thesis, London Schoool of Economics and Political Science, London.

Matthiasson, T. 1992. Principles for distribution of rent from a "commons". *Marine Policy.* 16 (3) pp 210-231.

McGoodwin, J.R. 1990. *Crisis in the World's Fisheries: People, Problems, and Policies.* Stanford University Press, Stanford.

O'Meara, D. 1996. *Forty lost years. The apartheid state and the politics of the National party, 1948-94.* Ohio University Press, Athens.

Republic of South Africa (RSA) 1997a. *A Marine Fisheries Policy for South Africa. WhitePaper.* Minister of Environmental Affairs and Tourism, Pretoria.

Republic of South Africa (RSA) 1997b, *Marine Living Resources Bill (as introduced).* Minister of Environmental Affairs and Tourism, Pretoria.

Republic of South Africa (RSA) 1998. *Marine Living Resources Bill (as amended by the Portfolio Committee on Environmental Affairs and Tourism (National Assembly).* Minister of Environmental Affairs and Tourism, Pretoria.

Sea Fisheries 1996a. *Report of the Access Rights Panel.* Sea Fisheries, Cape Town.

Sea Fisheries 1996b. *Report of the Legal Task, Team.* Sea Fisheries, Cape Town.

Streeck, W. and P.C. Schmitter 1988. Community, Market, State - and Associations? The Prospective Contribution of Interest Governance to Social Order. *In* W. Streeck and P.C. Schmitter (eds.) *Private Interest Government: Beyond Market and State.* Sage, London.

Stuttaford, M. 1994-98. *Fishing Industry Handbook.* Marine Information cc, Cape Town.

Townsend, R.E. 1997. Fisheries management implications of Alaskan communitydevelopment quotas, in G. Palsson and G. Petursdottir: *Social Implications of Quota Systems in Fisheries.* Tema Nord 1997:593. Nordic Council of Ministers, Copenhagen.

Weber, M. 1924/1978. *Economy and Society.* University of California Press, Berkeley.

PROPERTY RIGHTS AND RECREATIONAL FISHING: NEVER THE TWAIN SHALL MEET?

J. McMurran
Ministry of Fisheries
ASB Bank House, 101 – 103 The Terrace, P.O. Box 1020, Wellington New Zealand
<mcmurranj@fish.govt.nz>

1. INTRODUCTION[1]

There is sometimes a tendency to associate property rights solely with commercial fishing. However, there is a wider place for property rights in fisheries management. Explicit property rights can be vested in other groups, such as communities, recreational fishers, and indigenous fishers.

This paper examines ongoing work to improve the management of marine recreational fishing in New Zealand using a property-rights approach.

2. HOW IS RECREATIONAL FISHING MANAGED IN NEW ZEALAND?

A simplified version of the way fisheries are managed can be described as follows. For each fish stock, the fisheries Minister sets a total allowable catch (TAC), based on scientific advice as to the sustainable level of harvest from the fishery. The TAC is then allocated to the recreational, customary and commercial sectors. Customary and recreational-take is provided for when setting the annual commercial catch limit. There is no specific guidance for the Minister in setting recreational-take. Essentially the Minister weighs up competing interests and decides what is a reasonable share.

In a collective sense then recreational fishers have a right to a share of TAC. Individually, recreational fishers also have rights. Anyone, including an overseas tourist, is free to fish in the sea, provided they do not sell their catch and they comply with the amateur fishing regulations. New Zealand has world-class recreational fishing. Not surprisingly fishing is a very popular pursuit. About one in five New Zealanders fish recreationally in the sea in any one year and many overseas tourists join them.

The amateur fishing regulations include controls such as daily bag-limits, minimum fish sizes, closed areas, closed seasons, and method and gear restrictions. The regulations serve a range of functions including:

i. managing recreational take so as the TAC is not overshot
ii. enabling all recreational fishers to have a "fair go" rather than having high individual limits which could result in the majority of the (collective) recreational share going to a relatively small proportion of recreational fishers and
iii. fisheries compliance purposes - the commercial compliance regime applies when a person is found in possession of fish at a specified level well above the amateur bag limits.

The Ministry of Fisheries uses telephone and diary surveys, and boat-ramp interviews to monitor recreational catches.

3. HOW DO RECREATIONAL FISHERS' RIGHTS STACK UP WITH COMMERCIAL AND CUSTOMARY RIGHTS?

3.1 Status of rights

Underlying the brief description above about how recreational fishing is managed are some fundamental problems for the recreational sector, and indeed for fisheries management generally. The recreational fishing sector is in fact in a less-advantaged position in terms of how their rights stack up against the customary and commercial sectors. This is not a good position to be in as the fishery is a shared resource.

3.2 Commercial fishing rights

Over the past 15 years, commercial fishers have worked with government to implement clearly-defined, appropriately-specified and enforceable property-rights. In 1986 government introduced the quota management system (QMS) to manage commercial fishing in the marine environment, using individual transferable quota. The QMS has evolved over time, with a number of changes made to improve the system. *The Fisheries Amendment Act 1999* heralded the most recent changes. Amongst other things, the Act:

i. allows responsibility for the operation of the quota registry to be devolved to the fishing industry, and
ii. enables research, compliance and other services required by government to be directly purchased by the fishing industry.

Concurrent with the evolution of the QMS has been a change in behaviour of many quota-holders. Through representative organisations, quota holders are continuing to seek more direct responsibility and control over their fishing activities. Together these changes have strengthened the rights of the commercial fishing sector.

3.3 Customary fishing rights

The QMS provided private rights to harvest fisheries (shares of fishstocks) without first determining who owned the resource. Understandably the indigenous Māori population saw this as an affront to their rights under the Treaty of Waitangi signed with the Crown in 1840. The development of the QMS therefore triggered addressing customary fishing grievances.

The *Treaty of Waitangi (Fisheries Claims) Settlement Act 1992* split the commercial and non-commercial components of the customary fishing right and provided

---

[1] The views expressed in this paper are those of the author and do not necessarily represent the views of the Ministry of Fisheries.

for each in a different way. The commercial part of the Settlement provided for quota, cash and other assets to be deeded to Māori.

The non-commercial component of the customary fishing right continues to place Treaty obligations on the Crown. The Settlement requires the Minister, acting in accordance with the principles of the Treaty of Waitangi, to consult with *tangata whenua*[2] and develop policies to help recognise the use and management practices of Māori in the exercise of their non-commercial fishing rights.

Customary fishing regulations have been enacted for the management of customary (non-commercial) fishing. The regulations devolve responsibility for the management of customary fishing to Māori. A rigorous framework, involving authorisations (permits) issued by authorised individuals (*kaitiaki*)[3], and reporting of take is included. The regulations clearly signal the expanding role of Māori in managing their fishing rights and interests. One example of this greater role is the contract that the Ngai Tahu tribe has with the Crown for the delivery of non-criminal compliance services for customary fishing over most of the South Island.

3.4 Recreational fishing rights

Unfortunately for recreational fishers, their collective rights to a share of the fishery are not well defined, relative to customary and commercial fishers who share in the same resource.

When allocating the available catch, the Minister provides for customary and recreational take and then sets the annual commercial catch limit. Recreational fishers do not have any priority in law over commercial fishers, or vice versa. The fisheries Minister simply needs to make an allowance that he or she considers reasonable. If recreational fishers think the allowance the Minster sets for them is unfair, it is hard for them to take action for two reasons:

i. their right is loosely defined and decisions are hard to overturn unless the Minister acted unreasonably, and
ii. unlike commercial fishers, recreational fishers do not have ample funds to take legal action to defend their rights.

Rights that are not well defined are difficult to protect and, or, enhance. Population growth in many regions popular for fishing; environmental pressures such as algal blooms; and competing demands for coastal space (*e.g.* from marine farming and marine protected areas) are likely to put increased demand on available fisheries resources. The risk for the recreational sector is they may not be well-placed to protect their interests as these pressure continue to grow. There is a danger that the recreational sector could shoulder a disproportionate burden relative to commercial sector in any adjustment that is necessary, and that the quality of recreational fishing may decline over time.

A related problem with recreational fishing rights is how the rights are managed. Recreational fishers largely rely on the government to give effect to, and manage, their rights. Until recently there has been little discussion about whether this is the best way to manage recreational fishing.

Management by the Ministry of Fisheries tends towards something of a "one size fits all" approach. For example, the amateur fishing regulations are similar in approach around the country. However, New Zealand's coastline and coastal communities are diverse with different needs and local conditions. Many of the frustrations that recreational fishers have are local concerns. Commonly there are concerns expressed about the impact of commercial fishing on recreational fishing in particular areas. The Ministry does not have the detailed knowledge, and more importantly the resources, to become heavily involved in local disputes. The Ministry's primary role is to ensure the sustainability of fishstocks, rather than advocating the cause of one sector (*e.g.* recreational) at the expense of another or mediating disputes. It is not surprising that recreational fishers sometimes express frustration about the lack of response when they have made the effort to influence fisheries management decisions. These factors, and a concern by many recreational fishers that the quality of fishing has declined, suggest it is unlikely that recreational fishers can rely on government to fully meet their needs and aspirations.

4. POLICY REFORM

Over the past year the Ministry of Fisheries has been working collaboratively with the New Zealand Recreational Fishing Council Inc. (NZRFC). The NZRFC is the main national body for recreational fishers, representing a range of individuals and clubs throughout the country. The joint Ministry/NZRFC working group is preparing a public consultation document on improving recreational fishing. The intention is for the consultation to be managed and undertaken jointly by the working group.

Two key areas being examined in order to better position the recreational sector are:

i. better defining the recreational right by introducing a proportional-share arrangement, and
ii. enhancing recreational fishers' rights to directly manage their share.

5. BETTER DEFINING THE RIGHT - PROPORTIONAL SHARE

Recreational shares in key fisheries would be set as an on-going proportion of the available catch, rather than being subject to the Minister's discretion each time a stock is reviewed. The proportion would be set as a percentage of the available catch. For example, if the recreational share was set at 40% in a particular fishery and the available catch for the year was 100t, 40t would be allocated to the recreational sector. In subsequent years the proportion would remain in place, with the ton-

---

[2] This literally means people of the land and refers to the Māori population.
[3] The Māori people local to the area.

nage allocated varying in line with changes in the available catch.

A benefit of an on-going proportional share for the recreational sector is protection of their share from erosion. There is also the potential that in some fisheries, the recreational sector could make a case for a higher share than at present when shares are first set.

Perhaps more importantly however, having a share known in advance provides the recreational sector with greater status to sit around the table with customary and commercial fishers in the area and work out how they can manage the fishery so they all benefit. A proportional share would remove the current incentives for both the commercial and recreational sectors to lobby the Minister to increase their collective share. Such behaviour is time-consuming and not a productive use of resources — a good example of a zero-sum game because an increase for one sector results in a decrease for the others. And, position taking, or gaming, is encouraged which contributes to tensions between sector groups and diverts attention from opportunities to work together constructively.

However, a proportional-share arrangement would mean that the obvious way to improve fishing for all three harvest groups would be by working together to increase yields, or coming to agreements over use of particular areas within fisheries. The following sorts of agreements might be possible:

i. commercial fishers stay out of a particular area at particular times (*e.g.* a harbour over the summer holiday period) or cease to use particular methods in certain areas, in return for the recreational fishers supporting a commercial harvest strategy
ii. fewer fish be harvested in order to generate larger fish and better catch rates in the fishery, and
iii. different areas be set aside for commercial and non-commercial shellfish harvesting.

For agreements like these to be enforceable, they would need to be reflected in regulations. If doing so, the Crown would need to look at compliance-costs and the degree to which the individuals who negotiated the agreement are representative. Mandate is a particular issue for recreational fishers, as customary and commercial fishers tend to be more readily identifiable and are often affiliated with representative groups.

There are a number of issues to consider in better-defining the recreational right with a proportional share, including:

i. which stocks would be subject to the proportional option
ii. how the proportional shares would be set
iii. how recreational fishing would be managed within the share
iv. what response could be made if recreational demand significantly increased after shares were set
v. whether there are any circumstances when the level of the shares could be reviewed, and
vi. what if recreational shares were at a higher level than currently set - how would any costs be managed?

If the benefits of a proportional share are kept in mind, none of these issues are insurmountable. For example, data analysis and consultation could be used to identify those fisheries where the recreational take is significant. As to the level of the shares, it need not be the current share. The overall objective would be to give the recreational sector access to a "fair" share of the available catch. Criteria such as the value of the particular fishstock to each sector, historical catch-rates, and the degree to which commercial fishing is restricted in the fishery could be used as a basis for negotiating the level of the share. However, matters such as these can never be an exact science because there is imperfect information. As such, there would need to be a process involving the Crown and stakeholders in the particular fishery to work the issues through.

6. SHARED MANAGEMENT

The concept would see a legislative framework to enable mandated regionally-based recreational management groups (RMGs) to be established to:

i. manage recreational fishing with the Crown
ii. work with commercial and customary fishers to develop plans to manage harvesting.

The role and functions of RMGs would be clearly specified in statute. The bodies would need to be representative of regional recreational fishers and accountable to the government and fishers. An RMG would give recreational fishers a stronger voice to act for recreational interests at the local and national level.

In managing recreational fishing, an RMG would need to develop some form of a plan including the following sorts of matters:

i. the objectives for recreational fishing in the fishery or area
ii. fisheries management controls to give effect to those objectives
iii. governance rules for decision making by the RMG
iv. supporting services – compliance, research and education
v. specification of how the environmental obligations in the Fisheries Act would be met, and
vi. funding.

An important role in managing recreational fishing would be recommending[4] management controls (*e.g.* closed areas, daily bag-limits, etc.) for recreational fishing. The controls would be set with reference to the collective share and compliance costs would need to be considered. The government would also need to be satisfied that the resource's sustainability and Treaty of Waitangi obligations were not put at risk. The RMG would have flexibility to customise controls to suit the needs of the fishers they represent. For example, there might be a wish to allow use of scuba divers for obtaining paua (abalone), something that the rules do not currently

---

[4] Controls would probably need to be gazetted by government so they can be effectively enforced.

allow. RMGs could also help enforce the controls. For example, they could be responsible for operating the Honorary Fishery Officer network currently co-ordinated by the Ministry of Fisheries.

Having a recognised mandated recreational body would also facilitate all three harvest groups, customary, recreational and commercial, in coming to agreement about how best to manage the fishery they share. The sorts of agreements outlined earlier in discussions about the proportional share concept would be easier to implement with a mandated RMG. The three harvest groups could also undertake other work to promote their shared interests. They might for example:

i. make representations to local councils seeking more sustainable land-management practices if important fish nursery areas are being adversely affected by run-off and pollution, and
ii. investigate technologies to reduce mortality of undersized fish, reduce capture of unwanted bycatch, and improve detection of blackmarket shellfish.

RMGs would be managing shares of fishstocks of considerable value, both monetary and non-monetary. A number of issues would need to be resolved more fully before RMGs could be established. These would include:

i. How a mandate would be established
ii. Role and functions of RMGs
iii. How to ensuring the Crown continues to deliver on Treaty and sustainability obligations
iv. How RMGs should be funded, and
v. Whether trading (or leasing) of shares between sectors should be permitted.

None of these issues is insurmountable. Indeed there is already one model of shared fisheries management in New Zealand, that of the regionally-based Fish and Game Councils which manages trout fishing. However, if shared management of marine recreational fishing does happen, it will not happen overnight as issues like the ones above will need to be worked through. The intention is that if the recreational sector is interested in having a much greater say in how recreational fishing in managed, the *Fisheries Act 1996* would be amended to enable RMGs to be established over time to assume management rights.

7. CONCLUSION

There is a place for the use of property-rights approaches in the management of recreational fishing. In an ideal world, rights for all harvest groups would be better defined at the same time. However, New Zealand's situation suggests there is potential for formalised recreational property-rights even when individual transferable quota already exists for commercial fishers and the rights of customary fishers are also well-defined.

Introducing a formalised property-rights regime for the recreational fishing is not a task for the faint hearted, and will be something that takes considerable time. There are some major challenges that need to be resolved in a calm way with a longer-term strategic perspective. However, within every challenge lie opportunities. The potential benefits of better defined recreational rights are not limited to protecting the recreational share from reduction. The benefits extend to better fisheries outcomes through more responsive management, more collaboration and much greater participation in fisheries management decisions. The potential benefits suggest that progress will be made in New Zealand and that the challenge of using property- rights to improve recreational fishing is one worth investigating.

NEGOTIATING THE ESTABLISHMENT AND MANAGEMENT OF INDIGENOUS COASTAL AND MARINE RESOURCES

D. Campbell
DCafe
P.O. Box 228, Kippax, ACT 2615 Australia
<dcampbell.fish@bigpond.com>

1. INTRODUCTION[1]

The 1992 Mabo no 2 decision recognised the indigenous rights of Aborigines and Torres Strait Islanders to their country. However, the extent of the decision was limited to that part of the indigenous estate above the mean high-water mark. The *Native Title Act 1993*, in providing the legislative basis of the Mabo decision, also leaves the question of offshore native title rights and interests in abeyance.

Justice Olney in the *Croker Island 1998* Federal Court ruled that communal native title exists in relation to the sea 'which washes the shores of the relevant land masses', and sea-bed within the claimed area. In making this decision, he found native title sea rights might exist co-jointly with non-indigenous rights, but only to the extent that native title rights yield to inconsistencies with other legal rights and interests. The possibility of exclusive native title, however, was extended from the landward side of the mean high-water mark to the landward side of the jurisdictional limits of the State or Territory - as set by the coastal baseline. In most cases this will be to the landward side of the mean low-water mark, although exclusive native title rights can also apply to enclosed waters, such as Mission Bay in the Northern Territory (*Croker Island 1998* s.51).

The Croker Island native title claimants and the Commonwealth government have appealed the Croker Island decision to the Federal Court. It is expected that any decision by the Federal Court will eventually be appealed to the High Court, with the possibility of a ruling by 2001.

In addition, the High Court, on 7 October 1999 found Queensland's State conservation laws did not extinguish the native title rights of Aborigines and Torres Strait Islanders to carry out traditional hunting. While the full implications of the decision are yet to be fully understood the decision is likely to impact on fisheries management and conservation law in all Australian government jurisdictions[2].

Such uncertainty in defining coastal and marine native title rights works to the loss of indigenous and non-indigenous people and society as a whole. Those with an interest in marine and coastal resources can mitigate such losses by negotiating institutional structures to overcome poorly defined rights.

This paper, deals with the rights, issues and institutional structures by which local groups may set up and negotiate agreements. The nature of the problem, and how inadequate rights may lead to the demand for new institutional structures are reviewed in Section 2. Section 3 covers the nature of rights, including legal and economic rights, how different bundles of rights might affect what can be done with an asset and questions regarding private, community and government holding of rights. The nature of legal and economic rights are discussed in Section 3. In Section 4 some of the issues requiring consideration when developing new institutional structures are discussed. These include the costs in transacting an agreement, the importance of information and issues of compliance. In Section 5 a number of different possible compliance procedures, including Indigenous Land (Sea) Right Agreements and Indigenous Protected Areas, are reviewed. The conclusion and possible future directions are presented in Section 6.

2. THE NATURE OF THE PROBLEM: THE EBB AND FLOW OF CHANGING RIGHTS

2.1 Establishment of native title rights

With the passing of the *Native Title Act 1993*, Aboriginal and Torres Strait Islander peoples are establishing native title rights to land and coastal areas where they can show an ongoing indigenous connection to the area[3]. Such changes have altered the expectancies of non-indigenous as well as indigenous peoples over future resource access

---

[1] This project received funding from the Fish Resources Research Fund

[2] The extent of this decision is still to be defined. The Canadian Supreme Court decision, *Regina v Sparrow 1990*, gave that customary rights to fish were only required to give way to conservation requirements. This is in line with the Law Reform Commission's report (1986) on the recognition of Aboriginal customary laws, where it found (p. 200):

'As a matter of general principle, Aboriginal traditional hunting and fishing should take priority over non-traditional activities, including commercial and recreational activities, where the traditional activities are carried on for subsistence purposes. Once this principle is established the precise allocation is a matter for the appropriate licensing and management authorities acting in consultation with Aboriginal and other user groups'.

[3] By April 1998 the National Native Title Tribunal had received native title applications from Aborigines and Torres Strait Islander people to 140 locations that included areas of sea. Of these, 73 were in Queensland, 35 were in Western Australia, 5 were in South Australia, 11 in the Northern Territory, 11 in New South Wales, 3 in Victoria, 1 in Tasmania, and 1 (Jervis Bay) in the Commonwealth.

and the associated cultural and economic benefits. In addition, uses by non-indigenous users of marine and coastal resources can affect indigenous cultural and economic uses of coastal and marine resources.

All marine activities have the potential to invade the privacy of marine and coastal land-owning groups, including: commercial and recreational fishing, aquaculture[4], recreational boating, tourism, marine park zoning and management, port operations and shipping (Smyth 1997). While there is no quantitative data, coastal tourism including recreational fishing is likely to have a substantial and ongoing long-term impact on indigenous coastal communities. The geographical extent, the large numbers involved and difficulties in identifying and monitoring individual behaviour means the impact will likely be insidious and difficult to control. Indirect effects due to activities beyond a community's indigenous estate are also likely to be important. This includes changes in fisheries management, effects to the coast or sea bottom from coastal and offshore construction, bottom trawling and changes in fishing pressure. All such events will increasingly impact the cultural and economic relationship of Aboriginal and Torres Strait Islander peoples with the sea and foreshore.

Access to marine and coastal resources is also important to recreational and commercial fishers, aquaculturalists, recreationalists and other non-indigenous users while native title to sea and foreshore areas is also likely to be important to fisheries management and in setting fishing regulations. Uncertainty over outcomes for existing and future claims can place a level of uncertainty on investment decisions of commercial fishers fishery managers, and commercial tourist operators. For instance, does the increasing use by recreationalists and commercial fishers to the sea and foreshore areas qualify for compensation to holders of rights under the *Native Title Act 1993*?

Except for infrastructure that had been constructed, there is no requirement under the *Native Title Act 1993* to negotiate with native title holders concerning acts in the waters and seabed to the seaward side of the mean high-water mark. Such infrastructure includes the establishment of ports and jetties, but does not explicitly exclude aquaculture infrastructure, such as for oyster production. As a result, aquaculturalists may still need to negotiate access in those locations in which exclusive native title rights apply. If native title rights are found to apply to the sea, any diminution of rights or benefits from the sea estate is likely to require compensation to the native title holders. Whether compensation payment is the responsibility of the perpetuator of the act, or the Commonwealth government, depends on the nature of the act incurred.

A decision in support of s.51 of the *Croker Island 1998* decision concerning the extension of exclusive native title rights to the mean low-water mark, will affect s.26 (3) of the *Native Title Act 1993* – which restricts the seaward extension of exclusive native title to the mean high-water mark. As well as affecting on those wishing to construct coastal structures, such as aquaculture ponds, confirmation of the original Croker Island decision will also affect fishing operations in the intertidal zone, as for barramundi and mud crab. Also, questions exist on the differential effect native title decisions can have on different fishers operating in the same fishery, even when entitlements are the same (see Loveday 1998 pp. 2-3). A decision in favour of co-joint native title rights might remove some ambiguity, although the inadequacy of property rights to migratory resources, such as fish, will continue.

Reference is made in judicial decisions such as the 1999 *Yorta Yorta case*, and in the literature on native title rights, to a tide that washes over the rights of indigenous people. This tide, which changes the institutional nature of rights to the seas and coastal landscape, ebbs as well as flows. The problem is to manage these currents so that the foreshore is not eroded, the waters are not polluted and the potential benefits from marine and coastal resources are not washed away.

2.2 Social costs and the dissipation of benefits

Ambiguity and uncertain rights over marine and coastal resources places many of these assets in the public domain where there is little control over who has access to the benefits obtainable. This uncertainty over rights is likely to result in a race by resource users for the benefits obtainable from these assets; a situation common to fisheries management and marine resources in general.

The nature of the losses from poorly defined rights include the loss of customary indigenous benefits and resource rent[5], through the effect on:

i. indigenous people, when they are not fully compensated for their losses[6]
ii. those responsible for payment of compensation for future acts and
iii. society at large when:

    a) indigenous assets are not used for their highest valued use[7]

---

[4] For example, plans for an aquaculture project in Darwin were withdrawn due to uncertainty with native title and the perceived inability of the Northern Territory government and Aboriginal representatives to be able to handle applications on land which may be subject to native title claims AIATSIS (1997 p. 11).

[5] The application of resource rent to fisheries is discussed in Campbell and Haynes (1990).
[6] For instance, the requirement in s.51A(1) of the *Native Title Act 1993*, for compensation to be constrained to an amount no greater than freehold value can result compensation for specific sites of high cultural value being a pittance.
[7] In response to the High Court's 1996 *Wik* decision, the Commonwealth government in 1998 amended the *Native Title Act 1993* such that the Commonwealth would meet the compensation costs for some compensatable acts. Circumstances in which those committing and benefiting from acts resulting in the loss of native title rights do not pay compensation are likely to compound the social loss from such acts. That is, while losses

b) excess resources are used in the race to capture the benefits obtainable from poorly defined marine and coastal resources
c) there are costs of transacting new institutional structures, including obtaining information on the nature of assets and ensuring compliance under these new institutional structures; and
d) resources are lost to future use.

Such losses are but part of the conditions leading to the setting up of new institutional structures which are summarized in Table 1.

probable participants, lack of communication, and a lack of the necessary institutional conditions to establish binding agreements, may need to be overcome. She concludes that successfully generated new institutional structures included:

i. a definition of those who will be authorised to use the resource
ii. the relating of the specific attributes of resources to the characteristics of identified resource users
iii. the need for new institutional structures to be at least partly designed by local resource users
iv. a need for the application and compliance with the

Table 1
Conditions leading to establishing new institutional arrangements

The conditions leading to the demand for new institutional structures	The conditions for local groups to successfully set up new institutional arrangements	The structure of the institutional arrangement for the given conditions

2.3 The demand for new institutional structures

Barzel (1997), North (1990) and Ostrom (1990) discuss the entry of assets into the public arena as a result of poorly defined rights and the conditions for interested groups to address the loss of social benefits[8]. Formally, incentives for self-generated institutional change are likely to occur among groups with an interest in marine and coastal areas when:

i. groups are interested in minimizing cost
ii. choices are constrained by budget constraints
iii. interested parties can be separated into definable groups and
iv. market imperfections, such as inadequate property rights, exist (see Hayami and Rutlan 1985).

While these conditions may create a demand for new institutional structures, they do not set out the conditions in which local groups will enter negotiations and for which contracts will be agreed to. In particular, the institutional conditions required for local groups to develop contracts to new institutional structures and to defining new or altered rights and responsibilities, setting rules, developing behavioural constraints (norms of behaviour and convention) and enforcement characteristics.

2.4 Requirements for self-generated negotiation

Ostrom (1990) suggests that the necessary factors to predict participation are poorly understood. In general, though, myopic behaviour, lack of mutual trust among

new institutional requirements to be monitored by individuals who are accountable to local resource users and
v. the development of a schedule of graduated punishments for non-compliance which needs to be sanctioned (Ostrom 1990, pp. 185-6).

Some of the factors likely to affect commercial fisher participation in changing to institutional rules and rights are discussed by Libecap 1990 Ch. 6). In particular, cooperation among fishers is likely to decrease, when the impact of any changes among individual fishers is uneven. Such differences could occur as the result of differences in the type of fishing gear used, the amount of catch taken and differences in the location fished. Loveday (1998) discusses the importance of relative differences in the effect of native title on Queensland commercial fishers.

3. THE NATURE OF RIGHTS
3.1 Legal and economic rights

Legal rights involve what has been assigned to a person, group, organization or jurisdiction by the state, or marine and coastal land-owning group through legislation, custom, indigenous law or other means. Provision of legal rights occur as a result of formal arrangements, including constitutional, statutory, judicial rulings or as part of an organised system of indigenous law, and informal conventions and custom. The nature of property rights will affect the decisions made in regard to how resources are used, to the net social benefit enjoyed by indigenous and non-indigenous people and by society as a whole from fish and other marine and coastal resources.

Economic rights depend on the enforcement of legal rights and relate to the right-holder's ability to enjoy benefits from a piece of property and the assets contained therein. That is, economic rights include the ability to

---

exceed zero, beneficiaries may be better off by as little as zero, thus resulting in a net social loss.

[8] Using the prisoner's dilemma paradigm, Baland and Plateau (1996), Campbell (1995) and Ostrom (1990), show that, in spite of an overwhelming social benefit, incentives exist for private individuals and different groups to not cooperate. These authors also discuss the conditions in which cooperation might be improved.

enjoy benefits either directly through consumption and cultural appreciation, or indirectly through exchange, including barter, sale, rent, inheritance and gift-giving.

Realisation of the benefits of economic rights depends on the nature of the legal rights and, in this sense, legal rights provide a means to an end. However, the existence of legal rights is not enough as the ongoing enjoyment of benefits also depends on the effective power of an individual, group or community to assert control over the different attributes of their rights.

The ability to assert control over assets will be affected by knowledge on what the attributes of a resource are. Without this knowledge, control is likely to be incomplete, 'unknown' attributes are likely to fall into the public domain and the value or benefits of the attributes will be lost through the costly races for possession (Barzel 1997).

Granted and recognised rights will define the range of privileges and responsibilities of right holders to specific assets, such as possible parcels of water 'as far as the eye can see', intertidal zones, reefs, and fish. Although the legal basis of native title rights to coastal resources is known, the nature of sea-rights is not.

3.2 Different bundles of rights

Discussions in regard to rights in fisheries usually focus on privately held rights to fish and the use of input controls and individual transferable catch quota. The rights to fish and marine and coastal resources are more extensive and concern rights in addition to private rights to fish. For instance, rights of access, removal, management, exclusion and alienation (Table 2) need to be considered. The characteristics of each of these bundles of rights are important when considering who should hold these rights and whether they are held privately, by a defined group or community, by government, or be held co-jointly (Schlager and Ostrom 1992).

The mix of customary rights and responsibilities of Aboriginal and Torres Strait Islander communities means they too, are concerned with rights in addition to private or community-held catch and removal rights. Many of the speakers at the 1999 National Indigenous Sea Rights Conference in Hobart spoke of the importance for Aboriginal and Torres Strait Islander peoples to exercise indigenous rights to marine resources[9]. In particular, they spoke of the importance of their involvement in negotiating the establishment of national parks in the seas and foreshores of their indigenous estates, their management of access to marine resources, and the need to achieve and maintain cultural and economic benefits from their sea estate (for example, J. Caristopherson[10], 28 September 1999).

Differences in rights will affect the uses and the manner in which assets are utilised. However, benefits also depend on the attempts by others to capture benefits and the attempts by owners, non-government organizations and government to protect these rights. In addition, the value of an asset is unlikely to be fully realised if access to asset attributes is restricted to a single individual or group. That is, those placing the highest value on an attribute are capable of paying the highest price for access.

What is shown in this paper is that negotiation and the development of new institutional structures can be used to:

i. remove the uncertainty and social loss due to poorly defined rights
ii. establish governance structures to shore up and maintain compliance to rights
iii. ensure those able to make the best use of the assets have access to them and
iv. ensure that right holders are able to achieve the greatest benefits obtainable from their rights.

4. CONSIDERATIONS IN NEGOTIATING NEW INSTITUTIONAL STRUCTURES

4.1 Relevant factors

Optimal achievement of benefits from new institutional structures will depend upon consideration of a number of factors, including:

i. what the objective is in developing new institutional structures
ii. the amount of information on the nature and attributes of the assets involved
iii. the nature of the new property rights or institutional structures achieved as a result of negotiation
iv. the level of compliance with the negotiated agreement
v. the transaction costs incurred in negotiation and enforcing the new institutional structure and
vi. how well the negotiated agreement meets the objective.

From an economic perspective, an expected optimal outcome would be when the expected benefit that might be obtained through additional negotiation equals the expected additional transaction cost.

4.2 Transaction costs

Transaction costs are those costs associated with the transfer, capture and protection of rights, or, 'the costs of measuring and enforcing agreement' (North 1990, p. 362). That is, transaction costs are the costs incurred identifying performance requirements or outcomes, obtaining necessary information, establishing and sustaining new property rights and ensuring compliance.

4.3 Information

Information on the nature of the assets held by native title rights is important in establishing and enforcing rights and in ensuring benefits are realised. For instance, because of limited knowledge, rights-holders may be

---

[9] See, for instance, the Declaration of the National Indigenous Sea rights Conference (Anon. 1999).
[10] Mr Caristopherson is a claimant in the Croker Island case and an executive member of the Northern Territory Northern Land Council.

Table 2
The different types of rights to be considered

Type of right	Definition
Access	The right to enter a defined area or location
Removal	The right to obtain the products or a resource, such as taking fish
Management	The right to regulate internal use patterns and transform the resource by making improvements
Exclusion	The right to determine who will have an access right, and how that right may be transferred.
Alienation	The right to sell or lease either or both of the above collective choice rights

From Schlager and Ostrom (1992).

ignorant of all the benefits from the use of their assets, the benefits obtainable from exchange, and the benefits from their joint-use with others. Poorly identified benefits can therefore result in inappropriate rights to the assets, poor governance structures and the loss of benefits.

The likely wide scope and complexity due to the large geographical area covered, variability in the number and nature of assets and the large number of interest groups will place a high demand for information. The veil of ignorance that overlies future events, including future judicial decisions and legislative acts, further compounds this potential loss of benefits[11].

Likely shortages of information means that negotiations will need to be structured to proceed within the boundaries of available information, while allowing time and future resources to collect additional information. Agreements can also be structured to allow reassessment on the basis of future events and judicial and legislative change. As a result, information considerations are likely to affect the type of structure or processes used to carry out negotiation[12]. However, re-entering, or maintaining, the negotiation process throughout the life of an agreement is not costless and the expected benefits need to be weighted against the additional cost.

4.4 Compliance: establishing and enforcing rights

Compliance can be obtained through enforcement and by setting up institutional structures to provide incentives for compliance. With a simple model, the level of compliance can be described as a function of the probability of being found out in not complying, times the resulting penalties compared to the benefits of non-compliance. That is, compliance could be expected to occur when the expected penalty exceeds the benefits of non-compliance. The costs of monitoring, however, are likely to be so high that in many situations, socially unacceptably high penalties would be required.

Compliance can be improved through the integration of compliance and monitoring structures within an agreement. A more complex model of compliance includes social influence and moral obligation as factors in explaining compliance behaviour, where such factors can be enhanced through education, persuasion and the development of shared social links (Sutinen 1996). Persuasion and the development of such social links might be built into an agreement. The rules and requirements built into an agreement can be used to change the incentives faced by participants in the agreement through changes in the institutional governance structures so as to:

i. lower monitoring costs
ii. increase the probability of being found out and
iii. change the expected benefits of not complying.

This might be achieved by building a strategic alliance within an agreement by linking current actions with previous actions. An example would be to link continued access by municipal staff to collect sand in an area of significant indigenous cultural interest to their preventing all others from entering the area. Linking compliance with ongoing benefits and requiring the council to monitor and prevent entry by others, changes the council's benefit and cost profile and leaves monitoring to the council, who are likely to be better placed to prevent entry to the area. It is then a matter for the resource owner to monitor the council compliance. An alternative could be different forms of triggering events, which will require the identification of what actions to take such as renegotiate some part of the agreement (Campbell et al. 1996). An example of a triggering event might be when the judicial response to the appeals against the *1998 Croker Island decision* is handed down[13]. Integration of compli-

---

[11] The implications of many of these judicial and legislative decisions have not been fully thought through. A case in point is the 1998 amendment to the *Native Title Act 1993* limiting compensation to no more than a freehold value (The *Native Title Amendment Bill 1997 Explanatory Memorandum (House of Representatives 1997)* develops this discussion further (p. 248). It is questionable whether this meets the requirement of s51 of the Australian constitution. Aside from this, questions remain as to how to measure indigenous rights (Campbell 1999a). A methodology based on the choices made by indigenous people has been suggested by Campbell (1999b), although there are shortcomings with this approach that require further consideration.

[12] Jones (1999) discusses this approach in comparison with other options.

[13] Triggering events would be best defined according to generic characteristics, such as the type of judicial or legislative decision, rather than in regard to particular decisions. Other acts, such as changes in recreational participation, or changes in

ance can also include the development of different forms of performance/compliance indicators.

4.5 Structural complexity

The preceding discussions on information needs and on compliance included the use of different approaches to circumvent information shortfalls and to obtain compliance. Consistent with these observations, Campbell (1995, pp. 221-233) shows non-compliance and opportunistic behaviour is likely to increase with short-term agreements and simple models involving "one-off" rather than ongoing negotiation and agreement. That is, simple agreements that have a brief time duration provide little or no incentive for ongoing cooperation and compliance. A complex hierarchical system, requiring ongoing review and assessment between the parties to an agreement, is more likely to ensure ongoing cooperation between the members of the agreement. This conclusion is supported by research carried out by Oxley (1997) in which she found conditions that led to poorly defined rights, increasing contractual risk and moral hazard led to an increasing reliance on hierarchic institutional structures. Consistent with this, she also found that the trade-off between transaction costs and moral hazard led to increasingly complex governance structures.

Complex hierarchical institutional structures may have the capacity to ensure that all parties to an agreement have an incentive to make the agreement work to their mutual advantage. Parties may, however, work against an agreement and towards its eventual breakdown if the agreement has disadvantaged them and they are better off outside the agreement.

5. USE OF INDIGENOUS LAND USE AND INDIGENOUS PROTECTED AREA AGREEMENTS

5.1 Requirements for agreement

A number of approaches can be taken to resolve poorly defined rights, including the use of costly judicial appeals. For instance, resolution of the Miriuwong-Gajerrong case in the Kimberly region cost approximately $A4.66 million, while, by comparison, the Cape York Heads of Agreement cost in the order of $A20 000 (McCann 1999)[14]. In addition, judicial decisions are inflexible while net gains can be increased if only those subsets and particular commodity attributes required by the other person are transferred. The courts have, and continue to play an important role in the recognition and resolution of questions relating to indigenous rights. In particular, they are important in the resolution of questions of law, in the enforcement of agreements and the provision of contract security. The courts, therefore, do provide an important basis of support to the establishment, settlement and maintenance of agreements, and, in the long term, the development of mutual trust and confidence[15].

The approaches available to indigenous and non-indigenous people outside of direct use of the courts include indigenous land use agreements (ILUA) and indigenous protected area agreements (IPA). It is important to note that IPA agreements do not have the flexibility or the extent of ILUA and can not be used to establish native title rights. The availability of these choices does not preclude the use of the same processes that are available to the general public. One approach put forward as an option to resolve issues concerning native title rights is the use of a Coasian solution. The applicability of this option is critiqued in Appendix 1.

5.2 Indigenous land use agreements

The 1998 amendments to the *Native Title Act 1993* provided important institutional changes to the original Act in the provision of indigenous land use agreements to facilitate local and regional agreements (McCann 1999). In particular, the registration requirement for ILUA is important to monitoring and enforcement of agreements and the provision of security. In addition, the constraints on what can be agreed to under an ILUA protect other members of indigenous communities, including grounds by which removal of an ILUA from the National Native Title Tribunal (NNTT) register can be ordered by the Federal Court.

In an environment of uncertain rights, an Indigenous Land Use Agreement can help people who are unsure on how to proceed in an area where native title is held or is claimed. As a result, they provide an important framework in which much of the uncertainty over future rights can be removed and the ebb and flow of rights along the coast and seas can be managed. A critical characteristic of ILUA, highlighted by Smith (1998), is they are 'instruments of consent'. While this can leave the process open to strategic behaviour, a 'shared commitment to negotiated outcomes will arguably facilitate better post-agreement relations between indigenous people and the wider community than do judicial or arbitrated agreements'. There are several strengths of the ILUA process:

i. Its flexibility allows the inclusion of all interest groups and negotiation of agreements without final

---

recreational or social; conditions might also be used as triggering events.

[14] Justice Olney's comments regarding use of adversarial litigation in the Yorta Yorta case are noteworthy: 'The time and expense expended in the preparation and presentation of a large part of the evidence has proved to be unproductive, a circumstance which calls into question the suitability of the process of adversary litigation for the purpose of determining matters relating to native title (from Neate 1999, p. 9)

[15] Decisions on the use of the courts relative to other options should be the same as choices made in regard to any other option., that is, by comparing the marginal transaction costs incurred to the marginal benefits obtained. The problem is that many court decisions will set legal precedents that are important to others who are not involved in the court case. As a result, because all costs and benefits are not accounted for, some cases that on a public good basis should have been taken to court are not.

resolution of individual claimants within an area. Therefore:
   a) in an environment in which parties to an agreement may be uncertain of their legal rights, ILUA offer an opportunity for interested parties to come to an understanding in those areas in which there is certainty, while withholding action in regard to those areas in which there are uncertain legal rights (Smith 1999, p. 10) and
   b) it can provide a process that is sympathetic to indigenous norms and processes.
ii. Indigenous people can come to agreement with non-indigenous people over an area without necessarily resolving questions in regard to intra-group ownership or title over particular areas (see Jones 1999).
iii. They provide the legal certainty and security all parties require, as agreements are made binding on their being placed on the NNTT registrar.

There are three types of ILUA:
i. Body corporate agreements, that involve
ii. Area agreements and
iii. Alternative procedure agreements.

The primary characteristics separating the three types of ILUA is shown in Table 3. While all three types of ILUA may apply to onshore sites, pending the final decision on the appeals to the *Croker Island 1993* decision, *only area agreements and alternative procedure agreements are applicable to sea rights*.

Two examples of indigenous land use agreements involving marine or coastal estate are the Quandamooka community and Redlands Shire Council agreement (Anon. 1997) and the Sea Forum alliance of traditional owners in the Southern Great Barrier Reef (Muir 1999). Both instances provide examples where participants take a long-term view of ten or more years and in which there is a strong emphasis on establishing and maintaining a process.

The Quandamooka Land Council, representing the three clans with traditional links to the area, and the Redlands Shire agreement, initiated in 1994 with the lodgement by the QLC to the National Native Title Tribunal, notice of an application for a determination of native title. The purpose of the agreement was to establish an understanding between the parties with an interest in the area in a process leading to an agreement on native title. The agreement is focused on North Stradbroke Island/Minjerribah, and its surrounding seas located southeast of the city of Brisbane. The project has eleven guiding principles including:
i. recognition of the interdependence between cultural and natural landscapes
ii. recognition of Quandamooka's environmental systems in their local, regional, national and global context
iii. respect and incorporation of the custodial obligations of the traditional owners

iv. promotion of sustainable economic development opportunities for both indigenous and non-indigenous people recognising the relationship between economic sustainability, community development and cultural resource management initiatives and
v. the use of broad definitions of natural, built, economic resilience and sustainability as limits to human activity (Anon. 1997).

The Sea Forum was initiated in 1997 and has a broader scope. It involves a larger number of indigenous communities with sea country estates, several Commonwealth and State government agencies, a number of Shire councils and other interest groups such as the Queensland Commercial Fisherman's Organisation. The geographical area is the southern Great Barrier Reef and consists of the three southern zones of the Great Barrier Reef Marine Park, including Fraser Island. It is a community-based alliance of traditional owner groups who have interests in sea country and who have come together to facilitate the development of sea estate agreements within a regional framework. The primary purpose of the agreement is to assist Aboriginal people with sea country estates within the area to achieve their aspirations for resource management (Muir 1999).

The Sea Forum is structured to address those elements common to all indigenous people within the region. The intention is to develop protocols that have procedural integrity as a basis for ongoing negotiations. The process is designed to accommodate those issues that relate to the region, while leaving local Aboriginal communities to speak to those issues that are specific to their own country.

Both organizations provide examples of local indigenous groups coming together to resolve questions of rights and future use of coastal and marine areas, although the nature of indigenous rights to marine areas remains uncertain. The process is advantageous to the indigenous and to the non-indigenous signatories to the respective agreements. The agreements appear to be hierarchical in nature, requiring ongoing consultation and cooperation over a number of years. At least, the process has saved participants from expensive court actions, and the availability of this option may help to ensure a commitment by all of the parties concerned. While the structure of the agreements provides incentives for ongoing compliance, it would be useful to know how monitoring of compliance is built into the procedures and institutional structures.

5.3 Indigenous protected areas

The indigenous protected area (IPA) programme, administered by Environment Australia, is part of the national reserve system (NRS) programme. The NRS was established as a means of coordinating the cooperation with the States, Territories and the wider community to

Table 3
Indigenous land use agreements (ILUAs) and indigenous protected areas (IPAs)

Type of agreement	Primary characteristics
Indigenous land use agreements[a]	The statutory basis of ILUA is the *Native Title Act 1993*, as amended in 1998. Agreements can be given by native title groups for any consideration and subject to any conditions; any person may request assistance from the National Native Title Tribunal (NNTT) in making agreements; an application for registration of each type can be made in writing by any of the parties to the registrar of ILUAs; ILUAs are registered with the NNTT and are legally binding.
Body corporate agreements	Can be made where there <u>has been a determination/s on who</u> holds native title over. Agreement covers <u>the whole of the area;</u> Can be made if <u>there is a registered native title body/s corporate</u> for the whole of the agreement area.
Area agreements	Can be made between persons who <u>claim</u> to hold native title over a particular area and other people or organizations about the use <u>of land and waters</u> in that area. Can be made if there is <u>no registered native title body/s corporate</u> for the whole of the agreement area.
Alternative procedures	Can be made between people who <u>claim</u> to hold native title over a particular area and other people or organizations about the use of <u>land and waters</u> in that area. Can be made if there <u>is no registered body/s corporate for the whole</u> of the agreement area, <u>but requires a body/s corporate for part</u> of the agreement area.
Indigenous protected areas[b]	These are an approach by which Aboriginal and Torres Strait Islander peoples can care for and protect lands and waters for present and future generations. It may include land over which Aboriginal and Torres Strait Islanders are custodians and which will be managed for cultural biodiversity and conservation, permitting customary sustainable resource use and sharing of benefits. IPAs are part of broader National Reserve System established and coordinated by Environment Australia.

[a] Native Title Act 1993, Smith (1999), National Native Title Tribunal 1998a,b,c,d.
[b] Leitch (1999).

develop a national system of protected[16] areas. The incorporation of the IPA programme within the NRS is to include establishing and managing protected areas on indigenous estates and establishing cooperative or joint management with indigenous people over government owned protected areas.

The agreements are intended to apply for a period of three to five years and are monitored in consultation with other agencies by the landowner. Beside emphasising cooperative or joint management, the IPA programme places limitations on the management agreement. Funding for the programme is supplied from the National Heritage Trust.

The traditional Yolngu owners gave the Parks and Wildlife Commission of the Northern Territory 'in principal' support for the establishment of an indigenous protected area in the vicinity of Nhulunby, in the northeast of the 'top end'. The area proposed includes coastal areas currently managed by the Dhimurra Land Management Aboriginal Corporation. In giving this support, the Commission were advised of their wish that the area not be administered jointly, but be continued to be administered by the traditional owners. The initial informal response from the Commission was that this option is unlikely to be acceptable.

The nature of such agreements, might provide indigenous people an opportunity to place parts of their coastal estate in hold until more information is obtained on the nature of their estate and the nature of rights are further resolved. It also provides an option by which indigenous people can obtain financial and administrative support in managing their estate. The example highlights concern that can occur in regard to control of traditional estates. Again, questions of compliance and maintaining ongoing commitment exist. It may be interesting in the future to compare both approaches given that indigenous protected areas appear to have a shorter duration and do not necessarily require as complex a structure.

6. CONCLUSION

The ebb and flow in coastal and marine rights and the poor definition of these rights create uncertainty and the loss of benefits Aborigines, Torres Strait Islanders and non-indigenous people might enjoy. Those with an interest in marine and coastal resources can mitigate such losses by negotiating institutional structures to overcome

---

[16] A protected area is defined as *'An area of land or sea specially dedicated to the protection and maintenance of biodiversity and associated cultural resources and management through legal and/or other effective means'* (Environment Australia 1998). It is important to note the place given to cultural resources in the context of conserving indigenous estate.

poorly defined rights. This paper dealt with the necessary conditions, rights, issues and institutional structures by which local groups may set up and negotiate agreements. Indigenous land use agreements and indigenous protected areas are discussed as low transaction-cost options applicable to situations involving native title right.

A useful direction for future research would be to examine how different institional structures might affect the decisions made on whether cooperation continues or whether there is failure in such cooperation. Such work should have possible effects of different performance or agreement compliance indicators as part of any new institutional structure.

7. LITERATURE CITED

AIATSIS 1997. *Native Title Newsletter*, no 1 Canberra (http:/www.aiatsis.gov.au/archive/ntru1-97.htm).

Anon. 1999. Declaration by the National Sea Rights Conference, Hobart, Tasmania, 28-30 September, ATSIC, Hobart.

Anon. 1997. *Native Title Process Agreement: Quandamooka Land Council Aboriginal Corporation and Redland Shire Council* (http./www.nntt.gov.au/).

Baland, J.M. and J-P. Platteau 1996. *Halting Degradation of Natural Resources: Is there a Role for Rural Communities?*, Food and Agricultural Organization of the United Nations, Rome, pp 1-439.

Barzel, Y. 1997. *Economic Analysis of Property Rights*, Cambridge University Press.

Campbell, D. 1999a. Valuation of indigenous fisheries, paper presented at the 43rd Annual Conference Australian Agricultural and resource Economics Society, Inc. Christchurch, 20-22 January.

Campbell, D. 1999b. Economic valuation of indigenous rights, paper presented at the 1999 International Symposium Society and Resource Management: 'Application of Social Sciences to resource management in the Asia-Pacific Region', Brisbane July 7-10.

Campbell, D. and J. Haynes. 1990. *Resource Rent in Fisheries*, ABARE Discussion Paper 90.10, Australian Government Publishing Service, Canberra.

Campbell, D., A. Stokes and D. Brown 1996. Issues in monitoring fisheries management performance, ABARE paper no 96.2, VIII International Fisheries Economics and Trade Conference, Marrakech, Morocco, 1-4 July.

Campbell, D.E. 1995. *Incentives: Motivation and the Economics of Information*, Cambridge University Press, Cambridge, New York Oakleigh.

Coase, R.D. 1960. The problem of social cost, *Journal of Law and Economics*. 17(2), 357-76.

Environment Australia 1998. *Indigenous Protected Area Programme: Advice to Applicants 1998/99*, Biodiversity Group, Environment Australia, ACT 1960.

Godden, D. 1999. Attenuating indigenous property rights: Land policy after the *Wik* decision, *The Australian Journal of Agricultural and Resource Economics*, vol. 43(1), 1-33.

Hayami, Y. and V. Ruttan. 1985. Toward a theory of technical and institutional change, *in* Koppel, Bruce M. (ed) *Agricultural Development: An International Perspective*, The John Hopkins University Press, Baltimore.

Jones, C. 1999. Native title and natural resource management: A way forward, paper presented to The 1999 International Symposium, Society and resource Management, University of Queensland, Brisbane July 7-10, 1985.

Law Reform Commission 1986. *The Recognition of Aboriginal Customary Laws*, Report 31(2) Australian Government Publishing Service, Canberra.

Leitch, K. 1999. Traditional owners support establishment of an indigenous protected area, *Waves: Newsheet of the Marine and Coastal Community Network*, Environment Australia, 6(1), 2.

Libecap, G.D. 1990. *Contracting for Property Rights*, Cambridge University Press, Cambridge, New York and Melbourne.

Loveday, T.D. 1998. 'Native title focuses on fisheries', *The Queensland Fisherman*, 16(8), 2-3.

McCann, L. 1999. Induced institutional innovation in Response to Transaction Costs: The Case of the National Native Title Tribunal, paper presented at the 43rd Annual Conference of the Australian Agricultural and Resource Economics Society, Christchurch, New Zealand, 20-22 January.

Muir, B. 1999. Sea Forum: A regional agreement process for Aboriginal traditional owner involvement in sea country management in the southern Great Barrier Reef, presentation notes for a talk given at the National Indigenous Sea Rights Conference: What the Sea Means to Indigenous People, Hobart Tasmania, 28-30 September.

National Native Title Tribunal 1998a. *Indigenous Land Use Agreements (ILUAs): Short Guide to ILA Registration*, (http://www.nntt.gov.au/nntt/publicatn.nsf/).

National Native Title Tribunal 1998b. *Indigenous Land Use Agreements (ILUAs): Body Corporate Agreements*, (http://www.nntt.gov.au/nntt/publicatn.nsf/).

National Native Title Tribunal 1998c. *Indigenous Land Use Agreements (ILUAs): Area Agreements*, (http://www.nntt.gov.au/nntt/publicatn.nsf/.

National Native Title Tribunal 1998d. *Indigenous Land Use Agreements (ILUAs): Alternative Procedure Agreements*, (http://www.nntt.gov.au/nntt/publicatn.nsf/).

Neate, G. 1999. Future directions in native title, a National Native Title Tribunal paper presented at a seminar organised by the Centre for Energy and Resource Law, University of Melbourne, March 1999.

North, D. 1990. A transaction cost theory of politics, *Journal of Theoretical Politics*, 2(4), 355-367.

Ostrom, E. 1990. *Governing the Commons: The Evolution of Institutions for Collective Action*, Cambridge University Press, Cambridge, New York, Oakleigh.

Oxley, J.E. 1997. Appropriability hazards and governance in strategic alliances: A transaction cost approach, *The Journal of Law, Economics, and Organization*, 13(2), 387-409.

Schlager, E. and E. Ostrom 1992. Property-Rights Regimes and Natural Resources, *Land Economics*, 68(3), 249-262.

Shapley, L.S. and M. Shubik, 1969. On the Core of an Economic System with Externalities. *American Economic Review* 59(4) pp 678-684.

Smith, D.E. 1999. Indigenous Land Use Agreements: The Opportunities, Challenges and Policy Implications of the Amended Native Title Act. Centre for Aboriginal Economic Policy Research, Discussion Paper 163/1998, Australian National University, Canberra.

Smyth, D. 1997. *Australia's Ocean Policy: Socio-cultural Considerations*, Issues paper no. 6. A report commissioned by Environment Australia.

Sutinen, J.G. 1996. *Fisheries Compliance & Management: Assessing Performance*, a report to the Australian Fisheries Management Authority, Canberra.

Whipple, R.T.M. 1997. *Assessing Compensation Under the Provisions of the Native Title Act 1993*, Working Paper Series, no. 97.10, Business School, Curtin University, Perth.

Appendix 1
A Coasian diversion

It has been suggested that a Coasian solution might be used to resolve questions regarding access to 'land' rights. Whipple (1997, p. 31) suggests that once native title rights and those entitled to them are defined, the Coase theorem may be applied. More recently, Godden (1999, p. 19), suggested that '[a]ll that is required is that the parties to the bargain feel that both have gained or, at worst, none has lost'. There is nothing particularly Coasian about the conditions given in either of the papers. A fundamental point in Coase's (1960) paper is that the distribution of rights is irrelevant. This is not the case for indigenous people because their budget depends on the allocation of rights to their estate and because so much that is of value to a community location is specific. Indeed, reference is made in this paper to a Coasian solution because its use would likely work against the interests of Aboriginal and Torres Strait Islanders and against a fair and sustainable solution to the overcoming poorly defined rights.

According to the 'Coase theorem', "*voluntary negotiation will lead to a fully efficient outcome, provided* (a) rights are well defined, (b) transaction are costless, and (c) there are no income effects." The outcomes of this, it is suggested, are '1. if markets are incomplete[17], people will negotiate and the efficient outcome will result; 2. there is no need for government intervention; and 3. the outcome is independent of the initial assignment of rights. The question is whether these necessary conditions are met.' (Farrell 1987, cited in Baland and Platteau 1996, p. 50).

First, the transfer of rights will result in the transfer of budgetary power and therefore the ability to pay for rights (Campbell 1999b). Therefore, in the unlikely event of an efficient outcome, a negative effect on economic welfare will occur. Second, a large part native title concerns the costly process of defining rights and their ownership. The assuming away of this problem through use of a Coasian solution does not remove the problem. Further, Shapley and Shubik (1969) argue that when more than two parties are involved, an economically efficient outcome depends on the initial distribution of rights (Baland and Platteau 1996 pp. 51-2).

---

[17] An incomplete market exists when the social opportunity costs or social benefits derived from the supply of a good or service are not fully accounted for and the good or service is correspondingly over supplied or under supplied

RECOGNITION OF AND PROVISION FOR INDIGENOUS AND COASTAL COMMUNITY FISHING RIGHTS USING PROPERTY RIGHTS INSTRUMENTS

M. Hooper and T. Lynch
New Zealand Ministry of Fisheries
PO Box 1020, Wellington, New Zealand
<hooperm@fish.govt.nz>

1. INTRODUCTION

Throughout the world, state management of fisheries using regulatory instruments has left indigenous and coastal communities subject to the changing national aspirations of governments. The recent recognition of indigenous rights in some Western post-colonial countries has occurred largely through the courts. Rather than embracing the indigenous communities' interests as valid in their own right, governments have sought to protect the interests of the dominant culture from the disruption that could be caused by the recognition of indigenous rights.

Using regulatory mechanisms alone to attempt to recognise the broad range of cultural and economic interests of indigenous communities will, in our view, inevitably fail. Regulatory mechanisms devised and controlled by the state will largely reflect the values and aspirations of the dominant culture as represented by that government. No matter how liberal, democratic and egalitarian the state may be, the final result is likely to further erode the ability of indigenous communities to manage, harvest and use natural resources in ways that are consistent with their cultural needs.

The management of fisheries through the use of property-rights is often perceived as being anathema to both the recognition of indigenous fishing rights and provision for the interests of coastal fishing communities. We argue that the opposite is in fact the case. Not only are indigenous fishing rights, and the rights and interests of coastal communities, compatible with a property-rights approach to fisheries management, such an approach can be used to settle claims involving indigenous fishing rights, to preserve those rights for future generations, and to integrate such rights within a wider fisheries management framework. A property rights-based system provides a far more robust and flexible mechanism to ensure the sustainable utilisation of a fishery, while providing for indigenous and other users of a fishery to exercise their often divergent social and economic aspirations.

The creation of an artificial fisheries property-right and the allocation of that right, or a proportion of that right, to the indigenous or coastal community rights-holders provides the basis for a more equitable relationship between the state and the relevant community. Property-rights are difficult to extinguish and provide a strong disincentive for the coercive use of state authority. Equally they can create equity between the different users of the fisheries resource if all rights are derived from the same legal base. In this paper, the fisheries property-right referred to is essentially a share of the total allowable catch (TAC) in a fishery. The right to a fixed proportion of the resource negates the need to compete for that resource and can provide the catalyst for co-operative management.

Central to the ideas expressed in this paper is the argument that property-rights cannot exist in isolation from rules as to how that property is to be managed and rules governing interactions between property rights-holders. The creation and allocation of an artificial fisheries property-right is not an end in itself. Rather, it establishes an equitable basis for the application of management tools that can provide for the different social, cultural and economic aspirations of the various rights-holders within the fishery.

Indigenous and coastal fishing communities usually have their own internal regulatory mechanisms for management of their fishing activities. Such regulatory mechanisms are integral to the nature of their fishing-rights. Recognising and providing for indigenous and coastal community fishing-rights means empowering the communities concerned to use those mechanisms and integrating them within the wider fisheries management framework. Property-rights instruments enable this to occur.

This paper outlines a basic four-step process for recognition and provision for indigenous and coastal community fishing-rights through the use of property-rights instruments. The four steps are: (a) define the nature of the fisheries right to be provided for; (b) quantify the right; (c) provide instruments for exercising the right; and (d) create incentives for co-operative management of fisheries. Each of these steps is described in general terms and then illustrated by reference to recent events in New Zealand resulting from the settlement of Māori fishing-rights claims.

2. THEORETICAL FRAMEWORK
2.1 Define the right

Many countries recognise the existence of indigenous rights in some form. However, there is often conflict between different sections of society as to the nature and extent of those rights. Indigenous rights, based on legal tenets such as aboriginal title and customary common law often exist without being properly defined in statute and without any reference to their relationship with other more tangible rights that may exist under state law.

Before such rights can be recognised and provided for they must be clearly defined. If there is conflict as to the nature and extent of those rights the defining process may take place in the courts where the nature of customary rights is drawn from historical accounts of behaviour and practice. However, the defining of the right could be a more deliberate process carried out by the state in order to recognise and preserve the socio-economic characteristics of a particular community or sector of society.

There is an important distinction between indigenous rights that are recognised by the courts and the state, and the rights or interests of coastal communities which may have no legal basis in common law or aboriginal title. Indigenous rights will be defined in accordance with customary practices and, at least in theory, can not be attenuated without the agreement of the rights-holders. Coastal communities may have no inherent rights to fisheries, in which case the decision to recognise their interests and define rights relating to those interests will be a purely political one.

When defining the right it is necessary to identify the behaviours that are to be provided for and the outcome that is sought. Ideally, this should be a participatory process involving the communities concerned and the lawmakers. In simple terms, it is a matter of working out who goes fishing, where they fish, what methods and equipment they use, what management techniques they employ, and what impacts they have on both the fishery, and other users of the resource. The objectives of both the state and the community for participation in the fishery, now and in the future, are central to this process. The expectations of the rights-holders may well extend beyond their current practices. It should then be possible to define the exact nature of the right to be provided for, which may be a greater or a lesser right than the one currently being exercised, depending on the objectives for the future of both the fishery and the communities concerned.

2.2 Quantify the right

Having defined the nature of the indigenous or coastal community fishing rights that are to be recognised and provided for, the next step is to determine the quantity of fish that these rights represent. It is at this stage of the process that the pre-existing rights or interests are translated into the artificial fisheries property right. In a single-user fishery, sustainability may well be the only factor that limits the quantity of fish that can be removed from the fishery in accordance with these rights. If there is more than one gear sector in the fishery then the fishery will need to be allocated between the different rights-holders.

A unique feature of property rights systems of fisheries management is that the rights of each stakeholder can be quantified, relative to the rights of others, usually as a proportion of the TAC in a particular fishery. The fisheries property right is the share of the TAC. Systems can then be put in place to enable each sector to exercise and manage its rights within the wider framework.

In most indigenous societies the rights to use a resource are not absolute. They are limited by concerns of resource sustainability, the common nature of community rights to the resource that might lie with more than one group and cultural or spiritual prohibitions on fishing activity. A perpetual access right to a share of a TAC, set according to the sustainability of the resource, provides a better expression of the autonomous nature of aboriginal title than regulatory mechanisms that can be more easily amended or diluted by the state.

The property rights approach provides guaranteed access to a share of the fishery, the opportunity for autonomous or semi-autonomous management of that share, the potential for communal ownership and management, and the ability to participate in compliance monitoring. However, the introduction of a property rights system does not negate the need for voluntary, or enforceable rules, to regulate internal community use of their right or to manage interactions between the community and other users of the fishery.

Setting of a TAC for a set period, and the proportional allocation of that catch between users, will establish an environmental boundary for harvests from the fishery. In itself, this will not necessarily enable indigenous groups to manage the taking of fish according to traditional practices relating to where and when the fish are caught, how they are caught, and the size or sex of the fish that are caught. Many indigenous or traditional coastal communities have well developed internal regulatory systems governing whom may utilise a fishery and how they may fish. In a modern context, statutory mechanisms may be necessary to give legal expression to the use of such traditional regulatory systems.

2.3 Provide instruments for exercising the right

The sorts of instruments necessary to provide for indigenous or community rights-holders to exercise their rights fall into two categories. First, instruments to enable the communities themselves to manage their own fishing activity within the parameters of the right as defined and quantified above. Second, instruments for managing the interaction between the indigenous or community rights-holders and other users of the fishery.

Essential to the first set of instruments is information as to how many of a particular species are harvested in a particular area. Such instruments need to perform three key functions: (a) establish who has the mandate to manage the fishery on behalf of the community concerned; (b) generate information about removals from a fishery; and (c) enable removals from a fishery to be stopped once the share of the TAC has been caught and the property-right fulfilled.

Obtaining basic information about the harvest from a fishery requires catches to be reported (and monitored). While reporting may seem onerous, the alternative is to rely on enforcement of the property-right through blunter input controls such as bag-limits, closed-seasons and method-restrictions. The better the information that is reported, the lesser degree to which the state needs to be involved in the management of the fishery, and the greater the flexibility for management of the right. The information provided enables the TAC to be refined to protect the fishery as management uncertainty, and hence risk is reduced.

As stated above, indigenous and coastal fishing communities usually have their own internal regulatory mechanisms for management of their fishing activity. The setting of a TAC and the allocation of property-rights to that TAC provide a basis for fishers to use their share in ways that are culturally appropriate to them and to regulate their internal activities to achieve those goals. The allocation of property-rights can be used to empower communities to use their own mechanisms, and provide for their integration within the wider fisheries management framework, as long as the three key functions are incorporated – establishment of mandate, information on removals, and the ability to stop fishing when removals reach a certain point.

The second set of instruments relates to shared fisheries, where the state must play additional roles in allocating access between sector groups. The state has a number of roles. In consultation with sector groups the state has a duty to set the TAC so as to protect the fishery. It must allocate proportions of the fishery between sectors. It also must provide instruments for resolving disputes between the indigenous or community rights holders and other participants in the fishery, and tools to allow indigenous or community rights to be exercised within the wider framework, in terms of both taking fish and managing fisheries. This latter component may require the incorporation of a spatial element into the rights framework, where the activities of other sectors may need to be curtailed so that the indigenous or community fishing rights can be realised.

2.4 Incentives for co-operative decision-making

The management of fisheries by regulatory mechanisms controlled by the state leaves the regulation-making process vulnerable to capture by the sector groups that wield the most influence on the regulatory authority. These lobby-groups may vary from those with political power (the dominant cultural group or local governments), those with economic or organisational strength (recreational or commercial fishing organisations) or those who simply share the values of the government. It is often difficult for indigenous groups or artisanal coastal communities to successfully participate in such an environment.

In such situations, it is likely that regulatory systems will result that reflect the values and aspirations of the dominant parts of society, submerging the values of other groups. Forced to compete under a set of rules designed to reflect the goals of the dominant group, marginalised groups will have little commitment to compliance with those laws, especially if they do not address their own cultural or socio-economic aspirations.

In contrast, when a TAC is established and a proportional share of that TAC is allocated among fishery users, the ability to lobby the state for preferential benefit is removed. The rights of each group to participate in, and benefit from, the fishery are derived from the same legal source. The alteration by the state of any group's rights becomes a threat to the rights of all users. From the point of allocation, there are strong incentives for sector groups to co-operate to maintain the integrity of the allocation mechanism and the property-rights of each group.

The tonnage of fish allocated becomes the principal limitation on deriving social or economic benefit from the fishery. This limitation provides strong incentives for commercial fishers to minimise costs of management, compliance and fishing as these expenses become the sole variables in the success of their fishing operations. In the case of indigenous and coastal communities, where cost and efficiency may not be the primary incentive for use of the resource, those groups have security of access to a share of the fishery, which they can catch without direct competition from other sector groups.

In many circumstances, indigenous and coastal communities are more limited to the area they fish than other users. The types and size of fish they require may also be different from those of other sector groups. The simple allocation of a proportional right to a quantum of fish will not resolve these qualitative issues. However, the allocation of a right to a share of a fishery enables the participants to set out operating rules and make trade-offs to reduce conflict and enable each sector-group to take the types of fish they need in the areas they fish. When the sector groups develop and agree to their own rules to achieve these goals, management and compliance costs are likely to be minimised.

There is equally a strong incentive for all sector groups to manage and enhance the fishery to increase the TAC in the knowledge that they will benefit from the fishery in proportion to the rights they have been allocated. The costs of enhancement can therefore be shared proportionally as well. Where a TAC is set for a specified period, there is an incentive for fishers to develop plans to re-build stocks and distribute the benefits in a manner that enables the specific concerns of each group to be addressed. Because each sector-group will benefit, the incentive in such long-term planning is to enable each group to modify its operations over time to achieve its own aspirations.

3. WHAT HAPPENED IN NEW ZEALAND?
3.1 Talking past each other – definition by the Courts

As the indigenous people of New Zealand, Māori held customary fishing rights under British common law. These rights were guaranteed by the *Treaty of Waitangi* in 1840, and were exempted from the rules and regulations in fisheries legislation made after the signing of the Treaty. While fisheries law in New Zealand always contained a clause that exempted Māori fishing rights from its ambit, the exact nature of these rights was never defined.

As a result these rights were negated by those same statutes through the application of the egalitarian principles of the dominant pakeha culture. The statutory provisions protecting Māori customary fishing-rights were worthless without the conjoint ability to self-define the nature of the right. Instead the right was left to be misinterpreted by faceless bureaucrats or a biased judiciary.

In New Zealand the task of defining the nature of Māori customary fishing-rights fell to the Courts and to the Waitangi Tribunal. The Waitangi Tribunal is a permanent commission of inquiry set up in 1975 to investigate claims regarding breaches of the Treaty of Waitangi. In an important test case in 1986 a Māori individual was found not guilty of taking undersized *paua* on the grounds that he was exercising a customary fishing-right. He had fished in accordance with customary practices by obtaining permission from the *kaitiaki*, or guardian, of the *tangata whenua* from the area where the fishing occurred, and acted in accordance with the instructions of the *kaitiaki*.

The concept of *tangata whenua*, or "people of the land", is crucial to the definition of Māori customary fishing-rights. *Tangata whenua* in a particular area are the *iwi* (tribe) or *hapu* (sub-tribe) that hold customary authority over that area. Rather than being general Māori rights, customary rights belong to *tangata whenua* and can only be exercised within their area. So in New Zealand, customary fishing-rights are held by *tangata whenua* in relation to their area of traditional authority.

The full nature and extent of customary fishing-rights was elucidated by the Waitangi Tribunal as a result of extensive research into tribal claims to fisheries. Māori customary fishing rights were found to contain both a commercial and a non-commercial component (based on evidence that Māori were trading seafood widely prior to the signing of the *Treaty of Waitangi*); the fisheries they exploited were extensive; and the methods they used to catch fish were advanced compared to those of their European counterparts. There was also a developmental component to the customary right, which meant that Māori had a right to a share of the deep-sea fisheries off the coast of New Zealand, even if they were not being fished at the time the Treaty was signed.

Most importantly, Māori customary fishing-rights pertained not only to the use of fisheries, but also to the management of the resource. Rather than being Māori rights, customary fishing-rights belong to *tangata whenua* – the tribe or sub-tribe that hold traditional authority over a particular area. While fishing practices differed between the different tribes, customary fisheries were always actively managed by individuals known as *kaitiaki* or guardians. Fishing outside of the rules set by the *kaitiaki* could make the fisher subject to severe penalties.

In the mid-1980s, New Zealand was moving to introduce a quota management system based on individual transferable quota (ITQ) for major commercial fish stocks. At the time, the Waitangi Tribunal observed that the ITQ right was analogous to the nature of the rights guaranteed to Māori under the Treaty of Waitangi – it guaranteed access, it was perpetual, it was tradable, and it allowed for autonomous management. It was this move to create an artificial property right to take fish for commercial purposes, and then allocate that right to existing commercial fishers, that drove Māori to injunct the Crown on the basis that their customary fishing-rights had not been taken into account.

The introduction of a property-rights system for fisheries in New Zealand not only gave rise to the largest indigenous rights claim in New Zealand's history, it also provided the means for that claim to be settled and for customary rights to be recognised and provided for within the wider legislative framework.

3.2 Quantifying Māori customary rights – the fisheries settlement

In New Zealand, the process of quantifying the indigenous fishing right occurred through the negotiated settlement of Māori claims in respect of their fishing-rights. In 1986 when the Crown was injuncted from proceeding with the introduction of the quota management system, the Courts advised the Ministry of Fisheries that they had no issue with the quota management system itself. The aims of the Crown in introducing the quota system were laudable. The problem was that indigenous rights had not been recognised or provided for in the allocation of commercial fishing quota.

An interim settlement of Māori fisheries claims was negotiated and legislated for in 1989 with a full and final settlement signed in 1992 and legislated for in the *Treaty of Waitangi (Fisheries Claims) Settlement Act 1992*. The principal effect of the settlement on the customary fishing-rights of Māori was to effect a split between the commercial and non-commercial components of those rights.

The commercial component of the customary right was settled through the provision of ITQ. The 1989 interim settlement provided for 10% of all existing quota to be bought back from fishers and provided to Māori. The 1992 Settlement provided Māori with a half-share in Sealord Products Ltd which owned a further 23% of existing quota. The Settlement also provided for 20% of quota for new species to be transferred to Māori on the introduction of those species into the quota management system.

The non-commercial component of the customary right was provided for through provision for the making of regulations that recognise and provide for the use and management practices of Māori in respect of both the taking of fish for non-commercial purposes and the management of traditional fishing grounds. The non-commercial needs of Māori have a *de facto* priority in the process for allocation of the TAC – the needs of Māori are to be provided for first, to the extent that they are not commercial.

3.3 Instruments for exercising the right – ITQs and customary fishing regulations

As outlined above, the fisheries settlement split the customary fishing-rights of Māori into commercial and non-commercial components. The commercial component was provided for through ITQ, while the non-commercial component was provided for through regulations. The Settlement legislation creates instruments for Māori to exercise their fishing-rights within a framework that also provides for the rights of commercial and recreational fishers. The intervening years have seen Māori and the Crown both working to implement that framework in what has been an evolutionary process.

The quota that Māori received as a result of the fisheries settlement is held by a central Commission, Te Ohu Kai Moana, that manages that quota on behalf of Māori. The quota held by the Commission is no different from other ITQ generated under the Quota Management System and the State interacts with all commercial rights-holders (quota-owners) on the same basis. The Commission currently leases quota to tribes on an annual basis. In time, the quota will be allocated to tribes, giving them all the benefits and obligations associated with quota ownership.

The commercial interests and objectives of Māori in respect of fisheries may differ from tribe to tribe. They may also be different from the interests of other commercial fishers in their area. Property-rights instruments such as ITQ allow the different priorities and interests of different groups to be realised within the same framework, while minimising the opportunity or need for the state to interfere with those interests.

The fisheries settlement stated that the non-commercial fishing-rights of Māori were to be provided for through regulations which recognised and provided for customary food gathering by Māori, and the special relationship between *tangata whenua* and their traditional fishing grounds. A modern regulatory framework provides an effective way of recognising and providing for the traditional fisheries management practices of Māori. The defining and quantifying of the non-commercial customary right made it possible for such regulatory instruments to be developed within the constraints of that right. Most important is that the regulatory framework is highly flexible, with regard to the way *tangata whenua* manage their fishing activity, but prescriptive in terms of mandate' issues - recording of catch and accountability mechanisms.

New Zealand's customary fishing regulations are based on a number of key underlying principles. The first principle centres on mandate. Before *tangata whenua* can take responsibility for managing their non-commercial fishing activity, they must establish mandated representatives for their area. The first part of the customary fishing regulations provides for *tangata whenua* to appoint *kaitiaki*, or guardians, who will be responsible for managing customary fishing in their area with the full support of the law. Disputes over who should be *kaitiaki* over tribal boundaries must be resolved by *tangata whenua* themselves. There is no role for the state in this process.

The second key principle is the ability to manage fishing activity and fisheries. *Kaitiaki* manage customary fishing through an authorisation system which requires them to specify the exact nature of the fishing activity that is being authorised, including species, quantities, areas, size limits, methods, purpose for which the fish will be used and instructions for the disposal of any bycatch. Each of these factors is entirely at the discretion of the *kaitiaki*, who must act within the bounds of sustainability and with due regard for the environment. The regulations also provide for the establishment of areas known as "*mataitai* reserves" over traditional fishing grounds. There is no commercial fishing permitted within these reserves and all non-commercial fishers, including non-*tangata whenua*, must act in accordance with bylaws made by the *kaitiaki* when fishing within the reserve area.

The third principle relates to the recording of accurate information on catch by a fishery. Fishers must report their actual catches to the *kaitiaki* who record the information for fisheries management and compliance purposes. *Kaitiaki* must report quarterly to the Ministry of Fisheries on how many of each species were taken from each quota management area within their traditional boundaries. That information is then used in the setting of sustainability measures and allocating the TAC amongst sector groups.

There is no legislative ability for the state to cap customary non-commercial removals from a fishery. An allowance is made for customary fishing in the annual process of allocating the TAC and this allowance is designed to reflect the estimated removals from that fishery in the coming year. The allowance is based largely on customary removals in the previous year although other information may influence the final quantum. The customary allowance is often referred to as being "uncapped" but the use of this term is misleading. Customary fishing is restricted – but by *kaitiaki*, not by the State. In reality a number of factors combine to limit customary take including the restrictions placed by the *kaitiaki*, the balancing of their non-commercial and commercial interests, and the harvest capacity of the area over which they have management authority.

Customary non-commercial fishing-rights have a *de facto* priority in the allocation of TACs for New Zealand fisheries. The right equates to a share of the TAC but it is not a fixed or proportional share. Any move to a fixed proportional share of the TAC is unlikely as long as recreational fishing continues to be managed by the state through blunt input controls with no recording of catch. The benefits and security offered by a fixed share can not be realised until all non-commercial fishers are able to demonstrate the same levels of accountability and control over removals exhibited by customary fishers and commercial ITQ holders.

The fourth key principle, and perhaps the most important one, is accountability. Individual customary fishers are accountable to the *kaitiaki* who authorises their activity. *Kaitiaki* are primarily accountable to the *tangata whenua* who appoint them, and to the Ministry of Fisheries, for the sustainable management of fisheries and for the maintenance of effective records for both management and compliance purposes. The state is still ultimately responsible for the overall sustainability of fisheries and for the provision of necessary information and assistance to *kaitiaki* to enable the effective implementation of the regulations.

3.4 The Chatham islands – allocation of fishing rights to a coastal community

As an aside from our examination of the indigenous fisheries settlement, it is relevant to note that the New Zealand government has also allocated ITQ in recognition of the social and economic needs of an isolated community. The Chatham Islands lie far off the east coast of New Zealand. The population of approximately 750 people is largely dependent on fishing for its economy. While some members of the population have received fishing-rights through the settlement of indigenous fishing-rights claims, others did not share these rights. The population base is not sufficient to maintain the infrastructure of the Chatham Islands through local government taxation.

When allocating ITQ at the time the Quota Management System was introduced, the government acknowledged the specific socio-economic circumstances of the Chatham Islands. The government recognised that while the principal purpose of the ITQ system was to provide for the economic rationalisation of commercial fishing and the sustainability of fisheries, the socio-economic needs of the Chathams were a special circumstance and special issue. The government allocated 2000t of ITQ to an Islands' Community Trust to enable it to generate income to fund social and infrastructural needs of the Islands.

3.5 Integration of fisheries rights – talking to each other

Māori, while often mistakenly considered a distinct stakeholder group, belong to all the key stakeholder groups in the New Zealand fisheries – commercial, recreational, customary and environmental. Their stake in both the commercial and non-commercial fishery means that Māori are well aware of the trade-offs that can be made to improve the returns to both commercial and non-commercial stakeholders. It is the clearly defined nature of the property-rights of each group that makes this possible.

In New Zealand, the government has tested a number of options for community participation in management. An example of how stakeholders can operate when presented with the incentives of a property-rights system is the joint management of the eel fishery in the South Island by Māori, commercial and recreational fishers, and the government. The eel fishery is of particular cultural importance to *tangata whenua* in the South Island. It is also a significant commercial fishery. The sustainability of the fishery itself is under threat from habitat degradation and erratic recruitment, which is exacerbated by commercial fishing.

The fishery required a change in management and, in particular, a cap on the harvest to ensure its future. The government offered participants in the fishery the opportunity to manage the fishery in conjunction with the Government. The broad terms of reference for management were that the fishery had to be managed sustainably, the customary fishing rights of *tangata whenua* had to be recognised and provided for in any management scheme, and the rights of commercial and recreational fishers had to be accommodated.

A Working Group of *tangata whenua* and commercial fishers considered all the management systems available and concluded that only a property-rights system would achieve the range of outcomes they sought. The property-rights system which has been developed includes a comprehensive planning process and regulatory supports to enable spatial allocation of fishing sites, and seasonal, size and method restrictions which reflect local realities and the biological complexities of the fishery.

The Working Group also recommended proportional allocation of the TAC. Twenty percent will be allocated to customary non-commercial fishing, a proportion to recreational interests to be held by the Crown pending reform of the recreational fishing-rights system, and 20% to Māori commercial interests via the Treaty of Waitangi Fisheries Commission. The balance will be allocated to existing commercial fishers.

The Working Group has established regional planning and management teams that develop and recommend rules for their area. Each group has been appointed as an advisory committee to the Minister of Fisheries with the authority to recommend management measures on all matters affecting the use and management of the fishery in their region. The management committees are now the primary source of advice to the Minister on the management of the eel fishery in the South Island. They have successfully developed complex rules to manage the Lake Waihora commercial eel fishery, which contains a large stock of migratory male eels, which do not grow to the normal legal size. Balanced with that is a customary fishery based on larger eels that migrate through the same area to spawn.

The local management focus has enabled the participants to set rules over the timing of fishing that enable most commercial activity to be focussed on male fish that will be lost to the fishery, while preserving larger eels for customary use. The end result has been an increase in the total allowable commercial catch in the fishery and the first successful customary harvests for decades.

4. CONCLUSION

The four steps outlined in this paper provide a framework for applying property-rights instruments to the task of recognising and providing for indigenous and coastal community fishing-rights. The four steps are to define the right, to quantify the right, to provide tools for exercising the right, and to create incentives for co-operative decision-making.

As the second half of the paper illustrates, this framework provides a context for examining recent events in New Zealand with regard to the recognition of Māori customary fishing-rights. In New Zealand, it was the introduction of ITQ property-rights for the commercial fishery that sparked the settlement of indigenous fisheries claims, not the other way around. The theoretical framework helps identify the progress made in providing for Māori customary fishing-rights, and where need go in the future.

There are several debates taking place in New Zealand about the implementation of the settlement of indigenous fishing-rights, including the allocation of commercial quota and assets among tribes and the acceptance of non-commercial customary fishing regulations. There are also a number of wider debates regarding the need to better define the rights and responsibilities of other sector groups, especially those of recreational fishers, to provide for the integration of all rights within a single framework. Again, the theoretical framework outlined in this paper provides a context in which those debates can be located.

The primary reason for introducing property-rights into fisheries management in New Zealand in the 1980s was the need to rationalise the fishing industry. For the then Government, the settlement of indigenous rights claims to fisheries was a necessary consequence of the introduction of property rights to the commercial sector, rather than an end in itself. It is only recently that the rights of non-commercial fishing sectors in New Zealand have been discussed in the context of property-rights instruments, and with social and cultural outcomes in mind rather than purely economic ones.

Many countries have not progressed down the property-rights track for management of their commercial fisheries, or taken steps to recognise and provide for the rights of their indigenous or coastal communities. To our minds, those countries have the luxury of a clean slate. We hope that the ideas expressed in this paper will aid in the consideration of property-rights instruments for achieving a range of outcomes beyond the economic and sustainability objectives often associated with fisheries property-rights.

THE IMPLEMENTATION OF AN ITALIAN NETWORK OF MARINE PROTECTED AREAS: RIGHTS-BASED STRATEGIES FOR COASTAL FISHERIES MANAGEMENT

F. Andaloro and L. Tunesi
Central Sea Research Institute (ICRAM)
Via di Casalotti 300 - 00166 Rome, Italy
<andalorf@tin.it> and <letunesi@tin.it>

1. A MEDITERRANEAN FISHERY OVERVIEW

The Mediterranean Sea represents only 0.8% of the planet's sea surface but is currently exposed to the pressure exerted by the presence of approximately 300 million people living along its coasts. This figure is expected to increase rapidly in the near future and poses serious concerns for the strains that will be exerted on the Mediterranean-basin marine ecosystem. The Mediterranean Sea has been one of the oldest and most important sources of food for the Mediterranean people and currently supports the employment of half million people in fisheries and in related activities.

With a total catch of 8 million tonnes the European Union (EU) is the third largest "country" in fisheries production, after China and Peru, but only 1 million tonnes of fish are caught in the Mediterranean Sea by the EU fishing fleets. In total 1 330 470t of fish (FAO 1997) are caught in the whole Mediterranean Sea including the large pelagic fishes captured by the Japanese fleet.

During the last thirty years the governments of the EU's coastal Mediterranean countries have provided incentives only for the industrial fishing fleets, excluding the small-scale fishery from any significant benefits. Thus, after the second world-war, the growing development of industrial fisheries caused a rapid over-exploitation of the main fish resources, a phenomenon that was largely tied to the characteristics of the basin of the semi-enclosed sea which exhibits high biodiversity, but lacks the large mono-specific fish stocks inhabiting the Atlantic Ocean. Today, despite modern fleets of small industrial vessels, the artisanal fishermen and their local communities still play a role in management and supply specialized fish-products to high-priced local outlets (Caddy 1999).

The main problems that limit the development of this artisanal activity are related to:

i. the power of the industrial fishery, representing a strong lobby, able to change the view of governments
ii. divisions within the artisanal category (fishermen are divided and sometimes also in contradiction between themselves)
iii. the greater interest that the EU/EEC fishery policy shows for the fisheries in the North Sea than the Mediterranean Sea
iv. the important effort exerted by the recreational fishery in Mediterranean coastal waters
v. the relevance of the tourist areas in the coastal zone and
vi. the relevance of environmental pollution in the coastal areas.

In addition to the above factors, there are the problems related to the relationships between the four EU coastal countries, which take approximately 80% of the Mediterranean catch, and the other 16 Mediterranean non-EU countries which fish in this sea and whose growing fishing fleets are acquiring fishing gears now forbidden by EU regulations. This phenomenon is happening at present in the sword-fish industry, because the drift-nets now banned by the EU countries, have since this prohibition been bought by other Mediterranean countries, which are also offering flags-of-convenience to EU vessels that have chosen to continue this type of fishing.

A common fishery-policy to solve this problem for the Mediterranean had been proposed in 1997 through the mandate given in this domain to the General Fisheries Council for the Mediterranean (GFCM-FAO). However, at the moment it appears that this authority is only theoretical. The only case of resource management in the wider Mediterranean was the introduction of a TAC in the tuna fishery in 1997, but the catch limit is valid only for the EEC fishing vessels for whom paradoxically, it is also difficult to determine a realistic, and verified, estimate of their catch.

2. A MEDITERRANEAN ENVIRONMENTAL PROFILE

The human impact on the marine environment is a function of the size of the surrounding population and of its level of socio-economic development. In the Mediterranean Sea the coastal zones currently support the bulk of the population and the growth of this human pressure results in increasing pollution of the Sea. Uncontrolled industrialisation of coastal areas, the discharge of industrial by-products in these areas, as well as the development of industrial-scale agriculture, have led to a great increase of adverse impacts on the Sea's marine ecosystems. Other important sources of pollution in the Mediterranean Sea are the dumping of chemical waste, oil and gas extraction, coastal aquaculture and shipping. Thirty percent of the world's marine oil and gas transportation occurs in the Mediterranean Sea and accounts for an estimated yearly spill in the sea of 600 000t of oil, catastrophes excluded.

Another important environmental problem is the deliberate, or accidental introduction, of exotic species. Some alien tropical species have immigrated from the Red Sea and Atlantic Ocean, as a probable consequence of global weather change, and are rapidly modifying Mediterranean biodiversity with unexpected consequences on the ecosystems.

3. THE ITALIAN FISHERY

The Italian fishery employs more than 45 000 fishermen and provides a total catch of 750 000t per year. The Italian fishing fleet is composed of:

1864 bottom trawlers
162 pelagic trawlers
285 small pelagic fish purse-seiners
850 dredges
9575 artisanal fishing vessels
3557 multigear fishing vessels
17 tuna purse-seiners
33 oceanic fishing vessels.

The artisanal sector is thus the most important Italian fishery activity in terms of number of fishermen and boats but they take only 12% of the total Italian catch while the trawlers provide another 30% and the multigear fishing vessels 35%.

The existing overexploitation has resulted in rapidly-decreasing catches per unit effort (CPUE) and the trawlers, needing to increase their fishing effort to maintain the catches, sometimes now are also fishing in shallow water, thus conflicting with the artisanal fishery. From an economics point of view, the artisanal fishery has low costs, obtains high quality products and does not produce discards (since highly selective gears such as trammel nets, long-lines seines and pots are used).

The artisanal fishery, notwithstanding the high number of fishermen it employs, is rapidly losing its culture and tradition, which are the most important aspects of an activity requiring a fundamental knowledge of fish behaviour and of the environment. Only an Integrated Coastal Zone Management scheme can enable the recovery of the artisanal fishery in Italy by permitting the resolution of conflicts and a reduction of pollution. Italian Law, at present, does not permit the direct management of national marine waters for fisheries, where the resources can be exploited by professional and recreational fishermen everywhere but fishery associations or co-operatives cannot propose any self-management projects. Other problems for the Italian fishery to adopt-property rights management strategies are social issues and the power of the recreational-fishery groups. Moreover, artisanal fishermen are divided and often in conflict with each other, notwithstanding this they are often associated in co-operatives.

4. ITALIAN MARINE PROTECTED AREAS NETWORK AS A PROPERTY RIGHTS-INSTRUMENT

Marine Protected Areas (MPAs) have a strategic role in the framework of integrated coastal management (ICM). An effective ICM strategy cannot occur in the absence of a solid scientific basis and activities. Natural science is vital for understanding the functioning of ecosystems, while social sciences are essential for comprehending patterns of human behaviour that cause ecological damage and for finding effective solutions.

The Central Institute for Applied Marine Research (ICRAM), is the Italian government body charged with providing scientific support to the public administration on matters related to marine conservation and its management policy. In particular, ICRAM has legal obligations in terms of providing guidance on the institution of marine protected areas and was recently charged by the Minister of the Environment to co-ordinate all research and monitoring activities in Italian marine protected areas.

ICRAM activities in support of the national objective for Marine Protected Areas in Italy are:

i. to guarantee scientific support to the Italian Ministry of the Environment by providing standards for the institution of new Marine Protected Areas (MPAs) and management of all the MPAs
ii. to support the managing bodies established for MPAs by co-ordining monitoring activities and research activities
iii. to indicate modalities for the training of personnel
iv. to provide adequate documentation and information.

Italian legislation identifies 48 marine sites deserving special protection due to their great natural interest. To date, 15 marine protected areas have been instituted and a limited number have fully-operational management schemes. The relations between MPAs and artisanal fishery in Italy are important, because about 10% of all the Italian fishermen work in the areas covered by the MPAs and many of the 48 MPAs are small islands with existing fishery activities.

Approximately 3 years ago, an important relationship was begun between Italian fishermen's associations and the environmental NGOs (previously in strong conflict among themselves). In this context the fishermen's associations acknowledged the negative ecological role brought about by illegal fishing and began to address the issues proposed by The Code of Conduct for Responsible Fisheries (FAO 1995).

The Italian national approach to the management of MPAs specifies where artisanal fishing activities are allowed (zones B and C), and for local fishermen only. Consequently MPAs have become the principal means in Italy to: apply fishery management strategies that are environment-friendly, adopt property-rights models, study the sustainable development of marine living resource and, apply the precautionary approach in the fishery.

The creation of the network of 48 MPAs may provide the means to meet the challenge of conservation and fishery management in Italy and the possibility of obtaining regulations for these areas that can be used as an instrument of property-rights for resident fishermen.

5. LITERATURE CITED

Caddy, J.F. 1999. Fisheries management in the twenty-first century: will new paradigms apply? *Reviews in Fish Biology and Fisheries*, **9**: 1-43.

FAO 1995. *Code of Conduct for Responsible Fisheries.* FAO. Rome, 41 pp.

FAO 1997. *The State of the World Fisheries and Aquaculture 1996.* FAO, Rome, 125 pp.

Tunesi, L. and G. Diviacco 1993. Environmental and socio-economic criteria for the establishment of marine coastal parks. *Intern. J. Environmental Studies*, **43**: 253-259.

ESTUARIES IN WESTERN AUSTRALIA – AN INTEGRATED APPROACH TO MANAGEMENT

J. Borg
Fisheries Western Australia
168-170 St Georges Terrace Perth WA 6000, Australia
<nborg@fish.wa.gov.au>

1. INTRODUCTION

While there are a large number of estuaries in Western Australia, only a few of these are open to commercial fishing. Those near Perth have among the oldest fisheries in the State and have been fished commercially and recreationally since the founding of Western Australia in 1829. Conflict between different sectors is recorded as far back as 1840.

Estuaries occur at the mouth of rivers where the saline water and tides of the ocean influence the riverine environment. The river brings with it sediment and nutrients which produce ideal nursery environments for for many types of animals - prawns, crabs, fish, birds. Although productivity is generally high, there is not a wide range of species in these environments and their abundance varies greatly throughout the year. This productivity, and the easy accessibility of estuaries, has meant that man has traditionally used the estuaries as a food source. These same factors have also made them attractive for recreational purposes.

Estuaries therefore present a particular challenge to fisheries managers - there is a highly-variable natural environment independent of fishing activity; this is further influenced by man-made interference in the environment which is complicated further by extractive uses such as commercial and recreational fishing. In the absence of a rights-based management regime, it is even more difficult to maintain fishing levels so that sustainability is assured.

As the management of each estuarine and marine embayment fishery in the State has been based on the same philosophy, this paper concentrates on the estuaries of the lower west coast of Western Australia to illustrate what has happened, why, and where Western Australia hopes to go from here in terms of rights-based management of these fisheries.

2. DESCRIPTION OF FISHERIES

In the mid 1970s, the lower west coast estuarine fisheries (Figure 1) were grouped as follows for commercial fisheries management purposes: Hardy Inlet Estuarine Fishery; Leschenault Estuarine Fishery; Mandurah Estuarine Fishery (Peel Harvey estuary); and Swan-Canning Estuarine Fishery. Other estuaries in this region were closed to commercial fishing. Commercial netting in all the major estuaries provides a valuable source of fresh fish for local and metropolitan Perth consumption. These fisheries were once the key source of fish for the State but now have more significance as a supply of local fresh fish and a lifestyle for participants.

The West Coast estuaries are also a principal focus of the State's recreational fishing and boating activity. This extends right across the range of fishing pursuits, including crabbing and prawning, as well as line fishing for species such as whiting, flathead, flounder, garfish, trevally, black bream, mulloway and tailor. In some estuaries such as the Peel Harvey and Leschenault, recreational netters also take mullet.

Much of the State's coastal urban development has focused on these estuaries, particularly the major estuaries. They are all widely used for recreation, including fishing, and being close to metropolitan Perth and the major southern tourist destinations, pressure has been applied by recreational fishing groups to have commercial fishing removed, or substantially decreased, in these estuaries.

3. HISTORY OF MANAGEMENT

Western Australia's estuaries have been fished for as long as there has been human habitation nearby. By the late 1840s there was already concern over the overexploitation of stocks in Western Australia's estuarine fisheries and closed-waters regulations were introduced. There was also conflict between commercial fishermen and amateur fishermen and other recreational users with regard to commercial fishing in the Swan-Canning estuary.

By the 20th century, a number of commercial estuarine fisheries had been established, however there were many problems, largely to do with marketing and competition for use of the estuaries. During the early 20th century, the activities of amateur fishermen had become intense, particularly in the Peel-Harvey and Swan-Canning estuaries, where in some years the amateurs caught as much as the commercial fishermen. Amateur fishing also intensified in the Leschenault estuary, which at this time was totally closed to all forms of netting.

By the late 1920s and early 1930s commercial catches from estuaries in south Western Australia had begun to increase. Growth in estuarine fishing continued from the 1930s through to the 1960s when an export market developed for rock lobster and later prawns, scallops and abalone. This resulted in a move by fishermen away from finfish fisheries. Markets for the small estuarine fish shrank considerably and demand for high quality species such as whiting and cobbler exceeded supply from estuaries. These fish were not in sufficient quantities to support the large number of estuarine fishermen operating in the state.

Figure 1
Lower west coast estuarine fisheries

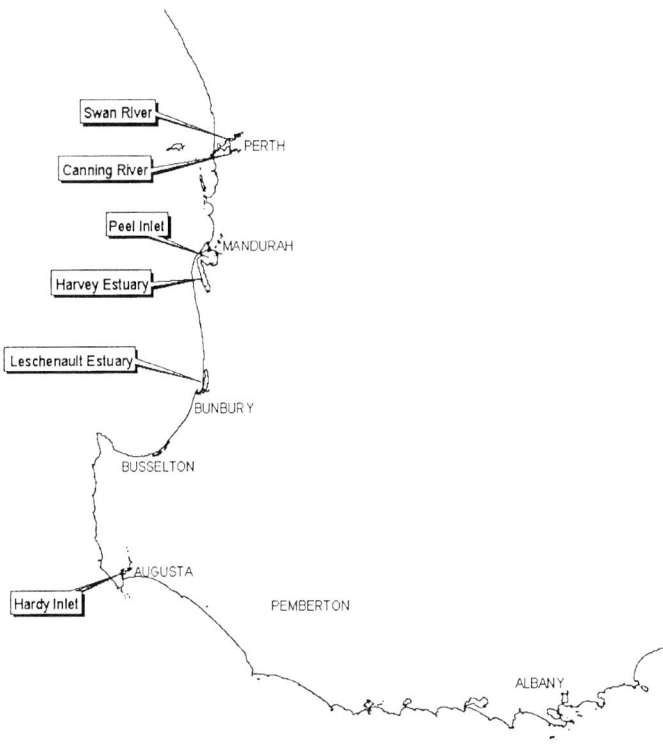

In 1969, the Government commissioned a study into the wholesale and retail marketing of wetfish in Western Australia (WD Scott & Co 1969). The study concluded that there were far more commercial fishermen than the estuaries could support and using economic indicators of the time, indicated estuarine fishing offered "full-time employment for only 20-30 fishermen.". It also noted that the estuaries were becoming increasingly used for recreational purposes. It recommended that "estuarine fishermen be given no grounds to expect any improvement in their economic position and be encouraged to leave the industry".

The then Department of Fisheries and Wildlife accepted the recommendations of the Scott report and put in management measures aimed at reducing the number of commercial fishermen to a level where each received a worthwhile economic return. The measures implemented were: no issue of any new estuarine licences and no renewal of estuarine fishing licences that had expired, either as a result of the death of the fisherman or the desire of the fisherman to leave the fishery.

These measures were not set within a strategic framework, nor legislated as management plans, rather they were managed through a series of Fisheries Notices and Ministerial Directions which restricted access, area of fishing and gear types and quantity. Such measures afforded fishermen no legal long-term rights; nor did they address the impact on the resource of fishing by recreational fishermen - it was not a multi-sectoral approach.

Lenanton (1984) noted two problems with the commercial fishing management arrangements for these fisheries: (a) management of the level of commercial take was not sufficient and (b) amateur fishermen were capable of fishing stocks to a low level as they were not concerned with the economic return from their efforts. He noted "sustained intensive amateur effort may seasonally reduce the abundance of certain species to levels which would make commercial exploitation unacceptable economically. Therefore it would be useful, and perhaps in the very near future necessary, from the point of view of responsible management of our estuarine fisheries, to determine the relative proportions of the estuarine fish resource taken annually by the licensed and unlicensed components of the amateur fishery".

Despite 150 years of commercial and recreational fishing conflict and the scientific backing for a multi-sectoral approach to management, there was not the political will to address management of the ever-growing recreational sector. Management continued to concentrate, and restrict, the commercial sector.

3.1 Adjustment over time

Although there have been occasions in the past where management measures in the estuarine and marine embayment fisheries have been introduced for reasons of both resource-sustainability and resource-sharing, the majority of management issues in these fisheries now revolve around resource-sharing.

It has been recognised by both industry and government over the years that there are competing uses for the state's estuaries. Various regulatory tools and adjustment mechanisms have been applied to the commercial fisheries in an attempt to further control the commercial take in the estuaries. Table 1 shows the reduction in commercial fishing licences in these estuaries over the last two decades. However, it is only in recent years that initial attempts have been made to quantify the recreational fishing effort in these estuaries and to start to identify the relative shares of the two fishing sectors.

Table 1
Number of commercial fishing units in the lower west coast estuaries

Fishery	Number of fishing units			
	1979	1987	1/1998	10/1998
Hardy Inlet estuarine	6	6	4	2
Leschenault estuarine	18	14	7	6
Mandurah estuarine	45	41	24	14
Swan-Canning estuarine	32	17	8	6
Total	**103**	**78**	**43**	**28**

In the past, the Government has responded to the needs of user-groups by introducing controls on commercial fishing through regulatory mechanisms - particularly temporal and area closures and non-transferability of licences - and through buyback schemes. However, it has become increasingly obvious that controls on commercial fishing are not sufficient; management needs to integrate, and apply to, all users of the resource.

Regulatory mechanisms

Associated with each estuary or marine embayment is a series of management arrangements, initially made up of a combination of legislation and policies. Legislative tools are found in the *Fish Resources Management Act 1994* (FRMA) and the *Fish Resources Management Regulations 1995*. The legislation and policies put in place restriction the number of participants, the time the fishery is open for commercial fishing, the areas which are closed to commercial fishing, the type and amount of gear that can be used, and in some cases, the species which can be taken or kept from these commercial operations. One of the most successful regulatory tools was the imposition of non-transferability, tempered in some fisheries by the 'grandfather' clause (*i.e.* allowing transfer to family members).

Fisheries WA is currently developing legislative options for repackaging the current management arrangements, along with any changes that may be negotiated with industry. Despite this, regulatory mechanisms have had little success in restructuring fisheries to meet management objectives unless they are supported by economic incentives.

3.2 Buyback strategies

Although a legislative solution, the *Fisheries Adjustment Schemes Act 1987* (as amended) provides the means to reduce commercial fishing effort in open access fisheries, inshore/coastal fisheries and in estuarine and marine embayment fisheries through buyback of fishing licences. This scheme is to be closed in 1999.

The resource sharing initiative

In 1996 the Government committed $A8 million over four years for voluntary resource-sharing. The initiative had two facets: (a) the Resource Reallocation and (b), Licence Buyout Initiative and the Guidelines for Voluntary Resource Sharing. The *Guidelines for Voluntary Resource Sharing* initiative provided Fisheries WA with a process for addressing resource-sharing issues. The objective is to provide an administrative means for achieving defensible voluntary resource-sharing arrangements among interested sectors. To date, two rounds of this process have been initiated, with negotiated agreement reached on one fishery.

In relation to the Voluntary Fishery Adjustment Schemes targetted schemes for fisheries were nominated by the Voluntary Resource Reallocation and Buy-out Committee of Management or through the *Guidelines for Voluntary Resource Sharing* process. Between January and May 1998, Voluntary Fisheries Adjustment Schemes operated for each estuarine fishery. These schemes were successful in reducing the number of fishing units in estuaries on the lower west coast from 43 in January 1998 to 28 in October 1998.

3.3 Strengths and weaknesses of approach at the time

The management approach taken in these estuarine fisheries was consistent with fisheries management theory throughout the world - single sector management, commencing with the fishing activities that were quantifiable and controllable. Although in retrospect, a rights-based, integrated approach from day 1 may find us in a better position today, neither the concept of allocating rights, nor of integrating management was envisaged in the mid 20th century when fisheries management commenced in Western Australia.

The following strengths and weaknesses of this approach have been identified:

Strengths
i. it dealt with an activity that was quantifiable - commercial fishing
ii. it reduced numbers with minimal pain to individual operators
iii. it was cost-effective in terms of the drafting and administering of formal management plans and
iv. it was politically acceptable and achievable.

Weaknesses
i. it avoided the controversial activity of recreational fishing
ii. it did not impart responsibility in the minds of recreational fishermen

iii. it was not based on a strategic framework so there was always the debate about what rights each fisherman held to the fish resources
iv. it was expensive in that the lack of clearly-defined rights meant constant conflict, consultations negotiations and reworking of management arrangements and
v. in some instances, government had to buy out fishermen for far more than the licences were worth.

4. A NEW APPROACH - RIGHTS-BASED MANAGEMENT

4.1 Long-term objectives

Fisheries management needs to take a long-term view and the question is: where do we want these estuarine fisheries to be in the next 20 years? Although the push from some sectors of the community is for the total removal of commercial fishing from estuarine and marine embayment fisheries, commercial fishing in these fisheries plays an important role. These fishermen:

i. collect data for research purposes - long-term data series on catches and variations in stocks is essential to the management of these systems (as it is for recreational fishermen)
ii. are custodians of the resource, the "eyes and ears" to report any potential problems and information on the environment and health of estuaries (as are recreational fishermen)
iii. supply fish to the local community, tourists and the metropolitan markets
iv. take a range of fish not exploited by recreational fishers, such as yelloweye mullet and Perth herring
v. collect fish and provide knowledge for aquaculture ventures.
vi. have the potential to participate in local tourism, not only through direct contact, but through ventures such as developing regional cuisine based on local fish.

Commercial fishermen, because of their consistent day-to-day and long-term involvement in fisheries of the estuaries, provide both quantitative and qualitative data. As commercial fishing becomes a yet smaller component of fishing activities in estuarine and marine embayment fisheries, supplementary research data will be needed, collected on a structured basis, from recreational fishermen, but no funds to do this are yet available.

4.2 The major management issues that affect rights-based management of estuarine fisheries

4.2.1 Fish stocks

Fisheries WA (1998) identified a number of finfish stocks that were fully over-exploited. They included some locally-depleted cobbler and black bream stocks in the temperate estuaries, however, for the most part, fish resources in West Australia's estuarine fisheries are fished commercially at sustainable levels. Although, at present, sustainability of most fish stocks is not the main issue in the these fisheries, there is still a need to identify target catch and effort reference points and to establish the relative catch share between the commercial and recreational fishing sectors.

Given the current lack of recreational fishing data, the short-term objective is to identify commercial catch trends that may indicate the commercial fisheries are taking up catch that would have previously been taken by fishermen who had left the fisheries under the Voluntary Fisheries Adjustment Program. It is also important that fishing catch and effort for species that are of concern in terms of sustainability be monitored. Collecting long term catch and effort data is also one of the most cost-effective ways of amassing data for setting target effort levels.

4.2.2 Environmental factors and resource variability

There are a number of natural environmental factors which affect production in an estuary, salinity, temperature, oxygen levels, and turbidity. Those species that adapt to varying levels of these factors, especially salinity, will be those which appear regularly in the annual commercial catch (Lenanton 1984).

Human interference in this already highly-variable environment further influences commercial production in the estuarine and marine embayment fisheries. Such effects include clearing vegetation in the catchment, area, agricultural runoff of silt and pesticides, discharge from adjacent industrial developments, clearing of seagrass and dredging. Mining in the 1970s adjacent to the Blackwood River was brought to an end because of the consequences in the river; and pesticide runoff in the Swan River recently caused a massive kill of bream.

Eutrophication is another problem in nearly every estuarine and marine embayment fishery along the coast - elevated nutrient levels result from clearing catchments and applying fertilisers. It was so severe in the Peel-Harvey Inlet that the Dawesville cut had to be made to clear out the affected water. As population spreads, so does the urbanisation of the foreshores. The shallow-water feeding grounds for fish are probably the most vulnerable to degradation as vegetation is cleared and silt enters the estuarine systems.

Urbanisation has also made some positive contributions to the environment. Canal developments, although they alter the habitat and therefore contribute to environmental problems, have in some instances provided an increase in fish numbers. Cobbler, for example, burrow under the walls of the canal which provides a safe habitat for them when they breed.

Fishing also has an impact on the environment. The removal of fish from the estuarine fisheries not only affects the numbers of target fish-species but can affect the food chain of other marine animals and the bird life that feed on them.

4.2.3 Resource-sharing

Resource-sharing is the key factor in the restructuring of Western Australia's estuarine fisheries. Historically, there has always been competition for the estuarine resources and their environment. As the population spreads from the city and major regional towns, the pressure to prevent or resolve conflict increases.

There are a large number of stakeholders in the estuarine and marine embayment fisheries along Western Australia's extensive coastline commercial fishermen, recreational fishermen, Aboriginal communities, aquaculturalists, conservationists, aquatic charter and fishing tour operators, those involved in the tourism industry, and the many downstream businesses that support each of them. There is also the Australian public who gains satisfaction from knowing that the State's estuarine and marine embayment fisheries are well managed and hence that the fish resources are sustainable.

4.2.4 Latent effort

Although various management tools have been successful in reducing the level of latent commercial fishing effort in the estuarine fisheries, there is still latent effort in some of the fisheries. This is largely because:

i. some participants are involved in diverse fishing operations - estuarine fishing is only part of their operations
ii. variation in market prices affect their profitability from time to time and, or
iii. low-level fishing activity by part-time operators, semi-retired operators or fishermen approaching retirement.

One of the tools that has assisted in control of latent effort is the prohibition on transferability of licences in these fisheries. Without transferability, there is little danger of existing latent effort being reactivated and so there is little need to further reduce commercial fishing effort. With transferability, experience has shown that new owners tend to operate differently to previous owners and in some cases increase fishing effort. Under these conditions, target numbers and revised management arrangements would have to allow for possible re-activation of latent effort.

4.2.5 Target levels

The determination of an acceptable number of fishing units for each fishery is required to address concerns over the re-activation of latent effort, fishing sustainability and economic viability of the remaining operators.

Fisheries WA is about to commence negotiations with industry over appropriate target-levels for key species in each estuarine fishery. Preliminary target-levels have been developed by Fisheries WA as a starting point for these negotiations. The targets are based on qualitative judgments, past activity and catches in the fisheries, the size of the estuaries, the need for coverage in terms of collecting research data, and the demand for use of the estuaries by other stakeholders. They are also premised on no change to existing management arrangements. Any change to fishing habits, such as more intense fishing by new entrants if transferability is introduced, would affect target harvests.

Although there will be temptation to yield to community pressure wanting to severely restrict commercial fishing in these estuarine fisheries, consideration needs to be given to the affect target harvest-levels will have on the commercial viability of the fishery - will resource supply be sufficient for processors and service industries, and for the market so that demand for these fisheries can be maintained?

The harvest-levels will determine the resource shares in these fisheries and the intention would then be for minimal cost-effective management. The resource-shares would provide an upper limit on the commercial harvest from estuarine fisheries. This system is not without problems. In addition to the influences mentioned above, it also requires consideration of the total catch, recreational and commercial, as well as the catch composition in the relevant years. It must also recognise that the catch in any year is affected by factors such as the environment, market demand, and recreational fishing.

4.2.6 Transferability

As a result of the recommendations of the Scott report, transferability of licences to operate in most estuarine fisheries has been essentially prohibited. However, in some fisheries, and subject to certain conditions, direct descendants of licensees have been granted various types of licences as assistants and trainees, eventually resulting in full access on application when the father/grandfather left the estuarine fishery.

The current transfer policies in the south west estuarine fisheries are:

i. Swan-Canning - no transfers, even between family members
ii. Peel-Harvey - transfers permitted between family members after consideration of individual circumstances
iii. Leschenault - transfers permitted between family members after consideration of individual circumstances and
iv. Hardy Inlet - no transfers, even between family members.

Despite these limitations, this personal access has had some of the nature of a property-right as fishers were able to surrender their access to the estuary, together with their associated fishing dinghy licences, to the General Fisheries Adjustment Scheme — a joint industry/Government funded initiative and, more recently, to a specific Government-funded series of schemes.

Transferability is a major issue in the management of most of these fisheries. Its introduction is likely to be tied to a series of trade-offs, not only in terms of reaching fishing-effort target-numbers, but also recreational fishing concessions. It should also be noted that full transfer would have the potential to increase the number of active participants and release latent effort. To account for this, management would need to incorporate measures to reduce the commercial effort to agreed target-levels, and hence restore the resource-shares between the sectors.

However, it may also be possible to use the market to restore resource-shares.

5. TOWARDS A STRATEGY

To a large extent, many resource-sharing issues have been settled for the estuarine fisheries on the west coast through the use of Voluntary Adjustment Schemes and the *Guidelines for Voluntary Resource Sharing* process. Some of the remaining resource-sharing issues could be resolved by moving commercial fishing effort on the species of interest to the recreational sector. The debate is now largely on the level of activity - are the target-levels proposed appropriate? In entering this debate, all parties are having to realise that reducing commercial fishing below the levels proposed raises questions as to the logic in managing the commercial fisheries at all. Their contribution to the total take from the fisheries will be small and the research data collected of little use - these fisheries may even cease to be commercially profitable. In addition, there would be the temptation for the market-demand for locally caught fresh fish to be filled by illegal sales of recreationally- caught fish. The debate must now consider how we reach these targets, how the fisheries are managed once the targets are attained, how we manage the overall sustainability of individual species in this scenario, and must take into account the large cost associated with complex management arrangements. Complex management is not required.

Licence numbers in these fisheries would be moved towards power levels and transferability introduced once these targets are reached. However, as transferability often results in an increase in fishing efficiency, an adjustment mechanism would be an essential part of any management arrangements. One option for adjustment is unitisation of effort. The system of individual transferable effort units (ITEs) involves allocating each fisherman a number of days each season/year in which particular amounts of gear can be used in a fishery. But, whatever tools are chosen for the future management of these fisheries, management must be simple and cost-effective.

A number of specific proposals have been posed and will be the basis of consultation with industry groups over the next few months, including:

i. possible removal of commercial fishing from one estuary

ii. retention of low-level commercial fishing in *representative* estuaries and perhaps further reduction in others, especially if commercial licences become transferable

iii. target fishing-levels set for each estuary

iv. transferability of licences, but only when effective controls on total commercial exploitation are in place, and fishing is subject to the provision of detailed research records and

v. commercial effort limited and managed, taking into account fishing and gear efficiency changes over time as a result of transferability and the capacity of each estuary.

6. CONCLUSIONS

Fisheries Western Australia has come to see the importance of developing management within a strategic framework that incorporates all user-groups into the management framework and consultative processes. Having the vision, however, does not guarantee success. The setting of long-term target levels and resource shares, and eventually the allocation of property-rights, requires political to accept the concept of integrated resource management, especially in estuaries where the conflict between commercial and recreational fishermen has been long-standing and public. There is still some work to achieve this political will in Western Australia, however, it is hoped that the extensive consultative process that commenced this month, will educate the community and turn the political tide.

7. LITERATURE CITED

Fisheries WA (Western Australia) 1998. State of the Fisheries Report 1997/1998.

Fisheries WA, 1999. Management Paper No 131 - Management Directions for Western Australia's Estuarine and Marine Embayment Fisheries.

Lenanton, R.J.C. 1984. Report No 2: The Commercial Fisheries of Temperate Western Australian Estuaries: Early Settlement to 1975, Department of Fisheries and Wildlife, Western Australia.

W.D. Scott & Co Pty Ltd 1969. A Report to the Honourable, the Minister for Fisheries and Fauna on a Pilot Study into the Economic Future of the Wet Fish Industry in Western Australia.

UNITED STATES' FISHERY COOPERATIVES: RATIONALIZING FISHERIES THROUGH PRIVATELY NEGOTIATED CONTRACTS

J. Leblanc
National Fisheries Institute
1901 N. Fort Myer Drive, Suite 700 Arlington, VA 22209 USA
<jleblanc@nfi.org>

1. INTRODUCTION

In the *Sustainable Fisheries Act 1996*, a reauthorization of the *Magnuson-Stevens Fishery Conservation and Management Act*, the United States' Congress imposed a moratorium on new Individual Fishing Quota (IFQs) programs in federally-managed fisheries. This moratorium was imposed after considerable congressional debate over IFQs and their use as a fishery management tool to address two issues in particular, overcapacity and the "race for fish". The focal point for this debate was the Pacific Northwest. United States' Senators from Washington State were at loggerheads with the Senators from Alaska, the Washington State members supported the ability of fishery managers to use IFQs and the Alaska members wished to impose a moratorium. The Alaskan Senators won.

The Washington state constituencies supporting IFQs included the Seattle, Washington-based catcher-processor fleet of approximately 34 vessels engaged primarily in the harvest of North Pacific pollock. This fishery, under the jurisdiction of the North Pacific Fishery Management Council, was severely overcapitalized. Capital stuffing in both the offshore (catcher-processor) and onshore (catcher vessels delivering to shoreside plants) during the 1990s had added three times the capacity needed to harvest the annual pollock total allowable catch (TAC). As a result, the length of the fishery was reduced from 12 months in 1990 to 3 months in 1998.

The catcher-processor fleet, harvesting and processing approximately 55% of the two billion pound annual pollock harvest, sought IFQs as a means to rationalize their fishery and thereby reduce the number of vessels engaged in the fishery and lengthen the fishing season. However, in the face of the congressional moratorium on IFQs until 2000 (with the political climate making an extension of the moratorium probable), the North Pacific catcher-processor fleet sought an alternative solution.

2. THE PACIFIC WHITING FISHERY

Four of the companies engaged in the North Pacific pollock fishery also operate ten catcher-processor vessels in the considerably smaller Pacific whiting fishery off the coasts of Washington, Oregon, and California; these operations are under the jurisdiction of the Pacific Fishery Management Council (PFMC). They faced similar challenges in this fishery. The ten vessels represented significantly more capacity than necessary to harvest the allowable TAC and the companies were engaged in a race for fish against one another.

The annual whiting quota in recent years has been approximately 210 000t. Under the PFMC, this quota is divided among three sectors of the fishery; 42% of the harvest is reserved for vessels delivering their catch to shoreside processing facilities, 24% is reserved for catcher vessels delivering to at-sea processing vessels, and 34% reserved for catcher-processors. A sector-specific federal limited-entry licence must be purchased to operate in the fishery.

As with pollock, whiting is primarily processed into surimi. Whiting is harvested using mid-water trawls. It is one of the cleanest fisheries in the world, with a bycatch rate of approximately 1%. Nonetheless, the fishery does have an incidental take of particularly sensitive species, yellowtail rockfish and endangered Chinook salmon.

Halfway through the 1997 season, four companies operating the ten catcher-processors vessels in this fishery formed the Pacific Whiting Conservation Cooperative (PWCC) which is organized as a nonprofit corporation under the laws of the state of Washington. Its purposes are:

i. To promote, through mutual cooperation of its members, the intelligent and orderly harvest of whiting in the federal Pacific coast whiting fishery
ii. To reduce waste and improve resource utilization and
iii. To reduce incidental catch of nontarget species.

From an operational standpoint, the PWCC members believed that the Cooperative would improve the fishery and established the following goals:

i. Elimination of the race for fish and removal of incentives to catch as much fish as possible, as fast as possible, and to substitute as the primary incentive increased efficiency, by allowing vessels to concentrate on product quality, recovery, and bycatch avoidance
ii. Cooperation to improve the efficiency of the harvest by using an independent monitoring service and sharing catch and bycatch information
iii. Conducting and funding research for resource conservation, including catch and bycatch monitoring, observers, stock assessment, and other scientific research.

To achieve these goals, the members of the PWCC entered into a legal contract that apportioned the whiting harvest among those qualified under federal regulations to participate in the catcher-processor sector of the fishery. This contractual agreement among members to harvest no more than a specific percentage of the sector allocation removes the incentive to race to catch as much fish as possible. No matter how fast or slow cooperative members run their operations, they are guaranteed a certain and specified amount of the whiting quota. To ensure compliance, the contract contains substantial financial penalties for members exceeding their share of the quota or violating other conditions of the contract.

It is important to point out that the PWCC is not involved in matters relating to pricing or marketing of whiting products. To ensure that the PWCC was not in violation of United States anti-trust laws, the PWCC requested that the U.S. Department of Justice's Anti-Trust Division review the proposed cooperative agreement. The DOJ provided a 'letter ruling' on May 20, 1997 which stated:

"The Department of Justice has previously stated that reliance on an olympic (sic) race system to gather a fixed quota of fish 'is both inefficient and wasteful' because it is likely to generate 'inefficient overinvestment in fishing and processing capacity'...To the extent that the proposed agreement allows for more efficient processing that increases the usable yield (output) of the processed Pacific whiting and/or reduces the inadvertent catching of other fish species whose preservation is also a matter of regulatory concern, it could have procompetitive effects." -- Acting Assistant Attorney General, Joel Klein.

With this affirmation from the U.S. Department of Justice, the PWCC was initiated. The certainty provided by the Cooperative allows the member companies to optimize the amount of capital they place in the fishery instead of maximizing it. Members no longer need to catch as much fish as possible as quickly as possible to outcompete the other members. The most competitive firm is no longer the one the catches the most fish, rather, it is the firm that makes the most from each fish caught. This has allowed the members of the Cooperative to reduce capacity and increase efficiency.

Since the inception of the Cooperative, the number of catcher-processors engaged in the fishery has been reduced by 30% from ten to seven vessels. In addition, vessel operations have shifted from the harvesting rate controlling the plant processing rate to situations in which optimizing the processing process now defines the harvesting rate. The fishery has completely shifted from input-controlled to output-controlled.

In 1997, prior to the Cooperative, the catcher-processor fleet had a recovery rate for the production of surimi from whiting of 17.2%. After the implementation of the Cooperative, the recovery rate increased to 20.6% while motherships (processing-only vessels that take fish over the side from catcher vessels and remained in an open-access fishery) had an average recovery rate of 17% for the season. This dramatic increase in utilization of the resource resulted in the production of an additional 5 269 435lb of food from the same amount of fish!

In 1998, the Cooperative members increased the recovery rate to 24% while motherships remained at 17%.

In addition, there was a significant shift by some members away from surimi production and into fillet and mince block production. The President of the PWCC, John Bundy of Glacier Fish Company, considers this shift to be an important benefit of the Cooperative as well. In the past, some catcher-processors had attempted to make a block product from whiting but because of the nature of the fish and adverse consequences to quality caused by the race for fish, such product was of poor quality and difficult to produce. The Cooperative, through slower fishing and processing, has allowed some catcher-processors to make a good, high quality block that has sold for significantly higher prices in the U.S. and Europe than if the whiting had been used for surimi for sales to Asian export markets.

In addition, by eliminating the race for fish, members of the Cooperative are able to take bycatch avoidance measures without suffering adverse competitive impacts. The certainty of a fixed percentage of the harvest allows vessels experiencing a relatively high encounter-rate with prohibited species to cease fishing operations and move to areas with a lower incidental catch rate without losing any competitive advantage. Before analyzing the bycatch rates experienced by the whiting catcher-processor fleet, it is important to point out that the baseline bycatch rates of this fishery are extremely low.

In 1997, prior to the Cooperative, catcher-processors caught 2.47kg of yellowtail rockfish and 0.009 individual Chinook salmon per tonne of whiting. After the implementation of the Cooperative, bycatch rates declined to 0.99kg of yellowtail and 0.008 individual Chinook salmon, compared with 3.43kg of yellowtail rockfish and 0.017 individual Chinook salmon per tonne for catcher vessels delivering to motherships.

In 1998, the catcher-processor fishery had bycatch rates of 0.96 kg of yellowtail rockfish and 0.008 individual Chinook salmon per tonne, while the catcher vessels delivering to motherships had rates of 6.51kg/t and 0.02/t, respectively.

In 1999, the Cooperative experienced somewhat higher bycatch rates of yellowtail rockfish resulting from a shift in fishing effort from south of the Columbia River to more northern waters with a higher abundance of yellowtail rockfish. Fishing patterns were altered because changes in environmental conditions (primarily higher ocean temperatures) affected the distribution of Pacific whiting stocks.

The PWCC has demonstrated that a cooperative approach to fishery operation can end the race for fish, reduce capacity, increase utilization, and decrease bycatch. It is also apparent that in order for a cooperative to be possible, certain prerequisites must be met. They include:

i. a defined harvest opportunity, or hard TAC
ii. a defined class of participants
iii. a closed system or limited access and
iv. strong, if not unanimous support among participants in the closed class.

In the absence of these conditions, there is no way to determine what is being divided up or among whom. In the absence of universal support for the cooperative, the entire effort falls to pieces. If a single, qualified participant in the fishery continued to engage in the race for fish, it would likely undermine the certainty of harvest opportunity for the cooperative members upon which the whole enterprise depends.

As impressive an accomplishment as the Pacific Whiting Conservation Cooperative is, one must confess that coordinating agreement among 4 companies with ten vessels is relatively easy. How could a Cooperative work in the North Pacific pollock catcher-processor fishery with twice as many companies and three times as many vessels?

3. NORTH PACIFIC POLLOCK

3.1 Antecedents

In 1998, the U.S. Congress passed the *American Fisheries Act* (AFA) and paved the way for fishery cooperatives in the North Pacific pollock fishery by fulfilling the prerequisites necessary for a successful cooperative:

3.2 A defined harvest opportunity, or hard TAC

The only condition already in place without passage of the AFA was the establishment of defined total allowable catch. The pollock fishery, which occurs within the U.S. 200 mile Exclusive Economic Zone off Alaska, is the largest U.S. fishery by volume. Over two billion pounds of pollock are landed annually. The North Pacific pollock resource is healthy and the fishery is managed conservatively to promote sustainable use. The groundfish fisheries are healthy because federal fishery scientists and managers set an allowable harvest level each year and fishing stops when the quota is reached. The TAC is set at, or below, the allowable biological catch (ABC) level, which is the amount of pollock that fishery scientists and managers determine can be harvested on a sustainable level. Accurate catch measurement is assured by a comprehensive set of federal rules, which provide for extensive federal fishery observer coverage and strict catch reporting requirements. All fish caught counts against the annual quota.

The pollock resource is harvested using mid-water trawl nets. The United Nations' Food and Agriculture Organization (FAO) identifies this pollock fishery as one of the "cleanest" fisheries in the world. Pollock swim in enormous, tightly packed schools and do not co-mingle with other fish species. In a typical tow, pollock comprise 98% of the catch. To further minimize incidental catch of non-pollock species, federal fishery regulations prohibit bottom trawling for pollock. Federal rules also require the retention of all pollock (and cod). Discards of those species are prohibited except in limited circumstances.

3.3 A defined class of participants

The AFA created a three-sector allocation system for the North Pacific pollock fishery. Federal fishery managers set the TAC of pollock annually. A certain percentage of the TAC is held in reserve to account for the anticipated bycatch of pollock by non-pollock fishermen. This "set aside" amounted to 6% of the TAC in 1999. Another 10% of the TAC is allocated to the Western Alaska CDQ program. The remainder of the TAC, which is referred to as the directed fishing allowance, is statutorily allocated among three user groups by the AFA. The inshore sector (catcher vessels delivering to shoreside processing plants) receives 50%. The catcher-processor sector is allocated 40%. The remaining 10% is available to the mothership sector.

3.4 A closed system or limited access

The AFA statutorily defines the eligible vessels in each sector. The legislation actually lists the vessels by name. Additional participants cannot enter the fishery in the absence of congressional action, which is considered highly unlikely, particularly in the catcher-processor sector. Twenty vessels are listed in the AFA in the catcher-processor sector. The act established a buy-back program for the remaining nine vessels, achieving a 31% reduction in capacity in this sector of the fishery.

3.5 Unanimous support among a manageable number of participants in the closed class

It is extremely important to state that in the absence of the ability to form a cooperative, the catcher-processor fleet would not have agreed to the capacity-reducing buy-back program or other requirements of the AFA. In the absence of a means to end the race for fish, the buy-back program would have eliminated the competitive advantage these firms had over one another, catching as much fish as quickly as possible. With the number of vessels in the catcher-processor sector reduced to 20, the companies felt it possible for an effective Cooperative to be negotiated, creating the political will to allow the AFA to become law.

In December 1998, the nine companies owning the 20 catcher-processors in the North Pacific pollock fishery formed the Pollock Conservation Cooperative (PCC). Similar to the PWCC, the PCC is a contractual agreement designed to apportion specified amounts of the catcher-processor sector allocation of the annual pollock quota among the parties to the agreement, with no consideration of pricing or marketing activities. Although the PCC is still awaiting a Business Review Letter from the U.S. Department of Justice, the previous letter ruling regarding the PWCC suggests the Department of Justice will approve the PCC as well.

Interestingly, there are actually two Cooperatives within the catcher-processor sector. The AFA reserved 8.5% of the catcher-processor pollock allocation for seven catcher vessels that traditionally delivered to the catcher-processors. These catcher vessels established the High Seas Catcher's Cooperative (HSCC). The PCC agreement provides assurances that the catcher vessels will not be disadvantaged in marketing their allotted catch among the PCC members. HSCC members report that prices received by the catcher vessels were the highest in the fishery this year and were, in fact, the highest prices ever received by catcher vessels.

In addition to the capacity reduction from the AFA made possible by the ability to form a cooperative, the operation of the PCC has resulted in even further capacity reduction. During the winter season, only 16 of the 20 eligible catcher-processors fished. In the summer/fall season, only 14 vessels fished. On average, this equates to an additional 20% reduction in capacity.

Although data on utilization rates and bycatch avoidance has yet to be compiled, the fishery was certainly operated in a much slower fashion. In 1998 the winter season lasted 26 days and the summer/fall season, 49 days. In 1999, the winter season lasted 57 days, and the summer/fall season lasted 92 days. These shifts suggest that improved product-recovery and decreased incidental-catch can be anticipated.

4. CONCLUSION

In the absence of a regulatory framework to address the race for fish and overcapitalization, one segment of the U.S. industry took the initiative to rationalize its own operations through fishery cooperatives, privately-negotiated contractual agreements. While operated within existing federal fishery management rules and regulations, the cooperatives provide the framework for ownership- or IFQ-like behavior and ensure compliance through contractual penalties. Cooperatives can achieve the benefits of IFQ systems without government regulation at the quota-share level.

5. ACKNOWLEDGEMENTS

Special thanks to Jim Gilmore of the At-Sea Processors Association and John Bundy of the Glacier Fish Company for assistance in preparing this paper.

DETERMINING THE IMPACTS OF ADOPTING PROPERTY RIGHTS AS A FISHERY MANAGEMENT TOOL IN REGULATED OPEN ACCESS FISHERIES

J.M. Ward
U.S. Department of Commerce
National Oceanic and Atmospheric Administration
National Marine Fisheries Service, F/ST1, 1315 East-West Highway
Silver Spring, Maryland, USA 20910
<Deborah.Hogans@noaa.gov>
and
W. Keithly
Coastal Fisheries Institute, Center for Wetland Resources
Louisiana State University, Baton Rouge, Louisiana
<WalterK@unix1.sncc.lsu.edu>

1. INTRODUCTION

The "problem of the commons" plagues marine fisheries managers since no-one conserves a resource that belongs to everyone. Management programmes designed to control inputs in the harvesting process have generally been unsuccessful if property rights for the *in situ* resource do not exist. In recent years, rights-based management measures have been developed to give fishermen partial property rights or access rights to fish-in-the-sea as an alternative approach to achieving fisheries rationalization. However, this alternative approach has also been criticized for its shortcomings. Copes (1986) presents many sound arguments against the use of individual transferable quotas (ITQs) as a fishery management instrument citing the results of actual applications. These include quota-busting, data-fouling, residual catch management, unstable stocks, short lived species, flash fisheries, real-time management, high-grading, multi-species fisheries, seasonal variations, spatial distribution of effort, TAC-setting, transitional gains trap, and lack of industry acceptance. In addition, impacts of transferability of individual quotas on allocation of income and equity have been identified. Finally, the question of whether ITQs are preferable to the common property or open access fishery scenario remains unanswered.

While many of these concerns have been addressed in the economics literature, some have not; including unstable stocks and short-lived species. For example, Anderson (1993), Arnason (1994), Townsend (1995), and Turner (1996) have addressed high grading and bycatch in ITQ fisheries. The incentives to bust quota were addressed by Sutinen (1987) and Muse (1991). Unstable fish stocks prevent fishery managers from setting total catch limits (TCLs) for their firm at the beginning of the season. The resulting fear of a TCL re-adjustment reduces the fishers' confidence since they may not have the entire season to harvest their quota share without causing them to lose a portion of their quota. This uncertainty causes a race-for-fish at the beginning of the season. In some short-lived species, no discernable relationship exists between the size of the catch and the subsequent recruitment. As in the Gulf of Mexico shrimp fishery, the high fecundity of even a small number of spawners given favorable environmental conditions is sufficient to fully restock the fishery in each season. The rationale for not implementing a rights-based management programme is that fish left unharvested are wasted. However, the cost of harvesting the last fish is also a concern, as are the benefits, since the distribution of fish among size classes could have an effect on ex-vessel prices. Last, excessive capacity levels needed to harvest the last fish could have negative impacts on other fish stocks and fishing operations as a result of the discarding of incidentally caught fish; *i.e.* bycatch.

An empirical model of the Gulf of Mexico shrimp fishery is used to investigate the possible effects of introducing a rights-based fishery management instrument into a fishery characterized by both a short-lived species and highly variable recruitment. The approach is to integrate empirical analyses of fleet dynamics, vessel operating costs, and market demand for shrimp in a dynamic optimization model of the Gulf of Mexico shrimp fishery. Random recruitment into the fishery is based on a random number generator with the same mean and variance as a fishery-independent survey of brown shrimp abundance conducted by the Galveston Laboratory, National Marine Fisheries Service (NMFS)[1]. Shadow prices that approximate the shrimp resource-rent are used to determine an annual lease value. This model can be used to compare and contrast deterministic recruitment with variable recruitment under different TCLs that accurately predict landings-levels versus those that are fixed over time at higher or lower than optimal levels. Although a simulation, the advantages of this approach are that the actual behaviour of a group of individual fishermen is the basis for the model assuming that the model is correctly specified. Various scenarios can be compared based on the same set of initial assumptions. Transition paths can be compared to long-run equilibrium conditions. Finally, an index based on the present value of net benefits can be generated for open-access, common-property, and rights-based fishery resource management.

The model is presented in a flow-chart and various scenarios will be compared using quasi-phase diagrams[2].

---

[1] Brown shrimp represents about 60% of total shrimp landings in the Gulf of Mexico.
[2] Multiple variables are allowed to adjust to their equilibrium values in these quasi-phase diagrams instead of the two variables represented on the axis. As a result, transition paths

The derivation of the shadow price for the harvest, access-right, management instrument is explained. Results of the computer simulation model are presented next and a summary of the results concludes the paper.

2. THE MODEL
2.1 Model structure

A simplified flow-chart of the empirical model for the Gulf of Mexico Shrimp Fishery is presented in Figure 1 (Keithly, Roberts and Ward 1993, Ward and Sutinen 1994, and War, Ozuna and Griffin 1995). The model is based on a multinomial logit model that predicts the probability that an individual fishing firm will enter or exit the fishing fleet (Ward and Sutinen 1994). Based on this probability of entry and exit and a known universe of firms in the fishing fleet, a new fleet size can be calculated for each time period. This new fleet size can be used to determine a change in fleet production levels that result in a new ex-vessel price for shrimp. This ex-vessel shrimp price is then used to calculate a new operating cost for an individual vessel that results in a change in its production level. This change in production level, fleet size, and ex-vessel price then affects the probability of entry and exit into the fishing fleet through a set of feed back loops in the programme.

The simulation model goes through a series of iterations until an equilibrium fleet size is found for the existing biological and market conditions initially set to describe the fishery at a point in time. Once an initial equilibrium point is found, the initial values are modified to reflect a change due to a management regulation. Then, the model allows firms to enter or exit the shrimp fishery until a new equilibrium fleet size is found. The present value of net benefits are calculated for the initial equilibrium point and compared to the present value of net benefits generated along the transition path to the new equilibrium point plus those generated at the new equilibrium point.

2.2 Shadow price[3]

The resource rent generated by the shrimp fishery is based on the assumption that firms maximize profits subject to a resource constraint (equation 1).

$$\Pi = N[Ph - Ch] \quad (1)$$

such that $Nh \leq TAC$

where N is the number of firms in the fishing fleet;
P is the ex-vessel price for shrimp harvested;
h is the level of shrimp harvested;
C is the unit cost of harvesting shrimp; and
TAC is the total allowable catch level.

From equation (1), a Lagrangian function (equation 2) can be set up,

$$L = N[Ph - Ch] + \lambda[TAC - Nh] \quad (2)$$

and first order conditions can be taken from which λ can be solved. The value of λ represents the change in profits that would occur with a one-unit relaxation of the constraint; i.e. the resource shadow price. Equation (3) calculates the value of the shadow price based on the analyses underlying the computer simulation model. Equation (3) varies endogenously with changes in fleet size (N), TAC, and crew share. TAC can be set as a fixed constant or allowed to vary depending upon the level of abundance in the fishery.

$$\lambda = -0.0186(TAC) - 7.659(TAC/Fleet\ Size) \times \quad (3)$$
$$1.07(length)^{0.6} \times (fuel\ price)^{0.04} \times (vessel\ age)^{-0.22}$$
$$\times (abundance)^{-0.08} \times (crew\ share)^{0.52}$$

2.3 Random recruitment

Abundance can be either an exogenous constant or can be allowed to vary randomly reflecting annual recruitment in the shrimp fishery. Random abundance levels are generated using a random number generator with a mean of 53.3 and a variance of 50 to generate the range of values found in the brown shrimp abundance index maintained by the NMFS's Galveston Laboratory.

3. RESULTS

Figures 2 through 9 demonstrate graphically the results from running the simulation model over time. These results are based on the assumption that the TAC can be set accurately, based on the fishery independent survey conducted by the NMFS Galveston Laboratory[4]. Figures 2 and 3 indicate the effect random abundance can have on the equilibrium fleet size over time. In Figure 2, the model approaches a stable equilibrium fleet size of approximately 3030 vessels with constant abundance by the seventy-fifth year. At this point in time, the harvest rights instrument is implemented in the fishery and the fleet size declines asymptotically to about 2884 vessels by year 100; a reduction of 5%. With the introduction of random abundance in Figure 3, a similar pattern with more variation in fleet size results. Fleet size initially reaches an equilibrium size of 3082 and then declines to approximately 2900 vessels; a reduction of 6%. However, the exogenous shocks caused by the random fluctuations in abundance do not allow the fishery to maintain a stable, long-run equilibrium.

Figures 4 and 5 compare the effect of constant and random abundance on the shadow price for the shrimp resource stock. The positive value of the shadow price after the imposition of the access-right management regulation in the 75th year varies by less than 1 cent/lb under constant abundance in Figure 4. The shadow price also appears to approach an equilibrium value of 25.5 cents/lb as the fleet approaches its new equilibrium level. Under variable abundance, the shadow price varies by slightly more (almost 2 cents/lb) and fluctuates around its mean value.

---

between long-run equilibrium points can intersect. This is especially true in the diagrams where abundance randomly shocks the system of equations that determine the transition path to the long-run equilibrium.
[3] Personal communication, Jon Sutinen, Dept. of Natural Resource Economics, University of Rhode Island, Kingston, RI.

[4] Results from simulations run with TACs fixed over time will be briefly discussed in Section 4.

Figure 1
Flow-chart of computer simulation model for the Gulf of Mexico Shrimp Fishery

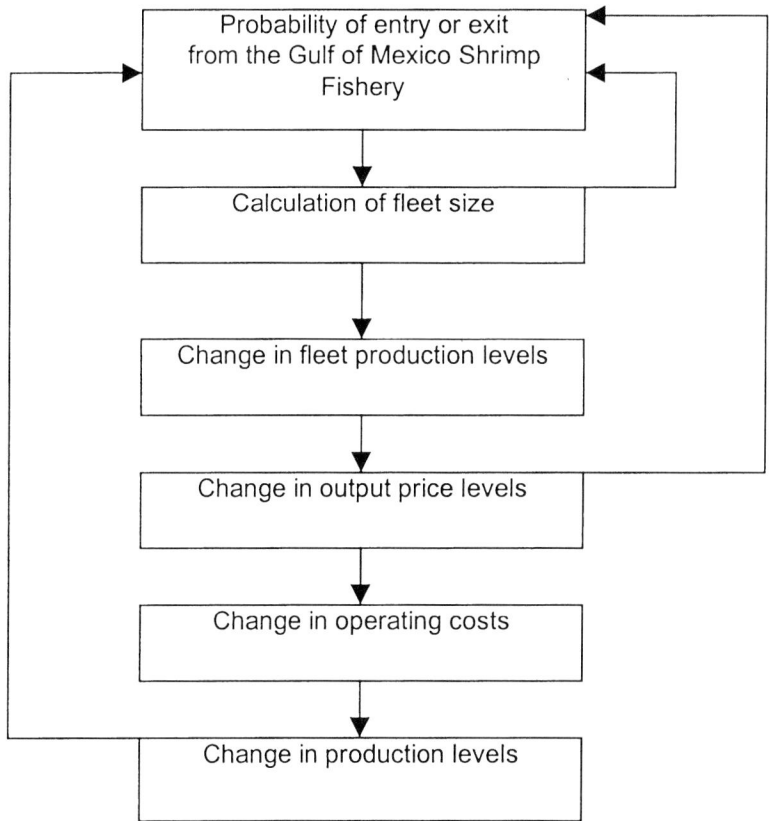

Figures 6 and 7 compare the effect of price changes on fleet size under constant and random recruitment. In Figure 6, the fleet oscillates toward an equilibrium fleet size of 3030 vessels with a price of $1.64 ranging from $1.59 to $1.68/lb. After the imposition of the access-right management instrument, fleet size declines toward 2880 vessels and price rises to $1.69 within a range of $1.68 to $1.70/lb. This pattern, of a price increase, a decline in its variance and a fleet size decrease, is less clear in Figure 7 because abundance is random, but is similar in direction and magnitude. Price rises from about $1.60 with a fleet size of between 3000 and 3100 vessels to a price of $1.76 with a fleet size of about 2800 vessels. The output from the simulation model indicates that prices rise because total landings fell as a result of the decline in fleet size even though landings per vessel have increased.

The increase in landings per vessel as price increases can be seen in Figures 8 and 9. With constant abundance (Figure 8), production for the individual vessel in the fishery spirals toward an equilibrium of approximately 67 000lb of shrimp landed per year at a price of a $1.63/lb. With the adoption of the access rights, management instrument, the individual vessel spirals toward a new equilibrium of 68 000lb with a price of $1.70/lb. With random abundance, the pattern is not as clear, but a close inspection indicates that a similar pattern in terms of direction and magnitude of change exists in Figure 9.

4. SUMMARY

The ability to accurately predict a TAC level based on fishery-independent surveys of abundance in this particular model of the Gulf of Mexico shrimp fishery suggests that random recruitment does not significantly affect the direction or magnitude of change induced by the access-rights instrument. Fewer firms each harvest more shrimp at a higher price, suggesting that net benefits to the nation will increase under a management programme of rationalization for this fishery. In fact, the computer simulation model results suggest that the benefits-to-costs ratio of adopting an access-rights management programme would be 1.99 with constant abundance and 2.03 with random abundance (the change in producer and consumer surplus relative to the status quo fishery scenario); probably not a significant difference.

The computer simulation model was also run using a constant TAC set at the highest and the lowest reported landings level to determine if model results would change substantially. With a low, constraining TAC, the shadow price for the shrimp resource was generally higher than under a high, less constraining, TAC and under a variable TAC set each year based on abundance. This resulted in a lower equilibrium size fleet, higher prices and higher landings for individual vessels in the fishery. The opposite result was true for a high constant TAC. Fewer

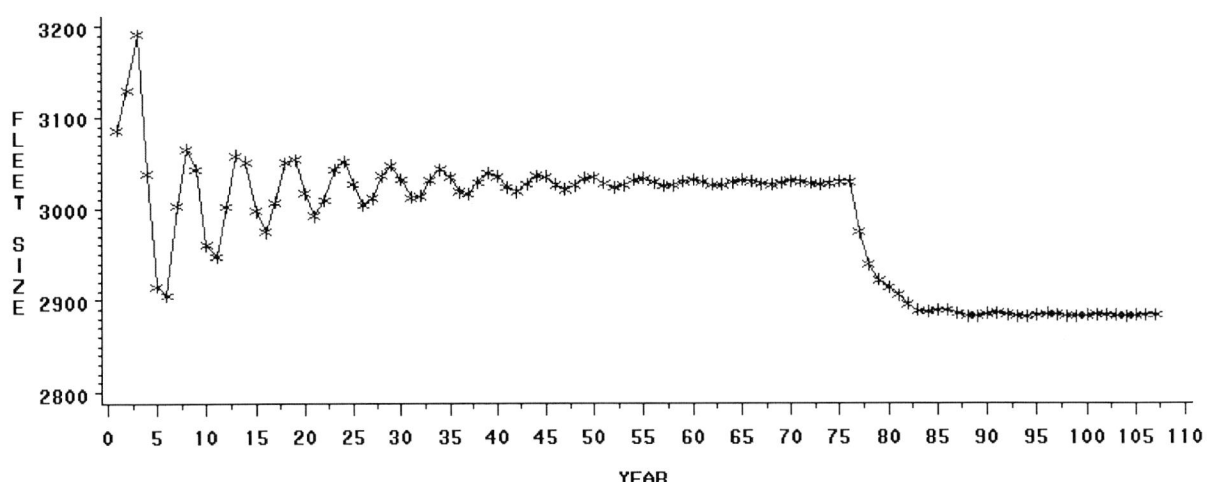

FIGURE 2
Projected Impact of Access Rights
On Fleet Size in the
Gulf of Mexico Shrimp Fishery
With Constant Abundance

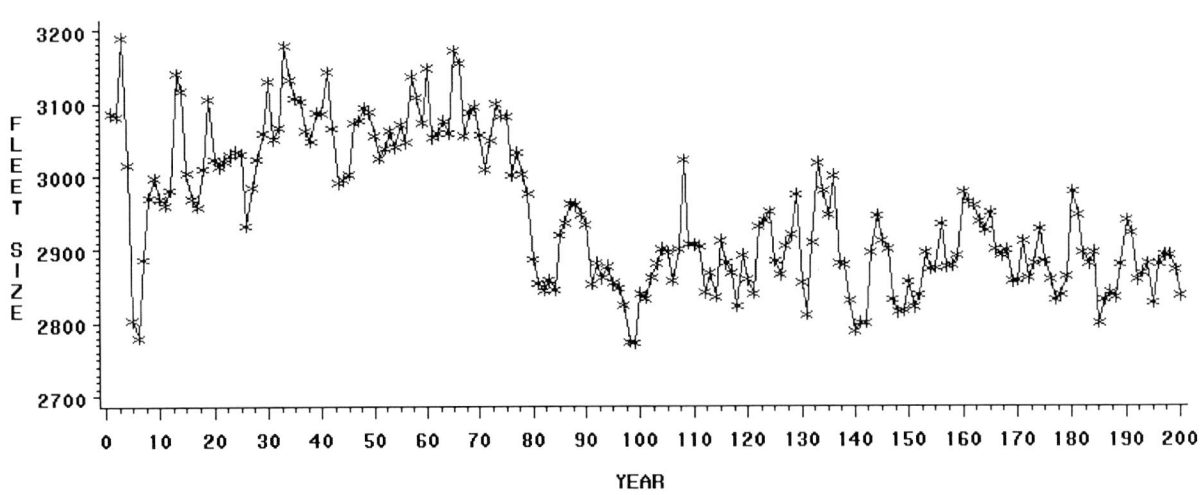

FIGURE 3
Projected Impact of Access Rights
On Fleet Size in the
Gulf of Mexico Shrimp Fishery
With Random Abundance

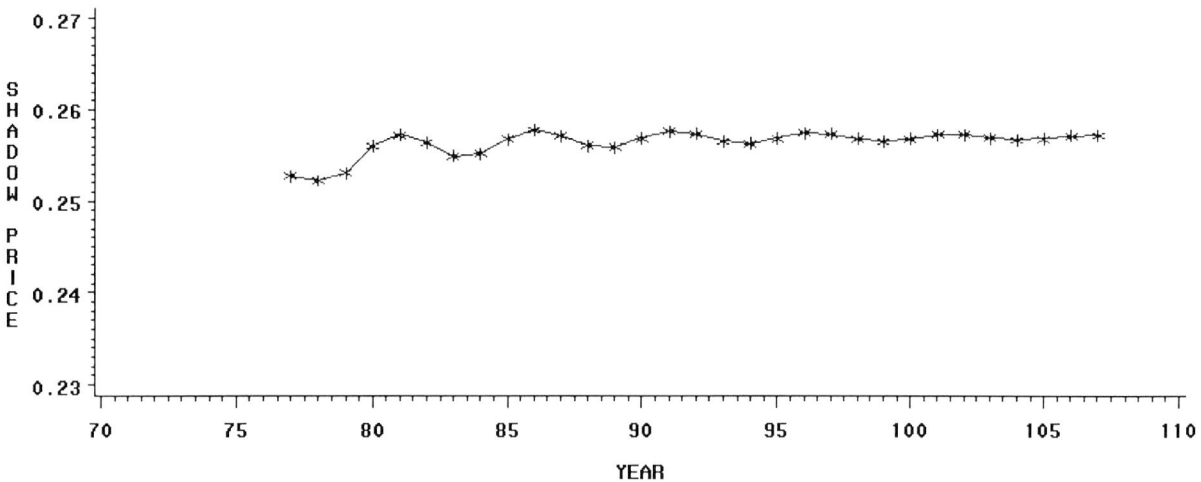

FIGURE 4
Projected Impact of Access Rights
On Fleet Size in the
Gulf of Mexico Shrimp Fishery
With Constant Abundance

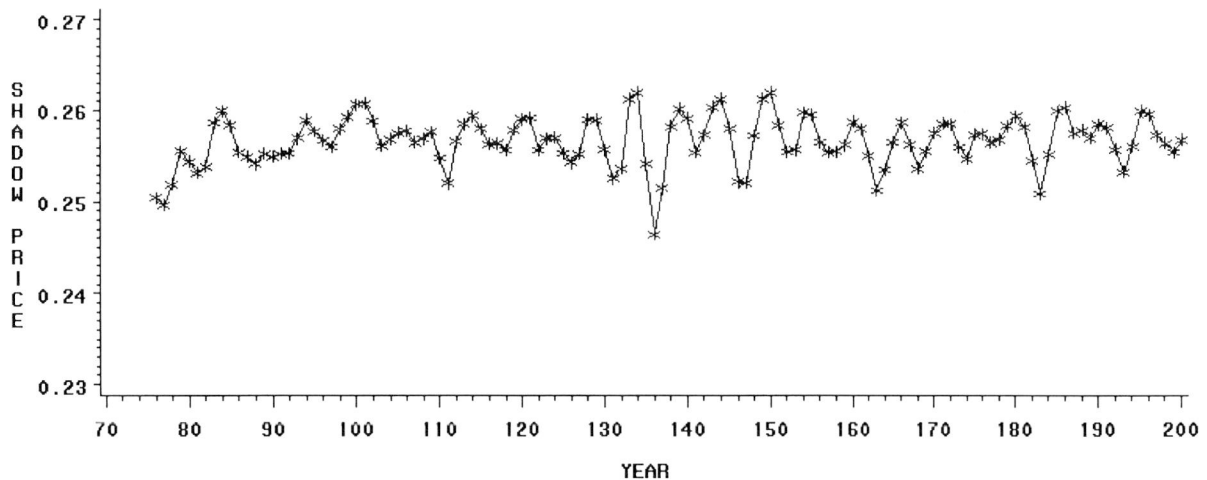

FIGURE 5
Projected Impact of Access Rights
On Fleet Size in the
Gulf of Mexico Shrimp Fishery
With Random Abundance

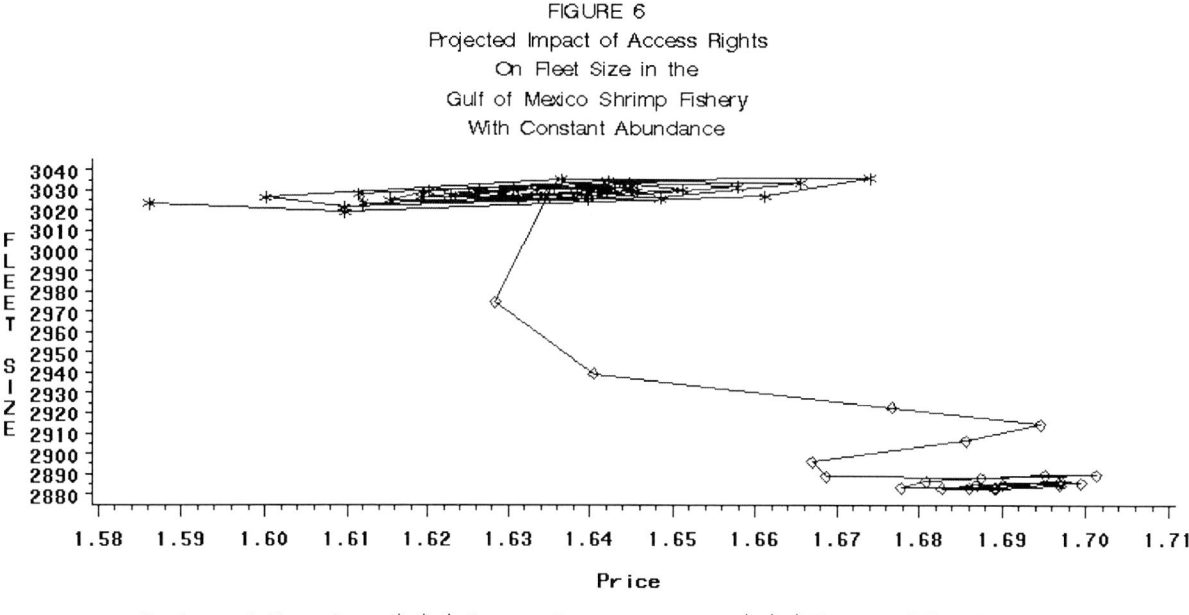

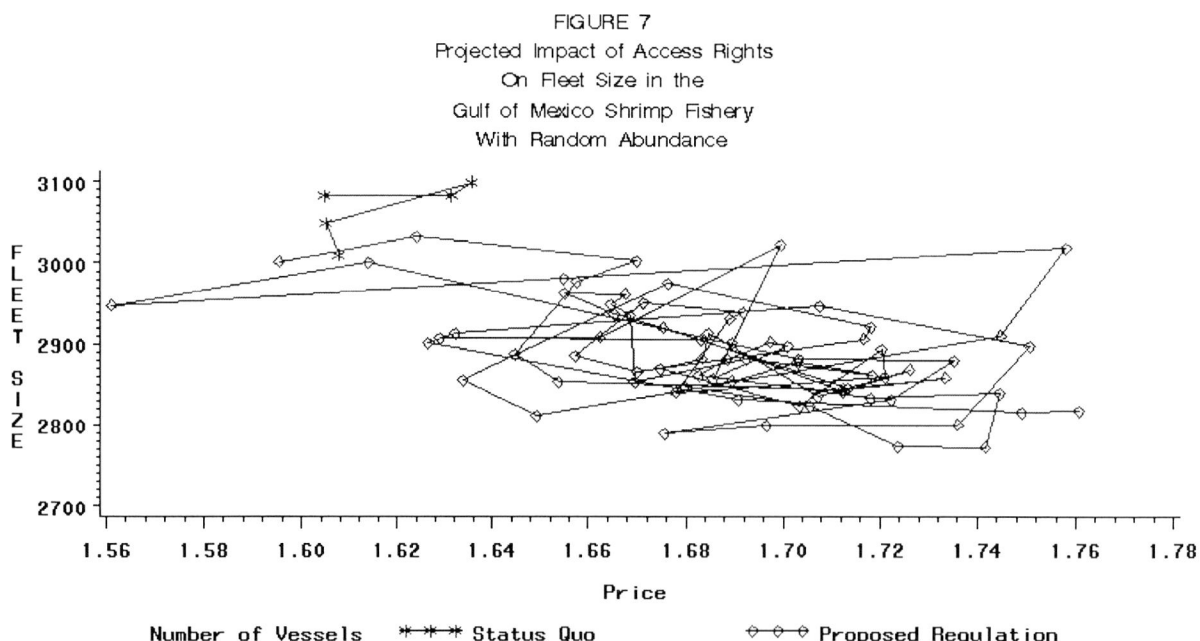

vessels exited the fishery, prices did not rise as much and landings per vessel while higher relative to the *status quo* fishery, did not increase as much as under the constraining or variable TAC scenarios.

In the political reality that is fisheries management, all access rights management programmes need to be evaluated in terms of the individual fisheries for which they are being designed to ensure they meet all management programme goals and objectives; not just economic efficiency criteria. The concern that stocks characterized by random recruitment and short lived species may not be suitable for these management programmes does not seem to be a problem in this particular fishery. In addition, the costs of harvesting the last shrimp do seem to outweigh the benefits. The costs to society associated with the reduction in shrimp harvest are more than out weighted by the benefits derived from reductions in harvesting costs. However, this result may not generalize to other fisheries characterized by short lived species and random recruitment. In addition, the Gulf of Mexico shrimp fishery is extremely complex and this computer simulation model while empirically based on actual fishers behaviour is by necessity a simplification of the real world and may not accurately predict change.

5. LITERATURE CITED

Anderson, L.G. 1993. Some Preliminary Thoughts on Discards, By-Catch, and Highgrading. Presented at the International Conference on Fisheries Economics, Os, Norway, May 26-28.

Arnason, R. 1994. On Catch Discarding in Fisheries. *Marine Resource Economics*, 9(3):189-207.

Copes, P. 1986. A Critical Review of the Individual Quota as a Device in Fisheries Management. *Land Economics*, 62(3), pp. 278-291.

Keithly, W.R., K.J. Roberts and J.M. Ward 1993. Effects of Shrimp Aquaculture on the U.S. Market: An Econometric Analysis. Chapter 8 *in* Upton Hatch and Henry Kinnucan (eds.). Aquaculture, Models and Economics. Westview Press, Boulder, Colorado.

Muse, B. 1991. Survey of Individual Quota Programs. Draft report, Alaska Commercial Fisheries Entry Commission, June. 32pp.

Sutinen, J.G. 1987. Enforcement of the MFCMA: An Economist's Perspective. Mar. Fish. **49**(3):36-43.

Townsend, R.E. 1995. Transferable Dynamic Stock Rights. Marine Policy, 19(2)153-158.

Turner, M.A. 1996. Value-based ITQs. *Marine Resource Economics*, 11(2):59-69.

Ward, J.M. and J.G. Sutinen 1994. Vessel Entry-Exit Behavior in the Gulf of Mexico Shrimp Fishery. *American Journal of Agricultural Economics*, 76(4):916-923.

Ward, J.M., T. Ozuna and W.L. Griffin 1995. Cost and Revenues in the Gulf of Mexico Shrimp Fishery. NOAA Technical Memorandum NMFS-SEFSC-371, National Marine Fisheries Service, Southeast Regional Office, Economics and Trade Analysis Division, 9721 Executive Center Drive, North, St. Petersburg, FL, May, 76 pp.

THE COMMERCIAL GEODUCK (*Panopea abrupta*) FISHERY IN BRITISH COLUMBIA, CANADA – AN OPERATIONAL PERSPECTIVE OF A LIMITED ENTRY FISHERY WITH INDIVIDUAL QUOTAS

S. Heizer
Operations Branch, Fisheries and Oceans Canada, Coastal British Columbia South
3225 Stephenson Point Road, Nanaimo, British Columbia, Canada V9T 1K3
<heizers@pac.dfo-mpo.gc.ca>

1. INTRODUCTION

The fishery for geoduck clams, *Panopea abrupta* (Conrad 1849), in British Columbia (BC) is one of several Canadian fisheries managed by limited entry and individual vessel quotas (IVQs). By Canadian standards, individual vessel quotas provide a 'right' of access to the harvest of these valuable clams. This opportunity is constrained by the fact that the Minister of Fisheries retains the absolute right to allocate licences and could take away a fisher's access to the fishery at his discretion (Canada, *Fisheries Act* Sec. 7). In 11 years of this programme, this has never happened and the IVQ programme has helped to render this fishery one of the most valuable in Canada.

Figure 1
Geoduck clam, *Panopea abrupta*

The great value of this fishery has allowed the fishers to not only pay for the incremental costs of the IVQ programme, such as monitoring, but also to contribute to a large share of the cost of managing the fishery. Fishers provide support for data collection and analysis, for scientific and other research as well as for personnel for programme operations.

This paper details, from an operational point of view, the benefits and costs of managing the BC geoduck fishery after Fisheries and Oceans Canada (the Department) provided licence holders certain 'rights of access'. The benefits have been in cost-recovery, advances in knowledge, manageability and control of the fishery and public health and safety. The costs have been increases in poaching and high-grading due to the high value of the product, a reduction in employment in the fishery and the high cost of entry into the fishery.

2. DESCRIPTION OF THE FISHERY

The geoduck is a large hiatellid clam which occurs from Alaska to Baja California. The name 'geoduck' comes from a Nisqually (native American) word for 'dig deep' (Quayle 1978). The clam lives buried up to 1m deep in sand and mud substrates (Goodwin and Pease 1989) from the lower intertidal zone to depths of at least 110m (Jamison *et al.* 1984). Once dug in, geoducks remain in the same spot for the duration of their lifespan, sometimes well in excess of 100 years (Goodwin and Pease 1989). If removed, they are unable to rebury themselves and quickly perish. The oldest clam on record was an estimated 146 years old (Harbo *et al.* 1983). Clams reach sizes of 4.5kg, but generally average about 1kg (Cox and Charman 1980).

Divers use surface-supplied air ('hookah' systems) and harvest geoducks from the substrate using water under pressure, delivered through a hose and nozzle system or 'stinger'. The diver grabs the clam by the siphon as he inserts the stinger, which delivers a stream of high-pressure water into the ground near the clam. As the ground liquefies, the diver is able to remove the clam alive. He then places each clam in a bag attached to his waist. At intervals, the diver returns to a line reaching to the bottom below the vessel and the dive-tender hoists the full bag to the deck. The crew places animals in containers and keeps them alive, moist and covered. The vessel usually delivers the clams on the day of harvest.

In the early period of the fishery, markets were for processed siphons in Japan and to a lesser extent to the United States (US) and Hong Kong. The body meat was marketed at a low value domestically and in the US. Lesser amounts of live product went to the US and Hong Kong. In the late 1980s, live geoducks became the primary product, and markets shifted from Japan to the People's Republic of China. The market in China required a year-round supply of clams. This enhanced the value of geoducks and resulted in the industry funding a pilot IVQ programme.

3. HISTORY OF THE FISHERY
3.1 The unlimited fishery

Geoducks have been harvested in British Columbia since 1976, when fishers seeing the success of fishers in neighbouring Washington State, began harvesting these clams in the southern regions of the province (Cox 1979). Prior to the commercial fishery, recreational and aboriginal fishers may have had some limited intertidal harvest. In 1976 and in early 1977 the Department issued seven special permits giving each permit-holder an opportunity to harvest geoducks from specific assigned areas in the Strait of Georgia. There was no harvest limit, but regulations required fishers to fish below a certain depth (3m below chart datum) and limited the 'stinger' design. In 1976, the seven permit holders using five vessels, harvested about 44 000kg of geoducks (Harbo *et al.* 1992). Figure 2 shows catches from 1976 through 1998.

Figure 2
BC geoduck catches from 1976 through 1998

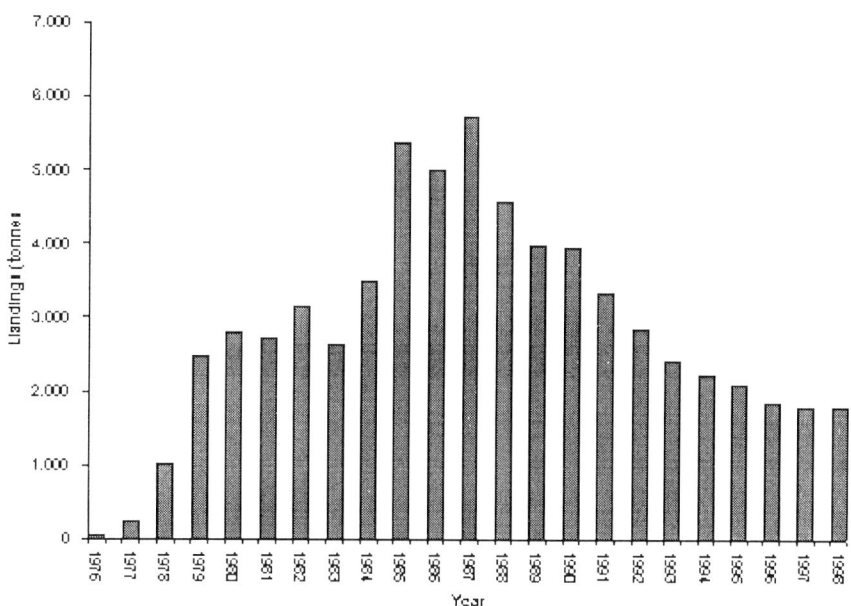

In July of 1977 the Department instituted a licensing system. For extensive discussions of the licensing and entry limitation in this fishery, see Muse (1998). The Department did not limit the number of licences issued and eliminated the exclusive harvesting rights which had accompanied the special permits. The holder of the licence, who could be a person or incorporated company, need not be present at the fishing operation. As well, the Department did not require that the licence be fished in order to guarantee continued entitlement. Licence- holders were required to submit sales slips, and to return harvest logs and charts of harvest locations. There was no total allowable catch (TAC) limiting harvest.

The number of active vessels and harvest levels rose dramatically during the early period between 1976 and 1979. For extensive discussion of the catch and effort in this fishery, see Campbell et al. (1998). In 1977 the Department issued thirty licences and 14 vessels landed about 245 400kg of geoducks. In 1978, 54 licences were issued, and 14 vessels landed about 1.0 million kg. In 1979 101 licences were issued, and 72 vessels landed 2.5 million kg. During the same period, the mean ex-vessel price also rose from $Can0.37/kg in 1977 to $Can0.68/kg in 1979.

3.2 The limited entry fishery

In mid-1979, due to concerns about the increasing effort and harvest levels, the Department imposed a moratorium on the issuance on new licences. In 1980 the Department issued 95 licences, and 63 vessels landed 2.8 million kg of geoducks. In the next two years, a legal framework and selection criteria were developed for limiting entry to the fishery. In 1981 and 1982, licences were issued to persons who landed more than 13 600kg of geoducks in either 1978 or 1979. The effect of this was to limit licences to 52 in both 1981 and 1982. In 1981, 49 vessels landed about 2.7 million kg, and in 1982, 53 vessels (one licence was fished on two vessels) landed 3.1 million kg.

In 1983, the current licence limitation programme came into force. A new vessel licence (called a 'G' licence) was issued to persons who met the landing criteria. The licences were vessel-based licences and carried limitations in transferability between vessels. Fifty-four licences were issued in 1983 and 1984 and increased to the current 55 in 1985 after a successful appeal.

3.3 Establishment of Total Allowable Catches

Managers instituted Total Allowable Catches (TACs) in 1979, one for the North Coast (1.59 million kg) and one for the South Coast (2.04 million kg) (Harbo et al. 1992). These initial TACs were arbitrary since available information on stock sizes was limited. Over the next 10 years, TACs fluctuated between 2.8 and 4.2 million kg, and reflected the Department's assessment of the ability of stocks to absorb heavy harvests.

The TACs were exceeded in 1979 1982 and 1984 to 1988, and in the period 1984 to 1988, the TACs were exceeded by an average of 34%! Catches peaked in 1987 with 7.5 million kg landed, exceeding the TAC by 1.5 million kg. There are several reasons why TACs were exceeded. The primary reason was that catch information did not come in quickly enough for managers to close the fishery in time. As well, effort was increasing because of the rising value of the fishery. Between 1979 and 1987, ex-vessel prices fluctuated between $Can68/kg and $Can88/kg then jumped to $Can08/kg, in 1987and reached $Can2.14/kg in 1989.

The TAC is currently set at 1% of the estimated virgin biomass. The biomass is calculated by applying the estimated virgin densities (in kg/m^2) times the estimated bed area (Hand *et al.* 1998a 1998b). The Department calculates the TAC annually to adjust for advances in understanding of bed size and geoduck densities.

3.4 The individual vessel quota fishery
3.4.1 IVQs

By 1988 it was obvious that limited entry was not able to control effort or harvest levels. Fishers referred to the fishery as a 'shotgun' fishery where high effort was expended as areas opened. In 1988, the fisher's association (the Underwater Harvesters Association or UHA) requested that an individual quota system be set up. They pointed to a number of problems including loss of profitability due to erratic product supply, difficulties in product handling and transport caused by periodic gluts which followed openings, TAC overruns, safety concerns created by the race-to-fish and economic losses possible from missing a 'shotgun' opening. Under the shotgun-style fishery, fishers could not properly service the profitable live market, opening in China, that demanded a year-round supply.

Estimates of incremental programme costs for an IVQ fishery were high, primarily for catch monitoring and the Department was not prepared to commit this expense. Fishers proposed to fund the necessary programmes, and a two-year pilot programme was started in 1989. This programme, with minor changes, has continued until today. The essential elements are described below.

3.4.2 Initial allocation

Often, the biggest problem in the development of an IVQ programme is the initial allocation of quotas. The determination of who is to get what portion of the TAC can be contentious. In the case of the geoduck IVQs, this was not a problem. Fishers approached the Department with the proposal that initial IVQs be equal. In this programme, the 55 licence holders each have access to 1/55 of the TAC. In the first year of the programme, several licence holders refused to join the UHA and pay the catch monitoring fee. But, in the second year of the programme, these fishers apparently realized the benefits and there was full participation.

3.4.3 Transferability

Although licences can be transferred, the quota may not be split and sold or traded. However, up to three licences can be fished from a single vessel. Unharvested quotas may not be carried over into the next fishing year. Small quota overages (less than 91kg) may be transferred to another vessel which has not harvested its whole quota. Larger quota overages are sold and the proceeds relinquished to the Crown.

3.4.4 Area licensing

Area licensing was instituted at the same time as IVQs. The coast was divided into three areas (a) the north coast, (b) the west coast of Vancouver Island and (c), the waters between Vancouver Island and mainland BC (See Figure 3). Licences are distributed to the three areas so that the TAC from each area can be taken exactly by the licences present. The number of licences assigned to the north has increased over time, as TACs increased. The UHA assigns specific vessels to areas through an internal process.

3.4.5 Rotational fishing

Each licence area is further subdivided into management areas, each with an assigned quota. One-third of the management areas are fished each year, at three times the annual 1% harvest rate. Thus each management area is unfished for two years and fished for one year. The benefits of this rotational fishery are the reduction of monitoring and operating costs as only one-third of the coast is fished each year.

3.4.6 Fishery monitoring

The Department and the UHA have currently developed a 5-year collaborative agreement that details the programme requirements and standards. Licence conditions also define many of the requirements. The Department requires licence holders to validate their catch when landed. The UHA hires a private firm to undertake this activity at a cost (in 1999) of $Can525 000. Fishers must notify the validation company when they move into or out of an area. They must also provide the company with the time and place of every landing and the method of transport to and from the landing port. The validation company weighs every load of geoducks as landed. The harvest-logs, first required under the limited entry fishery and also required under the IVQ fishery, have recently been incorporated into the validation form. The Department requires that information for each dive be recorded on this form, including time, place and duration of dive, name of diver and amount of product harvested. Precise data such as these allow for a much more refined understanding of the effects of harvest on a bed, with improvements in stock assessment activities and micro-management.

On the remote north coast, the UHA hires an on-grounds observer and charters a vessel that moves with the fleet and is present for every day of every opening. Besides monitoring all movements and landings, the observer gathers a great deal of site-specific catch and effort information and provides unique observations possible only from an observer who is not also concentrating on the job of harvesting. The cost of this patrolman is around $Can140 000, which is included in the costs of the validation programme.

4. EFFECTS OF THE LIMITED ENTRY/IVQ PROGRAMME
4.1 The value of the fishery

Since the institution of IVQs and the associated area licensing and rotational fisheries, the ex-vessel value of the fishery has ranged from $Can9.6 million, in 1991, to a maximum of $Can43.3 million in 1995, and had a value of $Can30 million in 1998. The wholesale value is twice these values. The drop in ex-vessel value, between 1996 and 1998, is attributed to the recent decline in prices for most fish and shellfish exported to Asia and to the continuing market presence of large quantities of product from Washington State. At the same time, quotas dropped

Figure 3
Map of coastal BC

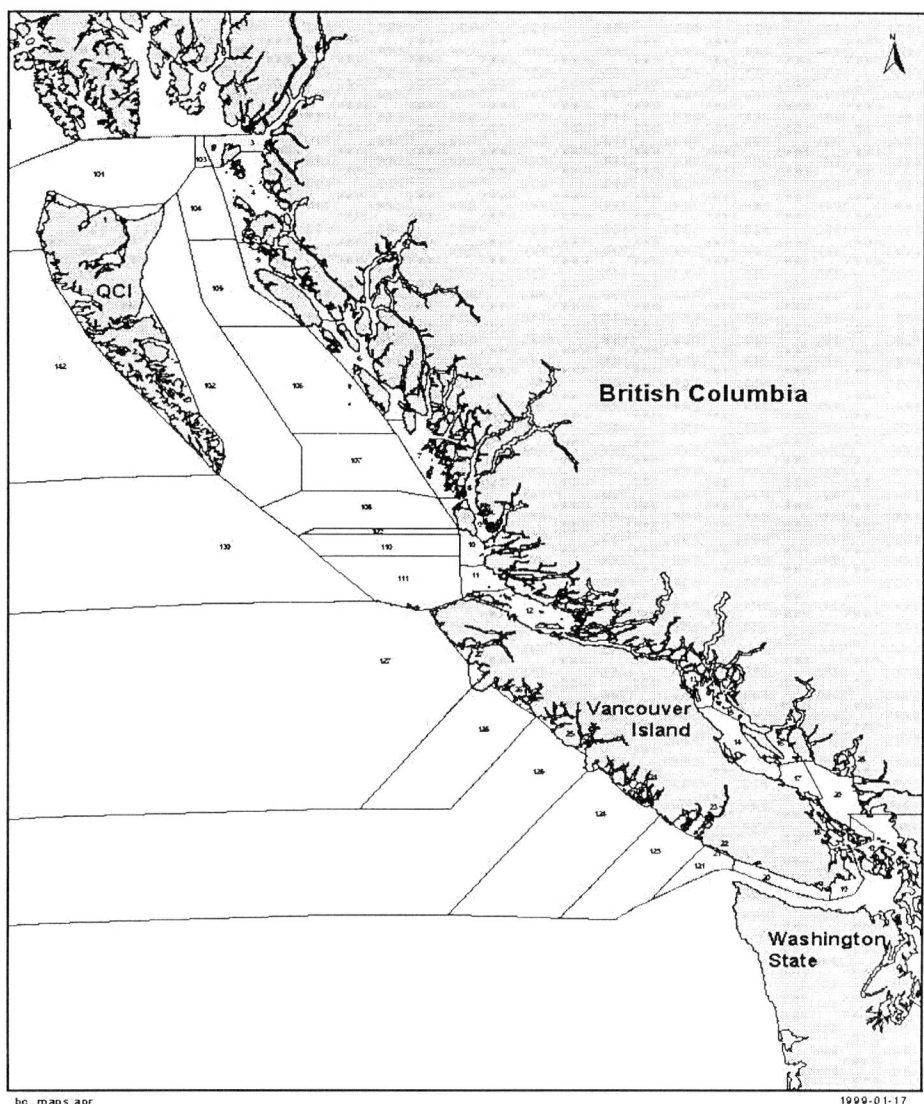

from 3.63 million kg to 1.80 million kg due stock assessments and a precautionary TAC-setting approach.

As Figure 4 shows the obvious result of the programme is a profitable fishery where IVQs have resulted in a high value based on quality and flexibility in supply to compensate for the significant decreases in quotas (Figure 4).

4.2 On management of the fishery
4.2.1 Cost recovery

The profitability of this fishery is tracked by a concomitant mechanism for cost recovery for the Department. Fishers currently pay $Can1.16 million (not including licence fees), which covers a significant proportion of costs incurred in the prosecution and management of the fishery. These payments can be divided into two types: (a) those required for the IVQ programme (catch validation, patrolman, water quality and biotoxin monitoring), and (b), those which fund other activities such as stock assessment and management costs for the programme.

Licence fees (currently known as 'access fees') have risen, as well, from $Can10 between 1985 and 1995, to $Can3615 in 1996, to $Can3520.80 in 1997 and 1998, to $Can7215.20 in 1999. Benefits go beyond the recovery of programme costs from fishers. Managers enjoy a rare ability to have improved fishery data collection and management, more operational flexibility to deal with real-time biological, enforcement and logistical issues, greater confidence in essential control functions such as fishery monitoring, enforcement, and biotoxin and contamination monitoring.

4.2.2 Catch validation

Perhaps the greatest benefit for the Department from the IVQ fishery is the industry-funded catch validation programme. Data obtained by this programme allow the Department to closely control the fishery, to be flexible in

Figure 4
Trends in value and quota size for the BC geoduck fishery

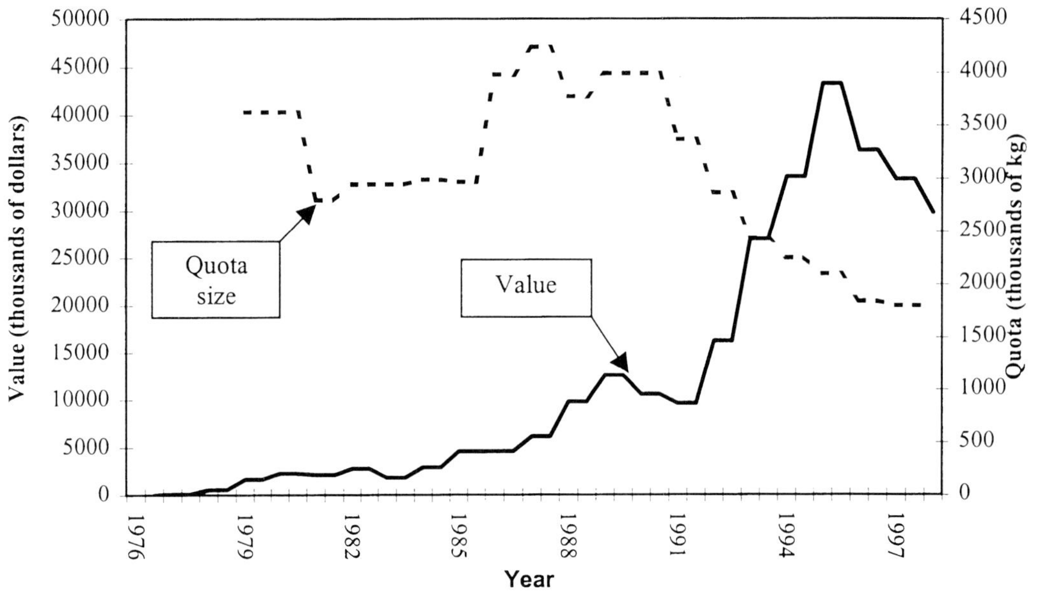

adjusting times and areas of openings and to have improved data for stock assessment purposes. Fiscal restraints on the Department would never otherwise allow the collection of such precise and timely data.

Improved fishery information collection and management has reduced over-harvesting. Prior to the IVQ programme, coastwide quotas were often exceeded by as much as 81%. In 1998, the quota overage was 342kg, or slightly less than 0.02%. This improved information is gradually leading towards bed-by-bed management as well. With bed-by-bed management, effort can be precisely targeted on specific areas, thus reducing over-harvesting on particularly productive areas while avoiding less desirable areas. Local over-harvesting can be masked when larger areas, including several beds, are open.

4.2.3 Public health and safety

The UHA has contributed to the costs of an extensive biotoxin monitoring programme ($Can187 000 in 1999), and for a growing water-testing programme ($Can100 000 in 1999). These programmes are essential to a molluscan fishery which exports the vast majority of its product live. These programmes allow the Department to confidently open remote, or problematical, areas while still complying with various international agreements regarding product safety. Again, fiscal restraints would not allow the Department to provide this level of service, and without industry support, a considerable fishable area would not be approved for harvesting.

4.2.4 Enforcement

Catch validation has introduced control structures which have made illegal activities (poaching, harvesting without a licence) more difficult. All shipments of geoducks are validated and labelled. This product tracking has reduced the necessity of routine enforcement. The UHA has also made financial and in-kind contributions towards enforcement activities.

The Department views the validation programme as providing effective enforcement controls. The Conservation and Protection Division (C&P) of the Department confirms this by maintaining, in the draft 2000 Integrated Fishery Management Plan for Geoducks[1], that the geoduck fishery is a '...low (enforcement) priority except where human health and safety issues are identified.' and that C&P will '...pursue opportunities to monitor and enforce issues and problems related to these fisheries in conjunction with the monitoring and enforcement activities dedicated to....identified priority fisheries....'.

4.2.5 Science/Research

Licence holders have supported stock assessment activities in response to reductions in quotas. The UHA spends approximately $Can150 000 annually on survey activities related to better understand the biology of geoducks. Surveys provide information needed for estimates of biomass (bed sizes, densities, and sizes of clams), for studies of recruitment and for population age-structure. Interestingly, data from these studies have been partially responsible for the drop in quotas, but the UHA continues to fund these studies. The UHA has also provided funds for analytical and reporting activities that have allowed the science sector to do stock assessment and research in support of the fishery that would otherwise be impossible.

4.2.6 Management support

The UHA provides the Department with the funding for a management biologist. The biologist (the author) works for the Department, but, in return for the funding, is assigned full time to the management of the geoduck fishery. The Department can now service the commercial

---

[1] *Pacific Region Integrated Fisheries Management Plan – Geoduck & Horse Clam 2000*. Available from the Communications Branch, Fisheries and Oceans Canada, Pacific Region, Vancouver, BC, Canada.

fishery better with a dedicated biologist, which was impossible in the past as management biologists oversaw several fisheries concurrently and often had other priorities at critical times.

4.2.7 Geoduck enhancement

The UHA has embarked on an enhancement programme. A biologist in the direct employ of the UHA is working with a local hatchery to spawn and rear geoducks and to plant juveniles in the wild. The UHA claims no proprietary or ownership rights to the planted seed and both the Department and the UHA currently consider them part of the common property resource.

Their studies provide stock assessment information as they help address how planting density and size affects survival, whether these planted animals will survive to make a significant impact on stock size and whether they might increase the local spawning stock and affect recruitment.

4.2.8 Consultation and co-management

Another major benefit of the IVQ programme is improved consultation with the industry. Commercial fishers and the Department have an ongoing formal and informal dialogue regarding the prosecution of the fishery. This has led to vast improvements in the relationship between the Department and fishers. There are still differences in philosophy and opinion, but there is greater trust and less confrontation. It is an excellent example of government/industry co-management.

Two agreements govern the co-management, one covering the provision of the dockside monitoring programme, the other covering the funding for the fishery manager. The current term of these contracts is five years. A number of projects are not covered by specific agreements (*i.e.* enforcement contributions, science and research commitments, geoduck enhancement). Instead, fishers and DFO prioritize these UHA funded initiatives through the consultation process.

4.3 On the industry
4.3.1 Financial benefits

Fishers enjoy a higher and more reliable income stream for several reasons. First, the ability to fish as the market requires has removed the periodic market gluts and has allowed fishers to develop and supply lucrative markets. Second, the higher-quality product being landed, a result of improved harvesting techniques and product handling, has increased the value of the product. Third, savings in vessel operating-costs have also had a marked effect on profitability.

Fishers have a greater stake in the industry, created in part, by increased profitability and the enhanced access to the fishery created by the limited entry and IVQ programme. Consequently, fishers have become more interested in the sustainablity of the fishery and less in developing strategies to better their individual competitive position. This has led to fishers requesting unchanging quotas to avoid both high and low fluctuations that often accompany fixed harvest rates. These fluctuations are often viewed by the market as a sign of instability in the fishery. Fishers were willing to take slightly less product each year in return for the stabilized quotas.

4.3.2 Health and safety

The industry believes in general that the number of incidents with vessels (sinkings, breakdowns, *etc.*) has decreased since the rush to fish has ended and vessels are better equipped and maintained. There is little data on this subject and the data that exist are incomplete as fishers do not necessarily report such incidents. Still, the industry is firmly convinced that it is a much safer fishery (J. Austin[2], pers. comm. 1999).

It is also believed that the number of accidents involving divers has also decreased as fishers feel much less pressure to dive deeper, to make decompression dives or to violate dive protocols. Not all diving incidents are reported. Nonetheless, the Medical Director for Hyperbaric Medicine at Vancouver General Hospital in Vancouver, BC reports that "[There is] ... a definite trend towards less decompression sickness amongst diving geoduck harvesters and a trend towards decreased fatalities in commercial divers in general, since the start of more aggressive Workers' Compensation Board of British Columbia involvement in the mid 1980's" (Lepawski[3], pers. comm. 1999). The reduction in the competitiveness in the fishery, with the start of IVQs in 1989, is another significant factor that allows fishers to adhere to safety guidelines.

4.3.3 Commercial geoduck crews

Although crew sizes and numbers of divers participating were initially reduced with the institution of IVQs, job security and safety conditions for those remaining in the industry has improved greatly. The crew on a geoduck harvesting vessel usually consists of a vessel master, a tender, who looks after the diver in the water, and two divers. In 1997, there were 86 divers fishing off 42 vessels. This is a significant change from the record number of divers fishing in 1988, the year just prior to IVQ implementation. In that year, 233 divers fished from 56 vessels for an average of just over four divers per vessel (Muse 1998). The decline in crew size is a consequence of efficiencies introduced through the IVQ programme and the declines in quotas. Nonetheless, crews are smaller and there is little turnover. Crew earnings have increased along with the increased value of the product (J. Austin[2], pers. comm.), employment is more secure and the industry is safer.

5. OTHER ISSUES
5.1 Aquaculture

A private aquaculture firm is spawning geoducks and rearing the seed in tenured tracts in the Strait of Georgia. This firm and the UHA are working cooperatively and jointly fund some disease studies. Manag-

---

[2] President British Columbia Underwater Harvesters Association. Qualicum Beach, British Columbia, Canada.

[3] Associate Clinical Professor, Faculty of Medicine, University of British Columbia and Medical Director, Hyperbaric Medicine, Vancouver General Hospital, Vancouver, British Columbia.

ers hope that geoducks harvested off these tenures will have marking protocols similar to those in the commercial fishery. In this way tracking in the wild fishery validation programme would not be compromised by quantities of unmarked cultured product in the marketplace.

5.2 First Nations

There is currently no involvement of First Nations in the geoduck fishery except for opportunities to harvest for food, social or ceremonial purposes. No harvest is reported. Treaty negotiations are now underway with First Nations and will certainly affect all fisheries. If First Nations obtain access to geoducks, managers hope that the control structures similar to those in place in the current commercial fishery would be implemented. Similarly, managers hope First Nations' involvement in aquaculture will also have appropriate control structures in place.

6. THE FUTURE

Although there are several uncertainties for the geoduck fishery, such as the ultimate effect of First Nation treaty negotiations, aquaculture development and Asian seafood markets, the future of the commercial fishery appears fairly robust. The understanding of the resource has improved substantially over the last ten years and allows a degree of comfort about both the sustainability of the fishery and co-operation between managers and the commercial industry

For the future, both industry and government desire a more formal co-management agreement. The UHA would like more secure, longer term access to the resource while, from a managers perspective, the Department would like a longer term agreement on the roles in the fishery, in particular, to secure programme funding. The will and positions of the respective stakeholders appear suitable for negotiating a successful co-management agreement.

7. CONCLUSIONS

The geoduck fishery is an example of a co-management success story. The fishery is very profitable. The fishers, because of the high value, are capable of making considerable investments in the future of the fishery by making significant contributions towards the cost of managing the fishery through fees and voluntary contributions.

Operational personnel involved in the fishery enjoy having enhanced tools and information at hand such as willing survey teams and timely catch and effort information. Consultation is largely lacking the strife found in many other fisheries. Fishers enjoy a greater responsibility and say in the fishery. And, there are advances in understanding of geoduck biology allowing a more scientific approach to the management of the fishery. And longer-term studies are in place due to commitments of funding from the fishers.

Control of the fishery is highly efficient due to the dock-side validation programme. Catch and effort information is timely enough to allow managers greater flexibility in making in-season course corrections. And the Department has more eyes and ears out on the grounds.

There has been some displacement of workers from the fishery due to efficiencies possible through the IVQ programme. As in other such fisheries, high-grading has become a problem of some magnitude. Because the fishery is not causing conservation concerns, there is little routine enforcement and poaching and other infractions have received little attention.

The Government of Canada may, at some time, be required to buy licences for First Nations in settlements of treaty negotiations. But, the high value of geoduck licences will be a big problem for this process. Government, the public and stakeholders can learn from the successes of the fishery and modify and apply them to new and existing fisheries.

8. LITERATURE CITED

Campbell, A., R.M. Harbo and C.M. Hand 1998. Harvesting and distribution of Pacific geoduck clams, *Panopea abrupta*, in British Columbia. In *Proceedings of the North Pacific Symposium on Invertebrate Stock Assessment and Management*. Edited by G.S. Jamieson and A. Campbell. *Spec. Publ. Fish. Aquat. Sci.* 125. pp349-358.

Canada 1985. Fisheries Act. R.S.C. 1985, c. F-14. Government of Canada, Ottawa, Canada.

Cox, R. 1979. The geoduck, *Panope generosa*: some general information on distribution, life history, harvesting, marketing and management in British Columbia. *Fisheries Development Report No. 15.* B.C. Mar. Res. Br., Ministry of Environment, Province of British Columbia. 25pp.

Cox, R. K. and E.M. Charman 1980. A survey of abundance and distribution (1977) of the geoduck clam *'Panope generosa'* in Queen Charlotte, Johnstone and Georgia Straits, British Columbia. *Fisheries Development Report No. 16.* B.C. Mar. Res. Br., Ministry of Environment, Province of British Columbia. 122pp.

Goodwin, C.L. and B.C. Pease. 1989. Species profiles, life histories and environmental requirements of coastal fish and invertebrates (Pacific Northwest) - Pacific geoduck clam. *U.S. Wildl. Serv. Biol. Rep.* 82(11.120). U.S. Army Corps of Engineers, TR EL-82-4. 15pp.

Hand, C.M., B.G. Vaughn and S. Heizer 1998a. Quota options and recommendations for the 1999 and 2000 geoduck clam fisheries. *Canadian Stock Assessment Secretariat Research Document* 98/146. 52pp.

Hand, C.M., K. Marcus, S. Heizer and R. Harbo 1998b. Quota options and recommendations for the 1997 and 1998 geoduck clam fisheries. *Can. Tech. Rpt. Fish. Aq. Sci.* 2221. pp71-159.

Harbo, R.M., B.G. Adkins, P.A. Breen and K.L. Hobbs 1983. Age and size in market samples of geoduck clams (*Panope generosa*). *Can Ms Rep. Fish. Aqaut. Sci.* No. 1714. 77pp.

Harbo, R., S. Farlinger, K. Hobbs and G. Thomas 1992. A review of quota management in the geoduck clam fishery in British Columbia, 1976 to 1990, and quota options for the 1991 fishery. *Can. MS. Rep. Fish. Aq. Sci.* 2178. 135pp.

Jamison, D., R. Heggen, and J. Lukes 1984. Underwater video in a regional benthos survey. *Proc. Pac. Cong. on Mar. Tech., Mar. Tech. Soc.*, Honolulu, Hawaii.

Muse, B. 1998. Management of the British Columbia Geoduck Fishery. *Alaska Commercial Fisheries Entry Commission.* CFEC 98-3N. 23pp.

Quayle, F.B. 1978. The intertidal bivalves of British Columbia. *Handbook* No. 17. B.C. Prov. Mus., Victoria, B.C. 104pp.

ABALONE AND THE IMPLEMENTATION OF A SHARE-BASED PROPERTY RIGHT IN NEW SOUTH WALES, AUSTRALIA

D. Watkins
NSW Fisheries
P.O. Box 21, Cronulla, NSW 2230 Australia
<watkinsd@fisheries.nsw.gov.au>

1. BACKGROUND

The commercial fishery for abalone commenced in the 1960s with access to the fishery by way of a commercial fishers licence only. Most abalone were dried or salted and sold on the local market. A market collapse in 1976 led to the establishment of a Parliamentary Select Committee of Inquiry in 1978. Result of this inquiry was the subsequent introduction of a restricted access regime for the abalone and the closely associated sea urchin fishery. The fishery was called the Abalone Restricted Fishery. Access to the fishery was limited to those with a demonstrated catch history of abalone or sea urchin, or to a class of individual who was an aborigine within the meaning of the *Aborigines Act 1969*. Turban shell was added to the definition of the restricted fishery in 1981.

Prior to the introduction of this management regime, an economic study was conducted to determine what a reasonable income would be for an abalone diver to receive if restricted access was introduced. The study concluded that about $A26 000 was a reasonable return at that time.

The results of the restriction process were that from over 100 applicants, 59 divers were granted access to the fishery. These divers were issued with an annually-renewable, non-transferable permit to operate in the abalone restricted fishery. It is worthwhile noting that 2 permits were issued to indigenous applicants at this time. In addition, 2 permits were surrendered during the first few years of restricted access.

Even though diver numbers had been restricted, developing markets and increasing returns resulted in greater pressure on the abalone resource and an industry initiative to reduce diver number was adopted in 1985. Under this initiative, which became commonly known as the "2 for 1", a new entrant could enter the fishery provided they secured the surrender of 2 of the permits that had originally been issued. A new permit that became known as consolidated permit could then be transferred on a one for one basis. The 2 permits that had been issued to indigenous fishers were surrendered under the "2 for 1" scheme.

By 1989 fourteen permits had been removed from the fishery under the "2 for 1" scheme. However, new entrants paid considerable sums of money to enter the fishery and, in most instances applied greater pressure to the resource than those they had replaced, in an attempt to service the substantive loans taken out to enter the fishery.

In a further attempt to reduce effort, a quota was introduced in 1989 and total output was capped. The decision was made, with majority industry support, to allocate equal quota to all permit holders. This decision was taken so that new entrants who had in most instances bought permits with relatively low catch history were not disadvantaged and had the ability to meet their financial commitments. Quota, of 10t per annum was initially allocated but reduced to 9t in 1992. The "2 for 1" and the quota scheme continued to operate until the declaration of the commencement of the share management fishery in 1996.

2. THE SHARE MANAGEMENT SCHEME

Abalone licences became "transferable" in 1985 under the " 2 for 1" scheme. Prices achieved for these licences rose steadily and increased from $A120 000 to $A850 000 by 1992. Industry and lending institutions were concerned with the lack of security and statutory right, *i.e.* licences were still a personal commodity and still annually renewable with quota allocated as a condition of the fishing licence. In addition to the perceived lack of security, partnerships or company ownership were not recognised and operators who wished to employ another person to fish on their behalf had to do so outside existing fisheries legislation through complex legal agreements.

In June 1993, the Minister at the time, recognising that the fisheries resources needed protection and that fishers needed increased security, established a Property Rights Working Group to investigate the feasibility of a management regime based on a transferable fishing right. This Group recommended a system whereby statutory transferable shares, similar to a Torrens[1] title, were issued to fishers based on past entitlements or catch history. This scheme has formed the basis for the *Fisheries Management Act 1994*. Share management was included in this Act following extensive research and consultation with industry.

The commencement of the 1994 Act provided for the new type of management regime based on a share property right and in February 1995 all of the State's commercial fisheries were included in Schedule 1 of the Act as share management fisheries.

---

[1] Introduced in South Australia in 1858: a system whereby title to land is evidenced by one document issued by a Government department

A change in Government in March 1995 resulted in a review into the implementation of the share management regime. The terms of reference of this review were:

"To review the implementation of share management fisheries to ensure that the concept is effectively applied as part of an integrated approach to fisheries management in NSW".

Following this review, all fisheries except rock lobster and abalone were removed from Schedule 1 and have, in the first instance, proceeded down a restricted fishery path. The decision for the abalone and rock lobster fisheries to progress to share management was taken because the participants had already been identified and management strategies were underway. The abalone fishery commenced as a limited access share management fishery on 1 February 1996 and the rock lobster fishery in July 1996.

3. LEGISLATION

The implementation of share management is a 4-stage process:

i. Stage 1 (Consultation) - where the Minister consults relevant industry bodies about which fisheries should become share management fisheries.
ii. Stage 2 (Identification of fishery and shareholders) - when a fishery is identified as a share management fishery by the inclusion of a description of the fishery in Schedule 1. During this second stage, an interim Management Advisory Committee for the fishery is established, the criteria for the allocation of shares in the fishery are determined, eligible persons are invited to apply for shares and shares are issued provisionally.
iii. Stage 3 (Access to the fishery limited to shareholders) - when access to the fishery is limited to provisional shareholders (and also to any person claiming to be eligible to receive shares). During this stage, appeals against the provisional issue of shares are determined and a draft management plan for the fishery is prepared.
iv. Stage 4 (Full implementation) - when the management plan for the fishery commences and the fishing, share transfer and other rights of shareholders are fully identified and exercisable and subject to review.

4. WHAT DOES A SHARE MANAGEMENT SYSTEM PROVIDE?

The share management regime provides fishers with the security of a transferable statutory property right that is initially allocated for a 10-year period, but is automatically renewed if the management plan for the fishery is amended. The shares are also automatically renewed at the end of the 10-year period. It provides fishers with a long-term access right on which business decisions can be based and it provides flexibility within the fishing business itself, *i.e.* the ability to nominate another person to fish on one's behalf and the ability to adjust the size of their operation to best suit one's needs. The size of a business can be adjusted on a permanent basis through share-trading or an annual basis through quota-trading if provided for in the management plan. Part of the rationale of this system is that by providing fishers with a greater property right, they will demonstrate greater husbandry of the resource.

5. WHAT DOES IT COST?

It is NSW Government policy that the costs to the Government for managing the fishery are fully recoverable. In addition, the legislation requires that shareholders make a periodic payment to the community for their privilege of right of access to the resource. Matters relating to this payment must have the Treasurer's concurrence.

6. CALCULATION OF MANAGEMENT CHARGES

The costs of managing the NSW abalone fishery are calculated based on the resources required to deliver an acceptable level of service to industry. Industry has chosen over the years to dedicate funds towards specific research and compliance. In 1997 the Independent Pricing and Regulatory Tribunal (IPART) examined the cost to NSW Fisheries of managing the State's commercial fisheries and whether there were other beneficiaries of that management. IPART recommended that fishers should only pay the efficient costs of management and that a discount on management costs should be provided to commercial fishers for other beneficiaries of that management, such as recreational fishers.

Salaries and operating costs are calculated based on identified resource requirements and then savings based on efficiency and other beneficiaries are subtracted from the total amount. This new total is then divided by the number of shares in the fishery to provide a per share cost. This fee is then payable by shareholders in quarterly installments. In 1999/2000 abalone shareholders paid a management fee of $A242 per share.

7. DETERMINING THE COMMUNITY CONTRIBUTION

A community contribution is payable by shareholders in a share management fishery for their privileged access to what is ostensibly a community-owned resource. The Treasurer's concurrence is required on the level of this charge and any other matters relating to the charge.

To assess the amount of economic rent available in the abalone fishery and to determine an appropriate pricing mechanism for this charge, an expression of interest was sought from an independent consultant. This independent economic assessment recommended a community contribution based on 10% of the gross value of the catch at an average beach price of $A33/kg. It further recommended that the rate of community contribution should be on a sliding scale to take into account significant rises and falls in beach price. Industry was provided with the opportunity to make submissions on this report and an amount of 6% of the gross value of the catch phased in over a three year period was agreed by the Minister and Treasurer to be specified in the draft management plan.

Share management provides fishers with the opportunity for compensation, based on the market of those shares should the fishery definition be removed from Schedule 1. In reality however, this is unlikely to occur for a fishery such as abalone as structural adjustment has already occurred. Compensation is not payable if the total allowable catch is determined to be zero.

8. TRANSFER OF SHARES DURING STAGE 3

Transfer of shares during the limited-access stage of share management is controlled by general regulatory provisions that require a fisher's whole fishing business to transfer. This regulation was implemented to prevent speculation prior to a statutory management plan being made. Once full shares have been issued, transfer will be subject to the management plan and will provide greater flexibility.

9. ABALONE FISHERY IMPLEMENTATION OF SHARE MANAGEMENT

As noted above the implementation of share management is a four-stage process, requiring extensive and exhaustive consultation with stakeholders. Section 50(1) of the Act states that:

> "If a restricted fishery becomes a share management fishery, the persons entitled to shares in the fishery are the persons who, immediately before it became a share management fishery, were entitled to take fish for sale in the restricted fishery. The allocation of shares to any such persons may be made having regard to existing entitlements in the restricted fishery."

The second stage of the share management process was critical. It was clear who was entitled to shares in the fishery, however how those shares were to be allocated was not. The legislation says based on past entitlements, but the industry were divided about the definition of an "entitlement". Those who entered the fishery under the "2 for 1" claimed that their method of access was the entitlement *i.e.* they should be allocated double the number of shares as they had "bought" 2 permits. The original divers however claimed that the entitlement was the right to take an equal amount of quota. And Section 78(3) of the Act requires:

> "An allocation (of quota) among shareholders in a particular fishery is to be made in proportion to the shareholdings of the persons concerned."

Therefore, if unequal numbers of shares were allocated to operators, different quota holdings would result. As equal quota had been allocated since the implementation of the quota scheme, regardless of entry, some saw this as inequitable.

The Department consulted extensively with industry with regard to the allocation of shares. All divers were personally interviewed to seek their views. They were provided with a questionnaire about how the share management system, particularly the allocation of shares, could be implemented and were requested to provide copies of any documentation that may have indicated that the "2 for 1" system would continue until all licences had been consolidated or, that they had a greater right of access to the fishery. In addition, the Department sought detailed legal advice on what constituted an entitlement in the abalone restricted fishery.

Based on the facts that: (a) no written assurance had ever been given to new entrants that the "2 for 1" would continue, or that they would receive a greater right of access to the fishery; (b) quota had always been allocated equally to all divers; (c) all divers had contributed equally to management charges and to a buy-back fund, on legal advice, the Minister decided to allocate 100 equal shares to the 37 participants in the abalone restricted fishery. Applicants were notified of this decision and the appeal process that was available to them. The third stage of share management commenced on 9 February 1996.

10. REGULATORY REQUIREMENTS FOR THE ABALONE SHARE MANAGEMENT FISHERY

The abalone restricted fishery had specific regulations that provided for new entrants ("2 for 1"), transfer of quota, collection of management fees and other administrative requirements. When it commenced as a share management fishery, the restricted fishery regulations ceased to have effect for the abalone share management fishery. They did however, still have effect for the sea urchin and turban shell (SUTS) fishery that had remained as a restricted fishery. To ensure that there was a statutory base for administering the fisheries and that fishing business transfer regulations were consistent, a regulatory amendment was made that took effect on 9 February 1996.

As sea urchin and turban shell were included in the definition of "abalone" for the restricted fishery, all restricted fishery regulations also applied to these species, including the transfer of permits and endorsements. When the abalone fishery progressed to share management, sea urchin and turban shell were still bound by the "2 for 1" entry criteria, even though this scheme was implemented to reduce abalone diver numbers.

The *Fisheries Management (General) Regulation 1995,* and not the amended abalone share management regulation, provides for transfer of shares while the fishery is in the limited access stage of share management. This regulation requires a shareholder to transfer the whole of their fishing business. As shares were allocated to those who had operated in the restricted fishery, a fishing business now comprised abalone shares and an endorsement to take sea urchin and turban shell. The "2 for 1" entry criteria for the SUTS fishery was now inconsistent with the regulations that had been made specifically for those fisheries and thus the businesses that proceeded to share management.

11. JUDICIAL AND POLITICAL CHALLENGES
11.1 Implementation of the system

The implementation of share management for the abalone fishery was challenged at both the judicial and political level. The underlying reason for the challenges being the demise of the "2 for 1" entry scheme as a result of the decision to allocate equal shares and not by a mechanism that reflected the "2 for 1". The amended Regulation was disallowed by the Upper House of NSW Parliament on 30 April 1996. Their intent was that the fishery would revert to a restricted fishery and that the "2 for 1" would become effective for abalone once again. The disallowance did not result in the reversion of the fishery, but merely resulted in a fishery that had to be managed by policy and regulation rather than on a legislative basis.

A challenge was mounted in the Supreme Court of NSW by those divers (Consolidated Divers Group) who had entered the fishery under the "2 for 1" scheme with the initial summons being served on 14 March 1996. As the disallowance did not have the desired effect of restoring the restricted fishery, the summons was amended in July 1996.

Following the 3-month statutory period required before a disallowed regulation can be resubmitted, and on the basis that the disallowance did not resurrect the "2 for 1", another regulatory amendment was made in December 1996 that was ostensibly the same as the previous amendment. This Regulation was disallowed in March 1997, this time on the basis that if it were not disallowed the Supreme Court process would be compromised. In addition, it was believed if the amendment was not disallowed, the "2 for 1' could not be implemented again. In December 1997, the Supreme Court ruled that there had been no invalidity or error of law in the process (of implementation of share management for the abalone fishery) so far.

A regulation amendment was again made in March 1999 to provide a statutory basis for the management of the abalone share management fishery and the associated sea urchin and turban shell restricted fishery. A further challenge to this regulation was withdrawn in September 1999.

The decision of the Supreme Court was challenged in the Court of Appeal and in December 1998 this Court comprehensively dismissed the Appeal. An application for special leave to appeal to the High Court was made following this decision, but withdrawn in July 1999, one week before it was due to be heard.

11.2 Share appeals

Any applicant for shares in a share management fishery may appeal against the allocation decision, regardless of whether they were actually issued shares. Seven appeals were lodged with the Panel in relation to the issue of shares and all were based on the decision to allocate equal shares. Some claimed that they should have been allocated twice the number of shares as "original" divers while others believed different classes of shares should have been allocated. It was not within the power of the Panel to allocate different classes of shares, and no additional shares could be allocated. Any additional allocations to a shareholder will result in a redistribution of shares within the fishery.

The Panel upheld the decision of the Minister and the Department in the allocation of provisional shares. They ruled that no evidence was supplied by the appellants that supported their claim that NSW Fisheries did not at all times comply with the directions of the Minister.

11.3 Importance of process

The process, particularly the administrative procedures proved to be critical in ensuring the successful implementation of share management for the abalone fishery. The process required the collection of the appropriate data, *i.e.* survey information of industry, legal advice and analysis of previous decisions to ensure that when the decision on the allocation of shares was eventually made, the data were accurate and there was evidence that all concerned had been consulted.

The process included exhaustive consultation, documentation of meetings, approval (*i.e.* sign-off of decisions, filing of relevant material and record keeping, appropriate wording of documents, timely gazettal of orders and proclamations and distribution through the relevant channels. The steps, tasks and interactions during the implementation of share management for the abalone fishery withstood the judicial and political challenge reinforcing the views that the Department was committed providing to quality customer service.

11.4 Allocation of the statutory property right

Fishers are keen to see the implementation of a management plan and the allocation of shares that have a statutory basis. They believe this will provide them with greater security and stability and that financial institutions will be more willing to lend on a property right that goes into perpetuity rather than an annually renewable licence.

The system will provide shareholders with a 10-year statutory property right that is automatically renewed after this period. This access right is separate to the fishing right. It will provide shareholders with a statutory management plan that contains objectives, performance indicators and trigger points, and thus a sustainable resource and economically viable businesses into the future.

12. FUTURE OF THE COMMERCIAL ABALONE FISHERY IN NSW

The management plan will provide the basis for future management of the commercial fishery. Resources will be channelled into those areas of most need and as the management advisory committee becomes increasingly more mature, a devolution of the management of the commercial fishery to a management committee should occur. The Government's role as custodian of the resource will however not diminish and there will still be a requirement for the Government to be involved in the monitoring, compliance and management of the fishery.

CANADIAN SCALLOP FISHERY MANAGEMENT: A CASE HISTORY AND COMPARISON OF PROPERTY RIGHTS *VS.* COMPETITIVE APPROACHES

F. G. Peacock[1], J. Nelson[2], E. Kenchington[3] and G. Stevens[1]

[1]Resource Management Branch
[2]Policy and Economics Branch
[3]Science Branch
Department of Fisheries and Oceans
Dartmouth, Nova Scotia, Canada
< PeacockG@mar.dfo-mpo.gc.a>

1. INTRODUCTION

Modern fishery management in Canada began in the late 1960s with the introduction of limited entry licensing in inshore lobster fisheries and it relatively quickly developed into a complex of licence, catch and effort controls covering most fisheries. These were extensively used to manage the industry after 1977 when extended jurisdiction by Canada to 200 miles promised prosperity to many fleets (Parsons 1993). Today, each fishery operates under an annually-developed conservation-oriented harvest plan established to control the composition of the catch and the amount of fishing. Catch composition is controlled by mesh-size regulations, minimum size limits, closed areas and closed seasons. Catch quotas and effort controls control the amount of fishing. The elements of the plans vary from fishery to fishery in response to the individual circumstances of the resource and the industry.

The Canadian sea scallop (*Placopecten magellanicus*) fisheries offer the opportunity to evaluate the effectiveness of some of these management practices as two contrasting management approaches were in place for a decade; one being applied to offshore resources, the other to those inshore and in particular, in the Bay of Fundy (Figure 1). Until 1986, the management of both inshore and offshore scallop fisheries was largely identical with both operating under competitive formats, although largely in different locations with only a small amount of overlap. In 1986, an agreement was reached between the two fleets and accepted by the Government of Canada, which altered the management and operations of these two fisheries. This agreement, which withstood a legal challenge by inshore participants in 1996, separated the fleets' operating areas, confining the inshore fleet to waters north of latitude 43°40'N in the Bay of Fundy and the offshore fleet to waters 12 miles from shore to the south. This agreement provided a basis for a divergence in management approaches following its introduction, *i.e.* a rights-based management plan in the offshore and a competitive fishery in the inshore.

The management system within the offshore fishery has been largely unchanged from 1986 through to the present. However, the competitive format pursued by the inshore fishery was indirectly affected by management decisions in the groundfish fishery, in particular by the introduction of ITQs to the groundfish fleet in 1991 as many fishermen were dual licence holders. The inshore scallop management plan was changed to an area management and ITQ system in 1997 (Kenchington *et al.* 1997).

This analysis reviews the period 1986 to 1996 in order to assess the effects of the inshore and offshore management approaches on the performances and profitability of the fleets, the fishermen and associated fishing communities and the resources being harvested. It extends an earlier review (Brander and Burke 1995) of the impacts of rights-based *versus* competitive fishing of Canadian sea scallops.

2. RESOURCE BIOLOGY

The biology of the resource is well understood and the pertinent aspects to fisheries management have been conveyed to both industry and management to assist with the development of harvesting plans. The sea scallop is found only in the northwest Atlantic Ocean from Virginia to Labrador. Within this range, scallops are concentrated in persistent areas or "beds", many of which support valuable commercial fisheries. In the Canadian region, there are major concentrations on Georges Bank, Scotian Shelf (Middle Grounds, Sable Island Bank, Western Bank, Browns Bank and German Bank), in the Bay of Fundy and on St. Pierre Bank (Figure 1). Different beds, and areas within beds, have different growth and yield characteristics dependent on temperature and food availability. Studies on movement show that although the scallop is a capable swimmer, and there is individual movement within a bed, the beds themselves do not shift markedly.

This species is a broadcast spawner (fertilization takes place in the sea) with separate sexes. To ensure successful spawning, adult scallops must live in close proximity to one another with both males and females in the population. The formation of dense beds is therefore critical to spawning success. Overfishing of beds has caused stock collapse in many parts of the world (Kenchington and Lundy 1996). These collapses may be due to destruction of the bed formation as well as reduction of broodstock abundance (Kenchington and Lundy 1996). Spawning occurs from August to September and the larvae are planktonic for five to six weeks before settling in October. With so many uncertainties it is not

Figure 1
Location of major scallop fishing grounds in Atlantic Canada
NAFO areas, the International Court of Justice line and place names referred to in the text are indicated.

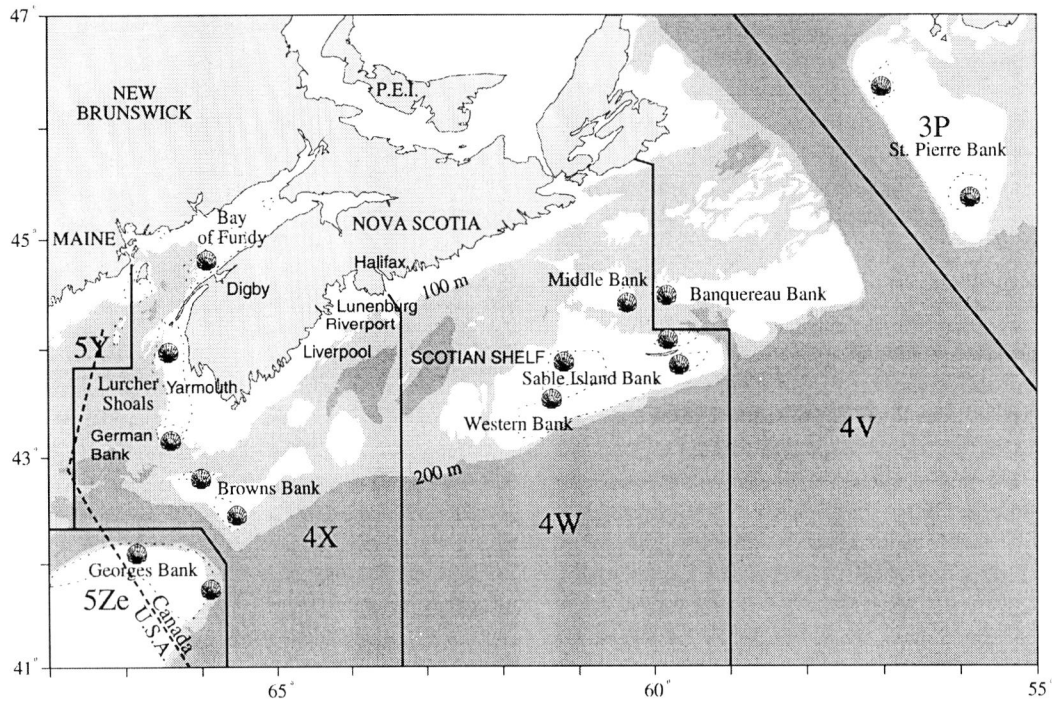

surprising to find that recruitment is highly variable and unpredictable. No stock/recruitment relationship has been clearly demonstrated.

The quality and weight of the scallop meat (adductor muscle) is directly influenced by this reproductive cycle. After spawning there is a 30-40% increase in adductor muscle wet-weight in the Bay of Fundy (Kenchington et al. 1994) and at least a 15-20% increase on Georges Bank (Serchuk 1983). Scallops can live to at least 20 years old, and animals up to 17 years old are found at low densities throughout the Bay of Fundy. The gear generally captures them first at age three or four. This lifespan offers the potential to manage year- classes so that catches can be stabilized through periods of poor recruitment.

3. THE FISHERY
3.1 Fleets and vessels: history and current status

Many diverse fleets fish scallops in Atlantic Canadian waters. Most are small, exploit only local, near-shore beds and involve fishermen, boats and communities for which scallops are strictly a seasonally exploited resource. In some cases local lobster boats are fitted with light scallop gear at the appropriate time of year. Elsewhere, slightly larger groundfish boats are used. There are such fisheries in the Gulf of St. Lawrence and at various points along the Atlantic coast. Within the Bay of Fundy, similar fisheries operate along the western (New Brunswick) shore and in the upper parts of the Bay.

Two more substantial fleets, primarily based in the Province of Nova Scotia, harvest scallops on a nearly year-round basis as their primary resource. One fleet fishes offshore and is owned by a small number of relatively large companies, some of which are traded publicly. These vessels are based along the south shore of Nova Scotia, primarily in Lunenburg, but also in Riverport, Liverpool and Yarmouth. The other fleet is based in the Bay of Fundy; many of its boats operate out of Digby, Nova Scotia. It is referred to as the "Full-Bay" fleet because the boats are licensed to fish scallops throughout the Bay of Fundy, including the traditionally-lucrative beds off Digby to which this fleet enjoys exclusive access. Elsewhere in the Bay, it shares the beds with other, local fleets of smaller boats. This Full-Bay fleet is owned by a mixture of small companies, some with a single boat and some with multi-boat fleets, with individual owners, some of whom are active captains.

In 1987 the licensing system recognized three distinct scallop fleets in the Bay of Fundy: The Full-Bay fleet, an "Upper-Bay" fleet that was confined to the upper reaches of the Bay, and a "Mid-Bay" fleet that was licensed to fish from the New Brunswick shore out to a "mid-bay" line. There were 99 licensed boats in the Full-Bay fleet, 16 in the Upper-Bay and 210 in the Mid-Bay, though not all of these boats are active in the scallop fisheries every year. The Full-Bay fleet regularly harvests over 80% of the total Bay of Fundy catch.

The inshore scallop fisheries are the oldest, having begun by the late nineteenth century in various parts of the Bay and on the south coast of Nova Scotia. The commercial dragger fishery in the Bay of Fundy began in 1920 in Digby. Scallops were fished from a 12m sloop, 4.6m in the beam with 11 horsepower and equipped with one drum and a head for hauling a single drag and hoisting it on deck. The drag allowed for the exploitation of scallops from greater depths than could be taken with tongs. The adoption of powered draggers and improved gear design allowed an expansion in the 1940s. Thereafter the Full-Bay fleet worked particularly the beds off Digby, and elsewhere in the Bay. This fleet also fished on the offshore banks, when the nearshore and Bay of Fundy beds were depleted and when particularly plentiful populations of scallops were present on the banks. Today, Full-Bay vessels are 15 to 19.8m in length.

The Canadian offshore scallop fishery developed after 1945 in response to an increased market demand for scallops. Canadian boats then competed with American ones on the outer banks from Newfoundland waters south to Virginia and fished on a continuous basis regardless of the status of the inshore stocks. This development brought with it the need for larger (27 to 45.7m overall length), all-weather vessels capable of harvesting scallops in the severe conditions of the winter months in the northwest Atlantic. The typical Canadian offshore scallop dragger came to be a wooden boat of about 30m in length and, in shape much like a North Sea side-trawler. They were fitted to tow two rakes simultaneously, one from either bow. Wooden boats were gradually replaced with steel boats through the 1970s. This offshore fleet dominated scalloping on the outer banks after 1950 and became the sole scallop fleet operating there after 1988.

Fishing methods have changed relatively little since dragging replaced tonging. The offshore fleet uses large steel rakes of the "New Bedford" type 4 to 4.9m in width, while the inshore boats use "Digby drags" – up to nine individual "buckets" (chain bags attached to steel frames) flexibly linked to a rigid bar. This arrangement is thought to be more efficient on the irregular bottom of the Bay of Fundy.

The offshore boats typically make trips of up to 12 days duration, working round the clock when on the scallop beds. The inshore fleets formerly went to sea for only one day at a time but they have come to undertake three- or four-day trips, sometimes with sufficient crew for 24-hour operations. Both the offshore and the Full-Bay fleets take scallops throughout the year.

3.2 Products and processing

Almost the only product of these fisheries is the adductor muscles, or "meats", of the scallops, which are separated from the rest of the animals ("shucked") at sea. Small scallops are sorted from the catch and returned to the sea alive and other processing is largely limited to washing and packing for market, which takes place at plants on shore. There has been a small "roe-on" fishery, using boats of the offshore fleet, in the spring of recent years. They produce whole scallops for markets requiring that product. However, Canadian scallops can accumulate dangerous quantities of Paralytic Shellfish Poisoning (PSP) toxins in most of their tissues other than the adductor muscle. Thus, the roe-on product must be carefully inspected and this specialist fishery is susceptible to closure due to unacceptable levels of toxins.

3.3 Markets

Apart from the small market in Europe for scallops with roe, scallops are sold as either fresh or frozen scallop meats in Canada and the United States. Normally, the price received for scallops at the wholesale level depends on supply and demand conditions in the US, the major market. Resource conditions on major fishing banks such as Georges Bank influence supply and demand is based on general economic factors, which are fairly stable. Prices for scallops have been extremely high for the last two or three years because of a shortage of scallops in the markets in relation to demand. The size of scallops also influences price - larger sizes get a higher price.

4. FLEET SEPARATION
4.1 The 1986 Agreement

Although the 1984 International Court of Justice (ICJ) decision delimited Canada's jurisdiction over the various scallop beds in the Northwest Atlantic, it left the different Canadian fleets sharing access to the national resource. That access was partitioned between the fleets in 1986, through an agreement between the various parties concerned, both government and industry. This "1986 Agreement" allowed different management regimes to develop in the Bay of Fundy and on the offshore banks, while it also influenced the different developments in the two areas.

4.2 Process and history leading up to the agreement

The first scallop management plans appeared in the early 1970s and as with other fisheries the development and operation of the offshore and inshore scallop fleets was done with few rules or regulations in place. Most accounts suggest that the two fleets operated harmoniously with only a small overlap of fishing areas even though there were no spatial restrictions on either fleet. By 1978 this had changed as declining resources and catches in the inshore created increased fishing activity on the outer banks. The implementation in this year of the "2.9% rule" whereby 2.9% of the previous year's catches on Georges Bank were allocated to the inshore fleet, paralleled a decline in catches of Bay of Fundy scallops necessitating greater dependence by some on the Georges Bank harvest.

By 1984 Canadian scallop landings from Georges Bank had declined to less than 2000t of meats. Both fleets fished to the same meat-count (maximum number of scallops per 500g sample) on Georges, but in other areas, only the offshore fleet was bound by strict meat-count regulations. The offshore fleet was not permitted in the Bay of Fundy or within the territorial sea (to 12 nautical miles). Prior to 1986 the Bay of Fundy was defined as a line from Digby Neck, N.S. to Grand Manan, N.B. Other issues, such as "shellstocking" (holding whole animals on board), conducted by the inshore fleet during peak years, were additional irritants. The two fleet sectors had divergent points of view to settle before economic stability could be restored and the fleets rationalized.

In 1986, the entire 2.9% inshore share was already taken by July and the inshore fleet became aggressively interested in increasing their share on Georges Bank. The added problem of non-complementary regulations for the two fleets harvesting Georges further convinced the offshore fleet of the necessity to change management course and to separate the fleets. Interest in an Enterprise Allocation (EA), (company quota) property-right plan grew following their introduction into the groundfish industry and logical extension to offshore scallops, which involved many of the same players (enterprises). Despite numerous discussions and a series of industry - government seminars in 1985-86, mutual agreement on a common management strategy for all scallop fleets could not be reached. However, agreement was reached between the inshore and offshore groups on the concept of exclusive zones for each fleet.

4.3 Terms of the Agreement

The 1986 Agreement dealt with a number of issues, including, among others:

i. Separation of the grounds open to the offshore and inshore fleets at the 43°40' N parallel of Latitude
ii. Phasing-out the Bay of Fundy fleet from Georges Bank (with 8% of the TAC in 1987, 4% in 1988 and none thereafter)
iii. Extension of the zone open to the New Brunswick "seven-mile" licences out to a new "Mid-Bay" line
iv. Cancellation of never-used licences
v. A voluntary licence-reduction programme
vi. Establishment of new closed areas
vii. Introduction of a meat-count limit in the Bay of Fundy and
viii. Stricter penalties for violators.

The 43°40'N line excluded the offshore fleet from portions of German Bank and Lurcher Shoal that it had previously fished, while shutting Browns and Georges Banks to the inshore fleets.

Following this agreement, which was accepted by the Minister of Fisheries, a permanent separation of the two fleets at the 43°40'N line was announced in October 1986. The offshore sector moved to a trial EA programme in June 1986 with nine corporations holding allocations. In 1986 a TAC on Georges Bank was established before the offshore companies completed sharing arrangements. This resulted in 50% of the TAC being harvested in a six-week period. The fishery was closed while companies determined shares then resumed under an EA programme and the race for the resource was over. The programme was made permanent in 1989. The 1986 Agreement is considered to be a milestone decision for the management of commercial scallop fishing off Nova Scotia.

5. EFFECT OF THE GROUNDFISH FISHERY AND ITQ SYSTEM ON THE INSHORE SCALLOP FISHERY

From 1981 to 1985, 50 to 60% of the dual (groundfish and scallop) licence holders among the Full-Bay scallop licence holders (Table 1) fished scallops exclusively (Figure 2) with no prominent monthly trend in activity (Kenchington et al. 1995). Mixed fishing within a year (Figure 2) was practiced by 15 to 25% of dual licence holders. During this period the groundfish TACs were high, although the catches of both cod (Campana and Hamel 1992) and haddock (Hurley et al. 1992) were declining.

Table 1
Number of Full-Bay of Fundy scallop licences and additional groundfish licences carried from 1981-95 Prior to 1986 the number of licences included transfers during the year and reflect the number of licences used during that year and not the absolute number of licences

Year	No. scallop licences	No. groundfish licences
1981	99	81
1982	107	86
1983	115	91
1984	106	79
1985	96	73
1986	96	74
1987	95	73
1988	98	68
1989	98	61
1990	99	61
1991	99	64
1992	99	61
1993	99	59
1994	99	58
1995	99	54

In 1986 this pattern changed (Figure 2) due to a number of factors. The 1986 separation agreement confined the inshore fleet to the Bay of Fundy at a time when scallop landings were at a low level. Government surveys that year detected the presence of a large number of scallop pre-recruits. As a result, an area off Digby, Nova Scotia referred to as the Inside Zone was extended to 8 miles from shore and closed to protect the incoming year-classes causing further spatial restriction. For the only period in this time (Figure 2) exclusive groundfish fishing was more common than exclusive scallop fishing among dual licence holders on a monthly (Jan. 1986 to June 1987, excluding May 1986) and yearly basis (Kenchington et al. 1995). Coincident with the decline in the scallop fishery, the groundfish fishery was at a peak in 1986 (Digou 1994), further increasing the incentive for dual licence holders to spend more time in that fishery (Figure 2). In 1986, and continuing through to 1994, a strong mixed fishery appeared with vessels spending a part of each year fishing both scallop and groundfish. The

Figure 2
Annual trends in active licences amongst the dual (groundfish-GF, scallop-SC) licence holders in the inshore Full-Bay fleet

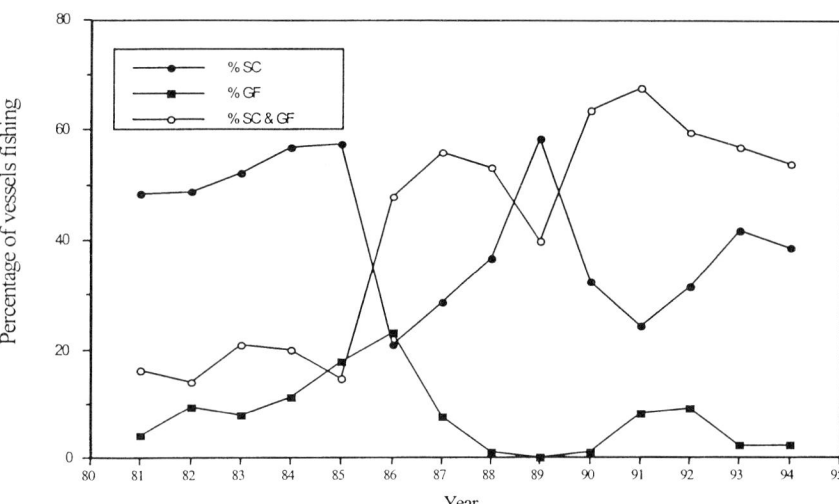

number of dual licence holders declined from 74 in 1986 to 54 in 1995 (Table 1).

Exploitation of the large scallop year-classes off Digby began with the opening of the seasonally restricted "Inside Zone" off Digby, Nova Scotia in October 1987, resulting in an increase in directed scallop fishing. In both 1987 and 1988 increased directed scallop activity was associated with the opening of the Inside Zone each year. In 1988 and 1989 all dual scallop licence holders fished from October to December (Kenchington et al. 1995). Exclusive demersal fishing activity fell dramatically from August of 1988 through to the summer of 1990 and total landings by this gear sector show a decline from 1986 to 1990 (Digou 1994). In 1987 and 1988 the groundfish fishery was closed or restricted several times during the year.

From 1989 to 1994 exclusive scallop fishing was high all year (generally over 60%; Figures 2, 3) because in 1989 the groundfish stocks had largely disappeared and the mobile gear sector (under 19.8m in length) exceeded their cod/haddock/pollock (CHP) quotas (introduced in 1989 - c.f. Hurley et al. 1991) and were tied up in June. By the second half of 1990 exclusive demersal fishing picked up and was regulated by CHP trip limit (Hurley et al. 1991). ITQs were introduced to that fleet in 1991 (Apostle et al. 1997). The decline in groundfish activity associated with the ITQ cuts of September 1993 can readily be seen in Figure 2. Groundfish fishing ceased for the next three months and has not returned to any significant degree due to low quotas. In 1994 there was more scallop-only fishing than in any previous year with over 90% of the dual licence holders targeting scallops (Figure 3). Thus, the consequence of small groundfish quotas has been a movement of the fleet into directed scallop fishing.

Multiple groundfish quotas have been transferred to a single boat allowing that boat to fish the quota economically, while the original owners of the quota free their boats up to fish scallop exclusively.

In conclusion, the imbalance between the scallop resource and capacity was further exacerbated by the introduction of groundfish quotas. However, the activation of the latent capacity in the fleet was also driven by the increased scallop abundance in the Bay. With a high price for scallops, unprecedented high landings and no meaningful effort controls, the increase in capacity would likely have arisen regardless of the groundfish ITQ plan. For a short while, and for the first time, the resource was greater than the available fleet capacity. Ultimately, the introduction of groundfish ITQs may only have facilitated an inevitable process.

6. FISHERY MANAGEMENT APPROACHES
6.1 Inshore fishery management

During most of the period under review (1986-1995) Bay of Fundy management plans were not conservation-oriented, reflecting the wish of the fishing and processing sector within the Inshore Scallop Advisory Committee (ISAC) for an uncontrolled harvest (i.e. no Total Allowable Catch or TAC). Minimum shell sizes were set too small (76mm) and meat counts too high (72 and 55 meats/500g) to be effective at controlling the composition of the catch to biologically-based target sizes. Closed areas, the only other tool addressing conservation, were partly effective. Other measures included limited-entry of vessels to the fishery, gear-restrictions, vessel-replacement rules and banning the splitting of scallop licences from groundfish licences. Throughout this period there was a major imbalance between resource and fleet capacity.

Figure 3
Relationship between the % of vessels fishing scallop-only
and those fishing groundfish and scallop by year

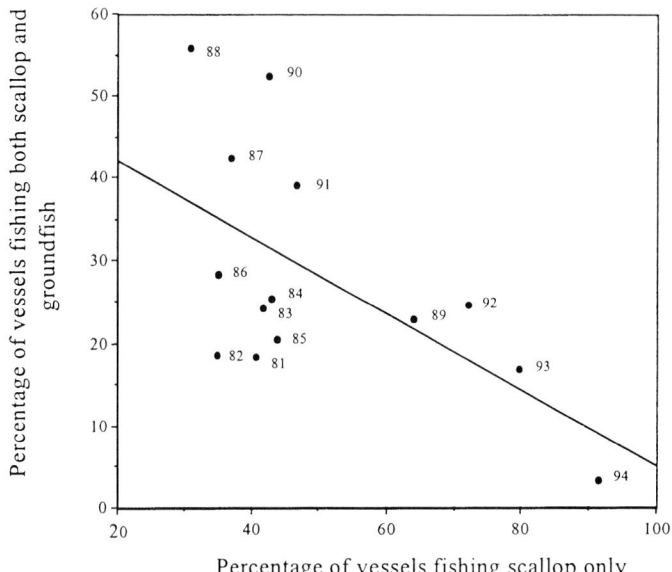

Significant progress was made in the 1995 and 1996 harvest plans. These had several initiatives that seriously addressed biological/conservation concerns, including lowering meat counts substantially to 45 meats/500g over large areas and maintaining area-closures put in place to protect broodstock. Distribution/effort controls which focused on issues such as crew size and weekday fishing[1] (introduced at industry's request) appear to have had little impact on the stocks. There was no limitation on overall harvest levels and no harvest targets were put in place.

The 1997 Bay of Fundy harvest plan was a milestone in this fishery. The Bay was divided into 7 biologically defined fishing areas (Kenchington *et al.* 1997). Each area was managed with a TAC, a biologically based meat count, shell height and voluntary minimum meat size, and in some cases, seasonal restrictions. The intent of the plan was to eventually manage each area individually with differing regulations adapted to the biology of the resource. This approach would allow for the protection and management of good year-classes and may facilitate further initiatives such as rotational fishing and enhancement projects. However, at present, enforcement is inadequate to allow for the opening of areas with different fishing regulations. Industry-funded dockside monitoring of the catch and vessel monitoring "black-boxes" will be introduced in 2000 at which time the full potential of the plan may become reality.

6.2 Offshore fishery management

The offshore fishery is managed through the Offshore Scallop Advisory Committee (OSAC), which is composed of the seven enterprises as well as federal and provincial representatives and a representative of the crew unions. Management methods include a Total Allowable Catch (TAC) to control harvest levels. Industry is provided with biologically-based TACs for different levels of exploitation and the final TAC is set through government/industry agreement.

To provide the needed stability for investment decisions on vessels, as well as the cost-savings from a rationalized fleet, the offshore corporations decided on an Enterprise Allocation (EA) option where the TAC is divided into quotas. An EA is a fixed percentage of the fleet quota that is allocated to an enterprise or company that holds licences for the fishery. This system provides the enterprise with a mechanism for assigning a vessel or vessels to catch the enterprise quota as required. EAs are not intended to convey ownership of the resource to enterprises in the industry. Rather, the enterprises hold valid licences to harvest, within the fishery, a specified quantity, in an organized and deliberate fashion without interference by the performance of others within the fleet. Under the competitive scenario, the majority of owners felt that the process of replacing the aging fleet would be disorderly and costly compared to the more orderly and stable environment under the EA programme. Scallop EAs were based on each firm's dependence on the fishery determined by the numbers of licences held (50%) and historical catch (50%).

No one enterprise may hold more than 50% of any specific scallop stock. The EA allocations are not permanently transferable although temporary transfers are permitted. Except under exceptional circumstances an enterprise cannot transfer in excess of 25% of its EA for more than 2 consecutive years. Temporary transfers of

---

[1] This term refers to fishing only during daylight hours, specifically 05.00 - 20.00.

quota are permitted within the year. Permanent transfers of quota can only be made if a Canadian purchases the entire company. All or part of any allocation that cannot be harvested must be offered to the remaining active enterprises. The EA plan includes provision for allocating licence and quota in the event of the collapse of the programme or of the bankruptcy of one of the enterprise holding licences.

With an EA process in place, enterprises could focus on maximizing efficiency by assigning appropriate vessels in appropriate circumstances and integrating firm's operations from sea to market. Previously, the lack of individual-catch controls led to costly competitive races for scallops and an unstable environment for planning business investments and operations. In addition to the TAC, there are minimum size provisions and complete hail and landing industry monitoring applications in place (Appendix 1). The meat counts are set at biologically-based levels and are different for different stocks. Industry also assists in stock surveys. Fleet capacity is voluntarily adjusted by enterprises as necessary and ongoing reviews of catches and catch rates allow for in-season adjustments.

7. IMPACTS OF THE 1986 AGREEMENT ON THE INSHORE FISHERY

7.1 Resource conservation

The scallop fishery in the Bay of Fundy continues to be strongly driven by the variability in recruitment as a consequence of the high levels of fishing mortality (Kenchington et al. 1997). During the past decade three exceptional year-classes have settled in the Bay, two in the Digby area and one on the beds in the lower Bay of Fundy. None of these year-classes were effectively managed. In the case of the Digby scallops a mass mortality event coincided with the annual closure of the beds such that the extent of the losses were not realized until it was too late. Coupled with extremely high fishing mortality ($F>1.0$) the resource was devastated to the point where densities were so low that large areas were closed to prevent further fishing of the broodstock in 1995. This event created two mindsets that have been detrimental to the development of conservation schemes. One is that there is no link between stock abundance and recruitment, since the 1984 and 1985 boom year-classes came from the lowest stock-abundance on record to that date. Secondly, that if scallops are too plentiful they will die and so they must be thinned out when they are small. These views are widely held by the majority of ISAC members, including the non-fishing representatives. Thus when the strong 1990 year-class recruited to the fishery in the lower Bay of Fundy it was heavily fished, resulting in low yields and there was no interest in trying to extend the life of the cohort.

As the scallop beds returned to more average densities the fishing effort remained high, bolstered by good prices and a fully dependent active fleet, resulting in the depletion of most inshore scallop beds and widespread recruitment failure on the traditional beds off Digby. Landings have been reduced dramatically, and coupled with the new more restrictive management measures, operational and economic difficulties existed. Seasonal closures for large parts of the year (fishing restricted to 1 to 3 months) were effective in maintaining the Grand Manan and Annapolis Basin stocks. Both of those areas show regular recruitment with a range of ages in the populations. The seasonal closure of a portion of the beds off Digby did not benefit recruitment, although the timing of the closure maximizes yield. The fishing season was prolonged in that area (October to May) and occurred during the settling period of the larvae. The season-length did not generally restrict the catch as in the other areas. Meat count regulations were not effective in protecting broodstock or strong year-classes. Changes to the meat counts toward biological-based recommendations have been too recent to see an effect on the resource.

7.2 Fishing capacity

Inshore scallop licence holders may have less incentive to over-invest in the fishery than others in the inshore groundfish dragger fleet where technical innovation provides a competitive advantage. In the scallop fleet, all vessels use the same gear and electronic equipment and since horsepower is not as much of an issue as in other fisheries this suggests that equipment-upgrading is not a major concern. Similarly, vessel-size is not an issue. Most fishermen use a vessel 16-17m in length and while replacement rules allow for increases, none occur, which suggests an ideal design exists for current conditions. However, during the late 1980s a major boat-building campaign resulted in many new vessels, adding a considerable debt burden to individuals within the fleet.

While capacity growth in this fleet did not come from investing in the "new-bigger" phenomenon so common in competitive fisheries, more than half the 99 scallop licence holders also had licences for groundfish otter trawling (Table 1). Consequently their vessels were larger and more powerful than those of the single (scallop) licence holders. Throughout most of the 1980s this represented latent scallop capacity with most dual licence holders targeting in groundfish for at least 50% of the fishing year.

7.3 Fishing effort

With the introduction of ITQs into the inshore groundfish dragger fleet in 1991, the ability of groundfish draggers to harvest individual quotas quickly and under an individual schedule, or to sell off the quota, freed up latent capacity to participate in the scallop fishery. This opportunity increased the number of full-time scallop operators and, while this number has fluctuated in recent years, the corresponding decrease in groundfish quotas has resulted in a major increase in committed scallop-effort in the 1990s (Kenchington et al. 1995). The number of vessels specializing in scallops, (landing more than 80% of their gross revenue from scallops) increased from 68 in 1991 to 90 by 1995. This was largely attributed to differences in management and allocation of resource procedures, nonetheless a major effort increase did occur in scallop fishing. The major downswing in the scallop resource (Figure 4), the impending tight controls of the Individual Transferable Quota on catch and effort on the scallop fleet for 1997, suggested that the dedicated scallop fleet would again decline and a reduction in the number of vessels active in the fishery did occur, and at present (1999) only 52 vessels are active.

Catch per unit effort (kg/hr) is at low levels on both the Digby and Brier Island/Lurcher Shoal beds and has been declining in recent years. The input-based competitive fishery for Bay of Fundy scallops had no mechanism to rationalize fleet capacity, other than limited entry, in the face of dwindling scallop resources. Costs of fishing increased and incomes of captains, crews and returns on vessel operations fell as vessel owners struggled to make ends meet.

7.4 Vessel earnings

By 1996 the scallop catch by the Bay of Fundy fleet had fallen to around 700t of meats for the first time since 1986 (Figure 4). Many inshore scallop vessels were suffering losses due to a scarcity of scallops and higher harvesting costs (Figure 5). Average gross revenue per vessel of scallop specialists (those who earned greater than 80% of their fishing income from scallops) based on a sample survey was $C187 000, slightly higher than the $C176 000 grossed in 1986, but only half the $C381 000 attained in 1989. There was a net average loss of $C4400 per vessel in 1996 after all expenses including labour and depreciation are considered. Average net vessel earnings were $C21 562 in 1986 and $C74 319 in 1989. This amounted to a 10.6% return on a vessel investment of $C202 708 in 1986 and 29.9% of $C248 336 in 1989 compared to a negative return in 1996.

Despite near record ex-vessel prices of $C8-$C9/lb in 1996, lower catches kept revenues down. Harvesting costs were higher in 1996 compared to 1986 due mainly to the fuel and fixed costs per vessel almost doubling. Increased fees for book-keeping, legal fees, etc. along with new fees for a Dockside Monitoring Programme and a $C6500 licence fee were responsible for the fixed cost increase in 1996. Operating, maintenance and repair costs were about $C20 000 per vessel higher in 1996 than in 1986 partly due to more days fishing and partly because of price increases in supplies and other inputs.

7.5 Crew shares

Labour costs for captain and crew, based on a share system, were down from $C76 000 in 1986 to $C66 000 for 1996. In 1989, the average labour bill was $C162 000 with the captain's share $C45,000 and each crew-member earning $C28 000. By 1996, the average crew-member's earnings had fallen to $C15 300 with the captain earning $C24 800. Labour cost as a percentage of gross revenue has fallen from 43% in 1986 and 42% in 1989 to 35% in 1996.

There has been an increase in the variety of crew-sharing throughout the fishery since 1986. Some of the vessels have hired captains while others are owner-operators. Some crews take a larger percentage than others for boat share and on some boats, crews share different operating expenses than others. Lay arrangements are not publicly announced, but among fleet members this information is known. The standard crew sharing arrangement is for a set percentage of the gross stock to be deducted for the boat, the crew pays certain operating costs (*e.g.* for food) and the remainder is split equally among the crew.

7.6 Employment

Employment in this fleet has not changed significantly since 1986, although it has declined from 1989/1990 when landings of scallops were at their highest. Employment has closely tracked the cycle of the scallop resource, increasing in good times and decreasing when scallops were scarce. The average vessel crew size in 1986 was 2.2 men, excluding the captain, which is similar to 1996 as most vessels carried a crew of two to three. During the peak year of 1989, the average crew size was 4.2 and numerous shuckers were hired to deal with scallops that had to be shucked at the wharf. Therefore, the downturn from 1989 to 1996 created a reduction in crew employment of approximately 200 people, and a further 300 part-time shuckers. Most of these workers have been absorbed into other industries but further reductions, necessitated by resource depletion, will not be so easily dealt with. The current ITQ programme implemented in the inshore fishery is expected to reduce fleet capacity by up to 50%. This will result in a decrease of 200 persons employed (4 crew including captain x 50 vessels) over the time it takes the fleet to rationalize.

The downturn in the inshore fishery has also affected the employment in the fishery, resulting in protests to acquire new fishing areas. Also, illegal fishing has occurred: in 1996, 11 incidents of illegal fishing were documented (fishing in more lucrative offshore zones).

7.7 Local employment effects

Most inshore vessels land in the Southwest Nova Scotia area with the largest landings occurring in Digby, followed by Yarmouth and Meteghan. Recent protests (1996) led by crews of Full-Bay vessels created a major write-in campaign to government including letters from local businesses hit hard by this inshore downturn. Inshore fishers, both crews and captains/owners, tend to spend money locally. A 50% reduction in revenue over the last three years has had a major negative impact on local businesses, towns and suppliers and ancillary industries supporting the fishery. The Full-Bay fishery is the largest employer in the Digby area today and its ill-health is affecting local centres significantly (F. MacIntosh, pers. comm., Mayor of Digby, Nova Scotia). Concurrent declines in the groundfish fishery have further exacerbated the situation. In April 1997, a number of scallop fishermen occupied Federal Government offices in Digby and Yarmouth as a protest to draw attention to their inability to make a living in the Bay of Fundy scallop fishery.

Figure 4
Inshore Full-Bay Scallop fleet landings and number of vessels active (1983-1998)

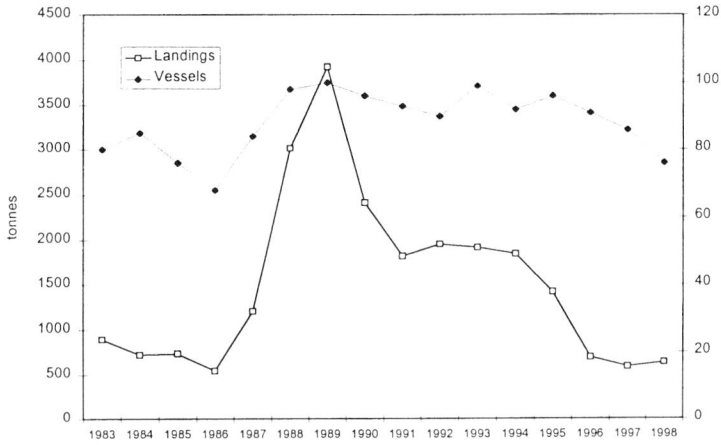

Figure 5
Inshore Full-Bay Scallop fleet revenue per vessel and number of vessels active (1983-1998)

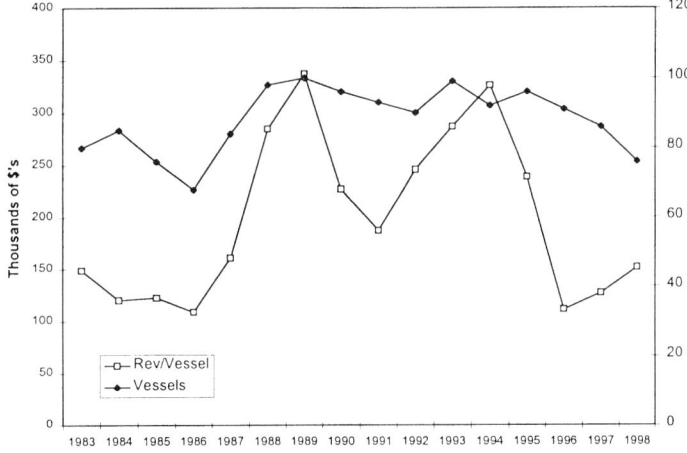

8. IMPACTS OF THE 1986 AGREEMENT ON THE OFFSHORE FISHERY

8.1 Resource conservation

During the 1980s the offshore fishery on Georges Bank was highly competitive and fishing mortality was high; consequently, the scallop stock biomass was reduced to low levels. After 1986, three strong year-classes (1986, 1988 and 1989) caused the stock to increase to peak levels in 1993. However, both the 1990 and 1991 year-classes were poor and the stock biomass dropped sharply to levels experienced during the 1980s (Robert and Butler 1995a). The 1992 year-class strength was close to the long-term average and was responsible for the increase in biomass available to the 1996 fishery. The TACs for Georges Bank have ranged from 2000t to 6850t since 1986 and was set at 3000t in 1996. This range is considerably smaller than that recorded from the landings during the period from 1957 to 1986 (732 to 11 126t). Low meat counts (e.g. 33/500g on Georges Bank) which have been in place for several years have not been completely effective in preventing the harvesting of small scallops. Industry has put in place a self-monitoring programme based on minimum meat sizes in the catch to try to resolve this problem.

The scallop fishery on Browns Bank was largely restricted to the southern part and along the edge of the Bank, at depths over 100m, during the years of the competitive fishery. The northern part of Browns Bank is now also exploited, with landings first seen in 1988 (Robert and Butler 1995b). The TAC has ranged from 220t in

Figure 6
Offshore scallop fleet landings and number of vessels active (1983-1998)

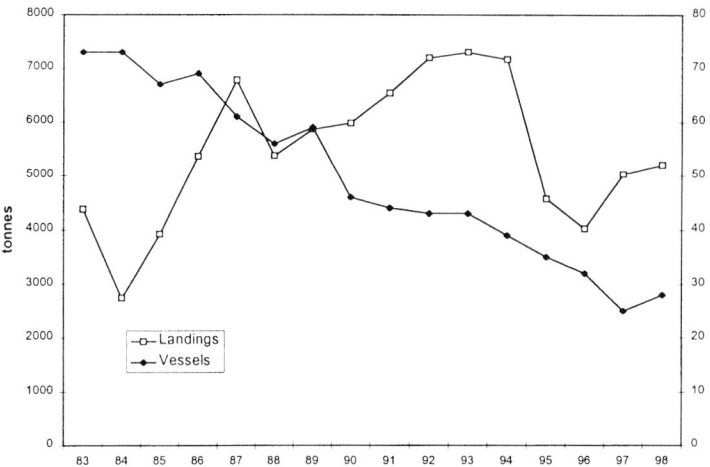

1990 to 2000t in 1995. The recent good catches (post 1993) are largely due to the exploitation of new fishing areas further to the east than previously fished. The area fished continued to expand through 1996 but the opportunities for further expansion are now small.

The scallop stocks on the eastern Scotian Shelf (Banquereau Bank, Middle Grounds, Western and Sable Island Banks) historically have been a relatively minor component of the offshore catch and are currently at very low levels. Prior to 1995 these beds were fished competitively; since then they have been grouped under one allocation management plan with a quota and meat count restrictions.

Thus, the scallop fishery on the offshore banks continues to be strongly driven by the variability in recruitment (Figure 6). High fishing mortality has created a fishery largely dependent on two age groups (4 and 5 year olds) on the major beds on Georges Bank which furthers the dependence on good recruitment. However, reduction in fishing effort and capacity has had a positive influence on landing variability as the range of variability has been reduced considerably.

8.2 Fishing capacity

Fishing capacity has been reduced under the EA programme (Figure 6). In 1986, there were nine companies with 76 licences operating 69 vessels. Forty-six of these were older wooden vessels of 28-30 m average length and the remainder were made of steel. Rules respecting numbers of participants, vessel replacement, and concentration were defined and approved within the management plan structure, and today there are seven companies operating 26 vessels of 25 to 45m length in operation. Most of these are older steel vessels, increased fishing capacity per vessel to some extent, however the reduction in the fleet more than compensates for the increased vessel capacity (Brander and Burke 1995). Some of these vessels will need to be replaced soon, as their average age is around 25 years.

8.3 Fishing effort

Fishing effort on Georges Bank initially increased from 1986 to 1988 but has declined steadily from 1989 to the present. Fishing effort on Browns Bank increased sharply in 1995 but returned to more average levels in 1996. Catch per unit effort (CPUE) tracked effort-levels in these cases except for 1996 when CPUE was high relative to effort on Georges Bank. This evidence supports the arguments which favoured the introduction of EAs, *i.e.* firms with individual quota allocations would deploy just enough fishing capacity to efficiently harvest their allocations (Anderson 1986) in a time pattern dictated by catch rates, prices, markets etc.

8.4 Vessel earnings

Revenues of the offshore scallop fleet have ranged between $C50 to $C80 million annually from 1986 to 1995. Revenue rose steadily from 1988 to 1994 reaching $C77 million before falling to $C50 million in 1995. Over this period, fishing effort in terms of both vessels and days at sea fell steadily, which resulted in an increase in revenue per vessel from $C800 000 in 1986 to almost $C2 million in 1994. The earnings level per vessel fell in 1995 to below $C1.5 million and was just below $C1.4 million in 1996 (Figure 7). From 1986 to the present, the number of vessels active in the offshore scallop fishery was reduced by half, *i.e.* from 69 in 1986 to 32 in 1996. The access fees are paid based upon the allocations to the company and in 1996 totaled approximately $C2.5 million. The formula for determining access fees is $C547.50/t of quota meat. These fees go directly to the Receiver General for Canada.

8.5 Crew shares

Lay arrangements are established by collective agreements for unionized offshore scallop crew-members and are similar for the non-unionized vessels. Crew shares have remained the same proportions of gross vessel revenues less operating expenses from 1986 to 1996 - crews share 60% and vessels take 40%.

8.6 Employment

There has been a negative impact on employment through the vessels removed from the offshore fishery. However, for the remaining crews opportunities for work have improved. The offshore reduction in vessels, initially targeted at 50%, has been surpassed. The reduction was gradual with the largest portion occurring between the years 1987 and 1992. Even today, the fleet size continues to adjust to current economic conditions. In several cases, down-sizing paralleled vessel replacement, but in some companies, a more direct aggressive approach was adopted.

The total number of crew affected over the 10-year period approximates 700 through eliminating 42 vessels since 1986 with an average crew size of 17. Seniority dictates employment in the offshore fishery that has unionized crews. Job-sharing is not a feature except in one company. Anecdotal evidence suggests that during the 1987-1992 period, approximately 300 crew members moved from the offshore to the inshore scallop fishery, which at the time was lucrative. It is estimated that 20% of inshore captains came from offshore operations and that as many as 200 crew from the offshore displacement remain in the inshore fishery. Discussions with employment centres suggest that most displaced crew-members have continued their attachment to the fishery; others have found work in construction and forestry. Some have left to work in the United States and other parts of Canada. It is understandable that the inshore fishery became an employment opportunity for offshore individuals with specific fishery skills. The slow and steady vessel removal in the offshore fishery, coupled with the ability of the local labour market at the time to absorb these workers, suggests a net benefit to the economy.

The earnings of captains and crew has improved since 1986, partialy due to improved fishing conditions, *i.e.* available quotas as well as an overall reduction in crew size. The increased wealth is shared among fewer crew as enterprises attempt to maximize profits. Thus, the ability of enterprises within this fishery to ensure a more stable harvest environment has improved productivity, not only at the corporate level, but also from the individual's perspective.

8.7 Local employment effects

Community-impacts appear to be minimal as the offshore fleet has traditionally landed in only five or six ports. In Yarmouth, N.S. the transfer of one complete offshore company to Lunenburg reduced the overall landings but, given the major port activities in the fishery in inshore scallops, groundfish, herring and lobster, the effect of this closure was not considered to be major. As well, the new activity in Lunenburg was as an employment offset. The total number of vessel landings is down to less than 500 from approximately 800 in the early 1990s. Offshore revenues are distributed locally, but crews tend to be more varied in residence creating a less concentrated impact.

9. CONCLUSIONS

The primary conclusion from this comparison of the two management approaches is that a property-rights re-

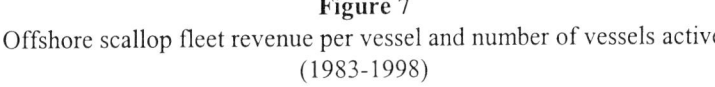

Figure 7
Offshore scallop fleet revenue per vessel and number of vessels active
(1983-1998)

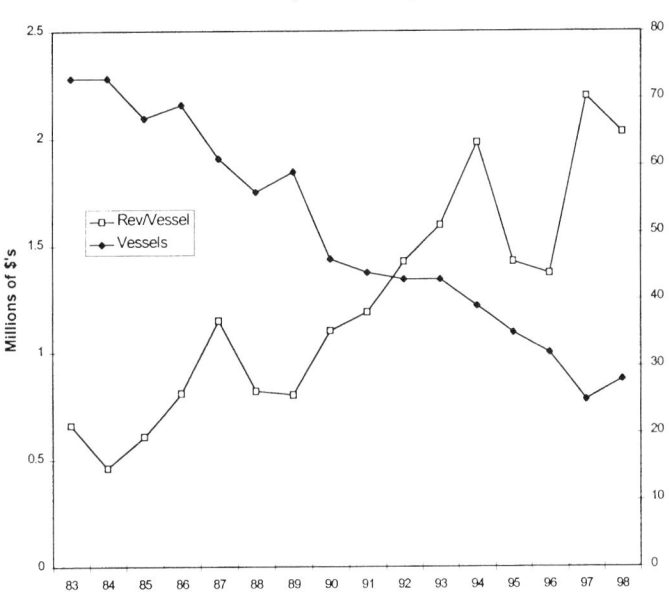

gime is superior to a competitive-fishery approach. Under an EA programme the offshore fleet was successful at matching capacity to the resource. Because fleet-reduction occurred gradually, the impacts on employees and associated communities were reduced. However, this reduction took place at a time when displaced workers could be absorbed into the inshore scallop, lobster and groundfish fisheries, which were then in a growth phase. With this balancing influence it is difficult to say what the impact would have been if reduction had taken place during a downturn in other fisheries and in the economy in general. Nevertheless, the EA programme has been highly successful at reducing fishing capacity in an orderly and efficient manner (as predicted by theory: *e.g.* Arnason 1996, Townsend 1998). From an economic viewpoint, the expected benefits of the EA programme have generally been realized. Firms with their own allocations are motivated to harvest their catch as efficiently as possible because they reap the financial gains of doing so. Since 1986 offshore scallop firms have adjusted their fleets to deploy just enough fishing capacity and fishing effort to efficiently harvest their allocations. Not only has the available scallop resource been harvested with less than half the number of vessels and crew that were engaged prior to the implementation of the EA programme, the amount of effort in terms of sea-days used to harvest it has been significantly reduced as well. This has resulted in increased economic returns per vessel as the fleet has been rationalized (Figure 7) and since the available catch can be harvested with less effort and cost, savings are generated by the fleet as a whole. This contrasts with quota-fisheries without individual shares, or competitive-fisheries where the race for available resource causes inefficiencies and waste in the production of effort. In the case of the inshore scallop fleet this has resulted in higher harvesting costs and subsequent lower vessel earnings (Figure 5).

The uncontrolled competitive inshore fishery failed to rationalize capacity, resulting in growth and recruitment over-fishing of all the major stocks (again, as predicted by theory: *e.g.* Arnason 1996). Confining a large dedicated fishing fleet to a restricted fishing area (as determined in the 1986 agreement between fleets), during a period of high and increasing effort, has proven to be detrimental to this resource. The need for a "fallow" time for heavily-fished scallop beds to recover is extremely difficult to maintain in situations where the financial and social demands of a competitive fishing dominate industry activities. The use of closed areas and time-periods have proven insufficient to stem over-fishing of scallop beds and indicates the need for far more extensive and restrictive controls.

The regulatory framework supporting the competitive fishery was insufficient to match fleet over-capacity and excess fishing effort on the resource. Further, there was a failure of the fleet to self-regulate and implement the voluntary licence reduction programme agreed to in 1986. Over-capacity was further exacerbated by fleet adjustments in 1991 with the creation of groundfish ITQs. Had recruitment remained moderately high and had the groundfish fishery flourished through this period, it is unlikely that scallop stock collapse would have occurred, even with the relatively high fishing on the beds. Thus, the failure of the inshore scallop competitive fishery is specifically a failure to adjust fishing effort to the optimal catch and size composition of the available resource. The resulting impact on workers and their communities has been great, and economic conditions are such that opportunities for employment in other sectors are almost non-existent.

The result of the long-term over-fishing and the associated decline in fleet viability in the inshore fishery has been an increase in demand for government programmes and services during the transition period; this has not been necessary within the offshore scenario. In the offshore fishery, dispute resolution was internalized, minimizing the need for government intervention.

And last, neither management approach has removed the dependence of these fisheries on scallop year-class strength. In a relatively long-lived species such as the sea scallop, which sees several good recruitment events within the lifetime of a cohort, has relatively low natural mortality in the recruited year-classes, and increases meat and gonad-yield occur throughout its lifetime, it should be possible to further stabilize landings. This will be one of the challenges of the next decade.

10. ACKNOWLEDGEMENTS

We thank Dr. T.J. Kenchington (Gadus Associates, Musquodoboit Hbr., N.S.) for his valuable comments and discussion, Dr. G. Robert, Mr. M. Lundy (Science Branch, Dept. Fisheries and Oceans, Dartmouth, N.S.) and two anonymous reviewers for providing comments on this paper.

11. LITERATURE CITED

Anderson, L.G. 1986. The Economics of Fisheries Management, The Johns Hopkins University Press, Baltimore and London, 296pp.

Apostle, R., B. McCay and K.H. Mikalsen 1997. The political construction of an IQ management system: The mobile gear ITQ experiment in the Scotia Fundy region of Canada, Social Implications of Quota Systems in Fisheries, G. Palsson and G. Petursdottir, Copenhagen, Nordic Council of Ministers, 27-49.

Arnason, R. 1996. Property rights as an organizational framework in fisheries: The cases of six fishing nations, Taking ownership: Property rights and fishery management on the Atlantic coast, B.L. Crowley, Halifax Nova Scotia, *Atlantic Institute for Market Studies*, 99-144.

Brander, L. and D.L. Burke 1995. Rights-based vs. competitive fishing of sea scallops *Placopecten magellanicus* in Nova Scotia, *Aquatic Living Resources*, 8, 279-288.

Campana, S, and J. Hamel 1992. Status of the 1991 4X cod fishery, *Canadian Atlantic Scientific Advisory Committee Research Document,* 92/46, 42pp.

Digou, D. 1994. Scotia-Fundy region harvesting sector overview. 1986-1993, *Economics Comm. Ann. Rep.,* 144, 39pp.

Hurley, P.C.F, J. Simon and K.T. 1992. Frank, Assessment of 4X haddock in 1991, *Canadian Atlantic Scientific Advisory Committee Research Document* 92/63, 40pp.

Kenchington, E. and M.J. Lundy 1996. An assessment of areas for scallop broodstock protection in the approaches to the Bay of Fundy, *DFO Atlantic Fisheries Research Document,* 96/13, 21 pp.

Kenchington, E., M.J. Lundy and V. Hazelton 1994. Seasonal changes in somatic and reproductive tissue weights in wild populations of *Placopecten magellanicus* from the Bay of Fundy, Canada, *In*: Bourne, N.F., Bunting, B.L. and Townsend, L.D., eds, Proceedings of the 9th International Pectinid Workshop, Nanaimo, B.C., Canada, April 22-27. 1993, Volume 2, *Canadian Technical Report of Fisheries and Aquatic Science,* 154-162.

Kenchington, E., M.J. Lundy and D.L. Roddick 1995. An overview of the scallop fishery in the Bay of Fundy 1986 to 1994 with a report on fishing activity trends amongst the dual licence holders in the Full-Bay fleet, *DFO Atlantic Fisheries Research Document,* 95/126, 40pp.

Kenchington, E., M.J. Lundy and S.J. Smith 1997. Bay of Fundy Scallop Stock Assessment: Areas 2, 3, 4, 5, 7, *Canadian Stock Assessment Secretariat Research Document,* 97/63, 98pp.

Parsons L.S. 1993. Management of marine fisheries in Canada, *Canadian Bulletin of Fisheries and Aquatic Science,* 225.

Robert, G. and M.A.E. Butler 1995a. Georges Bank scallop stock assessment-1994, *DFO Atlantic Fisheries Research Document,* 95/140, 37pp.

Robert, G. and M.A.E. Butler 1995b. Activity report for 1994-Scotian Shelf scallop fishing grounds, *DFO Atlantic Fisheries Research Document,* 95/141, 28 pp.

Serchuk, F.M. 1983. Seasonality in sea scallop shell height-meat weight relationships: review and analysis of temporal and spatial variability and implications for management measures based on meat count, *Woods Hole Laboratory Research Document,* 83-35, 30pp.

Townsend, R.E. 1998. Beyond ITQs: property rights as a management tool, *Fisheries Research,* 37, 203-210.

Appendix 1
Regulations

The following paraphrases and summarizes the regulations applicable to the inshore and offshore scallop fleets from 1986 to 1996. Shaded areas indicate regulations applicable to both fleet sectors. This summary does not consider variation orders issued annually to adjust seasons and meat counts, or to establish specific closures for conservation reasons. On 8 January 1986, the Atlantic Fishery Regulations, 1985 came into effect. These regulations were the result of a consolidation of the Atlantic Coast Marine Plant Regulations, the Atlantic Crab Fishery Regulations, the Atlantic Fishery Regulations, the Atlantic Fishing Registration and Licensing Regulations, the Atlantic Herring Fishery Regulations, the Fishing Gear Marking Regulations, and the Lobster Fishery Regulations.

1986
Regulations applicable to the offshore scallop fleet
1. Scallop Fishing Areas (SFAs) 1 to 28 were created and closed times established.
2. Offshore scallop vessels (>65') were prohibited from fishing in SFA 28 (Bay of Fundy) and from the Territorial Sea in SFAs 25 and 26 (4VWX).
3. A 33 per 500g average meat count was in effect for all SFAs except 21, 22 and 24 (Gulf of St. Lawrence). (Variation orders were used annually to adjust this count in certain SFAs.)
4. The average count was to be determined on the basis of eight or more samples of meats, each sample weighing 500 grams or more.
5. Offshore vessels were restricted
   - to trip limits of 13,700 kg (30,000 lb.), and
   - quarterly limits not to exceed 82,200 kg (181,000 lb.)
6. Offshore vessels could not fish for more than 12 consecutive 24 hour periods.
7. It was prohibited in any SFA to have scallop drags onboard a vessel unless that vessel was authorized to fish for scallops in that area at that time, or the scallop drags had to be unshackled and stowed.

Regulations applicable to the inshore scallop fleet
1. Scallop Fishing Areas (SFAs) 1 to 28 were created and closed times established.
2. That portion of SFA 28 from Parkers Cove to Sandy Cove within 6 nautical miles from shore was closed from May 1 to September 30 (Inside Fishing Zone).
3. The waters of Digby Gut and Annapolis Basin were closed from May 1 to November 30.
4. Inshore vessels (<65') were exempted from meat count regulations except in the SFAs 21, 22, 24 and 27 (Gulf of St. Lawrence and Georges Bank).
5. Inshore vessels only permitted to fish in SFA 27 (Georges Bank) under a written authorization issued by a fishery officer.
6. Written authorizations were valid only for the period specified therein.
7. Scallops caught and retained by holders of a written authorization were deemed to have been caught in SFA 27 (Georges Bank).
8. It was prohibited in any SFA to have scallop drags onboard a vessel unless that vessel was authorized to fish for scallops in that area at that time, or the scallop drags had to be unshackled and stowed.
9. Scallop Fishing in SFA 28 (Bay of Fundy) was prohibited using a drag with rings less than 82 mm inside diameter.

1987
Amendments to Regulations applicable to the offshore scallop fleet
1. Vessel classes for the nine offshore scallop license holders were introduced and closed times established for each vessel class in each of the SFAs.
2. The regulation which prohibited offshore scallop vessels (>65') from fishing in SFA 28 (Bay of Fundy) and from the Territorial Sea in SFAs 25 and 26 (4VWX) was revoked but the same rules were implemented as a condition of the offshore scallop fishing licenses.
3. License fees were increased (essentially doubled).
4. A definition for "shell height" was introduced.
5. A 45 per 500 gram average meat count was specifically implemented for SFA 25 (Eastern Scotian Shelf).
6. Any scallops caught and retained or found on board a vessel were deemed to have been caught in the SFA area in which the vessel was authorized to fish.
7. It was prohibited to have on board a vessel any scallops caught in SFA 27 (Georges Bank) unless the shell height was 105 mm or greater.
8. Offshore scallop vessels were required to hail to a fishery officer 12 hours before a vessel arrived at port
   - the port where the scallops would be landed, and
   - the time when scallops would be landed.
8. Created an offense to land at a port or time different than that hails unless by permission of a fishery officer.
9. A fishery officer could direct that scallops not be landed until they were first inspected and made it an offense not to comply with the fishery officer's direction.
10. Trans-shipping of scallops to another vessel was prohibited.
11. Offshore scallop license holders were required to weigh all scallops caught in SFAs 26 and 27 (Browns/German banks and Georges Bank) at the time of landing.

Amendments to Regulations applicable to the inshore scallop fleet
1. Vessel classes were introduced and closed times established for in each of the SFAs.
2. A definition for "shell height" was introduced.
3. That portion of SFA 28 from Parkers Cove to Sandy Cove within 8 nautical miles from shore was closed from May 1 to September 30.
4. A 72 per 500 gram average meat count was specifically implemented for SFA 28 (Bay of Fundy)

from May 1 to September 30 and 55 per 500 gram average meat count from October 1 to April 30.
5. Any scallops caught and retained or found on board a vessel were deemed to have been caught in the SFA area in which the vessel was authorized to fish (also applied to inshore vessels fishing Georges Bank under written authorizations).
6. It was prohibited to catch and retain or have on board a vessel in SFA 28 (Bay of Fundy) any scallops unless the shell height was 76 mm or greater.
7. Inshore scallop vessels fishing in SFA 27 (Georges Bank) under a written authorization were required to hail to a fishery officer 12 hours before a vessel arrived at port
    - the port where the scallops would be landed, and
    - the time when scallops would be landed.
8. Created an offense for inshore vessels operating under a written authorization to land at a port or time different than that hails unless by permission of a fishery officer.
9. A fishery officer could direct that scallops not be landed from inshore vessels operating under a written authorization until they were first inspected and made it an offense not to comply with the fishery officer's direction.
10. Trans-shipping of scallops from inshore vessels, operating under a written authorization, to another vessel was prohibited.
11. Inshore scallop license holders were required to weigh all scallops caught in SFA 27 (Georges Bank) at the time of landing.
12. It was prohibited for inshore vessels in SFA 28 (Bay of Fundy) to fish with
    - an offshore scallop drag;
    - a Green sweep scallop drag;
    - a scallop drag or drags greater than 5.5 m in total length;
    - a scallop drag with a bag that has rings of less than 82 mm inside diameter.

1989
Amendments to Regulations applicable to the offshore scallop fleet
1. Offshore vessel classes were revoked from the regulations but implemented as a condition of the fishing license.

1993
Amendments to Regulations applicable to the offshore scallop fleet
1. It was prohibited in any SFA, other than SFA 26 and 27 (Browns and German banks and Georges Bank), to have scallop drags onboard a vessel unless that vessel was authorized to fish for scallops in that area at that time, or, the scallop drags had to be unshackled and stowed.

2. In SFA 27 (Georges Bank) it was prohibited to have scallop drags on board a vessel unless the vessel was authorized to fish for scallops in that area at that time even if the drags were unshackled and stowed.

Amendments to Regulations applicable to the inshore scallop fleet
1. SFA 28 (Bay of Fundy) was subdivided into four SFAs (28A, 28B, 28C, and 28D.
2. The seasonal scallop fishing closure around Grand Manan Island in SFA 28B (from April 1 to 08:00 on the second Tuesday in January next following), previously implemented as a license condition, was formally implemented as a regulation.
3. The scallop fishing closure for waters within approximately 2 miles of the shore from the Canada/US boundary to Joes Point in SFA 28B, previously implemented as a license condition, was formally implemented as a regulation.
4. It was prohibited in any SFA, other than SFA 26 and 27 (Browns and German banks and Georges Bank), to have scallop drags onboard a vessel unless that vessel was authorized to fish for scallops in that area at that time, or, the scallop drags had to be unshackled and stowed.
5. In SFA 27 (Georges Bank) it was prohibited to have scallop drags on board a vessel unless the vessel was authorized to fish for scallops in that area at that time even if the drags were unshackled and stowed.
6. In SFA 26 (Browns and German banks) it was prohibited to have scallop drags on board a vessel unless the vessel was authorized to fish for scallops in that area at that time, or, the vessel was transiting SFA 26 under a written authorization from fishery officer, and, the scallop drags were unshackled and stowed.
7. Written authorizations for vessels transiting SFA 26 (Browns and German banks) were to be issued at the request of the master of the vessel if the home port was in SFA 29 (Territorial Sea) and that vessel was authorized to fish in SFA 28A, 28B, 28C or 28D, at that time.
8. Written authorizations must state the period for which the authorization is valid.

1996
Amendments to Regulations applicable to the offshore scallop fleet
1. License fees were amended and the new fee based on $547.50 per tonne of scallop meat allocated (previous fees were based on the number of vessels eligible to be licensed by each company).

Amendments to Regulations applicable to the inshore scallop fleet
1. License fees were amended and the new flat rate of $6,500 applied (previous fee was $200).

THE ORANGE ROUGHY MANAGEMENT COMPANY LIMITED – A POSITIVE EXAMPLE OF FISH RIGHTS IN ACTION

G. Clement
Clement and Associates Ltd
PO Box 2145, Tauranga, New Zealand
<clement@fishinfo.co.nz>

1. INTRODUCTION

New Zealand's deepwater fisheries for orange roughy and oreos earn $US67 million annually. Within New Zealand's 200 mile Exclusive Economic Zone (EEZ), these fisheries are predominantly found in depths between 800 and 1200 metres. New Zealand first developed these fisheries in the late 1970s, principally through joint venture arrangements with trawler operators from other countries. Since the late 1980s, almost all the catches have been taken by New Zealand domestic vessels.

In 1983, a quota system was introduced in the deepwater fisheries for orange roughy, squid, oreos, silver warehou, hake, ling and hoki. Initial allocations of quota were based on an assessment of catch history, investment in vessels and commitment to processing. These deepwater quotas were initially non-transferable and were converted into Individual Transferable Quotas (ITQs) in 1986 when the comprehensive Quota Management System (QMS) was introduced, principally to conserve and restructure inshore fisheries.

The objective of the QMS is sustainable fisheries management by maintaining fishstocks at or above the size that will produce the maximum sustainable yield (MSY). The setting and enforcement of conservative total allowable commercial catch (TACC) levels ensure resource conservation. A comprehensive scientific stock assessment programme underpins determination of the appropriate TACC levels. Each TACC is allocated as ITQs, which grant the right to catch a specified proportion of the TACC in perpetuity. Thus the property right, by ITQ, is a permanent harvest access right to a prescribed fishery for an annually determined tonnage of catch.

Quota-owners pay for the full costs of management and enforcement of their fisheries. It is therefore in their interests to act co-operatively, both to ensure that the appropriate information and management decisions are made and that value is obtained from their investment in the research and management costs of these fisheries. Their objective is the same as the Government's - sustainable fisheries.

2. THE ORANGE ROUGHY MANAGEMENT COMPANY LIMITED

In 1991, a consortium of orange roughy and oreo dory quota-owners united to form **The Orange Roughy Management Company Limited** (ORMC). Their objective was to maximise the value of the deepwater fisheries through sustainable management.

All fisheries are market driven – without markets there is no basis for fishermen and seafood companies to risk the substantial investments required to develop and to maintain fisheries. The Orange Roughy Management Company's vision is to maximise the long-term value of orange roughy and oreo fisheries in the world's best markets. This value can only be maximised by ensuring continuity of supply of consistent, high-quality products to niche markets that demand them. Long-term consistency of supply only results from sustainably managed fisheries.

Fish rights, through ITQs, provide the incentives for individual seafood companies, who would be competing for resources in an open access fishery, to work together allowing co-operation to replace competition. Co-operation can best happen through a quota-owners' company, such as ORMC, which provides a vehicle for the combination of their independent expertise and resources to the common purpose of improving the sustainable management and utilisation of their fisheries. ORMC, represents 99% of orange roughy and oreo quota-owners, and acts on their behalf to:

i. add value to shareholders' businesses through a direct involvement in improved management of these fisheries
ii. provide a united and credible voice on all matters concerning the sustainable management and utilisation of New Zealand's deepwater fisheries
iii. provide professional capability to undertake a range of projects to improve the management of these resources including fisheries research, strategic and fisheries planning, dispute resolution and relations with other stakeholders who have an interest in these fisheries and
iv. provide and maintain a direct dialogue with the Government and, in particular, the Minister of Fisheries, who has a statutory role as the fisheries manager within the New Zealand legislature, and his officials within the Ministry of Fisheries.

Fish rights, through ITQs, have provided the incentives for improved cooperation both amongst quota-owners and between quota-owners and the government.

3. BENEFITS ACHIEVED THROUGH ITQs
3.1 Range of benefits

The achievements of deepwater quota-owners through The Orange Roughy Management Company cover a broad range of activities highlighted below:

i. Management benefits
ii. Economic benefits
iii. Research benefits and
iv. Development benefits

3.2 Management benefits - Fish rights secure sustainability

Within the New Zealand deepwater fisheries, cooperative action by quota-owners has enhanced sustainable management measures – Fish rights secure sustainability.

Agreements, through civil contracts between ORMC, quota-owners and the Government, have implemented self-regulatory management controls that include:

i. closing areas to fishing
ii. establishing and maintaining catch limits for separate sub-areas within the TACC set for the large quota management areas
iii. voluntarily reducing catches through the setting aside of quota and supporting TACC reductions and
iv. spreading catches amongst discrete geographic locations, such as seamounts, to provide a spread of fishing effort.

Deepwater fisheries in New Zealand have been a relatively recent development, particularly in waters deeper than 500 metres, with new fisheries being developed continually over the last 15 – 20 years. During this period knowledge of the location and size of oreo and orange roughy populations has unfolded as both commercial fishermen and research scientists have learnt of the size and extent of these resources and their life history parameters. Fisheries for orange roughy and oreo are managed within a series of quota management areas (QMAs) within the New Zealand 200 mile Exclusive Economic Zone – a TACC being set for each management area.

There are eight QMAs for orange roughy within the quota management system. The main fishing grounds are distributed around the New Zealand EEZ, along the 1000 metre contour and on topographical features such as ridges, seamounts and canyons. Experience has demonstrated that fisheries in relatively localised areas appear to behave independently and fishing pressure can act to reduce population sizes. For orange roughy, the biomass which will support the MSY is estimated to be 30% of the size of the unfished biomass. On this basis the management strategy is to: fish the population size down by 70%, set a sustainable catch level and monitor the ongoing management.

Through the ORMC, industry has identified separate fisheries within a number of these large quota management areas and has reached agreement with the Minister of Fisheries to manage these fisheries separately, within the Quota Management System – effectively setting up paddocks, or delimited management areas, within each QMA. These measures allow for moderation of fishing pressure on a localised fishery basis.

Quota-owners cooperate to ensure that catches are within agreed limits on separate fisheries within the main QMAs. For example, the large ORH3B fishery has separate catch limits set in five areas, determined after assessment of the information that is available and in consultation with the Minister and then set in place through civil contract among the quota-owners. Within this process, where necessary, catch limits have been set at zero to provide for increased rates of recovery where the populations of localised fisheries have been assessed to be below the level that will produce the MSY. The Sub-Antarctic area, the southern portion of New Zealand's 200 mile zone, remains substantially unexplored. These unsheltered waters are exposed to the roaring forties and are notorious for their storms and high sea-states. Exploration of deepwater fisheries is expensive, difficult and dangerous. Quota-owners are progressively and cooperatively exploring these areas. An agreement with the Minister of Fisheries provides that quota-owners will voluntarily limit catches to 500 tonnes within a 12 nautical mile radius of any geographic feature. This is to both ensure spreading of the catch, to explore as much area as possible and to reduce the fishing impact on any new resource before its long term productive capacity can be assessed.

Quota-based management systems are information-intensive. The management target, to maintain the stock at or above the biomass level that will produce the MSY, requires detailed knowledge on a fishery-specific basis of a range of life-history parameters, including age, growth, stock boundaries, stock sizes and available yields. Collection of a range of biological data from the fishing fleet will increasingly become an important component in the assessment and management of these fisheries. Within the deepwater fisheries, ORMC has successfully established a scheme to collect a range of biological information on the new and developing fisheries using a mix of independent expertise, trained industry personnel and quality assured processes to assist with the management of these fisheries.

The oreo fishery currently comprises three species: smooth, black and spiky oreo, which are managed under a single TACC in each of four Quota Management Areas. There is agreement to manage fisheries for the main two species (smooth and black oreo) as separate species but the legislative and administrative changes required cannot be completed until 2001. In the meantime, ORMC has taken the initiative to develop a comprehensive fisheries plan for the management of oreo and set in place measures to separately manage smooth and black oreo in the main fishery on the Western Chatham Rise.

3.3 Economic benefits – Rights create rent

Fish rights also provide economic benefits. Open access fisheries not only result in poor resource conservation, but they also result in dissipation of resource rents. Fish rights create rents. ITQs have provided the incentives for the more efficient operators to invest in additional quota, new vessels, improved harvesting and processing capabilities and market development. Cooperation in management has improved dialogue between quota-owners and resulted in a broad range of initiatives that have increased cooperation in harvesting. For example, as management has become increasingly more complex with smaller areas and lower catch limits, the larger quota-owners have undertaken to lease, or to catch quota that is owned by smaller operators on their behalf. In these cases the contracting harvester may only charge for his additional direct marginal costs. This results in improved utilisation of existing investments in vessels, rationalisation of the fleet and processing capabilities and substantially less fishing pressure on the grounds. ITQs enable the rationalisation and reduction of harvesting costs.

ITQs also enable higher market returns. Security of resource-access allows quota-owners to focus on market needs. This has resulted in substantial increases in fish quality and a move away from bulk fishing during spawning seasons to fishing for small catches on a year-round basis to optimise fish quality and to ensure year round market supply. For example, during the late 1980s over 90% of the orange roughy catch from the Chatham Rise was taken during the spawning season, where 20 or more large trawlers operated in a very small area. Catch utilisation was sub-optimal with fish and product quality loss resulting from burst cod-end bags, crushing of the catch, spoilage and the need to reprocess on shore.

Today less than 10% of the Chatham Rise catch is taken during the spawning season. The majority is taken throughout the year by a fleet of less than 10 vessels, with fish quality being optimised through small catches, targeting of non-spawning fish and, in many cases, processing onboard to frozen-at-sea consumer-ready products. As a result, New Zealand companies have been able to obtain price premiums over competitors, not only because of quality and consistency of supply, but also in recognition of access to ongoing supplies into the future.

The security of resource-access provided by fishing rights within New Zealand has resulted in the most modern and efficient fishing fleet in the world. ITQs enable quota-owners to make both long-term investment decisions and to optimise the use of technology and capital on a year-by-year basis. There are now 15 modern factory trawlers in the New Zealand fleet each of which annually spend in the order of 320 days at sea. This is a direct result of a successful rights-based fishery providing secure ongoing access to resources and the environment for the optimum investment and use of technology and the creation of real economic rents.

3.4 Improved research – Rights require responsibility

Quota-owners, more than any other stakeholder in the fishery, have a need to know the productive limits of the fishery. It is in their interests to ensure sustainable management more than any other stakeholder. Their whole investment is underpinned by resource-access and by sustainable management. Management by ITQ requires significant levels of information. It is the responsibility of quota-owners to ensure this information, through fishery research, is available. A successful Fish rights regime requires that this responsibility be met.

In New Zealand quota-owners pay all the costs of fisheries management, research and enforcement. Thus the need to know the productive limits is coupled with the need to ensure that the substantial investments made in scientific are value-driven and can directly assist in the management and conservation of these resources.

Quota-owners have not only insisted on quality research but also understand that this work needs to be undertaken by independent organisations with international expertise and standing. ORMC shareholders have invested an estimated $NZ30 million in research over the past 10 years both through direct purchase and through government levies.

The ORMC has recognized that improved research is required over a wide range of issues and ORMC has directly contracted research in the following areas:

Stock discrimination

Current techniques in genetics have a relatively low power to determine stock differences and industry has funded research into the development and assessment of new techniques. ORMC negotiates and manages this research on behalf of quota-owners.

Age and growth

There remains uncertainty about the age, growth and recruitment parameters in deepwater fisheries. For management purposes it is critically important to have good estimates of these parameters to assess the productivity, and hence annual sustainable yields from these resources. Assessment of New Zealand's deepwater fisheries are based on the conservative assumption that orange roughy and oreo are slow-growing and long-lived fish. These assumptions are based on the interpretations of rings and otoliths as annual marks. However the periodicity of deposition of these rings remains unvalidated and thus unknown.

Orange roughy quota-owners, through ORMC, have funded studies to investigate alternative techniques and to validate existing techniques, including studies using radioactive isotopes, C_{14}, and endolymph chemistry. This has proven a difficult field to make positive progress but work is ongoing

Biological sampling

A range of biological parameters is routinely measured from commercial catches in exploratory areas. Funding and coordination of this project is managed through ORMC.

Environmental studies

Increasing awareness on the need to consider the possible impacts of fishing activities on the marine environment require quota-owners to take a broader view of the relevant factors in developing fisheries plans. Throughout the first 15 years of the Quota Management System the focus has been on determining and setting sustainable catch limits for target species. Quota-owners now accept the need for a broader focus and are working with Ministry of Fisheries officials and other stakeholders to develop a strategy for the management of non-target benthic fauna in areas such as deep-water seamounts.

Biomass surveys

A challenge has been to count orange roughy at depths of 1000 metres or more, in dense aggregations often close to the bottom and over sloping ground. Earlier techniques such as research trawl-surveys, egg-surveys and CPUE indices have proven useful during the fishing-down phase, where the management objectives have been to harvest the population size to the B_{MSY} level. These techniques provide relative indices of biomass and are useful to track changes in the resource side. Once "fishing down" to the B_{MSY} level is completed the sizes of stocks managed at B_{MSY} are not expected to change significantly. This leads to the need for precise estimates of biomass and preferably estimates of absolute biomass rather than relative indices. This is a challenge that scientists have yet to meet. To date ORMC has invested in excess of $NZ14 million in the development of acoustic technology for the biomass-assessment of deepwater stocks, and research and development are ongoing.

Stock assessment

Assessment of deepwater stocks through modeling techniques is at an advanced stage within New Zealand and ORMC has commissioned the University of Washington to develop Bayesian models applicable to these fisheries.

Risk analyses

Unfortunately the nature of fisheries science and stock assessment, particularly in these deepwater species, is such that the results are often imprecise. ORMC has promoted the use of a range of risk analyses to assist managers to make informed decisions.

The transition from research purchased by Government to research purchased by industry is an inevitable and desirable outcome of Fish rights. The responsibilities of the quota-owners require them to be informed about the state of their fisheries and to use this information to improve management.

As New Zealand moves towards the quota-owners directly purchasing their required research, the focus will continue to be on the quality of research and not just the quantity. The results must also be relevant to the application of fisheries management and not driven by the service providers' interests or capabilities.

In the direct purchase of research information, quota-owners also need to ensure that they can satisfy the concerns of outside commentators and stakeholders at large. Thus research must be of the highest quality and of internationally accepted standards and independent from the purchaser. The costs need to be relevant to the size of the fishery, research outcomes need to be relevant to the management of the fishery and provide value for the investment.

The commercial fleet operates in fisheries on a year-round basis and provides the optimum basis for data-gathering supplemented by the judicious application of fishery independent research. Quota-owners have the unique incentive to ensure that the research is both relevant and provides value.

3.5 Development benefits - Cooperation replaces competition

The development of improved fisheries management processes is another positive outcome of Fish rights. In New Zealand, management focus is shifting to the development of comprehensive fisheries plans, led by quota-owners. These plans will encapsulate the vision, strategies and processes for the ongoing management and development of each fishery, including important issues that need to be addressed, the information requirements and the management responses at pre-agreed trigger points or in response to other outcomes. Co-operation among quota-owners and between them and with the government has increased.

Deepwater quota-owners have supported all recent TACC changes, which have been based on scientifically-obtained information. In several fisheries, including Chatham Rise, Puysegur and ORH7A, quota-owners have set in place catch-limits below those recommended by scientists in order to increase the rate of rebuilding of these fisheries.

Fisheries do not exist in themselves – they must first be discovered and developed commercially to determine both their commercial and resource viability. The development of deepwater fisheries has proven risky and costly. These fisheries have all too often been found in remote, localized and inhospitable areas and require specialist expertise and technology to develop. Development of deepwater fisheries is not too dissimilar to looking for a needle in a haystack but at considerably greater commercial expense.

Security of access through quota has enabled the investment in leading-edge technologies such as swath-mapping which acoustically maps the ocean floor in swaths or strips up to 12 kilometres wide at orange roughy depths. The data can be digitally-enhanced to produce a range of products including acoustic images of the seafloor, the underwater equivalent of aerial photographs.

Swath-mapping also enables a much clearer understanding of these deepwater habitats and their nature through refined bathymetric outputs. This information will be critical in further improving the management of these fisheries, particular when we look for the assessment and management of possible environmental issues.

New Zealand leads the world in this field for deepwater fisheries and through the ORMC, has now mapped more of New Zealand's EEZ than any other group.

4. FISH RIGHTS IN ACTION

In summary, Fish rights have proven invaluable in the conservation and sustainable economic exploitation of resources within the New Zealand Quota Management System. The ORMC experience is that:

i. Fish rights secure sustainability
ii. Fish rights create rent
iii. Fish rights require responsibility and
iv. Under fish rights, co-operation replaces competition.

THE EFFECTS OF TRANSFERABLE PROPERTY RIGHTS ON THE FLEET CAPACITY AND OWNERSHIP OF HARVESTING RIGHTS IN THE DUTCH DEMERSAL NORTH SEA FISHERIES

W. P. Davidse
Agricultural Economics Research Institute, LEI - Burgemeester Patijnlaan 19
2502 LS The Hague, Netherlands
<w.p.davidse@lei.dlo.nl>

1. INTRODUCTION

This case study considers the development of fleet capacity and harvesting rights in the Dutch demersal North Sea fishery[1] since 1983. The Common Fisheries Policy (CFP) of the European Union was implemented in that year, which meant for this fishery a growing importance of harvesting rights. Individual vessel quota IQ for sole and plaice had already been introduced in 1976 within the framework of the North East Atlantic Fishery Convention (NEAFC)[2].

In the period 1976-1984 these IQs were perceived by the vessel owners as limitations rather than as rights and enforcement of these quota was rather weak so that they were not much more than 'a piece of paper'. Transferability of the IQs was officially allowed from 1985. This, and intensification of enforcement, gradually brought about a transition in attitudes from individual limitations towards valuable property-rights for sole and plaice.

The CFP of the European Union (EU) requires setting annual Total Allowable Catches (TACs) for almost all commercial species landed by vessels of the member states. The Council of Ministers of the EU decides annually on these TACs, which are proposed by the European Commission. Each country has its own management system to fullfil the TAC obligations. The Dutch fishing sector is so far the only one within the EU that operates under an individual transferable quota (ITQ) system.

The Dutch demersal North Sea fishery consisted by the end of 1983 of 595 vessels, owned by some 500 firms. The fishery is composed by four main segments:

i. Beam trawlers, targetting sole and plaice, they are by far the most important segment. Most of these vessels are equipped with an engine whose power exceeds 800kW
ii. Roundfish trawlers, concentrated in the 225-810kW engine-power range
iii. Vessels with a 221kW engine, mostly operating in different fisheries (beam trawling for flatfish, demersal trawl for cod and whiting and shrimp fishing) and
iv. Vessels under 221kW, generally specialised shrimp trawlers. A part of these vessels operate in the Wadden Sea, in the north of the Netherlands.

Together, these four segments are known in the Netherlands as the 'cutter' fishery.

Table 1 shows some major characteristics of the demersal North Sea fishery. Recent figures have been added to demonstrate the important changes. The following sections of this paper explain how transferable property rights have influenced the changes in fleet-capacity and ownership of rights.

Table 1
Characteristics of the Dutch demersal North Sea Fishery in 1983 and 1998

Annual quota (tonnes)	1983	1998
Sole	15 400	14 600
Plaice	53700	35 300
Cod	22 900	14900
Financial results		
Proceeds (mln NLG, deflated)[1]	840	607
Net profit (mln NLG, deflated)[2]	-44	39
Number of vessels	595	407
Value of harvesting rights per vessel (on average, NLG)[1][2]	150 000	5 000 000

[1] Deflated for 1983 on the basis of the NLG purchase power in 1998.
1 NLG= 0.45 EURO or 0.49 US$.
[2] Estimated on the basis of market prices.
Source: Ministry of Agriculture, Nature management and Fisheries; Shipping Inspection;
Dutch Agricultural Economics Research Institute, LEI.

2. THE NATURE OF THE HARVESTING RIGHT

Transferability of the IQs for sole and plaice was officially allowed in 1985 by the Ministry of Agriculture and Fisheries since informal trade of these documents had occurred more frequently and more in the early 1980s. An extension of rights-based fishing came in 1994 when ITQs for cod were introduced and then in 1996 with the implementation of herring and mackerel rights. As a result, all quota species have been brought under an ITQ regime nowadays.

Co-management groups have pooled the ITQs of their members since 1993. This results in a group-quota for eight different management groups whereby the board of each group is responsible for compliance with this group-quota. The ownership of the rights remains with the individual holders. The groups facilitate trade, hiring

---
[1] In the Netherlands known as 'Cutter fishery'.
[2] The case study of W. Smit, "Dutch Demersal North Sea Fisheries, Initial Allocation of Flatfish ITQs" describes the initial allocation of these individual vessel quota (FAO press).

and renting of the ITQs between their members, which makes the system far more flexible. The rights can be used as a collateral for a loan; in fact, the ITQs always serve as a security for the bank when a loan is acquired, for example to finance a new vessel.

Investments in ITQs used to be encouraged by a fiscal allowance for depreciation. This included a 12.5% annual depreciation from the purchase price of the right. Trade in the 1990s has led to high prices for the ITQs. In fact they have become an important production factor for the firms (as the high value of the harvesting rights in Table 1 indicates). The sole and plaice ITQs are responsible for the major part of this value.

Apart from the ITQs the Dutch rights-based fisheries management nowadays consists of a number of other individual rights:

i. Licences, expressed in quantities of horsepower-per-vessel, introduced in 1984. These transferable rights aim to limit the total engine power of the sea-going fleet and give an entitlement to fish on quota species. This licence scheme resulted from the first Multi-annual Guidance Programme (MAGP 1), implemented in 1985 within the framework of the CFP. The target of the subsequent MAGPs has been the limitation of the capacity of fishing fleets in European Union (EU) waters.
ii. Transferable entitlements for shrimp fishing in the North Sea and in the Wadden Sea area.
iii. Entitlements to fish in the coastal zone, the so-called List 1 and 11 documents, which may also be transferred.
iv. Limitation of gross tonnage (GT) per vessel, implemented in 1998, which has led to rising values for transferable GTs. This measure results from the Dutch obligations in MAGP IV, running from 1997-2001.

3. MEASUREMENT OF FLEET CAPACITY
3.1 Characterizing fleet capacity

As stated in Section 1, the development of the fleet capacity will only be considered here for the period 1983-1998 since EU's Common Fisheries Policy started in 1983. Individual quota changed gradually from limitations, towards valuable property-rights in the early 1980s and the transferability of these rights, officially allowed in 1985.

Specialised beam-trawlers, equipped with an engine exceeding 810kW (1100 horsepower) took the most important part (65%) of the total fleet capacity in terms of engine-power in 1983. These vessels target sole and plaice, taking turbot, cod and whiting as bycatch species. Their crew varies from 6-8 people.

The medium-size trawlers, with engine-powers ranging from 222-810kW operate in different fisheries such as otter-trawling and pair-trawling on cod and whiting, herring pair-trawling and also beam-trawling. This segment consisted of 173 vessels in 1983 counting for 24% of the total engine-power of the fleet. The 221kW vessels mostly operate in the beam-trawl and shrimp fishery, whereas most of the smallest vessels are specialised shrimp vessels. Engine-power is mostly used to express the capacity of the Dutch demersal fleet since this parameter is likely to have the main influence on the catches of the vessels. In particular for the beam-trawlers this relationship is rather clear. Table 2 gives an overview of the fleet capacity and its development.

Table 2
Dutch demersal North Sea fleet, number of vessels and total engine-power

	1983	1998
Total number of vessels	595	407
Number of vessels as to engine-power		
0 - 190kW	141	82
191 - 221 kW	78	142
222 - 1104kW	295	32
>1104 kW	81	151
Total engine-power (kW)	367 000	319 000

Source: Ministry of Agriculture, Nature management and Fisheries;
Shipping Inspection; LEI.

The fishing effort of the fleet is composed of capacity and time. It is usually expressed as horsepower times days-at-sea for the Dutch demersal fleet. Table 3 gives this effort for the different types of gear. Beam trawling counted for 77% of the total effort in 1983, followed by otter- and pair-trawling on roundfish (15%).

Table 3
Fishing effort of the Dutch demersal North Sea fleet
(*100 000 horse-power-days)

Fishing method	1983	1998
Beam trawl	656	703
Otter trawl and pair trawl, roundfish	126	32
Pair trawl, herring	35	6
Shrimp trawl	33	38
Other	5	12
Total	855	791

Source: LEI.

The major part of the fleet was rather young in 1983, having an age of ten years or less. This was caused by an investment wave in the period 1979-1983, resulting in an addition of 126 new vessels to the fleet (Figure 4).

3.2 Changes in fleet capacity over the period 1983-1998

The number of vessels has decreased significantly in this fifteen year period and the fleet composition changed dramatically (Tables 2 and 5). The mid-size vessels (222-1104 kW) almost disappeared and two other classes became far more important in 1998. These two segments, the 'Euro-cutters' (191-221 kW) and the bigger beamers (more than 1104kW) nowadays count for about 90% of the engine-power of the fleet. In terms of engine-power the capacity of the fleet diminished by 13%, whereas the fishing effort was at a 7% lower level in 1998. Thus, the average number of days-at-sea per vessel increased since 1983.

Table 4
Dutch demersal North Sea fleet,
age profile of the vessels

Age	1983	1998
0-10 years	231	94
11-20 years	170	148
> 20 years	194	165
Total number of vessels	595	407

Source: Ministry of Agriculture, Nature management and Fisheries; Shipping Inspection; LEI.

Table 5
Dutch demersal North Sea fishery,
changes in fleet capacity
1983-1998 (1983=100)

	Index 1998
Total number of vessels	68
Number of smallest vessels (0-90kW)	58
Number of 'Euro-cutters' (191-221kW)	182
Number of mid-size vessels (222-1104kW)	11
Number of bigger vessels (>1104kW)	186
Total engine-power (kW)	87
Total engine-power (standard kWs)[1]	77
Fishing effort (in horse power/days):	
Beam trawl	107
Otter/pair trawl	25
Shrimp trawl	115
Total fishing effort	93

[1] Explanation see Section 3.2.
Source: Ministry of Agriculture, Nature management and Fisheries; Shipping Inspection; LEI.

Another important change regards the age-composition of the fleet. The number of newer vessels (less than ten years old) decreased from 39% in 1983 to 23% in 1998. On the other side, the proportion of older vessels (more than 20 years) rose from 33% in 1983 to 41% in 1998.

The change in vessel numbers has been analyzed further in Table 6. It appears that different subsequent decommissioning schemes[3] have had an important impact on the fleet capacity. The decommissioned vessels had to be scrapped or sold to third countries, *i.e.* countries outside the EU.

A part of the vessels under 'other withdrawals' in Table 6 have been re-flagged to other EU countries. This means that the fleet under Dutch ownership is in fact bigger than the previous tables suggest. These re-flagged vessels operate in European waters and they are entitled to British, German and Belgian flatfish and cod quota. The re-flagged fleet counts for about 20% of the demersal North Sea fishery under Dutch flag (in 1998), in terms of vessel number, engine-power and fishing effort. Taking this into account the demersal North Sea fleet under Dutch ownership has stabilized more or less in the period 1983-1998 from the view of total engine-power and fishing-effort.

Table 6
Dutch demersal North Sea fishery, additions to and
withdrawals
from the fleet in the period 1988-1998

	Number of vessels
Fleet at 31 December 1987	611
Period 1988-1998:	
Newbuildings	+111
Second-hand, bought abroad	+22
Decommissioned	-161
Other withdrawals[1]	-176
Fleet at 31December 1998	407

[1] Sold to other countries, re-flagged, changed to other activities, scrapped etc.
Source: Fisheries Directorate; Shipping Inspection; LEI.

The changes in fleet capacity have been caused by a chain of several factors which are described below. It has to be kept in mind that there are no simple cause-effect relationships in these changes. Causes may be effects from other points of view. Transferable harvesting rights have played a role amongst other factors.

3.3 Common fisheries policy

The establishment of the CFP in 1983 was the first and main influencing factor through the implementation of TACs in the framework of the conservation policy and the introduction of MAGPs resulting from EU structural policy. The CFP has led to several national measures which have caused major changes in the structure and scope of the Dutch demersal North Sea fleet.

The national quota levels in the 1980s for sole, plaice, cod and whiting caused a big imbalance between the capacity of many cutters and their fishing rights. A study by LEI in 1988 (Pavel 1988)[4] pointed out that 70 000 - 100 000 horsepower-units of the operating fleet would face liquidity problems in the next 2-4 years, due to this disproportion. To comply with the EU TACs allocated to the Netherlands a number of measures have been implemented, such as distribution of the national quota through ITQs, days-at-sea regulations, decommissioning and heavy enforcement of the quota.

To fulfill the obligations resulting from the first MAGP the Dutch ministry of Agriculture, Nature

---

[3] The first decommissiong scheme started in 1988.

[4] Pavel, S. *et al.* "Vooruitzichten voor de Nederlandse plat- en rondvissector op korte en middellange termijn", LEI-DLO report nr 5.79, August 1988.

Management and Fisheries implemented a licence scheme in 1984 which led to a horsepower ceiling for the fleet. The total horsepower of the active fleet could increase until 1988, due to orders for new vessels which were in the pipeline when the licence scheme came into force in 1984. The decrease of this total horsepower in the period 1983-1998 (Table 5) demonstrates the effectiveness of the horsepower scheme, since it prevented an expansion after the profitable years of 1991 and 1992.

3.4 Decommissioning schemes

The first scheme started in 1988 and this was followed by subsequent programmes so that decommissioning grants could be obtained nearly throughout the whole period 1988-1998. Quota limitations for cod and whiting have forced most of the owners of otter- and pair trawlers to apply for decommissioning. This has been the main cause of the decline of the cutter fleet after 1988, in particular the dramatic decrease of the number of mid-size vessels. A total of 183 000 horsepower (135 000kW) from 161 vessels was withdrawn from the fleet in the period 1988-1998. The majority of these decommissioned vessels (120) belonged to the medium size group (222-1104kW).

A major intensification of enforcement of ITQs through monitoring of landings in 1988, which meant systematic control of landings carried out by some 100 inspectors, made the overcapacity of the fleet visible. This has contributed significantly to the effectiveness of the decommissioning schemes.

A maximum limit on the number of vessels entitled to fish within the 12-mile limit exists. This is a EU measure (Regulation nr. 55/87) whereby the concerned vessels are registered in two separate files. The power of the coastal vessels was permitted up to 300 HP (221kW). These entitlements have been the main cause of the increase of vessel power up to the 300 HP limit.

3.5 Economic performance and prices of fishing rights

The heavy enforcement, mentioned earlier, led to a sharp rise of prices for flat fish ITQs in 1988. The good profitability of the cutters in 1991 and 1992 kept these prices at a high level and even resulted in further price increases. The decommissioning process contributed importantly to the trade in ITQs in the period 1988-1998. This enabled those who remained in the industry to adjust their fishing rights to the available capacity of the vessel, by buying additional ITQs. The high Dutch prices for rights have stimulated purchases of different types of rights in other countries in the early 1990s, which led to re-flagging of vessels.

A special law for investment promotion (for all industries) was introduced in 1978. This allowed a deduction of a certain percentage (12% at a minimum) of the amount invested from the taxable income. In fact, it meant a diminishing amount of the income-tax, or corporate-tax and this stimulated new construction of fishing vessels in the period 1979-1988. This contributed to an increase of total fleet horsepower up to 1988. The investment allowance was abolished in 1988.

A good level of profitability in the years 1985-1987 and 1991-1992 stimulated the construction of new vessels, in combination (in the first period) with the investment allowances mentioned before. The existence of a second-hand market for vessels abroad enabled the investors in new vessels to sell their 'old' one at a rather high price and to transfer the horsepower licence from the vessel sold to the new one. In cases of expansion additional horsepower could be bought from those who withdrew their vessel from the Dutch fleet, apart from through the decommissioning scheme. However, this mechanism stopped nearly completely in the early 1990s, mainly due to the tightening of the licence schemes in the UK. This has caused a major fall in the demand for second hand vessels.

3.6 The role of transferable property-rights in changes to fleet capacity

As noted in Section 3.3, there has been a complex variety of causes for the changes in the capacity of the Dutch demersal North Sea fleet and it is difficult to assess separately the impact of transferability of rights separately. But it can be stated that the advent of input and output rights have contributed to many factors:

i. Withdrawals from the fleet, apart from decommissioning, to realise high earnings from selling of the ITQs.

ii. Decommissioning of vessels. Vessel owners who have left the fishery had to hand in their horsepower licence but they could keep their ITQ. The high earnings from these rights have stimulated decisions to decommission in a number of cases.

iii. Concentration of rights amongst the owners of the bigger beam trawlers, which has led to their dominating part of the fleet by these vessels in terms of total horsepower.

iv. The absence of a high level of new constructions in the 1990s after profitable years. The effective horse power 'ceiling' has prevented further fleet expansion. This constraint led to a shift in investments from vessels towards ITQs in the early 1990s. These investments in ITQs absorbed more or less the depreciation funds of the firms so that future new vessel-constructions will also be at a lower level than in the 1980s.

v. Re-flagging of vessels. The Dutch vessel owners have acquired much experience in the market for harvesting-rights. High prices for ITQs in the early 1990s prompted them to look at the situation abroad. Low prices for such rights in the UK and other countries have prompted operations to buy rights abroad by purchasing foreign firms. A number of Dutch cutters has been re-flagged to these foreign subsidary firms since they could not operate profitably at that time.

vi. In particular the possibility for hiring and renting of ITQs has contributed to a better adaptation of rules to business practices and enabled a better utilisation of the vessels and a more efficient uptake of quota. The co-management sytem, established in 1993, has created an important condition for this improved efficiency.

The Dutch experience demonstrates that co-management can secure for the ITQ-right benefits by sound management of group quota. This includes monitoring of landings and measures (warnings, not to land abroad etc.) when a group member has almost caught his ITQ[5]. Such group management guarantees that the individual holder he can fully take his own ITQ. The threat that colleagues will take a part of his ITQ by over-fishing their own quota has been removed in that case[6].

3.7 Consequences of changes in fleet capacity

The lower capacity of the demersal North Sea fleet, shown in Table 5, has had several consequences:

i. Improvement of the profitability level of the cutters. The sector has been profitable or at break-even level since 1991. This is a rather long period of good economic results in view of developments in the 1970s and the 1980s. Profitable years were followed by years with adverse results in that period. Fleet expansion through investments in new cutters after good years used to dissipate potential profitability. This is impossible now because of the effective engine-power licence scheme.

ii. Decrease in the level of employment from 2750 crew members in 1983 to 1920 by the end of 1997. Generally, those who have left the fishery could find a job ashore, in particular in the past few years. It is now even difficult to find enough capable crew-members for the cutters, due to sometimes good economic development in the Netherlands and ageing of the labour force.

iii. Decline of fishing communities. The industry fears that the 'critical mass' of some communities may be too small for sustainability in the longer run.

iv. A much larger proportion of the bigger beam trawlers in the fleet which has had consequences for the productivity of the sector. Such vessels show a decreasing yield per kW/day (Smit 1998: 47-53) so that the capacity has in fact diminished more than the 'nominal' figure for engine-power indicates. Therefore Table 5 also shows the capacity expressed in standard kWs. This measure corrects for lower yields per kW/day for the bigger beamers so that a better estimate for the real capacity is obtained. In the same way, real fishing effort is in fact lower than the index shows in Table 5. Taking into account this lower productivity among the bigger vessels the real fishing-effort decreased by some 20% in the period 1983-1998, instead of the 'nominal' 7%.

v. Difficulties to fully take the quota. The cutters are also limited by days-at-sea, apart from the engine-power limitation. The current MAGP intends further reduction of days-at-sea per vessel which may make it impossible for the fleet to land all the quota[7].

4. CONCENTRATION OF OWNERSHIP

4.1 Status prior to the programme

The demersal North Sea fleet consisted of 530 enterprises in 1985, the year when transferability of flatfish IQs was allowed officially by the Fisheries Directorate. These firms are family enterprises, employing a majority of family members in many cases. Ownership may rest with the father alone or together with several sons or brothers. In particular the situation of these being many sons of the same owner has led to expansion of the enterprise, when the objective was that each son would become skipper on a vessel. This kind of expansion was possible before rights-based fishing became effective, in the second half of the 1980s. It has resulted in a number of bigger firms owning more than one vessel. In fact, a concentration process was going on already before 1985, leading to more engine-power being exploited by fewer enterprises. About half of the total engine-power was concentrated in 1987 in these 'multi-vessel' companies, whereas their number counted for only 13% of the total number of firms[8].

The flatfish IQs, introduced in 1976, were not much more than 'a piece of paper' up to the mid-1980s. Informal trade in them at that time and the introduction of official transferability demonstrates their growing importance around 1985. Trade in subsequent years has led to a concentration of rights within the bigger firms. This accelerated after 1988, because strict enforcement of the ITQs forced the owners of the bigger cutters to acquire enough rights to ensure their vessel's operations. The introduction of ITQs for cod and whiting in 1994 and those for herring and mackerel in 1996 gradually strengthened this concentration process.

4.2 Restrictions for transfer of ownership

A number of more detailed quota regulations were in force in 1998:

i. A continuous individual quota regulation, whereby annual changes in the Dutch part of the EU TACs and resulting changes in ITQs were included as an appendix of the regulation. This continuity of rights replaced the annual allocation process in 1997.

ii. Related species are associated, which mean that there should always be an ITQ for sole and plaice, just as for cod and whiting.

---

[5] One of the co-management groups expelled three members in October 1999 and held under arrest one vessel because of ITQ over-fishing, Visserijnieuws 29 October 1999.

[6] This advantage of co-management in an ITQ fishery has been emphasized by Dick Langstraat, Chairman of the Dutch Fish Board. Transfer of some competence from the individual right holder to the collective of the management group is necessary in that case (pers.comm.).

---

[7] Conclusion in " Ondernemend vissen", p. 54.

[8] "Visserij in Cijfers 1987", p. 22, LEI 1988.

iii. Transfers of the ITQs have to be registered by the Fisheries Directorate of the Ministry of Agriculture, Nature management and Fisheries.
iv. The ITQs should be attached to a principal vessel, with the exemption that the rights may be reserved separately for a five year maximum (from 1 January 1998). This only applies for ITQs that have been included in a total quota of the group. This reservation-term enables right holders, which have *e.g.* sold their vessel, to hire out their ITQ temporarily while a new vessel is being built.

Some rules limit the transfer of ITQs explicitly:

i. Selling of a part of the quantity sole or plaice to vessels not having such ITQs is not allowed.
ii. A quantity of both sole and plaice should remain after such a sale; the same applies for cod and whiting ITQs.
iii. ITQ holders are not free to withdraw their ITQ from the group quota in the course of the year unless the group board agrees and 90% of the group quota has not been taken. Sale of the vessel and bankruptcy are two other cases in which the ITQ may be separated from the group quota.

A regulation stipulates a time-schedule for a number of requests to the Ministry. This regards mainly:

i. Formation of group-quota for the main species before 1 February
ii. Requests for transfer of a sole quota into a plaice quota, or vice versa, before 1 March and
iii. Requests for lease or rental transactions of ITQs between groups before 1 December. For fishermen who are non-group members this date is 1 March.

4.3 Prices received

The prices of ITQs are not publicly recorded but the co-management groups have a good overview of these prices through their involvement in the trade of rights. Table 7 contains price indications for flat-fish rights (Davidse *et al.* 1997: 105). These prices were obtained by the LEI cost-and-earnings panel and from interviews with representatives of co-management groups.

Rather high prices have also been paid for other entitlements, *e.g.* cod/whiting NLG 14-17/kg since 1998 and some NLG 300 000 for shrimp permits for the Wadden Sea. Further, horsepower-licences were priced at NLG 800-1500 in 1998 and 1999.

Prices of flat-fish ITQs increased sharply in 1987/1988 as Table 7 shows. This reflects the fact that control measures became very stringent in 1988. In that year systematic control of landings was implemented, carried out by about 100 inspectors. In 1993/1994 prices of sole and plaice quotas dropped, due to the high level of the national sole quota and diminished catches of plaice. The catches were low in many cases compared with the available plaice quota so that there was no need, in general, to buy plaice quota.

Table 7
Price indications of flat-fish ITQs

Year	Flat fish ITQ	
	Sole/plaice (NLG/kg)[1]	Sole only (NLG/kg)
1986	10-15	10-15
1987	70-85	.
1988	100-120	70-80
1989	100-120	70-80
1990	100-120	70-80
1991	130-150	90-95
1992	130-150	90-95
1993	70-95	55-75
1994	65-90	50-70
1995	. a)	60-80
1996	. a)	75-85
1997	. a)	70-90

[1] 1 NLG=0.45EURO or 0.49US$.
a) Plaice has been traded more and more separately, at higher prices: NLG 9-13 in 1996 and NLG 10-18 in 1997.
Source: LEI; Co-management groups.

With respect to price developments of Dutch ITQs the following major influencing factors can be distinguished in summary:

i. Enforcement of quotas: A major improvement of enforcement in 1988 caused a sharp price increase.
ii. Profitability of the fishery: Better profitability in 1991 also caused higher quota prices. Formerly, investments in fishing vessels used to increase sharply in such situations but in 1991 investments in vessels were to a major extent diverted to investments in flat-fish ITQs.
iii. Potential harvest of the fish relative to the quota level: In 1993/1994 the national plaice quota was rather high in view of the catch potential for this species. This contributed to a downward price trend for plaice quotas.

4.4 Effectiveness of regulations governing ownership of rights

ITQs should be attached to a vessel as they are owned by the owners of that vessel. A dis-association from the cutter-vessel is allowed for up to five years because of new constructions, or other reasons, if the vessel is included in a group quota. Loopholes have been the attachments to small boats to the extent that there is a disconnection from a real commercial vessel. Conditions for linkage with working vessels have been strenghthened by requirements for commercial exploitation. But the phenomenon of fishermen ashore who try to make a living from hiring out ITQs still exists and this a matter of concern, though not a major one for the industry. The possession of valuable rights has gradually resulted in rather complicated financial arrangements to facilitate, for example, succession of ownership to the son(s) of the right-holder. Nowadays the ITQs have the same fiscal status as agricultural-rights such as milk-quotas.

4.5 Affects of the programme

The major intensification of enforcement in 1988, accompanied by vessel decommissioning, prompted more and more transfers of rights. Vessel owners who had to adjust their flat-fish rights to the capacity of their vessel were prepared to pay high prices for sole and plaice ITQs. The extra proceeds from the additional quota held had only to cover the marginal cost for catching and landing the extra fish. Moreover, the possibility to avoid heavy fines was an important condition for this willingness to pay high prices.

Table 8 shows that the owners of the bigger cutters (over 1104kW) possessed 86% of the sole rights, whereas six months, in mid 1988, this percentage was 56%[9] (Salz 1996). The change in the rights situation followed the (an annual landing of 1.18t of sole on the basis of the 1994 quota) to 1.5-2.5% (354-590t) for the biggest ITQs.

Table 9 shows that some concentration of sole ITQs occurred in the period 1988-1994. Holders of bigger sole ITQs, owning 1% or more of the national sole-quota, had a higher share in the total ITQ in 1994 (8%) compared with 1988 (4.7%). On the other hand, the percentage of holders owning smaller ITQs (up to 0.5%) has decreased since 1988. The distribution of ownership with respect to the plaice ITQs (not shown in the table) followed the same development, though holders of the biggest plaice ITQs (1.5-2.5%) were somewhat less in number compared with the sole ITQs.

Table 9 also shows the ITQ distribution according to size of holding for the 1997 allocations. This shows that

Table 8
Concentration of fishing rights according to engine-power

Fleet Segment (kW-group)	Mid 1988			January 1998		
	Number of vessels	Total power 1000kW	% sole quota	Number of vessels	Total power 1000kW	% sole quota
0- 190	141	19	1.0	87	12	0.3
191- 221	125	27	4.7	143	32	9.6
222-1104	201	151	38.5	30	18	4.1
>1104	139	236	55.8	156	264	86.0
Total	**606**	**433**	**100**	**416**	**326**	**100**

Source: Ministry of Agriculture, Nature management and Fisheries; Shipping Inspection; LEI.

trend towards bigger beamers in the fleet. But the share of this segment in the total of flatfish rights has increased somewhat more than its contribution in the total fleet engine-power (an increase of 54% against 49% since 1988).

In addition to these output-rights the input-rights in the form of horsepower-licences became more important in the late 1980s and in the 1990s. Vessel owners who wished to expand the engine-power of their vessel or intended to build new cutters, had to buy additional horsepower-rights on the market. In this way a trade in horsepower-rights also has arisen, in particular in the 1990s[10].

In Table 9 the distribution of individual sole-quota according to the size of the ITQ is considered (Davidse et al. 1997: 184). This size is expressed as a percentage of the total national sole-quota. The level ranges from the 'mini' ITQs, representing 0.005% of the total sole-quota

the trend in concentration did not continue clearly in the period 1994-1997. The number of holders of the smallest ITQs decreased on the one hand, but also the number of bigger ITQ holders (category >1%) decreased somewhat. The underlying factor seems to be less trade in ITQs, since the number of holders remained rather constant between 1994 and 1997.

A significant decrease in the total number of ITQ holders (by 25%) in the period 1988-1994 was caused by selling of sole/plaice ITQs in combination with decommissioning, closing of enterprises for other reasons, or only stopping fishing with beam trawls. The share in the total Dutch sole/plaice-quotas of the holders of the 20% biggest ITQs is another measure of the level of concentration. In 1994 this group owned almost 60% of total Dutch sole quota and 56% of the plaice-quota. By 1997 it was 58% and 56% respectively. Nearly all of these ITQ holders are companies owning more than one vessel.

The regional concentration of flat-fish rights was also been considered in the property-rights study mentioned before. The conclusion from the 1997 situation was that the share of the national quota in Urk, the main Dutch fishing port, had decreased somewhat, whereas the Den Helder/Texel region had expanded their share. However, there was no major concentration of flat-fish

---

[9] The concentration level has been measured in sole rights and the same conclusions can be drawn for the ITQs for plaice and cod.

[10] In the first years after the introduction of horsepower licences a quantity of 'floating' licences existed because of extra orders for new constructions just before this licence scheme was put in place in 1985. Therefore, trade of these rights mainly began some five years after the introduction.

Table 9
Distribution of ITQ holders according to size of the ITQ, expressed as percentage share
in total allocated Dutch sole-quotas in 1988, 1994 and 1997

Percentage share of ITQ in total sole quota	Percentage of ITQ holders		
	1988 (n=387)	1994 (n=289)	1997 (n=276)
0.005 (mini ITQ)	20.2	17.3	14.9
0.005-0.5	65.3	57.7	59.7
0.5-1.0	9.8	17.0	17.8
1.0-1.5	3.9	4.5	3.6
1.5-2.5	0.8	3.5	3.6
>2.5	0.0	0.0	0.4
Total	100.0	100.0	100.0

ITQs in a limited number of regions in the period 1988-1997.

5. DISCUSSION
5.1 Reduction in fleet capacity

The Common Fisheries Policy (CFP) of the European Union (EU) has two main goals: limiting the catches by fixing annual TACs, and reduction of the fishing capacity by implementing multi-annual guidance programmes (MAGPs). The MAGP objectives for the Dutch demersal North Sea fishery have not been met so far, although an engine power licence scheme has been in place since 1985 and subsequent decommissioning schemes have been implemented. The owners of the cutters cannot be forced to leave the fishery so that the actual fleet reduction depends on the profitability of the fishery and also on the ownership situation, *i.e.* the presence of a successor for the current vessel's owner. In the past eight years most of the cutters have operated on, or above, break-even in relation to their costs, so that many skipper owners have not had an urgent reason to stop fishing.

The Dutch Ministry of Agriculture, Nature Management and Fisheries has been indeed successful in preventing expansion of the capacity after profitable fishing years. This kind of expansion occurred regularly before the licence scheme became effective in the early 1990s.

Looking at the developments of the past fifteen years two kinds of reactions by vessel owners were contrary to the expectations of policy-makers:

i. Extra orders for new constructions, just before the engine-power licence scheme was put into place in 1985. In response the Ministry has limited the validity of the 'floating' licences and has strengthened the condition for attachment to a vessel.
ii. Re-flagging of cutters to other countries, to get more harvesting rights for the remaining ones in the Dutch fishery. In fact this has not been a problem for the Dutch fisheries management since it contributed to compliance with the Dutch TAC and MAGP obligations.

The capacity of the cutter fleet was 8% above the MAGP requirements by the end of 1997 (LEI 1998: 52). This target has been expressed in effort (days-at-sea) for the current MAGP which runs from 1998-2002. A further reduction by 17% for the segment of bigger beam trawlers compared with the situation in 1998 [11] has occurred in this programme.

The industry heavily opposes these effort reductions since the quota allocations cannot be taken up with a lower fishing effort. The industry representatives state that the quota should count first, and not the capacity limitations. The co-management system has been able to comply with the quota-limit since 1993 so that further substantial reduction of capacity is not necessary. In fact, two kinds of management are now conflicting in the Dutch demersal North Sea fishery: (a) management by command and control, aiming at reducing the fleet within the framework of EU's CFP and (b), a type of co-management characterized by responsibility for compliance with national quota by groups of quota-holders. Centralized targets for fleet capacity levels thwart these decentralized quota responsibilities. Therefore, giving priority to fleet reduction above that needed to comply with national quota limits would heavily undermine the Dutch co-management system. This conflict between input- and output-targets is being discussed now between the Dutch government and the European Commission.

5.2 Concentration of ownership

ITQs for all quota species have now become an important production factor for fishing enterprises. It is a major intangible asset on the balance sheet of many firms. The rights can be 'banked' and fiscal allowances for depreciation are in force, as for other assets. The co-management system facilitates transfers and hiring of rights via the group boards. This combined management of ITQs has brought advantages in the past five years in

---
[11] Visserijnieuws 21 May 1999.

the form of price increases for plaice in particular, through compliance with much the lower quota.

Transferability has led to a continuing concentration of rights since the mid 1980s, although this did not result in a few companies owning a major part of the rights so far. It is reasonable to assume that this process will continue in future. Newcomers cannot enter the fishery since prices of rights are too high to permit a new firm to be profitable (Davidse *et al.* 1997:201). Hence the number of enterprises will diminish.

It is necessary to understand the nature of quota trade to explore future developments. ITQs have mainly been purchased by vessel owners who had already vessels with harvesting rights. Thus high prices could be paid since marginal revenues had only to cover the marginal costs. These high price levels will hamper concentration of rights because of finance limitations. Nowadays the economic depreciation of vessels has been absorbed more or less by quota investments so that vessel replacements may be difficult in a number of cases. However, the quota market determines the future price of rights. In this respect the concentration process has not been hampered, apart from some restrictions described in Section 4.2 above.

An important aspect of this issue concerning the concentration of rights is the review of the CFP due in 2001. Quota-hopping by Dutch enterprises in the past has made clear that concentration of rights has gone further than the national situation alone reveals. A question is whether new possibilities will arise to acquire harvesting rights in other countries from 2001 onwards. If so, the concentration of rights may accelerate in a more open, international market. That would bring the fishing sector more in line with other branches of the economy in the common EU market.

6. LITERATURE CITED

Davidse, W.P. *et al.* 1997. Property rights in fishing, LEI-DLO report 159. 328pp.

FAO in preparation. Case studies on the allocation of Transferable Quota Rights in Fisheries. *Fish.Tech. Paper,* FAO, Rome

Salz, P. 1996. ITQs in the Netherlands: twenty years of experience, ICES paper. 17pp.

Salz, P. *et al.* 1988. Vooruitzichten voor de Nederlandse plat- en rondvissector op korte en middellange termijn, LEI-DLO report nr 5.79. 64pp.

Smit, W. *et al.* 1998. Ondernemend vissen, Toekomstperspektief van de kottervisserij. Report 1.98.001. LEI. 100pp.

Smit W. *et al.* 1998. Visserij in cijfers, annual edition, LEI.

TRENDS IN FISHING CAPACITY AND AGGREGATION OF FISHING RIGHTS IN NEW ZEALAND UNDER INDIVIDUAL TRANSFERABLE QUOTA

R. Connor
Centre for Resource and Environmental Studies,
The Australian National University, Canberra, ACT 0200, Australia.
<rconnor@cres.anu.edu.au>

1. INTRODUCTION

The introduction of individual transferable quota (ITQ) could be expected to result in reduced ratios of fleet fishing capacity proxies to catch, due to increases in technical efficiency and greater utilisation of existing capacity with the exit of some vessels from the fleet. A concomitant expectation of the process of effort and capacity rationalisation is increased concentration of quota ownership. This paper is a progress report on an FAO-supported study (FAO in press) of quota concentration and fleet change under the New Zealand quota management system (QMS). Vessel capacity is assessed through the indicator of gross registered tonnage (GRT). Quota ownership, end of year holdings and catches are assessed through the use of a range of indices of market concentration and the Gini Index. A preliminary interpretation of the results is made here.

2. THE NEW ZEALAND QMS

New Zealand's quota management system (QMS), based on strongly-defined individual transferable quota (ITQ), was introduced in October 1986 for all of the significant fin-fish species. This followed four years of enterprise allocations (EA), with very limited transferability, applied to the deepwater species. The fishing of the large deepwater resource was being taken over by the New Zealand domestic industry following the declaration of the EEZ in 1978. These stocks had been discovered and fished by several foreign fleets, particularly those of the USSR, Japan and the Republic of Korea ("South Korea"). The EA scheme was used as a means of allocating the resources among the domestic companies investing in fishing and processing deepwater species. The NZ inshore fisheries were in decline in the early 1980s after 15 years of unrestricted fishing and government-supported development of capacity (Clark and Duncan 1986). A full moratorium on fishing permits was effected from 1980, part-time fishers were excluded in 1983, and after grandfathering quota to remaining fishers at average levels of their catch history for the best two out of three years, a quota buyback was undertaken for many inshore species for which TAC reductions were required (*ibid.*).

New Zealand fisheries may be divided into several sectors. The inshore fin-fisheries, of which 17 species or species groups were introduced to the QMS in 1986, produced a total catch of 34 000t (QMS species only) in the first year of quota management. A range of catching methods is used, including trawling, bottom long-line and set-netting. The most significant species by value is snapper, which is mainly exported to Japan. The deepwater trawl fisheries for orange roughy and oreo species operate in depths to 1500m, and in 1986-7 produced some 66 000t of catch. The further seven mid-depth species comprise the largest proportion of the total catch. In 1986-87 catch of these species was around 308 000t. Some 98 000t of this total was from squid and jack mackerel, which were brought into the quota system in its second year. In this study, these are excluded from the mid-water figures. This sector is dominated by hoki, which has a (currently fully caught) TACC of 250 000t. In 1987 the hoki catch reported within the quota system was 158 000t. Although the vast bulk of this fish is caught by large off-shore trawlers, the species is found in most fished demersal habitats, making hoki a by-catch species for many fishers. The QMS shellfish were all introduced in later years (paua/abalone 1987; rock lobster 1989; scallops 1993; oysters 1998). Rock lobster is included in this report, but problems with paua data are yet to be resolved. Scallops and oysters are not considered.

3. QUOTA CONCENTRATION
3.1 Background to the debate

A concern in the debate over ITQs is the concentration in ownership of quota shares. From an economic point of view there are two reasons for monitoring concentration. The first is that in fisheries where production is inefficient due to over-capitalisation, likely accompanied by stock-depletion and higher than necessary variable catching costs, the introduction of quota management often has as the explicit objective of reducing the number of vessels in the fishery. Assuming grandfathering is used in allocation, if rationalisation of the fleet is to occur, then some degree of quota concentration is to be expected as a result. On the other hand, there are limits beyond which concentration can be negative. Should a small number of owners control the large majority of quota, monopoly-type market-power effects may occur. These can include the manipulation of prices for both fish and quota to capture rents and to facilitate further accumulation of quota, and undue dominance in the labour market affecting wages and conditions of fishers (National Research Council 1999). The development of such monopoly-power may be avoided by specifying aggregation limits for quota. In practice such limits have been set at levels ranging from 0.5% to 35% of a fish stock. Otherwise, antitrust-type legislation may be relied upon to protect against monopoly power developing, without having to specify and arbitrary limit on individual holdings. However, some means to estimate concentration is required. There has been few published studies to date estimating quota concentration in fisheries (for examples see Gauvin, Ward and Burgess 1994; Hogan, Thorpe and Timcke 1999).

3.2 Methods and data

Standard measurements of concentration of such factors as market-share, assets, physical output, or employment among competing industrial companies are considered by Scherer (1973). Should data on marginal costs of production be available, which is usually not the case, a direct comparison with price can estimate the degree of monopoly-power present in a market. The most straightforward index for commonly available data is the Concentration Ratio (CR), which is the proportion of the factor chosen represented by a selected number of the largest firms. The top four firms are commonly used (CR4), and often a table is presented with a range of values (*e.g.* CR4, 8, 20 and 50).

An extension of this comparison is to use percentiles, or to construct a Lorenz curve, which is a plot of the cumulative total proportion of the factor represented against the proportion of the firms represented, sorted in rank order for the factor. For example, if fish quota owning firms are ordered by amount of quota held, the proportion of the total held by all up to each firm is plotted. In the case that all participants hold the same amount (for instance 100 firms with 1 per cent of the quota each) a 45 degree straight line is the result. Where there are differences in holdings a curve will sag below the 45 degree line but start and finish at the same points (0 and 100% of quota owned by 0 and 100% of owners). The Lorenz curve can be converted to a single number, the Gini Index (GI), by comparing the area between the 45 degree line and the curve with the total area under the 45 degree line. For any number of firms all with identical shares, the curve will be the line, and the GI is zero. As the distribution of shares becomes less equal, the index approaches unity. The GI is a measure of inequality in shares that does not take into account the number of participants. It is therefore not ideal for measuring competitiveness where firms are likely to have similar sized shares, as it will give the same answer for 2 firms with 50% of quota each and for 100 firms with 1% each. However, it may be useful in indicating inequality when used with other indices.

The Gini Index (GI) has a range from zero to one and is calculated as follows:

$$GI = 1 - \frac{\sum_{i=1}^{N}\left(\sum_{j=1}^{i} q_j / Q\right)}{(N+1)/2}$$

where q = quota or catch amount
Q = total of all quotas or catches
N = number of quota holdings or catches
and all q are ordered, 1 to N from smallest to largest.

An index more useful for estimating potential market power, the Herfindahl-Hirschman Index (HHI), sums the squared proportionate shares of all firms. This takes account of both the number of firms and inequality in quota shares, weighting the larger firms quadratically. The HHI is calculated as follows:

$$HHI = \sum_{i=1}^{N} (q_i / Q)^2$$

where q = quota or catch amount
Q = total of all quotas or catches
N = number of quota holdings or catches.

The index as calculated has a range of zero to one. In this study, both the GI and the HHI are expressed as percentages. Hence a calculated index of 0.75 is written as 75%.

With the New Zealand quota and catch data, the GI tends to be quite high for all stocks, in the 75% to 97% range. However, for the same GI, the HHI can range widely. For example:

IS1988   GI = 82.9%,   HHI = 1.6%
DW1995   GI = 82.7%,   HHI = 14.8%

In many years there were large numbers of small-quota holdings that were less than the minimum for a fishing permit to be issued. If these are eliminated from the index calculations, the GI decreases slightly as the degree of inequality has decreased, but the HHI increases as the number of firms has decreased and thereby the concentration of ownership has increased. This illustrates the utility of the two indexes. In fact, neither is sensitive to large numbers of small holdings being eliminated. On the other hand, all the indices are sensitive to the presence of single large owners at the top of the scale. In New Zealand, Te Ohu Kai Moana (TOKM) [the Treaty of Waitangi Fisheries Commission] holds approximately 10% of all quota species, and this makes them significantly larger owners than any other in fisheries such as paua (abalone) and rock lobster. This distorts some of the figures as this quota is essentially being held in trust and will (eventually) be distributed in small holdings to the 78 tribal groups (*Maori iwi*) recognised by TOKM.

Other data-issues that may affect results at this early stage in the study are government ownership of large holdings for some deepwater species in the first years of the quota system, and multiple quota accounts being maintained by fishing companies. Cross ownership by larger interests in the industry also effectively concentrates control over quota, but this is even more difficult to take into account.

3.3 Results of concentration analysis

The New Zealand data for quota ownership, holdings at the end of the season, and catch were processed separately to generate a range of indices:

i. HHI
ii. GI
iii. CR1, CR3, CR4, CR10
iv. Number and percentage of owners with 95% share
v. Percentage share controlled by top 5% of owners and
vi. Number with less than minimum holdings for the class.

The fisheries were split into three general classes of finfish species: inshore, mid-depth, and deep-water, and jack mackerel, squid, paua, and rock lobster were treated separately. Results are presented as Tables 1 to 3, and in Figures 1 to 4.

4. INSHORE SPECIES

The seventeen inshore finfish species (or species groups) are blue cod, blue nose, alfonsino, elephant fish, flatfish (group of 8 species), grey mullet, red gurnard, hapuka/bass, john dory, blue moki, red cod, school shark, snapper, rig, stargazer, tarakihi and trevally.

While total allowable commercial catches (TACCs) and therefore quota owned for these species increased by 15% between 1987 and 1998, the number of quota owners

Table 1
Quota owned - by sector

Fishery	Year	Total quota owned (t)	No. owning quota	HHI	CR1	CR3	CR4	CR10	No. owning 95%	% owning 95%	% owned by top 5%	No. owning < 5t	Gini index
Inshore	1987	59 015	1309	3.1%	12%	26%	30%	44%	563	43%	69%	454	84.1%
	1988	61 198	1289	3.2%	12%	25%	29%	45%	533	41%	71%	445	84.8%
	1989	64 715	1312	3.5%	12%	28%	32%	46%	525	40%	72%	452	85.3%
	1990	67 026	1320	3.4%	13%	27%	30%	44%	512	39%	71%	459	85.6%
	1991	66 666	1291	2.9%	10%	23%	28%	44%	491	38%	73%	439	86.1%
	1992	68 391	1244	2.8%	7%	21%	27%	46%	457	37%	74%	411	86.7%
	1993	67 694	1187	2.8%	9%	23%	28%	44%	426	36%	74%	398	86.9%
	1994	67 858	1161	2.8%	10%	23%	28%	44%	411	35%	74%	386	86.9%
	1995	68 057	1110	3.3%	11%	27%	31%	45%	391	35%	74%	364	86.9%
	1996	68 444	1077	3.0%	10%	24%	29%	46%	366	34%	75%	350	87.3%
	1997	69 071	1023	3.2%	10%	24%	29%	49%	343	34%	75%	323	87.5%
	1998	67 958	963	3.3%	10%	24%	30%	49%	322	33%	76%	292	87.6%
Mid-depth	1987	325 175	493	30.2%	53%	72%	76%	91%	15	3%	98%	327	98.2%
	1988	326 081	478	29.9%	52%	72%	77%	90%	17	4%	98%	317	98.1%
	1989	329 119	475	28.9%	51%	71%	78%	91%	16	3%	97%	312	98.1%
	1990	333 569	466	26.6%	49%	68%	74%	89%	18	4%	97%	291	97.9%
	1991	308 161	457	19.0%	40%	59%	66%	84%	23	5%	95%	290	97.3%
	1992	290 266	449	14.8%	31%	58%	65%	86%	20	4%	96%	278	97.3%
	1993	290 694	424	14.1%	31%	55%	62%	85%	22	5%	95%	259	97.0%
	1994	290 770	417	14.6%	31%	55%	65%	87%	22	5%	95%	247	97.0%
	1995	311 864	409	11.4%	21%	50%	60%	88%	21	5%	95%	236	96.9%
	1996	331 514	396	13.6%	29%	53%	63%	89%	18	5%	96%	224	97.0%
	1997	340 668	378	13.5%	28%	53%	63%	90%	17	4%	96%	217	97.0%
	1998	338 242	360	13.6%	28%	53%	63%	90%	17	5%	96%	210	96.9%
Deep-water	1987	83 010	44	12.4%	22%	55%	64%	88%	19	43%	40%	1	76.3%
	1988	85 216	38	13.8%	23%	56%	70%	90%	16	42%	41%	1	76.0%
	1989	86 623	43	13.6%	23%	56%	70%	90%	16	37%	41%	3	78.2%
	1990	70 353	42	10.9%	20%	50%	61%	85%	20	48%	36%	2	72.7%
	1991	64 110	41	15.8%	28%	64%	72%	87%	18	44%	50%	4	76.9%
	1992	63 076	49	15.5%	29%	60%	66%	84%	19	39%	52%	2	78.3%
	1993	61 700	47	15.9%	30%	61%	68%	84%	18	38%	52%	4	78.1%
	1994	61 534	49	16.1%	30%	62%	70%	85%	18	37%	52%	6	79.8%
	1995	56 233	47	15.6%	27%	61%	69%	86%	18	38%	52%	5	79.1%
	1996	50 050	41	16.5%	29%	62%	71%	89%	14	34%	53%	6	79.3%
	1997	50 474	39	16.2%	28%	62%	71%	92%	13	33%	50%	7	79.9%
	1998	50 474	40	16.1%	28%	62%	72%	92%	13	33%	50%	8	80.2%
Rock lobster	1987												
	1988												
	1989												
	1990	3 726	686	0.6%	6%	8%	9%	13%	503	73%	23%	411	48.7%
	1991	3 597	656	0.6%	4%	9%	10%	16%	479	73%	27%	418	51.2%
	1992	3 286	598	0.7%	6%	9%	11%	18%	438	73%	29%	382	52.6%
	1993	2 936	554	1.1%	8%	12%	14%	21%	404	73%	32%	379	54.5%
	1994	2 932	525	1.1%	8%	13%	15%	21%	383	73%	32%	350	55.0%
	1995	2 915	517	1.2%	8%	14%	15%	22%	372	72%	33%	342	55.5%
	1996	2 968	512	1.2%	8%	14%	16%	23%	358	70%	35%	336	57.5%
	1997	2 894	490	1.3%	8%	15%	17%	24%	344	70%	34%	315	57.1%
	1998	2 954	470	1.4%	8%	15%	17%	26%	329	70%	37%	293	58.1%

Table 2
Quota owned - All stocks combined

Fishery	Year	Total quota owned (t)	No. Owning Quota	HHI	CR1	CR3	CR4	CR10	No. owning 95%	% owning 95%	% owned by top 5%	No. owning < 5t	Gini index
All data	1987	467 201	1357	20.7%	42%	63%	68%	83%	62	5%	95%	457	97.2%
	1988	594 734	1459	17.4%	37%	58%	66%	83%	45	3%	96%	497	97.6%
	1989	671 927	1495	12.8%	29%	50%	59%	82%	45	3%	96%	520	97.6%
	1990	673 180	1899	11.3%	27%	48%	56%	78%	67	4%	96%	682	97.5%
	1991	568 021	1873	10.5%	24%	49%	56%	74%	82	4%	95%	694	97.2%
	1992	577 180	1809	9.5%	24%	43%	52%	76%	69	4%	96%	660	97.3%
	1993	581 436	1769	9.9%	26%	44%	51%	74%	79	4%	95%	641	97.2%
	1994	581 838	1764	10.3%	25%	45%	54%	76%	75	4%	96%	622	97.2%
	1995	595 312	1738	9.6%	20%	45%	55%	78%	72	4%	96%	616	97.3%
	1996	609 352	1718	10.6%	25%	45%	55%	80%	64	4%	96%	615	97.5%
	1997	646 610	1755	10.2%	23%	44%	55%	81%	57	3%	96%	650	97.6%
	1998	661 395	1673	9.8%	22%	43%	54%	79%	66	4%	96%	599	97.5%

Table 3
Quota held - all stocks combined

Fishery	Year	Total quota held (t)	No. holding quota	HHI	CR1	CR3	CR4	CR10	No. holding 95%	% holding 95%	% held by top 5%	No. holding < 5t	Gini Index
All data	1987	473 657	1461	8.5%	21%	43%	49%	74%	97	7%	94%	450	96.5%
	1988	618 676	1650	6.9%	15%	37%	47%	70%	74	4%	95%	487	97.0%
	1989	702 421	1695	6.2%	17%	35%	41%	63%	75	4%	96%	497	97.0%
	1990	683 091	2100	5.3%	15%	33%	39%	58%	97	5%	95%	658	96.9%
	1991	573 075	2233	4.5%	9%	26%	33%	59%	135	6%	94%	656	96.4%
	1992	582 929	2243	6.3%	17%	35%	43%	66%	113	5%	95%	668	96.7%
	1993	581 765	2151	5.5%	13%	32%	40%	64%	127	6%	94%	639	96.4%
	1994	581 879	2132	6.7%	17%	36%	44%	67%	132	6%	94%	587	96.4%
	1995	595 320	2052	10.6%	22%	51%	56%	70%	111	5%	95%	595	96.8%
	1996	609 353	1940	8.3%	18%	39%	49%	79%	71	4%	96%	639	97.4%
	1997	646 610	1947	7.9%	19%	40%	50%	73%	89	5%	95%	634	97.1%
	1998	661 395	1889	7.4%	17%	39%	48%	71%	91	5%	95%	578	97.0%

decreased by 26% from 1309 to 963 (Table 1). The HHI has increased marginally, which would be predicted by falling numbers of owners, but there is more going on here. While the top ten owners have steadily increased their share of the quota from 43% to just under 50%, some jockeying has been going on among the top three or four owners, with TOKM building its holdings to take the number one position with just under 10%. Given that there are almost one thousand owners (1998) having 50% owned by 10 interests (1% of owners) might seem concentrated, but it represents a low concentration relative to other sectors. The top 5% of owners has 75% of the inshore, up a little from 69%, and the proportion of all owners holding 95% of all quota has dropped from 43 to 33%. The HHI is low at around 3.3%, the GI has moved from 84% to 88%, indicating a moderately high degree of inequality in holding sizes, and the number with less than minimum holdings has dropped by a third.

The figures for end of year (EOY) holdings (Table 2), which take into account effective redistribution of access through leasing, show a decrease in concentration with respect to the ownership figures, systematically expressed in all indicators. Numbers of holders are about 10% up on numbers of owners, HHI is down slightly, all CRs are down and so on. The numbers with less than the minimum holdings are up slightly, presumably because some have leased almost all their quota.

For catch, the HHI and GI indicators are lower again than for holding and ownership. The CR1 and CR3 are the same as holdings, indicating the top owners are getting all their fish, but CR4 and CR10 drop off the pace slightly. The proportion caught by the top 5% in 1998 was two thirds the total catch, whereas the top 5% of owners owned three quarters of all quota. This indicates a small shift in effective share down the line to smaller operators.

Figure 1
Percentage of total inshore finfish quota owned by top 1, 3, 4 and 10 owners and by top 5% of owners

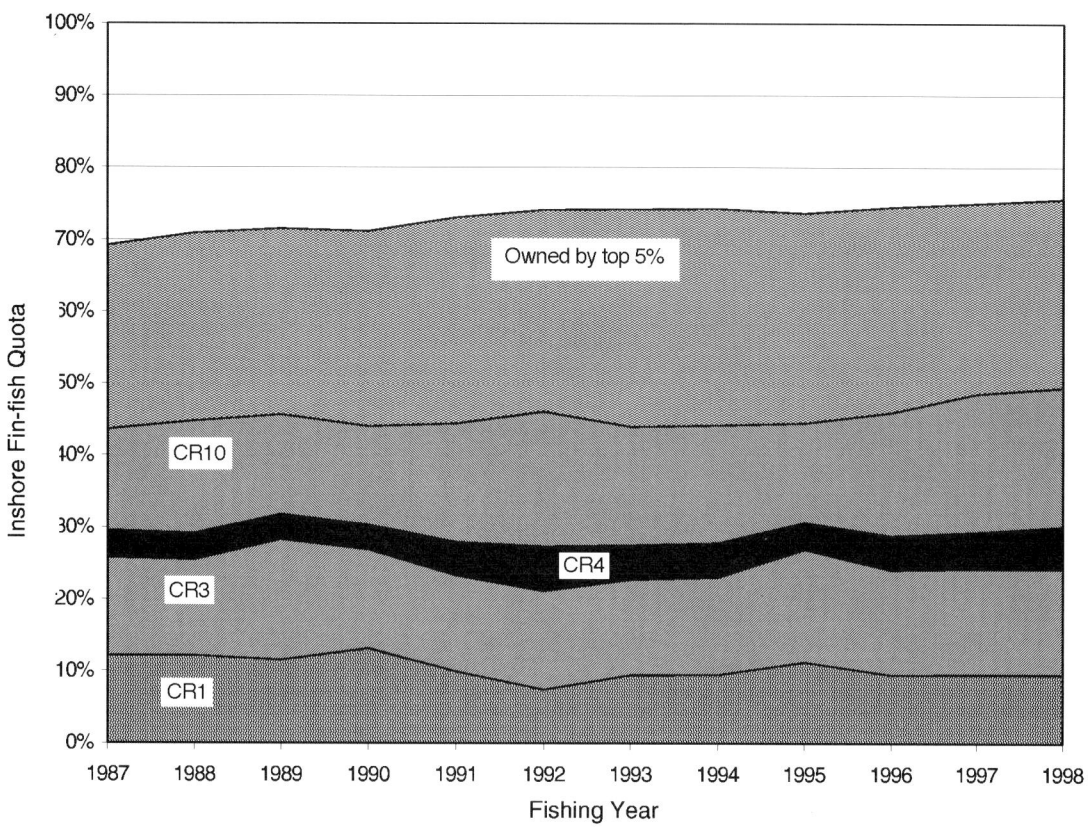

Figure 2
Percentage of total mid-depth finfish quota owned by top 1, 3, 4 and 10 owners and by top 5% of owners

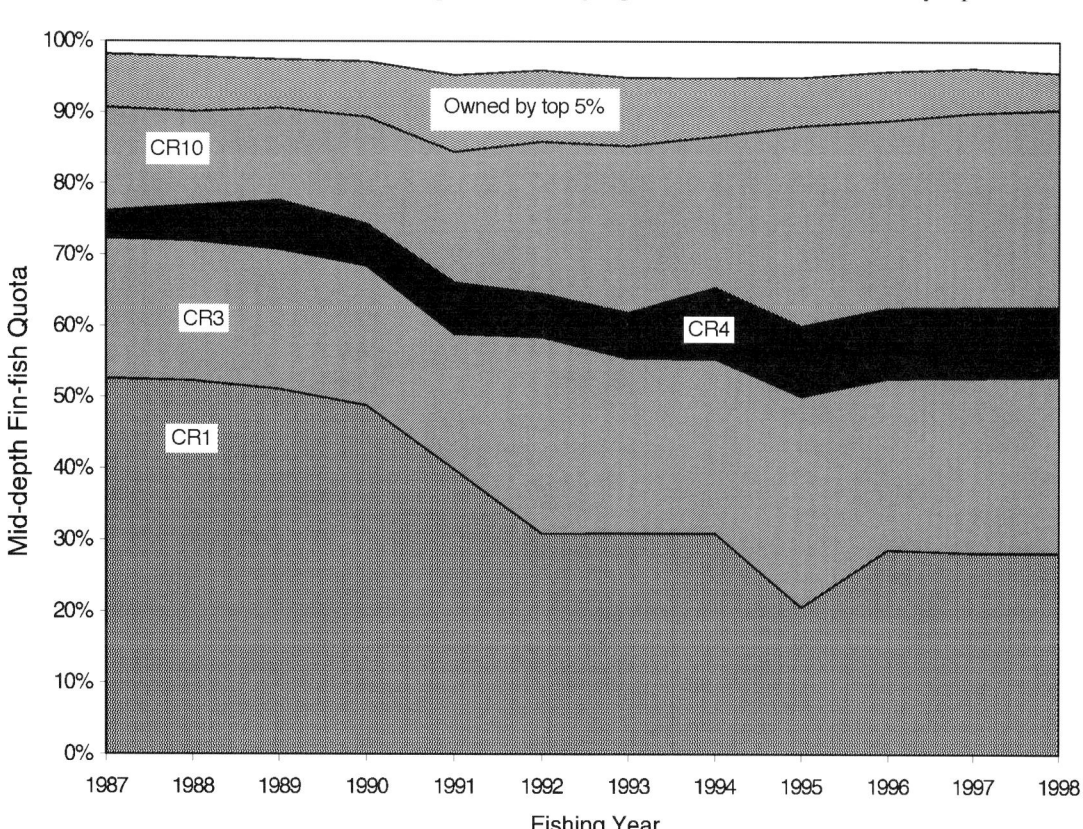

Figure 3
Percentage of total deep-water finfish quota owned by top 1, 3, 4 and 10 owners and by top 5% of owners

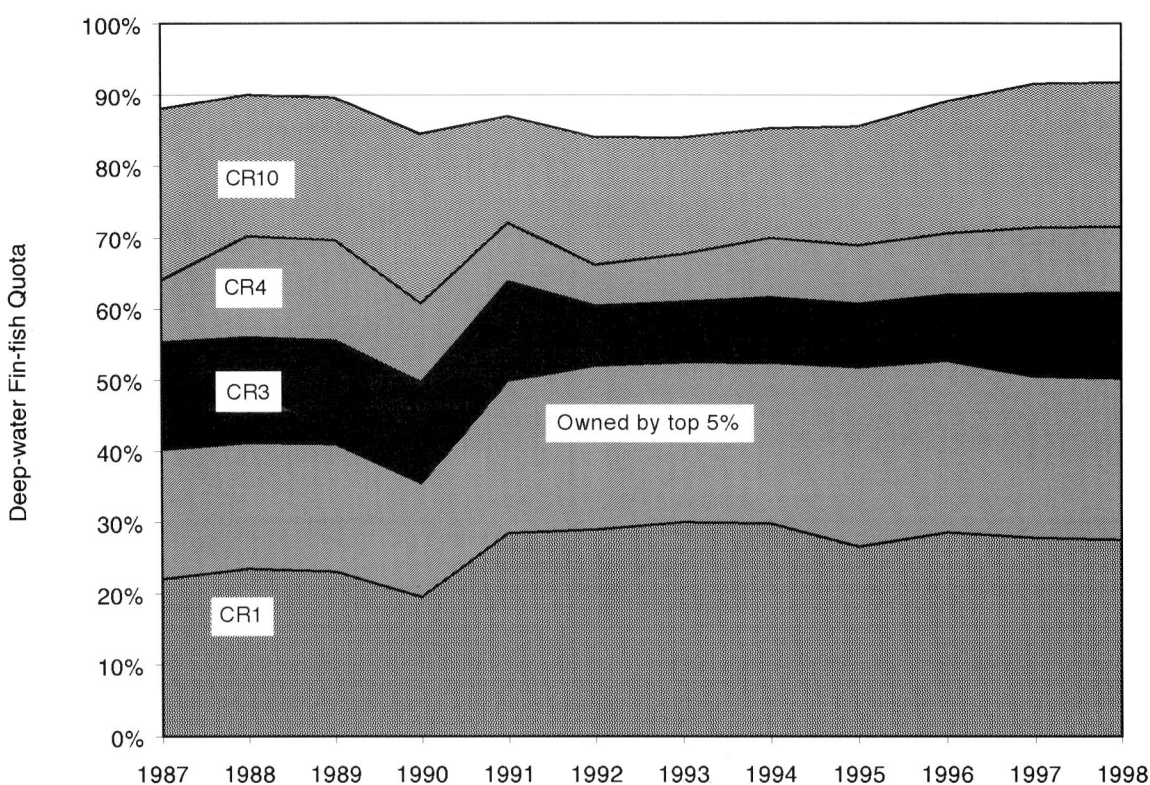

Figure 4
Percentage of total rock lobster quota owned by top 1, 3, 4 and 10 owners and by top 5% of owners

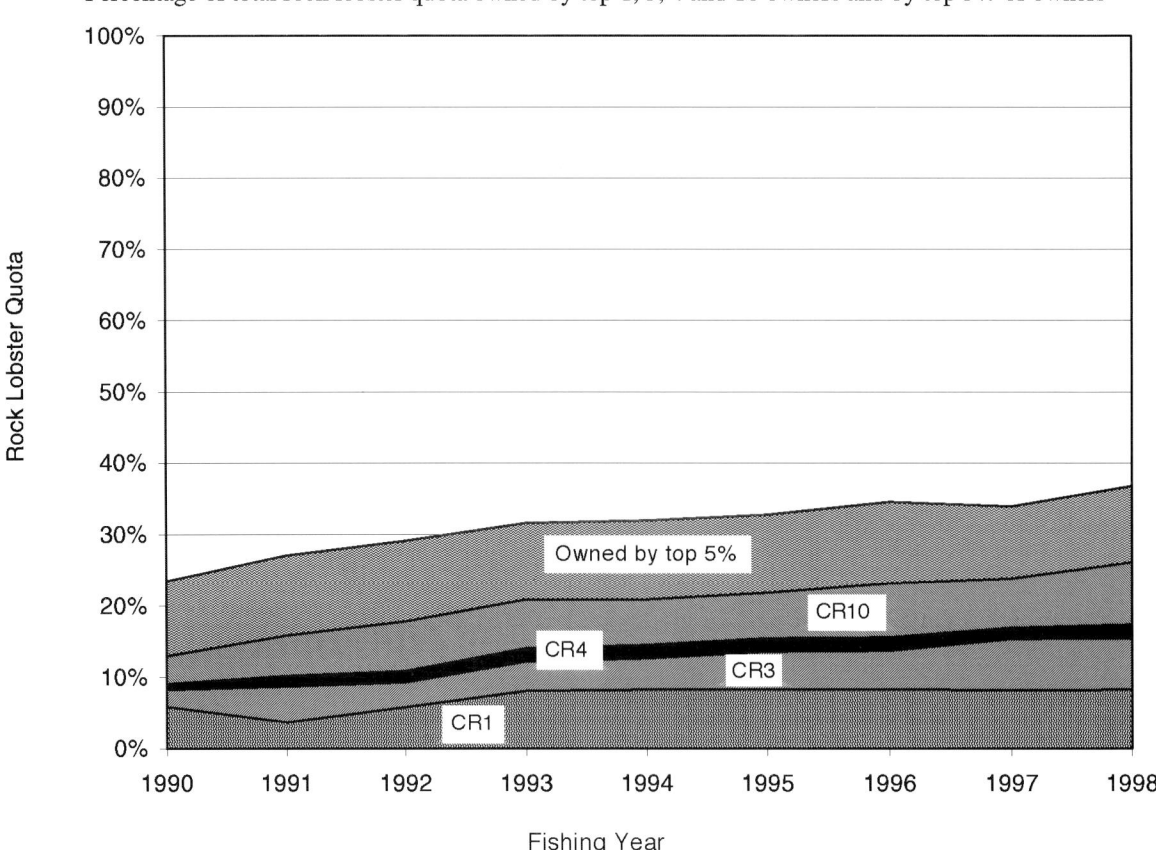

5. MID-DEPTH SPECIES

This group of species includes: barracouta, hake, hoki, ling, gemfish, silver warehou, and blue warehou.

The ownership of quota in this group is highly concentrated in comparison with the inshore species, reflecting the small number of companies with sufficient capital to participate in the bulk hoki fishery. The figures for the first four years of the quota system are confounded by the fact that the Crown (New Zealand Government) owned large amounts of quota that was leased out. Hence in this early period the CR1 is 53% which would have been illegal for any owner other than the Crown under the prevailing aggregation limits of 20% for inshore fin-fish and 35% for deepwater and mid-depth species. In 1992, the CR1 reached 31%, as the government got out of quota ownership with the move to proportional quota, and has stayed about there since. The number of quota owners has dropped by 25% over the period and those with less than minimum holdings (5t) have dropped by a third, accounting for most of the overall reduction. The HHI is about 14% and the GI is 97%, both remained stable since the exit of the Crown from ownership. The CRs are all high with 17 of 360 quota owners having 95% of the quota in 1998.

The relativity of end-of-year holdings to owned quota is similar to that of the inshore. Numbers of holders are about 10% up on numbers of owners, and all the indicators show that the quota is spread around a little more, with CR10 dropping from 90 to 82% indicating about 30 000t of quota is been leased out from this group of owners. The HHI drops from 14% for owned to 10% for "held". The effect of Crown ownership on the indices can be seen in the contrast between owned and "held" figures for the early years. The HHI for "held" in 1987 was only 11%, almost the same as in 1998, as are all the concentration ratios.

The catch indices show a further dilution of concentration with respect to holdings and ownership, with the biggest holders taking the losses. However, the top 5% of those reporting catch controlled 90% of the catch.

6. DEEPWATER SPECIES

This group comprises orange roughy and the oreo species.

A similar small number of companies controls 95% of these fisheries as is the case in the mid-depth stocks. The nature of the fishing limits the participation to large vessels and there are a total of around 40 owners, a number that has remained relatively static over the period. However, ownership has become more concentrated among the top four, with CR4 moving from 64 to 72% while CR10 has increased less than 4 percentage points. Of deepwater quota 95% is held in 13 accounts. The HHI has increased from 12.5 to 16% over the period. Despite this high and increasing concentration, the deepwater quota is more evenly spread among the owners than the other categories according to the GI, due to the limited scope for small players. The top 5% of owners (2) had 50% of the quota in 1998 with a GI ranging 72 to 80% over the period.

Holdings show somewhat different patterns to the other groups relative to owned quota. Numbers of participants are 25-30% higher, but the HHI barely moves, and the CRs are all about the same as for ownership. This pattern is likely to reflect the leasing of small amounts of quota to cover incidental catches in mid-depth fisheries, rather than any attempt by non-owners to target these species. Both catches and TACCs have fallen by about 40% over the period. Thirteen accounts reported 95% of the catch in 1998, and the other indicators are almost identical to ownership.

7. ROCK LOBSTER

This fishery has a relatively large number of participants with small tonnage holdings as it is a high-value, small-boat, near-shore fishery. Rock lobster was introduced into the QMS in 1989-90, and TACCs had fallen 20% in total by 1998. Total participants fell by a third from 686 to 470 over the 9-year period. This has doubled the HHI, but it is still very low at 1.4%. TOKM is the biggest owner, but the CR10 was only 26% in 1998. However, this has also doubled since 1989-90. The proportion of all quota owned by the top 5% has increased from 23 to 37% and the GI has ranged from 49% in 1990 to 58% in 1998, indicating a relatively low, but increasing, inequality in parcel sizes among quota owners. The average holding for rock lobster is around 6t.

The number of participants holding quota at the end of year is almost identical to ownership, but indices of holdings show large amounts of leasing by the big owners. This reflects both the fact that TOKM is the top owner and the nature of the fishery; it would be a busy lobsterman that brought in 250t in a season. The holding concentration ratios are all about half the ownership values and have remained stable over the period. A full 78% of holders are included in the group with 95% of the quota at year's end.

8. SUMMARY OF QUOTA CONCENTRATION

Most sectors assessed showed increasing concentration in quota ownership, but large shifts in distribution were not noted. All three fin-fish sectors began with large proportions held by the top few owners, which did not change. The highest concentration of quota ownership at the start of the system was in the mid-water species-group where the large hoki fishery dominates. The concentration indices for this sector decreased with time due to the exit of the government from quota ownership, but figures of EOY holdings were static over the period, as were ownership indices after 1992. The lowest concentration found was in the rock lobster fishery, where small parcels of quota are comparatively evenly distributed, reflecting the practicalities of the fishery. A couple of large holders, including TOKM which holds quota in trust for Māori, distort the indices somewhat. The deepwater quota is held by a small number of large interests, and this is reflected in the high HHI and a GI which is relatively low compared to that for the mid-depth species. The deep-water species had the highest HHI in 1998.

Consistently, EOY holdings and catch were less concentrated than ownership, with the exception of deep-water species. The major holders in the inshore and mid-depth sectors are leasing out about 10% of their quota to others and where there are shortfalls in catch these tend to be borne by the larger holders. In rock lobster half the quota of the top ten owners is leased out of the group.

9. FLEET CAPACITY

9.1 Methods and data

New Zealand fleet data for the period from the implementation of the quota management system (QMS) has been examined for indications of trends in capacity.

From data supplied by the Ministry of Fisheries, vessels are categorised as domestic, chartered, or foreign. Vessel dimensions include gross registered tonnage (GRT), which is used here as a proxy for fishing capacity[1]. GRT by length-class is summarised for domestic vessels from 1987 to 1998, and GRT by flag-state for charters (foreign vessel data are not yet assessed). The data are somewhat error-ridden with some significant gaps, although this improved in later years. For vessels with no recorded GRT, the averages of recorded GRT for the relevant length-class from the same year were used as estimates. After tracking many gross entry errors (most in the under-10m classes where they are more obvious), these averages are remarkably consistent, with any changes being smooth trends. Pre-QMS data is still under assessment and present more problems. Changes are likely in the fleet capacity across the boundary where the QMS was introduced, particularly for the inshore where total allowable catches were reduced considerably through a quota buy-back scheme.

9.2 Results of capacity assessment

Preliminary results are presented in Figures 5 and 6 for capacity, and in Figures 7 and 8 for catch. Total GRT in the domestic fleet in 1998 was up 43% on 1987. The big increases are in the larger classes, and these are described below.

Numbers of vessels less-than-10m LOA dropped substantially from 1850 in 1987 to 1050 in 1998, some 43%. This shed 2200 from 4000t for these classes, with most coming from the largest (8.5-10m) vessels. These 800 exiting vessels (under-10m) account for virtually all (93%) of the net exits from the fleet and represent nearly 30% of total number of vessels in the domestic fleet in 1987. However, the tonnage lost in these classes amounts to only 4% of 1987 totals. The number of vessels (10-15m) has been fairly static for the first 8 years and lost about 20% of their tonnage since 1994, another 4% of the total. Numbers and total tonnage of vessels in the 15 to 25m range have changed little. Vessels between 25 and 40m LOA have increased over 70% in tonnage and 62% in number. In the 30-35m class the average vessel tonnage has increased markedly.

The vessel classes with the greatest growth for the period were the 40-45m and 60-70m. In the 40-4m class, there were 4 vessels in 1987 and 18 in 1998. Average tonnage of these vessels has also increased by 50%, so total GRT has increased from 1600 to over 11 000. This distortion in the fleet is caused by a licensing rule that excludes vessels over 43m in length from many inshore areas. Vessels in the 60-70m class are now the largest vessels in the domestic fleet: there was 1 vessel in 1987 and 12 in 1998. Again, average tonnage has increased by 50% and total GRT has increased from about 1400 to 23 000. The data recorded up to ten vessels in length classes greater than 70m in the years since the implementation of the QMS, but all have now gone. Assuming these were part of the domestic fleet and are not entry errors, this capacity has been more than accounted for by the expansion in the two classes described.

The charter fleet is still important to the New Zealand fishing industry, with 125 000GRT active during 1998. This compares with less than 80 000GRT for the total domestic fleet, but charter vessels would not generally spend all year fishing in New Zealand waters. The 1998 charter tonnage is within 4% of the total in 1987, with 1997 being the lowest total since the start of the QMS. In the interim, a huge peak of 288 000GRT was registered in 1990. The majority of this (176 000GRT) was Russian (possibly reflecting chaos in the administration of the fleet following the collapse of the Soviet Union). The Japanese charter fleet was already declining off its peak by the previous year. From a traditional base in the Russian, Korean, and Japanese distant water fleets that have fished New Zealand waters since the 1950s and 1960s, the flag status of the charter capacity has diversified substantially since 1992, with some 20 nations now represented. Russian and Ukrainian flagged vessels still provide some 45% of charter tonnage.

9.3 Catch

Catch figures for the period have been reviewed to give some perspective on fleet changes. Total catches of quota species have increased over the period by some 30%. Inshore catches have increased by 60%, about 21 000t, while catches of mid-depth and deep-water species have risen 27%, or just over 100 000t. Of the inshore increase, a 43% is accounted for by one species, red cod, for which the catch has increased two and a half times over 1987. This species is subject to large variation in recruitment, which accounts for much of this seemingly dramatic increase. However, catches of all inshore quota species have increased over the period, many by substantial proportions.

In deeper waters, the net change is more than accounted for by the 100 000t increase in the hoki catch. The jack mackerel and ling fisheries have developed strongly, with new long-liners in the ling fishery accounting for some of the expansion in capacity. Big declines have been posted for gemfish, especially in the South Island where the catch has dropped from over 5000t in the early eighties to the point where there is no

---

[1] Discussion of methods for capacity estimation and the suitability of GRT as a proxy is not undertaken here due to time constraints and the preliminary nature of this report.

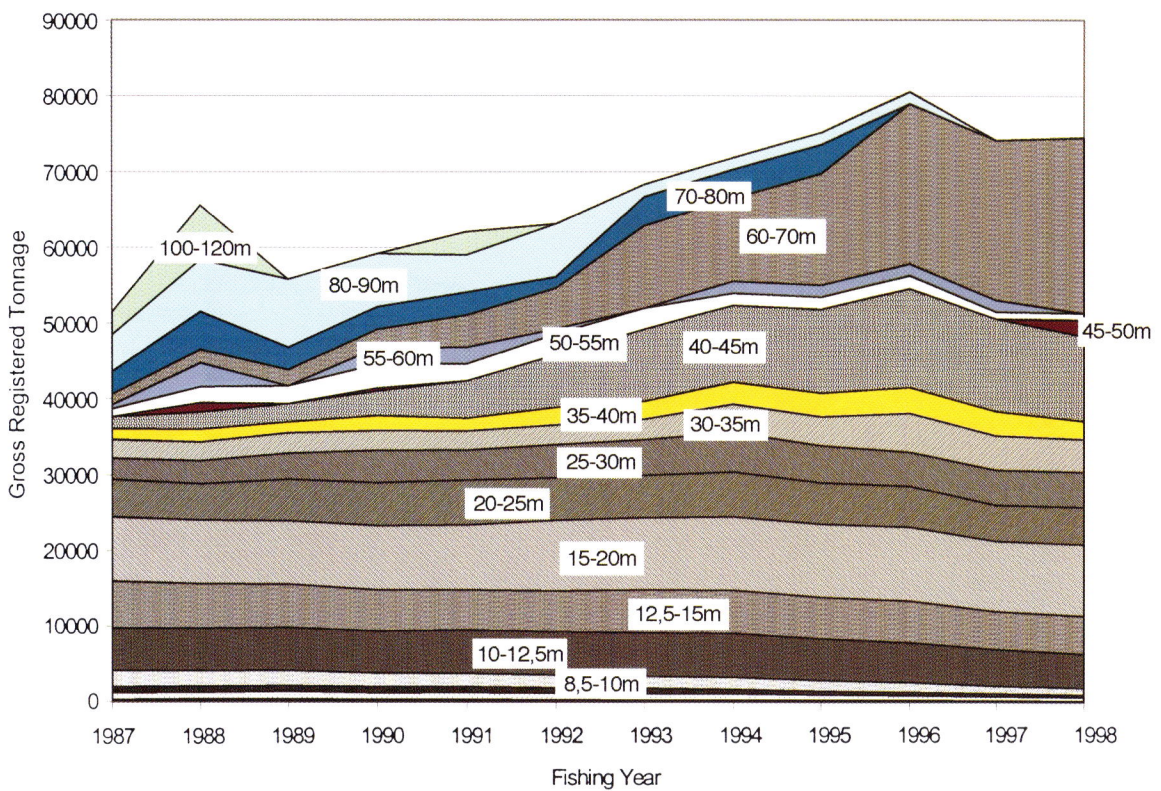

Figure 5
New Zealand domestic fleet: gross registered tonnes by size class (m)

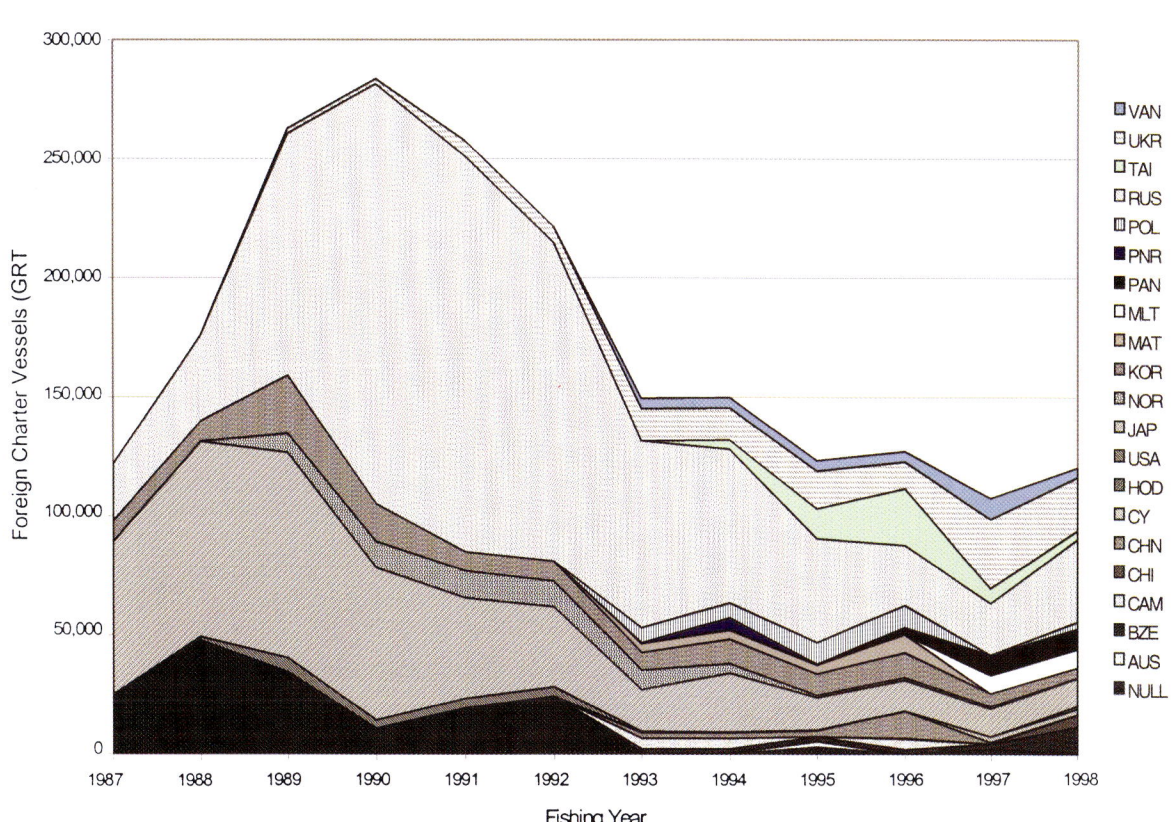

Figure 6
New Zeland foreign charter fleet: gross registered tonnes by flag state

Applying Rights-based Management 276

Figure 7
New Zealand ITQ Fisheries: Allowable and Actual Catch (t)

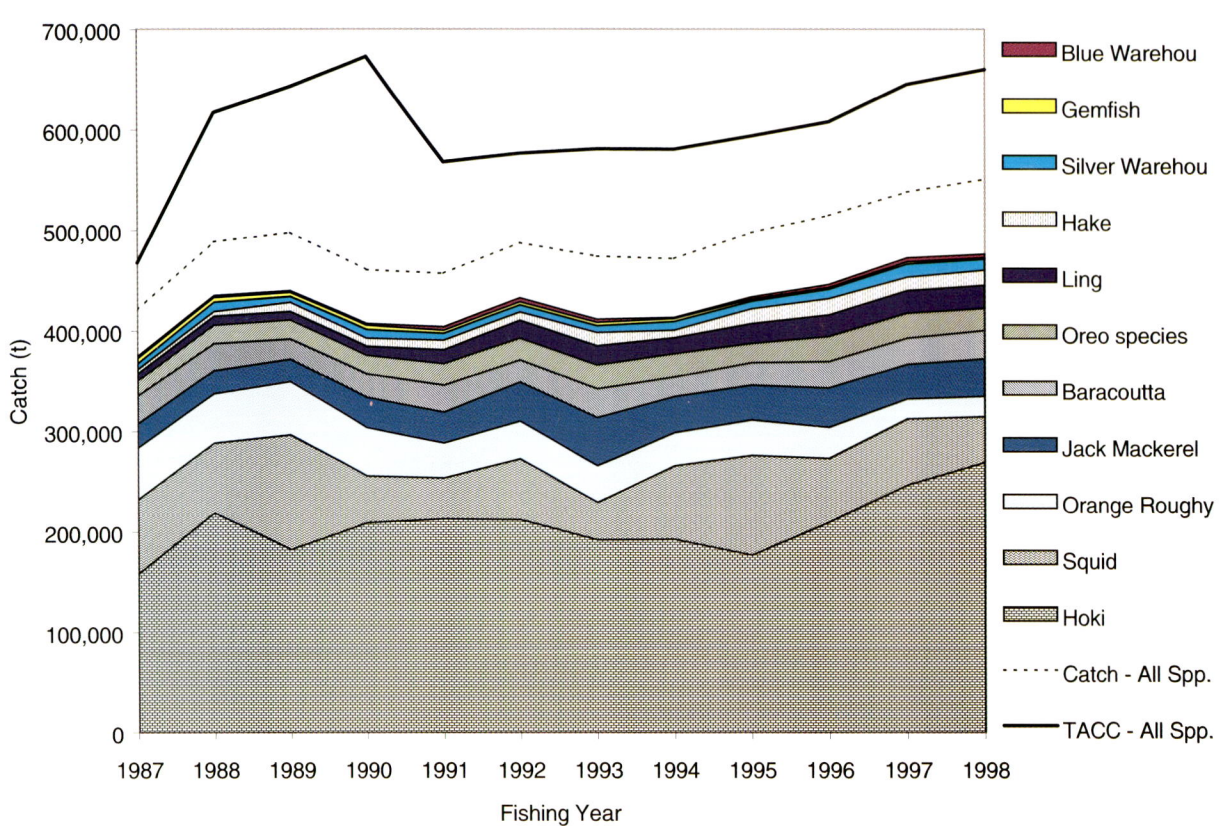

Figure 8
New Zealand inshore ITQ fisheries: allowable and actual catch (t)

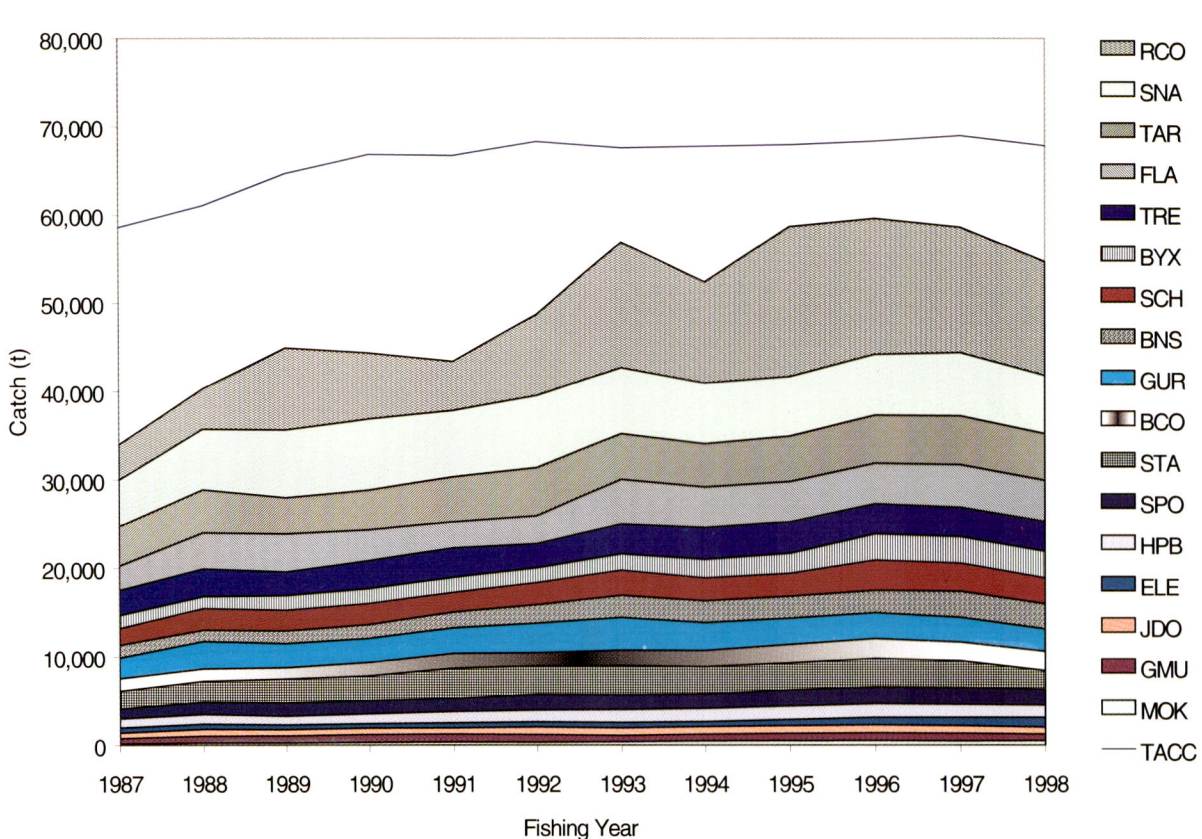

longer a target fishery. Orange roughy catches have also declined steadily, despite the serial discovery of new grounds. Catches of roughy in 1998 were 39% of 1987 totals, at just over 20 000t. The other main variable in the total catch over the period has been squid, which is subject to high annual variability in abundance.

9.4 Summary of fleet-capacity trends

Overall, the data indicates steady patterns of growth in both catches and in capacity of particular length classes in the fleet. The decline of the boats under 10m, and big increases in the 43m and 65m classes are the most conspicuous changes in fleet structure, with vessel tonnage in the 25 to 40m range growing more moderately over the period. A few companies are responsible for the increase in numbers of larger vessels. The 43m boats work in the Cook Strait and Hokitika Canyon hoki fisheries and other inshore areas, with some of these, and the larger entrants, replacing charter capacity. Some very large vessels, probably leased on an annual basis, were counted in the domestic fleet in the early years of the quota system, but now the 60-70m boats are the largest in the fleet.

Insufficient data are available at this stage to accurately separate the impact of charter capacity. Charter vessels are not in New Zealand waters all year round and a fuller assessment of the relationships and trends in capacity and catch will require data on how long the vessels are in-country. The expansion of the domestic fleet in the larger length classes indicates that some charter capacity is being domesticated on an ongoing basis. It has also proved difficult so far to obtain data for foreign licensed fishing, on either catch or vessel dimensions. Numbers of foreign vessels visiting New Zealand waters have declined steeply, and it is known from published catch data for important species that foreign catching has dwindled to insignificant levels in most fisheries (Annala and Sullivan 1997).

In the inshore fisheries where charter vessels do not operate, catches have been steadily expanding. While the under-15m fleet has contracted, the 25-45m fleet has experienced a 66% increase in numbers (27 vessels), and a 274% increase in GRT (14 300GRT), ably demonstrating the perverse effect that regulations such as length limits can have on investment and shipbuilding practice. This increase represents 27% of the total fleet tonnage in 1987, 19% of the 1998 total, and 62% of the net increase in tonnage over the period. As many of these vessels will be fishing hoki and other non-inshore species such as ling, it is impossible to get a true picture of inshore trends from the current data. However, while total domestic tonnage under 45m length is up 27%, the inshore catch (which does not include hoki) is up by 60%. In general this indicates, by the proxy of GRT, that capacity is not expanding as fast as catch, and it is likely that significant operational efficiencies have been achieved under quota management.

From preliminary analysis of available data from before the QMS, there seems to have been a major jump in vessel capacity with the introduction of the system, despite the buyback scheme and reduction of TACs. The under 40m fleet GRT figures increase by a third between 1984 and 1987 data. There are some problems with data between the two systems of recording that were in place at the time of the change. However, it seems likely that there was some rush to get (back) into fishing with the introduction of the new management arrangements. Since then total numbers participating have been steadily dropping again. This issue of the impact of the buyback on participation warrants further investigation.

10. CONCLUSIONS

The study, at this preliminary stage, is proving interesting from several points of view. The assessment of concentration of quota shows slow but steady reduction in the numbers of quota-holders in each sector except the deep-water. The HHI and GI indices are useful but more experience with these is needed through comparison across fisheries to get a better feel for what they really mean. This would be assisted by the adoption of standard indices across studies. The concentration ratios are easy to generate and understand and are the most useful for indicating the basic situation, especially where experience with the other idices is lacking.

Although the changes in concentration are not particularly alarming, the effect on individuals of exit from fishing can be profound. It is not possible in this type of assessment to understand what is happening at the local level. Even disaggregation of the data to the stock level may give a better indication of potential for adverse effects. Assessment of the impacts of concentration in particular fisheries will require more intensive techniques including economic modeling using price and cost data, inclusion of corporate structures and cross-ownership in the analysis, surveys of stakeholders and application of other social science methods.

Rationalisation of effort in the New Zealand fleet has been occurring, if slowly, over the period of operation of the quota system. Smaller boats, particularly in the 10m class have been disappearing, while capacity of the fleet vessels between 10 and 25m has remained static. At the same time the inshore catch has increased steadily, indicating better utilisation of existing assets. Vessels in the 25-43m range have increased markedly and the bottleneck created by the 43m rule prohibiting large boats from inshore and other key areas has distorted vessel design. At the top end, a few large vessels have moved out of the domestic fleet and capacity has expanded greatly in the 60-70m class. Chartering is still important to total effort and diversification in flag states of charter vessels indicates increasing competitiveness in the supply of capacity.

Overall from this assessment, the QMS appears to be living up to the promise of rationalisation, albeit at a somewhat more sedate pace in aggregate than some might have imagined in their enthusiasm for the concept. It is likely that in reality events have been more dynamic in particular areas than is indicated here. Much more detailed work should be done in estimating change and the effects of concentration of quota, but diminishing returns are likely to set in fairly quickly given the data available. A much richer understanding of the whole process of industry restructuring and change in the New Zealand fishing industry as a result of the quota system would be available through the involvement of industry in the research process, and could fill a substantial volume with useful and interesting lessons for managers and industry worldwide. In the interim, this study will attempt to address the gaps in the preliminary assessment of quota concentration and fleet change reported here.

11. LITERATURE CITED

Annala, J.H. and K.J. Sullivan 1997. Report from the Fishery Assessment Plenary, May 1997: stock assessments and yield estimates. Wellington, Ministry of Fisheries.

Clark, I.N. and A.J. Duncan 1986. New Zealand's Fisheries Management Policies - Past, Present and Future: The Implementation of an ITQ-Based Management System. Fisheries Access Control Programs Worldwide: Proceedings of the Workshop on Management Options for the North Pacific Longline Fisheries, Orcas Island, Washington, April 21-25, 1986, Alaska Sea Grant College Program, University of Alaska.

FAO in preparation. Case studies on the effect of Transferable Quota Rights on fishing fleet capacity and concentration of catch-quota ownership. Fisheries Technical Paper. FAO Rome.

Gauvin, J.R., J.M. Ward and E.E. Burgess 1994. "Description and evaluation of the Wreckfish (Polyprion americanus) fishery under Individual Transferable Quotas." Marine Resource Economics 9: 99-118.

Hogan, L., S. Thorpe and D. Timcke 1999. Tradable quotas in fisheries management: implications for Australia's south east fishery. Canberra, ABARE.

National Research Council 1999. Sharing the fish: toward a national policy on individual fishing quotas, National Academy Press.

Scherer, F.M. 1973. Industrial Market Structure and Economic Performance. Chicago, Rand McNally and Co.

MEASUREMENT OF CONCENTRATION IN CANADA'S SCOTIA-FUNDY INSHORE GROUNDFISH FISHERY

D.S.K. Liew
Policy and Economics, Department of Fisheries and Oceans Maritimes Region
P.O.Box 1035 - Dartmouth, Nova Scotia, Canada B2Y 4T3.
<LiewD@mar.dfo-mpc.gc.ca>

1. INTRODUCTION

With the increasing use of individual transferable quotas (ITQs) worldwide as a management regime in recent years, there has been increasing interest in the level of concentration of quota ownership in these fisheries. This paper looks at the concentration of ownership of two Scotia Fundy Inshore Groundfish fleets operating under different management regimes, the Mobile Gear fleet[1], vessels under 65 ft in length, that has used ITQs since 1991, and the Fixed Gear fleet of vessels under 65 ft that has used competitive quotas for most of this period. This report summarises the work in progress on what has happened eight years after the implementation of ITQs.

Concentration or quota ownership ('concentration') is viewed from three perspectives, at the individual vessel level, at the buyer level and at the geographical or port level of fish-landings. For the ITQ fishery, an analysis is also made on whether the quotas are being accumulated permanently or temporarily.

The Scotia-Fundy Inshore Groundfish fishery consists of approximately 3200 licenced vessels of which 438 are for mobile gear. As explained in Section 2.1 below, some mobile gear licences also have fixed gear designations and are currently fishing in the Fixed Gear[2] fishery. The Mobile Gear fleet has largely been managed by ITQs since 1991, with the initial allocation based on six stocks and six more added over the years since then. The Fixed Gear fleet was managed using competitive quotas until 1997 when the majority of the fleet (vessels under 45 ft in length) switched to community quotas and a small portion of the fleet (vessels 45-64 ft in length) changed to ITQs. These two fleets fished a number of common groundfish species and stocks.

Since 1992 there has been a major down-turn in the groundfish resource and most of the eastern Scotian Shelf demersal fisheries were closed completely. Between 1991 and 1998 the quota available to the Inshore Groundfish fleets declined by over 55%. As a result of the low quota levels, additional access-restrictions were implemented following the resource decline and these restrictions were still in existence in 1998.

2. QUOTA HOLDINGS BY LICENCE AND LANDINGS BY VESSEL

2.1 Mobile gear fishery

ITQs were initially allocated to 455 groundfish-mobile-gear-licences for six groundfish stocks. In brief, licence holders were given three choices: (a) to fish the IQs as allocated to them, (b) to fish from a competitive generalist pool made up of the sum of the IQs of pool members with the option to go back to the ITQ programme at a later date, or (c) to give up their IQs and fish in the competitive Fixed Gear fishery. All licence holders retained their mobile gear licences regardless of the choice they made and can come back to the ITQ programme at any time by purchasing quota. Between 1991 and 1998, 17 licences were cancelled for various reasons and as a result, 438 valid mobile gear licences remained in 1998. Of these, 329 licence holders had chosen to fish in the ITQ programme. The remaining 109 consisted of those who had chosen the Fixed Gear or Generalist option and are eligible to enter into the ITQ programme anytime by purchasing quota.

Table 1 shows the distribution of the 1998 quota-holdings of the 329 ITQ licences; of these, 74 had no permanent quota by 1998. Another 16 licences had permanent quota of less than one tonne, bringing the total number of licences with less than one tonne of permanent quota to 90, or 27%. After temporary quota transfers were included, the number of licences with less than one tonne of quota almost doubled by 53% to 175. As some of the small quota-holdings may be due to rounding when quotas were transferred, inactivity here is defined as quota-holdings of less than one tonne of groundfish.

The data showed that temporary quota-accumulation is an important factor in concentration. About 27% of licence holders have made "permanent exits" and another 26% have exited only temporarily, suggesting that the latter group may find it uneconomical to fish at low quota-levels but could potentially re-enter the fishery when the resource improves.

Figure 1 shows the cumulative distribution of quota holdings when quotas are ranked from the highest to the lowest. There are three curves; the curve on the far right shows the initial 1991 distribution, the middle curve depicts the 1998 permanent quota distribution and the curve on the left depicts the 1998 permanent plus temporary quota distribution. Note the shift from the 1998(P) curve to the 1998(P+T) curve after temporary quotas were included. At the start of the ITQ programme in 1991, 162 licencees held 80% of the quota. By 1998, the same

---

[1] The majority of the Mobile Gear fleet employed otter trawl gear, but a few vessels also fished with midwater trawls, danish and scottish seines. But, in Canada, the term "mobile gear" usually refers to demersal trawls.
[2] The Fixed Gear fleet fished using longline, gillnet, handline and automatic jigger gears.

Table 1
Distribution of 1998 quota holdings, groundfish ITQ fleet

Quota holdings (tonnes)	Permanent			Permanent and temporary		
	No. of licences	Group quota	% of total quota	No. of licences	Group quota	% of total quota
Chose ITQ						
0	74	0	0	126	0	0
< 1	16	6	0.0	49	13	0.1
1 - 10	31	129	0.6	26	100	0.5
10 - 30	54	1 102	5.1	15	308	1.4
30 - 50	30	1 200	5.6	16	614	2.8
50 - 100	44	3 189	14.7	20	1 538	7.0
100 - 200	54	7 715	35.5	26	3 780	17.2
200 - 300	18	4 579	21.1	22	5 476	24.9
> 300	8	3 797	17.5	29	10 160	46.2
Sub-total	329	21 718[1]	100.0	329	21 989[1]	100.0
Remaining	109			109		
TOTAL	**438**			**438**		

[1] The total quotas in these two columns are not equal due to inter-fleet transfers.

Figure 1
Cumulative quota holdings of groundfish ITQ licences, 1991 and 1998

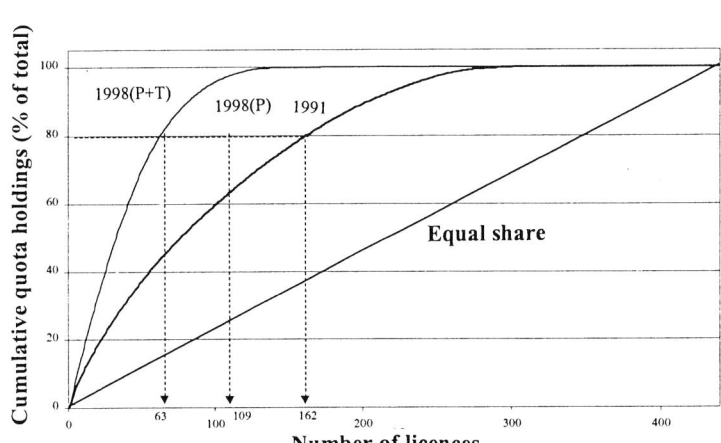

percentage was held permanently by 109 licencees and when temporary quotas were added, the number of licences was further reduced to 63.

Quotas are attached to the licence in the ITQ fishery, but multiple licences can be fished by a single vessel throughout the year by transferring the new licence with quota to the vessel after the quota on the previous licence has been caught. Besides quota transfers from licence to licence, stacking of multiple licences on to a vessel can result in fewer active vessels than licences with quota. Thus an important perspective is catch by vessel, instead of quota holdings by licence. Figure 2 shows the cumulative catch of the mobile gear vessels in 1998 compared to 1990. As the licence holders who opted to fish in the Generalist Pool or Fixed Gear sector can come back to the ITQ programme at any time, they have been included in the total licensed population. The Generalist licences fished using mobile gear, so their catches were included in both 1990 and 1998. Vessels that fished in the Fixed Gear Programme would not have any mobile gear activities in 1998 but would have recorded mobile gear landings in 1990, the year before the ITQ programme. These catches were included in the 1990 catch data and as a result, some of the observed increase in concentration was also caused by the choice made by these licence holders to go to the Fixed Gear category.

As expected, the figure shows that concentration of catches from fewer vessels occurred between 1990 and 1998. The number of active vessels (i.e. groundfish landings over 1t) declined from 343 to 146, a decline of 57%, and the number of vessels that accounted for 80% of the

Figure 2
Cumulative catch of groundfish mobile gear vessels under 65 ft, 1990 and 1998

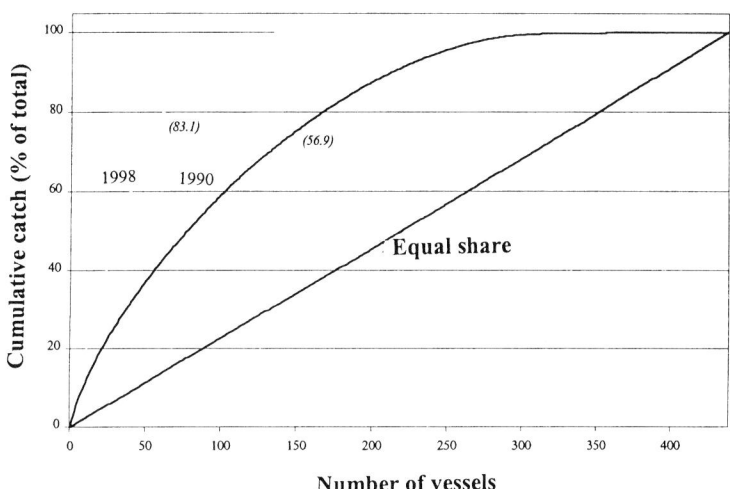

fish caught by mobile gear decreased from 166 to 61, a reduction of 63%.

2.2 Fixed gear fishery

Figure 3 shows the cumulative catch graph for the Fixed Gear fleet. It is evident that there had also been an increase in concentration in the Fixed Gear fishery between 1990 and 1998. The number of active Fixed Gear vessels decreased from 1660 in 1990 to 795 in 1998, a decline of 52%, and the number of vessels that accounted for 80% of the fish decreased from 445 in 1990 to 253 in 1998, a decrease of 43%.

3. METHODOLOGY TO MEASURE CONCENTRATION

3.1 The Gini Index of Concentration

The Gini concentration index is a measure widely used in the measurement of income inequalities. According to Needleman (1978), the most-frequently used summary measure of the degree of inequality of income distributions is the Gini coefficient of concentration. Examples of the applications of this index have included the measurement of income distributions in Canada (Needleman 1979), in Barbados (Holder and Prescod 1989) and

Figure 3
Cumulative catch of groundfish fixed gear vessels under 65 ft, 1990 and 1998

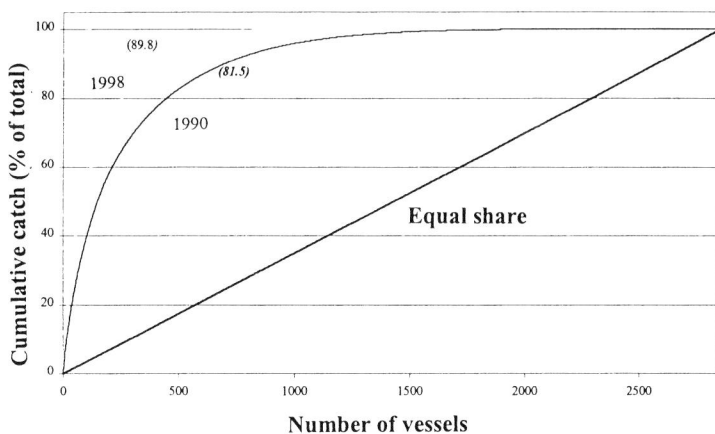

in a 12-country study (Berrebi *et al.* 1987). The literature on the Gini index has included studies on the characteristics, variations and different methods of computation of the index.

Before defining what the Gini index is it is useful to first consider the Lorenz curve. The Lorenz curve shows the proportion of total income received by a given (bottom) proportion of the population, *i.e.* it is the cumulative-income curve when incomes are ranked from lowest to highest. When incomes are equally distributed, the Lorenz curve will be a straight line across the diagonal. The further away the Lorenz curve is from the diagonal, the more inequality there is in the income distribution. The Gini coefficient (or index when multiplying by 100) is graphically represented by the area bounded by the Lorenz curve and the equal share line, divided by the total area below the equal share line.

In this application, I have inverted the Lorenz curve when the cumulative catch curves in Figures 1 to 3 plotted with catches ranked from highest to lowest, instead of from lowest to highest. The *Gini* index in this application is equivalent to the Area X divided by the total area of the upper triangle in Figure 4. This index has a scale of zero to 100; zero indicates no concentration and 100 indicates maximum concentration. When all vessels have equal catch, the cumulative catch curve is depicted by the diagonal line labeled A in Figure 4. As the level of concentration increases, the curve shifts from the right to the left. Curve C is more concentrated than B or A. In the limiting case when all the fish are caught by one vessel, the cumulative catch curve would be represented by the straight lines on the left and top side of the rectangle. The index in this case would be 100.

In practice, it may be impossible for one vessel to catch all the quota and the true maximum scenario would be the straight line (line D in Figure) defined by the minimum number of vessels required to harvest all the catch. The area bounded between this line and the equal-catch line would then be the true maximum. In this paper, no attempt is made to estimate the minimum number of vessels, and the one-vessel case is used as the maximum possible.

Note that this index only tracks concentration within the fleet, *i.e.* how vessels within the fleet stack up against each other. It does not take into account differences in the absolute level of catches by vessels across different fleet sectors.

3.2 Concentration indices of mobile gear and fixed gear fleets at the vessel level

The concentration indices for the Mobile and Fixed Gear fisheries were calculated using the above methodology and are shown in parenthesis in Table 2. Based on these indices, the Fixed Gear fishery is relatively more concentrated than the Mobile Gear fishery - the 1998 Fixed Gear index was 89.8 compared to 83.1 for the Mobile Gear. The concentration indices were much higher in the Fixed Gear fishery due to the higher percentages of in-active licences and also the more uneven distribution of catches among the active vessels. For example, in 1990, the top 10% of active vessels in Fixed Gear caught 53% of the fish while the top 10% in Mobile Gear only caught 29%. As noted in the previous section, the higher concentration index in the Fixed Gear fleet does not imply that the same number of Fixed Gear vessels caught more fish than Mobile Gear in absolute terms. It only

Figure 4
Cumulative catch at various concentration levels

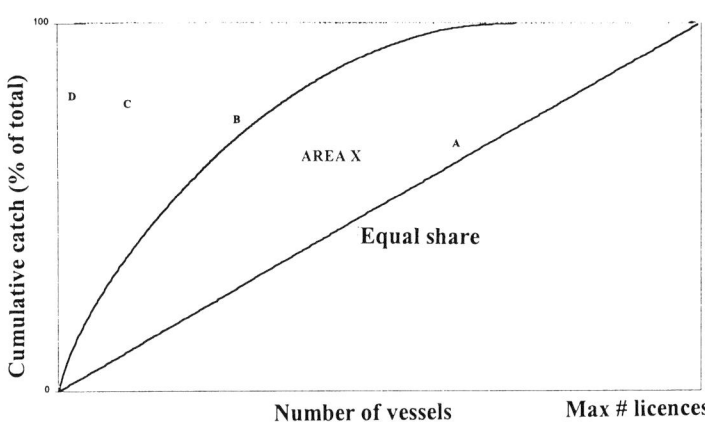

means that the same percentage of the Fixed Gear vessels caught a higher percentage of the catch compared to Mobile Gear.

3.3 Methodology to measure the change in concentration

The Mobile Gear index increased from 56.9 points in 1990 to 83.1 points in 1998, an increase of 26.2 points. Using conventional arithmetics, one could calculate the percentage change as an increase of 46%, which is the

26.2-point increase divided by the 1990-index of 56.9 points. Similarly, the increase in Fixed Gear would have been 10%, which is the difference between the 1998-index of 89.8 points and the 1990-index of 81.5 points, divided by the 1990-index of 81.5 points. But examination of the detailed data gives a different picture. The numbers in Table 2 help put this into perspective. The number of active vessels in the Mobile Gear fishery between 1990 and 1998 decreased by 57% - this com-

"room" to increase. This raises the need for another index, one that measures the change over time and is based on how much "room" there is left to increase. I call this proposed index the *Concentration-Change-Measurement* index.

As noted, the relative change between two points in time depends on the index at the starting-point. Higher starting-point indices do not have much "room" to in-

Table 2
Comparison of catch concentration indicators for groundfish mobile and fixed gear vessels under 65 ft

Cumulative catch	Mobile gear			Fixed gear		
%	No. vessels		%	No. vessels		%
	1990	1998		1990	1998	
50%	78	31	(60%)	148	83	(44%)
80%	166	61	(63%)	445	253	(43%)
100%[1]	343	146	(57%)	1660	795	(52%)

[1] Includes only vessels that caught over 1 tonne of groundfish.

pares to the 52% decrease for Fixed Gear. At the 50% cumulative catch level, the decrease for Mobile Gear was 60% compared to Fixed Gear's 44% and at the 80% level, the decline in Mobile Gear was 63% compared to 43% in Fixed Gear.

As illustrated above, calculating the changes using the 1990-indices as the bases does not give the true comparison of changes in the two fleets. This is because of the difference in the starting points of the index. In the case of Fixed Gear, a high starting index in 1990 means that the absolute concentration index does not have much

crease in absolute terms. For example in the Mobile Gear sector, the 1990 vessel concentration index was 56.9 and the index has to increase by 43.1 points to get to 100. By comparison, the Fixed Gear index of 81.5 only had to increase by 18.5 points to get to 100. If the starting-point index was used as the base, then the Mobile Gear index would have to increase by 76% (43.1/56.9) but Fixed

Table 3
Percentage of groundfish landings (mobile gear, vessels under 65 ft) purchased by buyer, 1990 and 1998

1990		1998	
Company	% of mobile gear landings	Company	% of mobile gear landings
A	7.0	1	8.9
1	5.0	2	7.8
5	4.4	3	6.8
6	3.5	4	6.6
B	3.1	5	6.4
7	2.8	6	6.2
2	2.8	7	6.1
C	2.8	8	6.0
D	2.6	9	4.9
E	2.5	10	4.3
Subtotal	36.5	Subtotal	64.5
Others (144 buyers)	63.5	Others (67 buyers)	36.0
Total (154 buyers)	100.0	Total (77 buyers)	100.0

Gear only has to increase by 23% (18.5/81.5) to get to maximum concentration of 100 points. This is not very meaningful for comparative purposes. A *Concentration-Change-Measurement* index could be based on a maximum increase of 100% regardless of the starting-index level. It would measure the actual increase against the "room" left to increase. For example, in the case of the Mobile Gear sector, the maximum "room" was 43.1 points and the increase between 1990 and 1998 was 26.2 points. Out of a maximum of 43.1 points, the 26.2-point increase represents 61%.

Note that this change-index tracks approximately the relative changes in the percentage decline in the number of vessels at various levels of cumulative catch. In the Mobile Gear sector example, declines at the 50%, 80% and 100% cumulative catch levels ranged from 57-63% and the *Concentration-Change-Measurement* index calculated for Mobile Gear of 61% lies within the range. Similarly, the declines at the 50%, 80% and 100% cumulative catch levels ranged from 43-52% in Fixed Gear sector and the *Concentration-Change-Measurement* index calculated for this fleet was 45%.

4. PURCHASES BY BUYERS
4.1 Mobile gear fishery

In 1990 there were 154 buyers for the fish caught by mobile gear and by 1998, there were only 77, a decline of 50%. At the 80% cumulative level, the decline was even higher – 47 buyers bought 80% of the fish in 1990 compared to 18 in 1998 or a decline of 62%. Applying the concentration methodology to buyers, the buyer concentration index was estimated to increase from 66.3 to 86.9

actual names of the buyers are not shown, instead, a coding system based on the 1998 Top-10 ranking was used. Rankings higher than 10 were not done. For example, Company 1 was the buyer that purchased the most fish caught by mobile gear in 1998 and Company 10 ranks number 10 in 1998. The 1990 list indicates the company's 1998 ranking numerically, but companies not in the 1998 Top-10 list were simply listed alphabetically in ascending order.

In 1990, the top-10 buyers purchased 36.5% of the mobile gear landings and by 1998 this percentage had increased to 64%. The largest buyer purchased about 9% of the total mobile gear landings in 1998, and seven out of the ten buyers were in the 6-8% range.

4.2 Fixed gear fishery

The number of buyers of fish caught by Fixed Gear decreased from 156 in 1990 to 119 in 1998, or a decrease of 24%. The buyer concentration index increased from 63.6 to 70.4 and the change-index was estimated to be 19%. Table 4 shows the distribution of fixed gear groundfish purchases by the top 10 buyers in 1990 and 1998. The top-10 companies purchased 36.4% of the fixed gear fish in 1998, up from 28.3% in 1990.

5. LANDINGS BY PORT
5.1 Mobile gear fishery

In 1990, 101 ports recorded landings from mobile gear and by 1998 this number had dropped to 53, a de-

Table 4
Percentage of groundfish length landings (fixed gear, vessels under 65 ft)
purchased by buyer, 1990 and 1998

1990		1998	
Company	% of fixed gear landings	Company	% of fixed gear landings
A	4.2	1	4.5
B	4.0	2	4.4
6	2.7	3	4.1
C	2.7	4	4.1
D	2.7	5	3.9
E	2.6	6	3.7
F	2.4	7	3.4
G	2.4	8	3.1
9	2.2	9	2.7
4	2.2	10	2.5
Subtotal	**28.3**	**Subtotal**	**36.4**
Others (146 buyers)	71.7	Others (109 buyers)	63.6
Total (156 buyers)	100.0	Total (119 buyers)	100.0

over this period and the change-index was estimated to be 61%.

The distribution of groundfish purchases mobile gear by the top-10 buyers in 1990 and 1998 are shown in Table 3. To protect the confidentiality of the buyers, the

cline of 48%. Eighty percent of the landings from mobile gear were made in the top-23 ports in 1990 and by 1998, that same percentage was landed in the top-9 ports. The geographical concentration index for Mobile Gear increased from 74.1 in 1990 to 91.0 in 1998. Based on the proposed change-index methodology, this was an increase of 65%.

The top-10 ports ranked by landing quantities in 1990 and 1998 are shown in Table 5. As it is possible that some of the ports may have only few buyers or vessels, actual names of the ports are not shown to protect the confidentiality of the players. Instead, a coding system similar to that used for buyers is used.

methodology to measure the relative change in the concentration levels between two time periods was proposed and applied to these fisheries. This index, called the *Concentration-Change-Measurement* index was based on the increase in the *Gini* index as a percentage of how much "room" there is left to increase. These indices were applied to vessels, buyers and ports for the years 1990 and 1998. The proposed methodology for measuring changes in concentration over time appears to provide a reason-

Table 5
Percentage of groundfish landings (mobile gear, vessels under 65 ft) by port, 1990 and 1998

1990		1998	
Port	% of mobile gear landings	Port	% of mobile gear landings
1	13.8	1	31.7
5	7.4	2	12.4
3	6.3	3	6.9
A	6.3	4	6.6
2	5.7	5	6.5
6	5.0	6	5.4
B	5.0	7	5.0
C	3.0	8	4.2
D	3.0	9	3.8
E	3.0	10	3.6
Subtotal	58.5	Subtotal	86.1
Others (91 buyers)	41.5	Others (43 buyers)	13.9
Total (101 buyers)	100.0	Total (53 buyers)	100.0

The top-port in 1998 accounted for 31.7% of all mobile gear landings. This port was also the top-port in 1990, but it accounted for only 13.8% of the landings then. Besides increased concentration among fewer ports, there had also been some shifts in the landings by port as some ports gain in importance while others lose out. Only 5 of the top-10 ports in 1990 made the top-10 list in 1998.

5.2 Fixed gear fishery

The number of ports with fixed gear landings decreased from 188 in 1990 to 151 in 1998, a decline of 20%. Eighty percent of fixed gear landings were accounted by the top-30 ports in 1998 compared to 39 in 1990. The Fixed Gear geographical concentration index increased from 75.4 to 80.7 and concentration was estimated to increase by 22%.

The top-10 ports accounted for 50.1% of all fixed gear landings in 1998. This was up marginally from 38.6% in 1990. The relative importance of the ports appear to be more stable between 1990 and 1998, 8 of the top-10 ports in 1990 were still in the top-10 ports in 1998.

6. SUMMARY AND CONCLUSIONS

Concentration indices based on the *Gini* concentration index were calculated for two Scotia Fundy Inshore Groundfish fleets, the Mobile Gear and the Fixed Gear. A

able measure of the relative changes in the number of participants for the two fisheries studied.

The concentration indices based on *Gini* and the *Concentration-Change-Measurement* methodology are summarized in Table 7. The number of vessels, buyers and ports at the 50%, 80% and 100% cumulative landings levels in 1990 and 1998 along with the relative changes between these two years are also shown.

At the vessel level, the number of active Mobile Gear vessels declined from 343 in 1990 to 146 in 1998, a reduction of 57%. Fixed Gear vessels also recorded a significant decline, from 1660 to 795, or a reduction of 52%. The concentration index increased from 56.9 in 1990 to 83.1 in 1998 in the Mobile Gear sector while in the Fixed Gear sector it increased from 81.5 to 89.8. Based on the *Concentration-Change-Measurement* methodology proposed, it is estimated that the relative increase in concentration between 1990 and 1998 was 61% for Mobile Gear and 45% for Fixed Gear.

The number of active buyers in the Mobile Gear fishery declined by half over the 1990 to 1998 period, from 154 to 77. By comparison, the number of buyers from Fixed Gear vessels declined by only 24%, from 156 to 119. A greater difference was noted in the changes in

Table 6
Percentage of groundfish fixed gear < 65' landings by port, 1990 and 1998

1990		1998	
Port	% of fixed gear landings	Port	% of fixed gear landings
5	6.3	1	6.7
1	5.0	2	6.6
3	4.3	3	6.1
4	3.7	4	5.4
6	3.7	5	4.9
9	3.5	6	4.6
8	3.3	7	4.2
A	3.3	8	4.2
B	3.0	9	4.1
10	2.5	10	3.3
Subtotal	**38.6**	**Subtotal**	**50.1**
Others (91 buyers)	61.4	Others (43 buyers)	49.9
Total (101 buyers)	**100.0**	**Total** (53 buyers)	**100.0**

Table 7
Concentration indices of groundfish mobile gear and fixed gear < 65' length, 1990-1998

Indicator	Mobile gear			Fixed gear		
	1990	1998	Relative change (%)	1990	1998	Relative change (%)
Vessel						
No. of active vessels	343	146	(57%)	1660	795	(52%)
No. of vessels that caught 50%	78	31	(60%)	148	83	(44%)
No. of vessels that caught 80%	166	61	(63%)	445	253	(43%)
Vessel concentration index	56.9	83.1	61%	81.5	89.8	45%
Buyer						
No. of active buyers	154	77	(50%)	156	119	(24%)
No. of buyers that bought 50%	17	8	(53%)	22	17	(23%)
No. of buyers that bought 80%	47	18	(62%)	47	41	(13%)
Buyer concentration index	66.3	86.9	61%	63.6	70.4	19%
Port						
No. of active ports	101	53	(48%)	188	151	(20%)
No. of ports that land 50%	8	3	(63%)	16	10	(37%)
No. of ports that land 80%	23	9	(61%)	39	30	(23%)
Port concentration index	74.1	91.0	65%	75.4	80.7	22%

the buyer concentration indices between 1990 and 1998, with Mobile Gear at 61% compared to Fixed Gear's 19%.

The number of ports receiving mobile gear landings declined from 101 in 1990 to 53 in 1998, or 48%. The decline for Fixed Gear landing sites was lower at 20%, from 188 to 151. The difference in the concentration indices in these two fisheries was even higher, with Mobile Gear at 65% and Fixed Gear at 22%.

For the Mobile Gear fleet, a significant amount of the quota accumulation was in the form of temporary quota-holdings. This could indicate that some of the vessels might have exited only temporarily and could conceivably return to the fishery when the resource conditions improve.

Finally, it should be noted that the study covers a period when there had been major resource-declines and the complete closure of fisheries for most eastern Scotian Shelf stocks. Part of the increased concentration would be due to the changes in the resource conditions. Some of the increased concentration in Mobile Gear was also due to the reduction in vessels as a result of licence holders opting for the Fixed Gear fishery. The analysis in this report basically compared the concentration levels in 1990 and 1998; the effects due to the resource decline or reduction in vessels that went to fish in the Fixed Gear fishery were not separated out. In the Mobile Gear fishery, a significant portion of the quota was accumulated on a temporary basis. It remains to be seen whether these concentration levels would prevail when the resource improves.

7. LITERATURE CITED

Atkinson, A.D. 1970. On the measurement of inequality, *Journal of Economic Theory, 2*, 244-263.

Berrebi, Z.M. and J. Silber 1987a. Interquantile differences, income inequality measurement and the Gini concentration index, *Mathematical Social Sciences, 13*, 67-72.

Berrebi, Z.M. and J. Silber 1987b. Dispersion, asymmetry and the Gini index of inequality, *International Economic Review, 28(2)*, 331-338.

Brown, J.A.C. and G. Mazzarino 1987b. Drawing the Lorenz curve and calculating the Gini concentration index from grouped data by computer, *Oxford Bulletin of Economics and Statistics, 46(3)*, 273-278, 1984.

Fisheries and Oceans Canada 1990. Atlantic Groundfish Management Plan.

Fisheries and Oceans Canada 1998. Integrated Fisheries Management Plan Atlantic Groundfish.

Holder, C. and R. Prescod 1989. The distribution of personal income in Barbados, *Social and Economic Studies, 38(1)*, 87-114.

Kondor, Y. 1975. The Gini coefficient of concentration and the Kuznets measure of inequality: A note, *Review of Income and Wealth, 21(3)*, 345.

Needleman, L. 1978. On the approximation of the Gini coefficient of concentration, *The Manchester School of Economic and Social Studies, 46(2)*, 105-122,

Needleman, L. 1979. Income distribution in Canada and the disaggregation of the Gini coefficient of concentration, *Canadian Public Policy, 4*, 497-505.

INDICATORS OF THE EFFECTIVENESS OF QUOTA MARKETS: THE SOUTH EAST TRAWL FISHERY OF AUSTRALIA

R. Connor
Centre for Resource and Environmental Studies
Australian National University, Canberra, ACT 0200, Australia
<rconnor@cres.anu.edu.au>
and
D. Alden
Australian Fisheries Management Authority
Box 7051, Canberra Mail Centre ACT 2610, Australia
<dave.alden@afma.gov.au>

This paper[1] presents some results of an investigation into potential indicators for the assessment of markets for individual transferable fishing quota where price data do not exist. The economic logic for implementing such markets and how they are expected to work is used as a basis for asking questions about how well they perform, and what might comprise evidence of problems. Given data on quota ownership, transfer and leasing, and associated catches, but no quota price data, indicators are suggested for monitoring and analysing market activity and are applied to data from the Australian South East Trawl Fishery (SETF). The focus is on aspects of the data that address issues of information asymmetry and transaction costs in particular, and briefly at the issue of the competitiveness. First we look at aspects of market participation as evidence that quota owners have sufficient information and knowledge to utilise trade when it is in their interests to do so. We also look for evidence here for the effect of transaction costs on participation. Then, the issues of quota "landlording" and quota concentration are examined for scale that might lead to the development of market power. Finally, we present an original analysis that looks at the match of catch against quota holdings and how this changes over time, again looking for evidence of possible asymmetries in transaction costs and information. The descriptive statistics and indicators show that the SETF quota market has contributed flexibility to the system. They also show that the market has assisted stakeholders to maximise their interests, given the constraints imposed by the annual total allowable catches, the allocations of quota species mix, the vicissitudes of the environment and the price of fish. The study provides an optimistic view of the health of the SETF quota market and our hope is that it can help separate the question of whether the market mechanism is working from the other issues of concern in the management of the fishery.

---

[1] This paper is to be published in *Marine and Freshwater Research*, 2001 Vol. 53.

WILL IMPROVING ACCESS RIGHTS LEAD TO BETTER MANAGEMENT – QUOTA MANAGEMENT IN THE TASMANIAN ROCK LOBSTER FISHERY

W. Ford
Tasmanian Department of Primary Industries, Water and Environment
GPO Box 44A Hobart, TAS 7001
<Wes.Ford@dpiwe.tas.gov.au>

1. INTRODUCTION

In March 1998 individual transferable quotas were introduced into the Tasmanian rock lobster fishery and this quota system has provided the industry with increased security and certainty of access. The move to quota management took place after nearly ten years of debate among industry groups and with the Government. The debate focused on what management arrangements should be used to ensure the long-term sustainability of the fishery. During the debate little attention was paid to providing better, more secure access rights and these only became an issue during the final stages of developing the quota management arrangements.

As a result of the changes to the management regime, licence holders now have certainty of licence renewal, security of future access to the commercial fishery and access rights which effectively exist in perpetuity. There is only limited scope for the Government to remove quota units from a licence holder. The improvements in the access rights were achieved indirectly through the quota-arrangement negotiations rather than through a structured design process where fishers negotiated particular outcomes.

One of the important questions now facing the Tasmanian rock lobster industry is whether more secure access rights will lead to better management of the rock lobster fishery, and if so how? While it may be too soon to answer these questions there are already some trends emerging that show the industry is changing, both in terms of ownership and long-term business planning. It should be noted that the industry is not united on the notion of strengthening the nature of the property right, many are happy with what they have and fear that investors will move in and control the fishery. A major consideration for the industry and the Government is whether or not more secure access rights lead to other management problems in the future, and if so, what might these be and how could they be avoided? These issues will need to be discussed and considered as part of any structured negotiations on the future nature of the property right associated with rock lobster quota units.

2. THE FISHERY

The Tasmanian rock lobster (*Jasus edwardsii*) fishery is managed under output controls, specifically individual transferable quotas. The current total allowable commercial catch (TACC) for the fishery is 1500t, while the recreational catch is about 5% of this. The TACC is allocated across 10 507 quota units. The allocation mechanism is phasing down to an equal allocation per quota unit and in March 2001 the quota allocations will be equal. The initial allocation saw about 9% of the TACC allocated on the basis of past catch history.

The fishery operates from 1 March to 23 February with a two month closure between mid-September and mid-November. This closure is to protect the spawning female rock lobsters and is also the time when many of the male rock lobster are moulting, and consequently are of poor commercial quality.

The fishery is managed as a single zone with size limits for male and female rock lobster, which are 110mm and 105mm carapace length respectively. There are also a number of input, or gear, controls that operate on the fishery. Fishers are limited to using 50 traps and each trap must be less than 1200mm x 1200mm x 750mm. The limit on the number of traps is to minimise the amount of octopus predation and reduce gear conflicts between fishers, and between the recreational sector. Historically fishers have been able to catch well in excess of 1500t using 10 500 traps, so there are few fishers who would wish to see the trap limits increased or removed.

The fishery extends from water less than 5m in depth out to depths of 120m and up to 20 nautical miles from the coast. The fishing grounds extend from the far north-west of the State around to the far north-east. Figure 1 shows the jurisdiction of the fishery (Anon. 1997).

3. WHY QUOTA MANAGEMENT WAS INTRODUCED

In the mid to late 1980s the industry and scientists had become increasingly concerned about the sustainability of the fishery following the doubling of fishing effort in a little over ten years, while catches were remaining relatively stable. Given the fishery had been exploited from the middle of last century and there had been heavy fishing pressure during the 1960s, there was general concern that the stocks were declining.

During the early 1990s various management options were explored and debated within the industry and with government. It was apparent that there were only two options, either a substantial reduction in the fishing effort (about 30%) or the introduction of a total allowable catch. Unfortunately the industry was divided on the issue with the majority recognising there needed to be reductions in the catch and effort, but no agreement about how to do it. Finally, in August 1996 the Government decided that the

Figure 1
The area of State waters for the Tasmanian rock lobster fishery and the approximate distribution of the fishing grounds

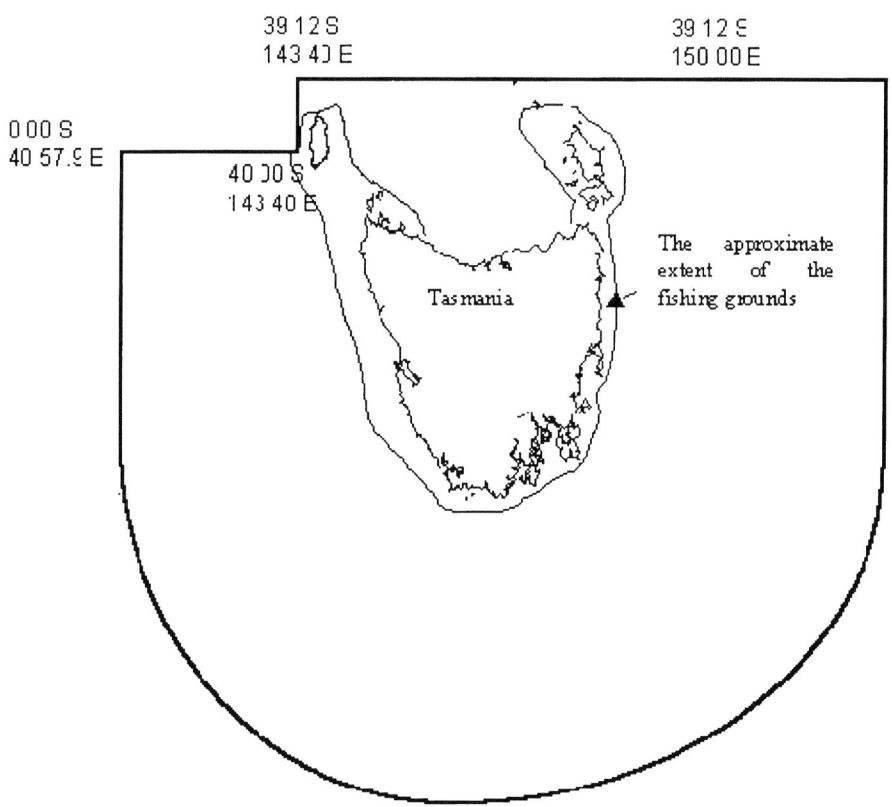

fishery would be managed by output controls and that individual transferable quotas would be introduced.

The Government had two objectives in mind, the first was to reduce the catch to a level which would be sustainable, and allow the biomass to rebuild over time. The second was to provide a mechanism whereby the industry could restructure and allow those who wished to leave the fishery to achieve a reasonable return on their access rights. It was recognised that whatever management option was adopted, reducing the catch would inevitably lead to less fishers participating in the fishery, and a more profitable industry. The Government, supported by a majority of the industry (about 75%), concluded that quota was the better option for achieving its two objectives.

The quota system has allowed the industry to commence restructuring. This process has seen the market value of the access rights increase from $A4000 per unit (rock lobster pot) in 1994 to $A10 000 in 1997, to about $A20 000 per unit (rock lobster quota unit) at present. While this has been beneficial to those leaving the fishery, those who have chosen to stay and invest in the fishery are paying considerable sums to buy back quota to achieve their previous catch levels.

In the first 18 months after quota was introduced it appears that the catch distribution of the fishery has changed little. It had been argued by some of the high catchers that the largely equal allocation of the quota (91% of the TACC) would mean that fishers would no longer be able to take large quantities, say in excess of 10t. Figure 2 shows clearly that this is not the case, in fact the number of licence holders taking more than 7t has remained the same, however there is a decrease in the number taking more than 11t. Given that 188t (11 %) less was taken for the November 1998 to September 1999 period, compared to the same period two years earlier, it is surprising that the number of fishers able to take in excess of 7t has not changed.

The similarity of the catch distributions shown in Figure 2 has occurred because fishers are buying or leasing additional quota units to maintain their historic catch levels. It would appear that so far the restructuring of the fleet has resulted a decline in the number of fishers who catch small quantities of rock lobster, many of these fishermen are from the older generation and took the opportunity to retire from the fishery.

Figure 2
A comparison of the distribution of the size of the catch for the last full fishing season prior to the introduction of quota and for the same period after quota management was introduced

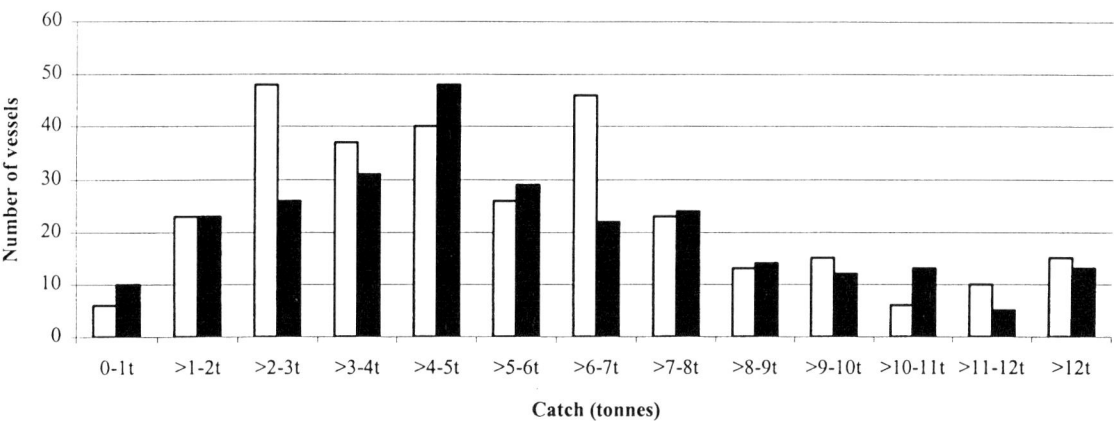

4. CHANGES TO OWNERSHIP OF THE ACCESS RIGHTS UNDER QUOTA MANAGEMENT

In October 1997 the rock lobster fishing fleet comprised 321 licences held by 294 licence holders and 308 vessels ranging in size from 6-26m in length. The majority of vessels are used primarily for rock lobster fishing but have the capacity to operate in other fisheries on a seasonal basis. The vessels are a mixture of wooden and steel displacement hull vessels, and a few fibreglass vessels. The average age of the fleet exceeds 15 years, with few new vessels operating in the fishery.

The industry is made up of fishers who own and operate licences, family operations, investors and lease holders. In January 1997, 188 licences were operated by the owner or by the nominated person if the holder was a company or partnership; 21 licences were operated by a family member of the owner, usually a son, brother, or husband. In some of these cases the operator may be purchasing the family business or has taken over the family business. At that time there were 112 licences that were leased or operated by someone other than the owner or the owner's family. New participants must buy an existing licence to enter the fishery and no additional licences are being issued. As of October 1997, 83.7% of the 321 licences were held by Tasmania-based owners. It is clear that the number of licences owned by Tasmanians has decreased under quota management, particularly as investors are now considering buying licences with less than 10 quota units.

The restructuring of the industry has seen the number of vessels drop from 308 in 1996/97 to 270 in 1998/99. Many of the 38 vessels that left the fishery were operated by skippers under lease arrangements. Unfortunately, such fishers are the most vulnerable particularly if the capacity of a fishery is reduced and such licence holders need to re-invest in their future.

The ownership of the access rights, in terms of the holder of the licence, can be determined from the licensing records. It is difficult to determine if the holder of the licence is in fact the beneficial owner, that is, the person who ultimately owns the "property" of the licence. At this stage the Government does not collect information on beneficial ownership for the rock lobster fishery. The licences can be held by individuals, partnerships or companies. Where the licence is held by a company, trust or partnership, the holder must nominate a natural person under Section 77 of the *Living Marine Resources Management Act 1995*.

In October 1997 the minimum holding was a licence with 15 rock lobster pots, the maximum was 50 pots per licence with no limits on the number of licences. Under quota management the minimum holding is 5 rock lobster quota units and the maximum is 200 quota units and no more than 100 can be held on one licence. The change in the distribution of the access units is shown in Figure 3. The nature of the holding of the access rights has changed under quota management, fishers must now hold, or lease, quota to cover their catching requirements. This has resulted in a doubling of the number of licence holders who hold more than 50 units, which has happened because these fishers are the ones who have historically taken the larger catches, so they have had to re-invest in access rights to allow them to continue their fishing operations at former levels.

Figure 3
A comparison of the distribution of fishing units 6 months prior to the introduction of quota management and 18 months after the quota commenced

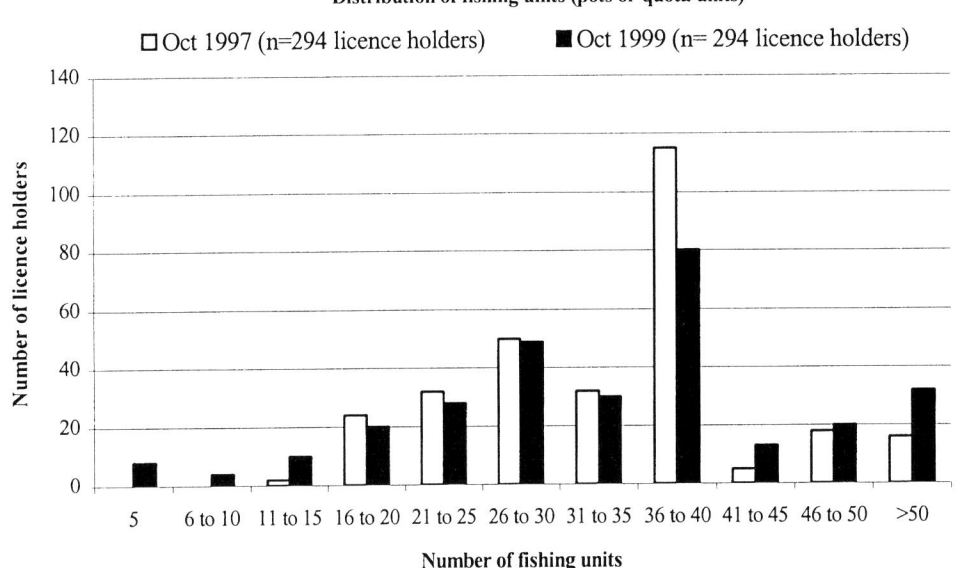

At the other end of the spectrum licence holders with small quota holdings increase. This group is made up of two types of licence holder: the first are vessel skippers who are trying to buy into the fishery, the second are investors looking for a sound investment with a good return. Currently the quota units are trading at between $A18 000 and $A20 000, with a return from leasing of $A1200 (6%-6.5%) after licence fees are deducted.

5. ROCK LOBSTER QUOTA UNITS AS AN ACCESS RIGHT

The rock lobster fishing licence and attached rock lobster quota units have the following characteristics:

i. The rock lobster quota units were allocated in perpetuity to the people who held commercial rock lobster licences by means of legislation (*Living Marine Resources Management Act 1995*). The quota units provide exclusive access to take 1/10,507 of the TACC. The quota units can be transfered free among the 315 licences. The licences and attached quota units are fully transferable.

ii. The licence is an annual licence which must be renewed by the Minister as long as (a) the licence fees are paid, and (b), the person has not been convicted of a relevant offence under a law of another State, Territory or the Commonwealth. If the licence renewal is refused under (b) the licence holder has a right of appeal to an independent appeal tribunal. Case history suggests that the refusal to renew a rock lobster licence on these grounds would require a serious offence in a rock lobster fishery. Therefore, this provision cannot be used for minor breaches of other fisheries laws.

iii. The Act provides that only the fishing licence is forfeited if 200 demerit points ($A20 000 in fines) are reached in any five year period and the rock lobster quota units are not forfeited. As the quota units are where the asset value lies, the licence holders investment is protected from forfeiture. While a licence holder who is convicted of the offences that resulted in the demerit points would be excluded from holding a licence for 5 years, they would be able to sell or transfer their quota units, thereby retaining the asset value of the quota. A licence holder who leased the licence to another fisher would not be prevented from buying another licence to transfer the quota units to.

iv. The Act requires that the holders of commercial fishing licences be allocated the rock lobster quota, regardless of the instrument that creates the licence. This means a new licence-type cannot be created to transfer the ownership of the quota units to a new group of licence holders.

The above three characteristics effectively mean the rock lobster quota units and rock lobster licences are issued to the licence holder in perpetuity, and they cannot forfeit the asset value if convicted under State law. However, as with any criminal proceeding, profits from crime can be seized by the Crown.

6. PROPERTY RIGHT OR ACCESS RIGHT

The concept of property rights applied to fisheries seems to mean different things to different people. It is increasingly apparent that the term property right is being applied to a spectrum of access rights. This spectrum ranges from access rights that can best be described as tenuous, through the range of licences to a private property right (ECS 1997). Simply put, the property right attached to any access right to a fishery represents the security of tenure and access, the tradeability and value associated with any particular regime.

A licence to fish which is effectively issued in perpetuity, provides security of tenure, and is tradeable, may well be a strong property right. This property right may be further strengthened by providing access to a fixed share of the annual harvest, by providing some form of instrument that goes beyond a licence (Deed, contract or statutory fishing right) (ECS 1997). It may also be strengthened by providing a mechanism whereby the access right can be used as collateral for financiers to lend against.

7. WHAT DOES THE INDUSTRY WANT

The rock lobster industry is seeking a strengthening of the property right nature of the access rights. However, at this stage it is unclear what is being sought aside from providing a register of ownership that allows interested parties to register their financial interest. What is not clear yet is how far the characteristics of such a register should go, that is, ranging from an ability to register an interest, to effectively providing a mortgage system, with a form of title that can be held by a mortgagor.

A simple register would allow a bank to register an interest in a licence that should then satisfy a common law right to have an existing claim over the licence should the licence holder try to sell the licence. A system whereby a bank could effectively take a mortgage over a licence would work in the same way banks mortgage houses or land.

The industry and the Government will be discussing this issue over the next year. The industry's main reason for wanting licence holders to be able to register an interest in rock lobster licences and their associated quota units is to ensure money can be borrowed against the licence. Historically, banks have only lent up to a percentage of the market value of the licence, often not more than 60%, and then only if there was considerable other collateral or guarantors. As the price of buying into this fishery, and other similarly well-managed Australian fisheries, increases it becomes harder for new entrants, particularly young fishers, to afford to buy a licence.

Providing a means whereby there is more incentive for financiers to lend funds equal to a greater portion of the value of the asset should make it easier for fishers to continue to buy licences and quota, rather than corporations. However, it can be argued that providing even more secure property rights will drive the price of access up further. In the long-term it is reasonable to expect that the price of the access rights in a sustainable, well-managed fishery, will reflect the investment potential with returns comparable to long-term bond rates.

At this stage the Tasmanian rock lobster industry is not seeking to replace their fishing licences with something that may have a greater property right value. Many in the industry feel they have a strong property right in their current licences and quota units. However, it is likely the industry will soon consider such options, following the Commonwealth's statutory fishing right process. There are those in the industry who are concerned that such changes will lead to investor-control and less scope for fishers to own their own access rights.

8. BENEFITS OF STRONG PROPERTY RIGHTS TO THE MANAGEMENT OF THE RESOURCE

As more fisheries look at strengthening the nature of the property right that governs access to a fishery, the benefit to the management of the resource should be considered. If the nature of the property right, or the ownership of those rights, actually puts more pressure on the stock, is strengthening the property right in the public benefit? Fisheries managers, whether they be government or industry, must accept that they have a responsibility to manage fish stocks for future generations, such that they have the same, if not a better, opportunity to harvest fish. In economic terms this means the net present value of the stock should be high, and remain the same over time.

Achieving long-term sustainability and inter-generational equity may be at odds with the nature of the property right, particularly if the property rights are held by corporations who must provide a high return to shareholders. There may be an economic disincentive to invest in the resources when growth rates are low, that is, when the grow rate is less than the economic discount rate it makes economic sense to catch the fish sooner, or in extreme cases 'mine' the stock. However, fishers, even those with large investments in a fishery, tend not to act in an economically rational way, as they are often concerned about leaving the fishery for their sons and grandsons.

Providing a stronger property right, which encourages investors, may lead to more 'rational' economic behaviour, which may increase the pressure to harvest a larger portion of the fishery now if growth rates are low.

It has been commonly suggested that improving the property rights will lead to better management as licence holders accept greater responsibility for management. However, is this always going to be the case or are there other factors that determine the success or otherwise of management? One function of good management must certainly be the number of participants compared to the catch and its value. The fewer the participants, the greater the earning capacity of individual fishers or licence holders. A fishery which has few players should result in the licence holders taking a strong interest in the sustainable management of the resource, and will generally be easier because each participant can take sufficient fish to ensure the viability of their business. Similarly, fisheries with few participants should be cheaper to manage as fewer resources are required to monitor and service the industry.

While it is difficult to compare fisheries because of factors relating to fishing costs and price differences, it is worth considering the Australian rock lobster fishery as an example. The Tasmanian Rock Lobster Fishery has an annual TACC of 1500t and is divided among 294 licence holders and must provide for a fleet of 270 vessels, their running costs and maintenance. The average gross return is about $A153 000 per licence holder, based on an average beach price of $A30/kg for *Jasus edwardsii*.

The Western Australian Rock Lobster Fishery (*Panulirus cygnus*) lands on average about 10 500t of Western rock lobster, at an average price of about $A20/kg. This means that the average gross income for the 596 licence holders is about $A350 000. The average number of rock lobster pots is about 116, which have a market value of at least $A20 000 per pot, or $A2.3 million (Donohue and Barker 1999).

The South Australian Southern Zone Rock Lobster Fishery, which takes the same species as the Tasmanian fishery, and competes in the same market, landed 1685t in 1997. This catch was divided among 184 licence holders (SASZRL-FMC 1999). The average gross return would have been in the order of $A275 000 per licence holder. These licences are worth about $A21 000 per quota unit, with an average of 65 units per licence, putting the value of a licence at about $A1.4M (Donohue and Barker 1999).

The comparison can be extended further when the South Australian Northern Zone Rock Lobster Fishery is considered. These fishers numbering 73 landed about 950t in 1997 (SANZRL-FMC 1999). This fishery is managed entirely by input controls and supposedly lacks the strength or property right that quota fisheries have. However in 1997 the fishery saw gross earnings of about $A390 000 per licence holder. This fishery appears to be managed sustainably, is producing good returns, has an asset value of about $A26 000 per fishing unit, but may be perceived by some as having as strong a property right as a quota fishery. What would the industry, the public or the resource gain from strengthening the nature of the property right in this fishery probably very little.

At the other extreme is the Victorian Eastern Zone Rock Lobster Fishery, again the same species and market. This fishery is in the process of considering quota management. However, the fishery is likely to have a TACC of about 75t divided among 76 licence holders (Anon - Victorian Fisheries 1996). This is likely to result in a gross return of about $A30 000 per licence holder. Is the strengthened property right that quota supposedly brings going to make it any easier to manage this fishery? Probably not. The problem is likely to be exacerbated as individuals have little to lose and may therefore be less willing to ensure any management strategy works.

Obviously, the South Australian and Western Australian fishers have a greater income, and therefore arguably a greater incentive as individuals to ensure their fishery remains well managed. One aspect of the strength of a property right is exclusivity (ECS 1997), meaning less participants in the fishery leads to a more exclusive, and therefore stronger, property right.

9. FUTURE PROBLEMS FOR MANAGEMENT

The strengthening of the nature of the property right in a well managed fishery has the tendency to increase the cost of access because it enables a higher long-term return on investment. If the return on investment is better than the long-term bond rate there will be a greater number of investors seeking Tasmanian rock lobster licences and quota to add to their investment portfolios. This inevitably will see an increase in the number of "absentee landlords" who lease out their licence or quota units. What problems may this lead to in the future?

In Tasmania we are starting to see such affects in the abalone fishery. Within five years it is likely that the Tasmanian Abalone Fishery will be owned entirely by investors who do not fish for abalone and already there are many contract divers who work for less than 15% of the beach price received. In the rock lobster fishery we already have fishers who lease licences and quota from investor licence holders.

One implication of having secure property rights for a licence or quota unit, which cannot be removed if convicted of serious fisheries offences, may be little incentive for the licence holder to ensure his lessee does the right thing. This may be more of a problem in fisheries where there are a large number of licence holders. In most instances, licence holders would have no idea what the fishers may, or may not, do. In fact, as the number of investors increases the direct contact with the actual fishers will decrease. We already see a profitable business of broking in fish quota – essentially matching quota holders with fishers and charging a fee for the service.

Under such a system, what incentive is there for lease holders to ensure they comply with the rules of the fishery? If they themselves have no licence to lose, and the quota they lease is secure from forfeiture, then the incentive to comply must be in deterrent value of the chances of being caught and the subsequent penalties. So if there is little incentive for some lease holders to only catch the quota they have leased, where they are being paid a fraction of the beach price, what are the implications for management and the resource?

If such illegal activity occurs, and even worse, increases, the pressure on the resource will increase, inevitably leading to cuts in total allowable catches. Such illegal activity undermines the asset value of the quota or licence. The response by government would probably be to increase the enforcement regime, which in turn will increase the licence fees. These are likely to be passed onto the fishers, reducing their profitability and increasing the incentive to operate illegally.

But how can the licence holder, or quota holder, control it? The answer seems to be increasing the level of enforcement, and the costs of management, which in turn reduce the nett returns to either the fishers or the quota holders. It is reasonable to conclude that the cost of enforcement in a fishery with strong property rights is likely to increase in proportion to the number of fishers, and decrease with their level of incentive to do the right thing. It is difficult to see easy solutions to some of these potential problems, however they should not be dismissed by governments or the industry, or both may suffer in the future. These issues need to be considered when management systems are being developed, as it is often easier to deal with such potential problems before they arise.

10. CONCLUSION

The Tasmanian Rock Lobster Fishery is managed by individual transferable quotas which provide a secure property right which entitles the holder to take 1/10 507 of the allowable commercial catch of rock lobster for each quota unit held. The property right of the licence is fully transferable while the property right of the rock lobster quota units are transferable between licence holders. The fact that quota units must be held on a licence, of which there are 315, means the property right has the benefit of exclusivity.

What is lacking in the property right of the licences and quota units is a suitable mechanism to allow financiers to register an interest in the licence. This issue is being addressed, with the Government who are committed to strengthening the nature of the property right in this way before January 2001.

The industry and the Government need to keep a watchful eye in case the industry changes to one where the fishery is dominated by lease holder fishers, rather than licence holder fishers. One of the significant downsides to providing more secure property rights may be the potential problems associated with absentee landlords with little incentive or ability to ensure their contracted fishers comply with the rules of the fishery. Such problems will be compounded where the fishers do not receive sufficient payment for catching the quota, and feel they need to take extra fish to make their operation pay.

11. LITERATURE CITED

Anon. 1996. Victorian Fisheries, Victorian rock lobster fishery future management position paper, Unpublished report – Victorian Fisheries, Department of Natural Resources and Environment, Victoria.

Anon. 1997. DPIF, Rock lobster fishery policy document, Unpublished report - Department of Primary Industry and Fisheries, Tasmania.

Donohue, K. and E. Barker 1999. (draft), Comparative study of quota management of rock lobster fisheries, Unpublished report - Rock Lobster Industry Advisory Committee, Western Australia.

Economic Consulting Services (ECS) 1997. Use of market mechanisms for allocation of commercial fishing access entitlements in Western Australia, Unpublished report for the Fisheries Department of Western Australia, Western Australia.

Living Marine Resources Management Act 1995, Part 4 - Division 6A, Tasmanian Statue number 25 of 1995.

SANZRL – FMC 1999. Annual report 1997-98, Unpublished report - South Australian Northern Zone Rock Lobster Fishery Management Committee, South Australia.

SASZRL – FMC 1999. Annual report 1997-98, Unpublished report - South Australian Southern Zone Rock Lobster Fishery Management Committee, South Australia.

INDIVIDUAL TRANSFERABLE CATCH QUOTA: AUSTRALIAN EXPERIENCE IN THE SOUTHERN BLUEFIN TUNA FISHERY

D.B. Campbell
David Campbell and Associates
PO Box 228 Kippax ACT 2615, Australia
<dcampbell@bigpond.au>
and
T. Battaglene
Canberra Wine Bureau
Winemakers Federation of Australia (INC)
Canberra, Australia
<tbattaglene@intereact.net.au>

ABSTRACT[1]

The use of individual transferable catch quota in Australia's southern bluefin tuna fishery led to a rapid reduction of the fishery with two thirds of the boats receiving quota leaving the fishery within two years of the management change. Although landings were reduced by more than a quarter, those remaining in the fishery enjoyed increased returns. These operator returns continued to increase, in spite of further reductions in allowable catch. In addition, when compared to that earned the year prior to the management changes, on average those who left the fishery enjoyed an increase in taxable income of over 20 per cent in New South Wales, and nearly 15 per cent in Western Australia. In addition, leavers from the fishery enjoyed an increase in capital value, which was estimated at 50 per cent in Western Australia.

Furthermore, the use of individual transferable catch quota created an institutional structure through which the Japanese gained access to Australian quota by a series of joint ventures. The cash flow resulting from tuna leasing under the joint venture was instrumental in maintaining viability for some Australian operators. In addition, the joint venture facilitated the transfer of Japanese longlining and farming technologies to Australian operators.

It is highly likely that many of these benefits would have been foregone had some other form of management, aside from individual transferable catch quota, been used to achieve the fishery restructuring and manage the fishery.

---

[1] This paper will be published in full with the same title in *Marine Policy,* **24**(2):09-117.

SHARK BAY PRAWN FISHERY - A SYNOPTIC HISTORY AND THE IMPORTANCE OF "PROPERTY RIGHTS" IN ITS ONGOING MANAGEMENT

P.P. Rogers and J.P. Penn
Fisheries WA
SGIO Atrium, Level 3, 170 St George's Terrace, Perth WA 6000, Australia
< progers@fish.wa.gov.au> <jpenn@fish.wa.gov.au>

1. INTRODUCTION

Shark Bay, a large hypersaline/marine embayment situated on the Western Australian coast between latitudes 24° 45' S and 26° 45' S and longitudes 112° 50' E and 114° 20'E (Figure 1), is the site of WA's most important penaeid prawn fishery, which has an annual value of production between $A30 million and $A40 million. Limited entry management of these prawn stocks was introduced in 1963 with initial access granted to 25 vessels. Over a period extending from 1963 to 1975 the number of vessels was gradually allowed to increase to 35 to maximize the social and economic benefits resulting from this fishery. Growth in fleet efficiency, *i.e.* increases in fishing power (Penn *et al.* 1997b) largely through improved technology, ultimately led to over-exploitation of the tiger prawn (*Penaeus esculentus*) stocks in the early 1980s. The king prawn (*Penaeus (Melicertus) plebejus*) stocks in contrast were not affected and increased production of this species compensated to some extent. The

Figure 1
Boundaries of the Shark Bay Prawn Managed Fishery

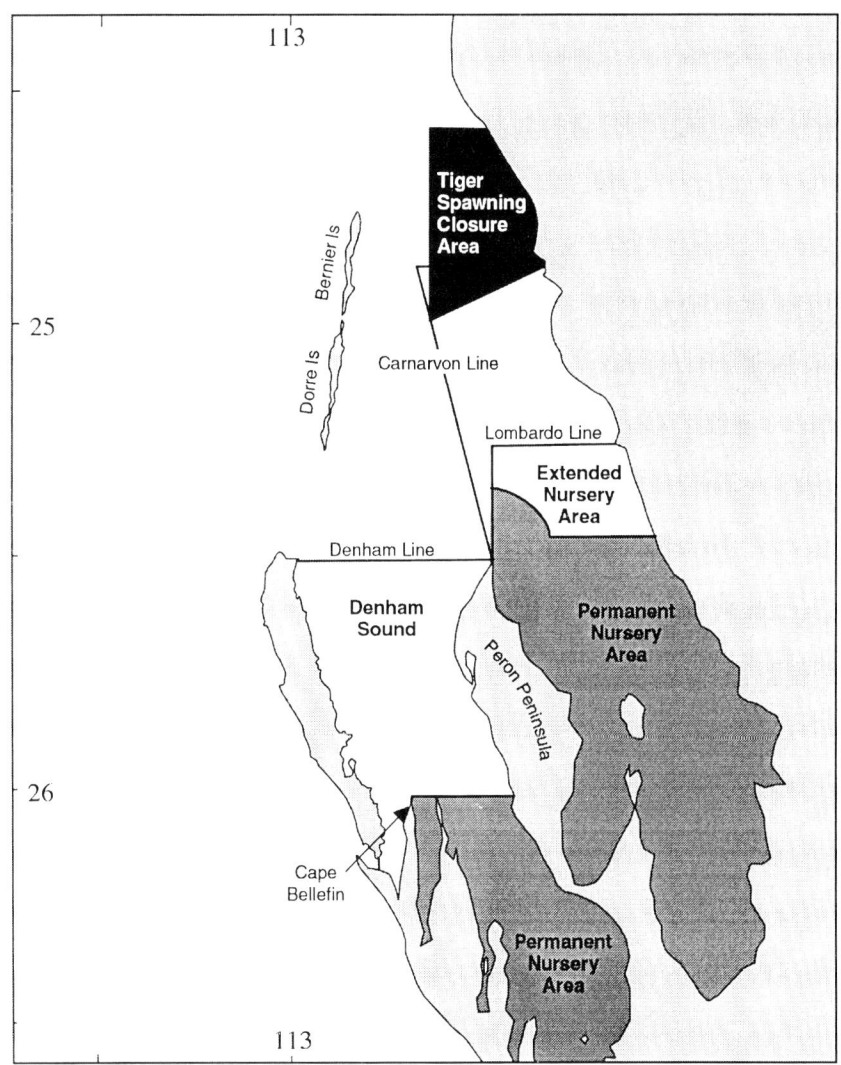

dual pressures of declining prices for prawns as world supply of aquaculture-produced prawns, expanded and lower tiger prawn catches ultimately led to a government-supported industry licence buyback scheme being implemented in 1990 at a cost of $A9.6 million. This reduced the fleet to 27 vessels for the 1990 and subsequent seasons. This buyback resulted in a substantial fishing effort reduction which, coupled with improved spatial closures, has resulted in the recovery of the tiger prawn stocks (Penn et al. 1997b) and improved returns to the industry. This paper traces some of the history regarding the management decision to vary the fleet numbers.

2. GRANTING OF ACCESS BETWEEN 1962 AND 1980

Although the presence of prawns in Shark Bay had been known since sailing ship surveys in 1905, the presence of commercially viable stocks was not confirmed until the late 1950s. Based on these latter surveys, commercial fishing began in 1962, when four vessels operated in the lower sections of the bay (Slack-Smith 1978). This delay in development was attributed to lack of processing capacity and the absence of experienced trawler skippers. This situation changed in 1962 when the Shark Bay whaling industry based at Carnarvon became uneconomic. The companies involved sought to diversify and began prawn trawling and processing using the infrastructure of the whaling industry.

The Nor'West Whaling Co. built two vessels and chartered a trawler from Queensland in 1962. They also commenced construction of a further five trawlers in late 1962 at a Fremantle shipyard. Because of the possibility of a rush of east coast vessels to Shark Bay, this company asked government for an exclusive licence to fish and process prawns in order to protect its capital investment. The discovery of large quantities of small prawns in the southern sections of the embayment also resulted in a request to close these prawn nursery areas to fishing.

The media interest in the developing prawn fishery and the likelihood of interstate vessels arriving in large numbers from Queensland resulted in the Minister for Fisheries announcing the following restrictions on 19 July 1962:

i. Fishing licences would not be issued to interstate trawlers unless they were purchased by, or under charter to, an approved local fisherman
ii. No local vessel was to fish for prawns in Shark Bay without the prior approval of the Department of Fisheries and Fauna and
iii. South of lines drawn east from Cape Bellefin access would be closed to all trawling.

In October 1962 the government further proclaimed that the number of vessels allowed to fish in Shark Bay during 1963 would be limited to 25. Incentives for the development of processing facilities were given to two companies already established in Shark Bay (Nor'West Whaling Co. and Planet Fisheries Co.) by excluding other companies from establishing shore-based processing plants. To ensure sufficient throughput to these processing plants, these companies were allocated 15 of the 25 prawning concessions, 10 to the Nor'West Whaling Co. and 5 to Planet Fisheries Co. The remainder were allocated to the rock lobster fishermen who had discovered and fished the Shark Bay prawn stocks prior to 1962. The limitation of licence access right in 1963 and later years provided an important catalytic role in encouraging risk-capital to be invested in new vessels and land-based prawn processing facilities.

The government received a number of objections to these restrictions in 1963 – in particular from fishermen already operating in the Shark Bay area. In February 1964, the government increased the number of concessions by five, three to Planet Fisheries and two to independent fishermen, one from Denham and the other from Carnarvon. This brought the total number of vessels and concessions to 30. Fishing access for these vessels was restricted to a three-year period.

Following Cabinet reviews of the Shark Bay prawn fishery in November 1966 and again in 1969 the government continued to support the retention of three-year licence periods (trienniums) but did not support an increase in vessel numbers. In 1971, the Nor'West Whaling Co. was permitted to have two additional "standby" vessels to cover for breakdowns. This situation was rationalised by issuing 32 full licences (Hancock 1975) for the 1972-74 triennium.

Following a scientific review in 1974 the then Minister approved three additional vessels on a trial basis for three years to "test" the potential for higher catches. These vessels, selected from the rock lobster fleet, were licensed for the 1975-77 triennium. These temporary licences were allocated through an administrative selection process based on criteria which partially took into account applicants' prior fishing history and access to vessels with trawling and freezing capacity. Following a review at the end of the 1975-77 triennium these vessels were issued with permanent licences and in exchange gave up their licences in the rock lobster fishery.

The approach and evaluations undertaken by fisheries administrators over this time were reported as being consistent with that subsequently recommended by Gulland (1984), i.e. to issue licences in some arbitrary number not expected to exceed the optimum number and to adjust the level of access as data became available, i.e. an adaptive management approach.

Further analysis of performance of the fishery past the 1975-77 triennium did not support any further increase in licensing as the prawn catches had levelled off over the period with changes in both nominal and effective effort. It was of note that over the period 1962-1981, while "goodwill" values emerged for vessels with Shark Bay trawl licences, these were not significant relative to the capital invested in boats and were rarely reported at that time. These goodwill payments associated with vessel transfers were not considered by the government, which only recorded the vessel transfer on its register.

The triennium review process around the granting of licences effectively lapsed at the end of the 1978-81 period when it became clear that ongoing uncontrolled efficiency-gains by the existing vessels were affecting on the stocks. Additional vessels were neither desirable nor sensible in the light of the severe decline in the tiger prawn stock in 1980 (Penn et al. 1995). Authorisations effectively became perpetual in succession.

3. THE SHARK BAY PRAWN FISHERY – 1981 to 1989

Until the 1980s, there was no documented spawning stock/recruitment relationship for penaeid prawns (Penn et al. 1995). This may be partly attributed to the lack of reliable catch and effort data in most fisheries, and to the obscuring effect of the well-documented environmentally-driven variations in prawn recruitment.

The king prawn stocks in Shark Bay continued to be fairly resilient to fishing pressure, but there was a significant decline in tiger prawns, the second most important and most valuable species, in the 1980s. Catches of tiger prawns fell from an average of about 650t/yr in the 1970s to about 300t/yr through the 1980s. The reduction in catch was attributed to heavy fishing on pre-spawning tiger prawns resulting from increases in vessel size and fishing power, and particularly to the targeting of localised areas of high tiger prawn abundance using radar (Penn et al. 1989).

The increasing importance of scallop fishing and the development of a dedicated scallop trawl fleet within Shark Bay in 1981 and 1982 further complicated management arrangements. The Shark Bay prawn management plan left the way open for a new trawl fleet to be established in Shark Bay using 100 mm mesh (not capable of retaining prawns) to target scallops. This resulted in 26 additional vessels targeting the Shark Bay scallop stocks in 1982 and the eventual establishment of a second managed trawl fishery of 14 dedicated scallop vessels under limited entry management arrangements within Shark Bay in 1983.

Resource-sharing of the scallop catch between the established prawn fleet and the specialised scallop fleet has however continued to be a matter of contention since then. All prawn vessels gained an ongoing entitlement to take scallops as part of their fishing entitlement under the Shark Bay scallop management plan. Catch-sharing between the two fleets has proved to be difficult to manage, particularly owing to the significant year-to-year variations in abundance of scallops (from 100 to 4000t meat weight). Limits on crew numbers, gear types and opening times for trawling have been crucial to achieving reasonable catch-share balance between the fleets; however, Ministerial decisions on these topics have often been required when the two sectors could not agree on the equity issues.

Spatial and temporal closures in this prawn fishery through the 1980s did not prove to be particularly successful in improving the recruitment in the tiger prawn stock, causing industry and government to re-examine the value of a prawn licence buyback programme. Spatial and temporal closures introduced during the 1980s, while not improving the tiger prawn recruitment, were however particularly successful in improving the average size and market value of king prawns. This improved value of the king prawn catch, together with improved profitability associated with greater value of the secondary catch of scallops, resulted in vessel licence goodwill values increasing from about $A100 000–200 000 in the early 1980s to about $A500 000 in 1985 and close to $A1 million by 1990.

4. THE SHARK BAY PRAWN FISHERY – 1990 to 1999

In the period of 1989 and 1990, government and the prawning industry examined the value of introducing a licence buyback scheme. There were a number of drivers for this approach:

i. The fleet, through technological and replacement vessel design improvements, had excess fishing power capacity and generated unproductive competition between vessels.
ii. Reduced fishing effort would increase prospects of a recovery of more valuable tiger prawn stocks and therefore recover catches to their historical level of about 600t.
iii. A reduced number of prawning vessels could improve catch per vessel by sharing the available king prawn and scallop catch amongst fewer vessels, thus improving industry profitability.
iv. A similar buyback arrangement in the nearby Exmouth Gulf fishery in the 1980s had generated a recovery in tiger prawn stocks and profitability.
v. There continued to be concern at the effects of world aquaculture and the recognised need for economic adjustment to offset an expected reduction in long-term prices for prawns.

One approach considered was to unitise gear entitlements and reduce gear usage; however, there was little industry support for an approach of this type owing to the relatively small number of vessels involved. Thinking among industry members at the time focused on the level of company ownership versus independent operators and considered the independents to be disadvantaged i.e. single vessel owners would have been unable to restructure their gear units to create more efficient vessel gear configurations.

In assessing the licence buyback proposal, government and the prawning industry used a financial model to ascertain a bid price for licences. The scheme was targeted at acquiring between four and eight licences under an industry-funded licence buyback scheme financed through government loan facilities. Offers by individual

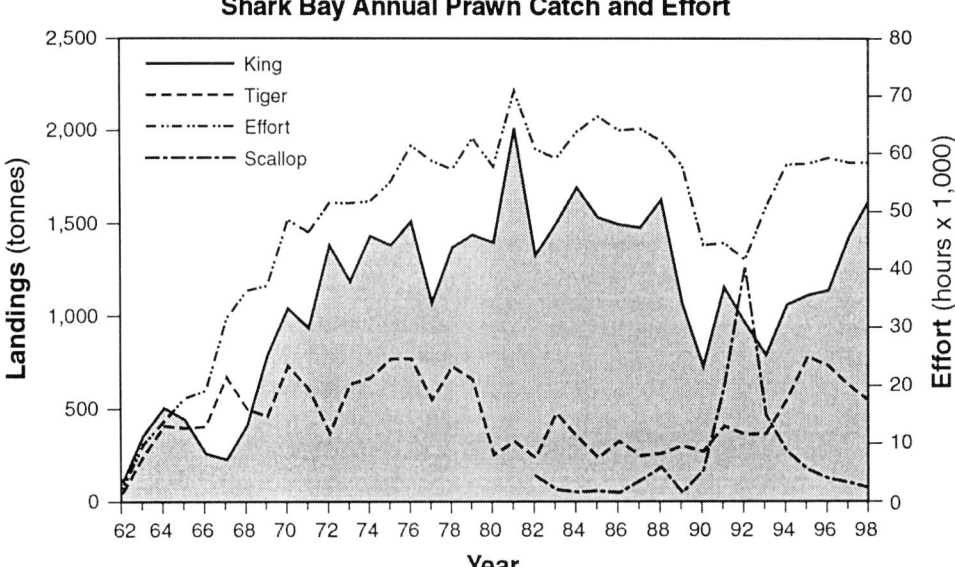

Figure 2
Shark Bay Annual Prawn Catch and Effort

industry members (vessels) to sell to the buyback scheme, *i.e.* to leave the fishery, had to be on a voluntary basis.

To gain an understanding of the financial model developed and the setting of a final offer price, all licensees with their accountants and financial advisers were invited to attend a briefing on the discounted cash flow financial model developed. Offers were then invited from all licensees wishing to leave the industry, while individual commitments by those wishing to remain and meet the financial costs of the scheme were obtained by the management agency. The members of the Licence Buyback Management Advisory Committee established at the time indicated they would not support the buyback proposal unless the clear majority of industry licensees supported the scheme. This was achieved after detailed consideration and debate among the 35 vessel owners.

In early 1990, the scheme was established with 8 licences being acquired and removed from the fishery at a total cost of $A9.6 million ($A1.2m per vessel removed) financed initially over a 15-year period. The loan arrangements were renegotiated after significant falls in interest rates in the mid-1990s, although the full costs at all times were met by industry through an annual levy on the remaining vessel licences. Full payment is expected to be completed by 2003/2004.

In the 1998 season, Shark Bay prawn catches totalled 2185t, including 1614t of king prawns, 538t of tiger prawns and small quantities of other prawns, with 75t of scallop meat taken by the prawn fleet. These levels of catches in aggregate are relatively typical of production in recent seasons, which have seen the recovery of the tiger prawn catch closer to its historical 500-600t. The value of a licence in 1999 is reported at about $A2.5 million, with the average prawn catch per vessel now exceeding 80t/boat compared with about 40t in the early 1990s.

Fishing effort targeting prawns (rather than scallops) fell substantially at the time of the buyback (Figure 2) when the abundance of scallops coincidentally increased apparently due to environmental effects on scallop recruitment (Caputi *et al.* 1998). This environmental effect also appeared to have a negative impact on king prawn catches in this period (Lenanton *et al.* 1991). However, as the fishery re-focused on prawns following a return to "normal" scallop abundance, the remaining fleet adjusted to a slightly lower nominal effort level than before the buyback. This has been due to improved catch rates encouraging the fleet to fully use all of the fishing days allowed and to improvements in replacement vessels permitting operations in all weather conditions.

The Shark Bay prawn fishery is now considered one of the more profitable prawn fisheries in Australia. With recent trends in the fishery toward further increases in vessel efficiency through GPS, better-designed replacement vessels and trawl-nets, it is not surprising that some industry members are again exploring the question as to whether further vessel reductions to improve economic performance ought to be pursued at the end of the current licence-buyback scheme.

5. LESSONS LEARNT FROM THE MANAGEMENT HISTORY

5.1 Benefits from early intervention

Licence limitation with controls on fleet capacity has provided fishing companies and individual fishermen with security to make long-term investments in better vessels and processing capacity. While this is not an unexpected outcome, experience in other, less-controlled fisheries points to excess fishing capacity quickly developing and undermining the profitability of the fishery. With early government intervention in the case of Shark Bay in limiting vessel numbers and allowing more orderly development, while not completely avoiding the problem of over-fishing, the task of stock-recovery and management to improve economic performance has been made much easier.

5.2 Increased value of licences

The goodwill values in the Shark Bay prawn fishery, tied to authorisations to fish did not become significant until the early 1980s. This was in part due to the shift from shorter-term tenures tied to ongoing triennium reviews, which later became long-term more secure access entitlements. Both profitability and security of tenure have proved to be important elements in the development of goodwill values tied to licences and industry support for management. Licence values have grown significantly in Shark Bay from about $A100 000–200 000 to $A2.5 million over a 20-year period. This growth in value appears to have resulted from improvements in profit performance as well as industry confidence in management relative to the success of other prawn fishery management arrangements across Australia.

5.3 Renewable and transferable access rights

Early exercise of control on growth in the exploitation of the Shark Bay fishery, through licence limitation and effective granting of renewable and transferable access rights, has led to more effective management capacity. Specifically, the success of the buyback initiative introduced in the early 1990s illustrates the benefits of adaptive co-management. The introduction of this arrangement could not have been achieved without the corporate involvement of industry and government in this issue. The long-term protection of access rights (licence values) and confidence in management arrangements enabled industry to collectively and responsibly consider management alternatives and create innovative solutions. Having an already-established and profitable fishery did not inhibit the Shark Bay prawning industry from considering long-term issues and innovative management responses. Nor did the high goodwill-values tied to licences at the time impede industry involvement or financing arrangements. To the contrary, other than more recent entrants to the fishery, industry had significant capacity to fund the licence-buyout arrangement.

5.4 Industry support (and voluntary participation)

The voluntary licence buyout scheme achieved wide industry support and has proven to be a workable tool for economic and biological adjustments. The voluntary Shark Bay prawn licence buyback scheme was successful due to four prime factors: (a) the scheme was voluntary; (b) had wide industry support (greater than 75%); (c) costs were fully met by industry and (d), the remaining licence holders were the ultimate beneficiaries. Consistent with other licence-buyback schemes in Western Australia, the business case for adjustment or licence-acquisition was able to be adequately demonstrated. As a result of these factors, there was virtually no political impediment to the scheme being introduced.

The financial elements of the system were as follows:

Shark Bay Prawn Fishery buy-back scheme

	Licence value	$A1.2m
	Number of licences acquired	8
	Total industry debt	$A9.6m
Licensee	Repayments (first 15 months)	$A74 464
	Repayments (next four years)	$A63 555/yr
	Annual interest rate	15.5%
	Terms-	interest fixed for 5 years and interest floated for next 10 years

5.5 Early and effective control of fishing effort

Early controls on maximum boat-sizes and maximum gear-size specifications were critical in preventing excessive vessel over-capitalisation. These controls were important elements in preventing rapid escalation of fishing effort in the prawn fishery, and allowed the technology-driven overshoot in the effort-levels relative to the tiger prawn stock to be addressed. Specific management controls on gear were introduced early in the history of the fishery and have been a feature of the fishery since the mid-1970s. The current controls under the management plan are as in Table 1.

The primary controls introduced in 1963 to protect nursery areas have essentially remained unchanged over 30 years. Other measures, *i.e.* gear controls, vessel sizes *etc.*, were introduced progressively as vessel competition increased and in the main kept fleet's catching-capacity in balance with the need to optimise exploitation to maximise catch values from larger prawns and regenerate tiger prawn spawning-stock levels.

5.6 Control on fleet fishing capacity

Fleet-capacity controls were the cornerstone to preventing a rapid blowout in fishing effort. Increases in fishing power and increased effectiveness of fishing effort as better designed boats and technology continued to improve have been adjusted for by the fleet-reduction programme. Importantly, these input management controls provided a workable basis from which industry and government could adjust fishing capacity within workable time frames. These controls, consistent with other input-based managed fisheries, have not prevented effective effort from continuing to increase, but the adaptability of the overall management system has allowed ongoing adjustments.

5.7 Use of input controls

Quotas (ITQs) and more specific statutory rights have not been considered to be a workable management tool for the Shark Bay prawn fishery either in early 1962 or today. Variable annual recruitment to the prawn stocks

Table 1
Details of regulations

Regulation	Shark Bay Fishery

Limited entry
    Number of licensed trawlers
    Vessel replacement limit (max. size)
	27
	375 vessel units

Fishing gear controls
    Trawl nets (number and max. head-rope length)
    Otter boards (max. dimensions)
    Ground chain (max. number and size)
	2 x 14.6m
	2.4 x 0.9m
	2 chains/net. 10mm link diameter

Closed seasons	November 1 to March 1 (adjusted annually to fit lunar cycles)
Within-season closures	3-5 days closed to fishing over each full moon

Closed areas
    Permanent nursery areas
    Temporary nursery areas
    Spawning stock closure (for tiger prawns)
	Closed at all times
	Closed August 1 to April 15
	Variable (July-November)

and lack of predictability due to the environment, even when stock/recruitment relationships are known, make ITQs inappropriate as a form of management. The individual transferable effort (ITE) system adopted in this fishery (Penn et al. 1997a) is a more reliable management arrangement which allows industry to accommodate variations in prawn abundance while leaving sufficient escapement to maintain breeding stocks.

5.8 Provision of information

Benefits from fishing rights arising out of access entitlements tied to licensing and management of prawn fisheries by input controls have been demonstrated in the Shark Bay prawn fishery. These have largely occurred as a consequence of early management by licence-limitation of the Shark Bay prawn fishery and the close involvement of industry and government in ongoing fishery management decision-making. The linkage of licensing to an obligation to provide detailed logbook information also facilitated the development of unique biological models (Penn et al. 1997b) covering breeding stock and recruitment that has enabled more effective management.

5.9 Industry responsibility (for the environment)

Industry responsibilities in the management system have extended beyond financial management and production outcomes of the fishery. The industry is now involved in development of fishing technology aimed at improving product quality and value while reducing by-catch wastage and minimising the risks of trawling on other non-fish species. At present, the Shark Bay prawning industry is, in partnership with government, developing new approaches to accommodate changing attitudes to natural resource use in Shark Bay, which has been listed as a World Heritage area. As a result of long-term and valuable access rights, the industry has had a significant incentive to participate in the process to develop environmentally responsible fishing practices through research programmes.

6. CONCLUSION

Tradeable access rights in the form of limited entry licences, together with other biologically-based input controls, have proven to be a successful long-term management approach for the Shark Bay prawn fishery. Individual licence values in excess of $A2.5 million are now being realised. The key benefit emerging from the use of such rights-based management is the incentive for industry to work with government to ensure long-term ecological sustainable fisheries production and deal with the short-term and longer-term fishery adjustments required to meet wider community expectations. Today, the Shark Bay prawn industry can be seen to be a successful and profitable fishery that contributes significantly to the economy of the State to the benefit of the wider community. However, neither industry nor government should remain complacent as technological changes will continue and will need to be taken into account by the management regime. Clearly, the basis for this success rests substantially on having an appropriate management based on access rights for the fishery. This will continue to engender the historical spirit of cooperation which has characterised this important fishery.

7. LITERATURE CITED

Caputi, N., J.W. Penn, L.M. Joll and C.F. Chubb 1998. Stock–recruitment–environment relationships for invertebrate species of Western Australia, in *Proceedings of the North Pacific Symposium on Invertebrate Stock Assessment and Management*, (eds) G.S. Jamieson and A. Campbell, *Canadian Special Publications in Fisheries and Aquatic Science* **125**: 247-55.

Gulland, J.A. 1984. Introductory guidelines to shrimp management: some further thoughts, in *Penaeid Shrimps, their Biology and Management*, (eds) J.A.

Gulland and B.J. Rothschild, Fishing News Books, Surrey, England, pp. 290-99.

Hancock, D.A. 1975. The basis for management of Western Australian prawn fisheries, in *First Australian National Prawn Seminar, Maroochydore, Queensland, November 1973*, (ed.) P.C. Young, Australian Government Publishing Service, Canberra, pp. 252-69.

Lenanton, R.C., L.M. Joll, J.W. Penn and K. Jones 1991. The influence of the Leeuwin Current on coastal fisheries in Western Australia, in *The Leeuwin Current: an influence on the coastal climate and marine life of Western Australia*, (eds) A.F. Pearce and D.I. Walker, *Journal of the Royal Society of Western Australia* **74**: 101-114.

Penn, J.W., N. Caputi and N.G. Hall 1995. Spawner–recruit relationships for the tiger prawn (*Penaeus esculentus*) stocks in Western Australia. *ICES Marine Science Symposia* **199**: 320-33.

Penn, J.W., N.G. Hall and N. Caputi 1989. Resource assessment and management perspectives of the prawn fisheries of Western Australia, in *Marine Invertebrate Fisheries: their assessment and management*, (ed.) J.F. Caddy, J. Wiley & Sons Inc., New York, pp. 115-40.

Penn, J.W., G.R. Morgan and P.J. Millington 1997a. Franchising fisheries resources, an alternative model for defining access rights in Western Australian fisheries, in *Developing and Sustaining World Fisheries Resources: Proceedings of the Second World Fisheries Congress*, (eds) D.A. Hancock, D.C. Smith, A. Grant and J.P. Beumer, CSIRO, Australia, pp. 383-90.

Penn, J.W., R.A. Watson, N. Caputi and N. Hall 1997b. Protecting vulnerable stocks in multi-species prawn fisheries, in *Developing and Sustaining World Fisheries Resources: Proceedings of the Second World Fisheries Congress*, (eds) D.A. Hancock, D.C. Smith, A. Grant and J.P. Beumer, CSIRO, Australia, pp.122-9.

Slack-Smith, R.J. 1978. Early history of the Shark Bay prawn fishery, Western Australia. *Fisheries Research Bulletin* (WA Department of Fisheries and Wildlife) **20**, 44 p.

EFFICIENT ACCESS RIGHT REGIMES FOR EXPLORATORY AND DEVELOPMENTAL FISHERIES

A. Cox and A. Kemp
ABARE - Fisheries Economics Section
PO Box 1563, Canberra, ACT 2601 Australia
< acox@abare.gov.au>

1. INTRODUCTION

The development of new (that is, undeveloped or previously non-commercial) fishery resources is best viewed as an evolutionary process from the initial exploration of a fish resource through to the full development of the fishery. While the various stages in the process are often considered as discrete steps, the process is in fact continuous. From initial exploration onwards, the fishery resource may be effectively managed, with the form of management varying considerably as the fishery evolves. Under current Australian Fisheries Management Authority (AFMA) policy, specified management arrangements are implemented during the exploratory phase as part of an exploratory fishing programme. Fishing concessions take the form of Fishing Permits, which are granted to eligible operators for a limited period. As knowledge is gained, and assuming that fishing activity can be sustained, there will be a shift from the exploratory phase towards a point where a statutory management plan, complete with Statutory Fishing Rights, is in place.

There is significant risk involved in exploring for new fish resources as there is uncertainty about the biological characteristics of potential stocks in prospective fishing grounds. The focus of exploratory fishing is to gather information about the likely commercial viability of any new resources. Commercial viability depends on a number of factors applying at a given time, including the market conditions, the available fishing technology, the size of the fish biomass and the management arrangements in place. A number of these factors may change over time, such that a resource that is not commercially viable at one time may become viable at another.

The problem facing fishers who are considering undertaking exploratory fishing is how to evaluate the expected net returns from exploratory fishing (accounting for the risk associated with exploration) and from any ongoing fishery. Society also has an interest in the relationship between the risks and reward from exploratory fishing. It is important that governments encourage fish resources to be developed (or conserved) in ways which maximise the benefits to the community at large. The incentive for fishers to act privately in ways that are consistent with maximisation of community benefits will depend heavily on the type of fisheries access- rights offered and the means by which they are allocated.

The issue of efficient exploration where there is uncertainty as to the existence and extent of a resource has been well developed in the literature for non-renewable resources such as for oil, gas and gold (*e.g.* Industry Commission 1991, Hogan *et al* 1996). In the case of renewable resources however such as fisheries, there has been little economic discussion, despite significant policy debate in recent years. The paper by Campbell *et al* (1993) is one of the few attempts to discuss the policy options for exploratory fishing from an economic perspective. But there is a considerable scientific literature on the development of biological criteria for the sustainable development of new fish resources (Walters 1998, McAllister and Kirkwood 1998).

The purpose in this paper is to examine the economic efficiency effects of the access-rights regime prevailing for exploratory fishing in waters managed by the Commonwealth government. Emphasis is given to the type of access-rights being offered to fishers and the mechanisms by which they are allocated. Alternative access-rights regimes that may improve economic efficiency in the exploration and exploitation of Australia's fish resources are also discussed. No attempt is made here to examine the impact of these alternative access-rights regimes on other policy goals of the government and the AFMA.

2. MANAGEMENT OF EXPLORATORY FISHING

Ideally, management of fish stocks should be based on accumulated information concerning the abundance of fish in the stock, the distribution of fish, the impact of fishing on the stock and other aspects of the marine environment and the effect of different harvesting strategies. In practice, at the exploratory stage of fishing, little information is available to enable fisheries managers to define a total allowable catch (TAC). In the initial stages of fishing a new stock, therefore, the major concern is to protect the integrity of the stock from overfishing.

The fish-down phase associated with exploiting a newly found resource provides a buffer against rapid overexploitation as it can withstand higher catch rates in the short term without unduly affecting the longer- term sustainable catch rates. The size of the buffer varies according to the size of the initial biomass and the biological characteristics of the species. The larger the biomass, the greater are the quantities that can be taken in the fish-down phase. For slow growing species (such as orange roughy), the maximum economic yield is lower and the fish-down phase is shorter.

However, the presence of this safety buffer does not mean that the initial catching phase should be free of management control. Given the lack of information about the stock size, the management focus is essentially short-term as information is gathered to enable an assessment of the sustainable catch level to take place. To facilitate this, the most appropriate form of short-term management may be to impose tight catch and operating restrictions on those vessels undertaking the exploratory fishing. Management arrangements involving more informed TAC levels may be introduced once additional information is collected. Indeed, this could occur quite quickly as it is not necessary to know the absolute stock size with a great degree of certainty in order to set a TAC (Arnason 1990). Experience with exploratory fishing in the Heard Island and the McDonald Islands fishery and the Macquarie Island fishery has demonstrated that it is possible to quickly establish a TAC for new fisheries.

3. EXPLORATORY FISHING IN AUSTRALIA

In recent years the policy debate in Australia has focussed on the exploratory and developmental fisheries around Heard Island and the McDonald Islands (HIMI) and Macquarie Island. There are also exploratory fisheries currently being considered in the waters surrounding Norfolk Island, Christmas Island and the Cocos and Keeling Islands. AFMA has also released a new policy on the exploration of fish resources (AFMA 1999a), which represents a substantial revision of the earlier exploratory fishing policy.

Commercial activity in the HIMI fishery commenced in March 1997 following a series of promising random stratified trawl surveys targeting Patagonian toothfish and Mackerel icefish in the early 1990s (AFMA 1998a, p. 9). An interim management policy for the exploratory fishery was in place from November 1996 to August 1997, during which time Scientific Permits were granted to two operators. The boats operated under a number of conditions designed to minimise the impact of fishing on the environment and to collect data on the fishery.

The HIMI Fishery Management Policy was issued in February 1998 as AFMA considered the fishery to have evolved beyond the exploratory stage (AFMA 1998a). Under the management plan, a limit of two boats was placed on the fishery, with a three year limit on the Fishing Permits (subject to annual review). A restricted individual transferable quota (ITQ) system was introduced. Each operator was granted 50% of the total allowable catch (TAC) for Patagonian toothfish and Mackerel icefish and is allowed to transfer quota units to the other operator, but not to third parties outside the fishery. It is expected that a statutory management plan will be developed for the fishery at the expiry of the current management policy.

Exploratory fishing around Macquarie Island was conducted by a single Fishing Permit holder between November 1994 and April 1996 (AFMA 1998b, p. 3). During this exploratory phase, the major target species was Patagonian toothfish. In November 1996, AFMA issued the Macquarie Island Developmental Fishery Management Policy which permitted access by a single boat until June 1999. As with the HIMI fishery, the boat operated under strict conditions regarding the environmental impact of fishing activity (due to the environmental significance of the area) and the collection of data for stock assessment. The Macquarie Island Interim Management Policy (AFMA 1999b) was released in June 1999 and essentially provided for a continuation of the previous policy regime until the introduction of a statutory management plan, expected in June 2001.

Under the interim management policy, only one Fishing Permit is granted and only one vessel is permitted to operate in the fishery. There is only a limited right to transfer the permit (subject to AFMA approval). The successful applicant was to be selected by a panel convened by AFMA according to selection criteria laid down in the interim management policy. The interim policy also stipulated that, in the event of the fishery operating beyond 2001 under a statutory management plan, a minimum of 50% of the ongoing rights in the fishery will be allocated to the operator(s) who participated in the fishery up to that point. The remainder of the rights will be allocated by an auction or competitive tender. Furthermore, AFMA may make available information on catches, effort and other information relevant to the fishery to prospective participants in the allocation process.

There are also a number of applications for exploratory fishing in the waters surrounding Norfolk Island, Cocos Islands and Christmas Island. At the time of preparing this paper, these applications had yet to be assessed and permits issued. The issue of exploratory fishing in these areas is complicated by the fact that there are established fisheries in the regions and the exploratory fishing is aimed at developing new species not currently under management arrangements.

4. CURRENT MANAGEMENT ARRANGEMENTS FOR EXPLORATORY FISHING

4.1 Development objectives

The management arrangements for exploratory fishing in Commonwealth waters are detailed in AFMA (1999a). The policy takes as its basis the legislated objectives contained in the *Fisheries Administration Act 1991* and *Fisheries Management Act 1991*. The key objectives are to pursue:

i. optimum utilisation of the living resources of the Australian Fishing Zone
ii. maximum economic efficiency and
iii. ecologically sustainable development.

Exploratory fishing is defined in the policy as a process of data-gathering, with 'the activities of fishers providing information on target and non-target species that can be used to determine what level, if any, of sustained harvesting of the resource can be supported in the longer term' (AFMA 1999a). Fishing activity is considered to be exploratory where research or anecdotal information suggests that a fish resource may exist and there is inadequate information on which to base a reliable stock assessment and set an appropriate level of long-term harvesting effort.

The following sections describe the AFMA exploratory fishing policy.

4.2 Exploratory management report

The AFMA will prepare an Exploratory Management Report as a prerequisite to the granting of a Fishing Permit(s). The report will be prepared if there are any applicants for exploratory fishing in an area or at the AFMA's initiative. The report will state the AFMA's policy and management arrangements in relation to the proposed exploratory fishing programme. It will detail any existing knowledge about the potential resource including, where known, the geographic area and a description of the supporting environment. It will also specify such things as: the expected number of participants in the exploratory programme; the fishing method(s) that participants will be authorised to use; any restrictions on the configuration and quantity of fishing gear; any known environmental issues affecting the programme; the conditions to be imposed on any permits granted; the data that participants will be required to provide; and, where possible, the quantity and nature of any ongoing rights that will be granted (should an ongoing fishery eventuate). Where there is to be a public call for applications, the report will also indicate the number of participants being sought for the exploratory fishing programme.

4.3 Allocation of fishing permits

Fishing Permits are a form of access-rights that the AFMA allocates for exploratory fishing. These permits are not generally transferable except under strict conditions set by the AFMA. Where there is a single applicant for a Fishing Permit, then that applicant will be given access-rights for the duration of the exploratory fishing programme. These rights are conditional on the fisher demonstrating that the conditions specified under the Exploratory Management Report can be met. This is not an exclusive access-right as the AFMA reserves the right to grant additional Fishing Permits allowing others to participate in the exploratory fishing programme. This may occur if the initial permit-holder is unable to meet the conditions of the Exploratory Management Report (for example, if they do not have sufficient capability to collect adequate data for stock assessment purposes). In this case, the AFMA may issue a public call for additional participants.

Where more than one application is received for the same exploratory fishing programme, access-rights are allocated on a 'first-come first-served' basis. Once again, the successful applicant must meet the conditions specified by the AFMA in the Exploratory Management Report. If it is deemed necessary to have more than one operator in the exploratory programme, the applicants will be ranked according to their suitability against criteria set out in the Exploratory Management Report.

It is also possible that the AFMA may initiate and prepare an Exploratory Management Report and publicly call for applications in relation to a specified exploratory programme. The allocation of access-rights is determined by the AFMA against criteria specified in the Exploratory Management Report.

4.4 Duration of access-rights

The AFMA may grant a Fishing Permit for a period of up to five years. A number of factors are to be taken into account when determining the period of access including: (a) the time likely to be required to obtain suitable data for stock assessment purposes; (b) the potential impacts on the environment over time; (c) the cost of specialised equipment needed to undertake the exploratory fishing; and (d), the degree of interest in the exploratory programme.

4.5 Ongoing access-rights

It is the AFMA's aim that more formal management arrangements be put in place for the end of the exploratory phase, should an ongoing fishery be identified. In the event that an ongoing fishery is identified, participants in the exploratory programme will be given first offer of participating in the ongoing fishery. A minimum percentage of the ongoing rights in the fishery will be divided and a proportion will be offered to each participant in the exploratory fishing phase. As was noted above, this minimum was 50% of the rights in the case of the Macquarie Island fishery. The minimum percentage to be offered will be decided by AFMA and specified in the Exploratory Management Report. The proportion offered to each participant will be based on an amount for each year or part year of participation and will take account of each participant's level of risk, cost and effort associated with undertaking the exploratory fishing programme. The proportion of ongoing rights granted will also depend on the need to provide suitable incentives to those contemplating investing in such programmes. Ongoing rights will be transferable unless otherwise determined by AFMA.

4.6 Cost recovery

The full costs of assessing applications for access to exploratory fishing programmes are charged to the applicants. The costs include an application fee, a non-refundable fee for service associated with assessing the applications (including the cost of preparing the Exploratory Management Report, consultation with external agencies and gathering of any additional information) and the management costs during the exploratory programme (including the costs of data collection and analysis, observer coverage, the grant of the Permit and licensing and compliance programmes). Cost-recovery will also apply to any ongoing fishery following the exploratory phase.

5. EVALUATION OF THE EXPLORATORY FISHING PROGRAMME

At the end of the exploratory fishing programme, the AFMA will prepare an evaluation that will provide the basis for determining the future direction of the fishery. The report will consider whether sufficient data are available for stock assessment, the environmental impact of fishing, whether the fishery is sustainable and whether ongoing management would be cost effective. There are options that the AFMA may pursue as a result of the evaluation: (a) extending the exploratory fishing phase; (b) ceasing exploratory (and all) fishing; or (c) move to a formally managed fishery with longer-term access-rights.

In summary, the current AFMA policy covering exploratory fishing can be characterised as a system allocating highly conditional access-rights for a limited duration on a first come first served basis. The degree of exclusivity of the access-rights is uncertain as the AFMA reserves the right to issue additional permits, and the right has only limited transferability. Moreover, holders of exploratory access-rights have only the guarantee of a minimum percentage of the ongoing rights in any fishery that eventuates. Such a policy regime provides only limited incentives to undertake exploratory fishing and cannot be regarded as an economically efficient mechanism for allocating access-rights.

6. ECONOMIC EFFICIENCY AND EXPLORATORY FISHING

The economic efficiency of exploratory fishing policy will be influenced by the type of access-rights granted to participants and may be affected by the way they are allocated. While these aspects are dependent, they can be considered separately from a conceptual perspective. It is necessary however, to first consider what is meant by economic efficiency in the context of exploratory and developmental fishing.

Efficiency in marine resource allocation arises when rights of access are allocated so that use provides the greatest value to the community. Value in this context refers to commercial, non-commercial (for example, recreational fishing) and non-use (for example, conservation concerns) values. Such decisions on allocation need to be made at several levels. At one level, there is competition for access to resources between commercial and recreational fishing sectors, and between conservation and other non-use interests. In the commercial sector, there may be competition between individual fishers for share of the resource. At another level, resource allocation has a temporal dimension — the amount of resources harvested in any particular period will influence the amount available in the future.

Fisheries managers therefore need to decide on the allocation of marine resources across sectors and use, and providing a framework that determines how individual fishing firms compete for the resource within the commercial sector. For example, managers have a range of options for regulating access by commercial operators to fish resources. In theory, instruments which establish tradable property-rights for the fish resource (such as a system of individual transferable quotas) provide the greatest likelihood for an efficient allocation of a given level of harvest (assuming that the TAC is optimally set with respect to the maximum economic yield). The ability of operators to buy and sell access-rights to the fish resource under such a system should facilitate the allocation of the resource to those who value it most. An efficient outcome can also be achieved through the granting of sole ownership rights to exploratory and ongoing fishing, where the single owner cannot influence the market price for either inputs or outputs. Such an option would also reduce management costs in many cases as these costs could be privately met by the operator. Alternative regulatory instruments, such as input controls or individual vessel quotas, all affect, to varying degrees, the efficiency of resource allocation within a commercial fishery.

The cost of fishing is also be an important determinant of the economic efficiency of resource use. Productive efficiency is achieved when a given amount of output (fish) is extracted at least cost given the available technology and price of productive inputs. Commercial fishermen face a range of choices about technologies to use in fishing and the amount of, and combination of, inputs such as fuel, labour and bait. The ability of fishing operations to achieve efficiency in production is influenced by the experience and skill of operators, the availability of inputs, the scale of their operation and so on. The regulations determining the conditions of access to the resource can also be important. Restrictions on the use of certain inputs, seasonal closure of fishing grounds and output restrictions all affect, to varying degrees, the ability of fishing operators to best utilise available inputs and technologies to produce a given amount of output.

The main objective underlying the pursuit of efficiency in resource allocation and productive efficiency is to maximise the economic-rent from the fishery for the benefit of society as a whole. Economic-rent can be defined as the returns over and above the costs of operations. These costs include all normal cash expenditure plus depreciation, a margin for the risks involved and a return for the investment in exploring and developing the fishery over time.

In the context of exploratory and developmental fisheries, pursuit of the economic efficiency objective is complicated by the considerable uncertainty that will surround the existence and extent of new fish resources. The objective for managers in establishing an exploratory fishing regime for a given area or species then becomes one of maximising the expected economic-rent from prospective fisheries. In this regard, the management regime for exploratory fishing has the same basic objective as a regime for an established fishery, except with a higher degree of biological uncertainty.

The decisions made by fisheries managers about the conditions to be met by prospective operators in an exploratory fishery will affect the decisions made by these operators in responding to the risks and the expected net returns of exploratory fishing. In principle, it is the biological uncertainty that will be the major influencing factor in private sector decisions about the risk-reward relationship — regulatory measures which affect the degree of uncertainty surrounding exploratory fishing will also influence the perceived riskiness of exploration. Therefore, one of the primary concerns in the management of exploratory fishing should be to provide a regulatory environment in which the process for determining and allocating access rights to exploratory and ongoing fisheries is clear and certain. It may, of course, be difficult to achieve as knowledge about new fish resources and the supporting environment gathered during exploration may alter the premises on which the exploratory regime was based at the outset.

7. EFFICIENT EXPLORATORY FISHING RIGHTS

7.1 Defining features

The extent to which improving the system of access-rights will encourage increased efficiency in exploratory fishing depends on how the rights are defined. There are five characteristics of an exploratory fishing right that define the rights held and facilitate productive efficiency in the fishery. This definition applies more generally to fisheries resource use, rather than solely to exploratory fishing. They are a right:

i. of access to the fishery and take fish during the exploratory phase (access and withdrawal right)
ii. of divisibility of the permit (divisibility right)
iii. to transfer the permit (transferability right) and
iv. to exclude others from access to the fishery and withdrawal of fish (exclusivity right).

7.2 Access and withdrawal

Access and withdrawal rights require a clear definition of the fishery for which exploration is about to commence. Fisheries can be defined in terms of species of fish and, or, geographic region. From an efficiency perspective, whether a fishery is defined in terms of fish species or geographic region is inconsequential.

The access and withdrawal rights can also be defined in terms of duration and timing for conservation reasons. These restrictions can be used to set the bounds of sustainability for the fishery. So long as rights of divisibility, transferability and exclusivity exist, these restrictions will not influence the ability of the system to lead to efficient exploration within sustainability constraints.

7.3 Divisibility

Divisibility allows an operator with an exploration permit to subcontract other operators to explore the fishery. For example, if an operator wished to use their exploratory entitlement at a certain time of year, they could hire another operator to engage in exploratory activity for the rest of the year. This could occur due to the seasonal nature of some fish species. Divisibility allows the operator to reduce the costs associated with exploratory fishing, which in turn leads to more efficient exploration.

7.4 Transferability

Transferability is an extension of divisibility and allows the firm to transfer or sell permanently all or part of the right to exploration to other operators. This right allows the permit to be owned by the most efficient operators for its duration. The lowest-cost explorer at any given point would be prepared to a pay greater may amount for the right to explore than any other operator. Restrictions on the transferability of the rights may reduce the efficiency with which exploration is done.

7.5 Exclusivity

The level of exclusivity of access and withdrawal will affect the level of investment by an operator in exploration of the fishery. From an efficiency perspective, there is no reason to allocate more than a single exploratory right, as long as the holder of the right is free to choose the number of boats in the exploration phase and to transfer the right when desired. These rights allow the market to determine the efficient number of boats engaged in exploration. While there may be efficiencies associated with collaborative action of operators in exploration, this decision is best left to market mechanisms rather than to central authorities to make. From a fisheries management perspective, fewer exploration rights issued may be more cost-effective as compliance may be easier to ensure and enforce.

7.6 Duration

The duration of the exploratory right needs to be sufficient to provide incentive to private operators to invest in exploration of the fishery. The duration should also be long enough not to distort the decisions of exploration operators. Short time-limits may encourage investment in exploration now, or result in no exploration occurring, whereas it may be efficient to explore at some other time in the future. Long-duration permits allow operators to decide the most efficient time to explore. With rights of transfer and divisibility, if other operators consider it more appropriate to explore now, there are opportunities to transfer the rights, which will lead to an efficient investment in exploration.

Short-term access-rights may give better control over fishing capacity than do access-rights for long periods, as it is possible to reduce capacity quickly by not renewing rights or increase it by allocating additional rights (Commonwealth of Australia 1989). However, this argument is weaker where the fishery is managed with output-controls such as a TAC. The allowable output can be increased or decreased in response to new information, leaving decisions on capacity and effort to the private sector. Moreover, limited tenure rights are likely to discourage operators from investing in exploratory activity. As in general operators are not prepared to invest when there is a low possibility of long-term benefits.

Another issue affecting the duration of the exploratory right is the relationship between the exploratory right and any ongoing rights in the fishery (should one eventuate). It is clear that the share of any future fishing development that is assigned to the exploratory right will have a major influence on the expected economic-rent associated with exploratory fishing. From an economic efficiency perspective, there is no reason why the rights to all the future catch should not be allocated with the exploratory fishing right, provided the other aspects of efficient access-rights (divisibility, transferability and so on) are incorporated. The distinction between exploratory fishing and fishing carried out under a statutory management plan is artificial given that there is effectively a continuum in the phases of a fishery's development. Indeed, creating two regimes — one for exploratory fishing and one for a managed fishery — may exacerbate efficiency concerns associated with the duration of access-rights and their allocation.

However, some sectors of the community may consider the allocation of 100% of a community-owned resource to one, or several, private operators to be inequitable. This is an equity question that may be taken into account by government together with the efficiency impacts of decisions.

7.7 Certainty of process

Significant uncertainty can be reduced through the definition and allocation of rights to exploration of a potential new fishery. Maintaining the stability of any property-rights regime is important for minimising uncertainty. The possibility of changes to the rules after the issuing of exploratory licences can have significant effects if risk premiums are built into the investment decisions of private operators (often referred to as 'sovereign risk'). Upon the issuing of an exploratory permit that defines these rights, no changes should be made for the duration of the permit.

8. ASSESSMENT OF ACCESS RIGHTS UNDER CURRENT AFMA POLICY

8.1 Conceptual background

When considered against the requirements for efficient access-rights for exploratory fishing, it is clear that there are some important shortcomings in the way the access-rights are defined under current AFMA policy on exploratory fishing.

First, the right of divisibility is not explicitly addressed in the general AFMA policy. However, under the management plans for the HIMI and Macquarie Island fisheries, the right of divisibility is explicitly ruled out. These plans do not permit two or more companies, with two or more boats, to jointly use a single Fishing Permit on a rotational basis. This lack of flexibility is partly due to the specialised nature of fishing in the environmentally sensitive regions and partly to the need for the AFMA to be sure that vessels operating in the areas meet environmental and safety guidelines. However, given that both the fisheries are managed under a TAC, it is not clear why more than one boat should not be permitted provided they meet environmental and safety standards. This would increase the ability to operate in the most efficient manner.

Second, while current AFMA policy states that exploratory rights are generally transferable, this has not been the case for the HIMI fishery. The rationale for this restriction is that the successful applicants for the permits have special experience, skills and reputation, as well as commitments made personally under the Fishing and Research Plans and these factors cannot be readily transferred. Similar restrictions apply in the Macquarie Island fishery, although the management plan does provide the AFMA with some discretion in allowing transfer but under restricted conditions.

Third, the fact that the AFMA reserves the right to admit other participants to an exploratory programme, primarily at its discretion, reduces the exclusivity attached to an exploratory right, and thereby reduces the efficiency of exploration.

Fourth, exploratory rights are currently short-term. The AFMA policy provides for a maximum of five years duration for an exploratory permit. The permit in the Macquarie Island fishery is issued on an annual basis, while the permits in the HIMI fishery are valid for three years, (the length of the management policy), subject to annual review. While these fisheries have ostensibly progressed beyond the exploratory phase, they are not yet managed under a statutory management plan, which provides for longer-term statutory rights. From an efficiency perspective however, there is no reason why access rights need be of such a short duration. Indeed, it is not clear that there needs to be a distinction between exploratory and ongoing rights in the fishery.

Last, one of the features of the AFMA exploratory policy is the considerable administrative discretion available to the fisheries managers. While a degree of flexibility is required for managers to be able to respond to changing circumstances and new information, it needs to be recognised that there is a cost associated with this discretion. This cost is an increase in the perceived risk associated with investment decisions. The major source of uncertainty in exploratory fishing should ideally relate to the characteristics of the fish stock, rather than to any current or future management arrangements. While the Exploratory Management Report is designed to provide such certainty with respect to current and future management arrangements, many of the discretionary powers reserved by the AFMA relate to management decisions which have a potential impact on economic efficiency. Examples include the limited rights of divisibility, exclusivity and transferability currently attached to exploratory permits, with the AFMA approval required before the exercise of the limited rights.

8.2 Economic efficiency and allocation mechanisms

The existence and transferability of properly-defined rights is sufficient to ensure efficiency in exploratory fishing (and, indeed, in any fishery). The method of allocation of the rights will have only a limited impact on economic efficiency, provided that transactions costs are negligible. The choice of allocation mechanisms will primarily have distributional effects — that is, determine who receives the economic rents from fishing — and therefore could be used to achieve equity goals of the government and the AFMA.

As a result, the most important criteria in choosing the most appropriate method for allocating rights are that transactions costs are kept to a minimum and that the costs of conducting the allocation process are recovered. A third criterion could also be added — to achieve an appropriate return to Australian society as owners of the fish resource for the right of access to the resource (Commonwealth of Australia 1989). The current Commonwealth government has a policy of not introducing a resource-rent tax on the fishing industry (Liberal Party 1999).

The allocation process must be able to deal with two issues: (a) the initial allocation of the exploratory right; and (b), the assignment of ongoing rights. To achieve these, two broad classes of allocation mechanisms can be identified. The first class is one-stage mechanisms in which all of the rights to both the exploratory phase and the managed phase are allocated at the same time. The second class can be termed two-stage mechanisms where there is an initial allocation of exploration rights and a subsequent allocation of managed-phase rights.

The objectives implicit in the use of each class of mechanism differ considerably. From an economic efficiency perspective, it is only necessary to allocate rights once, provided the rights are well defined in terms of transferability and so on. The rights then have value and may be traded through market mechanisms to those operators who value them the highest. Therefore, a properly designed one-stage allocation mechanism is all that is required to meet the efficiency criterion.

Two-stage mechanisms may also meet the efficiency criterion, although there are a greater number of considerations to be addressed in their design to ensure efficiency. Moreover, the transactions costs associated with such an allocation and subsequent reallocation of rights may be considerable. Two-stage mechanisms also have an implicit revenue-raising objective. The primary reason for conducting a second stage in the allocation process is to redistribute a greater share of the expected economic-rent in a fishery between the government and the private sector. Uncertainty about the biological characteristics of the resource will (hopefully) have been reduced as the result of exploration, and current and prospective operators will have more information on which to base their evaluations of the expected economic-rent available in the fishery. At this stage that the distribution of the economic-rent between the government and private operators does not affect economic efficiency.

The concept of maximising economic-rent is not new in fishery's management in Australia (*e.g.* Campbell and Haynes 1990). For example, the use of individual transferable quotas (ITQs) in existing fisheries ensures that the economic-rent is maximised. As the allocation of the initial quotas is usually on the basis of historical catch considerations, the economic-rent is given to the existing operators without returns being extracted by the government, except through normal taxation means. Irrespective of the allocation of economic-rent between government and private operators, the ITQ system will maximise the economic-rent for society as a whole.

8.3 One-stage mechanisms
8.3.1 Main objectives

The primary objective in a one-stage allocation process is to allocate the access-rights for a fishery between prospective operators. While considerable uncertainty may surround the extent and size of the potential fish stock at the stage of this allocation, this uncertainty does not affect the ability to achieve optimal investment in exploration, so long as the rights of divisibility, exclusivity and transferability exist. The risks associated with the potential fishery are borne exclusively by the private operator(s). The operator(s) is also in a position to secure a return from the information obtained about the characteristics of the fish stock and the commercial viability of fishing operations.

The two most common methods for allocating rights are — a first-come first-served system, and auction or tender. Other methods are used, such as administrative discretion and ballot, but they are not addressed as they can be quickly dismissed on efficiency and cost-effectiveness requirements.

8.3.2 First-come first-served

Under a first-come first-served (FCFS) system, the rights to fish in a particular area, or for a particular species, are allocated to the first operator who applies. Such rights are also referred to as finder's rights or pioneer rights. If the rights are transferable and divisible, then they may find their way to the most efficient operator. Such a system provides an incentive to fishers to undertake exploratory fishing, and so, would encourages greater exploratory fishing than might currently be the case.

It can be argued however that a FCFS system can be inefficient in that it can encourage excessive exploration. This may be the case if the rights are seen as having value. This is presently not the case as the exploratory rights under the current AFMA policy are heavily encumbered and of short duration. An FCFS system creates an incentive to acquire exploratory permits for areas which are considered promising before someone else does. Individual explorers would be expected to be willing to spend up to the expected value of the discounted future stream of rents from exploration.

As a result, there is an incentive to undertake preliminary exploration, even when there is only a small probability of positive returns from exploration when actual exploration is linked to the right to exploit. It can lead to situations such as in the minerals area of 'gold rushes', where the effort expended in exploration by many participants is greater than socially optimal (Industry Commission 1991, pp. 41–3). While the returns to successful explorers are significantly greater than their costs, this should be offset against the many unsuccessful exploration attempts and subsequent losses. This process - rent-dissipation - is where the expected gains from exploiting the fishery resources in an area are dissipated by losses associated with exploration in other areas, by the industry as a whole. Such dissipation of rent may be economically inefficient from society's perspective, while being quite rational from the viewpoint of individual fishers. This highlights the need to consider the incentives and losses in other areas of the industry, against the gains in any single exploration programme.

Against this, an advantage of the FCFS system is that the administrative costs may be lower than for other mechanisms discussed below in Section 8.3.3. These cost-advantages however, may be outweighed by potential efficiency-losses associated with rent-dissipation.

8.3.3 Auctions and tenders

Cash-bidding uses competitive market mechanisms to auction the rights to exploration of a potential fishery. The term 'auction' is used here to describe both public auction processes and sealed tenders. In terms of economic efficiency, the form of auction has no significance, though it can affect the administrative costs of the process. Auctions allows potential bidders in the market to assess the risks associated with exploration of the new fishery and incorporate these assessments into their bid price.

Auctions can be either a public process, where bids are taken from the floor, or a sealed-tender process, where bids assessed against a set of criteria that may include other operating considerations than price. As a one-stage mechanism, the auction may give the successful bidder 100% of the TAC if the fishery is developed to the managed phase. Alternatively, a number of rights to a potential fishery could be auctioned, *e.g.* two permits each granting 50% of the TAC.

In theory, where there is a competitive bidding system with a number of operators, the same information and no collusion, then the highest bid will reflect the best assessment of the potential value and risks associated with the fishery. The result will lead to efficient investment in exploration, given the risks. The operator becomes free to determine the pattern and timing of exploration and subsequent development given the sustainability conditions attached to the rights.

But, problems arise if operators collude on bid prices and violate the competitive nature of the system. This might still result in an efficient outcome although the rents would accrue to the colluders rather than to the community. Information asymmetries can also lead to inefficiencies in the auction process, although bids should reflect the information available to each operator. To avoid speculative bids sealed-bid auctions may be considered more appropriate. Further, small numbers of potential market participants can restrict the competitive nature of the auction process creating inefficiencies.

The rules associated with auctions need to be clear and nondiscretionary. All available information should be made available to all potential market participants when bids are called for. The process could be triggered either by an application or directly by the AFMA with the same rules applying for either trigger. A basic requirement of the auction is that the highest bid at least recovers the cost of conducting the auction (the reserve price).

One of the issues in the use of auctions is that the bids will be based on the expected economic-rent. If this valuation is based on little information about the commercial viability of the fishery, the bids offered may prove to be significantly below or above the true realised rent (Munro and Parkes 1999). The provision of additional information by government to improve the quality and quantity of the information available to prospective bidders is likely to reduce the distribution of bids, but may or may not increase the expected bids. This may be appropriate if the costs associated with obtaining the information are relatively small.

The major difference between a FCFS system and auctions lies in the distribution of rent in the allocation of the initial right. Under the FCFS system, the initial holder of the right receives the bulk of the economic-rent from the fishery and this will be incorporated into the price at which the rights are transferred in the future. Under an auction, the government will be the recipient of a share of the expected rents that would otherwise have gone to the private sector.

8.4 Two-stage mechanisms

The major objective underlying the use of two-stage mechanisms for rights-allocation is to extract a proportion of the economic-rents generated by a developing fishery. This is done in the second stage of the allocation process by using the information gained during the exploration. There are likely to be significant distributional issues associated with the design of any two-stage mechanism. The primary issue concerns the sharing of risk and economic-rent between the government and the private sector. In terms of efficiency of the system, however, the difference between the one-stage and two-stage mechanisms are minimal. Arguably, one-stage mechanisms are administratively easier and may therefore be less expensive to operate hence more efficient if the benefits are similar.

The relative efficiency of alternative two-stage mechanisms depends on how the information gathered during the exploratory phase on the location and abundance of fish stocks is treated. This information is potentially a valuable commodity and there are likely to be conflicting opinions on whether and how it should be disseminated prior to the second stage allocation. The incentives to invest in exploration will be influenced by the extent to which explorers can realise the benefits from the information they generate.

The current AFMA policy on the allocation of exploratory and ongoing rights is a two-stage mechanism. The initial allocation of exploratory rights is done on a first-come, first-served basis. The second stage comprises the allocation of a specified minimum proportion of the ongoing rights to the first stage participants. The remainder is allocated by an unspecified method (although the *Fisheries Management Act 1991* allows for the distribution of rights under a statutory management plan to take place by auction, ballot or tender).

This mechanism is being used in the HIMI and Macquarie Island fisheries where a portion of the managed phase rights are to be auctioned. Such a system may lead to rent-dissipation if explorers are not required to share information with other participants in the auction. The explorers will be able to make a more informed bid at the auction. This does not necessarily mean however that they will be the successful bidder.

Alternatively, operators in the exploratory phase may be required to provide any information regarding the commercial viability of fish stocks to the government, who then make it available to all prospective participants in the auction. In this case, there may or may not be private incentives to explore at the socially optimal level. Unlike minerals exploration, exploratory fishing generates a return from the fish caught during exploration. This return may offset the reduced benefit of information generated by exploration due to having to provide the information to all other prospective participants. If the returns offset the lost information benefit, then the private level of exploration will also be socially optimal. If not, then the private level of exploration may be socially sub optimal, hence there may be a role for government provision of information.

An alternative mechanism might involve two auctions — one for the exploratory rights, followed by one for the managed-phase rights. Such a mechanism may result in an economically-efficient outcome although there are more complexities that must be addressed in the optimal design of the process. These complexities relate primarily to generation and distribution of information and the timing of the two auctions. For example, to make the second auction competitive, it would be necessary to pool the information gathered during exploration.

It is beyond the scope of this paper to address these complexities in detail. It is worth noting however that an auction–auction process will provide government with a greater share of the economic-rent than other one-stage and two-stage mechanisms. If this is to be an explicit objective in the allocation of rights, then there are a number of alternatives *e.g.* combining an auction with some form of charge or tax on economic-rent.

9. CONCLUDING COMMENTS

It is unclear why there is a distinction between an exploratory fishery and a managed fishery. It is not feasible to draw a distinct line beyond which fishing is no longer exploratory and becomes an ongoing concern. Rather it is best represented as a continuum, along which fisheries managers exercise varying degrees of control over fishing activity. Indeed, it is possible to argue that exploratory fisheries have been managed more tightly than many of the established fisheries.

One solution, therefore, may be to abandon the distinction between exploratory and managed fishing. In terms of economic efficiency, it is only necessary to allocate access-rights at the initial stage, provided those rights are fully and efficiently defined (with respect to transferability, durability and exclusivity). The question of the allocation of ongoing rights then becomes irrelevant. The ability to transfer access-rights ensures that they go to the most efficient operators and the longer duration of the rights provides a better incentive to pursue an efficient exploration programme.

The choice of allocation mechanism is dependent on the requirements of the AFMA and the priorities of the Commonwealth government. If the primary objective is to allocate rights to those who value them most, then an auction system is preferable to the current system of first-come first-served. Auction processes reduce the potential for rent-dissipation that exists under a FCFS system. They also fulfil equity objectives, by giving all industry operators the opportunity to engage in the exploratory process. If there is however a desire to extract a return to the community for the use of publicly-owned resources, then some form of two-stage mechanism (involving either an auction–auction or an auction–tax process) has merit.

10. LITERATURE CITED

AFMA (Australian Fisheries Management Authority) 1998a. *Heard Island and McDonald Islands Fishery Policy*, Canberra, February.

AFMA (Australian Fisheries Management Authority) 1998b. *The Future of the Macquarie Island Developmental Fishery*, Canberra, December.

AFMA (Australian Fisheries Management Authority) 1999a. *Exploration of Fish Resources, Fisheries Management Paper Series*, FMP No. 5, Canberra, April.

AFMA (Australian Fisheries Management Authority) 1999b. *Macquarie Island Fishery Interim Management Policy*, Canberra, June.

Arnason, R. 1990. 'Minimum information management in fisheries', *Canadian Journal of Economics*, vol. 23, no. 3, pp. 630–53.

Campbell, D. and J. Haynes 1990. *Resource Rent in Fisheries*, ABARE Research Report 90.10, Canberra.

Campbell, D., A. Kingston and T. Battaglene 1993. *Policy Options for Exploratory Fishing*, ABARE report prepared for the Australian Fisheries Management Authority, March, Canberra.

Commonwealth of Australia 1989. *New Directions for Commonwealth Fisheries Management in the 1990s*, AGPS, Canberra.

Hogan, L., S. Thorpe, S. Zheng, L. Ho Trieu, G. Fok and K. Donaldson 1996. *Net Economic Benefits from Australia's Oil and Gas Resources: Exploration, Development and Production*, ABARE Research Report 96.4, Canberra.

Industry Commission 1991. *Mining and Minerals Processing in Australia*, vol. 3, *Issues in Detail*, Report No. 7, February, AGPS, Canberra

Liberal Party 1999. 'Wealth From The Sea', in Liberal Party of Australia 1998 Election Policy, www.liberal.org.au/ARCHIVES/election98/policy/fisheries/fisheries.html (accessed 26 October 1999).

McAllister, M.K. and G.P. Kirkwood 1998. 'Using Bayesian decision analysis to help achieve a precautionary approach for managing developing fisheries', *Canadian Journal of Fisheries and Aquatic Sciences*, vol. 55, pp. 2642–61.

Munro, G. and G. Parkes 1999. *Auction and Tender as Allocation Mechanisms in Fisheries and their Resource Rent Implications*, Report prepared by MRAG Americas Inc for Austral Fisheries Pty Ltd, February, Perth.

Walters, C. 1998. 'Evaluation of quota management policies for developing fisheries', *Canadian Journal of Fisheries and Aquatic Sciences*, vol. 55, pp. 691–705.

DEVELOPING NEW FISHERIES IN WESTERN AUSTRALIA

J. Borg
Fisheries Western Australia
168-170 St Georges Terrace, Perth WA 6000, Australia
<nborg@fish.wa.gov.au>
and
R. Sellers
Fisheries Division, Northern Territory Department of Primary Industry and Fisheries
GPO Box 990, Darwin NT 0801, Australia
<Richard.Sellers@DPIF.nt.gov.au>

1. INTRODUCTION

The Developing New Fisheries in Western Australia (DNF) process moves the management of new, exploratory and developing fisheries from an *ad hoc* system, where there was no certainty and no process for moving towards rights-based management, to a clear and transparent process where both fishermen and government have a strategic framework within which to work.

2. HOW FISHERIES MANAGEMENT IS ADMINISTERED IN WESTERN AUSTRALIA

2.1 Legislative base for granting access

The Minister for Fisheries, on behalf of the Western Australian (WA) community, has a stewardship responsible for the State's fisheries and their environment for the benefit of current and future generations. Fisheries Western Australia ensures the conservation, development and sharing of Western Australia's fish and other living aquatic resources on behalf of the Minister in accordance with all relevant laws enacted by the WA Parliament.

Prior to 1995, fisheries management in Western Australia was governed through the *Fisheries Act 1905*. This Act allowed for the gazetting of Fisheries Notices which either established the management rules for particular fisheries or set in place prohibitions covering particular fish resources, fishing methods or areas. It resulted in a series of limited entry fisheries (LEFs), restricted entry fisheries (REFs), and licence conditions that exempted people from existing prohibitions. A licence to fish in any of the fisheries covered by Fisheries Notices or licence conditions did not confer a perpetual right. However, rights were often assumed by licence holders unless serious breaches of the law caused the licence to be removed. On most cases, nothing the government did or said acted against this assumption.

A small number of fishing activities (mainly line fishing and hand netting) not covered by any limiting Notice remained outside of these Fisheries Notices, and hence any other formal management arrangements. These fisheries became known as the developing and exploratory fisheries and there was no policy or process to deal with applications to undertake such fishing activity.

In reality, all fishing activity in this category was prohibited by way of Fisheries Notices. However, fishermen who considered markets, technology and safer harvesting methods, now made these fishing activities potential fisheries of the future, by applying to the Executive Director for an exemption from the relevant prohibition. As little was usually known about fish resources subject to these applications, there was little way of refusing the application. This *ad hoc* process commonly resulted in many more authorisations being issued than was required for the sustainable exploration or development of new resources. This excess, and sometimes latent, effort proved to be legally difficult and administratively costly to remove.

The *Fish Resources Management Act 1994* (FRMA) was introduced in 1995. It is the predominant legislation governing the activities of Fisheries WA. The FRMA provides the framework for modern fisheries management and protection of the marine environment. The introduction of the FRMA did not change the types of fisheries that existed in the State - there are still different levels of management applied to fisheries at various stages of development. It does, however, provide additional powers to the Minister and Executive Director to regulate fishing activity. In addition to the declaration of managed fisheries, interim managed fisheries, and allowance for licence conditions, the Minister and Executive Director are now able to issue exemptions to certain provisions of the Act and Regulations. The extent of this power is listed in Section 7 of the FRMA. The other power given to the Executive Director, which is the legislative base for this process, is his ability to require an applicant to provide additional information in relation to an application to conduct a particular fishing activity.

The motivation for the DNF process comes from the objects of the *FRMA* (Section 3), that is:

> *"to ensure that the exploitation of fish resources is carried out in a sustainable manner...;*
> *to foster the development of commercial fishing...;*
> *to achieve the optimum economic, social and other benefits from the use of fish resources....; and*
> *to enable the allocation of fish resources between users of those resources..."*

2.2 Basis for access to fisheries not covered by management plans

Although management plans and prohibition notices covered most fishing activities off the coast, the policy of

Fisheries WA and its predecessors was to assist fishermen wherever possible in undertaking new fishing ventures. If a fisherman wanted to test or develop a new fishery that person would write and ask for an endorsement on their licence that would exempt them from the relevant prohibition (but not from a management plan). With some minor exceptions, it could safely be assumed that the appropriate endorsement would be granted.

What usually resulted from applications was -

i. considerable administrative work to ascertain whether such fishing was sustainable or desirable
ii. if it was decided that it was inappropriate to grant to exemption, considerable resources were used trying to show why an endorsement should not be issued
iii. often the resource was not commercially viable, the cost of harvesting too high, or the equipment not sufficient and fishing did not occur, or if it did, it led to no new viable commercial fisheries
iv. expectation of, or the realisation of, a permanent right because the condition or endorsement was never taken off the licence and
v. the frequency of individual applications created a constant unpredictable and unplanned workload for fisheries management staff.

It was also rare to get useful scientific or basic monitoring data from these fishing activities. Consequently, there could be no guarantee of sustainability of any resource fished under such arrangements.

In addition, new applicants would apply for access to resources that previous attempts had not proven viable and it was equally as difficult and expensive to refuse access. If refused, the applicant had recourse to the Minister to appeal under the 1905 Act. Under the FRMA, appealants may go to an Independent Tribunal.

There was no process and no policy to deal with these applications under either the 1905 or 1994 Acts. This lack of strategic management of exploratory and developmental fisheries meant that management of these fisheries was inefficient and costly and this was no longer acceptable in a cost-recovery environment.

3. WHAT IS A DEVELOPING FISHERY?

For the purposes of the DNF process, a developing fishery is defined as:

"a fishery within which there is little or no exploitation, there is potential for development and which is currently subject to a prohibition. There also may be little information regarding:

i. *the stock(s) under consideration,*
ii. *the role of said stock(s) in both local and larger marine ecosystems,*
iii. *the possible uses of the harvested materials,*
iv. *potential domestic and/or foreign markets, and*
v. *explicit management objectives, policies, and/or operating regulations."*

The process excludes new activities within existing managed fisheries, however, it does include fish resources already fished at exploratory or developmental levels at the time the 'new' fishery was created as well as those which are unexploited. Thus, the concept of a developing fishery applies to fisheries in which there is development potential but minimal strategic policy, management direction or guidance.

4. POLICY UNDERPINNING THE DEVELOPING FISHERIES MODEL

4.1 Principles behind the DNF

The principles outlined in this process reflect the fact that:

i. the Government, and hence Fisheries WA, takes a precautionary approach to ensure the sustainability of developing fisheries
ii. the objects of the *FRMA* require, among other things, achieving the optimum economic, social and other benefits from the use of fish resources
iii. Fisheries WA has responsibilities for conservation of the marine environment in general
iv. the risks of developing fisheries must be assumed by those participating in them
v. the assumption of risk by pioneer participants will be recognised and
vi. administrative, managerial simplicity and accountability are mandatory conditions in Fisheries WA's management of developing fisheries.

4.2 Ecologically sustainable development

One result of the sustainability mandate is that it is no longer appropriate for Fisheries WA to undertake a narrowly-based management strategy of a single resource species, rather it must take a broad approach that considers the ecosystems that support all marine species.

The goals of an ecosystem-based approach to fisheries management allows for development based on the equitable and the sustainable use of both species and ecosystems, the maintenance of essential ecological processes, and the preservation of biological diversity that aims at meeting present needs without compromising the ability of future generations to have the same privilege.

4.3 Precautionary principle

When information is uncertain, unreliable or inadequate, Fisheries WA must take a cautious approach to the development of fisheries to reduce the risk of development being to the detriment of resources, the environment or to other stake-holders. To foster sustainable development and minimise risk the DNF has been designed to make participants in each proposed venture fully aware of the biological, economic and social implications of the proposed activities by:

i. ensuring careful analysis of the proposed venture
ii. adopting a risk-averse approach regarding the exploitation of new resources and
iii. requiring consultation with other stake-holders as part of the application process.

4.4 Recognition of developers

In considering how to achieve the optimum benefits from developing fisheries, the DNF provides for recognition of the status of fisheries developers or "pioneers". It is recognised that a pioneer who develops a new commercial fishery that does not affect community-values associated with access to the community aquatic resources should receive some benefit. While the concept of differentiating between initial, or pioneering participants in a developing fishery, and parties who subsequently want to participate in the fishery after the developmental stage, is recognised by Fisheries WA in accordance with the process described in Section 71 of the *FRMA*:

s.71. *"(1) The fact that a person engaged in fishing, or used any boat for fishing, in a fishery before a management plan was determined for that fishery is not to be taken as conferring upon that person any right to the grant of an authorisation if a management plan is determined for that fishery.*

(2) Despite subsection (1), the Executive Director is to take into account a person's history of fishing in a fishery when determining whether or not to grant the person an authorisation."

Therefore, the Government reserves the right to consider a fisher's previous participation following the developmental stage and with Section 71 in mind. However, the Government does not provide any guarantee that initial participants (whether the participant be an individual or body corporate) in a developing fishery will be granted the "rights" to future access.

5. THE DEVELOPING NEW FISHERIES PROCESS

5.1 Rationale and objectives

One of the key benefits of the DNF process is that a developing fishery will not exist in a timeless vacuum with no formal beginning, middle and end. The framework within which fishermen in developing fisheries are operating is now explicit, as are the rights associated with each stage of development.

The developing fishery stage represents the first of three stages in the life of a fishery:

i. the *developing fishery* stage (or "*interim managed developmental fishery*" for the purposes of s.58 of the FRMA). This initial stage is of limited duration and ends when predetermined benchmarks are reached and management changes are triggered (generally a three year period)
ii. the *interim managed fishery* stage which is, at maximum, three years in duration and which may be less if participant-initiated triggers move it towards managed fishery status and finally
iii. the *managed fishery* stage which sees long-term access allocated through either a management plan, or other subsidiary legislation.

It could be that a fishery never progresses beyond the developing stage if the mechanism for triggering the change never occurs and, or, the fishery is closed. Similarly, the fishery may be managed through a series of Regulation Licences, with the management arrangements established within the *Fish Resources Management Regulations 1995* (the *Regulations*), or through modification to existing Section 43 Orders (prohibition notices). However, this overall perspective enables both Fisheries WA and applicants to better address planning, administrative and budgeting needs.

5.2 Establishing a developing fishery - the participants

The Government recognises that fish resources are public resources available to all and that participation in a developing fishery is open to all. However, for biological and administrative reasons participation in a developing fishery is subject to some criteria and rules.

The *Regulations* specify those people who can actively use fish resources for commercial purposes as those who hold a Commercial Fishing Licence and, where a boat is involved, a Western Australian Fishing Boat Licence.

Therefore, the minimum requirements to participate actively in a developing fishery are:

i. a Commercial Fishing Licence and
ii. a Western Australian Fishing Boat Licence (where a boat is involved).

5.3 Establishing a developing fishery - the process

There are seven steps to the process of establishing a developing fishery. This process is designed to provide a decision on the proposed developing fishery within three to six months and is intended to accommodate the maximum number of interested parties and their submissions and to facilitate efforts towards developing new fisheries. The application process will be initiated twice a year, but may move to once a year depending on demand.

The steps of the application process are:

i. *Expression of interest:* Following an advertisement placed in *The West Australian* by Fisheries WA calling for expressions of interest, people may express their interest in a potential developmental fishery
ii. *Ministerial decision:* The Minister determines the necessity of the regulation regarding the take of the resource described in the expression of interest (*i.e.* a prohibition)[1]

---

[1] Those fishing activities not covered by a formal legislated management plan are open-access, however, there may be prohibitions over certain fishing activities and, or, certain areas. These are by way of an Order issued under Section 43 of the *Fish Resources Management Act 1994*. This section allows the Minister to 'prohibit persons or any specified class of persons from engaging in any fishing activity of a specified class'. These Orders are gazetted in the Government Gazette and are subsidiary legislation to the main fisheries Act. Thus in these situations

iii. *Application:* Based on the Minister's decision regarding the development of a fishery for the resource, people who have expressed their interest will be notified by Fisheries WA of the opportunity to submit a formal application form and business plan to develop the fishery
iv. *Assessment:* The assessment of all applications and business plans by the Developing Fishery Assessment Committee (DFAC)
v. *Notification of status:* The stage at which the applicant is notified of the approval/refusal of an application
vi. *Implementation/participation*: The stage when successful applicants commence participation and bring the developing fishery to life and
vii. *Review, assessment and modification:* The review of performance against business plans annually, with a full review of the developmental fishery, its condition, and status occurring at the end of the developmental period. Both the review and status are to be assessed by the DFAC using data and information supplied by Fisheries WA, participants and independent observers.

A diagrammatic representation of the total assessment process is shown in Figure 1.

6. TRANSFERABILITY IN THE DEVELOPMENTAL PHASE

Authorisations issued for developing fisheries will be non-transferable during the initial developmental phase. However, if at any time a review of the developing fishery is conducted and it is determined that the transfer of authorisations in the fishery is appropriate, any new licensee that enters the fishery after a transfer will be required to operate under the same business plan or conditions imposed under the FRMA. If this is not to be the case, the modified business plan must be submitted to the DFAC for consideration and approval by the Executive Director.

7. BEYOND THE DEVELOPING FISHERY
7.1 Trigger to move to the next stage
At the end of the developmental fishery period, the Executive Director will recommend to the Minister whether the fishery ceases, remains a developmental fishery, is managed as a 'Regulation' or 'S 43 Order' fishery or moves to the next stage of an interim managed fishery.

7.2 Subsequent participation
The Government supports the concept of clearly differentiating between the initial or pioneer participants in a developing fishery and those who subsequently want to participate. Because of this, the concepts of access and allocating access are important ones. In essence, there are two situations in which subsequent participation is an issue *i.e.* when the fully transferable authorisations are transferred, and when the addition of participants in a fishery is an administrative matter rather than an automatic occurrence.

In fisheries managed by input-controls and with limits on the number of total participants the addition of new participants beyond the initially declared number will automatically trigger the mechanisms for moving to an interim managed fishery. Thus, the issue of authorisations for these additional participants becomes part of the management arrangements for the interim managed fishery.

For those developing fisheries managed under transferable output-control regimes, the issue of access and subsequent access will be strictly a function of the private purchase of the individually transferable units. Regardless of how access is allocated, the Government has adopted a strategy that allows it to reserve the right to choose and design the allocation mechanisms on a case-by-case basis. Further, regardless of the method of allocation, the process would be advertised in *The West Australian* newspaper and any funds received as a result of the process would be used to cover, or subsidise, costs of management (including compliance and enforcement) and research in the fishery.

Subsequent entrants, *i.e.* those who have not incurred the costs associated with discovering, developing, and marketing new fisheries, may pay a premium for making use of information gleaned by pioneers. Those who have committed considerable amounts of time, equipment and resources should be able to recoup their exploratory costs during the post-development phase of the fishery.

Having described the allowances for additional access to developing fisheries, the situation may arise that the level of effort during the developmental period is too large. Should this be the case, an interim managed fishery management plan would need to be developed, using criteria that result in an appropriate level of effort remaining in the fishery.

8. CONCLUSION
Fisheries WA advertised for the first round of the DNF process on 15 October 1999. This initial round will be used to refine the process and has attracted wide interest and support throughout WA, including from peak industry bodies. In the future, Fisheries WA proposes to review existing developing fisheries by requiring fishermen with current developmental fisheries authorisations to make applications through the DNF process. This will

---

there is not usually a specific fishery to be developed or explored, it is more likely that the species to be targeted or the method to be used is currently prohibited and it is up to the applicant to convince the government that they should be excepted from that prohibition in order to explore the potential of a new fishery.

Figure 1
The developing fishery process

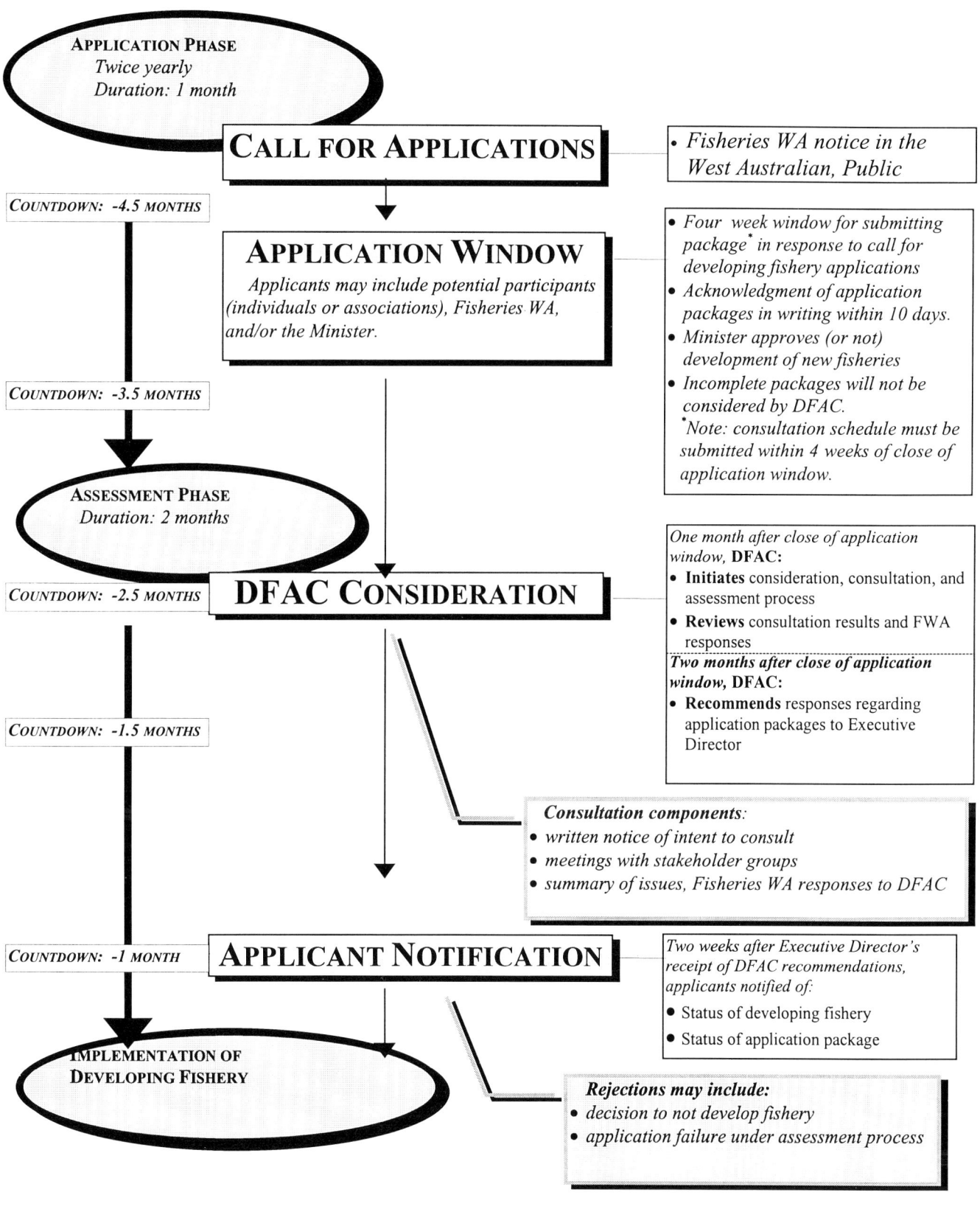

The developing fishery process *(continued)*

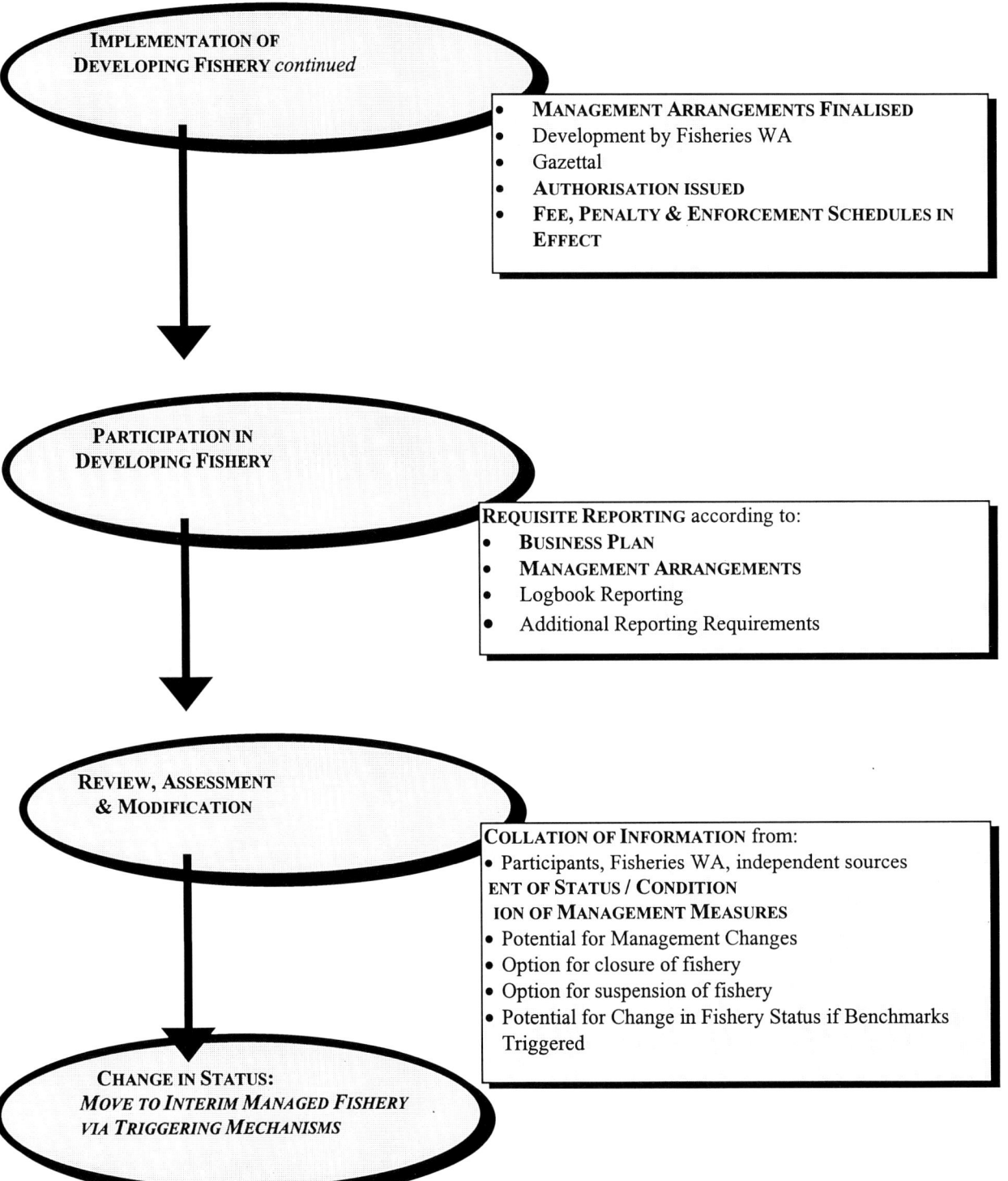

If a developing fishery is terminated, the licences will automatically be cancelled and there would be no continued right of access for those who bought into the fishery.

ensure that a consistent approach is applied to all developing fisheries in WA and that existing developmental fishery rights are clarified. The DNF process is an excellent example of co-management and is explicit in the delineation of rights of access to developmental fisheries.

9. LITERATURE CITED

Fisheries Western Australia 1999. Developing new fisheries in Western Australia. A guide to applicants for developing fisheries. Fisheries Management Paper No 130. 42pp.

INCORPORATION OF A FISH PROPERTY RIGHT: A HYPOTHETICAL EXAMPLE

W. Zacharin
Fisheries and Aquaculture, Primary Industries and Resources SA
GPO Box 1625, Adelaide, Australia 5001
< zacharin.will@saugov.sa.gov.au>
and
R. Edwards
Complete Fisheries Management
12 Greenhill Rd Wayville, Adelaide, Australia 5034
<redwards@gazebo.os.com.au>

1. INTRODUCTION

Few statutory fishing rights have been created in Australia or worldwide in relation to inshore marine resources. The majority of wild capture fisheries have access arrangements determined by one being the holder of a commercial fishing licence or permit, which is usually issued for a period of 12 months under the relevant fisheries legislation. All Australian States and Territories have formal consultative structures and mechanisms in place which provide advice to their Governments on the best management or access arrangements for specific fisheries. However, the management committees are in all cases advisory only and this advice may be accepted, amended or rejected by Government for a variety of reasons, including strong opposing views from other stakeholders who may be affected by a management decision.

This lack of fishing rights worldwide has been due to the inability to effectively overcome the common property nature of the resource which Hardin (1968) termed the "tragedy of the commons". It is becoming clear that 'limited entry' alone, which is the norm in Australian fisheries, does not necessarily overcome the 'tragedy of the commons'. We need only look at emerging demands for reallocation, or new allocations, of limited fish resources away from the commercial sector for 'use' in marine park exclusion zones, or for recreational fishers, tourists and charter operators, to understand that the ocean still is being treated as 'commons'.

It is incongruous that after many years of structural adjustment, intensive research and tightening of management controls aimed at achieving sustainability in the commercial sectors that insufficient notice is being taken of past experience and early warning signals of to effectively tackle the new 'tragedy of the commons'. These emerging trends, which can only occur in the absence of clearly defined property-rights for all user-groups, see a number of important impediments to development of fisheries. These derive primarily from lack of certainty.

The key impact on the commercial sector is the distortion of investment with an overriding imperative to take the short view. This results in sub-optimal decisions about development opportunities including:

i. investment in market assets such as brands, infrastructure and human resources
ii. investment in down-stream processing and value-adding
iii. investment in more environmentally friendly practices and facilities such as waste-disposal infrastructure
iv. voluntary stock research programmes and
v. catching capacity and related input supply services and infrastructure (*e.g.* fuel facilities).

The situation is further exacerbated by over-investment in:

i. financial and human resources to defend or capture the 'commons'
ii. gear and resources to ensure short-term profits
iii. methods to 'beat the system' and,
iv. government resources to deal with those competing for the 'commons'.

Similar costs and/or impediments to development in other sectors such as recreational and conservation are also evident and increasing.

If Government were to investigate moving forward in relation to independent day-to-day management of any marine resource, what would be the organisational implications of delegating responsibility for management of the resource and its habitat? Social researchers and fisheries managers have suggested that private 'ownership', or delegation of the stewardship role, is not appropriate for marine resources because of the common-property nature of the resource.

However, the agriculture, forestry and mining industries have gained legitimate long-term access to these crown resources. The Government has leased these resources to the private sector in return for an economic-rent from their exploitation. Can the same principles used to lease the access and management rights of mining and forestry resources be applied to the commercial fishing industry? What could motivate fishers to promote collective interests at the expense of individual interests? This paper describes the constraints, advantages and disadvantages of establishing a public company to manage a marine resource, and uses the Northern Zone rock lobster fishery in South Australia as an example. Other corporate models have been described by Townsend (1995a) and Townsend and Pooley (1995).

2. NORTHERN ZONE ROCK LOBSTER FISHERY

The Northern Zone rock lobster fishery[1] extends from the mouth of the River Murray west to the border with Western Australia and to 200 nautical miles offshore. There are 70 licence holders in the fishery with between 25 and 60 pots held per licence. The commercial catch in 1998-99 was 1016t and has averaged around 950t/reason over the past 10 years. Licences are issued on an annual basis.

The fishing season is from 1 November to 31 May. Fishing effort is controlled by input-controls, the main mechanisms being an innovative flexible time-closure system, restrictions on the number of pots, boat horse-power and a minimum lobster size (Zacharin 1997). Over the last 10 years virtually all latent effort has been removed from the fishery and ongoing adjustments are made by the industry to account for any potential effort-creep over time.

There is also a recreational fishery which is active in inshore waters and accounts for about 30t/yr. Recreational lobster are taken by diving, commercial pots and drop and hoop nets.

The fishery includes a number of marine protected areas, the most significant being the Great Australian Bight Marine Park (GABMP). The park includes a mix of multiple use and total exclusion zones, which traditionally have been commercial rock lobster fishing areas.

Advice about resource-management, allocation, cost-recovery and service-delivery is provided to the Minister by his appointed Fishery Management Committee. The committee follows a strategic plan and has a 5-year fishery management plan covering biological and economic objectives.

3. PUBLIC COMPANY CONCEPT

A public company in Australia is a company which has an unlimited number of members and may be listed on the Australian Stock Exchange (ASX). The Corporations Law in Australia considers the company to have 3 distinct elements: the legal/economic entity, the directors and the shareholders or members. The company must have a constitution (Memorandum of Association) and replaceable rules (Articles of Association).

In relation to the first element, the company as an economic entity has the ability to manage its own financial, physical and human resources to fulfil its primary objectives, which in most cases is profit. However, with the management of a marine resource this function would also include the determination of access arrangements, harvesting protocols, collection of licence fees from a variety of individuals or other companies harvesting the resource, and the responsibility for audits (both financial and physical) under the Corporations Law. This discussion primarily focuses on commercial access issues with some comments on access by other extractive and non-extractive users.

Assume at this point that the Government is able to lease the resource to the company for a period of 50 years (we discuss how this may be achieved later). The first question that must be addressed by the company is the selection of the directors by the primary shareholders. Under the Articles of Association, groups of shareholders may have a right to appoint one or more directors, and certainly in the first instance the Government would also wish to nominate a director. The requirement for a Government director (for example similar to the current arrangements with Telstra Corporation) would be necessary to enable the Government to fulfil its statutory obligations under the current legislation and common law.

Ownership of forests and mineral resources in all the instances we have investigated remains with the crown and it should be no different in this case. However, the Government in entering into a long-term lease of access agreement with the company would require the ability to nominate a director to the board to protect their ownership and interests in such issues as environmental management, monitoring sustainable resource use and equity issues.

Other different shareholders with a right to appoint directors should be current licence-holders and perhaps industry associations that represent a significant majority of licence-holders. The board may also consider representation from the rock lobster processing sector, recreational fishing interests and any traditional users of the resource.

An appropriate initial board structure may be as follows:

Chairman selected by the board members
Directors (4) nominated by current licence-holders and recreational interests
Government nominee
Non-executive directors (2)
Executive director (Chief Executive Officer)

For many fishers being a director of a public company would be legally different from their current experience on fishery management advisory committees. The Corporations Laws in most countries stipulate that directors owe a 'fiduciary duty' to the company. A fiduciary duty has been defined by the High Court of Australia as the duty to act with fidelity and trust to others. That is, the director must act honestly, in good faith, and to the best of his, or her, ability in the interests of the company the courts have treated the company as being the shareholders

---

[1] The rock lobster fishery jurisdiction extends to 200 nautical miles under the Offshore Constitutional Settlement agreement between the Commonwealth Government and that of South Australia. South Australia has management responsibility for the species and licences boats. No bycatch of lobster is permitted by any other boat.

or the members). The courts have, in some circumstances, also extended this to include future shareholders[2].

An interesting legal argument would be what obligations would there be under the Corporations Law on this company to prevent degradation of the resource or other negative impacts on future shareholders? It is highly recommended that any directors investigate liability insurance.

4. SHAREHOLDERS

The key feature of the operating arrangements would be that the right to access the resource for any purpose including commercial, recreational, charter and conservation, would be linked to the holding of shares. This would set the basis for involvement in the company, managing the resource, contribution to the costs of management and shifts in access shares between competing shareholders.

How then could shares be allocated to existing licence holders, future licence holders, investors or other interested parties? There are numerous permutations one can develop, the most radical being that the company purchases all existing licences under an agreed pricing arrangement. But, this option would be expensive and of no benefit to current licence holders. We suggest as an initial allocation mechanism, the following strategy.

All current licence holders are issued shares that reflect their current access to the fishery. This could be determined by either a simple or complex calculation based on the number of pots which is the existing 'currency' of access. As there are 3950 pots permitted in the Northern Zone rock lobster fishery, it would be simple to issue shares based on the number of pots held and their valuation. For example, 3950 pots at a current market value of $A33 000/pot gives a total pot valuation of $A130.35 million or 130.35 million $A1 shares. A licence holder with 60 pots could be issued a total of 1.98 million shares.

Only those holders with a minimum number of shares (25 pots or 825 000 shares) would be issued with a harvesting licence by the company (stipulated in the Articles of Association). Other shareholders that subsequently purchase shares would have to lease those shares to harvesters, or accumulate a minimum share-parcel to qualify for a harvester's licence from the company. As the fishery is managed by input-controls, share-holdings must match pot-allocations to respective harvesters, with 33 000 shares representing an 'active pot' in the fishery.

The same principles could be applied to the recreational sector with the current catch estimate being converted to shares. The current catch of about 30t is equivalent to the catch from 120 commercial pots and hence at $A33 000/pot an initial allocation of 3.96 million shares to the recreational sector would be made. At first the recreational shares may be held by the South Australian Government on behalf of the recreational fishing community.

The concept could be extended to incorporate the share of the resource used for marine parks. It is estimated about 10t of catch is not accessible in the GAB MP total exclusion area. Thus 1.32 million shares could be issued and held by the South Australian Government on behalf of non-extractive (conservation) stakeholders.

It must be remembered that the shares provide an access-right to take rock lobster granted by the company under the lease contract with the South Australian Government. They do not provide for ownership of a proportion of the resource. However, the shares would be considered as 'property' and would have all the rights of an asset in relation to ownership and transferability.

Share-trading would be similar to current transfer arrangements, where investors (fishers, conservation interests, recreational and government) may hold, accumulate and direct the use of shares. This share-trading process might be simplified if the fishery were managed by output-controls in the form of individual transferable quotas (ITQs) as would any mechanism for allocating recreational catch (*e.g.* individual animal tags on a fee-per-tag basis).

The structure would allow new and/or current user-groups seeking variations in access to the resource, to participate in a formal transparent mechanism with a commercial basis. Governments, on behalf of the community, could participate in the market, adjusting shares in line with community expectations as they change over time. The benefits in terms of clearly-defined shares and a secure investment climate for all user-groups cannot be understated. Re-allocation decisions would be determined by the market, a fair and transparent process.

5. LEASE AGREEEMENT

The Government would have responsibilities in the areas of sustainability, environment, public accountability, economic development and equity. The nature of the lease would be performance-based reflecting these responsibilities, with the most important issue for the company being the legal obligations stipulated by the Government for management of the resource.

Not only would the Government (on behalf of the community) require that the company harvest the resource in an ecologically sustainable fashion, but that the company also demonstrate sustainable management by contracting, or directly employing, various professional staff to conduct scientific assessment, environmental assessment and other services as required. The lease would require that the fishery be managed for optimal utilisation, while maintaining the resource base at a sustainable level set using biological performance indicators, such as catch per unit effort and exploitation rate. The lease would be explicit in these matters of rights and

---

[2] Jeffree v The National Companies &Securities Commission (1989) in the Western Australian Full Court.

responsibilities of the Government, the company and the shareholders.

Further conditions in relation to biological and environmental performance would be required in a five- year fishery management plan, that would be explicit about:

i. the term of the plan
ii. achieving minimum standards - biological and environmental
iii. harvest strategies
iv. approvals processes
v. reporting procedures
vi. compliance procedures and
vii. audit and review processes.

The term of the lease would be a minimum of 50 years with an 'evergreen' extension of 5 years at the end of each management plan period. Extension would be subject to meeting the conditions of the lease and approval of the management plan.

Other lease conditions would relate to minimum and maximum access provided to commercial fishers, recreational fishers, traditional users of the fishery and conservation interests as well as any other requirements for shareholding and trading. A maximum shareholding of 15% would apply to any one individual or company.

A lease or rent payment would be expected in consideration of exclusivity to the resource. The amount of payment may be influenced by the significant rural and regional economic benefits that the rock lobster industry currently generates, but the Government could expect some rent for the access-right.

A requirement would be that shareholders, through an agreed licensing system, make licence-fee payments for services required to manage their access to the resource and their contribution to the lease. In the case of the commercial and recreational sectors, this would cover research, compliance and management costs, while conservation costs might include marine park research, compliance and management.

Appropriate penalties for breach of the lease would apply including confiscation of shares. Procedures for dispute resolution would be predetermined as would circumstances where compensation may be applicable.

6. COSTS OF MANAGEMENT

At present all commercial licence holders pay full management-cost recovery to the South Australian Government under a fee for service arrangement. The current fees for the Northern Zone rock lobster fishery are approximately $A700 000, of which $A236 267 funds fisheries-compliance activities. The company may be able to reduce the need for a high level of compliance if shareholders and harvesters are motivated to adhere loyally to the regulations imposed for management of the resource[3].

Loyalty may be built up if shareholders have a greater say in the management regulations and broader involvement in the decision-making processes and their implementation. However, the company would still require a public enforcement agency for when harvesters breach the regulations. This cost would be borne by Government, but would be considered in determining the lease fee. A significant saving may be made in this area if compliance of the regulations by the shareholders were increased by company practices and procedures.

In regard to a number of other management costs, we have made an arbitrary assessment of costs for comparison in Table 1. Additional savings could be achieved by promoting greater use of the harvesters in providing data on the fishery for scientific and management purposes. Further reduction in scientific research costs could be achieved by the direct employment of a research officer for the company.

A major cost increase would occur in administration costs. At present this represents costs for corporate services from Government which are minimal. With a company, the Chairman, directors and chief executive officer must be paid a salary and this has been set at appropriate market rates. The 'other services' represent costs for ancillary programmes such as community awareness or additional research, which may be significantly reduced in alternate years. Overall, the indicative budget suggests a potential saving of about $A50 000.

Possible sources of revenue for the company may include licence fees from recreational fishers. There are approximately 3600 pots used by recreational fishers in the Northern Zone fishery. Current pot registration fees set by Government at $A45/pot and revenues of $A162 000/yr are received. Some of this revenue may have to be shared with Government for providing compliance services for this fishery. An agreement on licence fees for the recreational sector would be determined as part of the contract conditions to ensure fair access by this sector. Additional revenue may be raised by the company by issuing additional recreational pots or by conducting a voluntary share buy-back over time and auctioning or leasing those shares to new or existing harvesters.

A new cost would be the annual lease payment to the South Australian Government for exclusive access and management rights to the resource. As suggested, this fee should cover costs to the Government for enforcing regulations. Whether additional rent should be paid would depend on the wider communities views on leasing of the resource.

7. REPORTING REQUIREMENTS

Under the requirements of the Corporations Law, a company must submit annual accounts and an audit of financial resources. However, it is the additional requirements of Government which are likely to be negotiated as

---

[3] Hardin (1968) stated that 'the only kind of coercion I recommend is mutual coercion mutually agreed up by the majority of the people affected'.

Table 1
A comparison of management costs for the current arrangements and the proposed public company

Service	Existing management committee ($A,000)	Company structure ($A,000)
Scientific research	166.2	150.0
Economic research	3.6	3.6
Policy & management	31.9	30.0
Regulatory/licensing	34.3	15.0
Compliance	236.3	100.0
Directorate	20.4	220.0
Operational management costs	49.0	20.0
Extension officer	31.5	15.0
FRDC[1] levy	62.6	62.6
Environmental programme	30.0	30.0
Other services	43.0	10.0
TOTAL	708.8	656.2

[1] Fisheries Research and Development Corporation

part of the lease contract that would occupy the company's attention.

An integral part of the lease arrangement would be conditions to ensure that the rock lobster resource was not over-exploited, degraded in any fashion or managed in a way that would harm the adjacent marine environment. These conditions would be difficult to quantify and monitor, but are critical for the company and Government to establish if long-term access rights to a resource are to be agreed upon. This issue is probably best handled by establishing measurable biological performance indicators and reference points in a formal management plan that form part of the contract specifications.

Biological performance indicators currently used in the fishery are:

i. Catch per unit effort (kg/pot lift)
ii. Exploitation rate (the fraction of the population harvested annually)
iii. Egg production (a derived index using legal-sized females)
iv. Pre-recruit abundance (under-size catch per unit of effort) and
v. Mean size of rock lobster landed across the fishery.

An audit of the biological and environmental status of the fishery would be presented to the Government on an annual basis. These reports would be subject to external review by appropriately qualified scientists approved by both Government and the company. If no agreement could be reached within a specified period, a reviewer may be chosen by the Australian Securities Commission. This is similar to cases where there is dispute on the financial reporting of public companies in Australia.

The lease contract would have to include substantial financial penalties for breach of contract to enforce performance. The penalty may be dealt with by placing a constraint on future catches by the company for a specified period, particularly if the resource was being over-exploited. Or a direct financial penalty could be imposed. Such a payment would have to be recouped from revenue sources which would mean the current shareholders.

8. SHAREHOLDERS BENEFITS

What benefits would the proposed corporate model provide to shareholders and Government? The proposed structure and operating lease do not establish a 'profit' generating activity as such, as the only source of income would be from shareholders in the form of licence fees to cover costs of management. Hence 'dividends' to stakeholders in the traditional sense would not accrue.

Rather, the 'dividends' would take the form of better defined access and a framework for managing changes in resource shares over time and the resultant investment confidence for all shareholders. An increase in rock lobster biomass may also lead to higher share prices.

It has been established in a number of countries that natural resource management can be improved through the strengthening of property rights. For wild fisheries, the challenge is to devise a system that will make the incentives of those who have exclusive access to the resource converge with the public interest in the conservation and efficient utilisation of the resource (Pearse 1994).

Any management system must reward fishers for their collective effort and motivate all harvesters to think locally but act globally in exploiting the resource. There is no doubt that a collective co-operative approach could improve the performance of the fishery and the costs of management. Jentoft and McCay (1995) suggest that *"ultimately, fishers control to what extent a management system will work or not; almost no matter how much Government spends on compliance and enforcement"*.

Benefits of providing a company with long-term management rights may be:

i. Greater control and flexibility in setting annual catch levels
ii. Greater certainty in access arrangements to the resource
iii. Improved flexibility in management decision making (*e.g.* changes to minimum size, market responsiveness)
iv. Greater compliance by harvesters
v. Reduced management costs to harvesters
vi. Flexibility in the choice of service providers, such as those for research, compliance and market information and
vii. Capacity to raise funds in the market for other opportunities.

Benefits to the South Australian Government may include:

i. Reduced management costs of the fishery
ii. Reduced vulnerability to political pressure and the power of vested interests and
iii. Allocation decisions between commercial, recreational fishers and other stakeholders would be determined by agreed negotiation and formalised in the lease contract for a specified period.

This is not to say that the company once established might not engage in commercial activity. Adoption of a development charter would seem inevitable and would see a corporate approach to decisions about:

i. Investment in market development and investment
ii. Investment in lobster on-growing and culture
iii. Development of recreational and tourism infrastructure
iv. Marketing of recreational and tourism opportunities
v. Service delivery (*e.g.* management, communications, development, marketing, research and conservation) and
vi. Acquisition of catching and processing capacity.

Each of these activities has potential to generate profit and would require the raising of capital and the conduct of operations on a commercial basis.

What are the disadvantages of transferring management to a public company and changing the equity of current licence-holders? Would the lease contract confer any stronger property-right to the resource than currently exists? We suggest there are significant benefits in formally determining resource shares. Would private investors move into the market and purchase significant shareholdings, thereby changing the small-business nature of the fishery and its contribution to regional economic growth? This is likely to occur if the value of shares rises and fishers decide to invest their capital elsewhere. Would the actual costs of management increase asbiological and environmental monitoring and performance require increasing resources to enable the company to fulfil audit requirements? This is uncertain at this stage, though. Greater participation by harvesters may result in the perceived cost-savings.

9. SUMMARY

As limits to the lobster resource have become better defined in recent years, the Government, commercial fishers, recreational fishers and conservationists, have turned their collective energies to methods of maximising the benefits to the South Australian economy and community from the limited resource. An economic, or development, focus dictates that a commercial approach to resource-sharing as opposed to government-driven prescriptions are inevitable if all the benefits and efficiencies on offer are to be captured.

It may well be however, that Australian society is not yet prepared to see ownership of marine resources transferred to private interests as has happened in Japan. Fisheries management remains a political issue in western society and the 'manager' needs to pursue multiple conflicting goals.

Co-management and leasing long-term access rights to the resource takes a middle road between overall Government concerns for efficient resource utilisation and conservation, and local concerns for equal opportunities, self-determination and self-control (Jentoft 1989). However, a move to corporate management over cooperative management is an alternative step and perhaps provides a better mechanism for dealing with the rapid structural changes and globalisation of markets in the seafood industry.

There are some other examples of privatisation described in Pomeroy and Berkes (1997) which use pooled quota-holdings between user-groups or auction mechanisms to determine who gains resource access. Gonzalez (1996) has suggested territorial use rights to control access and fishing effort in the presence of an open-access situation while Townsend (1995b) has described a transferable dynamic stock-rights system. However, all these mechanisms seem to display some continuing problems over ownership and equity of access. Our proposed corporate model would create incentives for cooperation and should promote more effective, efficient and equitable management regimes which would benefit the whole community.

Rock lobster fisheries are single species fisheries using single gear which make them conducive to private management. The proposed leasing of the northern zone rock lobster fishery to a public company may be achievable because of the corporate culture already prevalent in the licence-holders participating in this fishery. Without a collective corporate view being taken by licence-holders, the move to a public company model will not be achievable.

10. LITERATURE CITED

Gonzalez, E. 1996. Territorial Use Rights in Chilean Fisheries. *Marine Resource Economics* 11, 211-218.

Hardin, G. 1968. The tragedy of the commons. *Science* 162, 1243-1248.

Jentoft, S. 1989. Fisheries Co-management. *Marine Policy* 13, 137-154.

Jentoft, S. and B. McCay 1995. User participation in fisheries management. Lessons drawn from international experiences. *Marine Policy* 19(3), 227-246.

Pearse, H. 1994. Fishing rights and fishing policy: The development of property rights as instruments of fisheries management. Proceedings of the World Fisheries Congress, New Delhi pp 76-91.

Pomeroy, R.S. and F. Berkes 1997. Two to Tango: The role of Government in Fisheries Co-management. *Marine Policy* 21(5), 465-480.

Townsend, R.E. 1995a. Fisheries self-governance: corporate or cooperative structures? *Marine Policy* 19(1), 39-45.

Townsend, R.E. 1995b. Transferable dynamic stock rights. *Marine Policy* 19(2), 153-158.

Townsend, R.E. and S.G. Pooley 1995. Corporate management of the Northwestern Hawaiian Islands lobster fishery. *Ocean & Coastal Man.* 28(1-3), 63-83.

Zacharin, W.F. 1997. Management plan for the South Australian northern zone rock lobster fishery. *SA Fish. Man. Ser.* 28, 1-24.

CHALLENGES TO THE CO-EXISTENCE OF MARINE FARMING AND CAPTURE FISHERIES IN NEW ZEALAND

K. Drummond, P. Kirk and L. Nelson
Ministry of Fisheries
Private Bag 14, Nelson, New Zealand
<drummonk, kirkp or nelsonl@fish.govt.nz>

1. INTRODUCTION[1]

Since the mid-1980s the focus of New Zealand's fisheries policy has been to establish a rights-based framework for capture fisheries. The aim of this policy has been to encourage efficient resource allocation in the fisheries sector, while ensuring the catch is sustainable. The fisheries sector includes customary, recreational and commercial fishing interests. Since fisheries are a common pool resource, these groups frequently have rival demands for the limited yields available on a sustainable basis. There is also rivalry within the groups, particularly the commercial fishing interests, as different subgroups seek to maximise yield from particular species. The use of coastal space for marine farming development adds to this rivalry.

Faced with competing demands for a common pool resource, the government has established an institutional framework which has evolved whereby the government sets the environmental limits to harvesting and establishes rules for use and access to fisheries resources that enable optimal use to be achieved.

The government does not seek to determine or dictate the optimal use of fisheries resources. Its aim is to set the boundaries within which optimal use can emerge as a reflection of individual choices, which in turn reflect the individual values, both commercial and non-commercial that derive from harvesting fisheries resources.

The challenge this paper discusses is to extend this framework to better encompass the rivalry between marine farming and capture fisheries. Other papers presented to these proceedings discuss the challenge of improving the integration of recreational and customary fisheries into this framework in New Zealand.

The underlying principle is that optimal use of resources will emerge as a result of providing individuals with tools to enable them to achieve their own well-being. Such tools can only be exercised subject to not adversely affecting the environment (including not jeopardising the potential productivity for future generations). This principle is common across New Zealand's natural resource management legislation.

The tools provided to achieve optimal use involve the definition and allocation of fisheries rights. Specification of rights, underpinned by regulatory requirements to ensure sustainability, provides a mechanism for environmental protection to be achieved at least cost and for optimal use to emerge through individual choices.

Imperfect or incomplete specification and allocation of rights occurs in New Zealand. This is due either to incomplete development of the policy, such as is the case for recreational fishing, or to difficulty in establishing a complete set of rights that accurately reflects resource scarcity and the full range of benefits derived from fisheries resources and coastal space. In these circumstances, the rights-based approach to fisheries management must have a complementary regulatory environment which facilitates consideration of the trade-offs associated with transferring, or abridging, use rights from one group to another.

2. EXISTING FRAMEWORK
2.1 Capture fisheries

The existing institutional framework to manage fisheries resources has established rights of access for customary fishers, recreational fishers, commercial fishers, and marine farmers. These rights may be represented within coastal communities, but the extent to which this occurs in practice varies between communities.

For capture fisheries an annual total allowable catch is set which covers all harvesting from capture fisheries. The stock-specific total allowable catch is set with reference to achieving maximum sustainable yield over time. Modifications to the catch limit are made to take account of the needs of interdependent species, and ancillary restrictions are put in place to control adverse effects of fishing on the environment, including effects on marine biodiversity.

Customary fishing is managed within a regime implemented by guardians nominated by Māori. The guardians are responsible for sustainable management and authorising customary fishing within their area. This approach allows for area-based fisheries management rights to be established, and guardians are now being appointed around New Zealand.

Recreational fishing requires no authorisation. It takes place within an allowance determined by the government, and managed by the government using daily bag-limits, and controls on methods and seasons, all of which may be area specific.

Commercial fishing takes place within the constraint of an annual total allowable commercial catch (TACC)

---

[1] The views in this paper are the views of the authors, and not of the Ministry of Fisheries or the New Zealand Government.

which is set after allowing for customary and recreational use. It is area and stock specific. Commercial harvesting rights are allocated as individual transferable quota (ITQ), which represents a share of the TACC. Any variation in the TACC, for instance due to changes in the total catch limit or the allowance made for customary and recreational use, results in a change in the tonnage represented by a particular ITQ holding. Commercial fishing rights cannot be exercised in some areas set aside for customary or recreational use, but such areas can only be set up after taking into account the effect on commercial fishers.

2.2 Marine farming

Marine farming is authorised on a case-by-case basis. The framework requires two 'consents': one (a 'structure consent') from the local government and one (a 'farming consent) from the Ministry of Fisheries (a national government body). The consents are issued under separate pieces of legislation and deal with different aspects of marine farming.

The 'structure consent' authorises the placement of marine farming structures in the coastal area. In reaching a determination on this consent, the local government considers the impact of structures on the environment and on non-fishing interests in the coastal area. This structure consent is limited in duration, may be partly exclusive and is transferable. The local government decision can be appealed, in the first instance to an environment court.

The 'farming consent' authorises the possession, holding and growing of fish. In reaching a determination the Ministry of Fisheries is required to consider the impact of marine farming on fishing interests and on the sustainability of the fisheries resource. This farming consent can only be issued to the holder of a structure consent, is limited to the term of the structure consent, is exclusive in terms of ownership of the farmed product, and is transferable. The Ministry of Fisheries decision to either approve or decline a farming consent is not subject to appeal other than through judicial review. Given that this decision has the potential to affect existing fishing rights, it is unusual that there is no legal process available to challenge any encroachment of these rights. The farming consent also imposes requirements for farmers to record product flow, primarily to assist with detecting laundering of fish illegally obtained from capture fisheries through the marine farming system.

2.3 Coastal communities

The extent to which coastal communities exercise rights to capture fisheries or marine farming will depend on employment and leisure opportunities available to a particular community. In general, customary fishing has a strong community base and up to 20% of all New Zealanders are estimated to exercise their right to go recreational fishing in any one year. Direct community interest in commercial harvesting depends on factors such as the physical nature of the coast, the historical use of the area, recent demographic trends and proximity to processing capacity and freight services. However on the Chatham Islands a community trust has been established to manage certain ITQ rights.

Coastal communities also derive benefits from non-extractive use of the coastal marine area. The preferred form of use for a particular community may, for instance, focus on maximising the tourism benefit that can be derived from the maintenance of undeveloped seascapes, estuaries and wetland habitat. To manage the use of the coastal marine area, government has established a regional planning framework that allows the balance between resource development and production to be determined at a regional or local level via public participation in the creation of coastal management plans.

It is within the context of such plans that the opportunities for marine farming are effectively determined and applications for structure consents are considered by the local government authority. These plans are, however, prevented from establishing rules that are for the purpose of controlling the effects of fishing, the presumption being that any harvesting must occur within environmental limits. However, there remains potential for planning decisions to affect capture fishery rights-holders if the rules are determined for other reasons.

Coastal plans are at various stages of development in New Zealand. Decisions made under these plans have, in some cases, directed the potential for marine farming into an area where such development could undermine fisheries rights. This possibility has led to rights-holders in commercial capture fisheries taking legal action in an attempt to protect their rights under the coastal planning process. This action is occurring prior to the Ministry of Fisheries considering whether or not a farming consent can be issued.

2.4 Co-existence

In summary, within the existing framework harvesting rights provide customary, recreational and commercial fishing interest's access to a fishery provided the adverse effects of fishing on the environment are adequately controlled. Marine farming rights are a right to use a defined area for the purpose of marine farming provided the adverse effects on the environment are controlled and there is no undue adverse effect on fishing or other coastal interests. The poor fit of rights arises because evaluation of the impacts of marine farming on fishing interests is not integrated with the evaluation of marine farming impacts on other coastal interests. Opportunities to reach mutually agreeable outcomes are foreclosed by the sequential nature of the dual authorisation process.

This 'separation' in the decision-making process matters because both marine farming and fishing are significant contributors to New Zealand's social, cultural and economic well-being. New Zealand is predominantly a coastal society — fishing is a highly valued activity, both for its commercial and non-commercial benefits. Māori cultural ties with fishing are strong. Recreational fishing is a popular activity. ITQs have provided a base for significant investment in harvesting and processing technologies, and marketing. Marine farming is a substantial production sector already with scope for further development. Progress is also being made in establishing capacity within the fisheries sectors to assume collective responsibility for fisheries resource management.

Failure to integrate the evaluation of impacts on existing users, both fishing and non-fishing, when authorising the establishment of new marine farms, leads to poor resource management outcomes and poor allocation of resources. As the fishing and marine farming sectors move to maximise their access to resources (be they fisheries resources or coastal space), effort is spent in securing outcomes through political lobbying (at both the national and local government levels) rather than through working agreements between the various interests. This has sometimes resulted in costly appeals to the courts, as fishing interests seek to protect their rights and marine farming interests seek to establish theirs, within the local government authorisation process.

Expansion of marine farming often can only come at the expense of capture fisheries production. Points of tension between marine farming and capture fisheries arise over access to:

i. the productive capacity of the ecosystem, *e.g.* nutrient flows or habitat
ii. water space and
iii. spat, juveniles or broodstock required to stock marine farms.

At present there is no incentive for parties (marine farm applicants or existing fishing or farming interests) to enter into direct negotiation with each other. In addition, there is an absence of capability to assume the necessary collective responsibility to enter into such agreements. And, the absence of incentives to reach agreements is in itself discouraging the development of collective capacity. This situation will remain as long as government retains the authority to re-allocate resources.

Agreements for the co-existence of capture fisheries and marine farming need to be stable. Stability encourages negotiating in good faith, and discourages parties from seeking to undermine the agreement outside of the process. A stable agreement between capture fisheries and marine farmers will depend on the agreement also embracing the legitimate interests of other users of coastal space. These interests may well be driven by non-fishing values.

Under the existing framework, decisions by local or national government, whether or not supported by agreements, may establish precedents for future decisions. This possibility actively discourages parties with collective management capacity from direct negotiations with other interested parties, for fear of having any agreement interpreted more widely than was intended by interests not bound by a collective body.

3. POLICY REFORM

Policy reform is required to create a mechanism whereby capture fishing and marine farming interests can negotiate their own durable access agreements. A pure rights-based approach suggests the need to create a new 'development' right to coastal space or coastal productivity that could be freely traded between the interests (including community interests). Definition and allocation of such a set of rights would enable coastal development to occur in the absence of a planning regime. However optimal resource allocation would be highly dependent on whether the right was correctly defined (*i.e.* whether it embraced all the relevant development features) at the start, and on choosing an efficient allocation method since transaction costs will not be negligible.

In New Zealand collective management capacity is in the early stages of development within the capture fishery and marine farming sectors. While considerable progress has occurred with some groups, most notably within the commercial sector, the government must retain a role in evaluating agreements to ensure they are equitable. This role should continue until such time as the parties have established sufficient collective capacity to be representatives of particular interests.

We suggest adopting a pragmatic approach to policy development for the use of fisheries resources. In our view incremental reform can secure a fair process to redistribute access to coastal resources and enable optimum resource allocation. This could be achieved by ensuring that the marine farming planning framework complements the rights-based framework for fishing, by encouraging and recognising inter-sectoral access agreements. Such a reform encourages the use of a market mechanism without the need to create and allocate a new coastal development right.

If the best use of an area is for marine farming, it is desirable that market mechanisms reveal this best use. The costs to commercial and non-commercial fishing interests from restricted access to a particular area or from the downstream environmental impacts of marine farming (*e.g.* redistribution of nutrient flows or attraction of predator species) should be less than the benefits to marine farmers. In this situation there is a net gain to be made from changing the use of the area from fishing to marine farming. Agreement should be able to be reached between fishing and marine farming interests to enable marine farming to become established. Transaction costs associated with these agreements, and mechanisms that lower them, will be important considerations in designing a framework for managing the change in use.

The government should provide a framework for any re-allocation of resources between interests with collective management capacity to be dictated by individual choice. In time, that framework may be able to be extended to all fishing interests. This could be achieved through direct negotiation on behalf of an interest group, but more likely through a regulatory role by placing limits on the extent that agreements between parties with collective management capacity can affect other interests.

We consider that any agreements established between the marine farming and capture fishery rights-holders must also be considered alongside the interests of coastal communities. The government has already established a framework that has an underlying presumption that local government has knowledge of the resource management issues of its region, apart from those that relate to the management of capture fisheries – where it is presumed central government has that knowledge. This framework requires adjustment so that, where environmental limits permit, negotiated arrangements can be readily factored into the coastal planning framework. This approach would enhance the prospect of transparently considering the full range of costs and benefits of rules governing the use of the coastal marine area.

In situations where marine farming and capture fishery rights-holders are unable to reach agreements, adjustments to the coastal planning framework are still required to ensure the rights of existing capture fishers and marine farmers are explicitly considered in the context of the productive capacity of the ecosystem, including nutrient flows and habitat. This is to avoid local government inadvertently providing for marine farming in areas where it would have an undue adverse impact on production capacity for naturally occurring stocks or existing farmed stocks.

4. CONCLUSION

Since the mid-1980s the New Zealand government has progressively developed a rights-based approach to the management of natural resources. In the coastal marine area the approach adopted can be characterised as one of setting the environmental limits for development, while use has been enabled by a combination of:

i.   allocating rights for use and access to fisheries resources in a manner that will, in time, enable the optimum use to be determined for each class of rights-holders and
ii. establishing a coastal planning framework that is intended to allow for optimal use of the natural and physical resources (other than for fisheries).

The aim of government over this period has been to develop a resource management system that reflects both scarcity and the multitude of benefits derived from fisheries resources and coastal space. However the approach adopted to achieve the two forms of benefits has lead to a dislocation in the management of the coastal marine area. In our view this represents an institutional failure.

The main fault in the current approach is that it does not enable the full range of use-options that are available to capture fisheries interests, marine farmers and coastal communities to be considered concurrently. A regulatory environment that is complementary to the environmental limits and enabling goals is required, so that the full trade-off associated with transferring or abridging use-rights from one group to another, is explicitly recognised.

We suggest that the underlying principles that should guide the development of a revised framework are as follows:

i.   within the context of firm environmental limits, the framework should maximise opportunities for innovative solutions to resource use conflicts
ii. redistribution of access or use-rights to resources should depend, to the extent practical, on individuals' evaluation of the trade-offs involved
iii. when a decision is made to reallocate coastal resources of any sort, the full cost and benefits of that decision should be considered and
iv. to ensure the cost and benefits of reallocation are considered, the institutional framework should encourage negotiation.

Given the imperfect nature of rights so far allocated, which prevents free trade between the interests, we further suggest that the institutional framework provide for:

i.   evaluation of agreements to ensure they do not adversely affect other interests, or interests involved in the agreements who do not have collective management capacity and
ii. registering agreements struck through negotiation so that the coastal planning process can take into account the internalisation of costs and benefits which has occurred.

Given that the current environment of developing collective management capacity in New Zealand has, until recently, focused on the commercial sector, we consider that there remains a core role for government in representing non-commercial fishing interests in the coastal marine area. This includes recreational fishing interests and coastal communities. It will also include the interests of customary Māori until the customary management regime develops to the point where guardians prefer to manage the interface with marine farming in their area directly.

THE ROLE OF PROPERTY RIGHTS IN THE DEVELOPMENT OF NEW ZEALAND'S MARINE FARMING INDUSTRY

M. Harte
New Zealand Seafood Industry Council
Private Bag 24-901, Wellington, New Zealand
<michael@seafood.co.nz>
and
R. Bess
School of Business
Nelson Polytechnic, Private Bag 19, Nelson, New Zealand
<rbess@nelpoly.ac.nz>

1. INTRODUCTION[1]

The ecological and economic performance of the New Zealand seafood industry has improved dramatically in the last 15 years (Annala 1996). The implementation in 1986 of a property rights-based quota management system (QMS), based on individual transferable quota (ITQ) and permits for non-ITQ fisheries for wild fisheries, has been a key reason for this growth. Another reason has been the more recent rapid growth of marine farming. Although the QMS provides clear and appropriate property rights as the basis to managing wild fisheries, the development of a property rights framework for managing marine farming has been slow to come.

A recent independent review of fisheries legislation and its administration commissioned by the New Zealand Government found that legislation regulating marine farming in New Zealand is fragmented and outdated (PricewaterhouseCoopers 1998). These legislative arrangements and their historical development raise two critical issues for the marine farming industry:

i. Uncertainty and erosion of marine farming rights and
ii. Poor integration of marine farming rights with those of wild fisheries.

Failure to explore and adopt innovative policies to address the potential insecurity of marine farming tenure and inconsistencies between marine farming and the quota management system for wild fisheries could jeopardise the New Zealand marine farming industry's future growth and success in international markets. The integrity of the quota management system could also be undermined.

The first part of this paper describes the performance and management of New Zealand's marine farming and wild fisheries. The section contrasts the quality of existing rights for ITQ owners with the rights of marine farmers. The second part discusses the future management of New Zealand's fisheries and suggests two solutions to improve the security of marine farming property rights and resolve some inconsistencies in the rights and responsibilities for marine farming and wild fisheries. These solutions highlight the need for institutional changes to the management of wild fisheries and marine farming in order to strengthen the property rights basis for their economic and ecological success.

2. SUSTAINING THE SUCCESS OF NEW ZEALAND'S FISHERIES

2.1 Management and performance of the marine farming industry

Marine farming in New Zealand dates back to early this century when inter-tidal cultivation of rock oysters began. By the 1960s permitted marine farmers had developed more sophisticated catching and growing operations with the use of inter-tidal racks holding sticks and trays. In the early 1970s rock oyster farming accidentally introduced the pacific oyster, which continues to produce significant volumes for export. In the early 1970s the Greenshell™ mussel became another farmed species. Since the late 1970s, the Greenshell™ mussel has experienced substantial growth in the volume and value of exports. In the early 1980s the farming of salmon in seawater cages was established.

In the last decade the volume and value of marine farming exports have increased dramatically. Figure 1 shows the volume of exports from marine farming, and Figure 2 shows the value of exports. Most of the growth in marine farming has been driven by the outstanding success of the Greenshell™ mussel, which has surpassed the rate of growth of all other seafood exports. From 1988 to 1998 Greenshell™ mussel export volume increased by 473%, and export value increased by 413%. In 1998 Greenshell™ mussels became the second highest valued seafood export species. The export volume of New Zealand's farmed salmon also grew steadily from 1988 to 1996, but the export value grew at a lower rate. Salmon prices on world markets declined as record growth in farmed salmon, particularly from Norway and Chile, substantially increased availability (FIB 1996). Farmed oysters increased in export value from $NZ4 million in 1988 to $NZ8 million in 1992 and then remained fairly constant at $NZ10-$11 million from 1994 to 1998. The volume of farmed oyster exports increased steadily from 734t in 1988 to a high of 1436t in 1995.

---

[1] Views expressed are those of the authors alone and not those of their respective organisations.

Figure 1
Volume of exports from the New Zealand marine farming industry

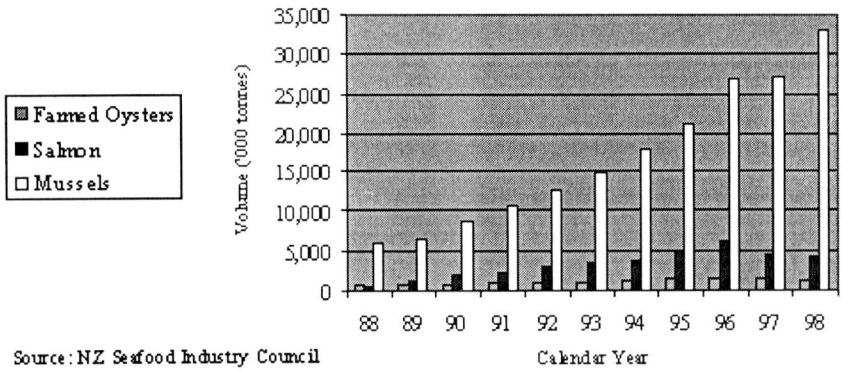

Source: NZ Seafood Industry Council

Figure 2
Value of exports from the New Zealand marine farming industry

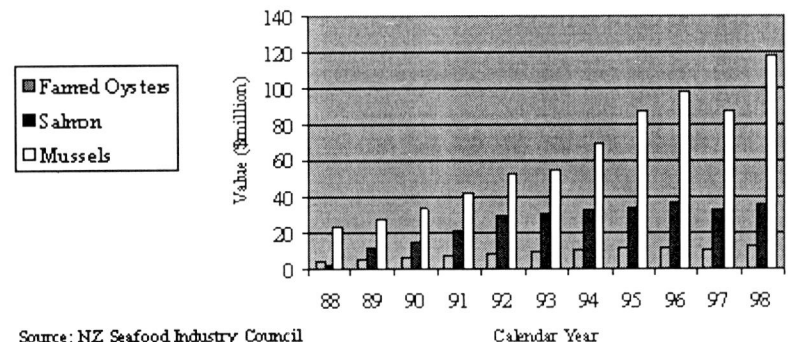

Source: NZ Seafood Industry Council

The increases in export volume and value are attributable to the investments made over the previous years and collective efforts that have brought about several innovations in harvesting, processing, marketing and farm management techniques. For example, the Greenshell™ mussel sector has co-ordinated collective effort on issues such as the development of:

i. an environmental policy and code of practice that help ensure a high level of environmental protection and sustainability
ii. export market studies that assist industry players to enter or expand overseas markets
iii. a portfolio of research projects that ensures relevant research is carried out with industry support and
iv. expansion of the Greenshell trademark.

Two different management regimes regulate marine farming:

i. the *Marine Farming Act (MFA) 1971*, which applies to around 680 farms that were established before enactment of the Resource Management Act (RMA) in 1991 and
ii. the joint *RMA and Fisheries Act 1983 (FA 1983)* regime, which applies to approximately 120 farms that have been established since 1991.

Until 1991 marine farm licences were issued under the MFA, which provided tenure of up to 14 years, with a preference right for an extension of 14 years, or less where considered appropriate. The MFA was not intended to create a perpetual right of coastal occupation. Conditions placed on MFA leases and licences including conditions relating to tenure, can be varied. In practice, however, applications for variation of conditions relating to the extension of tenure are routinely granted. Marine farms established up till 1991 by way of leases and licences under the MFA 1971 can continue operations indefinitely pursuant to the RMA, Section 426. The marine farming industry, therefore, considers the arrangement under the MFA to represent the granting of a perpetual right to occupation.

Since 1991, applications for marine farms have come under the joint regim RMA and FA 1983 e. Under this regime marine farmers must obtain (a) a RMA coastal permit for occupation of space and management of environmental effects and (b) a FA 1983 marine farming permit for the possession of stock. Under this joint regime, coastal permits and marine farming permits can be granted for up to 35 years. In practice, however, permits for coastal occupancy are generally granted for much less than 35 years, and the granting of marine farming permits are matched accordingly. The marine farming industry regard the security of tenure provided them under the joint regime to be significantly less than the security provided under the MFA.

The RMA, FA 1983 and the *Fisheries Act 1996* (FA 1996) set out obligations to manage the adverse effects of marine farm activities on the aquatic environment. The RMA resource-consent application for marine farming and spat-catching includes the obligation to avoid, remedy or mitigate any adverse effect on the environment arising from an activity carried out, by, or on behalf, of a person. The FA 1983 provides permits for those involved in marine farming or spat-catching who hold a resource-consent authorising the activity under the RMA. The Chief Executive of the Ministry of Fisheries (MFish) may only issue a marine farming permit if satisfied that the effects of the activity would not have undue adverse effects on fishing or the sustainability of any fisheries resource. The term 'sustainability' is not defined in the FA 1983, although it is defined in its successor, the FA 1996. However, 'fishery resource', according to the FA 1983, means any fishery, stock, species, habitat, or location of fish, aquatic life, or seaweed. This definition encompasses a range of environmental parameters that overlap with obligations under the RMA.

In line with the RMA, the FA 1996 contains an obligation to avoid, remedy or mitigate the adverse effects of, in this case, fishing on the aquatic environment. 'Fishing' includes the activity of harvesting, which is the end point of marine farming, but it does not refer to marine farming's other core element, growing the marine stock. Therefore, it is unclear whether the environmental considerations of the FA 1996 apply to marine farming, and so far this point has not been tested in the courts. Despite this lack of clarity, marine farming continues to be managed jointly under the FA 1983 and the RMA.

Problems with overlapping environmental regulations stem from the jurisdictional boundaries between fisheries law and the RMA. Provisions in the RMA and FA 1996 attempt to define these boundaries, but they, as well as associated compliance costs, remain unclear. Overlapping environmental regulations and compliance costs should be rectified by legislative changes that clearly state which environmental effects should be regulated by which legislation.

As demonstrated, the joint FA 1983 and RMA regime provides little clarity about the rights of marine farmers. The existence of two management regimes, the MFA and the FA 1983/RMA, for marine farming generates confusion and unnecessary uncertainty for the industry about which regime applies to a particular situation. These two regimes make it difficult for the Government to provide an integrated and consistent approach to the management of marine farming. The continuation of these two regimes increases the likelihood that local and central government and other interests could erode the rights, and hence economic security, of the marine farming industry. Understandably, under this legislative confusion, the marine farming industry has experienced difficulty in securing investment financing and this industry continues to work towards legislation reform. Until its property rights are defined and established, the industry's growth prospects will remain uncertain.

2.2 The quota management system and the performance of wild fisheries

The QMS and allocation of ITQ was introduced in New Zealand with the *Fisheries Amendment Act 1986* (FAA 1986). This represented a radical departure from previous fisheries management regimes. Detailed descriptions of the management of New Zealand's fisheries and of events leading up to the introduction of property rights-based management can be found in Sharp (1997), Gaffney (1997) and Harding (1991). The seven founding aims of the QMS (Luxton 1997) were to:

i. rebuild inshore fisheries where required
ii. ensure that catches were limited to levels that could be sustained over the long term
iii. ensure that catches were harvested efficiently with maximum benefit to the industry and to New Zealand
iv. allocate catch entitlements equitably based on individual permit holder's commitment to the fishery
v. integrate management of inshore and offshore fisheries
vi. develop a management system applicable on both national- and regional-bases and
vii. enhance the recreational fishery.

The QMS as introduced in the FAA 1986 was initially viewed by most in the industry as a relatively simple and workable management system. However, the FAA 1986 retained several aspects of the FA 1983, including Mfish's retention of power to impose various input controls such as restrictions on fishing gear, fishing methods, landings, fish size, fishing seasons and fishing areas. These input controls were still required for management of non-ITQ species. The traditional input controls implemented under Fisheries Management Plans (FMPs), as required under the FA 1983, contradicted the basis of ITQ where quota owners are able to determine the most efficient timing and means of catching their quota. Under FMPs, however, a total allowable commercial catch (TACC) managed in part with input controls could potentially impinge on quota owners' rights as created by the QMS. The need to run dual management systems leads to inconsistencies in management practice and makes it more difficult to achieve the QMS's intended degree of efficiency and co-ordination (Falloon 1993).

The QMS has been revised continually, requiring substantial time and effort by the MFish and the industry to implement legislation and policy changes. The relatively simple QMS has become complex, bureaucratic and expensive to manage while the industry requests that fisheries management be simplified. The industry contends that the increasing complexity and bureaucracy of the QMS has imposed unnecessary financial costs on individual fishers and fishing firms and has not provided the industry overall with corresponding benefits (Horton 1997).

Figure 3
New Zealand seafood exports by volume and value

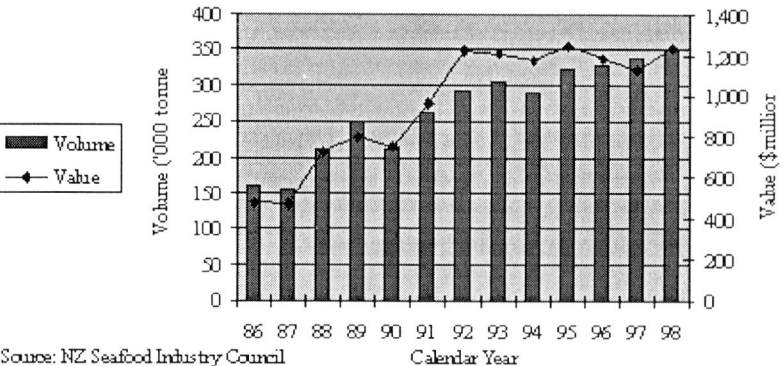

Source: NZ Seafood Industry Council

The strengths and weaknesses of New Zealand's QMS are well documented in fisheries management literature (Batstone and Sharp 1999, Clark et al. 1988, Dewees 1989, Memon and Cullen 1992, Sissenwine and Mace 1992). New Zealand's QMS stands out from those of other nations since New Zealand initially applied ITQ to the majority of commercially caught species. Iceland and New Zealand remain the only two nations that have implemented ITQ comprehensively. Today New Zealand's QMS has over 180 fish stocks present in 10 Quota Management Areas covering 40 species out of 100 species caught commercially. This represents over 85% of the total known fish catch in the 200-mile exclusive economic zone. It is the MFish's intention to bring more fish stocks under the control of the QMS over the next two to three years and eventually place all commercially caught species under the QMS. However, the QMS and allocation of ITQ is less suitable to marine farming since it does not include wild fisheries.

There is general recognition that the QMS has played a significant role in improving the biological status of the fisheries resource and commercial return to fishers (Annala 1996). Since implementation of the QMS, the industry has experienced steady and impressive growth in the volume and value of production. The surprising rates of growth experienced during the late 1980s and early 1990s was due primarily to the expansion of the deepwater fisheries. Between 1986 and 1989 the value of seafood exports increased by an astonishing rate of 69.3%. After a slight decline in the levels of production and value in 1990, the seafood industry again experienced dramatic growth. Between 1990 and 1992 the value of seafood exports increased 65.5%. Since 1992 overall seafood export value has remained fairly constant. In 1998 export value was $NZ1.237 billion, $NZ18 million higher than the 1992 export value. The gradual appreciation of the New Zealand dollar from 1992 to 1997 exacerbated poor international trading conditions and reduced returns to seafood firms. During this time catch-levels declined due to reductions in some TACCs (FIB 1996). Without the outstanding growth in marine farming exports, particularly Greenshell™ mussels, overall seafood export value and volume would have declined beginning 1992. Figure 3 shows New Zealand's seafood export value and volume from 1986 to 1998.

2.3 Comparison of property rights – ITQ owners and marine farmers

Scott's (1988) six characteristics – duration, flexibility, exclusivity, quality of title, transferability and divisibility – are useful in comparing the property rights granted under the QMS and the marine farming legislation. For simplicity, the rights of ITQ owners are compared to the rights of marine farmers under the joint RMA/FA 1983 regime. This comparison is an important step in better defining the marine farming and fishing rights. Each combination of characteristics can be shown by the six-pointed, star-shaped figure formed by joining the measured points on the six characteristic axes as illustrated in Figures 4 and 5. A property-rights regime that maximises all characteristics creates a large hexagon when the end points of each axis are linked. The mapping of characteristic scores helps reveal the differences in the specification of property rights for ITQ owners and marine farmers.

Duration refers to the time-frame that property rights are in effect. A short duration leads to costly or uncertain renewal and, or, extension of property rights. More permanent duration is valuable to property rights holders as it reduces renewal costs and uncertainty and raises incentives to invest for the long term. Similarly, the more likely property rights holders will be to invest in enhancing their fisheries. ITQ owners have durable property rights since ITQ is in perpetuity, subject to changes in TACC.

Flexibility refers to the ability of property rights holders to structure operations to achieve goals of their choice such as maximising profits by way of increasing the value of their catch rather than the volume. Flexibility in the exercise of rights and responsibilities is similar for both ITQ owners and marine farmers.

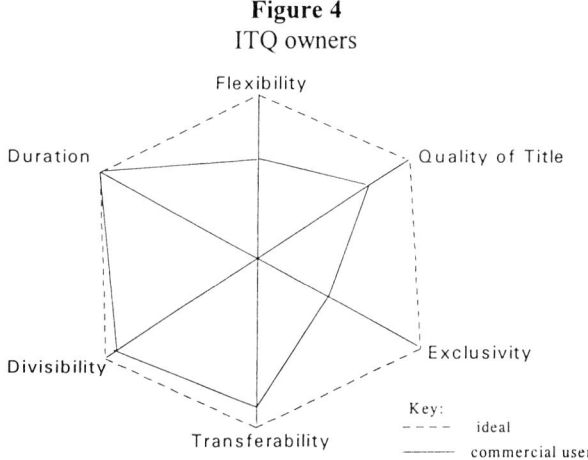

Figure 4
ITQ owners

Exclusivity refers to the extent that a person's property rights overlap with the rights of others. More exclusive rights are less likely to have operational clashes with other property rights holders and more likely that similar rights holders will co-ordinate their activities. Since ITQ owners compete to exercise their rights to a common fishery and, or, common fishing grounds, their harvest rights are less exclusive than the rights of marine farmers who have sole occupancy of a portion of coastal space.

Transferability refers to the ability to transfer title to property rights, thereby providing more efficient operators the option to buy rights from less efficient operators. ITQ is an instrument for transferability, which assists retirement from fisheries and reduces overcapitalisation. Only ITQ owners' rights are fully transferable.

Divisibility refers to the ability to divide (a) property rights more narrowly, producing new recognised rights specified perhaps by season, region, ground, species, age or other classification and (b), the amount of quota into smaller amounts and to transfer some quota to others. Only ITQ owners' rights are fully divisible.

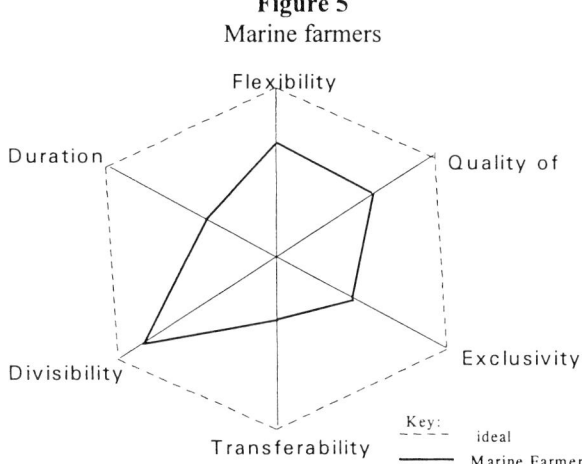

Figure 5
Marine farmers

Quality of title refers to certainty and security. The more predictable entitlement of the property rights the higher the quality of their title. If property rights holders can expect little change over time to their entitlements, the more certain and secure are their rights, which increases the likelihood that they will invest in the management of their fishery. ITQ owners' rights have considerably higher quality of title than marine farmers'.

Figures 4 and 5 show that the mapping of these six characteristics for both ITQ owners and marine farmers produces irregular shapes, signifying that there is still scope for better-defined property rights for both groups. However, Figures 4 and 5 demonstrate that ITQ owners' rights score better than marine farmers' rights on most characteristics. Without clearly defined, appropriate and enforceable rights and responsibilities for marine farmers there is no consistent basis for resolving durable solutions when conflicts arise between themselves and with ITQ owners.

3. MANAGING MARINE FARMING FOR THE FUTURE

3.1 Improving the marine farming property rights

A comparison of Figures 4 and 5 suggests the real difference in property rights regimes for ITQ owners and marine farmers is the duration of their rights. Extending rights in perpetuity, as is the case with ITQ, to marine farmers is a political issue rather than a policy problem. The seafood industry has proposed that fisheries legislation be amended to create a new marine farming regime that would specify the rights that all marine farms would have. These new marine farming rights would have the following characteristics:

i. Exclusive right to farmed stock, subject to the stock always being in the continuous and exclusive possession of the marine farmer
ii. The right to farm any species, subject to the environmental effects of farming the species being authorised, and not subject to any prohibition notice
iii. Perpetual first right of renewal and
iv. Ability to sub-lease and trade.

The Government is currently preparing to consult with the public on proposed legislation amendments that could be the first step in reforming the rights for marine farming. These new marine farming rights, as proposed, would allow farmers to possess, harvest and sell farmed stock, subject to holding RMA coastal permits for occupation of space and management of environmental effects. It is being proposed that all existing marine farms come under this new regime to ensure equivalency of title and consistency of rights among all marine farmers.

3.2 Integrating marine farming and commercial fishing rights

The challenge in integrating marine farming rights with those of commercial fishing, including rights granted through ITQ ownership and permits for non-ITQ fisheries, is competition between the two for coastal space. The QMS provides ITQ and non-ITQ fishers the right to fish almost anywhere in the relevant quota management area. In reality though, some commercial fishing is highly localised. The development of a new marine farm can, therefore, cause an unavoidable economic loss for some commercial fishing rights holders.

Present mechanisms available under the FA 1996 are inadequate for dealing with conflicts over coastal space. Though the RMA is often the vehicle used by commercial fishing rights holders to oppose the development of new marine farmers, the issue is largely about loss of income and rights to coastal space rather than environmental effects. The RMA is not designed to deal with economic and rights-based issues. Actions by commercial fishing rights holders to limit the impact of marine farms on their livelihood, unless conducted carefully and expansively, risk being dismissed as *ultra vires*. On the other hand, under current law there is no explicit requirement for new marine farming operations to compensate existing commercial fisheries rights holders for potentially adverse effects of their operations. There is little incentive, therefore, for those setting up new marine farming operations to seek non-regulatory solutions to conflicts over rights to coastal space and loss of income for existing commercial fishing rights holders. Such conflicts have the potential to stifle appropriate development, particularly where opportunities exist to move from lower value harvesting activity (per unit area) to higher value intensive marine farming. Resolving these rights-based problems means amending marine farming legislation to better integrate the rights of commercial fishing rights holders and marine farmers, which would provide greater investment certainty for both groups.

Better defined rights for marine farmers could encourage acceptance of voluntary or trade-off agreements for spatial use where both commercial fisheries rights holders and marine farmers can gain from the transaction. Commercial fisheries rights holders and marine farmers operating in the same discrete area are ideally suited to voluntary agreements, or decentralised management agreements (DMAs). DMAs could create fisheries where both commercial fisheries rights holders and marine farmers would have rights to fish and, or, farm in a defined and enforceable area governed by a series of rules over use and exploitation. These arrangements are suited to coastal areas with mixed fisheries and stocks of limited mobility such as occurs in many inshore fisheries.

However, DMAs are not encouraged under the current legislative framework since only the Chief Executive of MFish, or a delegated officer, can determine allocation of marine farming rights in coastal marine areas. Where adverse effects from marine farming could be demonstrated, trade-off agreements could be more attractive if the potential for compensation provisions were introduced. A DMA may also include transferable rights and the ability to expand or contract entitlements through purchases and disposal of rights. This way existing commercial fishing rights holders might not be as disadvantaged through the establishment of new marine farming activity in their fishery.

A DMA could alleviate pressure on a fisheries resource or marine area by direct restrictions on output (*e.g.* commercial catch limits, marine farming harvest) or by indirect restrictions on inputs (*e.g.* occupation of coastal space, methods and seasons for harvesting) and by modifying rules according to changing circumstances. How well a DMA may operate depends on the pressure placed on the resource and the effectiveness of institutional arrangements to respond to changes. The key characteristics of a successful DMA would be the robustness of its institutional arrangements, limitations on territorial use and the feasibility of exclusion.

Exclusivity is an especially critical issue for success. For example, ITQ owners are a readily identifiable community of interests better able to collectively negotiate a management agreement. The nature of the rights granted by the ITQ system allows ITQ owners, more or less, the ability to exclude and, or, bind new entrants to a fishery. In contrast, marine farmers do not become rights holders until they are granted harvesting and occupation rights under the joint RMA/FA 1983 regime. Marine farmers are, therefore, excluded from negotiating the potential impact of proposed marine farms prior to the granting of rights.

Negotiations could occur between the commercial fisheries rights holders and regulatory authorities. However, political and legal issues and power asymmetries limit the likelihood of this occurring. Instead, there could be a clear separation of the granting of occupation rights to coastal space from the authorisation of marine farming activity. This would allow a tendering or allocation process for a specific area of coastal space. A subsequent round of negotiations could then occur between potential marine farmers with a provisional occupation right and commercial fisheries rights holders. A successful outcome could be that marine farmers are granted a perpetual harvesting right having achieved a negotiated outcome with commercial fishers. Where agreement could not be reached an independent dispute-resolution process could facilitate access agreements between commercial fisheries rights holders and marine farmers. Where compensation to commercial fisheries rights holders could not resolve their loss of rights to coastal space, provisional occupation rights might be surrendered and costs incurred in applying for the occupation right could be refunded.

Provided commercial fisheries rights holders and marine farmers can protect their rights against external pressures, such as pollution, DMAs are likely to be effective in (a) eliminating wasteful effort, (b) establishing equity of rights between commercial fisheries rights holders and marine farmers, (c) providing incentives for innovative management, (d) resolving conflicts, and (e), achieving profitable and sustainable fisheries. The administrative feasibility of DMAs would be high while monitoring, policing and enforcement costs would be relatively low.

4. CONCLUSION

The decade since the introduction of a property rights-based management system has been accompanied by industry profitability, unparalleled levels of investment and generally improved fish availability due to developments in stock assessments and implementation of rebuilding strategies. In addition, the marine farming industry, particularly the Greenshell™ mussel sector, has experienced dramatic growth in volume and value. However, the full economic and biological potential of New Zealand's fisheries has yet to be realised. Failure to further develop the QMS and the management of marine farming would incline most harvesters to adopt short-term perspectives towards fisheries management. As Jentoft *et al.* (1998) suggest, fishers do not easily accept government intervention unless it makes sense in the way they see their problems, know their fishery and have learned to understand the marine environment.

The underlying mood within the seafood industry is one of confidence in the future and there is general agreement that more commercial species should be managed under the QMS. The main reason for this confidence is that at every major crisis point the property rights-based fisheries management system has emerged stronger and better specified. It has taken most of the last decade for ITQ to outgrow its experimental and tentative status, but now it is viewed as irreversible and secure. Although marine farming is less suitable to the QMS, there is growing recognition that its integration with wild fisheries must be improved. Recent government and stakeholder initiatives to better define and manage marine farming have been welcomed by the commercial sector, however, as outlined in this paper, many issues remain unresolved. Government must find ways to resolve the confusion caused by the current management of marine farming and its inevitable competition with commercial fisheries rights holders over use of coastal space.

Until security of tenure in marine farming is defined and established, including determination of current and proposed coastal space occupation and structures, its growth prospects will remain in jeopardy. Failure to explore and adopt innovative policies to address the potential insecurity of marine farming tenure could jeopardise the industry's future growth and success in international markets. The integrity of the QMS for wild fisheries could also be undermined.

5. LITERATURE CITED

Annala, J. 1996. New Zealand's ITQ system: have the first eight years been a success or failure? *Reviews in Fish Biology and Fisheries*, 6, 43-62.

Batstone, C.J. and B..M.H. Sharp 1999. New Zealand's quota management system: The first ten years. *Marine Policy*, 23, 177-190.

Clark, I. Major, P. and N. Mollet 1988. Development and implementation of New Zealand's ITQ management system. *Marine Resource Economics*, 5, 325-349.

Dewes, C. 1989. Assessment of the implementation of individual transferable quotas in New Zealand's inshore fisheries. *North American Journal of Fisheries Management*, 9(2), 131-139.

Falloon, R. 1993. Individual transferable quotas: The New Zealand case. In *The Use of Individual Quotas in Fisheries Management.* OECD, Paris, 44-62.

FIB 1996. *The New Zealand Seafood Industry Economic Review*. New Zealand Fishing Industry Board, Wellington.

Gaffney, K.R. 1997. *Property Right Based Fisheries Management: Lessons From New Zealand's Quota Management System*. Unpublished Masters Thesis, Victoria University of Wellington.

Harding, R.J. 1991. *New Zealand Fisheries Management: A Study in Bureaucratisation*. Unpublished PhD thesis, Victoria University of Wellington.

Horton, C. 1997. *Open Letter to Minister of Fisheries.* New Zealand Fishing Industry Board, Wellington.

Jentoft, S. B.J. McCay and D.C. Wilson 1998. Social theory and fisheries co-management. *Marine Policy*, 22 (4-5) 423-436.

Luxton, J. 1997. *Stakeholder Management of Recreational Fisheries*. Address to the Recreation Fishing Council Annual General Meeting, Bay of Islands, July 1997.

Memon, A.P. and R. Cullen 1992. Fisheries policies and their impact on the New Zealand Maori. *Marine Resource Economics*, 7, 153-167.

Pricewaterhouse Coopers 1998. *Fishing for the Future: Review of the Fisheries Act 1996*. Ministry of Fisheries, Wellington.

Scott, A. 1988. Development of property in the fishery. *Marine Resource Economics*, 5, 289-311.

Sharp, B.M.H. 1997. From regulated access to transferable harvesting rights: policy insights from New Zealand. *Marine Policy,* 21(6), 501-517.

Sissenwine, M.P. and P.M. Mace 1992. ITQs in New Zealand: The era of fixed quota in perpetuity. *Fisheries Bulletin*, 90, 147-160.

THE NATURE OF "RIGHTS" IN THE WESTERN AUSTRALIAN PEARLING INDUSTRY

H.G. Brayford
Fisheries Management Services Division, Fisheries Western Australia
<hbrayford@fish.wa.gov.au>
and
G. Paust
Locked Bag 39, Cloisters Square P.O., Perth, WA 6854
<gpaust@fish.wa.gov.au>

1. INTRODUCTION

Australian South Sea pearls are recognised as the purest, finest and most valuable pearls in the world. Pearls are produced from the pearl oyster *Pinctada maxima* in the pristine coastal waters of northern Australia. The farming of *P. maxima* for pearls and associated products is Australia's most valuable aquaculture sector. Western Australia is the major producer with 16 licencees participating in an industry worth in excess of $A260m in exports.

Currently, the industry is substantially based on the collection of pearl oysters from the wild under a quota system for subsequent pearl production. However to maintain its leading position in the production and sale of quality South Sea pearls on the world market, the industry has also developed hatchery technology for pearl oyster production. As a result, the industry is going through a gradual, quota controlled phase of growth and the value of pearl production may exceed $A500m by 2010.

Under the current legislative and management framework, three categories of "rights" have emerged:

i. The right to access wildstock pearl oysters for pearl production
ii. The right to seed hatchery produced pearl oysters for pearl production
iii. The right to occupy an area of coastal waters to conduct pearl farming or hatchery activities ("pearl farm lease rights").

This paper describes the nature of these rights including a description of allocation processes and emerging issues. Issues of particular relevance include the allocation of coastal water sites for pearl production, given competing use, native title and the objective to maintain Australia's leading position in the world pearl market.

2. BACKGROUND

Pearling in Western Australia comprises three main activities - the harvesting of wildstock oysters, the production and grow out of hatchery spat and the subsequent farming activity to produce pearls.

Prior to 1995, the wildstock pearl oyster fishery was managed as a Joint Authority Fishery by the State and Commonwealth under State law. Aspects relating to pearl production (*i.e.* farming) were managed solely by the State. Since 1995, following an arrangement between the State and the Commonwealth, all aspects of the industry have been regulated by the State under the *Pearling Act 1990*.

From the early 1970s to the early 1980s, there were major problems within the industry associated with the availability of shells and excessive mortalities of shell stocks. These problems led the Commonwealth and State Governments at the time to impose prohibitions and restrictions such as on the taking of pearl shell for Mother-of-Pearl and placing quota levels on the companies involved in the industry. The stock situation became so serious in the early 1980s that the Governments imposed a moratorium until December 1987 on the number of companies licensed to fish and farm shell.

In June 1987, an advisory committee, known as the Pearling Industry Review Committee (PIRC), was established to independently review and assess the structure and operations of the industry and to report and make recommendations with respect to the future development and management of the industry. The Review Committee was asked to study and review all aspects of the industry and present its recommendations for consideration prior to the end of the moratorium period. The final report was presented to the State and Federal Governments in February 1988.

The Review Committee made a number of recommendations. Those of relevance to this particular discussion included recommendations that the industry be quota-controlled with entry by licence transfer only. The licence holders at the time, plus some pending new applicants were recommended as the companies to comprise the initial licencees in the industry. The establishment of an advisory committee was also recommended.

The PIRC report was the forerunner to the preparation of new pearling legislation, the *Pearling Act 1990*, and the subsequent publication of Ministerial Policy Guidelines in 1992 which, together with the Act, provide the basis for the current management framework. It should be noted that *P maxima* is the only species declared for the purposes of the *Pearling Act 1990*. All other pearl oyster species are regulated under general fisheries legislation.

3. WILDSTOCK ACCESS RIGHTS

The industry is based mainly on pearl oysters from the north coast of Western Australia. Oysters are hand collected by divers from the wild shell grounds and following seeding and a period of rest in the same area (3 + months) are transported to pearl farms in coastal waters off north-western Australia or, in some instances, to farms in the Northern Territory.

The wildstock fishery is managed in a similar way to a number of other commercial fisheries in Western Australia. To ensure sustainability of the stock, catches are limited each year by way of a Total Allowable Catch (TAC). Individual pearling licencees are then allocated a portion of the TAC by way of Individual Transferable Quotas.

The report of the Pearling Industry Review Committee noted that the first major management measures related to quota were introduced in the fishery in the early 1980s in response to concern about lowered catch rates (although, it is understood that prior to this, approval was required from the Commonwealth and State, as joint managers, to take pearl shell). These arrangements included the commencement of a system of quotas aimed at setting limits to the quantity of pearl oysters taken for pearl culture. The quotas set were not specifically intended to reduce the numbers taken in previous years but was also structured to establish a system for discussion about future requirements and a mechanism whereby those requirements could be controlled.

The TAC throughout the 1980s varied from time to time, based principally on whether or not stocks were considered limiting the harvest. Individual quota levels within the TAC reflected the commitment of the various companies, their traditional catching ability and infrastructure investment. This approach to quota allocation was consistent with approaches taken in other fisheries at the time.

With the introduction of new pearling legislation in 1990, quota levels for existing licencees were based on historical catch levels or in some cases, it is understood, reduced. Two new licences had also been granted as a result of the Review Committee's findings on the basis that shell was collected in Zone 3 (*i.e.* north of the main fishing grounds at Eighty Mile Beach.

During 1993, the then Joint Authority identified an opportunity for new entrants in the Zone 1 sector of the fishery (south of Eighty Mile Beach). Following a public expression of interest process - a first for the Western Australian pearling industry - three new pearling licences were granted in Zone 1, subject to conditions. The licencees were each allocated what was considered at the time to be the economically viable minimum quota level of 15 quota units (at the time 15 000 shell).

Under current arrangements, the TAC is normally 572 000 shells with one quota unit having a par value of 1000 shells. The quota for individual licencees ranges from 15 units to 100 units. In recent years, the quotas have remained relatively stable although the TAC in Zones 2/3 was increased by 55 000 shell for the 1995, 1996 and 1997 seasons as a result of a temporary increase in recruitment. This increase was allocated on a pro-rata basis across Zone 2 and 3 licence-holders. The quota was returned to its par value in 1998.

Quotas values are noted on pearling licences. Both quotas and licences are transferable with approval of the Executive Director of Fisheries. Quota units must be transferred in minimum parcels of 15 quota units. At present, the way is open for an applicant to be considered for a pearling licence, on the basis that quota units are acquired by way of transfer from an existing licencee and subject to both the transferee and the transferor having a minimum holding of not less than 15 quota units after approval of the transfer.

Pearling licences are renewable annually. Licences are not issued as a right and, if the Executive Director thinks it would be in the better interests of the pearling industry to do so, the Executive Director may refuse to issue a licence.

4. HATCHERY RIGHTS

In the last few years, hatchery technology has developed for the propagation of pearl oysters for subsequent pearl production. Pearl producers acquire pearl spat from licensed hatcheries (the majority of which are owned and run by existing industry licencees) who grow out the spat to the minimum seeding size which, in the case of hatchery oysters, is 90mm.

Under current arrangements, if a hatchery proponent does not hold a pearling licence, then a hatchery licence will only be issued where there is an agreement to supply spat to the holder of a pearling licence, or alternatively, if the hatchery licence is issued jointly between the proponent and a pearling licencee. Hatcheries will only be licensed for the purposes of providing spat or pearl oysters to the Australian pearling industry.

While there are no restrictions on the production of spat from licensed hatcheries or subsequent grow-out, there is a limit on the number of hatchery oysters that pearl producers can use for pearl production each year. This policy was introduced in 1992 and each pearling licencee, at the time, was allocated 'hatchery options' which entitled that company to seed a certain number of hatchery oysters each year for pearl production. Recent new entrants to Zone 1 have also been issued options following a review of individual performance.

There are currently a total of 350 000 hatchery options within the industry (350 units) with each company having been allocated 20 000 annual options. One company was allocated an additional 30 000 options following a general understanding within industry when that particular company made a decision to establish the first hatchery in 1989. Options are transferable among licencees. Hatchery options have a term of 10 years unless extended for special reasons such as approved development plans that have a longer life.

Licencees may apply to convert options within the ten-year period (expiring in 2002 although some companies have an "extension" until 2005 based on approved development plans) to permanent hatchery quota subject to meeting certain development conditions. To be eligible to convert options to permanent hatchery quota, licencees must demonstrate, over a three-year period, the successful production of a minimum average of at least 1000 pearl oysters suitable for round pearl production. Once converted, for all intents and purposes, hatchery quota is treated in the same manner as wildstock quota.

Options were initially issued with a ten-year life span rather than on a permanent basis for public policy reasons. The objective was to encourage licencees to develop the technology for hatchery development and grow-out or lose the options. It was also to ensure that the right to use hatchery shell was not seen as gifting a valuable asset to existing licencees without suitable investment and allocation of resources to the development of technology. The hatchery policy is also a balance between the need for pearling licencees to gain experience in hatchery and grow-out technology and, given the sensitivity of the market to rapid increases in production, the need to discourage over-production to maintain the current value of the industry and revenue back to Australia.

5. PEARL FARM LEASE-RIGHTS

As outlined above, following collection, seeding and a period of rest, pearl oysters are transported to pearl farms in coastal waters of Western Australia between Exmouth Gulf and the Western Australia/Northern Territory border and in some cases to farms in the Northern Territory. Shell can also be acquired from hatcheries. The oysters spend the remainder of their culture life (up to 4-10 years) on the farm. Shells are held in panels attached to longlines and are subject to regular cleaning to remove fouling. Pearls are harvested at the farm site on an annual basis (2 years of pearl growth) and some can be re-operated up to 4 times for round pearl production.

Under the pearling legislation, approval for a pearl farm site is by way of a pearl oyster farm lease. A lease confers a right to occupy an area of coastal waters to conduct pearling (or hatchery) activities. It does not confer a general right of exclusivity apart from pearling (or hatchery) activities authorised by licence.

Leases are only granted to the holder of a pearling (or hatchery) licence and the main 'home' leases are issued for a period up to 21 years, subject to renewal.

In December 1997, following community concern about the granting of licences and leases for aquaculture and pearling development, a Ministerial Policy Guideline was issued on the assessment process for pearl farm lease and aquaculture licence applications in coastal waters of Western Australia. The guideline process includes full public consultation with relevant decision-making authorities and interest groups. It also specifies the timeframes to apply at each stage of the assessment process and important matters to be taken into account. The guideline is not intended to limit in any way the statutory discretion of the Executive Director. All applications must also meet the requirements of the Act and other relevant legislation such as environmental and Native Title legislation.

Decisions made on pearl farm lease applications are subject to a statutory appeals process under the Act with appeals being determined by the Minister for Fisheries.

6. CONSULTATION AND COMMITTEE PROCESSES

Fisheries Western Australia ('Fisheries WA'), operating under the Pearling Act 1990, is the lead government agency in Western Australia for the regulation and management of the pearl oyster fishery and associated pearling and hatchery activities. It is also responsible for leasing areas of coastal waters for use as pearl farm sites.

To assist in achieving its objectives, Fisheries WA works in co-operation, as relevant, with industry. The agency often consults with industry members and/or the Pearl Producers Association (PPA), either through formal channels or on an informal basis, on a range of policy and management issues. The PPA is the representative body for the pearling industry and its membership comprises all licencees engaged in the commercial harvesting of *P. maxima* and in pearl production in Western Australia. The PPA greatly assists the decision-making process. It provides an industry or corporate view on many issues which, while in the better interests of the industry, are not necessarily the same as individual company imperatives but are supported nevertheless.

The Pearling Industry Advisory Committee (PIAC) established by the Pearling Act 1990 is a statutory committee providing to the Minister for Fisheries and to the Executive Director of Fisheries WA advice on matters relating to policy, management and development of the pearl oyster fishery and pearling industry. PIAC members are appointed by the Minister and may include people directly involved in the industry, people external to the industry and people from Government. PIAC can seek additional technical and expert advice where necessary.

PIAC provides advice to the Minister and to the Executive Director on a range of issues including matters relating to the TAC for a particular zone or zones and quota levels. In providing such advice PIAC normally takes into account research advice on stock status.

7. ISSUES

The pearling industry is unique in some respects in that it incorporates both fishing and aquaculture components. As a result, a complex set of rights have evolved relating to the collection of wildstock oysters, the use of hatchery oysters for pearl production and pearl farm lease-rights. While collectively these form the basis of the management framework, a number of issues have emerged in each set of rights.

As outlined above, wildstock quota-rights in the fishery are managed in much the same way as quotas in other commercial fisheries principally on grounds of maintaining sustainability and currently TACs are only subject to minor changes. No significant change in the TAC or in quota levels is anticipated for the future. The fishery is zoned and, as with other fisheries, there are issues surrounding the potential diminution of rights and issues of equity within and across zones if, for example, boundary changes are suggested or changes in the TAC for a particular zone are proposed. In most cases, these issues are resolved in discussion between Fisheries WA, the PPA and PIAC providing useful examples of the co-operative approach to management and, in particular, the industry's corporate approach to dealing with industry-wide issues.

While pearling licences are renewed annually and quota is issued on an annual basis, there is an expectation of permanency about licences and quotas even those these 'ongoing' renewals are not stated in the Act as a 'right of renewal'. Since the new Act came into effect in 1990, there are no examples known to the authors of licences not being renewed. During 1995, 1996 and 1997, however, one licencee had 5 quota units suspended because of a breach of the Act. From an industry perspective, greater security over licences and quota would be preferable particularly in light of the large capital-investment required by licencees in the industry.

The pearling industry is market driven. The high value of South Sea pearls is based on their beauty, rarity and the image of luxury attached to them. To maintain market stability, pearl producers must be attuned to the sensitivity of the world market. Historically, Western Australian pearl production has been constrained by the availability of wildstock pearl oysters. However the development of hatchery technology within the industry provides a means for increased production without further exploitation of the wildstock. The current management framework, encompassed within Ministerial Policy Guidelines, encourages the development of hatchery and grow-out technology but also limits the use of hatchery oysters for round-pearl production with the objective of maintaining the price of Australian South Sea pearls in world markets.

Debate on the hatchery policy has, at times noted that the policy limits the use of hatchery oysters for pearl production to existing licensed pearl producers, although the way may be open for new entrants via the purchase and transfer of hatchery quota subject to minimum quota holdings.

The pearling legislation, including the hatchery policy, is currently subject to review under National Competition Policy (NCP) agreements. The review, which is being conducted by an independent consultant, is due for completion in 1999. It is focussed on the guiding principle of NCP, namely to ascertain whether:

i. the existing regulatory mechanisms restrict competition and, if so, whether the benefits of the restrictions to the community as a whole outweigh their costs
ii. the objectives of the legislation can only be achieved by restricting competition and
iii. the current regulations are the most efficient method of achieving these restrictions.

The hatchery policy will be particularly scrutinised and the outcome of the review, whilst uncertain, may see a shift in the current management framework.

The allocation of pearl farm leases is becoming an increasingly complex issue given competing demands for marine resources. Traditionally, issues concerning property rights in fisheries have focussed on the right to harvest fish stocks. However, more and more, issues concerning the use of marine waters are coming to the fore, particularly given developments in aquaculture technology and stock enhancement.

Coastal communities are increasingly concerned about any expansion of pearling or aquaculture activities. The major issues include:

i. the potential direct impact of pearling and aquaculture activities on existing activities such as recreation and tourism
ii. in high use marine areas, a desire for unfettered access to, and use of, coastal waters and no further alienation of what is a common resource
iii. an expectation that applications for leases will be considered in an open, transparent manner and the views of other users and interest groups will be taken into account in the lease assessment process
iv. the perception of potential environmental impact of pearling and aquaculture activities
v. an expansion of pearling and aquaculture activities in existing or proposed marine reserves and the lack of a co-ordinated approach to marine planning
vi. the potential impact of pearling and aquaculture development on native title rights and interests.

To assist in resolving some of these issues the Minister for Fisheries issued a Policy Guideline in December 1997 which outlines the process for public consultation and assessment of lease applications. The guideline process ensures that the views of relevant user and interest groups are taken into account in the decision-making process and provides greater certainty for applicants in relation to the time-frames and assessment processes to apply. The Guideline process also places a responsibility on those making submissions against a particular proposal to put their best case and to provide information or evidence in support of their claims.

The Guideline process does not preclude the requirement for applications to be assessed in accordance with the *Pearling Act 1990* and with other relevant legislation relating to, for example, marine parks, environmental impact and native title. Processes and practices have been implemented by Fisheries WA to ensure compliance with relevant legislation including assessment of applications by the Department of Environmental Protection.

Notwithstanding the Guideline process and procedures implemented by Fisheries WA, it is difficult for pearl producers to secure lease sites in high and/or multi-use coastal areas and to secure long-term leases. A co-ordinated approach to marine planning across Government, with a high level of industry and community involvement may assist, however this is a complex issue given the various jurisdictions and lack of a statutory base. In addition, the setting aside of particular areas for pearling development can only be done based on rather gross parameters of site suitability. Currently there is no technical method that is feasible for ensuring that a site will be productive and produce high quality pearls. Industry practice is to test a site for several years to establish its suitability as a quality pearl-growth site. The risk for government in a marine-planning approach is that if this commits industry to culturing pearls in areas specifically set aside for that purpose and they prove to be unproductive and uneconomical, who is accountable?

This issue of availability of sites has also been recognised by the PPA and is being addressed by way of technology development, particularly in relation to: offshore farming systems; possible research in the areas of carrying-capacity; and working with other user groups, such as charter operators, to address issues of common concern and minimise potential conflict. The PPA also recently commissioned an independent report on the environmental impact of pearling activities through the fishing and farming stages. The report concluded that, in general, the industry was environmentally benign, producing a high-value product with a minimum of environmental disruption.

Ultimately, the major issue for fisheries-management inherent in the increased conflict over marine area use is how to measure, or determine, what is the most productive use of an area, *i.e.* what returns the best 'value' to the community as a whole. It is a simple process for the commercial usage of an area to have a dollar value attributed to its use but how does one measure, say, the 'value' of a 'recreation use only' area? Often decisions about allocation of coastal waters are based on subjective evaluation of optimum return to the community from what are essentially unmeasurable and uncomparable 'values'.

8. CONCLUSION

The wildstock pearl oyster fishery is one of the most valuable commercial fisheries in Western Australia. The culture of quality pearls is also Australia's most valuable aquaculture sector. Hatchery development is also well advanced. The industry is considered to be highly-developed with a complex set of management arrangements administered under one Act, separate to the general fisheries legislation.

Three sets of rights have emerged in the industry - wildstock access rights, hatchery rights and pearl farm lease rights. Currently the 16 licencees in the industry have each of these rights to varying levels. The rights have been allocated on differing bases, which included biological sustainability and economic considerations. Many of the issues surrounding rights of access, particularly in the wildstock sector, have been largely resolved in co-operation with industry and the management framework is considered to be relatively stable.

Industry's ability to secure additional, and long-term, lease sites, particularly as hatchery production expands within the current industry, is an important issue. Government has implemented processes and guidelines to ensure that pearl farms develop in an ordered manner and potential conflicts with other user groups are minimised. Industry is also addressing this issue by way of technology development, consideration of new research areas and working with other interest groups to address issues of common concern. The main unknown for the future is the outcome of the NCP review, which may result in changes to the current management framework, most likely in the hatchery sector if there is to be any change.

A fundamental sequel to the creation of these rights from an industry perspective is the security of tenure over quota and marine leases. The considerable capital-investment that is required by licencees in the industry requires a high level of confidence in the security of tenure over the allocated rights. Financial institutions seem to have similar concerns. Industry and financial institutions would like to see more permanent types of tenure rather than annual renewal and short term leases that are not reflected as 'a right to renewal' in the Act.

This expectation requires Government to balance the need for security of tenure (which will encourage the level of investment to optimise returns to the industry and community) against the need for Government to be able to implement possible future changes in resource allocation of what is considered a 'community resource'.

9. LITERATURE CITED

Fisheries Western Australia 1988. Report of the Pearling Industry Advisory Committee.

Fisheries Western Australia 1997. Pearl Oyster Fishery Ministerial Policy Guidelines FWA.

Centre for International Economics 1998. Review of the Western Australian Pearling Act 1990 under National Competition Review of Legislative Restrictions on Competition (Discussion Paper).

10. ACKNOWLEDGEMENTS

The assistance of Mr Michael Buckley, Executive Officer of the Pearl Producers Association, is appreciated.

PROPERTY RIGHTS AS AN ALTERNATIVE TO SUBSIDISATION OF FISHING AND A KEY TO ELIMINATING INTERNATIONAL SEAFOOD TRADE DISTORTIONS

A. Macfarlane
New Zealand Seafood Industry Council Ltd
Private Bag 24-901, Wellington, New Zealand
<alastair@seafood.co.nz>

1. BACKGROUND

The key issue facing capture fishing globally is achieving and maintaining optimal, sustainable exploitation of fisheries resources. Fishing is the only significant food producing sector left dependent on harvesting wild resources. Failure to secure sustainable management of fisheries resources ultimately calls into question the economic sustainability for fishing industries. Fish and fish products are the most widely traded food commodity grouping. Failure to secure sustainable resource management will affect the capacity of fishing nations to supply, and importing countries to obtain supplies, of fish products.

In most managed fisheries, management related risk is primarily borne by the manager – generally Governments. In the absence of clearly defined and allocated rights to access fisheries resources, the manager is faced with the need to deal with the consequences of decisions relating to resource availability or access.

The human reaction to exploitation of scarce resources in the absence of legally defensible access rights is to maximise individual exploitation to the extent possible. This has been well described in the literature (Gordon 1954). Resource abundance can be reduced to a scarcity as a result. This process is illustrated in Figure 1. In the case of fisheries, a renewable natural resource can, in the terms of Article XX(g) of the GATT 1947 Agreement, be turned into an exhaustible natural resource.

It is inevitable that once the technical capacity exists to exploit fisheries at their maximum sustainable yield levels, it is only be a matter of adding more capacity in the race for fish to start down the road towards overexploitation. Pulling a fishery back to sustainability will entail reducing harvests to levels which allow stocks to recover and then ensuring that catch levels are constrained to long term and optimally sustainable levels. Addressing the issue of biological sustainability can be relatively straight forward through setting and enforcing catch limits. However unless fishing effort is also constrained, the result will be economic failure of fishing businesses unable to achieve profitable catch levels.

Limiting access to fisheries can take many forms – including reducing the number of fishers permitted to be in a fishery or reducing the opportunity to fish by seasonal closure or limiting days-at-sea. Many have the disadvantage of failing to address the issue of developing technology or innovation which dissipate profits generated by rationing of access in the race to fish competitively and maximise catch.

When fisheries access is limited by number of participants and the total catch available to each participant, such access rights, if tradable, can enable individuals or communities to negotiate reduced catch levels to meet sustainable management imperatives. Thus, pressure on governments to provide subsidies can be reduced – indeed eliminated altogether – through the creation of tradable access rights with a capital value. This provides a built-in mechanism to provide financial compensation to those who choose to exit the fishery through trading of their rights to those who choose to stay. The value of the rights so traded will be a direct result of the profit available in the fishery. As with land, the experience in New Zealand's comprehensive rights-based system is for the profit in a fishery to be directly reflected in lease and sales value of access rights to a fishery.

Governments find themselves under pressure from fishing communities and businesses to provide financial transfers to assist with reducing the allowable harvest levels required to achieve resource sustainability and, or, allow for stock recovery. Faced with the prospect of whole communities losing their means of economic livelihood, the political imperatives are obvious. Lack of individual or community property rights ensures that governments have few, if any, alternatives to providing direct financial assistance.

2. THE PROBLEM OF SUBSIDIES

Subsidies to fishing can be categorised into two general groups; (a) cost reduction and (b), revenue enhancing. Figure 2 illustrates the effect of cost reducing subsidies. Examples of cost reducing subsidies provided to fishing can include the provision of services such as discounted or free access to port and storage facilities and fuel subsidies. Such subsidies simply shift the cost/revenue equilibrium point to the right, *i.e.* to a level of higher fishing effort, and restore the prospect, if not the reality of profit. But unless access to the fishery is limited, the lure of unallocated profitability will see the profit dissipated in increased effort. In the absence of effective controls on catch, cost reducing subsidies also lead to increased total catch until costs again catch up with revenues. Or, there is a resource failure.

Figure 3 illustrates the effect of revenue enhancing subsidies. Examples include vessel construction subsidies and subsidies to shift effort to under-utilised fisheries.

Figure 1
The simple fisheries model
(WTO Committee on Trade Environment WT/CTE/W/111 11 March 1999)

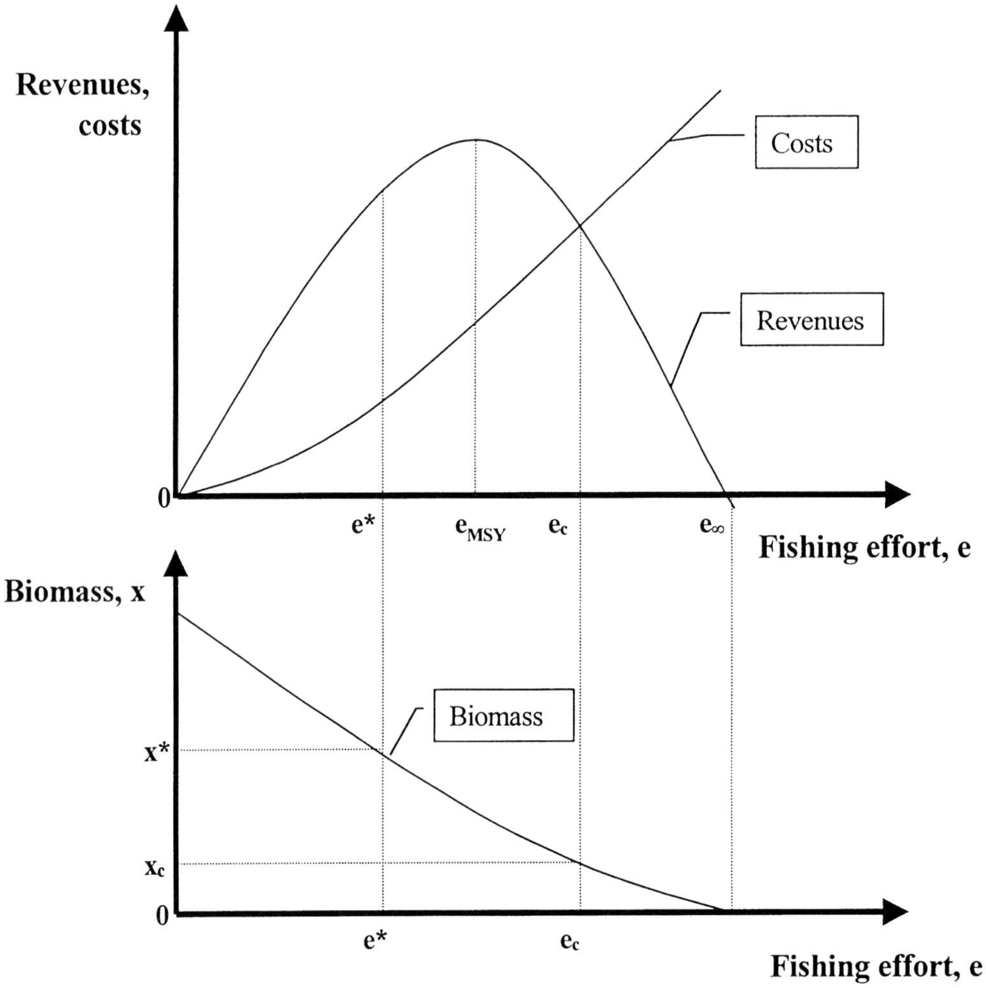

[Graph 1: Revenues, costs vs Fishing effort, e; showing Costs and Revenues curves with points e^*, e_{MSY}, e_c, e_∞]

[Graph 2: Biomass, x vs Fishing effort, e; showing Biomass curve with points x^*, x_c and e^*, e_c]

(WTO Committee on Trade Environment WT/CTE/111 11 March 1999)

Enhancing revenues has the same effect as reducing costs and can lead to increased effort and increased catch until the point of cost/price equilibrium has been reached.

The amount of subsidies provided to the world fisheries sector have been estimated to be on a par with the global assistance provided for beef and pork production (Table 1) (Milazzo 1998). It is unlikely that such an amount of assistance can be provided without distorting markets, affecting production and the international trade in seafood. Seafood exports predominantly originate from developing countries (Figure 4) while the leading markets are those of the developed countries - Japan, USA and the European Union (Figure 5).

Coincidentally, Governments in those leading markets manage fisheries largely without property rights and are the leading providers of subsidies to their fish harvesting sectors. The catching sectors of the key import markets countries are not generally major suppliers to international trade. But, the subsidies they provide to their industries to overcome losses arising from reduced catches as TACs are reduced, or are provided as cash

Table 1
Average global good subsidies
(Including trade barriers)

Product	Subsidy (%)
Wheat	48
Coarse grains	36
Rice	86
Oilseeds	24
Sugar	48
Beef and veal	35
Pork	22
Poultry	14
Lamb and mutton	45
Eggs	14
Fish	30-35

Figure 2
Effect of cost subsidies

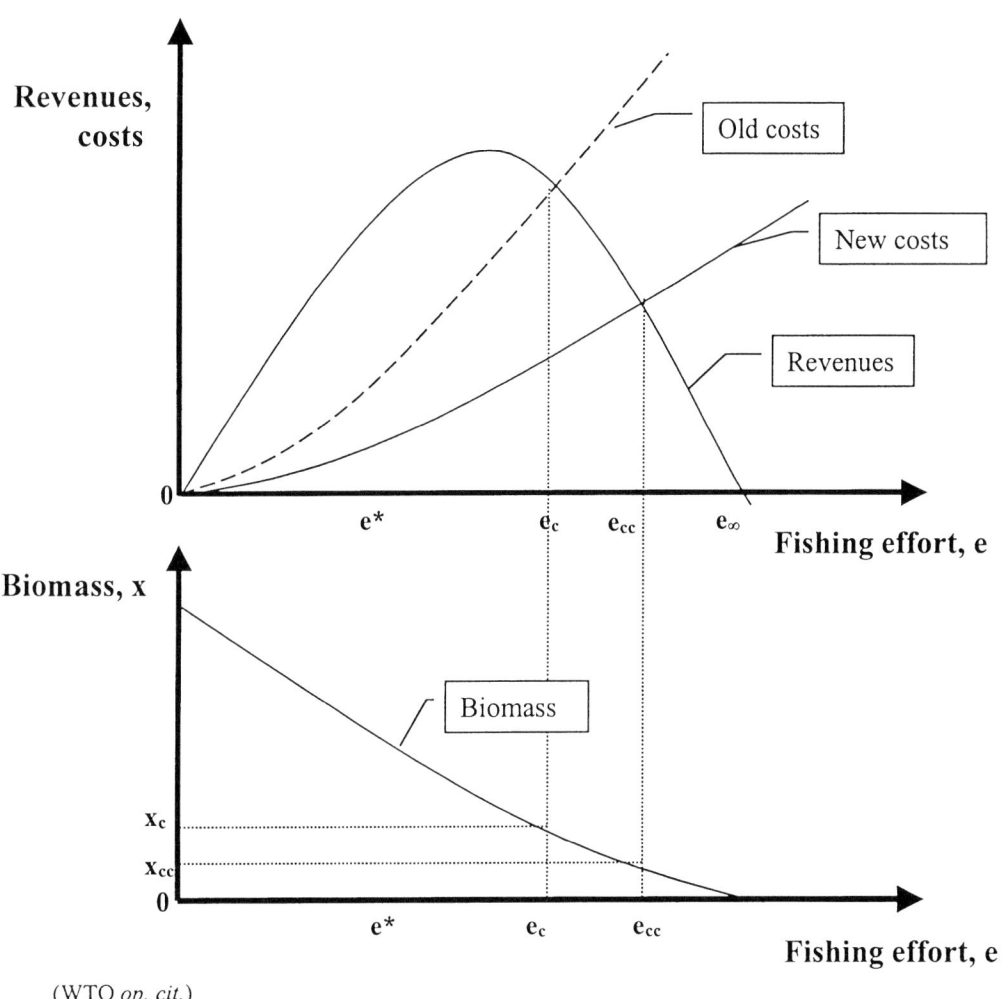

(WTO *op. cit.*)

adjustment assistance to participants to exit the fisheries and thus reduce capacity, often are used to facilitate their entry in to other less fully utilised fisheries. Here, the problems are created anew. The subsidies also assist their beneficiaries to compete against imports in their domestic markets. In that way subsidies contribute to distorting the international trade in seafood products.

There is little, or no, evidence of government subsidies to capture fisheries leading to surplus fish production that is then exported with the aid of subsidies. The reasons is that subsidies arise when fisheries are in trouble and governments attempt to minimise social dislocation. As a result, the arguably pernicious level of subsidisation to fishing in some countries, as estimated by Milazzo (1998), has failed to attract attention by way of use of dispute mechanisms of the World Trade Organisation.

Much of the fish in international trade originates from developing countries and is produced without the aid of subsidies. While prices in international trade in fish and fish products are generally established at levels that reflect available supply and competitive demand, competing domestic supplies arising from subsidies adds to available supply and it must be anticipated that the clearing prices which result are reduced accordingly. Thus, returns to unsubsidised exporters are reduced accordingly.

The WTO's Agreement on Subsidies and Countervailing Measures. Article 6, establishes grounds for determining the existence of serious prejudice to a Member and includes in Section 6.3 "the effect of a subsidy is to displace or impede the imports of a like product of another Member into the market of the subsidizing Member". However there is no dispute case history based on this provision – yet.

Figure 3
Effect of price supports

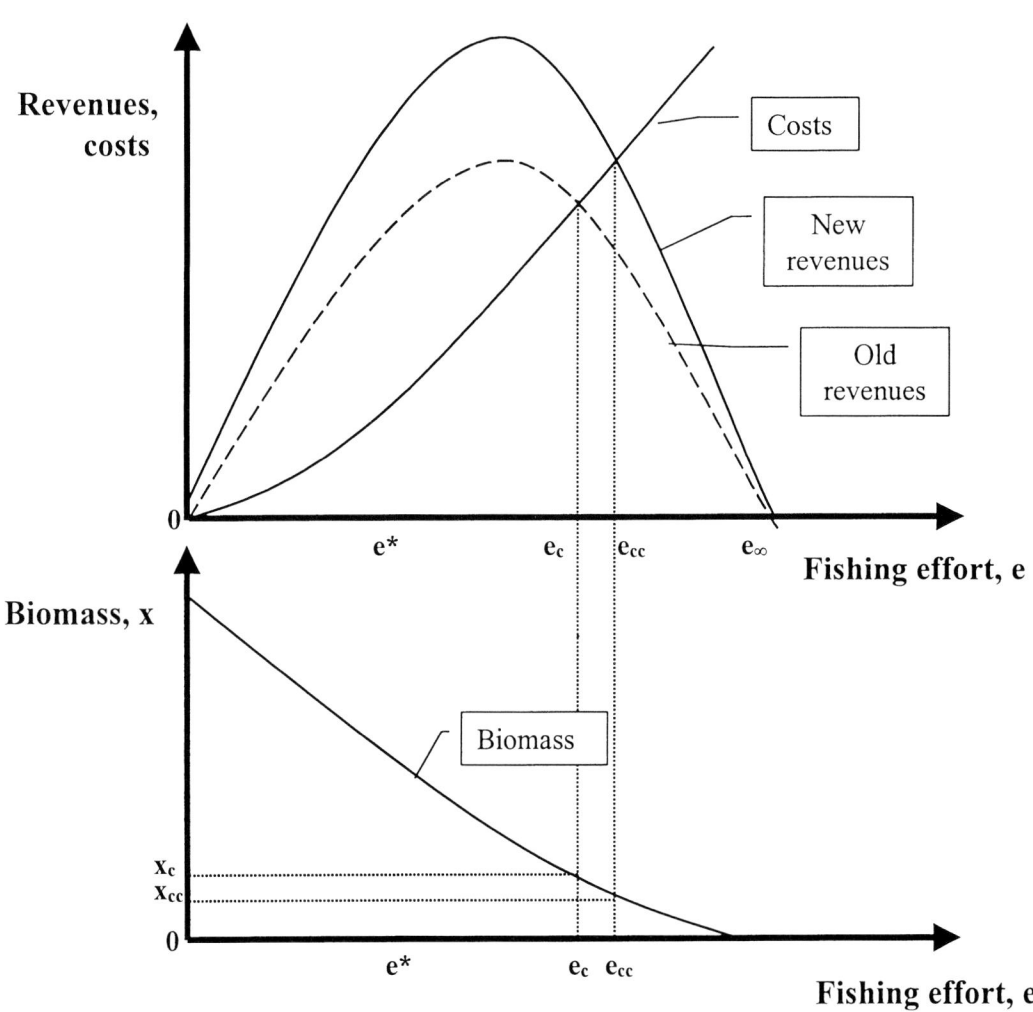

(WTO *op. cit.*)

The creation of tradable access rights provides Governments with an alternative to subsidising fishers through enabling time for the industry to adjust to operating at sustainable harvest levels. Tradable access rights also provide an opportunity for fishers to reduce the level of over-capitalization in fisheries through fishing effort buy-outs. In order for the forces of trading to have full effect, Governments need to allow fishing businesses to face up to the prospect of business failure as an alternative to selling their fishing rights and exiting the fishery. Governments must also accept the aggregation of fishing access rights into fewer hands if fleet over-capacity and excessive fleet capatilization is to be reduced.

So called 'multi-functionality' is difficult sustain if both the biological capacity of fisheries and their economic worth are such that large coastal communities cannot sustain incomes equivalent to those obtainable in cities. Fewer fish and fewer people in fishing communities can still result in economic and biological sustainability, but it may well be different from the result that can be achieved through subsidies with the underlying motive of achieving or maintaining social or political objectives.

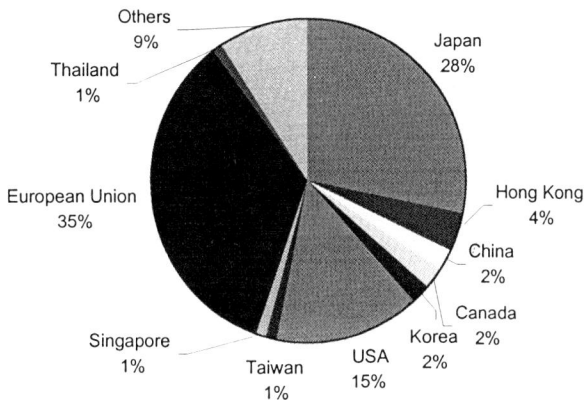

Figure 4
Major seafood importers (1997)

Source: FAO

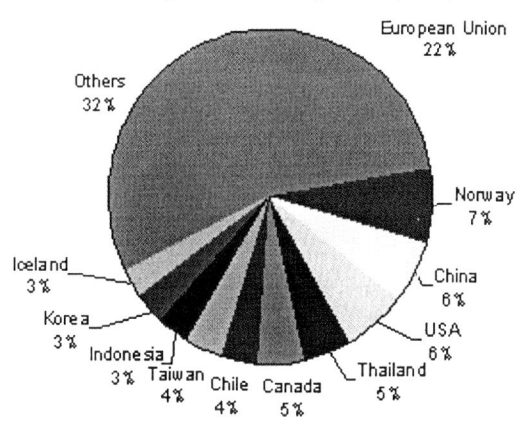

Figure 5
Major seafood exporters (1997)

Source: FAO

3. LITERATURE CITED

FAO 1999. FAO Yearbook Fishery Statistics Commodities Vol. 87 1997. Food and Agriculture Organisation of the United Nations, Rome

Gordon H.S. 1954. Economic Theory of a Common Property Resource: The Fishery, Journal of Political Economy **62**:124 – 42.

Milazzo, M. 1998. Subsidies in World Fisheries: A Re-examination. World Bank Technical Paper n. 406, Fisheries Series. World Bank, Washington DC

WTO 1999. Committee on Trade and the Environment. On the Environmental Impact of Fisheries Subsidies: A short report by the Icelandic Ministry of Fisheries. WT/CTE/W/111, 11 March 1999. World Trade Organisation, Geneva.

WTO 1994. Agreement on Subsidies and Countervailing Measures Pp 264 – 314 The Results of the Uruguay Round of Multilateral Trade Negotiations, The Legal Texts, The GATT Secretariat, Geneva.

WTO 1994. General Agreement on Tariffs and Trade (GATT 1947) Pp 485 – 537 The Results of the Uruguay Round of Multilateral Trade Negotiations, The Legal Texts, The GATT Secretariat, Geneva.

A PROPOSAL FOR COST RECOVERY IN THE ALASKA INDIVIDUAL FISHING QUOTA (IFQ) FISHERIES

P.J. Smith and J.T. Sproul
National Marine Fisheries Service
Alaska Region, P.O. Box 21668
Juneau, Alaska 99802, USA
<Phil.Smith@noaa.gov, John.Sproul@noaa.gov>

1. INTRODUCTION

Under the authority of the *Magnuson-Stevens Fishery Conservation and Management Act* (MSA or Act), marine fisheries in the Exclusive Economic Zone of the United States are managed by the Secretary of Commerce (Secretary), with operational responsibility delegated to the National Marine Fisheries Service (NMFS). Amendments to the Act, adopted in 1996, direct the Secretary to implement Federal regulations to recover actual costs associated with the management and enforcement of limited access programmes in U.S. fisheries that are managed under Individual Fishing Quota (IFQ) programmes [MSA, 1996; §304(d)]. The Act also mandates the collection of fees from participants in the Bering Sea Community Development Quota (CDQ) programme; however, the programmatic design to do so has not yet been developed and will not be discussed here. The Act limits cost-recovery fees to three percent of the ex-vessel value of fish harvested under any such programmes, and further requires that the fees be collected at the time of landing, the time of filing a landing report or during the final quarter of the year during which the fish were landed.

It is not clear why the Congress chose to limit agency discretion in this way; some argue that it would have been much easier, less confusing, and administratively more efficient to simply adopt the "Canadian model" – a system whereby holders of Individual Vessel Quota (IVQ) pay their fees before their annual IVQ permit is issued. Fees are adjusted annually (depending to some extent on the annual Total Allowable Catch, or TAC) and the annual rate is anticipated to be sufficient to cover the administrative and enforcement costs of the programme. The 1999 "up-front" amount paid by Canadian IVQ holders is CA$ 0.20/lb of IVQ halibut allocations (Best 1999).

In addition to the direct cost recovery anticipated by the Act, the Secretary is authorized to reserve up to 25% of the fees collected for use in an IFQ loan programme to aid in financing the purchase of IFQ or quota share (QS) by "entry-level" fishermen and "fishermen who fish from small vessels." The funds so reserved may then be assigned to the NMFS (Financial Services Division) under the Federal Credit Reform Act (FCRA) and "leveraged" (at a ratio of approximate 1:50) with funds loaned from the U.S. Treasury to capitalize the Federally-subsidized IFQ loan programme.

To implement the mandate in the Alaska halibut and sablefish Individual Fishing Quota (IFQ) programme, the NMFS is developing a Proposed Rule and will seek public comment before finalizing the cost-recovery programme. The proposal has not yet been published, and is therefore a "work in progress;" however, certain of its elements have been discussed in detail with members of the North Pacific Fishery Management Council (Council) and others, and are set out here.

2. BACKGROUND

The Alaska Region of the NMFS, through its Restricted Access Management (RAM) programme, administers the Alaska IFQ Programme. The IFQ programme is a limited access system authorized by section 303(b) of the *Magnuson-Stevens Act* and *the Northern Pacific Halibut Act of 1982*. The programme has been fully effective since its implementation in March 1995; regulations for the implementation of the programme can be found at 50 CFR part 679.

Approximately 5000 persons (individuals and companies) currently participate in the programme. As the following table shows, the IFQ system of managing access to the Alaska halibut and sablefish fisheries imposes additional costs on the public in an amount estimated at about US$ 2.8 million (Smith 2000*)*.

Table 1
Estimated annual costs of managing and enforcing the Alaska halibut and sablefish IFQ programme in Alaska

Expense category	Estimated annual costs
Restricted access management and sustainable fisheries	1 400 000
Administrative appeals	200 000
Alaska Enforcement Division	1 200 000
Total	**2 800 000**

To date, there has been no requirement that industry pay any of these costs. However, the amendment to the *Magnuson-Stevens Act* noted above requires the government to recover most, if not all, of these costs directly from IFQ programme participants.

The proposal discussed here was developed over a two-year period beginning in 1997. Staff from the NMFS Alaska Fisheries Science Center in Seattle and the Sustainable Fisheries Division in Juneau met on several occasions with the North Pacific Fishery Management Council (Council) and its committees, to address such issues as: Who should pay? How should payment be calculated? How should costs be calculated? How should the NMFS impose penalties for non-compliance? Where does the money go and what is done with it? and a variety of related questions. Taken together, the answers to these questions provide the framework for the proposed programme discussed herein.

3. PROGRAMME DESIGN

3.1 Who pays?

Under the IFQ programme, any IFQ permit holder who delivers IFQ halibut or sablefish must do so to a "Registered Buyer" (*i.e.* a processor or other buyer who holds a specific permit, issued by the NMFS, to purchase IFQ product). Because it is the transaction between the fisherman and the buyer that gives rise to the "ex-vessel" price of the fish (upon which the fee is to be premised), the payer under the programme could be either the IFQ permit holder or the buyer.

Some precedent exists for the buyer bearing the burden. The State of Alaska imposes a variety of taxes on the commercial fisheries it operates, including a "raw fish tax" (which is paid into the general fund of the state and relevant local governments), a marketing tax, and a salmon enhancement tax (Alaska statutes). All of these taxes are withheld by fish-buyers and paid annually to the State of Alaska's Department of Revenue. In developing this plan, initial thinking was that the Registered Buyers would likely be the most logical payer, especially as many of them are the same entities that withhold and pay the State of Alaska taxes.

However, upon reflection it was determined that Registered Buyers would not be the appropriate payer under the cost-recovery programme. First, not all Registered Buyers have the management infrastructure necessary to maintain records and funds and pay them on an annual basis to the NMFS. Second, because anyone who applies for a Registered Buyer permit may receive one, it would be difficult to enforce sanctions against those who failed to comply with the terms of the programme. And finally, the Registered Buyers resisted the role of "tax collector" noting that it was the IFQ holders, and not the buyers, who were benefitting the most from the IFQ programme. Accordingly, a decision was made to require each person who used their annual IFQ permit to make the payment to the NMFS at season's end.

3.2 Determining basis for payment

The Act requires that payment be based on the "ex vessel" price paid to fishermen for their catch. This mandate assumes that fishers are simply wholesalers of fish to a purchaser or processor, who then sells the fish at the retail level. However, a number of fishers are their own "buyers" and sell directly to the public from their vessels or after transporting their catch to a public market. In those cases, there is no distinction between the ex-vessel price and the wholesale price.

Accordingly, the decision was made to propose two separate approaches to determining just what the price should be. The first is the one envisioned in the Act; *i.e.* using the price paid to the fisherman (the "actual" price). The second option was a derived price calculated by the NMFS from reports submitted by Registered Buyers (the "standard" price). The standard price would be sensitive to species, time and area (port) of first sale and would be the "average" price paid to fishermen by buyers.

Those IFQ holders who bear the burden of calculating and paying the fee could either use the "standard" price compiled by the NMFS or they could base their fee on the actual prices they received from the sale of fish. If they choose the latter option, they would be required to keep appropriate documentation of their receipts to withstand any audits or other investigations by the NMFS.

3.3 Fee percent calculation

The Act envisions that the fees collected would be sufficient to cover the cost of managing and enforcing the halibut/sablefish IFQ programme. However, it also limits the percentage amount that can be collected to 3% of the total ex-vessel receipts and directs that 25% of the receipts be used for a different purpose (the loan programme). If the actual costs of management and enforcement are as reported (at $2.8 million), then it is apparent that, for the near term at least, the percentage charged will most likely remain somewhat below the maximum, or 3%. Simple arithmetic can be used to estimate the amount of fees that will be collected, as follows (estimates are based on projected 1999 harvest and estimated ex-vessel prices):

i. If the halibut harvest is 50 000 000lbs, and the ex-vessel value is ~$2.00/lb, the total ex-vessel value of the halibut IFQ fishery is $100 000 000
ii. If the sablefish harvest is 25 000 000lbs, and the ex-vessel value of sablefish is ~ $2.00/lb, the total ex-vessel value of the sablefish IFQ fishery is $50 000 000
iii. Three percent of $150 000 000 is $4 500 000 and
iv. After 25% ($1 125 000) is deducted for the loan programme, the balance ($3 375 000) is available to offset the actual costs of managing and enforcing the IFQ programme.

As we have seen, those actual costs amount to $2 800 000; so, it is probable that a fee of somewhat less than 3% may be charged. Because the two factors (total ex-vessel value of the fishery and actual cost of managing and enforcing the programme) that determine the percentage fee that will be charged are dynamic, the fee percentage is expected to vary from year to year. Accordingly, the plan is to assume that the maximum rate of 3% will be needed, but to allow the NMFS Regional Administrator to set a different (lower) rate if the value of the fishery unexpectedly increases, or if the annual cost of managing and enforcing the programme is adjusted.

3.4 Billing and payment

As noted above, the responsibility for paying the fee to the NMFS will lie with the IFQ permit holder. To facilitate payment, the NMFS will annually compute (for each permit holder), the total landings of IFQ halibut or sablefish, apply the "standard price" calculation to the number and location of lbs landed, multiply the totals by 0.03 and present the permit holder with a report and a bill. The IFQ season currently ends in mid-November; it is expected that the billing can be completed and mailed by no later than 1 December of any given year.

When the report and billing is received the IFQ holder will have the option of paying the amount billed (*i.e.* the amount that is based on the standard price computed by the NMFS from information submitted by Registered Buyers) or paying the fee based on the permit holder's "actual" receipts for fish sold.

Payment in full will be due to the NMFS by no later than 31 January of the year following the landings that cause the payment obligation to be incurred.

3.5 Underpayment, late payment, and non-payment

As IFQ permit holders who have landed fish against their IFQ permit, and who thereby have incurred an obligation to pay a fee under the cost recovery programme, will have until 31 January of the year following the landings that give rise to the obligation to make payment in full to the NMFS. The fisher has a choice: either pay the fee based on the "standard" price as computed by the NMFS from Registered Buyer reports, or pay the fee based on his/her actual receipts from the sale of fish.

If they choose the former, and pay in full, the obligation is met. However, if they choose to pay based on actual receipts, they bear the burden of demonstrating the veracity of the information provided and the accuracy of the calculations. It is anticipated that the NMFS will accept, without question, most such payments; however, some may vary so significantly from the payments that would have been due using the standard price calculations that the NMFS could make an inquiry. Further, the NMFS could randomly select certain returns for audit. In either case, the burden of demonstrating the validity of the payment would lie with the permit holder.

When the NMFS questions a payment, the payer would be so notified and given the opportunity to submit information and evidence in support of their position that sufficient payment had been made. In the case of those paying against "actual" receipts, the sort of evidence that would be expected would be contemporary records of fish sales that showed the time and place of sale, the amount sold and the total paid.

If, upon receipt of additional evidence from a permit holder, the NMFS still believes that insufficient payment has been made, an "Initial Administrative Determination" (IAD) to that effect would be produced. The IAD could be appealed to the NMFS Office of Administrative Appeals who would conduct an inquiry (perhaps ordering an evidentiary hearing) and produce a decision. The NMFS Regional Administrator would have 30 days during which to review the decision and could order its adoption, reverse it, or remand the matter for further work. Once a decision acceptable to the NMFS Regional Administrator had been produced, it would become the final agency action on the matter, subject only to further appeal through the US court system.

A permit holder who has incurred a fee obligation, but who does not file a return at all, would immediately receive an IAD from NMFS/RAM. If a permit holder does not respond, the appeal could result in a final agency action requiring payment.

During the pendency of any adverse administrative action (a determination that an insufficient fee, or no fee, has been filed, and that monies are due and owing to NMFS) on a fee obligation, any IFQ permit (including the underlying Quota Share from which the IFQ is derived) would be immediately designated non-transferable in the hands of the debtor; and the debtor would not be allowed to receive any additional Quota Share or IFQ by transfer. Once the matter was resolved, transferability would be restored; however, if the matter could not be administratively resolved, and a final agency action determines that payment is due, the use of any permit in the hands of the debtor would be suspended. If, after a period of 30 days, no payment had been received, the case could be referred within the Department of Commerce and subject to additional Federal collection procedures. In that event, continuing noncompliance with the fee requirement could lead to forfeiture of annual IFQ permits and/or permanent revocation of Quota Share.

4. LIMITED ACCESS SYSTEM ADMINISTRATIVE FUND

With the exception of the 25% of the fees that are diverted to support the IFQ loan programme, all fees collected from participants in the IFQ fisheries are to be deposited in the Limited Access System Administrative Fund (LASAF) established within the U.S. Treasury. Appropriations from that fund are intended to support the management and enforcement costs in the NMFS region from which they were collected. Additional deposits to the LASAF are expected to be derived from CDQ programme fees and, eventually, from payments to the Central Registry System for Limited Access System Permits established by the Act.

5. ANNUAL REPORTING

A final element in the proposed cost-recovery plan is the requirement that an Annual Report be prepared and distributed to IFQ programme participants and the general public. The Report would cover such items as the estimated total ex-vessel value of the fisheries (as derived from Registered Buyers reports), the numbers of participants paying fees, the amount of the fees, compliance with reporting and other fee collection requirements, and the use to which the fees were put (including a fully transparent budget for the actual cost of managing and enforcing the IFQ programme).

Although a decision on the question is yet to be made, it is envisioned that the required Report would be incorporated in the annual "IFQ Report to the Fleet" that is already prepared and distributed (and posted on the NMFS/Alaska Region Internet web site) each year.

6. CONCLUSION

Although the *Magnuson-Stevens Act* has long authorized NMFS to charge nominal fees for the issuance of permits [provided that "...the level of fees charged...shall not exceed the administrative costs incurred in issuing the permits..." -- MSA, Sec. 304(d)(1)], the inclusion of a specific fee for the IFQ programmes is an innovation that is yet to be tested. The programme described here is intended to meet the requirements of the law with respect to the level of fees that may be collected and the peculiar limitation on how that is to be accomplished further. It is intended to strike a proper balance between NMFS as the regulator and enforcer of tax collection (on the one hand) with the role of NMFS as a collaborator with industry in devising fisheries management programmes that are appropriate, responsive, and yet be acceptable to all participants. Although the regulatory framework for the programme is yet finalized, there are indications that, as a whole, the IFQ fleet is willing to pay the necessary price for the net benefits they experience from holding IFQ and fishing under the system.

In its important work, *Sharing the Fish, Toward a National Policy on Individual Fishing Quotas*, the National Research Council noted with approval that the *Magnuson-Stevens Act* now provides for a fee collection programme, but commented that "...in practice, the limit of 3% may well be too low for some IFQ programmes and should be increased..." Likewise, the NRC recommended: "The Magnuson-Stevens Act should be amended to authorize the capture of rent in excess of cost recovery."

Whether this latter recommendation will become reality is, of course, unknown. However, successful implementation of the existing statute would be an appropriate first step toward the realization of the broader goal of eliminating pernicious national subsidies.

7. LITERATURE CITED

Best, G. 1999. Commissioner (Canada), International Pacific Halibut Commission, personal communication.

Federal Fisheries Investment Task Force 1999. Study of Federal Investment.

National Marine Fisheries Service 1996. Magnuson-Stevens Fishery Conservation and Management Act as amended through October 11, 1996. *NOAA Technical Memorandum*, NMFS-F/SPO-23, 121 pp.

National Research Council 1999. Sharing the Fish: Toward a national policy on individual fishing quotas. National Research Council. National Academy Press, Washington, D.C.

National Institute of Economic and Industry Research 1998. Subsidies to use of the natural resources. Commonwealth of Australia, Paragon Printers.

Smith, P.J. #. How "privatization" can result in more government - the Alaska halibut and sablefish experience.

SOUTH AFRICAN PERSPECTIVES ON RIGHTS IN FISHING AND IMPLICATIONS FOR RESOURCE MANAGEMENT

D.J. Bailey
Bato Star Fishing (Pty) Ltd, Consultative Advisory Forum
P O Box 7251, Roggebaaai, Cape Town, 8012 South Africa
<dbailey@iafrica.com>

1. INTRODUCTION

Our policy development processes in fisheries management are necessary focused on our own domestic situation. It is, however, important to ensure that we keep up with developments in this area in the rest of the world. I am confident that we are on the right track with our policies, which may need some more definition in some instances, but our biggest challenge lies in achieving an equitable distribution of rights and successful implementation of a rights-based culture.

This paper presents a brief history and an objective perspective of the South African position regarding access rights at this point in time. This perspective is derived primarily from the outcome of policy consultations with stakeholders through the Fisheries Policy Development Committee (FPDC), the subsequent Government Policy White Paper and our new *Marine Living Resources Act* which came into effect in September 1998, just over a year ago.

The question of the nature of fishing rights is very pertinent to the South African situation at this time. We are currently in the process of restructuring our rights allocation regime and it is envisaged that long-term rights will be allocated over the next few years. This should bring about the conditions required for the South African industry to maintain their reputation for quality products and stability of supply into the global markets.

Our fisheries policy are based on the objectives of EQUITY, SUSTAINABILITY and STABILITY. It is important for the future of our fisheries that the nature of the rights and the outcome of the allocation process make a significant contribution to these policy objectives. Equity refers to the need to achieve an ownership profile in the industry which is representative of the South African population. Sustainability refers to the need to manage our resources responsibly for long-term benefit. Stability expresses the need for a stable industry in terms of resource levels and security of rights.

2. THE CONTEXT IN SOUTH AFRICA

It is necessary to reflect on the question of rights in the South African context. South Africa is a country where for a long time emphasis was placed on the differences between people of various racial groups through the policy of apartheid. While the proponents of apartheid tried to find philosophical justification for their policies, it came down to denial of basic rights to black South Africans. Fifty years of apartheid led to our current situation where the political, economic and social differences between groups are largely demarcated along racial lines.

An end to this madness was signalled by the radical change of direction by the ruling National Party in 1990 in the unbanning of all political parties and the release of political prisoners. There had, however, never been a level playing field for all races in South Africa. From the days of colonialism and slavery, the indigenous black people of South Africa were dispossessed, cheated and confined to a second class role in South African society. Our leaders were either forced into exile, imprisoned or killed by the regime. Some of the more high profile legislative and institutional arrangements were as follows:

i. *1913 Land Act* - confined over 80% of the population to 13% of the land
ii. the *Group Areas Act* defined the areas where black people could live and operate businesses and resulted in thousands of people being forcibly removed from their homes
iii. labour preference areas and Pass Laws restricted the free movement of black people
iv. labour laws entrenched a system of job reservation by defining certain jobs for 'whites only' and
v. each racial group had its own education department with huge disparities in resource allocation between the departments.

These apartheid policies were largely supported by industry, particularly those dependent on government contracts or quota allocations, in their employment and staff promotion policies.

This was where we found ourselves in April 1994 after our first democratic elections. It is clear, even after our second democratic elections earlier this year that we will have to work hard to sustain the 'miracle' which brought about a democratic political dispensation to ensure the normalisation of our racialised social and economic situation and the alleviation of poverty. Soon after it came to power, the new African National Congress government tabled the Reconstruction and Development Programme (RDP) to address the country's social and economic problems through the upliftment of previously marginalised groups. The RDP clearly envisaged addressing areas of unequal economic opportunities such as access to government contracts and the allocation of concessions including access to marine resources.

The RDP has been superseded by a new economic policy, but the basic objectives of the RDP, to normalise the distribution of economic opportunities, still guides government thinking on issues such as access to marine resources. In addition to this, our overall industrial policy strongly advocates the promotion of small and medium enterprises to serve as the engine of employment and economic growth through a more vibrant competitive environment.

The biggest challenge these policies are meant to address is the high level of unemployment in South Africa, estimated at over 30%, and the low levels of economic growth over the last few years. Again it is the largely unskilled black population which suffers the highest levels of unemployment. With a lack of alternative employment opportunities in coastal communities, the fishing industry is seen as a last resort by many people.

We also have a Constitution which, while recognising that certain imbalances need to be addressed, guarantees the rights of all South Africans. Important for fisheries management is the constitutional right to administrative justice. This constitutional provision will underpin a strong rights-based culture in our fisheries.

3. RIGHTS IN SOUTH AFRICA'S FISHERIES

The situation sketched in this brief background is clearly reflected in the pattern of quota distribution in the fishing industry in 1994:

Fishery	TAC (tonnes)	Number of quota holders	% of TAC held by 3 largest
Hake	148 300	31	80
W C Rock Lobster	1 500	99	30
S C Rock Lobster	427 (Tails)	6	82
Abalone	615	12	75
Pilchard	105 000	47	40
Anchovy	70 000	30	80
Sole	872	10	71

TAC: Total Allowable Catch.

These figures are made more stark if one considers that the black South African population, which constitutes 87% of the population, has virtually no interest in the ownership and management of these companies.

This is the situation which the FPDC, tasked with developing a national fisheries policy, was faced with when it started looking at policy initiatives for the transformation and management of the industry in December 1994, some seven months after our first democratic elections. The FPDC represented all stakeholders, *i.e.* government, industry representatives from the various fishing sectors, labour, recreational fishers, conservationists and coastal community representation through various regional Fishing Forums. The FPDC reached broad consensus on the important issues of transformation and the nature of fishing rights and published a final report after two and a half years of deliberations and negotiation.

It was clear that in terms of commercial fishing the major obstacle would be to achieve a more equitable distribution of fishing rights, a distribution that would more fairly reflect the demographics of South Africa and would be broadly accepted by stakeholders. This would be the initial allocation which many of the speakers at the FishRights99 Mini Course and Core Conference acknowledged as the necessary condition for the introduction of a more advanced ITQ system.

As the FPDC had strong representation from the existing industry as well as aspirant new entrants, its recommendations on the nature of the fishing right is a good yardstick of industry's views on the matter. The FPDC did not deal extensively with the characteristics of Security and Exclusivity as these were to a large degree entrenched in the existing quota rights. Rights emerging from a legitimate transformation process should also offer enhanced Security and Exclusivity. The FPDC recommended that rights should be transferable, subject to:

i. an initial payment on allocation and
ii. an initial moratorium on transfers to new entrants.

The FPDC concluded that long-term rights (in perpetuity) are more desirable because of the enhanced economic security for the rights-holder. It did envisage sanctions such as cancellation of the right for non-usage and compliance transgressions.

Apart from the initial payment for the right the FPDC also recommended annual tax and rental payments. The FPDC recommended that all users, including recreational fishers, should pay for the privilege of access in order to support sound management of marine resources. It further recommended that there should be a cap on both the maximum number of participants in the different sectors as well as the maximum allocation held by any one quota holder (30% of TAC)[1].

The *Marine Living Resources Act* emerged after the publication of the White Paper and further lengthy debate and trade-offs in Parliament. The Act, which became effective in September 1998, has now put in place a right which is Exclusive (determined as a portion of the TAC/TAE), Secure (in terms of legal processes), has limited Durability (up to 15 years) and regulated Transferability. Long-term rights will be leased by the state with an annual lease fee payable.

In recognition of the plight of impoverished coastal communities the FPDC recommended further investigation into ways of incorporating rights for unemployed coastal people and for providing immediate poverty relief. The Act makes provision for the granting of a subsistence fishing right to the unemployed and poor. This is a new fishing sector and an extensive consultation process has been launched to ensure the proper functioning and management of this sector. Unlike recreational fishers,

---

[1] This allocation was made by the Diemont Commission as it became known. Its full title was: Commission of Inquiry into the Allocation of Quotas for the Exploitation of Living Marine Resources on a Firm Basis, and it was chaired by Judge Diemont, appointed in 1985.

subsistence licence holders will be allowed to sell their catch.

The nature of the fishing right as it stands now makes it difficult for small and medium enterprises (SME's) to obtain finance as the banks are not prepared to accept the right as collateral. The FPDC did make reference to a dedicated development finance institution for the fishing industry. There are some development finance institutions and government credit guarantee schemes operating at the moment and I believe that these institutions will be able to extend their focus to the fishing industry to resolve the question of finance for SME's in the fishing industry.

With this fairly prolonged policy development process behind us everybody seems to be reasonably happy with the right as defined in the Act. This could be because the primary focus at this stage is on the allocation process, which is the major cause of uncertainty in the industry. The negative consequences of this uncertainty are a lack of compliance and weak stakeholder participation in resource management processes. Our current challenge is thus to stabilise our stakeholder base through the finalisation of the allocation process, building of strong co-management structures and entrenching a strong rights-based culture through our economic contribution to South Africa as an equitable, sustainable and stable industry.

MAKING FISHING RIGHTS WORTHWHILE – SUSTAINABLE FISHERIES

K.Truelove
Marine Group, Environment Australia
GPO 787, Canberra ACT 2601, Australia
<kerry.truelove@ea.gov.au>

1. INTRODUCTION

This paper is not about who should get property rights, nor about the principle upon which they should be granted or withdrawn. Instead, it deals with the issue of the value of a property right in terms of longterm sustainability. My point of view is simply that the value (to the owner and the community) of a right – any right – which is dependent upon the use of a natural resource is directly proportional to the long-term sustainability of that resource and the environment in which it exists. The key issue here is the phrase "long-term".

Two serious questions have been raised in the pursuit of property rights: (a) What are they? and (b) who should have them? I have no doubt that there is a plethora of explanations given as to what constitutes a property right. To take a relatively simple definition, property rights are "varying degrees of ownership of a resource by particular individuals or associations" (Cooke 1984, quoted in Harden Jones 1994). In the fisheries context they may include licences to fish, gear entitlements, individual transferable quotas (ITQs), and they basically are predicated upon a limited-access fishery since where there is no limit there is no preferential right of access.

"Who should have them?" is perhaps the most vexed question of them all. At a philosophical level, there is concern at the concept of property rights over a common property resource. Van der Elst et al. (1997) suggest that a consensus is emerging that the owner of the fisheries resources of a country is the community, with government charged with the task of allocating those resources. Fish and the environment in which they exist are a community resource to which commercial fishers obtain a form of preferential access, through such devices as licences. Conditions are placed upon those with the "right" to fish by the managers, in response to the community preference that the common resource be maintained through time.

It could be seen in terms of the community as a whole relinquishing a proportion of its right to the resource to the fishers. So in a sense the commercial fisher has an inalienable right, as a member of the community, to access the community resource plus that fragment of the overall community right which implicitly has been ceded to him through the tacit agreement that he be allowed to fish for profit. Meantime, every member of the community retains that inalienable right of access. To extend the analogy, the question arises: does the commercial fisher then have two rights to the resource – a commercial right (*i.e.* that proportion of the community right ceded to him) plus his community right? My argument is no: instead, he has an enhanced right of access through the tacit agreement of the community; not twice the rights of any other member of the community.

These are deep philosophical issues and it would take someone with a great understanding of human behaviour to unravel them. This paper, rather, concentrates on the fundamental principle that fisheries should be sustainable if properly managed in the ecological context and recognises that not even a natural, unfished system is inherently sustainable.

2. THE RIGHTS AND RESPONSIBILITIES OF STEWARDSHIP

2.1 Community property and the commercial fisher

Assume there is a community of 200 with access to a single species of fish. If rights are distributed evenly throughout the community, this would logically mean each person has right to $1/200^{th}$ of the resource that is available and can be harvested sustainably – note, not to the overall resource but only the portion that can be harvested sustainably. Of course, everyone's use of the resource will not be the same. For some it may be a matter of fishing for dinner once a week; others may prefer simply to go into the aquatic environment and observe it; still others may not use the resource in any physical way but "use" in the sense of having a warm feeling because the resource is there. These all are legitimate uses of a resource.

Now let 10 members of the community become commercial fishers, and to make a living they need $1/20^{th}$ of the available resource. To maintain the overall harvest at a sustainable level, the community must cede to each commercial fisher a proportion of the community right so that $1/20^{th}$ is available to each fisher. If each commercial fisher "ceded" the same proportion from his own community right to fish (but retained that right for his commercial operation), then ten operators would take up the entire resource available for sustainable harvest: 200/200 (I assume that the community does not wish to harvest the resource and is content to let the commercial fishers do that). If a commercial fisher did not cede that proportion of his right which the community as a whole has decided should be ceded, access rights reach 210/200.

This is not so sustainable. Still, it's easier to let the commercial operator continue making his profit, and keep the own recreational rights to fish or other forms of access. Everyone keeps their rights, and if the commercial fisher makes a profit, well good luck to him; everyone is happy. Besides, the commercial fishers have paid the community for the rights to exploit the resource for private commercial profit.

So things in the community of 200 can go along swimmingly until the numbers game changes: an eleventh commercial fisher starts operating, the community grows, the resource starts to fail. In all of these scenarios, the value of the right starts to fall. Eleven fishers means that each cannot have access to $1/20^{th}$ of the resource any more, yet $1/20^{th}$ is needed for their ventures to survive economically; a community of 201 means that the initial right of each member is reduced to 99.5% of the original (with consequent impacts on the proportions available for commercial operators); a decline in productivity of the resource means fewer fish make up the $1/200^{th}$ of the resource that is the initial right, with consequential economic and ecological implications.

This is a simplistic argument and the pundits will rightly argue that the whole issue is a great deal more complicated. My scenario does not take into account factors such as the cost to commercial fishers (licences, fuel, gear, etc.); the benefit to the fishers in making a living; the benefit to the community through employment generated by commercial operations; the benefit generated in terms of licence fees; the cost to the community in having the management authority; and so forth. These are all social and economic issues, and are critical in our socio-political world; yet in an environmental sense they are almost at the level of the *non-sequitur*. Our approach on the environmental front has been uncoordinated and leaves much to be done while we are at an unknown point on the sustainability curve.

Whatever the argument, it is hard to avoid from the reality that fish and the environment in which they live are finite; we cannot parcel them up into property rights to an infinite number of people without reducing the size, quality and value of those property rights. We can parcel up the resource explicitly, as part of a controlled management arrangement; or we can do it implicitly, by allowing pressure (fishing, pollution and degradation, etc.) to increase beyond the capacity of the aquatic environment and then deal with the social, economic and ecological decline that inevitably must follow.

2.2 For every right there is responsibility

I do not believe many would disagree that for every right there is a responsibility. If you get your driver's licence, you also get the responsibility of driving safely for yourself, your passengers, other users of the road and the community. Sometimes that responsibility will outweigh the benefits of the right; sometimes the responsibility is quite light; but the two go hand-in-hand.

In the community of 200, in the initial case, each person has the responsibility to ensure their use of the resource does not exceed a $1/200^{th}$ share of the resource. With the advent of the commercial fishers, each member of the community continues to have the same responsibility to ensure their use does not exceed their right, which in the 10-fisher scenario equates to nil because they have ceded that right to the commercial fishers. The community as a whole also has the responsibility for minimising its impact on the aquatic environment, *e.g.* sewage outfalls and other forms of pollution, as well as the usual forms of social, political and economic responsibilities such as voting, driving on the correct side of the street, and paying for what they obtain.

These responsibilities also lie on commercial fishers, yet they have additional responsibilities flowing from their preferential access to the resource. They must ensure their use does not exceed the $1/20^{th}$ that is their right. They must satisfy additional responsibilities brought about through mechanisms such as management arrangements – reporting their catch and providing the relevant information sought by the community as part of the cost of access to the $1/20^{th}$ of the resource and, because they are the ones working physically in the aquatic environment, a responsibility to minimise their impact on that environment. These impacts can be in terms of unintended catch, lost or discarded gear, excessive removal of prey or predator species, polluting discharges, translocation of pest species, and so forth.

2.3 The umpire's decision is final: but who is the umpire?

If a property right is not going to be handed out to fishers unfettered by regulation and community expectation, we must assume that there will be an umpire to control the exercise of the right and attendant responsibilities. Who is that umpire? The community as a whole, which has ceded to fishers' preferential access? The management agency, which has been created by the community to establish the degree of right and undertake the oversight role? Or is it, ultimately, the aquatic environment which must carry the overall brunt of the exercise of that right?

One cannot fish without having an impact. Thus, defining the point at which the community is willing to bear the cost associated with that impact and where it is not, becomes an exercise in clairvoyance. Management agencies with the assistance of scientists and with the increasing involvement of fishers and members of the general community, are engaged in developing estimates of this point. Sometimes these are target reference points, such as a total allowable catch, or an individual transferable quota; more critically it may be a limit reference point, the point in a decline in biomass where a management action is required.

Management agencies have the most onerous responsibility – they must be able to determine what proportion of the total resource can be harvested sustainably within the context of all impacts upon the resource and supporting aquatic environment. They do not do this without scientific help; as Garcia (1994a) notes, fishery scientists are expected to: (a) determine the theoretical potential production of a stock; (b) calculate as a benchmark the corresponding level of fishing effort; (c) determine appropriate size at first capture; (d) recommend means whereby these can be achieved and the trade-offs involved; and (e), assess the effects of fishing and forecast impacts of management options. Once that is done, management agencies must determine how the proportion available for fishing should be distributed and to whom; then they must ensure that each of the rights holders discharge their responsibilities for the use of the right. What is more, they must do this in a climate of chronic and continuing uncertainty about the state of the fish stocks and the aquatic environment. Whatever the approach, there always is a risk that the reference point is wrong and fisheries management is a process of risk management as much as anything else.

In an anthropocentric sense, once the community cedes preferential rights of access to fishers, the next level of umpiring is the community, through decisions taken by the government as the community representatives (tempting though it may be to equate the management agency with the government, it would not be accurate. The former is an instrument of the latter, but the two are not synonymous and governments legitimately may take decisions against the advice of management agencies). On a less anthropocentric scale, the final umpire, the one we avoid calling upon, is the environment itself.

3. WHAT IS THE WORTH OF A FISHING RIGHT?

3.1 What is sustainability?

The first question must be: what is sustainability? In our human time scales coal is an exhaustible resource, rather than a sustainable one and the most acceptable approach is to treat it as such by managing mining impacts so that they are ecologically sustainable on the surrounding environment – an approach consistent with the National Strategy on Ecologically Sustainable Development.

Fisheries are quite different. They are based upon a renewable biological resource. They have the potential, if properly managed, to be one of the most long-term sustainable industries; they provide a relatively swift turnover in biomass, faster than many forest resources and infinitely faster than mineral resources which take millennia to develop. Many aquatic ecosystems appear to be remarkably resilient to perturbations, and in comparison with many of our terrestrial systems are relatively undamaged.

In accepting fishing – be it recreational, artisanal, indigenous, commercial or heavy industrial – the community is implicitly accepting that the aquatic environment in which the fishing is taking place will be affected, the scale of which will depend upon the scale of fishing. In the case of full-scale commercial fishing or wide-spread, heavy recreational fishing, removal of a large number of select species from an ecosystem would have an impact on the ecosystem (small-scale, artisanal fishing may be a different case). Further, there may be a significant time delay before those effects become clear. At the same time one needs to accept that ecosystems are dynamic; with or without fishing, they continually change from one point of transition to another. Thus, the trick in

fisheries management is to ensure that the ecology, amended as it is by fishing (and other factors which may be beyond the control of the agency or of fishers), can continue to change over long periods of time, without a nett overall loss in quality.

I regard this as "sustainability". It is a definition as vexed with problems as any other – what, for instance, is "a nett overall loss in quality"? More importantly, how is it measured? These are questions fisheries agencies continue to grapple with them, as they battle with operationalising the principles of ecologically sustainable development (ESD). There are no neat explanations, no tidy definitions.

3.2 The strategic approach to sustainability – Australian style

Australia has a *National Strategy for Ecologically Sustainable Development* (Commonwealth of Australia 1992). The National Strategy recognises that there is no universally accepted definition of ESD and defines ESD as *"using, conserving and enhancing the community's resources so that ecological processes, on which life depends, are maintained, and the total quality of life, now and in the future, can be increased"*. The Strategy clearly recognises that there is no identifiable point at which ESD can be said to have been achieved and that some key changes to the way we think, act and make decisions will help ensure Australia's economic development is ecologically sustainable. Two main features are identified which distinguish an ecologically sustainable approach to development:

i. consideration, in an integrated way, of the wider economic, social and environmental implications of decisions and actions (nationally, internationally, and in terms of the biosphere) and

ii. taking a long-term rather than short-term view when taking those decisions and actions.

Within the definition of ESD are goals, core objectives and principles which need to be considered as a package with no objective or principle taking precedence over others. This requires a balanced approach to the goal of ESD, which is *"development that improves the total quality of life, both now and in the future, in a way that maintains the ecological processes on which life depends"*. The core objectives of the strategy can be summarised as: (a) enhancement of individual and community welfare through an economic development approach that safeguards the welfare of future generations; and (b) provision of inter- and intra-generation equity to protect biodiversity and maintain essential ecological processes and life-support systems. These core objectives imply the following guiding principles:

i. integration of short and long-term economic, environmental, social and equity considerations in decision making processes

ii. where there are threats of serious or irreversible environmental damage, lack of full scientific certainty

should not be used as a reason for postponing measures to prevent environmental degradation
iii. recognition and consideration of the global dimension of environmental impacts of actions and policies
iv. recognition of the need to develop a strong, growing and diversified economy which can enhance the capacity for environmental protection
v. recognition of the need to maintain and enhance international competitiveness in an environmentally sound manner
vi. adoption of cost-effective and flexible policy instruments (*e.g.* improved valuation, pricing and incentive mechanisms) and
vii. broad community involvement in decisions and actions on issues affecting the community.

The National Strategy also contains a specific challenge for the fishing sector which is to adopt an ecosystem-based management framework through satisfying objectives relating to: (a) a framework of resource stewardship; (b) state of the aquatic environment reporting; and (c) information dissemination.

The *Report on the Implementation of the National Strategy for Ecologically Sustainable Development 1993-1995* (Commonwealth of Australia 1996) indicated that a significant amount of work has been done since the finalisation of the National Strategy to achieve these objectives. Most action appears to have occurred in areas of resource stewardship, such as fisheries. One indication of this has been the promulgation of fisheries legislation embracing the basic principles of ESD in all States (other than South Australia) and the Northern Territory; South Australian fisheries legislation, promulgated in 1982, already encompassed the concepts of equity and sustainability, as does the 1991 Commonwealth fisheries legislation.

But new legislation and a national commitment merely provide the framework within which sustainability may be achieved. Further, after the adoption of the National Strategy there remained an emphasis in fisheries management upon the social and economic aspects contained in point i. above, with the environmental aspect addressed sporadically and in response to pressure. Dovers and Mobbs (1997) suggest that Australia's disjointed and *ad hoc* approach to environmental policy "*can only be addressed by institutionalising environmental and sustainability concerns on par with social and, especially, economic concerns*".

The draft report of a recent review of the adoption of ESD in Commonwealth departments suggested that it had varied widely across agencies, and stated that implementation has been best in areas of natural resource management: "*In the area of natural resource management and environment protection, the integration of economic, environmental and social considerations has been seen as a core policy concern. These areas provide the best examples of ESD implementation. A common model in these areas is various forms of partnerships among key stakeholders to achieve mutually agreed, integrated ESD outcomes. However, in some cases action has been taken in response to a looming problem*" (Productivity Commission 1999). Five main impediments to implementation of ESD are identified:

i. a lack of clarity or understanding as to what constitutes ESD-related policies
ii. the complex issues associated with the implementation of ESD and the information and data requirements
iii. failure to adopt good policy-making practices
iv. deficiencies in intra- and inter-governmental coordination in policy-making and
v. insufficient attention to longer-term sustainable development issues.

This is changing, but it takes time.

Biologically a fundamental error may have been made in assuming there should be no hierarchy between the economic, social and environmental implications of decisions, since if the environmental safety-net fails, economic and social considerations will collapse as well. I am not sure that the reverse is true, but if a society is so pressured by the need for short-term survival it is unlikely to pay attention to long-term issues such as environmental sustainability. I am equally uncertain whether Australians are faced with this particular conundrum; but it is an issue facing less fortunate nations.

3.3 Granting of property rights as a means of achieving sustainability

The argument has been raised that ill-defined or inappropriate property rights to resources can lead first to, overcapitalisation and economically inefficient exploitation, as fishers in open-access fisheries seek to increase their share by fishing more, and then to over-fishing of the resource (Commonwealth of Australia 1989; National Research Council 1999). The argument can be made that granting property rights should ensure the resource is looked after by its "owners" because they have a stake in the future of the resource. In the case of fisheries, this may be a measure for achieving sustainable exploitation.

But there are two main troubles with this assumption: (a) who are the "owners" of the resource, of the right, and who has the right to distribute the rights; and (b) whether many "owners" of the same property can collectively assure its sustainability.

On the first, Van der Elst *et al.* (1997) make it clear that an emerging consensus is that resource owners are the community as a whole.

Can the grant of a property right vest some form of ownership of the resource in the person receiving the right? I think it more likely that granting of the right merely gives ownership of the right (and its attendant responsibilities), not ownership of the resource *per se*. Admittedly, the right itself can be seen as that portion of the resource the community has granted to the fisher for his exclusive use, and as such there is a form of ownership of the resource involved. But it is not ownership of the total resource and should not be interpreted as such.

If the community is the owner of the resource, it follows that the community as a whole is the owner of the right to distribute rights. By creating management agencies vested with the responsibility for ensuring conservation and sustainable use of the resource, the community grants to the agencies the right to distribute rights. The management agency becomes the agent for the community, distributing rights on the community's behalf.

In the case of right-holders assuring sustainability of the resource, this may be relatively easy if only one individual or organisation exercised the right. Where a fisher had exclusive access to a fishery, he would fish at the level to maximise the profit. As the effort required to take maximum economic yield (MEY) is less than that required to take the maximum sustainable yield, exploitation at this level would not result in biological overexploitation (Commonwealth of Australia 1989). The fisher is happy – he is making a bomb – and the resource is not overly stressed. All going well, the aquatic environment in which the resource exists is also not overly stressed.

However, the situation is far less clear when you have a large number of individuals who have rights. A lot will depend on the nature of the property right. For example, the theory regarding ITQs is that because each fisher has a right to take a specified quantity of a certain species, he or she is under no obligation to race for fish and can exercise that right at a time which maximises his return. This may be an excellent theory for a species which is uniformly distributed across a fishery in time and space, but may not be so when a fishery depends on periodic aggregations of the fish unless there is a method of establishing total allowable catches (TACs) that takes account of the aggregating behaviour.

A classic example of this is eastern gemfish (*Rexea solandri*). The fishery for this occurs in early to mid-August, during migration to a spawning site off central and northern New South Wales. Although ITQs were introduced in 1989, concerns about recruitment led to successive reductions in the TAC and culminated in a zero allowable catch in 1993 that remained in place until 1996. Of course, the decline in TACs and the poor performance of the fishery even under a rebuilding strategy cannot be assumed to be the result of only ITQs and the preferred method of targeting the fish. There may be environmental factors, such as poor weather conditions hampering fishing, and scientists suspect that during the period of the zero TAC, discarding may have been a serious problem (Tilzey and Chesson 1998). But even under an ITQ system the catch rates of a stressed species continued to decline, apparently as a result of poor recruitment. Even now, ten years after ITQs were introduced, the species is considered overfished. The property right did not automatically mean the fishery was sustainable!

ITQs are perhaps an unfair choice – they may have excellent application in fisheries where there are a limited number of species with similar life histories and even distribution in space and time, and for species not prone to large natural fluctuations. But one of the main rationales for the use of ITQs is that they are an economically efficient means of distributing the fishing right. The argument is that the more-efficient operators buy out less-efficient operators, at prices exceeding the earning capacity of the less-efficient operators (Commonwealth of Australia 1989). Result: same number of ITQ in the hands of the most economically efficient fishers, less waste of inputs such as fuels and gears, ideally only the least polluting vessels in use, maximum economic return to the community (each ITQ being used to its fullest extent), etc.

This may work for fishers who are driven by economic considerations. But it may not be a good theory if fishing is a lifestyle choice or for subsistence purposes and not necessarily to make money. Social and cultural forces operate in communities, and it cannot be assumed that everyone will behave in an economically rational manner. And it is uncertain that a few highly efficient operators only is necessarily the community's choice – the issue may be either: (a) many fishers eking a living; or (b) a few efficient operators making a good living and the remainder, bought-out, surviving on social security or similar support mechanisms (assuming the alternative employment market is tight) – with consequent social and political implications. I am unaware of that question ever being put to the community, although studies on the sociological impacts of ITQs are becoming more in vogue.

ITQs have resulted in a few unexpected behavioural problems such high-grading. Scientific advice provided to management is usually from fishery-dependent sources, such as logbooks. Discards are unlikely to be recorded in logbooks, which makes accurate measurement of the impact of fishing on target stocks difficult. Accurate assessment of fishing mortality is a serious issue in ensuring sustainability. None the less, ITQs, because they purport to guarantee the holder the right to take a specified quantity of fish, are probably one of the strongest forms of property-rights currently in the fisheries manager's toolbox today. They provide a great deal more certainty for the fisher than property-rights based on gear controls, which offer little indication of the amount that might be taken and hence the likely economic return.

Another form of property-right is the delineation of geographic areas with exclusive rights to harvest. This approach may be useful for sessile organisms such as oysters where clear boundaries can be identified and policed (Commonwealth of Australia 1989). The approach also is being pursued in Queensland in the management of the coral fishery, where licences to take coral are tied to specific, identified sections of the reef and management arrangements limit the proportion of the licensee's area which can be harvested.

Certainly some form of rights-based allocation can provide incentives for better conservation and management of the resource. Open-access merely encourages a fishing free-for-all, in which the aquatic environment, the fish, and the fishers are the ultimate victims. A more rights-based allocation approach that removes or reduces the competition among participants and provides for access privileges is needed to encourage more sustainable fishing.

3.4 At what level do fisheries property rights exist?

While much of the debate seems to be at the individual fishery or fisher level, it goes all the way from the individual to the national or even global level. Article 2 of the *United Nations Convention on the Law of the Sea* (UNCLOS) establishes that the coastal state has sovereignty over its territorial sea (including, by inference with the use of the term "sovereignty", the living resources contained within the territorial sea); Article 56 establishes that the coastal state has sovereign rights "*...for the purposes of exploring and exploiting, conserving and managing the natural resources...*" of the exclusive economic zone (EEZ); and Article 61 vests in the coastal state the right to determine the total allowable catch of living resources within the EEZ. Are these property rights, on a national scale? – I argue that they are. UNCLOS also recognises that there should be responsibilities attached to the rights: within the EEZ, under Article 62 the coastal state has an obligation to identify surplus stocks and make them available to other fishing nations 'if' it is unable to harvest the stock, while Articles 63 and 64 place obligations on the coastal states to cooperate in the conservation and optimum use of straddling stocks and highly migratory stocks respectively.

Even on the high-seas UNCLOS attributes rights and responsibilities on fishers; Article 87 provides for the freedom to fish the high-seas, and at the same time stipulates that this freedom should be "*exercised by all States with due regard for the interests of other States in their exercise of the freedom of the high-seas*". More rights and responsibilities are associated in the freedom of the high-seas through Article 94 (the duties of the flag State), Article 116 (the right for nationals to fish on the high-seas, subject to various obligations), and Article 119, which outlines the responsibilities on States regarding the conservation of living resources of the high-seas.

Van der Elst *et al.* (1997) identify five broad categories of property: common, private, communal, State or national, and global or international. They recognise that current fisheries management practice is usually a mixture of these regimes. Within their framework, I argue that UNCLOS provides for the latter two categories while the first three are forms of property internal to a State; but most fisheries management is a mixture of all categories.

This means that in terms of sustainability the responsibility attached to the right of access, the ultimate responsibility of ensuring resource use for now and the future, lies at all levels.

3.5 What is needed to ensure long-term sustainability?
3.5.1 Management, uncertainty and related issues

Fishing rights and management are inseparable, because unless a fishery is sustainable the full value of the fishing right will not be realised. To get the sustainable fishing right, you need a sustainable environment - and that requires management. There needs to be an explicit mechanism to ensure that any fishery subject to fish-rights is sustainable in ecological, social and economic terms, and most importantly, in ecological terms. This mechanism needs to incorporate equity among users, setting of TACs (or similar management measures) for target and byproduct species which are environmentally sustainable; establishing the total allowable impact on non-target species and the aquatic environment, management rules for exercising of rights, protection for sensitive areas, cooperative approaches to management and time.

One theme running through this list is *management*. Sustainable property rights require environmentally sustainable management. Environmentally sustainable management requires active consideration of all who affect the resource and the aquatic environment: the commercial fisher, the recreational fisher, the artisanal/subsistence fisher, the tourist, transport systems, non-aquatic sources of pollution; targeted species, bycatch species, discards, physical impacts on the aquatic environment, etc. To address these issue requires a commitment to transparent precautionary risk assessment and management and genuine commitment by the stakeholders to abide by the management regime.

Since fisheries management exists in uncertainty, and since it is difficult politically, economically and socially to recall or reduce property rights, management needs to be *adaptable*. It must adjust for changing perceptions of the carrying capacity of both the resource and the aquatic environment. It must be ready to deal with unexpected environmental perturbations, including those that may not initially appear to affect the target species. Of course, since fishery managers cannot control all such factors, it needs to include some accommodation to deal with these aspects.

Where does this lead? – to me it looks like adaptive management – about which much has been written. According to Dovers and Mobbs (1997), adaptive environmental management is a learning process, "*management-as-experiment*", that moves managers away from crisis management, which frequently results in only partial solutions that tend to lock-up systems. As Johnson (1999) notes, "*if we hope to improve management, we must learn as we go*". The approach is to view management actions as experiments, design them to produce critical information to reduce uncertainty and provide broader knowledge and experience.

Dovers and Mobbs suggest it requires seven elements:

i. actively sought information (research, monitoring and communication) which is more appropriate that much which currently is obtained, through adequate resourcing so that the collection can persist in time
ii. intellectual collaboration across disciplinary and professional boundaries, which may need new methodologies, forums and modes of communication and requires widening the management advisory group to include multiple scientific disciplines, policy-makers, managers, etc.
iii. creative legislative reform, which creates effective institutions – rather than command-and-control approaches
iv. incorporation of ecology into management, *i.e.* factors such as temporal scales, functions of different components, potential "spare" ecological capacity
v. greater participation by the public and stakeholders, including valuing of local and traditional knowledge, over long periods of time
vi. commitment from politicians, policy makers, communities, companies, non-government organisations and scientists, even when the task is long and difficult
vii. appropriate and effective institutional arrangements that, have the life-span to address adaptive issues, integrate research/information and policy that operate through participatory processes, accommodate action in uncertainty and recognise the experimental nature of decisions, are properly resourced to deal with ecosystem-scale issues and have an appropriate statutory base.

Much argument has occurred as to whether adaptive management is useful in developing fisheries, where data are poor and increases in catch need to be managed cautiously to prevent overfishing. But adaptive management also has a role in developed and over-fished fisheries where it may be useful to manage reductions in catch in accordance with ecosystem issues. This requires the support of all stakeholders and acceptance that time scales are likely to be long.

3.5.2 Equity among users

To be sustainable, fisheries property rights have to take account of the total use of the resource – commercial, recreational, artisanal/subsistence, indigenous, and aquacultural. If those rights are not distributed equitably (not necessarily evenly), are not recognised and supported by the community, do not include rights to recreational, artisanal or subsistence and indigenous operators, *etc.*, it is hard to see how the existence of those rights can assure sustainability. Taylor-Moore (1997) provides a useful schematic of the allocation mechanisms available to fisheries managers. He notes that each mechanism has positive and negative aspects and that the current approach to fisheries allocation touches upon only some of these aspects.

But in our community of 200, who are the stakeholders? and are they all users? The answer depends on what is "use" and whether it should be confined to a physical interaction, or whether existence-value is also a use. I argue that it is, but that does not mean that those who value the resource for its existence-value need access to a part of the TAC.

As Van der Elst *et al.* (1997) note, whatever system of granting rights is used it must be "*legitimate and enjoy the support of all participants*". Further, the existence of fishing rights as a management tool does not necessarily denote exclusive rights of access (National Research Council 1999) – Ballantine (1997) notes that '*the public no longer accepts as an axiom that existing user groups should control whole ecosystems*'. The right may relate to specific components of the aquatic resource, but does not relate to the marine environment *per se*.

3.5.3 Total allowable catches for target and byproduct species

One of the most fundamental issues in ensuring that property-rights are set at levels so that long-term sustainability is assured, is sufficient knowledge of the biological resource to allow a TAC to be set. It is much easier for management agencies to increase or hold steady quotas or TACs than to reduce them and thus affect the short-term economic wellbeing of the fishery.

Qualitative stock assessments can be made for valuable, well-established fisheries because usually there is enough knowledge about the target species and the fishery. The status of the resource is usually assessed against biological reference points; a trend has been to focus on limit reference points (LRPs) or undesirably low levels of the resource. Assessments dealing with LRPs tend to assess the probability that the stock is, or may fall, below the reference point at current fishing levels. Alden (1999) notes that reliance on stock assessment approach leads to two major problems: the focus on removals does not take into account other measures related to sustainability, such as age at capture and critical habitats and the lack of political credibility because of continuing uncertainty about the stock.

More recently, fishery assessments have evaluated harvest strategies and assessed them against a range of performance measures which include sustainability of the resource. Harvest strategies generally comprise (a) data collection and monitoring, (b) stock assessment, and (c) the decision rules used to set management measures (*e.g.* quotas and effort restrictions). The extent to which different combinations of data, assessment and decision rules achieve the prescribed objectives of management are assessed, and adaptive strategies identified that should achieve management goals over time. This approach can be undertaken for fisheries where normal stock assessment is limited by lack of information, *e.g.* in newly developing or low value fisheries (Smith 1994).

Important components in both approaches to stock assessment include assessing the status of the resource relative to explicit reference points. Where this cannot be done directly, or inferred from stock assessment methods, proxies or sustainability indicators may be used to measure changes in resource status over time. Reference levels or trigger points, for such indicators often form an important component of harvesting strategies for species so managed.

Underlying whatever mechanism is used to allocate fishing-rights is the need for understanding the biology of the fishery. Much has been done on this aspect, for example Fisheries Assessment Groups provide advice to the Commonwealth Management Advisory Committees or Consultative Committees on the status of target stocks and similar mechanisms exist in State and Territory fisheries. These mechanisms are working reasonably well although in some cases the assessments seem to have missed the mark, as a result of the uncertainty issue. And, in many cases we are far from having the confidence to state categorically that a target catch is sustainable in the environmental context, *i.e.* it can be maintained without detriment to the aquatic environment and despite natural fluctuations.

Targeting itself is a vexed term. Some gears are quite species and/or size specific, while others are less discriminatory, *e.g.* in trawling the fisher draws his net across the floor after fish, any fish, while some hook and line methods can be size-selective, although they may take a range of species in that size. Perhaps the only truly targeted fishing method is hand collection. But without knowing what is targeted, can one say what is being taken is ecologically sustainable?

3.5.4 Total allowable impact on non-target species and the aquatic environment

Fundamental to the search for asustainable fishing-right is information on the environmental impact of fishing. Sadly, little effort has been directed at non-target species, the impact of fishing on the aquatic environment and the aquatic environment itself. There are no "standard" procedures for dealing with the wider ecological impacts of fishing. While some approaches dealing with threatened and endangered species have started to appear, agreed methods for assessments are generally lacking. Apart from limited research on the effects of spatial fishing closures and bycatch reduction devices, there has been little progress in dealing with threats to habitats and food chains, let alone whole ecosystems. Management will have to deal with higher levels of uncertainty about basic ecological processes, even less-adequate data than for target species, and the difficulties of differentiating effects of fishing from other impacts on marine ecological systems. One approach may be to use indicators for various components of marine ecological systems. Reference levels for the indicators could be chosen with regard to the acceptable levels of change to aquatic ecosystems (as defined by the community).

Fisheries management needs to recognise that bycatch is a management issue and in Australia, measures are being introduced to deal with bycatch mitigation. The *National Policy on Fisheries Bycatch* (Ministerial Council on Forestry, Fisheries and Aquaculture 1999) provides a framework for coordinating efforts between government, fishers and the community to tackle this problem. The policy will support continuing research into methods minimising bycatch and will foster cooperation between stakeholders to ensure complementary management practices across all fisheries. The policy is intended to balance short-term needs with long-term goals, so management practices are starting to deal with this aspect of the environmental equation.

But, bycatch is arguably the easiest part of the environmental equation to tackle; there is a growing body of research into effects on non-target species and bycatch mitigation measures seem to be proliferating at near-exponential rates. The challenge is to deal with the less tractable problems where a technological fix is less likely.

Ideally fisheries assessments should consider total removals in determining what proportion of the estimated removable biomass should be made available to fishers. Total removals should mean, on the human side, commercial, recreational, subsistence/artisanal, indigenous; and on the ecosystem side, predators and the food web as a whole. This is a tall order; most management agencies can calculate commercial removals and may have some concept of recreational take, and perhaps artisanal subsistence and indigenous removals. At a minimum, they can ask fishers, and use the results in stock assessments. The reality is that estimates of the ecosystem components require a level of knowledge bordering on god-hood, and proxies (or sustainability indicators) may be the only option. The development and use of indicators and associated reference levels would provide a formal basis for evaluating possible impacts of fishing, and a starting point for the evaluation and development of strategies to deal with such impacts.

But, sustainability indicators have their own problems, *e.g.* what are the right indicators? What do they actually indicate? Are they being used properly? Are they flexible enough to accommodate constantly shifting ecological balances? The Food and Agriculture Organisation of the United Nations is in the process of developing a number of technical guidelines for responsible fisheries. Among these are guidelines on the development and use of indicators for sustainable development of marine capture fisheries, initially drafted at the Technical Consultation on Sustainability Indicators in Marine Capture Fisheries held in Sydney, Australia, 18-22 January 1999 (FAO 1999).

3.5.5 Management rules for exercise of the right

Though fisheries management occurs in a climate of uncertainty, arrangements are needed to ensure that the fishery remains sustainable. This requires that property rights are appropriate to the resource and the aquatic environment and that there is a clear and agreed mechanism for dealing with problems. For example, should the species prove to be over-exploited, management arrangements should facilitate stock recovery by restraining fishing effort to a previously agreed level at which it is believed the stock may recover (Commonwealth of Australia 1989).

Another issue for sustainable fishing rights is the need to ensure that fishing capacity does not exceed the carrying capacity of the aquatic resource. The OECD (1997, quoted in National Research Council 1999) noted that if capacity is held at levels at or below that necessary to obtain the MSY or MEY, then potential for overfishing is reduced. Thus management based on input controls needs to take account of technological creep around rules aimed at controlling fishing effort, or contain a mechanism, for example a limit on net lengths, which is deliberately set low to ensure that technological creep does not undermine the sustainability of the fishery. As the National Research Council (1999) noted: *"fishers are ingenious in circumventing regulations to limit entry and fishing power."* One way for management rules to deal with this is to address the fisher's incentives and formulate a system which reduces, or removes, the short-term rewards arising from overcapitalisation and increase likely future returns.

One important element in adaptive management is the use of appropriate performance criteria. Performance criteria measure the success of management strategies against specific management objectives. For sustainability purposes they frequently use biological reference points. Indicators also can be used in the development and evaluation of different management strategy options. Smith (1997) outlines approaches used with performance criteria in three Commonwealth managed fisheries to: (a) establish targets, such as maintaining spawning biomass above a pre-determined level, with a series of management responses dependent on the probability of achieving targets or limits within a specified time; (b) achieve stock rebuilding; and (c) calculate effort-reduction to achieve a pre-determined rebuilding target. To be effective, performance indicators should include commitment to staged management responses.

In Figure 1, the unbroken curved line represents fishing mortality, the shaded lines represent indicators (used here as triggers), the optimal yield (within the ecosystem context) and the "biological bottom line" below which only draconian management options are available such as total or partial fishery closures or massive effort reductions.

3.5.6 The precautionary approach

To ensure sustainability (in the environmental sense) management must apply the Precautionary Principle: *"where there are threats of serious or irreversible environmental damage, lack of full scientific certainty should not be used as a reason for postponing measures to prevent environmental degradation"* (Commonwealth of Australia 1992). As noted, fisheries management proceeds in a situation of scientific, managerial and operational uncertainty. The Precautionary Principle relates to situations where there are serious threats of environmental damage, and an argument can be mounted that they may not necessarily apply to overall fisheries management.

The difficulty with pursuing this point of view is that those *"threats of serious or irreversible environmental damage"* are not likely to advertise themselves as such, and are unlikely to be known until the damage becomes clear. It may then be too late for management agencies to manoeuvre and may require draconian measures. So an application of the Precautionary Principle is to assume it relates to all situations, not just those under threat of serious environmental damage. The Principle should only aim at reducing impacts to a level acceptable to the community, not at the elimination of all impacts.

Garcia (1994a) notes that *"a major property of the Principle is that it inverses the course of action, requiring that measures are taken first and, subsequently, relaxed if research demonstrated convincingly that they are not necessary"*. He suggests that the Precautionary Principle could appear to be an attempt to fill in the gaps created by the UNCLOS regime guiding utilisation of marine resources, but argues that in fisheries where scientific data, management and monitoring regimes are available, the UNCLOS approach should prevail and decisions should be taken on the basis of the best scientific evidence available. Both Garcia (1994b) and Taylor-Moore (1997) suggest that an over-precautionary application of the Precautionary Principle may have serious social and economic ramifications. Taylor-Moore suggests that a precautionary approach to resource allocation *"places production within an acceptable level of ecological risk and uncertainty, with resource failure as the bottom line"*.

The difficulty with this approach is that the mechanism for accurately identifying an *"acceptable level of ecological risk"* seems to be missing, or at best is in slow development. So for the moment application of the Precautionary Principle is another balancing act. In light of the way fisheries have developed in the past and the emphasis placed upon the economic and social side of the equation, the application of the Principle may need to be revised to make it more rigorously precautionary and less 'in principle', to achieve long-term sustainability. One approach may be the philosophy of the ICES Advisory Committee on Fisheries Management for *"stocks where, at present, it is not possible to carry out any analytical assessment with an acceptable reliability...precautionary TACs to reduce the danger of excessive efforts being exerted on these stocks"* (Serchuk and Grainger 1992, quoted in Garcia 1994a).

More, the National Strategy for ESD may fail. The Precautionary Principle is one of the guiding principles, but the National Strategy states that the core objectives and guiding principles should not be interpreted so that one has precedence over the others. However, Garcia

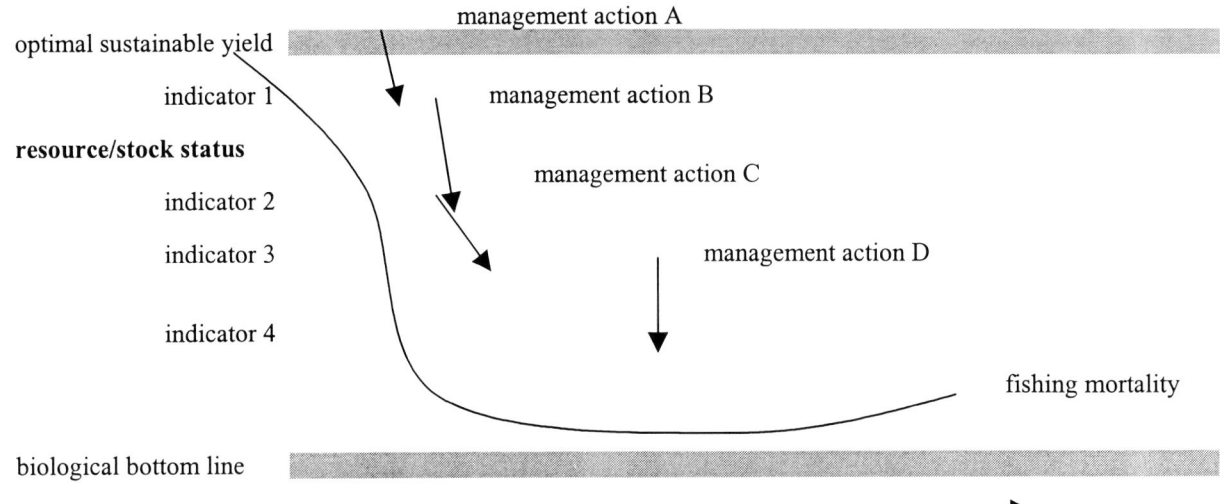

Figure 1
Diagrammatic on the use of triggers in management

(1994a) notes the following "preventative" management approaches (which I regard as precautionary) may be particularly relevant for developing fisheries:

i. staged development with impact monitoring
ii. early effort limitations
iii. institutional or financial controls to avoid explosive development
iv. precautionary quotas for species where proper assessments are not available
v. use of pessimistic models for stock where low resilience is suspected
vi. multispecies management
vii. experimental management to test systems responses
viii. development targets below MSY
ix. adoption of the concept of "safe biological limits"
x. modelling systems response across the range of uncertainties and
xi. cautious management thresholds (*e.g.* minium spawning biomass) and course of action to be taken before crises occur.

One avenue for bringing the concepts of sustainability and the Precautionary Principle into an operational framework, within finite financial and human resources, may be the development of sustainability indicators. Garcia (1994a) suggests that new indicators will be required to address species sustainability (*e.g.* minimum reproductive biomass, safe biological limits, maximum statistical probability of ecological or economic collapse) and ecosystem management (global stress indicators, resilience factors, habitat conditions). He suggests that criteria should be developed for assessing impacts of developments, taking into account reduction of target and associated species, levels of risk to species caused by combinations of fishing and environmental variability, and the degree of reversibility of observed or forecast impacts. The FAO work suggests that sustainability indicators enhance communication, transparency, effectiveness and accountability in natural resource management. In particular, indicators provide information about activity at a given scale and enable comparisons at other scales (*e.g.* local fishing community impact against overall pressure on the stock).

But, sustainability indicators are not an end in themselves. They are one tool that may be useful for developing fisheries where the costs of data collection and analysis may be prohibitive and where a carefully selected set of indicators may simplify evaluation and reporting. They would require intelligent use and the support of all stakeholders to achieve desired ends. They may also be a useful tool in monitoring the appropriateness of fishing rights.

3.5.7 Protection for sensitive areas

One mechanism for helping fisheries achieve environmental sustainability may be the use of protected areas. Fishery closures have existed for some time. They are used to protect spawning or nursery areas at specific times of year, to protecting vulnerable bycatch species throughout the year and to allow target species the opportunity to grow to an optimal marketing size.

Some Australian fisheries management agencies are declaring marine protected areas (MPAs) with a view to protect fish habitat in areas of special environmental importance. NSW Fisheries, for instance, has created eight MPAs; each is unique, and the type of protection varies – in some, only diving and observation are permitted while others may allow multiple use including fishing. Similarly, Queensland's Department of Primary Industries has declared Fish Habitat Areas, under the *Queensland Fisheries Act 1994*, which are established for the protection and management of fisheries resources and wetland habitats to ensure the continuation of productive recreational, commercial and traditional fisheries in the region.

MPAs declared by nature conservation agencies may be established for biodiversity conservation measures rather than fishery protection measures, and may prove useful for fisheries management as well. Ballantine (1997) suggests that 'no-take' marine reserves would be a practical, effective addition to the fisheries management toolbox. He suggests that as fishing may have serious and complex effects, it would be useful to minimise all of those effects in some places. In effect the protected area becomes an insurance policy. He postulates that a network of such reserves would be needed to provide sustainability. The National Research Council (1999) goes further in suggesting that a significant proportion of the total area could usefully be a protected area. They suggest 20% because the marine environment is more open than the terrestrial environment and has greater geographic exchange, thus a greater area is required for the same effect.

There is a growing trend towards engaging fishers and management agencies closely in the development of MPAs and seeking solutions which allow multiple use of those MPAs. The National Research Council (1999) suggest that, for sustainability, protected areas should be used as part of the management regime as they are less susceptible to error. Perhaps in future fishers and management agencies will look at "conservation" MPAs in view of their fishery benefits, rather than bodies of water excluding fishing operations. Certainly fishers have expressed fears that "conservation" MPAs may be declared over existing fishing grounds so that they will lose access to those grounds. But, it would be a strange "conservation" MPA that was declared over a stretch of water which has been subjected to heavy fishing operations – particularly trawling – for several decades.

MPAs may be a legitimate management tool in the search for environmental sustainability. However, as with any management tool, they have implications for the allocation of fishing rights and vice-versa. Most obvious is the situation where a protected area is declared in an area where there are pre-existing fishing rights. Even where the protected area is intended to be, by providing refuge, a fisheries management tool, it could be argued that the protected area inhibits the fisher in the exercise of his right.

3.5.8 Cooperative approach to management

Management of a common property resource, such as fisheries and the aquatic environment, requires cooperation between the commercial, recreational, artisanal/subsistence and indigenous fishers; the environmental movement; the fisheries management agencies; the scientific community; the environmental management agencies; and the community as a whole. Cooperation requires transparency in decision-making, so that all parties understand and support decisions which affect their community resource. To be effective, decision-making processes need to be transparent, efficient, robust and supported by all.

The trend in Australia is towards co-management arrangements, whereby commercial and recreational fishers and conservation groups are represented on management bodies such as Management Advisory Committees and Consultative Committees along side scientists and managers. Fishers, managers and fisheries scientists have a legitimate role in such arrangements, but they are not the only ones with such a role. Recreational fishers also may make substantial use of the resource, and conservation organisations are increasingly becoming involved in fisheries management either as part of management bodies or through the political arena. A relatively new trend on the part of the conservation movement is to seek the inclusion of ecologists, as well as fishery scientists, on management advisory bodies in order to provide a more ecological view of fisheries management.

These co-management arrangements work with varying degrees of success, hinging upon the confidence and support of each sector. One impediment to support for the outcomes may be a perception that the whole process is skewed toward the needs of commercial fishers, particularly where a significant proportion of management costs are recovered from them. While it is reasonable to assume that those making the greatest use (and profit) from the resource should pay for its management to ensure its sustainability, it is also reasonable to assume that management agencies, being dependent upon cost-recovery to operate, will find themselves between the rock of hard management decisions, and the hard place of disaffected stakeholders, the fishers.

Another issue in cooperative approaches is the peculiar role of scientists in fisheries management. Scientists form the backbone of fishery assessment work and their advice is critical to management decisions. And Australia's view is that management decisions should be based upon good science. Yet management decisions, while informed by scientific advice, are value judgements and can be taken for political or economic reasons. As a consequence they may not reflect the scientific advice – for example, the fishing industry often uses scientific uncertainty and/or natural variability to argue against a reduction in catch, even where a precautionary application of the Precautionary Principle requires reduction of fishing effort.

Property rights should not undermine these cooperative mechanisms; instead they need to acknowledge the legitimacy of issues raised by other sectors and accommodate management responses. The support of fishers is critical in ensuring that management arrangements work; the National Research Council (1999) noted that pressure on the part of fishers for liberal catch quotas is often strong and spills over into political arenas. Certainly adaptive management arrangements will have limited effect if they do not have the support of the rights-holders, *i.e.* fishers. To quote from FAO (1993), *"Implicit to the use of an LRP is that negotiation between the parties concerned in the fishery pre-establishes the response to be taken automatically once the fisheries assessments indicate that a "red zone" is being approached or has been entered, where pre-agreed long-term conservation objectives are endangered".* This need for pre-agreement is not confined to the operation of a limit reference point but holds for any management strategy and any stakeholder in that management strategy. Similarly, pressure on the part of conservation organisations in the past frequently has been expressed through the political arena, rather than directly at the level of management agencies, causing political "intervention". And, political decisions, which should be expressions of community desires, happen at the domestic levels and in international fisheries arenas alike.

Property rights also need to be established to accommodate situations where communities want certain species to be for recreational, artisanal/subsistence, or indigenous use. For example, Australia may need consider allocations in response to native title (though native title is not as clear-cut as in New Zealand, through the Treaty of Waitangi, because no such treaty was made during white settlement of Australia).

Legal recognition in Australia of customary (indigenous) fishing rights, or native title, to the seabed is unclear at the moment. However, in the opinion of some justices, a 'right-to-fish' based upon traditional laws and customs is a recognisable form of native title defended by the common law of Australia, but its evidentiary requirements are exacting. The Commonwealth *Native Title Act 1993* expressly permits native title holders to hunt, fish, gather, carry out a cultural or spiritual activity, or do any other kind of prescribed activity for their personal, domestic or non-commercial communal needs, and does not link these activities with native title to land. The Act also exempts native title holders from requiring a licence or permit to take game if it is taken for personal, community or other non-commercial and traditional purposes. Certainly indigenous peoples in Australia are becoming more active in pursuit of sea rights, as demonstrated by the recent convening in Hobart of the first National Indigenous Sea Rights Conference.

3.5.9 Time

Environmental sustainability is a long-term objective and our National Strategy for ESD clearly indicates that long-term considerations are valid. Property rights also should be long-term to allow the holder to develop a sense of ownership and stewardship for the resource (Van der Elst *et al.* 1997). Long-term access rights also encourage fishers to develop strategic approaches to the exercise of the right, whereas short-term property rights may encourage cut-throat fishing because the right to fish might be negated at the end of the year.

However, extending a long-term right to fishers places additional responsibilities upon management agencies and scientists to "get their sums right", because it may be difficult to reduce the access right if the initial allocation was too great. So, in granting property rights management agencies need to look at staged progressions, starting off short-term but on the understanding that longer-term access is the ultimate goal. When that longer-term access becomes available may be contingent upon increasingly effective adaptive management arrangements, so that the danger of over-estimating the carrying capacity of the aquatic environment is reduced incrementally.

4. RECENT COMMONWEALTH GOVERNMENT INITIATIVES, SUSTAINABILITY AND THE USE OF PROPERTY RIGHTS

The issue of sustainability in policy is not new – one of the first Commonwealth government policies to address the issue of sustainability was the 1989 fisheries policy *New Directions for Commonwealth Fisheries Management in the 1990s*, and a National Strategy for ESD had been published in 1992. However, there has been a resurgence of policy activity directed towards the sustainability question. The most substantial is *Australia's Oceans Policy*, released in December 1998. The policy is directed towards an integrated planning and management regime that will allow government and community to ensure the conservation of Australia's marine biodiversity while providing security for marine-based industries and other ocean users. At the core of this policy is a commitment to ecosystem-based management, to be implemented through a new Regional Marine Planning process. Based on large marine ecosystems, it will integrate sectoral commercial interests and conservation requirements.

Oceans Policy contains a number of commitments relevant to environmental sustainability in fisheries. Specifically, the policy seeks ecologically sustainable fishing practices through a variety of mechanisms including by-catch policies; a network of fisheries extension officers to promote environmentally sound fishing practices; plans to reduce the threat of fishing on protected marine life and enhance its recovery; and development of performance and sustainability indicators for fisheries. Other fisheries-specific activities include structural adjustment to remove excess capacity in over-capitalised fisheries, and measures to improve the management of recreational and charter fishing including the development of ESD mechanisms for recreational fishers.

Among other actions under the *Oceans Policy* is the creation of new marine reserves in Commonwealth waters. Since the release of *Oceans Policy*, MPAs have been declared for the Tasmanian seamounts and Macquarie Island, and are in development for a number of other sites.

Oceans Policy also commits the government to accelerated development of a National Representative System of Marine Protected Areas. This involves

Commonwealth, State and Territory governments working to expand existing marine parks and reserves. The intention is to set up a national system of protected areas to protect areas which represent all major ecological regions plus the communities of plants and animals they contain. In establishing the National Representative System of Marine Protected Areas, Australia is contributing to the development of the Global Representative System of Marine Protected Areas.

Another aspect of *Oceans Policy* is a revision of Schedule 4 of the Commonwealth *Wildlife Protection (Regulation of Exports and Imports) Act 1982*. Specifically, the Minister for Environment and Heritage is proposing to remove an existing exemption under this Act that allows native marine fish species to be exported from Australia without export controls. The proposal is that all fish species intended for export should be assessed to determine if they are managed in an ecologically sustainable manner before the exemption is applied. All management regimes in which the species is taken must be assessed, and exemption will only be granted if all regimes are assessed as being demonstrably sustainable. However, exports may continue under an export-authorisation regime. Work is underway on developing appropriate guidelines for undertaking those assessments, based upon those developed by the Marine Stewardship Council.

A second environmental action has been the promulgation of the *Environment Protection and Biodiversity Conservation Act 1999*, replacing the *Commonwealth Environment Protection (Impact of Conservation) Act 1974*; *National Parks and Wildlife Conservation Act 1975*; *Whale Protection Act 1980*; *World Heritage Properties Conservation Act 1983*; and *Endangered Species Protection Act 1992*. The new Act will come into effect in July 2000 and requires that management arrangements for all Commonwealth managed fisheries not previously assessed under the *Environmental Protection (Impact of Proposals) Act 1974* be strategically assessed by the Minister for Environment and Heritage. Where fisheries are not subject of formal management plans (as defined in the *Fisheries Management Act 1991*), assessment must start within five years. Work has commenced on implementation of the strategic assessment side of the Act. These last two initiatives, while arousing concern in fisheries circles, are the government's response to growing concern at the sustainability of Australia's fisheries and management arrangements. The assessments to be undertaken will provide an external review of fisheries sustainability.

At the same time external reviews may increase transparency in fisheries decision-making. A complaint, often made, is that fisheries management is a closed loop between fishers, fisheries management agencies and fisheries scientists, to which the community has limited access. While the trend towards co-management models with wider representation on management advisory bodies is a start, it does not appear to meet all community expectations. For example, some in the community feel that co-management bodies should more accurately reflect the numerical proportion of the community with an interest in the fishery.

The existence of an external body explicitly recognising a fishery as being demonstrably managed in a sustainable manner has potential benefits for fisher and fisheries management through recognition that the fishery is well-managed, as fishers may be able to extract market premiums while reducing conflict with community groups. A more explicit recognition would be through an arrangement such as the certification process being developed by the Marine Stewardship Council,

Finally, although it is not a Commonwealth initiative, during 1999 the Standing Committee on Fisheries and Aquaculture, comprising the heads of fisheries agencies throughout Australia, commenced a process to agree a set of national ESD criteria and indicators for use in Australian fisheries. In this will be the development of cost-effective and practical indicators for fisheries, including research programmes for those fisheries where data are not readily available. None of these initiatives are directed specifically towards fish rights, but they touch upon the issue of sustainability in fisheries, and so influence those rights if they are to be sustainable in the long-term.

5. CONCLUSION

To be worthwhile, a fisheries property right needs to be sustainable. There would be very little point in fishers paying for the right of access if the resource cannot support the level of exploitation. Short-term sustainability may suit operators who intend to move out of the fishery fairly swiftly, but may not suit the majority of fishers who are in the fishery for the long haul. Accordingly, fisheries property rights need to be directed towards the long-term sustainable user.

There are no neat fixes to achieving that sustainability, just hard work in a multiplicity of fields: management, science, technology, operations, public relations. While all are important, underpinning them is a need for effective and participatory management regimes that incorporate distinct management-decision rules and contingency plans, established through transparent and agreed processes. These rules need to be adaptive and accommodate natural and anthropogenic fluctuations, so that the resource and the aquatic environment do not become destabilised by the exercise of the right. Distinct performance and sustainability indicators need to be an integral part of the management rules, and triggers need to be used more widely to improve management response time.

Management arrangements need to be risk-averse and contain agreement on acceptable levels of risk. Ideally, management rules should be established before rights are issued, so that rights holders avoid unpleasant surprises. In circumstances where rights already exist, innovative ways are needed to bring about consensus without excessive compromise to the aquatic environment and minimum economic and social disruption.

Management rules should be based on an appropriate mix of good science and an effective monitoring regime, using fishery-independent and fishery-generated sources of information. In data-poor fisheries, management rules have to be precautionary, giving time for the science to develop to a robust level. Monitoring needs to address all sources of fishing-related mortality, whether it be target, bycatch, discarded or fish never hauled on board but killed none the less.

Wherever possible, management should avoid reliance on any one management arrangement and make full use of all management options. Most important, management rules need to be precautionary in approach and use existing "insurance policy" type tools, such as protected areas and closed seasons, to reduce fishing impacts. To achieve this fishers, management agencies and the community must have a common view on, and support for, property rights. In particular, developing fisheries may provide an ideal opportunity for the community to address the question of what socio-political and environmental, as well as economic, terms it wants of the fishery and the industry. Underlying this is a need for active information dissemination – Australia is far from a common view on the state of its fisheries, yet until there is such a view it is hard to see how the different sectors affected by fishing in general and property rights in particular can support those rights. If the rights are not supported, they will become a battlefield.

Adaptive management is a start, but it is not the whole answer. Johnson (1999) identifies a number of problems in the implementation of adaptive management which include: difficulties in developing acceptable predictive models; conflicts regarding management goals and ecological values; inadequate attention to non-scientific information; and unwillingness to implement long-term policies. One critical factor to be taken into account in the use of adaptive management approaches is the resilience of the natural system – it would be highly undesirable to start experiments on a system so stressed, either naturally or through anthropogenic forces, that it is already unstable. As Gunderson (1999) notes *"if there is no resilience in the ecological system, or flexibility among stakeholders in the coupled social system, one simply cannot manage adaptively"*.

The jury is still out on whether it is better for a fishery to fish selectively or to take a broad swathe through the ecological continuum. Ecologically, the latter may be more sustainable as it may not throw the bulk of the impact on a few of the organisms fulfilling ecological niches. Garrod (1973, discussed in Garcia 1994a) suggests that a reasonable approach in managing ecological impacts of a multi-species fishery would be to exploit all species proportionally to their abundance. Operationally that may be a challenge, particularly where multiple species are taken in each fishing operation and the mix of species varies in time and place. Economically it may be quite a different story, unless markets will accept a greater mix of species with lower tonnages of some species.

6. LITERATURE CITED

Alden, R. 1999. A troubled transition for fisheries. 25-28. *Environment* 41(4).

Ballantine, W.J. 1997. 'No-take' marine reserve networks support fisheries. In *Developing and sustaining world fisheries resources: the state of science and management; 2nd World Fisheries Congress proceedings*. Hancock, D.A., D.C. Smith, A. Grant, and J.P. Beumer (eds). CSIRO, Collingwood, Victoria. 702-706.

Commonwealth of Australia 1989. *New Directions for Commonwealth Fisheries Management in the 1990s*. Canberra.

Commonwealth of Australia 1992. *National Strategy for Ecologically Sustainable Development*. Canberra.

Commonwealth of Australia 1996. *Report on the Implementation of the National Strategy for Ecologically Sustainable Development (1993-1995)*. Canberra. 206 pp.

Commonwealth of Australia 1998. *Australia's Oceans Policy*. Canberra. 96 pp (in two parts).

Cooke, J.G. 1984. Glossary of technical terms. In *Exploitation of Marine Communities* edited by R.M. May. Rep. Dahlem Workshop. Springer-Verlag, Berlin. 241-48. 1984 (Quoted in Harden Jones, F.R. 1994).

Dovers, S.R. and C.D. Mobbs 1997. An alluring prospect? Ecology, and the requirements of adaptive management. In *Linking science, policy and people*. Elsevier Science Ltd. 39-57.

FAO 1993. Reference points for fishery management: their potential application to straddling and highly migratory resources. *FAO Fisheries Circular No. 864*. Rome. 52 pp.

FAO 1999. The development and use of indicators for sustainable development of marine capture fisheries. *FAO Technical Guidelines for Responsible Fisheries: No 8*. FAO, Rome.

Garcia, S.M. 1994a. The Precautionary Principle: its implications for capture fisheries management. *Ocean & Coastal Management* 22, Elsevier Science Ltd. 99-125.

Garcia, S.M. 1994b. The precautionary approach to fisheries with reference to straddling and highly migratory fish stocks. *FAO Fisheries Circular No. 871*. FAO, Rome. 76 pp.

Garrod, D.J. 1973. Management of multiple resources. *J. Fish. Res. Bd. Can.* 30. 1977. 85pp.

Gunderson, L. 1999. Resilience, flexibility and adaptive management – antidotes to spurious certitude? *Conservation Ecology* 3(1): 7.

Harden Jones, F.R. 1994. *Fisheries ecologically sustainable development: terms and concepts.* IASOS, University of Tasmania, Hobart, Tasmania. 205 pp.

Johnson, B.L. 1999. Introduction to the special feature: adaptive management – scientifically sound, socially challenged? *Conservation Ecology* 3(1): 10.

Ministerial Council on Forestry, Fisheries and Aquaculture 1999. *National Policy on Fisheries Bycatch.* Department of Agriculture, Fisheries and Forestry – Australia. 14 pp.

National Research Council 1999. *Sustaining marine fisheries.* National Academy Press, Washington DC. 164pp.

OECD 1997. *Towards sustainable fisheries: economic aspects of the management of living marine resources.* OECD Publications, Paris, France (Quoted in National Research Council 1999).

Productivity Commission 1999. *Implementation of Ecologically Sustainable Development by Commonwealth Departments and Agencies.* Draft Report. Canberra. 232 pp.

Serchuk, F. and R. Grainger 1992. Revised procedures for providing fishery management advice by the International Council for the Exploration o the Sea. The new form of ACFM advice. In *NMFS Second Annual National Stock Assessment Workshop 31/3-2/4/1992.* Southwest Fisheries Science Center, La Jolla, CA. 14pp.

Smith, A.D.M. 1994. Management strategy evaluation - the light on the hill. In *Population Dynamics for Fisheries Management. Australian Society for Fish Biology Workshop Proceedings, Perth, 24-25 August 1993.* Hancock, DA (ed.), Australian Society for Fish Biology, Perth. 249-253.

Smith, A.D.M. 1997. Quantification of objectives, strategies and performance criteria for fishery management plans – an Australian perspective. In *Developing and sustaining world fisheries resources: the state of science and management; 2nd World Fisheries Congress proceedings.* Hancock, DA., Smith, DC., Grant, A., and Beumer, JP. (eds). CSIRO, Collingwood, Victoria.

Taylor-Moore, N. 1997. Allocation of inshore marine fisheries resources. In *Developing and sustaining world fisheries resources: the state of science and management; 2nd World Fisheries Congress proceedings.* Hancock, D.A., D.C. Smith, A. Grant and J.P. Beumer (eds). CSIRO, Collingwood, Victoria. 352-357.

Tilzey, R. and J. Chesson 1998. South East Fishery – quota species. In *Fishery status reports 1998: Resource assessment of Australian Commonwealth fisheries.* Caton, A., McLoughlin, K. and Staples, D. (eds). Bureau of Rural Sciences, Canberra. 65-84.

Van der Elst, R., G. Branch, D. Butterworth, P. Wickens and K. Cochrane 1997. How can fisheries resources be allocated ... who owns the fish?. In *Developing and sustaining world fisheries resources: the state of science and management; 2nd World Fisheries Congress proceedings.* D.A.Hancock, D.C. Smith, A. Grant and J.P. Beumer. (eds). CSIRO, Collingwood, Victoria. 307-314.

EVOLUTION OF SELF-GOVERNANCE WITHIN A HARVESTING SYSTEM GOVERNED BY INDIVIDUAL TRANSFERABLE QUOTA

M. Arbuckle
Challenger Scallop Enhancement Company Limited
Private Bage 24901 - Wellington, New Zealand
<arbuckle@scallop.co.nz>
and
K. Drummond
Ministry of Fisheries
Private Bag 14 - Nelson, New Zealand
<drummonk@fish.govt.nz>

1. INTRODUCTION[1]

There is considerable discussion within the economics literature about the possibility of fishermen governing themselves under private fishery right management regimes (*e.g.* Scott 1998, Scott 1993 and Johnston 1995). Allocation of Individual Transferable Quota (ITQ) in a fishery can provide a framework for such activity to occur. Scott (1993), for example, argues that the allocation of ITQs in a fishery overcomes many of the obstacles to fishery self-governance:

"In many fisheries the ITQ will be less a new instrument of regulation, less a kind of individual property right, than a membership card in a self-governing fishery group. Compared to the old scattered voluntary inshore groups, this new type will have access to information, will indeed produce it itself. It need not be homogenous, for its distributional problems will be largely resolved by the prior distribution of ITQs."

Developing self-governance within particular fisheries only resolves some of the resource management problems faced in the oceans. Keen (1988) presents a convincing case for a completely integrated approach to the management of marine resources. He argues that under a common property regime a tragedy point is reached at the Maximum Sustainable Yield (MSY) of the fishery and that *full ownership* of fisheries resources, including the productive capacity of the "ocean pastures", is needed to redress the problems that are created. He notes:

"...three reasons stand out for change from a commons to a full owner framework once the tragedy point is reached.

They are:

i. The imperative to exploit the resource before someone else does.

ii. The imperative to take the most valuable species first.

iii. The imperative to forgo investment that would improve productivity of the resource."

The question often debated is to what extent ITQ's, or rights in individual fish stocks, can address these issues. Keen (1988) describes the underlying problem as follows:

"... marine fishery resources are almost all wild animal species, many of which range over wide areas. The habitat of any one resource species overlaps in part or in full the habitats of several others. Rights to the habitats of one wild stock or species cannot be assigned without creating a potential for conflict with owners of the habitats of other stocks or species."

The Challenger Scallop Enhancement Company (Challenger), is the most advanced self-governing organisation for fisheries management in New Zealand. Challenger operates as a cost centre for a range of fisheries managed under the Quota Management System (QMS) within defined geographical localities. These fisheries are centred at the top of the South Island of New Zealand and extend down a large extent of the west coast of the North Island and South Island.

Challenger undertakes a wide range of commercial and fisheries responsibilities for the Challenger group of companies (incorporating scallops (*Pecten novaezelandiae*) dredge oysters (*Tiostrea chilensis*) and 20 finfish stocks). Those responsibilities include the development of policy and management plans for fisheries, and the implementation of those plans with statutory force and effect. The plans are developed to integrate commercial rights between fisheries and with the interests of recreational and customary fishers. To facilitate that integration Challenger, along with Government, support and service a number of advisory groups.

This paper provides an analysis of the highly developed fishery management regime in the inshore fisheries in the Challenger area, that culminated in the establishment of the Challenger Group. Our analysis is not exhaustive and we acknowledge that there are a large number of individuals and organisations within government and industry that have influenced and, at times, driven the developmental steps we describe. Rather, we

---

[1] The views expressed in this paper are those of the authors and are not necessarily those of Challenger, or the Ministry of Fisheries, New Zealand.

have focussed on the main imperatives driving the development of fisheries management and self-governance in the Challenger fisheries. We have concentrated on the evolution of self-governance in the Challenger scallop fishery, which provides the foundation for the establishment of the wider Challenger group. We conclude by summarising the imperatives motivating the development of a self-governance capacity and we reflect on the problems and the opportunities available to further progress in self-governance initiatives in the Challenger area.

2. EVOLUTION OF SELF-GOVERNANCE IN CHALLENGER FISHERIES

2.1 Structure

Clarke and Clough (1998) have categorised the evolution of the fisheries management framework in New Zealand into three distinct phases; limited entry (1908-1963), regulated open entry (1963-1983), and the evolution of a property rights-based system (1983 onwards). For the purposes of describing initiatives within the Challenger Group of fisheries in this paper, the first two phases of Clark and Clough are combined and the final phase is classified according to four key themes: (a) effort restriction (1983-1986), (b) development of the QMS (1986-1990), (c) institutional respecification (1990-1996), and (d) self-governance under environmental standards (1996 onwards).

2.2 Limited and regulated open entry (up to 1983)

2.2.1 General environment

At the turn of the century the first substantive legislation governing the harvest of wild fisheries was passed. The 1908 legislation favoured regulatory control of fisheries (based on biological considerations) rather than rights-based management (Ackroyd et al. 1990). In the 1960s the Government concluded that capture fisheries had considerable development potential that was not being realised. As a consequence commercial coastal fisheries were de-licensed in 1964. The fishing industry, along with other primary industries in New Zealand, developed in the 1960s and 1970s in a climate of central government support, subsidy and investment. The statutory New Zealand Fishing Industry Board (NZFIB) was established to promote the development of the fishing industry in 1963. Fisheries licensing systems were deregulated and access protected by the establishment of an exclusive economic zone. Direct subsidies to the industry focused on building catching and processing capacity. Subsidy support ranged from government guaranteed mortgages for the purchase of new fishing vessels to subsidised vessel ownership savings schemes and the direct allocation of funds for development.

Just as subsidies for pastoral farming were leading to over-capitalisation on farms and development of sub-marginal land, so too they stimulated over-investment in fishing in New Zealand's inshore stocks and the status of a range of these stocks reached crisis level in the early 1980s. Government moved to re-regulate fisheries in 1977 by introducing fisheries licensing – initially into some shell-fisheries. At the same time development of deepwater fishery resources was becoming the focus of a number of New Zealand fishing companies. This interest was stimulated by the declaration of the 200 nm Exclusive Economic Zone (EEZ) in 1977 together with the prevailing national protection and support for domestic investment in fishing capacity. The Government responded to the deepwater development opportunities by designing and implementing a prototype for the QMS in the form of Deepwater Enterprise Allocations.

2.2.2 Challenger fisheries management

Under the open access regime, vessel numbers in the Challenger scallop fishery increased to about 200 by 1975. Catches then began to decline from their peak of 1246t (meatweight) reached in the same year. Successions of controls designed to limit fishing were introduced (often at the request of the commercial fishery) in an attempt to manage expanding effort. The Challenger scallop fishery was the first fishery to be licensed in New Zealand and the licensing system introduced in 1977 reduced the number of vessels to 136. In spite of these steps thescallop catches continued to decline. By 1980, the size limit had been removed and landings fell to 40t. In addition, the commercial fleet had moved into inshore areas not previously fished and this created tensions with the recreational sector. The scallop fishery was subsequently closed to commercial fishing for two years (1981 and 1982). When the fishery reopened for the 1983 season only 48 boats were issued with licences to harvest scallops.

2.3 Effort restriction (1983 to 1986)

2.3.1 General Environment

The QMS prototype was not necessarily seen to be the answer for the full range of problems in the inshore fishery. Instead new legislation was enacted in 1983 that provided for the establishment of government administered fishery management plans. The Act also provided for the administrative removal of a large number of commercial fishers who were considered not sufficiently reliant on fishing for their income. But, the difficulties in addressing the issue of over-capitalisation in the inshore fisheries within the context of a regulatory environment soon became apparent. In 1984 the Government agreed to fund a scheme to reduce capital and effort in the industry, along with the introduction of ITQ, into the inshore fisheries. The opportunities for Government subsidies for fisheries development subsequently ceased.

2.3.2 Challenger fisheries management

Management of the scallop fishery under the boat-licensing regime continued. Under this regime, the government allocated a defined, and equal, catch limit for each vessel on an annual basis. Recommendations for management of the fishery, prepared by the Ministry of Agriculture and Fisheries (MAF), were implemented by a Controlled Fisheries Authority established as a statutory body. Also in 1983, MAF, with the support of industry representatives, were successful in obtaining funding and assistance from the Overseas Fishery Co-operation Foundation of Japan to establish a scallop enhancement programme, using Japanese techniques.

The 48 boat licence holders assisted in providing advice on enhancement trials and management of the scallop fishery through an informal committee set up by the NZFIB. This form of co-operative approach was atypical of inshore fisheries management at that time. Some management rules were also adopted by this group, which included the maintenance of daily catch limits for each vessel. Before, such measures had only been implemented under regulation.

2.4 Development of the QMS (1986 to 1990)
2.4.1 General environment

On 1 October 1986 the Government introduced the QMS into both the deepwater and the inshore finfish fisheries, the former prior to substantive development, and the latter after many stocks had been over-fished. Crothers (1988) stated that the QMS was introduced to achieve two broad goals: (a) conservation – to limit catches to levels that will result in the maximum production of the stock; and (b) allocation – to maximise the net economic return of the nation. Luxton (1997) documents the following founding aims of the QMS for inshore fisheries;

i. Rebuild inshore fisheries where required and ensure that catches were limited to levels that could be sustained over the long term
ii. Ensure that catches are harvested efficiently with maximum benefit to the industry and to New Zealand
iii. Allocate catch entitlements equitably, based on individual permit holder's commitment to the fishery
iv. Integrate management of inshore and offshore fisheries
v. Develop a management system that can be applied both nationally and regionally and
vi. Enhance the recreational fishery.

Initially, ITQ's allocated under the QMS represented rights to harvest a fixed tonnage of the Total Allowable Commercial Catch (TACC) from a particular fishstock, within a Quota Management Area (QMA). If, due to concerns for the biological sustainability of the resource, there was a need to reduce the TACC, then the Government was required to purchase the amount of excess quota.

In 1987 Māori successfully sought an injunction against the Crown that prevented additional fisheries being introduced into the QMS. The basis of the injunction was that under the terms of New Zealand's founding document, the Treaty of Waitangi, Māori had not authorised the Crown to allocate individual rights to fishery resources. This led to uncertainty within the industry with regard to their existing property rights and their future under the QMS (Annala 1996). An interim solution to Māori commercial fishing claims was negotiated between the Crown and Māori in 1989. This provided a cash sum and guaranteed Māori 10% of any ITQ. Access to fisheries for customary purposes remained a right over and above ITQ allocation. The 1989 interim solution enabled rock lobsters to be added to the list of stocks managed under the QMS, but initially ITQ was only issued for a fixed term.

2.4.2 Challenger fisheries management

Twenty inshore finfish stocks were introduced into the QMS in 1996. A number of industry participants in the inshore finfish fishery chose to sell their catch-history entitlements to the Government and left the fishery. Management continued to be regulatory-based and subject to TACC constraints for those fishers who chose to stay.

The Challenger scallop fishery was retained under the separate restricted licensing structure. Initial seeding trials proved successful and the first seeded scallops were harvested in 1986 (Bull 1994). Over the period 1986 to 1989 the scale of scallop-seeding steadily increased. Scallops seeded onto the seafloor were harvested to fund the further development of the enhancement programme. Funds were voluntarily contributed by licence holders and were paid into a Trust Account, which was administered by MAF.

With the growing success of the scallop enhancement programme a rotational fishing regime was introduced into the fishery, under which the fishing grounds in Tasman and Golden Bay were divided into 9 sectors. Each year an agreed number of sectors were opened to commercial fishing and after being fished each sector was then seeded with scallop spat caught on longlines at spat-catching sites located within the fishery. The commercial size limit was reduced from 100 to 90mm to allow stocks in rotational open areas to be fished at economically optimal levels. At the same time the recreational daily limit was increased from 20 to 50 scallops per person per day to provide better access to recreational fishers. Industry participation in the management of the fishery remained in the form of an informal advisory group. Recreational fishing interests remained independent of this group.

Industry representatives reached agreement with MAF to continue to provide funds (by way of a voluntary levy on all scallops landed) for managing and implementing the enhancement programme in 1989. To support this arrangement, MAF agreed to recommend that the Government place the scallop fishery under quota management. The reform 'agreement' incorporated a proposal whereby any annual catch allocation above 576t (which represented a 12t allocation for each of the 48 boat-licences) would belong to the Government to be sold by public tender and/or be used to settle Māori quota claims. The 'agreement' was entered into at a time when Māori commercial fisheries claims under the Treaty of Waitangi were still to be finally settled and this uncertainty motivated Government to retain an interest in the yield of the fishery.

The 'agreement' also recorded the industry's desire to secure the right to manage the enhancement scheme if MAF was required to retire from direct involvement through a change in policy or government direction. Cost-recovery arrangements specified in the 'agreement' were implemented voluntarily before the fishery was introduced into the QMS and in effect the scallop fishery became the first fishery in New Zealand to operate under a form of cost-recovery although voluntary. This has continued. Early industry co-operation and involvement in the management of the scallop fishery was driven by the growing success of enhancement and the opportunities that quota rights provided to an industry stagnating under a fixed vessel-licensing regime.

2.5 Institutional re-specification (1990 to 1996)
2.5.1 General environment

In 1990 all quota rights were made proportional, with the risks and costs of quota reductions being borne by the industry. The revised system, prompted by the catch reductions required for deepwater stocks, had the effect of reducing the financial liability on Government. It also stimulated calls for greater involvement by the industry in the management of the harvesting rights. Subsequent 'fine-tuning' of the QMS regime and associated regulatory framework has had a similar effect and has lead to further specification of the ITQ right (Annala 1996, Batstene and Sharp 1999).

A more lasting settlement of Māori commercial claims was required, but this did not occur until 1992. At that time a full and final discharge of the Crown's commercial fisheries obligations under the Treaty of Waitangi was achieved. The settlement provided the following benefits:

i. NZ$150 million for the purchase of 50% of New Zealand's largest fishing company, which had 25% of total allocated fish quota and achieved NZ $247 million in sales in the year until March 1992
ii. The transfer of 20% of the quota for all new species entering the QMS to Māori and
iii. Regulations to recognise and provide for the customary food-gathering and the special relationship between Māori and those places which are of customary food-gathering importance to the extent that such food-gathering is not conducted in a commercial manner.

While the settlement of Māori claims to fisheries was being negotiated the Government initiated a review of fisheries legislation in 1991, with the view of simplifying the Government's approach to fisheries management. The conclusions reached at that time are set out in Pearce (1991). A new approach to fisheries management based on property rights and economic incentives was proposed. This system required an extension of the QMS, improvement of the terms and conditions of quota rights and harmonisation of other parts of the policy framework. Pearce proposed a greater role for resource users in the management of the marine resource:

"Within the limits of official conservation prescriptions, those who hold the rights to fish should be encouraged to manage resources and their fishing operations, taking account of all the costs and benefits of their actions. This will involve making collective decisions about fishing patterns and fishing rules, projects of enhancement, exploratory fishing and research, financing these activities and administering their arrangements with the Government, among other things.
To enable quota-holders to engage in this kind of collective action in an orderly fashion, they should have legal authority to organise themselves. ..."

The Southern Scallop Fishery was identified by Pearce as a prime candidate for this type of improvement and he noted the progress already made in enhancing the fishery.

The Government's response to Pearce's report was to establish a Ministerial Task Force to initiate the simplification of New Zealand's fisheries management framework. The improvements described by Pearce were ultimately carried through into the report submitted by the Fisheries Task Force (Wheeler et al. 1992). This report provided the foundations for the legislative framework to come into effect in 1996. In the interim, a number of amendments to existing legislation were required to implement the more urgent changes.

A further important feature of the period 1990 to 1996 was the establishment and implementation of a fisheries management cost-recovery regime. When ITQs were initially introduced, the Government established a system of resource rentals which were set for each quota stock on the basis of the quota held. Resource rentals were established, in principle, to provide a return to government of some of the economic surplus derived from ITQs. In practice the resource rentals collected were used to offset the costs of quota reductions implemented in a number of fisheries.

Resource rentals were one of the most contentious elements of the fisheries legislation. Batstene and Sharp (1999) recorded that industry vigorously opposed them, and noted that even if ITQs created an economic surplus in the fishery the problem of determining the Government's share of the surplus was not straight forward. They concluded that resource rentals contributed to commercial uncertainty in the fishery. The resource rental debate was ultimately resolved by a decision to replace that system with a regime that recovered the Government's cost of fishery management through a levy.

In addition to the overwhelming industry opposition to resource rentals, the decision to move to cost-recovery can be linked to a number of other factors. First, the settlement of Māori claims raised the question as to whether Māori should be expected to pay resource rental on ITQs; if this was to be avoided it would lead to difficulties in differentiating between Māori and non-Māori ITQs in a fully tradeable environment.

Legislation to provide for greater government efficiency and accountability in the form of *The State Sector Act 1988*, *The Public Finance Act 1989* and *The Fiscal Responsibility Act 1994* also provided strong imperatives on Government to recover costs of fisheries management from industry. The cost-recovery regime introduced in 1994 was applicable only to commercial fisheries, and the Government continued to meet the costs of managing non-commercial fisheries. The original cost-recovery regime was based on the principle of avoidable cost[2].

---

[2] The principle of *avoidable cost* implies that the costs of services required by Government should be recovered from those who necessitate the provision of those services. Since few fisheries services would be required if there was no fishing industry,

Cost-recovery also led to an increased expectation from the industry to have a greater role in the management of fisheries managed under the QMS. The tensions created progressively increased by the move towards the delivery of traditional government services by the private sector. To facilitate this process the Government initiated a reform of its fishery management agency, ultimately effected on 1 July 1995 the Ministry of Fisheries (MFish) was established under the provisions of new legislation.

The Fisheries Policy and MAF Fisheries (operational) arms of MAF, including fisheries enforcement, were split off and formed into Mfish which became responsible for purchasing research, using funds largely recovered from industry. The former research functions of MAF Fisheries were transferred into the National Institute of Water and Atmosphere (NIWA), a Crown-owned agency, in an effort to ensure better transparency and accountability in the new cost-recovery environment. MFish became a purchaser of research services rather than a provider.

2.5.2 Challenger fisheries management

Between 1989 and 1991 scallop seeded into rotationally harvested areas in the Challenger scallop fishery increased from 81 million to 630 million spat. Scallop catches increased from 240 to 672t (meat-weight) over the same period.

The general principles of the quota allocation and levy arrangements agreed to in 1989, including provision for allocation of funds for other enhancement activities, were carried through into legislation in 1992. The Challenger scallop fishery was introduced into a modified form of the QMS and the QMA for the fishery was defined in law as the Southern Scallop Fishery (SSF). Each of the 48 licence holders was provided with a fixed tonnage ITQ allocation of 12t (meat-weight). This quota was allocated in perpetuity but was not made proportional like quota in other fisheries. An additional 64t (10%) was allocated to Māori as an interim settlement of fishery claims made against the Crown.

Provision was also made to allocate any quota available in excess of 640t to the Crown. The Crown was required to lease 10% of this quota to Māori without charge. The remaining quota was required to be allocated by the Crown as leasehold rights at a market price with preference being given to existing quota holders. Any reductions below 640t were to be effected by way of a proportionate reduction across all quota holdings. The Government effectively legislated to capture the upside benefits of the enhancement programme. The downside risks and the costs of enhancement remained with quota holders. Under the legislation this arrangement was to be discontinued in 1997 at which time all quota in the fishery would become proportion-based on the catch limit set at that time.

A compulsory levy was established under the 1992 legislation to fully fund the enhancement programme in accordance with a plan determined by the Minister of Fisheries. The legislation required that the fishery be enhanced to achieve the maximum economic yield by 1997, make specific provision for enhancing recreational-only areas, and take into account impacts of the enhancement programme on other fisheries. A new statutory body, the Southern Scallop Fishery Advisory Committee (SSFAC), was established under the legislation. It replaced the informal NZFIB advisory group.

Faced with competing demands for marine farm space, the scallop industry, initially through a NZFIB administered scallop advisory committee, became actively involved in protecting their fishing and management rights. By October 1993 applications to place marine farm structures in areas fished and enhanced in the scallop fishery exceeded 4000 hectares. With the establishment of the levy, it became apparent to the scallop industry that they would need to provide an alternative funding and administrative structure to protect fishing and management rights. The quota holders established the Challenger Scallop Quota Holders Association for this purpose in late 1993.

Māori quota was allocated equally to eight *Iwi* (or tribal groups) located within the boundaries of the SSF and represented the first and, so far, the only commercial fisheries settlement made directly between the Crown and *Iwi*. The 1992 introduction of the fishery into the QMS coincided with the signing of the final Deed of Settlement with Māoridom over claims for commercial fishing rights. The Minister, on introducing the legislation into the House, acknowledged the SSF as a post-settlement fishery and one that would require a 20% quota allocation to Māoridom in terms of the Deed.

Over the next three years, catches in the SSF increased from 710t in 1992 to a high of 850t in 1994. The effort to catch spat using bags set in the water expanded from around 200 000 bags to over 500 000 and the number of spat caught and area seeded increased proportionately. High seeding-success rates experienced in the early 1990s did not, however, continue. In particular, very low survival rates were recorded from seeding efforts undertaken in 1992/93. Adult scallops taken in 1993 were also in poor condition. Other problems with enhancement such as over-settlement of other species on spat-catching equipment were also becoming significant. The industry and government were faced with a productivity crisis in the fishery and this stimulated investment in research to identify the extent and nature of the problem. The SSFAC, along with MAF and local government, commissioned a scientific review of the enhancement programme in the context of the total marine environment.

In 1993 the costs of the enhancement programme were also increasing markedly with operational expendi-

---

under *avoidable cost principle* the industry has had to pay the full cost of most of the fisheries management services the Government provided. More recently, the Government has moved to adopt the *attributable cost principle* for fisheries management. This states that the costs of services should be recovered from those who benefit from the services. The application of this principle has lead to a reduction in the costs recovered from the fishing industry, and a corresponding increase in the costs incurred by Government on behalf of the non-commercial sectors.

ture exceeding NZ$1.6 million in the 1993/94 financial year. Trust funds were being managed on a non-accrued basis consistent with financial accounting in the state sector as a whole. This cash-based budgeting system provided only a limited ability to manage levy and cash flow requirements between years. This approach was particularly risky given the prospects of poor future fishery performance (and potential levy income) resulting from the 1993 seeding failure. Also in 1993, the programme itself was reliant on vessels contracted by MAF from the aquaculture industry to undertake harvesting activities. This approach was found to be costly and it provided only limited ability to develop specialised equipment designed to improve spat-handling and seeding techniques.

Management of the scallop fishery itself during 1992 and 1993 was the subject of considerable contention and disagreement between government and the industry. This disagreement was stimulated by the complex and perverse set of management objectives and financial incentives established in the 1992 legislation. Government, for its part, was required to ensure that the fishery was being utilised in a sustainable manner, to provide adequately for non-commercial fishing interests and to ensure that impacts of enhancement on other fisheries were taken into account. It also needed to balance these roles with obtaining a return from selling leasehold rights to Crown quota by ensuring that the fishery was enhanced to its full potential within five years. In contrast, the scallop-quota owners had been disenfranchised from the benefits of any investment in enhancement. They were required to fund not only the development of the scallop fishery but also other enhancement programmes with funds levied from their scallop landings. They did, however, retain a right of preference to any quota fishing rights leased by the Crown when purchased at a market price.

Local Māori had also started proceedings at the start of the 1993 scallop season which claimed that they should be allocated an additional 10% of quota, in line with the Deed of Settlement and the Minister's Hansard address made in 1992. They were successful in an application to the Court, which stopped the Crown from leasing any additional quota until their claim for an additional 10% could be reviewed. By 1994 it also became apparent that further expansion of the enhancement programme was unlikely to provide any expanded catches by the end of the five-year transitional period.

Cost-recovery and the restructure of MAF were additional factors driving the establishment and operation of Challenger. The enhancement programme, cost-recovered through a specific levy and delivered by MAF, did not fit into the accountability structures and new role of the Ministry. A new way of delivering the enhancement programme, as well as funding programme activities, within the wider cost-recovery environment was therefore sought. Contracting the service out to an external provider was an option consistent with MAF's new, and wider, purchasing roles.

The scallop-quota owners, with the support of the NZFIB, set about designing an organisation that could implement such a contract. It was not a realistic option to contract the programme out to a scientific provider since it was now a significant, and specialist, commercial operation in its own right. Challenger was designed and established for that purpose. It was set up as a company given the need for strong commercial accountabilities in an organisation that would be holding significant assets and financial responsibilities. Company law provided protections for minor shareholder interests. Shareholding was linked to quota ownership to ensure that management of the fishery was directly accountable to long term investments in the fishery. Shareholding was allocated initially on a tonnage basis to reflect the fisheries framework in place at the time. The Company was incorporated in May 1994 and industry members on the SSFAC were appointed as establishment Directors to the Company.

Three years on, much of the legislative and institutional framework, and the rights needed to realise the opportunities identified by Pearce in 1991 were in place in the Southern Scallop Fishery. Only the legislative support, identified by Pearce as a precursor to collective action in a quota fishery, had not been carried through into law. This meant that any collective arrangements adopted and implemented amongst quota owners and other industry participants would require 100% voluntary participation.

In April 1994, the Minister of Fisheries and the SSFAC agreed to a set of reforms to address the range of issues and problems affecting the fishery. These reforms included the early introduction of proportional quota into the fishery at a catch level of 850t, allocation of an additional 10% of quota directly to *Iwi* and a direct role for the industry in the management of the fishery. The reforms were effected by civil contract and legislation.

In July 1994, Challenger signed an Agreement for Provision of Services (Agreement) with MAF to implement an Enhancement Plan (Plan) approved by the Minister of Fisheries. Levies collected by the Ministry, and held in Trust for the purpose of implementing the Plan, were used to fund the Agreement. The Agreement was in essence an operational plan for the enhancement of the scallop fishery but also included requirements to collect information for research, provide advice on management of the fishery and review the enhancement planning framework. It also specified consultation requirements with recreational interests to implement these responsibilities. Rights to catch spat in the scallop fishery to service the enhancement programme were transferred to Challenger.

The Agreement was entered into for a term of three years ending in June 1997. MAF had discretion to terminate the Agreement in the event that legislation passed that removed its ability to levy quota holders of the purpose of implementing the Plan. The Articles of Association of Challenger specifically provided quota holder-based funding mechanisms for the implementation of the Plan in the event that such a circumstance arose.

In early 1995 Challenger established a recreational advisory group for the scallop fishery served at its expense. All management proposals developed by

Challenger in consultation with this group were approved by the Minister for implementation. These decisions included a reduction of the TACC to 720t, which under the legislative amendment meant that the Government's interest in the TACC ceased. Low seeding successes experienced in 1993 resulted in a significant catch reduction to 521t in 1995, down from 850t (equal to the full TACC) the previous year.

In May 1995 Challenger took possession of the **Tasman Challenger** a purpose-designed vessel for the enhancement programme constructed at cost of about $NZ1 000 000. The first Annual General Meeting of the Company was held on 4 August 1995. Share certificates were issued to all quota owners at this meeting. Over the 1994/95 summer period, the enhancement programme continued to be implemented pursuant to the Agreement.

Between June and October 1995 legislation was introduced to change the SSF into a proportional quota management system and provide for the allocation of additional quota to Māori. Provisions allowing the catch limit in the SSF to be set at a level other than at the MSY, which applied to other fisheries, were established. Statutory imperatives to enhance the fishery to obtain the maximum economic yield were retained in the interim. The 1995 legislation also provided for the repeal of the Government-imposed levy supporting the scallop enhancement programme. As a consequence the Agreement funding the implementation of the programme through the administrations of Challenger became uncertain. This decision placed considerable financial risks on Challenger activities and threatened the future viability of the enhancement programme given that the only alternate source of funding was to resort to a voluntary framework (albeit supported to an extent by the constitutional provisions of Challenger).

The Government, however, provided a solution by reforming legislation enabling the establishment of a compulsory levy by majority vote (subject to various checks and balances) of quota holders of a fishery. This reform represented the final step in the process of change from government-managed funding to industry, or self, funding but was separate, both in principle and practice, from the wider cost-recovery regime applicable to the industry as a whole.

Challenger was nonetheless well placed within the new and wider cost-recovery environment. Recovery of research costs and the costs of enhancement were already internalised within Challenger's operations and this was recognised by Government. The main cost-recovery issues faced by Challenger in the wider regime were related costs of fisheries management advice, administration of quota management system (including catch registry returns), lack of provision for government-sharing of enforcement costs and, most critically, the recovery of enforcement costs. These were contentious issues not so much because of the actual costs incurred by Government, but rather the process used for distributing these costs amongst the industry. A formula linked to tonnage and landed value across a grouping of fisheries was used for allocation. This meant that high-valued fisheries attracted a greater share of the costs and that the cost-recovery regime was not sensitive to self-regulation and the cost savings that such initiatives created for Government.

In December 1995 the size limit of scallop for amateur fishers was reduced from 100 to 90mm. Recreational fishers were also able to fish in rotational sectors closed to the commercial fleet. In normal growing conditions the scallops that are sown or naturally settle into the fishery grow to 90mm in 18 months. This important decision, supported by Challenger, enabled fully integrated recreational fishing and indirectly, the aspirations of customary Māori fishers, into the rotational seeding and enhancement programme applying to the commercial sector.

The Agreement was retained through the 1995/96 spat-catching period. Around 660 million spat were caught and released over Golden and Tasman Bay. The **Tasman Challenger** had been fitted with on-board equipment designed to process spat and improve seeding operations. Trials using this equipment on a commercial scale were done during these spat-seeding operations. Over-settlements of mussel spat, however, continued to cause difficulties during harvesting operations. Catches from the fishery dropped further to 231t in 1996, placing additional pressures on levy and cash flow requirements to support the enhancement and management programme. Most catches were, however, from enhancement stock not natural settlements and the imperative to maintain and improve the programme remained.

2.6 Self-governance under environmental standards (1996 onwards)
2.6.1 General environment

New fisheries management legislation was introduced in 1996. The new Act provided a purpose and principles for fisheries management, which in turn set standards and specifications that had to be met by all agencies that wished to assume a role in managing fisheries. In addition to developing the new *Fisheries Act*, in its first year the newly formed MFish also developed a strategic direction for managing fisheries. This work included establishing the Ministry's role in the process of fisheries management and ways in which the new direction should be identified and achieved. The document 'Changing Course – towards fisheries 2010' set out the framework for developing the strategy to manage fisheries in the future, based around four major themes:

i. An ecosystem based management approach was required to effectively manage fisheries in the context of the environment in which they exist
ii. Long-term goals for the management of fisheries were required and there needed to be an agreed understanding of how to achieve those goals
iii. Effective management of fisheries was vital to the community, and successful management depended on the involvement, co-operation and support of all those with an interest in the fishery and
iv. Sustainable fisheries contribute to economic growth.

In 1998 the Ministry published a 5-year strategic plan (1998-2003) identifying the role of the Government in ensuring that the fisheries resource was utilised sustainably and describing the way in which MFish must operate. The core roles of MFish as an agency of the Government were broadly summarised as:

i. ensuring ecological sustainability
ii. meeting Treaty of Waitangi and international obligations
iii. enabling efficient resource use and
iv. ensuring the integrity of management systems.

On the national front, the New Zealand Seafood Industry also initiated a reform of its representative bodies. By 1997 over 20 rights-based fishery management organisations existed in New Zealand in various stages of development. Many were initially established with the assistance of the NZFIB and fishing industry representative organisations. In 1997, representatives of the fishing industry established a new organisation designed to provide generic services for the industry, and accountable through its shareholdings to the various fishery management organisations. Shareholdings in the newly established Seafood Industry Council Ltd (SeaFIC) were allocated to the fishery management organisations, including quota-based management companies and representative bodies of the aquaculture industry. SeaFIC is funded through an agency agreement with the NZFIB. Shareholder voting rights for SeaFIC are based on levy contributions made by shareholder interests to the NZFIB.

Growing concern about lack of progress and cost associated with implementing the new *Fisheries Act 1996* stimulated MFish and the Minister of Fisheries to commission an independent review of the Act in early 1998. The Independent Reviewer's Report completed in September 1998 recommended the following (Hartevelt 1998):

i. a fundamental realignment of the roles of Government and fisheries stakeholders and the implementation of transparent consultation and decision-making processes
ii. a simplified and less prescriptive operating regime than exists under the *Fisheries Act 1996* and
iii. devolving to fisheries-rights holders the responsibility for fisheries management at the discretion of the Minister.

During 1999 the Government passed legislation that gave further effect to the devolution of certain service functions that formerly rested with MFish. In August 1999 the responsibility for managing the QMS registries was transferred to a subsidiary of SeaFIC. This devolution of functions was a significant step, although it still retains the delivery of services such as catch-reporting and catch-balancing with a central body. The responsibility for ensuring that environmental standards are adhered to was retained by Government. Management of most QMS fisheries, including inshore fin-fisheries, continued to be managed centrally by regulation and TACC constraints.

By the end of 1999, the QMS was confirmed as a system that limited the total take and provided quota owners with an ongoing right to a share of the total catch in the fishery. Further, the QMS reforms had provided an opportunity for quota holders to have more control over their future, and change from maximising their individual share of the catch to managing the fishery for collective benefit.

Luxton (1999) noted that the Government viewed the marine environment as a precious and finite resource and, accordingly, the use of it needed to be as efficient as possible in order to generate maximum returns for the economy with minimum impact on the environment. Upton (1999) declared that the Government's quest was to establish a framework where laws set the framework for business to operate most efficiently, enabling outcomes that are both economically and environmentally sustainable.

2.6.2 Challenger fisheries management

In April 1996 Challenger carried out a ballot among scallop quota holders proposing that a compulsory levy be established by vote of quota holders. With the prospects of such a funding-framework Challenger prepared a comprehensive plan for the management of the 1996 commercial scallop season. That plan included an analysis of scallop enhancement and survey data, recommendations for management, projected budgets and funding requirements for management of the fishery for the next twelve-month period. The plan contained a draft civil contract that contained binding rules for the management of the fishery. All management arrangements, including a proposal to introduce a 20% levy and the compliance plan, were approved in principle by shareholders at an Extraordinary General Meeting of the Company. Under the plan Challenger established a daily catch-balancing system and a dockside-monitoring programme to ensure compliance. Penalties were established within the contract for any catch taken in excess of the daily limits.

The contract also provided for Challenger to exclude commercial scallop vessels from all, or parts, of the scallop fishery if scallop yields were poor or for purposes of managing biotoxin risks. The contract established penalties (subsequently, and more effectively, defined as "agreed damages") for breach of rotational fishing closures and area agreements implemented to provide better access to the recreational fishing sector. Catch reports provided under the contract were used to monitor compliance with total catch limits set for parts of the fishery.

The Minister of Fisheries and MFish approved, and where appropriate, implemented the annual management plan. Importantly for the self-governance of the fishery, the Minister agreed to retain the TACC at 720t even though available yields were much lower. The quota owners in turn agreed to lease excess quota (being 51.39% of the quota available) above estimated catch levels, to Challenger to be held and not fished. The Minister provided approvals to exempt Challenger and its Directors from aggregation limits to effect this management decision.

In July 1996 the compulsory levy order came into effect. This enabled Challenger to fund management activities directly and ultimately led to the dis-establishment of the Government levy as well as the Agreement. A new set of Articles of Association for the Company was developed, and adopted by shareholders, that changed the shareholding base of the Company to better reflect the proportional quota system. Under the new Articles all quota owners qualify for one share in the Company. Shareholder voting rights for appointment of Directors and management of other Company business are exercised in proportion to the total quota owned by shareholders. These provisions are consistent with the requirements for setting the levy, by majority vote at a general meeting, as specified under the compulsory levy order.

Industry participants invited to Challenger's Annual General Meeting (AGM) in August 1996 included fishermen, quota holders and processors as well as shareholders (the quota owners). All parties endorsed and signed the 1996 Compliance Contract, which specified management rules for the upcoming season which had been approved by the Minister. The successful introduction of this contract with full support demonstrated that it was feasible to establish collective management agreements in this fishery even without the statutory support envisaged by Pearce back in 1991.

A meeting of permit holders in the closely associated dredge oyster fishery was also held during the AGM to update permit holders on progress with introducing this fishery into the QMS. In consultation with dredge oyster permit holders, Māori interests, and MFish, Challenger facilitated a quota allocation proposal which was ultimately endorsed by those parties. Challenger supported this initiative because of the need to start integrating catching and management arrangements between the two fisheries. Until that time, scallop and oyster dredging activities proceeded in an uncoordinated manner so that oyster fishing often occurred across scallop beds within a month of the newseeding operations being completed.

The *Fisheries Act 1996* provided a framework for the development of standards and specifications for management of the enhancement programme (and ultimately the fishery) by Challenger. Rather than specify the goals of the enhancement programme in legislation as the previous statute had done, the programme was integrated into the wider fishery management framework. Importantly, the SSF remained exempt from normal sustainability criteria (based on MSY) on the basis that the purpose and principles of the legislation were better delivered by rotational fishing and enhancement in the fishery.

The Act also introduced the dredge oyster fishery into the QMS. This was the second fishery after the scallop fishery to be introduced since the settlement of Māori claims to commercial fishing. The Challenger Dredge Oyster Fisheries Management Company was subsequently incorporated in February 1997. Challenger was immediately contracted to provide management and fisheries services to this new Company. Challenger established itself as the Shellfish Quality Assurance Programme Delivery Centre for both the scallop and dredge oyster fisheries to manage and enforce biotoxin and sanitation requirements specified in legislation and detailed in government-approved Shellfish Quality Management Plans.

The role of the Challenger Scallop Recreational Advisory Group was expanded to include dredge oysters. This Group, along with the two Challenger companies, supported removal of the amateur season applying to the dredge oyster fishery at the same time that the commercial dredge oyster season was rationalised so that it opened during the scallop season.

Management recommendations for the 1997 scallop season had to be developed under the umbrella of the purpose and principles of the new *Fisheries Act 1996*. In anticipation of this Challenger commissioned a study in 1997 to explore the use of reference points or "environmental bottom lines" to underpin the sustainability framework for the SSF. This work modelled the effectiveness of rotational fishing and enhancement as compared to other, more traditional, fishing strategies designed to ensure sustainability. The modelling study helped confirm the statutory basis for the SSF exemption from sustainability measures applying to other fisheries under the Act. This information underpinned the management plan sent to the Minister by Challenger following the completion of the annual scallop abundance survey in 1997.

In 1997 the annual scallop survey was further expanded and refined. In particular, a new analysis was undertaken to obtain estimates of commercial yields. Challenger needed accurate estimates to set catch limits ando allow more effective utilisation of the fishery through the leaseback arrangements. The information was also critical for budgeting purposes. Research activities were expanded to include assessments of the population distribution (*i.e.* scallop numbers at different sizes). This information added greater certainty to the process of setting sustainability measures and gave Challenger and MFish added security in determining appropriate harvesting strategies.

The *Fisheries Act 1996* also placed different consultation requirements on Government and these were co-ordinated with Challenger's role in the development of harvesting plans for the fishery. Initial management proposals developed by the Board for consultation included a proposal to harvest a fixed tonnage from an area of the fishery, out of phase with the planned rotational fishing cycle. Challenger successfully proposed to establish a daily catch-balancing and reporting regime to ensure compliance with these provisions.

The Minister of Fisheries subsequently endorsed the rotational programme as the primary means of ensuring sustainability in the scallop fishery. He approved all management plan proposals for the 1997 season including the right to harvest outside of rotation in the circumstances described. That decision was important for future self-governance of the fishery for a number of reasons. First, it provided Challenger with the opportunity to demonstrate its ability to undertake micro-management roles that could not be effectively implemented by Government. Second, it demonstrated the value and effectiveness of the new legislation in providing for management under a self-governance framework. Third, it allowed Challenger to demonstrate to other sector groups, and in particular the recreational sector, its ability to manage the fishery, leading to a greater level of acceptance of the flexible management regime that had evolved.

Catches in 1997 increased to 300t. The levy was maintained at 20%, but even so it was insufficient to fund scallop seeding targets set in the previous year. Problems in programme operations were also encountered and in particular the seeding approach developed was leading to loss of equipment. Challenger determined to cut back seeding and harvesting operations to operate within the processing capacity of the **Tasman Challenger**. In taking this step Challenger was able to fully process all spat on board and remove the risk of loss of spat bags during seeding. This also enabled Challenger to improve the quality of processing operations thus increasing the chances of seeding successes. Even with the reduced spat-catching capacity, 403 million scallops were caught and seeded.

In January 1998, the Minister of Fisheries approved a revised Enhancement Plan developed by Challenger. This Plan established targets for enhancement but also identified constraints on areas and amounts seeded per unit area as well as information-reporting requirements designed to meet the purpose and principles of the legislation. The Minister's approval was subject to a range of reporting conditions addressing these issues.

Against the backdrop of its strategic direction, and its new structure and purpose, MFish took steps to formalise management arrangements in the scallop fishery. A comprehensive Memorandum of Understanding (MOU) was entered into with Challenger in March 1998 to enable better use of resources in the scallop fishery, within environmental constraints. Under the MOU, Challenger agreed to provide MFish with information and advice on enhancement and management of the scallop fishery. The MOU records, *inter alia*:

"The MOU is intended to contribute to the ability of [Challenger] to continue to develop opportunities for responsible self-management of the Southern Scallop Fishery. The relationship between [MFish] and [Challenger] has been characterised by the high quality of routine information provided to [MFish] by [Challenger]; the Chief Executive is authorised to enter into the MOU through his powers set out in Part III of the State Sector Act.

[MFish] records its objective of maximising the ability of stakeholders to act in a collective manner and develop opportunities for self-management. Providing such opportunities is said to be consistent with the Government's role in enabling efficient resource use by providing the framework to allow owners of harvesting rights to make decisions regarding the operation of those rights.

[MFish] seeks to safeguard its ability to deliver its core responsibilities: ensuring and fulfilling environmental principles, meeting Treaty of Waitangi and international obligations, enabling efficient resource use and ensuring the integrity of management systems."

The development of the management plan for the 1998 scallop season was again preceded by an annual stock abundance survey. The survey was further expanded and included a more detailed analysis of the population structure of juvenile and adult scallops. The survey also incorporated the dredge oyster fishery. Once again, the Minister of Fisheries approved the management plan. By 1998 the annual management plan for the fishery had become a complex document incorporating the following:

i. Financial results of the prior financial year as outlined in the 1998 Annual Report (pursuant to compulsory levies legislation the Minister was required to table this Report in the House of Representatives)
ii. The 1998/99 Business Plan and levy (set at 17 % for the 1998 season)
iii. A summary report on all scallop enhancement activities relevant to harvesting in 1998 and future years
iv. Information available from the annual scallop abundance survey
v. A report on consultations undertaken
vi. Other information relevant to the management of the fishery (including additional survey data)
vii. Recommendations for management of the scallop and dredge oyster fisheries and
viii. A draft 1998 Compliance Contract.

Scallop harvests in 1998 increased to 547t. A large proportion of the recruited scallop population was left for harvest in 1999 to allow the scallop size to improve to meet export market requirements. Recreational fishery access was considerably enhanced by this decision. For the 1998/99 spat-catching season further improvements were introduced to the spat-processing operation to keep spat protected during transportation. A research programme was also established to identify ways of improving survival of seeded spat. An extensive spat-monitoring programme is now undertaken to determine when to set gear to catch spat.

During 1999 over 325 million scallop spat were caught and seeded over sections of the SSF. Seeding activities were targeted at the most productive areas and away from areas with high numbers of naturally recruited scallops as identified from extensive pre-release surveys. Post-release surveys have shown very high survival rates. As a result of the research commissioned on seeding techniques Challenger is introducing processing operations this year to further improve survival of scallop spat when seeded. Challenger also monitors water quality to assess productivity and algal events and is in the process of establishing a more intensive monitoring programme to assess productivity on the seafloor as well as in the water column. Challenger has also commissioned a modelling study on behalf of the Challenger Dredge Oyster Fishery

Management Company to assess options for better management of that fishery.

A survey of scallop stocks was carried out in May and June 1999. Results of this survey were used for developing a management plan for the fishery, which was discussed with recreational fishers, *Iwi* and environmentalists. This year (1999), responsibility for management of rotationally closed areas was devolved to Challenger which ensures compliance with these closures, as well as a range of other management measures under a contract, once again signed by all industry participants (including those in the dredge oyster fishery). MFish has appropriately retained an audit role to ensure that the management outcomes established by Challenger and implemented by Challenger to achieve the purpose and principles of the Fisheries Act 1996, are realised. Scallop catches this year (1999) are predicted to reach 720t. Management measures adopted to meet quality targets (*i.e.* yield and colour) are proving successful and current information indicates that catches will further improve in the future.

In addition to managing shell fisheries, Challenger is now developing a collective structure for the management of inshore fin-fisheries in QMA 7 and 8. Challenger is in the process of incorporating the Challenger Finfisheries' Management Company (CFC) and allocating shares to quota-owners. In contrast to the scallop and dredge oyster companies, each of which has less than forty shareholders, CFC has 416 quota-owners who qualify for shareholdings covering 20 fish species. The costs of administering such a structure are much higher, but not insurmountable, given that 75% of the quota is held by as little as thirteen individual companies. In the interim, Challenger has already taken a number of steps to develop self-governance frameworks for this fishery. In 1998, with the agreement of the Minister, Challenger established a monitoring, reporting and effort-split regime[3] designed to reduce fishing effort in the rig (*Mustelus lenticulatus*) fishery within the boundaries of the SSF.

2.7 Challenger coastal marine area management

Part way through the process of fishery management integration, Challenger and Government are being faced with a more pervading crisis of adjustment. Various interests are seeking opportunities for spatial and productive use of the environment on a significant scale. New applications for marine farming within areas where the scallop fishery now operates exceed 10 000 hectares. The situation faced by Challenger and Government is how to place the rights and responsibilities allocated and devolved to Challenger and quota holders within the management of the Coastal Marine Area (*i.e.* within 12 nautical miles of the coast) as a whole.

The management of the environment (including fisheries) is controlled by both local and central government under different political accountabilities and within differing legislative, legal and spatial jurisdictions. The extent to which the roles of central and local government overlap in terms of their mandates under separate legislation and within the Coastal Marine Area (CMA) is an unresolved issue. From a resource management perspective, the debate is about the extent of external effects created by any new uses that are authorised and about how these should be managed.

Under legislation, the rights of commercial fishers are protected against certain undue adverse effect from other users of fisheries in the QMA. This legal constraint provides a threshold against which other claims for access to fisheries resources must be measured. It applies to the establishment of exclusive recreational fishing zones, marine farming operations, and no-take marine reserves. In each case the legislation provides, among other things, that such allocations cannot be made if there is an undue adverse effect on the ability of commercial fishers to harvest their entitlement. In the context of Challenger, the commercial fishing rights are expressed as ITQs, able to be exercised subject to a regulatory framework ensuring sustainability.

Ways to resolve these problems are simple in resource-economics theory. The potential externalities created by proposed new uses should, as far as possible, be internalised through better specification of the rights and management structures already established. Alternatively, Government will need to centralise the management of these effects within a regulatory environment. The latter approach potentially involves reallocation or attenuation of the rights already issued in the marine environment. This approach is a retrograde step from Challenger's perspective.

3. CONCLUSION

A complex set of biological, financial and institutional imperatives have driven increased self-governance by quota owners and other industry participants in the Challenger area. These include the following:

i. over-exploitation of fish stocks and the limitation of participants
ii. development of commercial enhancement technology and infrastructure demonstrating the presence of management opportunities
iii. establishment of perpetual quota rights allowing transfer and security for investment
iv. fishery-productivity failures and fish-quality demands requiring management attention
v. introduction of proportional quota and the resolution of Māori claims to fisheries resources providing opportunity for development
vi. enactment of government financial accountabilities and introduction of cost-recovery providing incentives for management efficiency
vii. restructure of fisheries law, and the government management agency providing opportunity for self-governance
viii. the establishment of funding streams and collective frameworks for self-governance and

---

[3] The term *effort-split regime* is best described as a scheme designed to ensure that a specified maximum catch of species within the TACC set by the Minister is taken from a sub-area of the wider QMA for this fishery. The purpose of the effort split is to reduce fishing pressure on one component of what is thought to be two stocks of a species within a single QMA.

ix. competition for use of coastal space and productive capacity in the marine environment stimulating collective action to protect investments and rights from reallocation.

Self-governance in the SSF has developed under these imperatives in two forms. On one hand industry participants have responded to the specific opportunities provided under the legislative framework for this fishery, which were developed and implemented by the Government. On the other, some key initiatives undertaken by industry interests have driven legislative and institutional reform.

The authority of Challenger to undertake fishery management responsibilities is devolved and delegated to Challenger within the context of a number of planning documents and permit authorities. For the scallop fishery, these include a statutorily-approved Enhancement Plan, an annually-reviewed fishery management plan (including a harvest strategy and compliance programme), a Memorandum of Understanding between Challenger and MFish outlining information needs for the fishery, Shellfish Quality Management Plans, and various permit and consent authorities to effect these responsibilities. Information needs and management for the dredge oyster fishery are integrated into these documents.

The pattern of development has occurred differently in the Challenger fin-fisheries. The centralised process of management has limited opportunities for self-governance in these fisheries. Devolution of the management of QMS registries to SeaFIC has provided opportunity for progress, but this depends on how responsive this new centrally-based system is to the needs of fisheries management. Further development of management in Challenger fin-fisheries is also faced with the difficult job of organising the interests of a much larger number of quota holders, fishers and processors without statutory support.

Challenger is now positioned as a regulator of fisheries-use in the marine environment under the watchful eye of Government. It is an organisation sensitised by management of fisheries within the constraints of that environment. It has the incentives, the capacity and the responsibility to preserve and improve its role and the interests of its shareholders within that environment.

Challenger has recently initiated Environment Court and High Court proceedings in opposition to local government planning proposals for the CMA which include provision for marine farm uses. Challenger is defending its management role and the rights of its shareholders within the context of spatial use and the life supporting capacity of the environment it operates within. The difficulties of allocating rights in fisheries in the marine environment identified by Keen (1988) are a reality in the Challenger CMA.

An underlying issue to be resolved is how to allocate any super-profits that might be created from allowing the marine environment (including its productive capacity) to be utilised in alternative ways. The politics of fairness, envy and environmental management that have pervaded the allocation of ITQs are being revisited on the marine environment as a whole. Challenger is being seen as the barrier to progress by many of the new aspirants for these benefits.

4. LITERATURE CITED

Ackroyd, P., R.P. Hide and B.M.H.Sharp 1990. New Zealand's ITQ system: prospects for the evolution of sole ownership corporations, *Report to MFish*, pp. 86.

Annala, J.H. 1996. New Zealand's ITQ system: have the first eight years been a success or failure? *Reviews in Fish Biology and Fisheries*, pp.43-62.

Batstene, C.J. and B.M.H. Sharp 1999. New Zealand's quota management system: the first ten years, *Marine Policy*, pp.177-190.

Bull, M.F. 1994. Enhancement and management of New Zealand's "southern scallop" fishery. *In* Bourne, N.F., Bunting, B.L., and Townsend, L.D. (eds), *Proceedings of the 9th International Pectinid Workshop*, Nanaimo, B. C. 1994. Canada, April 22-27, 1993, Volume 2, pp.131-136.

Clarke, M. and P. Clough 1998. New Zealand's fisheries: co-management and property rights, working paper 98/16, *New Zealand Institute of Economic Research*, pp. 50.

Crothers, S. 1988. Individual transferable quotas: the New Zealand experience, *Fisheries* pp.13(1), 10-12.

Hartevelt, T. 1998. Fishing for the Future: Review of the Fisheries Act 1996: Report of the Independent Reviewer of the Fisheries Act 1996 to the Minister of Food, Fibre, Biosecurity and Border Control, *Report to the Minister of Food Fibre Biosecurity and Border Control*, pp. 91.

Johnston, R. 1995. Implications of taxing Quota Value in an Individual Transferable Quota Fishery, *Marine Economics (10)*, pp.327-340 (see p.337).

Keen, E.A. 1988. Ownership and Productivity of Marine Fishery Resources: An Essay on the Resolution of Conflict in the Use of the Ocean Pastures, The McDonald and Woodward Publishing Company, Virginia, pp.122.

Luxton, J. 1997. (Minister of Food Fibre Biosecurity and Border Control), Stakeholder management of recreational fisheries, *Address to the New Zealand Recreational Fishing Council*, Annual General Meeting, Bay of Islands. July 1997.

Luxton, J. 1999. (Minister of Food Fibre Biosecurity and Border Control), Sustainable development of ocean resources, *Address to Our Oceans Conference*, Te Papa, Wellington. October 1999.

Ministry of Fisheries (New Zealand) 1996: Changing course – towards fisheries 2010, pp.24.

Ministry of Fisheries (New Zealand) 1998: Five year strategic plan 1998-2003, pp.24.

Pearce, P. H. 1991. Building on Progress: Fisheries Policy development in New Zealand, *A report prepared for the Minister of Fisheries*, pp.28.

Scott, A. 1998. Development of Property in the Fishery, *Marine Resource Economics (5)*, pp.289-311 (see pp.305-6).

Scott, A. 1993. Obstacles to fishery self-government, *Marine Resource Economics (8)*, 187-199.

Upton, S. 1999. (Minister of the Environment), Corporatisation and sustainable management, Address to resource management law association conference, Christchurch. October 1999.

Wheeler, B., J. Bradford, C. Collins, A. Duncan, B. Wilson and H. Young 1992. Sustainable Fisheries (Tiakina nga Taonga a Tangaroa): Report of the Fisheries Task Force to the Minister of Fisheries on the Review of Fisheries Legislation. pp 75.

STRONGER RIGHTS, HIGHER FEES, GREATER SAY: LINKAGES FOR THE PACIFIC HALIBUT FISHERY IN CANADA

G.S. Gislason
GSGislason & Associates Ltd.
PO Box 10321, Pacific Centre, 880 - 609 Granville Street
Vancouver, BC, Canada V7Y 1G5
<gsg@gsg.bc.ca>

1. INTRODUCTION

Commercial halibut licence holders in British Columbia, Canada gained much stronger rights under the Individual Vessel Quota (IVQ) fisheries management programme introduced in 1991. Licence holders have to pay substantial fees to fund a dockside monitoring programme (DMP) for offloading catches, a dedicated enforcement presence, and other fisheries management services. As well, the Halibut Advisory Board (HAB) comprised of elected licence holders, was formed to advise the Canada Department of Fisheries and Oceans (DFO) on the halibut industry and, by implication, to ensure accountability to industry.

This paper outlines the evolution and linkages between rights, fees and accountability for the Pacific Halibut Individual Vessel Quota (IVQ) Programme in British Columbia, Canada. The paper focuses on one aspect of the many changes in industry relationships that inevitably occur when licence holders attain stronger rights, in this case the new relationship between rights (licence) holders and government (DFO) fishery management.

2. BACKGROUND
2.1 History

Pacific halibut is the largest of all flatfish and among the largest fish in the world. Pacific halibut inhabits the continental shelf of the US and Canada, ranging from California north to the Bering Sea (Bell 1981). The International Pacific Halibut Commission (1998) outlines the development of the commercial halibut fishery along the Pacific Coast of North America. The fishery was pioneered by fishermen of Norwegian ancestry and started in the late 1880s.

The commercial halibut fishery started in 1888. Fishermen used sailing vessels to fish off Washington State, USA, landed the halibut in Tacoma, Washington and shipped the iced fish by the newly-completed transcontinental railway to Boston. After railway service to Vancouver and Prince Rupert in British Columbia connected the West Coast of Canada to Eastern Canada, these two locales became important West Coast landing ports. In fact, Prince Rupert became known as "The Halibut Capital of the World".

From the beginning in the 19th century, Canadian and US boats fished on common grounds as there were no international boundaries pertaining to fishing at the time. Both fleets gained port privileges to land halibut and take on supplies in the other country.

Initially, sailing vessels and then steam-powered vessels fished with several dories. Two men in each dory pulled the lines by hand. These were replaced in the 1920s by diesel-powered schooners designed to mechanically haul longline gear directly from the deck. Over the next 50 years, several other technological innovations were adopted including the replacement of natural fibre lines by nylon lines, the introduction of snap-on gear, and the conversion to circle-shaped hooks from J-shaped hooks (IPHC studies concluded that circle hooks were 2 to 3 times more efficient at catching halibut).

2.2 Fishing methods

The directed halibut fishery uses setline gear where a skate, the basic unit of gear, consists of groundline, gangions or branch lines, and baited hooks. The gear is set, left to 'soak' for several hours, and then retrieved by a power-driven gurdy. Bait used includes octopus, herring and other fish. Snap-on gear has gained favour on those boats that participate in other hook and line fisheries beside halibut (snap-on gear differs from traditional setline gear in that the gangions are attached to the groundline with metal snaps, rather than being tied to the groundline with wire).

Halibut are dressed on-board by removing the viscera and gills and the body cavity is filled with ice. The head is not removed until the catch is delivered at dockside and sold. The halibut then is beheaded in the plant and graded by weight according to trade categories.

Today, the vast majority of BC halibut is sold as fresh, headed and gutted whole fish into the US market. Halibut is a popular food fish as it can grow to 200kg[1] or more (although the average landed size would be in the 5 to 20kg range), is firm-textured and white-fleshed and has relatively few bones. The fish, if well iced, keeps for an extended period of time without spoiling. Figure 1 displays the catch in tonnes and the real, inflation-adjusted catch value over the past 35 years.

2.3 The International Pacific Halibut Commission (IPHC)

Appendix 1 summarizes the evolution of fisheries management for the Pacific halibut fishery of Canada. The halibut fishery initially was unregulated by season, catch or any other constraints. By 1910, the halibut fleets of both countries had expanded and overfishing became apparent. The industry asked the governments of both the

---

[1] In this paper, all halibut catches are reported in dressed, head-off weight.

Figure 1
British Columbia commercial halibut catch and landed value, 1965 to 1999

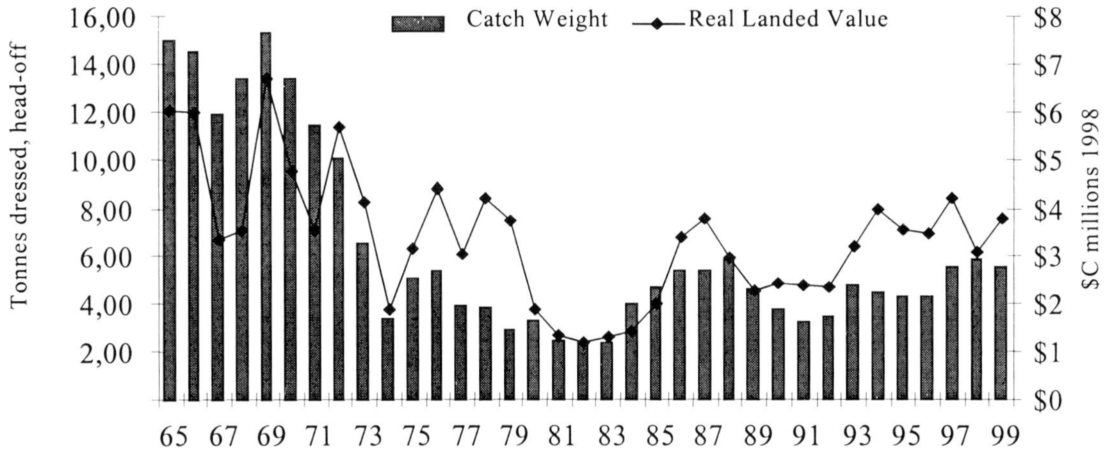

US and Canada for international management of the resource.

In 1923, the US and Canada signed a Convention under which the International Fisheries Commission was formed in 1924. The Commission was mandated to regulate by closed season alone, but it soon became clear that the three-month closure imposed was inadequate to protect the resource. A new convention was signed in 1930 under which the Commission could institute other conservation measures such as catch limits and gear regulations.

The convention was further modified in 1953 so separate fishing seasons by area could be established. The commission also changed its name to the International Pacific Halibut Commission (IPHC). The IPHC conducts basic research and stock assessment to develop and maintain stocks at a level which would permit maximum sustainable yield.

Halibut catches in both US and Canadian waters declined in the 1960s and the early 1970s due to a combination of factors including poor recruitment and increasing halibut bycatch by trawlers. In Canada, during the late 1960s and early 1970s many former halibut longline vessels were retrofitted to participate in the BC salmon fishery using seine gear. By 1974, the combined US-Canada catch had declined to under 10 000t, or less than one-third of the average catches in the mid 1960s (see Figure 1).

Both the US and Canada extended their coastal jurisdiction to 200nm in 1977. This resulted in an amendment to the 1953 Halibut Convention, termed a protocol, which was signed in 1979. The Protocol called for the phaseout of the fishing by one country's fleet in the other country's waters. At this time, the Canadian fleet caught much more halibut in US waters than did the US fleet in Canadian waters.

To this day, the IPHC continues to conduct Pacific halibut stock assessment work, set Total Allowable Catches (TACs) for both the US and Canada and enact other regulatory measures. The Federal Government of Canada paid about $C1.2 million to support the activities of the IPHC in 1998. Canada's Department of Fisheries and Oceans manages the Canadian portion of the fishery within IPHC parameters.

2.4 Limited entry in Canada 1979

The 1979 Protocol also enabled the individual governments to make regulations concerning their own fleets which did not interfere with 1979 Commission regulations. In 1979, Canada immediately imposed limited entry on the halibut fleet and those 435 vessel owners who had participated in the fishery in recent years received halibut "L" licences. The US fishery remained an open access fishery.

3. THE INDIVIDUAL VESSEL QUOTA SYSTEM: 1991 TO DATE

3.1 The setting

Turris and Sporer (1994) outline the development of the individual quota system for Pacific Halibut in Canada. Several problems emerged during the 1980s under the "derby" or competitive fishery format whereby licensed vessels competed for the available TAC. The fleet had become unmanageable and catches in most years exceeded the halibut TAC set by the IPHC. The length of the season became shorter each year with the result that by the late 1980s, the season was less than ten days long. In November 1988, a small group of licence holders approached DFO to explore the possibility of adopting an individual quota management system.

In 1989 DFO released a discussion paper of problems and prospects for the halibut industry. In 1990 DFO held a referendum of licence holders who voted in favour of implementing a two-year, trial Individual Vessel Quota (IVQ or IQ) programme. DFO adopted a system of non-transferable IVQs for a two-year trial period for the halibut fleet in 1991.

The commercial halibut fishing season under IVQs is usually mid-March to mid-November. Licence holders are free to catch their IVQ at any time during the season. Vessels must 'hail-out' their intention to fish and 'hail-in' their intention to land and unload their catch at designated landing ports.

3.2 Initial allocation

In 1991 each of the existing 435 "L" licence holders in 1991 received an Individual Vessel Quota where the quota level comprised a percentage of the TAC. DFO based the IVQ formula on a combination of recent vessel catch history and vessel length (70% catch history and 30% vessel length).

3.3 Transferability and new entry

For the initial two-year trial period, quota consolidation, or "stacking" of more than one quota on a single vessel, was not permitted (but a new industry entrant could buy a licence and quota from an existing licence holder). A 1992 review of the IVQ programme (EB Economics 1992) indicated industry support for the continuation of the individual quota system but that stacking, on a temporary basis, should be allowed. Industry endorsed the concept in a vote in the fall of 1992.

DFO made IVQs stackable on a temporary basis for the 1993 halibut fishery but at the end of the season any transferred IVQ reverted back to the original licence "L" licensed vessels). These rules persisted over the 1993 to 1998 period.

Starting in 1999, both permanent and temporary transfers were allowed. Any level of transfer was allowed so long as:

i. no one vessel had more than 1% of the TAC (unless it had fished greater than this amount from 1993 to 1998) and
ii. each "L" licensed vessel held a minimum amount of permanent IVQ set at 0.01149% of the TAC (5% of the average initial allocation for the 435 "L" licensed vessels). The minimum could be temporarily reallocated during the year.

The restriction i. above implies a minimum fleet size of 100 active vessels and room for more consolidation or stacking of quotas than occurred under the previous system. Table 1 (below) shows fishery parameters and fees paid since the introduction of IVQs in 1991.

3.4 Industry user fees

At the outset of the IVQ programme in 1991, the industry agreed to pay the incremental costs associated with the programme (prior to this, each of the 435 licence holders paid only $C10/ yr for a basic "L" licence).

New costs associated with the 1991 IVQ programme included: (a) a Dockside Monitoring Program (DMP) for

Table 1
Overview of British Columbia Halibut Fishery under individual vessel quota management

	Year								
	1991	1992	1993	1994	1995	1996	1997	1998	1999[a]
Fishing activity									
TAC (tonnes)	3357	3629	4763	4536	4318	4318	5670	5897	5489
Catch (tonnes)	3250	3459	4789	4490	4314	4312	5589	5847	5540
Value ($C million)	21.6	21.7	30.2	37.4	34.0	34.1	41.6	30.9	38.7
No. of licences	435	435	435	435	435	435	435	435	435
No. of licences fished	433	431	351	313	294	281	279	288	265
No. of offload events	1173	1150	1255	1148	1177	1094	1211	1335	1284
% TAC transferred[b]	0	0	19%	34%	39%	44%	49%	50%	61%
Fees paid ($C000)									
Flat rate fee[c]	109	109	109	109	109	109	109	109	109
Management levy[c]	652	652	697	620	625	625	820	853	794
Economic rent levy	0	0	0	0	0	951	1249	1298	1200
Total	761	761	806	729	734	1685	1878	2260	2103

Source: GSGislason & Associates (1999) and information from Archipelago Marine Research Ltd.
[a] Rent fee is estimated.
[b] No quota transfers were allowed from 1991 to 1992; only temporary quota transfers allowed from 1993 to 1998; temporary and permanent quota transfers allowed in 1999.
[c] Flat rate fee plus management levy to fund dockside monitoring, enforcement and other fisheries management activities (economic rent levy goes to general government revenue and does not fund specific fisheries management activities)

holder. To guard against the possible concentration of quotas in a few hands each initial halibut quota was split into two equal shares and quota shares could transfer freely so long as no more than four shares were held or fished by any one licensed halibut vessel. In essence this imposed a minimum fleet size of 218 vessels (half the 435 vessel offloading of halibut, (b) DFO enforcement officer salaries and expenses, (c) DFO management salaries and expenses and (d), other items. Halibut licence holders paid a total of $C761 000 in fees in 1991.

The initial cost recovery mechanism consisted of a two-part licence fee paid to DFO before the season – a flat fee of $C250 per licence holder plus a fee per-tonne of TAC (the per tonne fee has been set at $C144.70 since 1995). The intent was that the revenues realized would go to fund halibut fishery management activities.

In 1996, DFO started to collect, in addition to the two-part licence fee, a quota fee of $C220.30 per tonne of TAC that was not tied to fisheries management. This latter fee represented a "resource rent" that flowed into the federal government's central treasury, the Consolidated Revenue Fund (CRF). The federal government changed the resource rent fee to $C310 per tonne TAC in 1999 (less a 40% credit up to a maximum of $C1,000). The resource rent fee is pegged to a percentage of halibut landed value in a base period.

In 1999, the total of all fees paid comprised an estimated $C2.1 million, or about 5.4% of the estimated $C38 million in halibut fleet landed value.

3.5 The Halibut Advisory Board

The HAB was created in 1991 to provide "wide ranging advice to DFO to assist in the overall planning, management and enforcement of the Canadian Halibut Fishery". The Board presently consists of 21 individuals – one DFO non-voting chairman, 11 elected halibut licence holder members, and nine appointed members (representing processors, the provincial government, and aboriginal and recreational interests). The Board meets four or five times a year for one to two days each.

The HAB provides advice to DFO in three main areas (a) fisheries management regulations, (b) enforcement and (c) dockside monitoring/tagging. DFO provides fisheries management and enforcement services directly to the industry and a third party contractor provides dockside monitoring and halibut tagging services. The halibut management fees pay for these services.

4. LINKAGES – RIGHTS, FEES AND ACCOUNTABILITY

4.1 Stronger rights

The halibut IVQ programme gives licence holders predetermined shares of the available catch and, as a result, has strengthened and more clearly defined access rights to the resource. But these rights are not property rights *per se* as the rights do not entail all of the attributes of pure property such as security, durability, exclusivity, transferability, etc. Fish are subject to the "rule of capture" whereby a fisherman does not have ownership to individual fish until the fish are caught. Accordingly, the rights of halibut licence holders are access rights rather than strict property rights.

There is a continuum of rights regimes with "open access" at one end and "pure property" at the other. The 1979 limited entry programme moved the industry away from open access and further across the spectrum. The IVQ programme of 1991, and its refinements since then, most notably the move to transferability, pushed the bundle of rights closer to the pure property end of the spectrum.

Nevertheless, the position of the Canadian Department of Fisheries and Oceans has always been that a commercial fishing licence is a privilege, granted annually, not a property right. The absolute right to issue, suspend, cancel and refuse issuance or reissuance of any licence is at the sole discretion of the federal Minister of Fisheries and Oceans.

However, DFO's actual behaviour is at odds with this stance given that: (a) halibut licences and quotas do trade in the open market at substantial sums without objection by DFO and (b) DFO has purchased halibut licences and quotas and then reissued the licence and quota as communal halibut "F" licences to aboriginal bands under their Aboriginal Fishing Strategy. The evidence is compelling that the rights of halibut licence holders do entail some of the key attributes of property, *i.e.* certain segments are excluded from use (*i.e.* exclusivity) and the rights can be sold (*i.e.* transferability).

The nature of rights with respect to security of tenure can also be changed through legislation. For example, Gislason (1999) indicated that individual quota holders for the Lake Winnipeg commercial fishery in Central Canada, subject to provincial, not federal, management, had their property rights entrenched in legislation as "the allocation of an individual quota entitlement to a fisherman...constitutes a property interest of the fisherman in a right to fish the specified quota".

The value of access rights to the fishery depends on several factors including: (a) the revenue potential of the fishery (expected catch and prices); (b) the costs of harvesting (normal returns to capital and labour, the costs of purchased inputs); (c) levies for management fees and economic rent charges; (d) the strength of rights and security of access to the resource which affects the business planning time horizon; and (e) other factors (*e.g.* government taxation policy re capital gains).

Reviews of the halibut IVQ programme by EB Economics (1992) and Turris and Sporer (1994) show that revenues increased and costs decreased because of the programme. These two benefits plus the value of stronger rights have more than compensated for the increase in industry levies. The result has been a substantial increase in the market or trading value of halibut licences in the 1990s. Presently halibut quota may sell for up to $C40/kg and the aggregate value of quota rights may be $C200 million or more. Not all of this amount represents a potential capital gain to existing licence holders as many have bought into the industry at substantial cost since 1991.

4.2 Higher fees

These stronger rights of halibut licence holders under IVQs have come at a cost. A condition placed by DFO on the move to IVQs in 1991 was that the industry fund all incremental costs associated with managing the fishery. The new "user pay" costs were substantial as a monitoring programme for individual catches/offloads had to be incorporated, five enforcement positions staffed and a management team put in place.

These "cost-recovery" fees have increased from $C761 000 in 1991 to an estimated $C903 000 in 1999. These fees comprise the deemed revenue of the "Halibut Program" on which the Halibut Advisory Board provides advice to DFO as to how to spend. However, Gislason (1999) noted the following:

i. HAB can only advise DFO as to how to spend the money. The money is DFO's money.
ii. DFO does not have a separate bank account or accounting system for the "Halibut Program". It normally does not provide a final financial statement of revenues and expenditures for each fiscal year.
iii. DFO enforcement officers and DFO management charge substantial time and salaries against the "Halibut Program" but these individuals, up to the end of 1998, did not keep formal time sheets. The HAB could not determine if the allocated costs borne by industry were reasonable.
iv. Since 1996 and irrespective of the size of the TAC, the halibut management levy of $C144.70/t has been fixed in regulation. It is virtually impossible, therefore, to match revenue generation in any year to needed programme expenditures, *i.e.* to be revenue-neutral.
v. Until recently any surpluses from the "Halibut Program" in a given year were not carried forward to the next year (the "Halibut Program" has generated a surplus in every year since 1991). The surpluses remained in general government revenue.

The result is that the HAB has had serious concerns about the accountability of the industry's licence fee contributions. The HAB and DFO are in on-going discussions as to the accountability issue.

The DFO introduced a "rent recovery" fee in 1996 to capture some of the private benefits accruing under restricted access to a public resource (Gardner Pinfold Consulting Economists Ltd. and GSGislason & Associates Ltd. 1999). The fact that industry pays both cost-recovery and rent-recovery fees gives the halibut licence holders a substantial say in fisheries management as discussed below.

4.3 Greater say – the last link in the trinity

The Halibut Advisory Board process has given halibut licence holders substantial say in fisheries management. The HAB helped to usher in the initial halibut IVQ Programme and was instrumental in the move to making licences transferable. HAB and DFO, without a formal legal agreement, have successfully practised co-management.

The HAB and the halibut industry have taken a greater interest in fisheries management, have extended their planning horizon, have co-operated with DFO and with each other and have accepted greater responsibility for their future. And, the interests of DFO and halibut fishermen have been more closely aligned. In essence, halibut licence holders have become "shareholders" not merely "stakeholders".

A key underpinning of the ascension to "shareholder" status has been the fact that industry pays both cost recovery and economic rent fees. Industry pays the full costs of IVQ programme management and in addition pays a return to the general public purse. Gislason (1995) has argued that fishermen are willing to pay more in fees if they receive a say in how the money is spent, and if they receive increased tenure, security or rights. The halibut fishery in British Columbia exemplifies this view.

Just as there is a continuum of rights, there is a continuum of accountability. The problems noted previously with the lack of financial control over industry levies and the limitations associated with the advisory role of HAB indicate that the halibut industry is constrained in its ability to achieve greater accountability.

To address this issue, the halibut licence holders created the Pacific Halibut Management Association (PHMA) in 1997 as a registered non-profit organization. The PHMA, in theory, will provide the institutional structure to further empower the industry and have a greater say over management of the halibut industry. The intent is to enter into a formal co-management agreement with DFO in the future, to receive industry cost-recovery levies directly and to authorize co-management spending. In recent years, DFO has provided "Halibut Program" surpluses as start-up money to the PHMA.

This PHMA initiative is consistent with the broad trend across Canada to institute industry co-management or partnerships through formal agreements (There is, however, confusion about the definitions of co-management and partnerships and how, if at all, they differ). DFO has indicated in its presentations to the Senate of Canada (1998) that the proposed amendments to the federal *Fisheries Act*: (a) help formalize the role of industry in decision-making ("greater say"); (b) share the costs of management ("higher fees"); and (c) provide greater security of tenure ("stronger rights"). That is, the move to partnering, or co-management, is consistent with the theme of this paper. In addition, the move to co-management blunts the criticism by Savoie (1998), Chairman of the Partnering Panel, and others that micromanaging has created a culture of paternalism in fisheries management in Canada. Co-management and accountability give industry greater say and responsibility for their future.

5. CONCLUSIONS

The Canadian experience with the Pacific halibut individual vessel quota (IVQ) programme demonstrates that the trinity of stronger rights, higher fees and greater say or accountability are inextricably linked. The move to an IVQ management system created much stronger rights and industry value, but also created new demands for catch monitoring, a dedicated enforcement presence and fishery management structure. The halibut industry embraced the "user pay" philosophy and agreed to pay for all incremental management costs at IVQ programme inception in 1991. Since 1996 the industry has contributed a resource rent to the general federal treasury. In return, the industry, through an Advisory Board, received a say in fisheries management. And the Board has been instrumental in modifying the IVQ programme design over time.

Under IVQs the industry became "shareholders" and not merely "stakeholders". Shareholders have rights, pay the price of admission to the decision-making table and, in return, have a say in how the entity operates. The halibut industry also has shown much greater interest in the management of the fishery and the long-term health of the resource, co-operated with one another and with the Department of Fisheries and Oceans and assumed much greater responsibility for their future. These desirable outcomes are inevitable from the move to industry co-management.

The Pacific halibut example also demonstrates that new institutional structures may be needed to give industry the accountability that they will expect and demand. In particular, industry input or say that is purely advisory and does not entail direct control over spending of their fisheries management fees may be deemed inadequate.

6. ACKNOWLEDGEMENTS

The author has benefited from discussions and information from several people, namely: Greg Clapp, John Secord, Chris Sporer, and Bruce Turris. Edna Lam provided valuable comments on a draft of the paper. Notwithstanding this assistance, the author has final responsibility for the analysis and conclusions of the study.

7. LITERATURE CITED

Bell, F.H. 1981. *The Pacific Halibut: The Resource and the Fishery*, Alaska Northwest Publishing Company.

Canada Fisheries and Oceans 1999. Pacific region integrated fisheries management plan: halibut.

EB Economics 1992. Evaluation study of individual quota management in the halibut industry, Evaluation report prepared for Canada Department of Fisheries and Oceans and BC Ministry of Agriculture, Fisheries and Food, November.

Gardner Pinfold Consulting Economists Ltd. and GSGislason & Associates Ltd. 1999. Cumulative impact of federal user fees on the commercial fish harvesting sector, Report prepared for Canada Fisheries and Oceans, March 1999.

Gislason, G.S. 1995. You pay, you say: an assessment of DFO's proposed new licence fee structure, Report prepared for Canada Fisheries and Oceans, April 1995.

Gislason, G.S. 2000. From social thought to economic reality: the first 25 years of the Lake Winnipeg IQ management programme. *In* Use of Property Rights in Fisheries Management. Proceedings of the FishRights99 Conference. Fremantle, Western Australia, 11-19 November 1999. FAO Fisheries Technical Paper No. 404/2. pp.118-126, FAO, Rome.

GSGislason & Associates Ltd. 1999. Halibut Advisory Board review, Prepared for Halibut Advisory Board, February 8, 1999.

International Pacific Halibut Commission 1998. *The Pacific Halibut: Biology, Fisheries and Management*, Technical Report No. 40.

Savoie, D.J. 1998. Chair, Partnering the fishery: report of the panel studying partnering, Report prepared for Canada Department of Fisheries and Oceans, December 1998.

Senate of Canada 1998. Privatization and quota licensing in Canada's fisheries, Report of the Senate Standing Committee on Fisheries, December 1998.

Turris, B. and C. Sporer 1994. Halibut IVQ program, *in* Experience with individual quota and enterprise allocation (IQ/EA) management in Canadian fisheries 1972-1994, Canada Fisheries and Oceans, November 1994.

Appendix 1
The evolution of property rights in the Pacific Halibut Fishery of Canada

Era		Fishery management regime
The unregulated fishery	1867	Canada becomes country under the *British North America Act*
	1871	British Columbia joins Dominion of Canada
	1888	Commercial halibut fishery starts, is open access and is unregulated
Regulation under the IPHC	1923	Formation of International Fisheries Commission by US and Canada to manage halibut fishery (by closed season)
	1930	New Convention signed to extend management tools to catch limits, gear restrictions
	1953	Commission changes name to International Pacific Halibut Commission (IPHC) and gains mandate to have separate seasons by management area
	1977	Both the US and Canada extend their respective coastal jurisdictions to 200nm
	1979	Amendment or protocol to the 1953 Halibut Convention calls for phaseout of the fishing of one country's fleet in the other country's waters
Limited entry	1979	DFO implements limited entry through "grandfathering" of licence holders (435 new "L" halibut vessel licences created).
	1988	Discussions between industry and DFO commence on individual quotas
	1990	Licence holders vote in favour of two year, trial Individual Vessel Quota (IVQ) Programme
The IVQ Program	1991	IVQ programme introduced
		– each of existing 435 licence holders gets a percentage share of the TAC where share based on 70:30 rule of catch history: vessel length
		– non-stackable for two years
		– industry pays much higher licence fees to pay for dedicated enforcement officers, a dockside monitoring programme (DMP), DFO management etc.
		– licence fee set at $C250 flat rate plus variable rate per tonne TAC (variable rate stabilizes at $C144.70 per tonne TAC in 1995)
		Halibut Advisory Board (HAB) consisting of halibut licence holders created
	1992	Programme review indicates support to make the temporary programme permanent and to allow temporary quota transfers
	1993	IVQs made transferable on a temporary (1 year) basis
		– each initial halibut quota split into two equal shares
		– licence holder can transfer 1 or 2 shares for the season
		– a licensed halibut vessel can have a maximum of four shares
	1996	New economic rent fee of $C220.30 per tonne of TAC introduced
	1999	Both permanent and temporary transfers of IVQs allowed subject to:
		– no one vessel having more than 1% of total TAC
		– each "L" licensed vessel maintaining a minimum amount of permanent IVQ of 0.1149% of the TAC (5% of the average initial allocation for the 435 licensed vessels). But the minimum can be temporarily reallocated for the year
	1999	Economic rent fee changed to $C310 per tonne of TAC (less a 40% credit up to a maximum of $C1000)
	1999	Partial on-board observer coverage of fleet

PROPERTY RIGHTS AND THEIR ROLE IN SUSTAINING NEW ZEALAND SEAFOOD FIRMS' COMPETITIVENESS

R. Bess
School of Business, Nelson Polytechnic,
Private Bag 19, Nelson, New Zealand
<rbess@nelpoly.ac.nz>

1. INTRODUCTION

The primary concern of the field of strategy is to determine how firms can acquire superior performance and the challenge for strategy researchers is to develop normative prescriptions on how firms can enhance their performance (Montgomery 1995). Gaining a competitive advantage remains a fundamental prerequisite to acquiring superior performance, while its absence is seen as the precursor to a firm's ultimate failure (Porter 1980). According to Hall (1993:610) firms have a sustainable competitive advantage 'when they consistently produce products and/or delivery systems with attributes which correspond to the key buying criteria for the majority of the customers in their target market'. The resource-based view argues that in highly competitive environments, these attributes, and the ability to align them with customers' key buying criteria stem from enduring firm-specific tangible and intangible resources. Strategy then becomes the art of creating value by reconfiguring new roles and relationships for those resources that really matter to a firm (Normann and Ramirez 1993).

There is growing awareness among contributors to the resource-based view that the most theoretically interesting variables are the least identifiable and measurable, eliciting increased interest in intangible resources, such as knowledge and interaction among individuals and groups (Godfrey and Hill 1995; Spender and Grant 1996). However, in the case of natural resource based industries, tangible resources, particularly secure property rights, can be a source of competitive advantage. Security of tenure in rights to natural resources could well be the fundamental basis to the development of firm-specific resources that in combination sustain a firm's competitive advantage.

This paper argues that in the case of New Zealand's seafood industry, the implementation of the Individual Transferable Quota (ITQ) system has been an important first step in the sustainable management of fisheries and the establishment of secure property rights, which provide the basis for firms' success in international markets. The security of rights to the fisheries resources and potential involvement in some fisheries management services provide individual firms with opportunities to enhance their competitiveness by reconfiguring value chain activities from harvesting to marketing. It is important to place this discussion on firm-specific resources within New Zealand's historical context. Beginning in the mid-1980s significant and rapid changes occurred in New Zealand due to the implementation of economic reforms and the transformation of the fisheries management system.

For this reason, following the discussion in Section 2 on the resource-based view, Section 3 outlines New Zealand's economic reforms, which began in 1984. The late 1980s and early 1990s have been referred to as a period of radical change; some refer to it as a revolution. It was during this period that the ITQ system was implemented. Section 4 describes the implementation of the ITQ system, including the legislation that brought the ITQ system into effect, the initial ITQ allocations, changes in quota ownership and subsequent legislative changes. Section 5 synthesises the statements made by several top managers of seafood firms about their firms' development and deployment of tangible and intangible resources to reduce reliance on commodity trading and enhance international competitiveness by offering value-added products and superior customer service.

2. THE RESOURCE-BASED VIEW

According to Barney (1991, 1995), tangible resources are considered unique and have a constrained supply while intangible resources consist of tacit knowledge or know-how that is culturally based, and therefore embedded within a firm, creating barriers to competitors understanding the source of advantage. Tangible resources, however, are viewed as becoming increasingly difficult to use as a basis for competitive advantage. Very few tangible resources have the uniqueness and supply limitations required to sustain an advantage, since their origin is typically from outside the firm. Therefore, the focus on firm-specific resources as a competitive advantage has remained primarily on intangibles, specifically the knowledge held by individuals within the firm, and the firm's ability to create and integrate knowledge into its production of economically viable products and services. Barney's (1991) VRIO framework, outlined in Figure 1, is perhaps the best description of an ideal firm-specific resource as a source of advantage. Barney suggests that heterogeneous and immobile resources that are rare, valued and embedded within the firm create barriers that impede competitors' ability to acquire, imitate and substitute the source of the firm's competitive advantage. Intangible resources come closest to meeting Barney's characterisation of the ideal resource.

According to Barney, managers must address four important questions about their resources and capabilities to understand internal sources of competitive advantage.

Figure 1
The relationship between resource heterogeneity and immobility, value, rareness,
imperfect imitability, substitutability, and sustained competitive advantage (Source: Barney

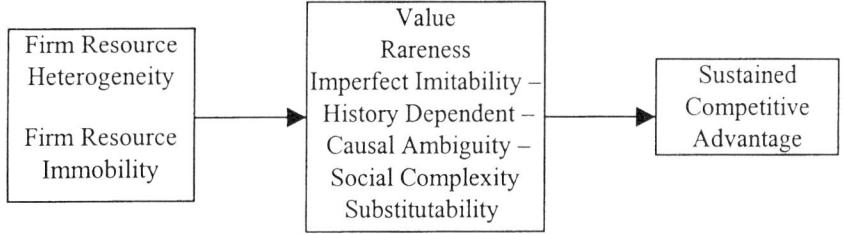

First, do a firm's resources and capabilities add value by enabling it to exploit opportunities and/or neutralise threats? The second question concerns rareness. A resource or capability is unlikely to provide a competitive advantage if numerous competing firms control it. And so the question is asked, how many competing firms already possess these valuable resources and capabilities? The third question addresses competitors' ability to imitate resources and capabilities that generate sustained competitive advantage. According to Barney, imitation can occur by duplication, where an imitating firm builds the same kinds of resources, and substitution of some resources for other resources. Imitation can be costly for competitors for three reasons: (a) the importance of history in creating firm resources; (b) the importance of numerous 'small decisions' in developing, nurturing, and exploiting resources; and (c) the importance of socially complex resources. A firm's competitive advantage not only depends on the value, rareness and inimitability of its resources and capabilities, but also on the firm's ability to exploit its resources and capabilities. The fourth question is then, is a firm organised to exploit the full competitive potential of its resources and capabilities? Barney's VRIO framework is addressed again in the last section.

3. NEW ZEALAND'S ECONOMIC REFORMS

The period of radical change, beginning 1984, should be seen in the context of New Zealand's early history and events leading up to the mid-1980s. In brief, New Zealand's economic and social environment began in the mid-1800s when New Zealand became a British settler state. In some ways, New Zealand's legacy of colonial dependence on Britain became permanent (Haworth 1994). New Zealand's economic links with Britain were the strongest and most lasting form of colonial dependence as Britain provided assured access to its markets for virtually all New Zealand export products, which historically have been primary product commodities. New Zealand's cultural and economic dependence on Britain remained in place until Britain's entry into the European Community in 1972. At that time New Zealand was thrust into the international trade arena while its economy remained strongly dependent on exports of primary product commodities. New Zealand's economic policies continued to subsidise primary product commodity exports and protect the growing domestic manufacturing sector from imports. High demand overseas for wool, meat and dairy products had produced euphoric dependence on agricultural exports (Carew 1987). New Zealand's exports had traditionally the highest concentration of commodities of all OECD nations, excluding Iceland (OECD 1983).

The unfavourable effects of New Zealand's prolonged reliance on a primary products-based economy began to surface as early as the 1960s. It became increasingly clear that New Zealand's continued reliance on agricultural exports, which were subsidised by the non-farming sectors, could not generate the earnings required to finance the nation's imports and the prices of protected domestic manufacturers' goods. The generous farm subsidies had translated into increased costs in land, equipment and services rather than increased rewards (Russell 1996). Low international prices for agricultural commodities resulted in the price of New Zealand's agricultural exports falling in real terms relative to the price of manufactured goods, causing the terms of trade to continually fall (OECD 1975). The New Zealand economy required significant structural changes before an economic recovery was possible and by the early 1980s the forces for change were strong. New Zealand's prolonged dependence on commodity exports and its continued use of central government controls prompted the Labour Government in 1984 to swiftly launch dramatic and sweeping economic reforms.

From 1984 to 1990 New Zealand experienced a redirection from its long history of centralised government control and an isolated economic system to a decentralised, market-based and outward-oriented economy. This period transformed New Zealand both economically and socially, and some refer to it as a time of revolution (Russell 1996). The major aim of the economic reforms was to revitalise the nation by removing subsidies and distortions while encouraging economic growth, efficiency and competition in a price stable environment (Carew 1987). All direct controls reintroduced by the National Government between 1982 and 1984 – plus previous policies on import quotas, subsidies and massive borrowing to sustain living standards – were reviewed by the Labour Government, which emphasised removing distortions and encouraging greater competition in the financial sector (OECD 1985). The Labour Government then implemented one of the most broad-based and rapid reforms of financial policy ever undertaken (Harper and Karacaoglu 1987). Conventional economic wisdom, however, holds that stabilisation of an economy should begin by reforms to the goods, trade and labour markets and continue with reform of the financial sector. Since financial markets adjust fairly quickly, when they are reformed first they reinforce distortions in those markets not yet reformed. Contrary to this conventional wisdom the Labour Government liberalised New Zealand's financial sector well ahead of other sectors and with as much speed as possible (McNelis and Bollard 1991). The Government took on a 'blitzkrieg' approach towards the reform process (Easton 1994), and with almost 'evangelical fervour', it set about redesigning the economic and social structure of New Zealand (Kelsey 1995). The swiftness with which the Government moved to change the economy reflected its view that the best solution was to aim straight for the cause of the problem rather than try to paste over the symptoms as had been done by the previous National Government (Carew 1987).

It was at this time that the Government approved implementation of the ITQ system. This climate of favouring market forces as the solution to economic and social issues strongly affected the options available for managing fisheries (Harding 1991). The mid to late 1980s was perhaps the optimal time period for the implementation of ITQ in New Zealand. Previous, and possibly subsequent, political and legal environments may not have approved the ITQ system. Beginning in 1990 the National Government acted swiftly to continue the momentum behind the economic reforms begun by the Labour Government. The National Government was resolved to address the welfare state from the start, acting quickly to introduce reforms to employment relations, social assistance, education and health care. However, the introduction of the Mixed Member Proportional (MMP) system in 1996 led to a slowdown in the reform process. The next national election was due in November 1999.

During the first phase of economic reforms only a minority of firms were able to emerge beyond the 'survival' mode of adaptation to the market-driven environment, while most firms required more time to respond effectively to the tougher market conditions (Campbell-Hunt, *et al.* 1993). Evidence of economic recovery and transformation beyond the survival stage of adaptation was not apparent until the early 1990s. By the mid-1990s some firms displayed evidence of specialising in aspects of the value chain where they had some competitive advantage and becoming internationally competitive (Campbell-Hunt and Corbett 1996). While the seafood industry experienced dramatic growth during the late 1980s and early 1990s, its export focus had to contend with the effects of finance sector liberalisation, such as severe fluctuations in interest rates and exchanges rates and elimination of import restrictions.

'There is quantitative evidence ... [that] New Zealand now has perhaps the least-interventionist or, equivalently, the most market-based economy in the OECD' (Lloyd 1997:118). New Zealand has been a world leader in implementing policies that have accelerated the closure of inefficient, uncompetitive firms, industries and government businesses. However, New Zealand has not had the same success in developing new economic growth. The Government's marginal changes to economic policies have not increased the nation's rate of innovation and investment necessary to create new growth industries (Alexander 1999). Arguably, the seafood industry is an exception. However, Section 4 demonstrates, the legislation surrounding the ITQ system continues to bring uncertainty and change that affects the industry's growth potential.

4. IMPLEMENTATION OF THE ITQ SYSTEM
4.1 Legislative effect to ITQ system

The New Zealand seafood industry experienced several problems initially in implementing the ITQ system. Initially, disagreement existed over the level of consultation required by the industry and the Ministry of Agriculture and Fisheries (MAF). Disagreement made the task of setting annual total allowable commercial catches (TACCs) a laborious process. Further, the legislative framework of the time rendered the new management system difficult to operationalise. Delays on critical issues, such as compensation to the industry for initial reductions in TACCs, led the industry to file a $NZ150 million lawsuit against the Government in October 1989. The lawsuit was later suspended as negotiations improved with a change in Government and a new Minister of Agriculture and Fisheries (FIB 1990).

Implementation of the ITQ system began with the *Fisheries Act 1983*, which first introduced significant changes to the fisheries management administrative system and statutory framework. The 1983 Act introduced individual quota allocated under regulations to participants in the seven main deepwater fisheries. The 1983 Act also outlined a framework for regional fisheries management to conserve the fish stocks, promote commercial and recreational fishing, limit access to fisheries and provide for optimum yields from fisheries (Cunningham 1983). The 1983 Act remained focused on regulations that limited access to fisheries to reduce catching effort, which had increased during the previous two decades due to Government implementing a regulated open entry system to encourage greater domestic participation. A permit scheme was implemented which led to a dramatic reduction in the numbers of part-time fishers, removing 2260 permits. This accounted for almost half of the commercial fishers in the early 1980s (Harding 1991). The 1983 Act did not provide any means of compensation for those exiting the industry. The permit scheme was not intended as a long-term management control. The 1983 Act was intended to implement a new regime that utilised long-term planning in controlling commercial fishing effort. The MAF divided the 200-mile Exclusive Economic Zone (EEZ) into ten fisheries management areas (FMA) with each area having its own set of controls. Long-term fisheries management control and planning was introduced by way of Fisheries Management Plans (FMP) for each FMA.

After consultation with the industry through public meetings, the National Fisheries Management Advisory Committee (NAFMAC) recommended to the Minister of Agriculture and Fisheries implementation of an ITQ system in combination with FMP (NAFMAC 1983). While some individuals and groups in the fishing industry pushed the Government to implement ITQ, initially there was not unanimous industry support. The Fishing Industry Board submitted a report (FIB 1984) outlining the industry's view on ITQ at that time, which emphasised that: ITQ would not be appropriate for some fisheries; that the administration system, including required documentation, should be kept as simple as possible; and ITQ should be allocated for a minimum of ten years so that investments could remain secure.

It was with some risk that the Labour Government, after lengthy consultation with the industry, considered the implementation of the ITQ system. It is important to recall that, to date, private property rights had been applied to fisheries management in theory only, so implementation of the ITQ system was an extreme departure from current fisheries management regimes throughout the world. The Fisheries Amendment Bill, which gave legislative effect to the ITQ system, encountered relatively little resistance in Parliament and became law on 25 July 1986. Perhaps the Fisheries Amendment Bill's radical nature was its attraction during this time of radical change.

The intended functions of the 1986 Act were: (a) to control the quantity of fish extracted from fisheries to sustainable levels by way of TACs (total allowable catch) and TACCs; (b) to maximise benefits from the fisheries to the nation by creating appropriate economic incentives for investment in fisheries, including the implementation of the ITQ system which would bring about rational industry restructuring; (c) to allocate ITQ to quota holders, and for quota to be a fully tradeable or leasable property right; (d) to maintain an efficient government-based monitoring system to keep track of catch against quota; and (e) to allow quota holders the right to catch up to their quota at any time during the fishing year, thus removing the 'race for fish' (Shallard 1997). However, the 1986 Act retained several aspects of the *Fisheries Act 1983*, including the Ministry of Agriculture and Fisheries, now the Ministry of Fisheries (MFish), retaining the power to impose various input controls such as restrictions on fishing gear, fishing methods, landings, fish size, fishing seasons, and fishing areas. These input controls were still required for management of non-ITQ species. The traditional input controls implemented under FMP, as required under the 1983 Act, contradicted the basis of ITQ where quota owners are able to determine the most efficient timing and means of catching their quota. Under FMP, however, a TACC managed in part with input controls could potentially impinge on quota owners' rights as created by the ITQ system. The need to run dual management systems has led to inconsistencies in management practice, and has made it more difficult to achieve the ITQ system's intended degree of efficiency and co-ordination (Falloon 1993).

4.2 Initial allocation of ITQ

Since the Government had already set in place an informal quota arrangement for some deepwater fisheries, beginning in the late 1970s, ITQ was first implemented for deepwater fisheries. In 1982 eleven firms were allocated quota which was then transferred to ITQ in 1986 (Clement and Pfahlert 1996). The industry supported the implementation of deepwater ITQs as there was concern that these fisheries could be quickly fished to destruction (Falloon 1993). Annual quota and harvest rights were allocated by means of 'the level of domestic investment, quantity of deepwater catch that had been supplied for onshore processing, onshore investment, and the extent to which this investment was committed to the processing of deepwater species' (Sharp 1997:510). The setting of initial TACs for the over-fished inshore fisheries was estimated conservatively, set between 25 to 75% lower than the 1983 catch levels (Sissenwine and Mace 1992). The quota for inshore fisheries were allocated with a provisional maximum assessment of quota based on each qualifying permit holder's catch history of the best two out of three years: 1981/82 1982/83 and 1983/84. However, in many cases the initial allocations of quota were significantly below fishers' catch histories resulting in substantial losses incurred by the industry (Clement and Pfahlert 1996). The initial allocation of quota was made in specific tonnage with the Government intending to buy and sell quota on the open market as a means of adjusting required changes in TACCs.

It is of interest to note that the Government retained ownership of substantial amounts of quota at the start of the ITQ system. In 1986 the TACCs totalled 520,901t: 60.8% of the TACCs, or 316 769t, was allocated to 1472 permit holders; 49% of the ITQs, or 255 241t, went to the 12 largest seafood firms. The Government retained the remaining 39.2% of ITQs, 204 132t. The Government has sold most of its quota by way of competitive tender on the open market. As early as December 1986, the Government sold 140 183t of ITQ, primarily hoki and orange roughy quota, for $76.6 million. In addition, the tender sale was conditional on the purchaser agreeing to lease the quota from the Government for five years and pay an additional annual lease payment to the Government, with ownership transferred at the end of the five years (Clement and Phahlert 1996).

Since the initial implementation of ITQ, the quota holding profile has changed considerably. The late 1980s and early 1990s was a period of consolidation in the industry. Some seafood firms with large quota holdings exited the industry while others purchased quota and other assets. These changes were a natural outcome of the large reductions in TACCs made during that time, particularly for the deepwater hoki and orange roughy fisheries (FIB 1990). Another reason given for these changes was the intention by some seafood firms to concentrate on 'core business activities' (Reorganisation, August 1991). Table 1 outlines the larger quota holdings at 1986 compared with ownership at 1991, 1996 and 1999. The seafood firms listed in Table 1 include their subsidiaries.

In 1991 Sealord Products Ltd held 136 180t of quota, 24.1% of the overall quota; Sanford Ltd held 93 972t of quota, 16.6%; and Amaltal Fishing Co Ltd/Talleys Fisheries Ltd. became the third largest non-government quota holder with 65 953t, 11.7%. The Government remained the largest quota holder with 185 420t, 32.7% of overall quota. By 1996 Sealord Products Ltd's overall quota holdings increased to 145 433t, 25.5%; Sanford Ltd's quota holdings increased to 115 298t, 20.2%; and Amaltal Fishing Co Ltd's quota holdings fell to 56 118t, 9.8%. This consolidation of ITQ holdings among the top three firms is mostly concentrated in the deepwater fisheries, which reflects the substantial investment necessary to efficiently harvest these fisheries (Sharp 1998). By 1996 the Government had reduced its quota holdings to only 874t, 0.2% of overall quota. The Government's substantial quota holdings were sold on the open market and allocated to the Treaty of Waitangi Fisheries Commission in accordance with the *Māori Fisheries Act 1989* and the *Treaty of Waitangi Settlement Act 1992*, which are described later. The Treaty of Waitangi Fisheries Commission became the third largest quota holder with 56 624t, 9.9% of overall quota. By 1999 the three largest quota holders remained unchanged. Sealord Products Ltd held 149 462t of quota, 22.1% of overall quota. Sanford Ltd held 141 243t of quota, 20.9%, and the Treaty of Waitangi Fisheries Commission's quota holdings increased to 72,235t, 10.7%. Amaltal Fishing Co Ltd increased its quota holdings to 62 333t, 9.2%.

Initial expansion of the deepwater fisheries relied heavily on joint venture partnerships and charter arrangements. The domestic fleet lacked the the larger vessels and technology to fish deepwater fisheries. The United Nations Law of the Sea Convention requires New Zealand to allow foreign licensed vessels to fish within the EEZ in the event New Zealand-controlled vessels cannot catch the annual TACC. In 1986, 18% of the catch from the EEZ was taken by foreign licensed vessels and by 1993 New Zealand-controlled vessels, including joint ventures and charters, took 99.8% of the catch (Dynamic Year 1993). As some New Zealand seafood firms expanded their efforts into the deepwater fisheries they purchased their own vessels. The 'New Zealandisation' of the fishing fleet was needed to meet firms' objective to further develop the deepwater fisheries and process product at sea to improve quality and add value.

Table 1
ITQ ownership at 1986 compared with 1991, 1996 and 1999
(Sources: FIB 1990; Clement and Pfahlert 1996; MFish 1999)

Quota Owners	December 1986		April 1991		August 1996		June 1999	
	tonnes	%	tonnes	%	tonnes	%	tonnes	%
Fletcher Fishing Ltd	56 675	10.9						
Sealord Products Ltd	55 796	10.7	136 180	24.1	145 433	25.5	149 462	22.1
Sanford Ltd	49 412	9.5	93 972	16.6	115 298	20.2	141 243	20.9
Amaltal Fishing Co Ltd	25 204	4.8	58 117	10.3	56 118	9.8	62 333	9.2
Skeggs Investments Ltd	19 432	3.7						
Independent Fisheries Ltd	13 622	2.6	19 032	3.3	27 815	4.9	37 224	5.5
Wanganui Trawlers Ltd	12 273	2.4	17 073	3.0			1 358	0.2
Wattie Fishing Ltd	8 887	1.7						
South Island Deepwater Fishing Ltd	6 207	1.2	7 343	1.3				
Southfish Co-Operative Ltd	4 101	0.8	9 968	1.8	8 836	1.5	685	0.1
United Fisheries Ltd	1 149	0.2			19 397	3.4	15 048	2.2
Talleys Fisheries Ltd	83	0	7 836	1.4	11 950	2.1	16 536	2.4
Moana Pacific Quota Holding Ltd					7 189	1.3	7 568	1.1
Simunovich Fisheries Ltd			5 106	0.9	6 553	1.1	7 284	1.1
Vela Fishing Ltd			14 782	2.6	27 863	4.9	31 839	4.7
Crown	204 132	39.2	185 420	32.7	874	0.2	9 303	1.4
Aotearoa Fisheries Ltd			3 956	0.7				
Treaty of Waitangi Fisheries Commission					56 624	9.9	72 235	10.7
Other	63 927	12.3	7 439	1.3	87 439	15.2	124 866	18.4
Total TACC	520 900	100	566 224	100	571 389	100	676 984	100

4.3 Subsequent changes to fisheries legislation

Although the New Zealand ITQ system is regarded as one of the most innovative and successful fisheries management options in the world, the industry overall views subsequent legislative changes as having resulted in a complex, bureaucratic administration system that causes the industry to incur expensive compliance costs. The industry contends that the increasing complexity and bureaucracy of the QMS has imposed unnecessary financial costs on individual fishers and fishing firms and has not provided the industry overall with corresponding benefits (Horton 1997). Recently, the fishing industry 'angry about the red tape and delays that they said were strangling their industry ... [joined in a] flare-waving protest ... before they came ashore in Lambton Harbour and marched on Parliament' (Anon. 1999:3). The level of frustration within the industry over fisheries management bureaucracy explains the industry's interest in adopting management alternatives that would simplify and preserve the integrity of the original QMS, as outlined in the 1986 Act. This section briefly outlines some of the legislative amendments which include addressing claims by indigenous Māori, an issue that remained unresolved until after the introduction of the ITQ system.

A significant number of part-time fishers were excluded from the initial allocation of quota, many of whom were Māori. The ITQ system also excluded any reference to Māori fishing rights, which Māori argue were secured under the Treaty of Waitangi 1840. It should be noted that, at the time the ITQ system was implemented, a growing resurgence in Māori culture and language and awareness of Māori rights under the Treaty was occuring. While the ITQ system initially prompted indigenous claims to large areas of fisheries, it also proved an effective means of resolving Māori fishing rights claims (Sullivan 1998).

Soon after the ITQ system was implemented, Māori obtained a series of injunctions issued by the High Court against further ITQ allocations. Following protracted disputes between Māori and the Government, the *Māori Fisheries Act 1989* was passed. This Act was considered an interim settlement, which required the Government to buy back and transfer 10% of TACCs to the Māori Fisheries Commission before 31 October 1992. The Commission would administer fishery assets on behalf of Māori. The *Treaty of Waitangi Settlement Act 1992* was intended to be the full and final settlement of all Māori fishing rights claims as secured under the *Treaty of Waitangi 1840*. The *Settlement Act*, otherwise known as the 'Sealord deal', resulted in substantial assets, primarily as quota holdings and half ownership of Sealord Products Ltd., being allocated to Māori. The *Settlement Act* also included the Crown recognising that (a) commercial fishing was important to Māori, (b) some coastal areas were of significance to Māori for customary food gathering and (c) Māori would participate in the Government's fisheries management processes. Henceforth, the taking or possession of fish by Māori was to be in accordance with the *Fisheries Act 1983* or any further regulations.

The Māori Fisheries Commission was reconstituted as the Treaty of Waitangi Fisheries Commission Te *Ohu Kai Moana* to administer both pre and post-settlement assets, and was empowered to devise, in consultation with Māori, a scheme to distribute its pre-settlement assets pursuant to the *Settlement Act 1992*. Since the *Settlement Act* also provides that Māori gain 20% of the quota holdings for all new species placed under the QMS, Māori will continue to have a major influence in the industry's development.

The *Fisheries Amendment Act 1990* brought about perhaps the most important change to the QMS, the basis of ITQ from a fixed tonnage to a proportional basis (Luxton 1997). During the late 1980s the MAF fisheries scientists became concerned about the level of some fish stocks, which led to large reductions in some TACCs, particularly for orange roughy. The inherent fluctuations in fish populations and uncertainty of stock assessments could not ensure a constant amount of quota by tonnage from one year to the next. Any reduction in TACCs would require that the Government repurchase quota on the open market. To avoid substantial outlays for quota repurchasing, the Government implemented a swap of quota from quota management areas (QMAs) where the fish stocks were stressed to QMAs where the fish stock remained plentiful. The MAF announced that effective, 1 October 1989, ITQ would change from a specified or fixed tonnage per year to a proportion of the TACC adjusted each year for sustainability measures. The implications of this change were that the MAF avoided substantial costs to repurchase quota from ITQ owners and from that point on could vary TACs and TACCs each year, with no compensation to ITQ owners.

The *Fisheries Amendment Act 1994* implemented a replacement for resource rentals, which were payments made by ITQ owners that went towards paying some of the costs of fisheries management. The Government implemented a full cost recovery levy on 1 October 1994 with the intention that resource users would pay the full cost of fisheries management and research. The cost recovery levy fit within the MFish's intention to simplify its administration by focusing on the delivery of core services and devolving non-core services to the private sector. The MFish's core services would include the allocation of harvest rights, liaising and disputing resolution, enforcement and prosecution, while all other services would in time become contestable. It was envisioned that relevant stakeholder groups, through consultation, would determine needed non-core services, and those who benefitted from the services would then pay for them. The Government proposed these changes with the view that it could not deliver the needed flexibility, responsiveness and diversity the industry required. The intended outcome would be lower costs for the MFish's services, which would then be paid in full by the industry by way of full cost recovery (Kidd 1994).

However, the cost recovery regime has imposed a cost of $NZ37 million on the industry for 1998/99, which includes $NZ7 million for the implementation of the *Fisheries Act 1996*. The Minister of Fisheries has called for a review of the 1996 Act to, in part, reduce business compliance costs.

The Fisheries Act 1996 brought about several significant changes. Several sections of the 1996 Act were in line with the 1992 Fisheries Task Force's recommendation that fisheries management adopt an 'ecosystem approach' to ensure the sustainability of the environment as well as fish stocks. This macro-management approach intends that the industry accept more responsibility for managing resource use while consulting with relevant stakeholders. The Minister of Fisheries' intentions were exemplified in his speech at the 1997 Fishing Industry Conference.

> 'It [Fisheries Act 1996] provides for more explicit environmental standards and gives further opportunities for the users of the fisheries to accept increasing responsibility for managing the resource... The new Fisheries Act makes considerable advances in issues relating to sustainability, expanding the opportunities for stakeholder participation in fisheries management and in better defining the role of Māori' (Luxton 1997:4).

Several sections in the 1996 Act clearly increase the Minister's authority to implement various regulations that could impact significantly on ITQ owners' ability to choose when, where and how they might catch their quota. While the MFish states that these 1996 Act sections are needed to ensure the sustainable management of fisheries and the environment and meet the Government's Waitangi Treaty and international obligations, the industry could argue that they 'attack at the heart of the security of tenure and the property rights created by the QMS' (Chapman, *et al.* 1997:8). Further, the Minister has urged the industry to accept more responsibility for managing fisheries by preparing for the devolution of some fisheries management services. The industry has responded by restructuring around property rights, with the resulting associations expecting to have a more direct involvement in and more responsibility for the management of their respective fisheries. At the 1997 Fishing Industry Conference the Minister stated:

> 'It is my clear view that we have reached a point in the development of fisheries management in New Zealand when it is vital that the fishing industry begin to assume a far greater level of responsibility to collectively manage fisheries within appropriate sustainability parameters. To progress co-management strategies it is necessary for you as an industry to begin to develop effective associations of users to assume the duties and responsibilities associated with property rights.

> My challenge to you is to continue to develop such associations so I can work with you to advance further the management models currently available to us and thus ensure a healthy future for your industry' (Luxton 1997:5).

The main reasons for the industry having responded favourably to this challenge were first, the industry acknowledged that the functions undertaken by the former Fishing Industry Board (FIB) and some trade associations duplicated efforts and led to unnecessary complications. Second, industry growth, recent investments and cost recovery have led quota owners to specialise their planning and operations along 'associations of users', as encouraged by the Minister. The industry has developed the Seafood Industry Council (SeaFIC) which took over several generic services and functions formerly undertaken by the FIB and some functions that had been the responsibility of various industry associations. It is the intention of SeaFIC to change its offering of services as associations, referred to as quota owning companies (QOCs), take increasing responsibility for providing their own administration, research, compliance, consultation, and development of management plans.

Despite the fishing industry's efforts to restructure into 'effective associations of users', the 1996 Act contains legislation that inhibits the industry from assuming 'the duties and responsibilities associated with property rights', as encouraged by the Minister (Luxton 1997:8). The industry view is that fisheries legislation must also reflect co-management principles before the industry can commit and invest further in the QOCs. The Minister agrees with this view:

> 'As it stands, the 1996 Act is very centralised and prescriptive... [and] the Act would have been expensive to implement and inflexible to manage, and may have led to poor management and environmental outcomes... In addition, the 1996 Act does not allow for the devolution of non-core Government fisheries services (approved in principle by Cabinet), it lacks a robust cost recovery scheme and provides few incentives for fisheries rights holders to take a constructive role in managing their share of the fisheries resource. Aquaculture rights are uncertain and recreational rights are poorly defined' (Luxton 1998:2).

The Minister initiated a review of the 1996 Act which concluded that there is an immediate need to amend aspects of the Act and its administration. The Reviewer recommended:

i. a fundamental realignment of the roles of Government and fisheries stakeholders and the implementation of transparent consultation and decision-making processes

ii. a simplified and less prescriptive operating regime than currently exists and

iii. devolving to fisheries rightholders the responsibility for fisheries management at the discretion of the Minister.

Some of the recommended amendments to the Act include:

i. enabling the Minister to devolve fisheries management functions to rightholder groups

ii. enabling the Minister to approve fish stock management plans developed by representative and accountable quota owner associations and

iii. providing for regulations designating appropriate specifications and standards for devolved management functions and the elements required in a fish stock management plan (Pricewaterhouse Coopers 1998).

As this section demonstrates, the transformation of New Zealand's fisheries management system to an ITQ system has not been without its contradictions to previous management regimes, which has rendered the ITQ system's intended degree of efficiency and co-ordination more difficult to achieve. Furthermore, subsequent legislative changes have, at times, conflicted with industry views, requiring both the industry and the MFish to spend considerable time and other resources in consultation to amend legislation. It should be noted that as seafood firms have focused on developing their international competitiveness, they have had to expend considerable resources adjusting to radical and contradictory changes in Government economic and fisheries management legislation. Section 5 below briefly outlines the firm-specific resources, tangible and intangible, that New Zealand seafood firms have developed and deployed to enhance and sustain their international competitiveness.

5. FIRM-SPECIFIC RESOURCES

There is general recognition that the QMS has improved the biological status of the fisheries resource and commercial return to fishers and fishing firms (Annala 1996). Overall, the seafood industry has experienced dramatic growth in the volume and value of production. However, since around 1993 the value of production has remained fairly constant. The gradual appreciation of the New Zealand dollar from 1992 to 1997 exacerbated poor international trading conditions and reduced returns to seafood firms. During this time catch levels declined due to reductions in some TACCs (FIB 1996). It is beyond the scope of this paper to analyse the financial performance of the industry and individual firms. Such analyses are difficult to undertake due to the lack of aggregated industry data, particularly during the last few years, and the commercial sensitivity of information on privately held firms. However, financial analysis based on historical data has little direct relevance when trying to assess a firm's strategic health. For example, low profitability measurements may, in fact, provide valid evidence of a strategically healthy firm seeking to make long-term investments. As mentioned, several seafood firms have made significant capital investments in vessel capability and quota holdings to enhance their ability to meet international market demand.

This section synthesises comments made by several top managers of seafood firms during interviews on the development and deployment of various resources that in combination act to enhance individual firms' competitiveness. As mentioned, researchers have been noting the increasing importance of intangible resources since tangible resources are viewed as becoming increasingly difficult to use as a basis for a sustained competitive advantage. Few tangible resources have the uniqueness and supply limitations required to sustain an advantage. However, in the case of some New Zealand seafood firms, several top managers have consistently stated that secure property rights by way of the ITQ system are the fundamental basis to their firms' international competitiveness. As mentioned, 49% of the initial ITQ was allocated to the 12 largest seafood firms. ITQ's transferability allows these firms the option to increase and alter the makeup of their quota holdings to better suit their strategic direction and international markets. Furthermore, firms' security of tenure in transferable property rights has provided incentives for investments in the catching and processing sectors. As the industry experienced dramatic growth, several firms made substantial capital investments in vessels and processing facilities, reducing their reliance on charter and joint venture arrangements. At the same time, several firms have focused on developing high-quality and high-value products and on improving customer relations in international markets.

By now most medium- and large-sized seafood firms are highly vertically integrated, having the majority if not all of their value chain activities within the boundaries of the firm. This ability to align virtually all activities in the value chain, (catching/harvesting, intermediate and final stage processing, sales and distribution and after-sales service) within the firm provides opportunities to focus on each activity and the best possible links between them. Some vertically integrated firms have developed these value chain activities and linkages between them, so they are, according to Barney's (1991) VRIO framework, valuable, rare and costly-to-imitate, factors which can enhance and sustain competitiveness when firms are organised to exploit the full competitive potential of their resources. These valuable, rare and costly-to-imitate intangibles are predominately the relationships firms have with their customers. It is understandable that since the initial allocation of ITQ, individual firms have concentrated first on their own value chain activities and how they are linked. Historically, there has been little evidence of inter-firm linking of value chain activities. However, there are now more reasons for some firms to consider inter-firm links: increased competitive pressures; acknowledged benefits of co-operative efforts, such as reducing costs and resource enhancement, as displayed by the Challenger Scallop Enhancement Company. And, the industry has already restructured along property rights, which is a first step towards co-operation. Co-operative efforts and more involvement in fisheries management by resource users is often referred to as co-management, defined by Jentoft, *et al.* (1998:426) as:

> '... a social system that changes the nature of the game, the relationships between players and what each of them strives for. Co-management means an ongoing collaborative and communicative process, where resource users and other actors are in an entrepreneurial and creative role'.

Assuming that a government has already converted fishers' catches into explicit annual numerical entitlements, such as ITQ, the first question to ask is, 'what aspects of fishing can self-governing groups actually control? The answer is: 'almost everything ... except measures or regulations to protect the size of the stock by restricting effort or the catch' (Scott 1993:189). According to Pinkerton (1989:8), 'sharing responsibility for enhancement is an excellent starting point for more comprehensive co-management'. Once new relationships are established it is easier to extend co-operation to other fisheries management functions, such as improving the quality of data and data analysis, reducing excessive investments by fishers in competitive gear, and reducing conflict between government and fishers and conflict among fishers' groups. From a strategic perspective, it is conceivable that groups of resources users in an 'entrepreneurial and creative role' could potentially develop sources of competitive advantage with, to use the VRIO framework, imperfect imitability due to their social complexity and causal ambiguity.

Figure 2 outlines the value chain and VRIO frameworks for a vertically integrated New Zealand seafood firm, which reflects the intention that firms and their respective associations have responsibility for some fisheries management services, including, to some extent, research and enhancement. Figure 2 shows the importance that secure tenure in tradeable property rights has for firms' establishing value chain activities and the links between them. According to the VRIO framework, the secure rights to fisheries resources are valuable, rare and costly for competitors to imitate, for two reasons. First, firms that received the initial ITQ allocations have higher barriers to imitation, since allocated quota was not purchased at market rates, as would be required of would-be competitors. Second, security of rights to fisheries resources remain rare elsewhere in the world, providing New Zealand firms with an advantage. Secure property rights provide firms with options for when, where and how to catch/harvest the fisheries resources, adjusted for sustainability measures and management of non-ITQ fisheries. Secure property rights provide ITQ owners with appropriate incentives to invest in specialised fishing vessels and gear that reflect the expected return on investment. In a vertically integrated firm, investment in vessels and gear are considered valuable for the purpose of ensuring supply and meeting quality standards, but these investments are not rare or costly to imitate because they generally lack uniqueness and supply limitations. Firms have also made substantial capital investments in processing facilities to ensure that product is processed quickly and reliably, and that processes incorporate innovations developed within the firm. The ability to process product quickly and reliably is valuable to firms, but it is not considered rare or costly to imitate. Some in-house innovations, however, have increased the value and speed of processing while increasing the difficulty and cost competitors have in imitating the process. At the same time, some firms have developed consistency of product and quality while improving flexibililty in meeting varied product specifications. Again, this ability is potentially of great value in meeting customer requirements, however, it is not rare since competitors can imitate this ability at similar cost.

Some firms' sales and distribution activities have developed speed and competence in meeting customer requirements, not only in terms of product requirements but also in terms of agreed delivery times and locations. This activity is considered valuable, rare and costly to imitate since it is linked back to secure rights to fisheries resources and the linking up of all downstream value chain activities. Upstream value chain activities are further enhanced with the application of appropriate and relevant research undertaken in collaboration with various research institutes both in New Zealand and abroad. The combining of these various tangible and intangible resources provide firms with the ability to build enduring customer relationships, which are valuable, rare and costly to imitate. Several top managers have noted that their customer bases have remained fairly constant as firms have been able to consistently meet their requirements for products, delivery and service and after-sales service. In so doing, seafood firm managers have built up relationships over the years that have been improved with reciprocal visits to each others' locations to understand better customer needs and how to meet them. Customer feedback has been vitally important in aligning all upstream actitves and has the potential to influence the start of the value chain, fisheries enhancement programmes. It is the combination of all the value chain activities and their links that have brought about the success of some New Zealand seafood firms, and this success is based on the secure rights to the fisheries resources. Further, a potentially highly valued source of competitiveness could come about with legislative changes that provide legitimacy to the co-operative efforts of resource user associations. These efforts could enhance the sustainability of fish stocks and result in developing innovative sources of advantage that sustain firms' international competitiveness.

Figure 2
Value chain and VRIO frameworks applied to a vertically integrated New Zealand seafood

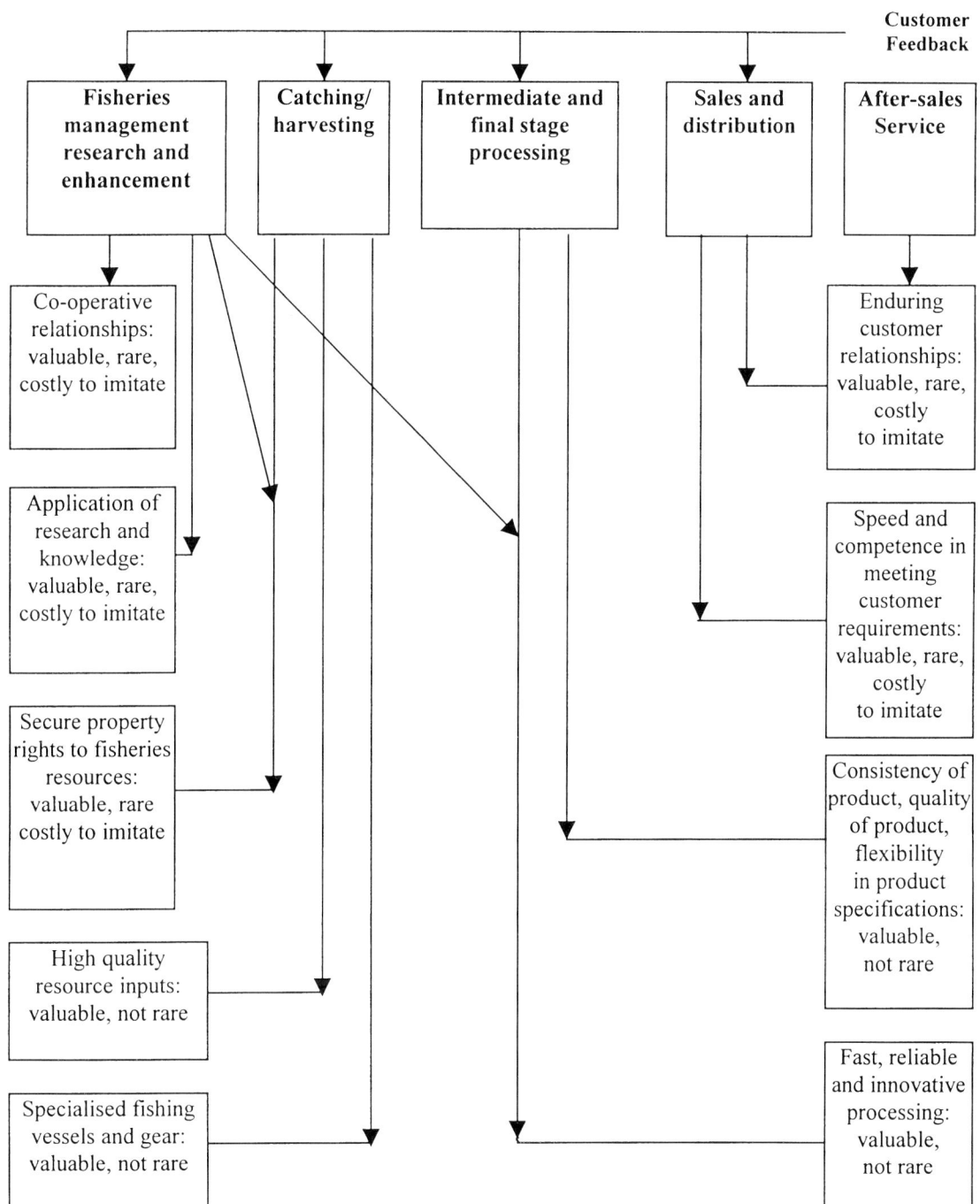

6. CONCLUSION

As this paper demonstrates, the history of New Zealand's seafood industry shows continuous challenge and change. Now that the ITQ system has been operative for 13 years, the property rights it established remain secure and irreversible. The ITQ system clearly has the support of Government which has frequently held it up as the principal example of successful application of property rights to fisheries management and one of the most innovative approaches to managing wild fish stocks. Government support was expressed by the Minister of Fisheries in his speech to the United Nations Food and Agriculture Organization in Rome on 11 March 1999. 'The most advanced form of fisheries property rights so far developed is the individual transferable quota, or ITQ' (Luxton 1999:1). The consultative process concerning the next Fisheries Act will be carried into the new millenium. The industry anticipates that the new Act will take steps toward resolving the cost to the industry for fisheries management and research services, reaching agreement on issues surrounding Māori, recreational and marine farming rights, and changing legislation to reflect co-management principles. It appears that the time is right for further development of the New Zealand fisheries management system by the adoption of co-management principles. If upcoming legislation fails to progress the implementation of co-management principles, the industry remains intent on simplifying the QMS by reducing administrative and compliance costs and providing fishers with appropriate economic incentives, while ensuring the sustainability of the fisheries. If upcoming legislative changes establish co-management principles, the next significant change to the QMS could well be the development of innovative and co-operative efforts among ITQ owners and other property rights holders and stakeholders. Although several New Zealand's seafood firms have already demonstrated enhancement of value chain activities, co-operative efforts could benefit individual firms by developing innovative and creative ways of reconfiguring value chain activities. As resource users organise to exploit the full competitive potential of their property rights, other value chain activities can be developed to decrease costs, increase quality, improve product development, marketing and distribution. Other benefits may arise through resource users' involvement in some fisheries management services that help ensure the sustainability of the fisheries through collaborative research and stock assessment efforts and enhancement programmes. The combination of these efforts could assist the industry to build further on the greater certainty and security of supply that is provided them by the ITQ system, more than is the case in most other nations.

7. LITERATURE CITED

Alexander, T. 1999. We need to take more risks. *The Dominion*, 8, 4 August 1999.

Annala, J.H. 1996. New Zealand's ITQ system: have the first eight years been a success or failure? *Reviews in FishBiology and Fisheries*, 6, 43-62.

Anon. 1991. Reorganisation in New Zealand. *Seafood International*, 13, August 1991.

Anon. 1999. Fishing flotilla brings protesters to parliament. *The Dominion*, 3, 26 August 1999.

Barney, J.B. 1991. Firm resources and sustained competitive advantage. *Journal of Management*, 17 (1), 99-120.

Barney, J.B. 1995. Looking inside for competitive advantage. *Academy of Management Executive*, 9 (4), 49-61.

Campbell-Hunt, C., D.A. Harper and R.T. Hamilton 1993. *Islands of Excellence? A Study of Management in New Zealand.* Research monograph 59, New Zealand Institute of Economic Research: Wellington.

Campbell-Hunt, C. and L.M. Corbett 1996. *A Season of Excellence? An Overview of New Zealand Enterprise in the Nineties.* Research monograph 65, New Zealand Institute of Economic Research: Wellington.

Carew, E. 1987. *New Zealand's Money Revolution*. Allen and Unwin: Wellington.

Chapman, Tripp, Sheffield and Young (Barristers'& Solicitors' Office 1997. *Overview of the Provisions of the Fisheries Act 1996.* Report prepared for the New Zealand Fishing Industry Board: Wellington.

Clement, G. and J. Pfahlert 1996. Changes in quota ownership 1982-1996. *Seafood New Zealand*, 4 (10), 18-21, Nov. 1996.

Cunningham, B.T. 1983. Regional management. In J.L. Taylor and G.G. Baird (eds.) *New Zealand Finfish Fisheries: The Resources and Their Management*, Trade Publications Ltd.: Auckland, 67-69.

Dynamic Year 1992 – A dynamic year for the seafood industry. *New Zealand Professional Fishermen*, 7-13, July 1993.

Easton, B. 1994. How did the health reforms blitzkrieg fail? *Political Science*, 46 (2), 215.

Falloon, R. 1993. Individual transferable quotas: The New Zealand case. In *The Use of Individual Quotas in Fisheries Management*, 44-62, OECD: Paris.

FIB 1984. *Individual Transferable Quotas (ITQs): Their Feasibility and Possible Administration in the Fishing Industry.* New Zealand Fishing Industry Board: Wellington.

FIB 1990. *New Zealand Fishing Industry Economic Review*. New Zealand Fishing Industry Board: Wellington.

FIB 1996. *The New Zealand Seafood Industry Economic Review*. New Zealand Fishing Industry Board: Wellington.

Godfrey, P.C. and C.W.L. Hill 1995. The problem of unobservables in strategic management research. *Strategic Management Journal*, 16, 519-533.

Harding, R.J. 1991. *New Zealand Fisheries Management: A Study in Bureaucratisation,* Unpublished PhD thesis: Victoria University of Wellington.

Harper, D. and G. Karacaoglu 1987. Financial policy reform in New Zealand. *In* A. Bollard and R. Buckle (eds.) *Economic Liberalisation in New Zealand.* Allen and Unwin/Port Nicholson, Press: Wellington.

Hall, R. 1993. A framework linking intangible resources and capabilities to sustainable competitive advantage. *Strategic Management Journal*, 14, 607-618.

Haworth, N. 1994. Neo-liberalism, economic internationalisation and the contemporary state in New Zealand, *in* B. Sharp. (ed.) *Leap into the Dark: The Changing Role of the State in New Zealand Since 1984.* Auckland University Press: Auckland

Horton, C. 1997. *Open Letter to Minister of Fisheries.* The New Zealand Fishing Industry Board: Wellington.

Jentoft, S., B.J. McCay and D.C. Wilson 1998. Social theory and fisheries co-management. *Marine Policy*, 22 (4-5), 423-436.

Kelsey, J. 1995. *The New Zealand Experiment: A World Model for Structural Adjustment?* Auckland University Press: Auckland.

Kidd, D.L. 1994. Speech introducing the 1994 Fisheries Bill to Parliament. *In* D.C. Sharp: speech given at the 1997 Fishing Industry Association Conference. Plaza International Hotel: Wellington.

Lloyd, P. 1997. [Review of the book A Study of Economic Reform: The Case of New Zealand, (eds.) B. Silverstone, A. Bollard and R. Lattimore, Amsterdam: North-Holland 1996]. *New Zealand Economic Papers*, 31 (1), 115-131, June 1997.

Luxton, J. 1997. Speech given at the 1997 Seafood Industry Conference, Plaza International Hotel: Wellington, 14-17 May 1997.

Luxton, J. 1998. Media Release. Office of Hon. John Luxton, Parliament Buildings: Wellington, November 1998.

Luxton, J. 1999. Fisheries speech given at the United Nations Food and Agriculture Organisation: Rome, 10 March 1999.

McNelis, P. and A. Bollard 1991. *From Financial Indulgence to Fiscal Repentence: Chile, Ireland, and New Zealand in the 1980's*. Working Paper 91/14, New Zealand Institute of Economic Research: Wellington.

MFish 1999. The New Zealand Ministry of Fisheries: Wellington.

Montgomery, C.A. 1995. Of diamonds and rust: A new look at resources. *In* (ed.) C. A. Montgomery *Resource-Based and Evolutionary Theories of the Firm: Towards a Synthesis*. Kluwer Academic Publishers: Boston. pp. 251-268.

NAFMAC 1983. *Future Policy for the Inshore Fishery: Discussion Paper*. New Zealand Ministry of Agriculture and Fisheries: Wellington.

Normann, R. and R. Ramirez 1993. From value chain to value constellation: Designing interactive strategy. *Harvard Business Review*, 71 (4), 65-77.

OECD 1975. *Economic Surveys: New Zealand*. Organisation for Economic Co-operation and Development: Paris.

OECD 1983. *Economic Surveys: New Zealand*. Organisation for Economic Co-operation and Development: Paris.

OECD 1985. *Economic Surveys: New Zealand*. Organisation for Economic Co-operation and Development: Paris.

Pinkerton, E. 1989. Introduction: Attaining better fisheries management through co-management – prospects, problems, and propositions. *In* E. Pinkerton (ed.) *Co-operative Management of Local Fisheries: New Directions for Improved Management and Community Development*. University of British Columbia Press: Vancouver. pp. 3-33.

Porter, M. 1980. *Competitive Strategy*. Free Press: New York.

PricewaterhouseCoopers 1998. *Fishing For the Future: Review of the Fisheries Act 1996*, Report of the Independent Reviewer of the Fisheries Act 1996 to the Minister of Food, Fibre, Biosecurity and Border Control: Wellington, September 1998.

Russell, M. 1996. *Revolution: New Zealand From Fortress to Free Market*. Hodder Moa Beckett: Auckland.

Scott, A. 1993. Obstacles to fishing self-government. *Marine Resources Economics*, 8, 187-199.

Shallard, B.D. 1997. Concepts and practice of individual transferable quotas for the management of fisheries – An overview. *Proceedings of the 2nd World Fisheries Congress* 1996 at pp. 391-395, CSIRO.

Sharp, B. 1997. From regulated access to transferable harvesting rights: Policy insights from New Zealand. *Marine Policy*, 21 (6), 501-517.

Sharp, B. 1998. Fishing. *In* (eds.) M. Pickford and A. Bollard *The Structure and Dynamics of New Zealand Industries*. The Dunmore Press: Palmerston North. pp. 53-85

Sissenwine, M.P. and P.M. Mace 1992. ITQs in New Zealand: The era of fixed quota in perpetuity. *Fisheries Bulletin US*, 90, 147-160.

Spender, J.C. and R.M. Grant 1996. Knowledge and the firm: Overview. *Strategic Management Journal*, 17 (Winter Special Issue), 5-9.

Sullivan, M.S. 1998. *Introduction to Brooker's Fisheries Law*. Brooker's Ltd.: Wellington.

WHY RECOVER COSTS? COST RECOVERY AND PROPERTY RIGHTS IN NEW ZEALAND

N. Wyatt
Ministry of Fisheries
PO Box 1020, Wellington, New Zealand
<wyattn@fish.govt.nz>

1. INTRODUCTION[1]

A number of countries have suffered fisheries management failures, while their governments were supposedly managing them. Some observers (*e.g.* Kaufmann and Geen 1997) have speculated that one reason for these failures is that governments do not face the correct incentives to provide or purchase the services that will lead to efficient fisheries management outcomes. They suggest that introducing a cost-recovery regime that recovers the costs of fisheries services from commercial rights-holders, will make the fishing industry take a greater interest in the quality of the service provided and exert pressure to ensure more efficient fisheries management.

My thesis is that the outcome will not depend on the existence of cost-recovery so much as the principles underlying the regime. States may use cost-recovery to achieve a number of objectives:

i. Efficiency (*e.g.* to motivate rights-holders to demand cost effective services)
ii. Equity (*e.g.* to reduce subsidies to the fishing industry)
iii. Fiscal (*e.g.* to help balance the Government's books)
iv. Social (*e.g.* to maintain employment levels in coastal communities) and
v. Compensation (*e.g.* to obtain a portion of the value of the resource for the owners).

The last of these objectives a resource rental issue rather than one of cost-recovery, as charges should reflect the value of the resource, not the cost of managing it. This list does demonstrate that each objective is likely to have a different solution for optimising the achievement of that objective. Trying to achieve several objectives at once is likely to result in a sub-optimal solution, which may be worse than having no cost-recovery regime at all.

2. A HISTORY OF COST RECOVERY

New Zealand introduced a cost-recovery regime in 1994. Prior to that only a "resource-rental" had been charged generating approximately $NZ20 million per annum. Two central components of the Government's decision to increase the returns from fisheries were a desire to ensure that the fisheries industry fairly bore the costs imposed by commercial exploitation of the fishery and a desire for the community to receive a return from the use of the fishery resource.

The first of these desires was to be achieved by recovering from the industry all expenditure that arose as a consequence of the existence of commercial fishing. In addition, costs that were jointly shared between the commercial and non-commercial sectors would be fully recovered from the industry "in recognition of the degradation of non-commercial values as a result of commercial exploitation". (New Zealand Officials Committee 1993) This became known as the "avoidable cost" principle. It seems most closely related to the "equity" objective noted above, with the recovery of joint costs containing a measure of the "compensation" objective. It was estimated at the time that this would recover $NZ53 million. The second desire would be achieved by retaining the resource-rentals of $NZ20 million although these were subsequently dropped.

Under the current *Fisheries Act 1996*, decisions on both fisheries management measures (*e.g.* setting the Total Allowable Catch) and the nature and extent of fisheries services (*e.g.* stock assessment research) are made by the Minister on an annual basis. Costs are recovered according to a regulation setting out the proportion of the costs of each activity carried out by the Ministry of Fisheries that is to be met by the Crown. There is no rationale given for the proportions in the regulation; they appear to be the result of a "gentlemen's agreement" between the Government and the industry made some years ago.

In practice, the nature and extent of fisheries services are generally determined so as to keep the total cost to both taxpayers and the fishing industry within reasonable bounds, current costs are close to $NZ23 million and $NZ39 million respectively. In other words, fisheries services and cost-recovery decisions are now made primarily to meet fiscal constraints. Further, the fisheries management decisions are made through a statutory process that was separate from decisions on fisheries services. Thus, there was little association between fisheries management decisions and the costs recovered from commercial rights-holders.

There has been little evidence to suggest a marked improvement in the cost-effectiveness of fisheries management in New Zealand resulting from cost-recovery. The industry assures the Government that it is not doing any better, while the Ministry does not collect the sort of information that might be used to judge how effectively it is managing. And alas, there has been an increase in hostility between the Government and the industry. Both

---

[1] The views in this paper are those of the author and not necessarily those of the Ministry of Fisheries or the New Zealand Government.

sides have also incurred high costs in the consultation and decision-making processes. Consultation on the nature and extent of fisheries services is increasingly driven by

cost-recovery rather than by fisheries-management considerations as the industry lobbies for more taxpayer-funded services and fewer industry-funded services. The industry, after all, does not see any close link between the costs faced by rights-holders and the benefits they receive from the services provided, which are still determined by the needs of the Government rather than the needs of rights-holders.

3. REVIEWING THE COST RECOVERY REGIME

There have been several reviews of the cost-recovery regime in recent years. These have focussed more on such issues as whether the "avoidable cost" principle is better or worse than the "attributable cost" principle, under which rights-holders are charged according to the benefits that the Government considers they have received from fisheries services. The reviews have focussed less on how the services to support property rights might be more cost-effectively managed.

A recent amendment (*Fisheries Amendment Act 1999*) to the *Fisheries Act 1996* has improved some aspects of the cost-recovery regime. It removed the separate statutory process for determining fisheries services and removed the recovery of costs of services already purchased by rights-holders. It also provided a number of principles that were to underpin any future rules for the recovery of costs, which could then be set under regulation.

The principles contained in the amended Fisheries Act are (slightly paraphrased):

i. if a service is provided at the request of an identifiable person, that person must pay a fee for the service
ii. the costs of services provided in the general public interest, rather than in the interest of an identifiable person or class or person, may not be recovered
iii. the costs of services provided to manage or administer the harvesting or farming of fisheries resources must, so far as is practicable, be attributed to those who benefit from harvesting or farming the resources
iv. the costs of services provided to avoid, remedy or mitigate the risk to, or an adverse effect on, the aquatic environment or its biological diversity, must so far as practicable, be attributed to those who cause the risk or adverse effect and
v. the Government may not recover the costs of services provided by another organisation to which it has delegated responsibility for the service.

These principles provide some guidance, but are still flexible enough to allow considerable variation in the amounts that are recovered. For example, the first principle does not specify if the fee for a requested service should recover the full cost, only the marginal cost, or the provision of the service. The "attribution" of costs in the third or fourth principles does not mean that the persons identified must pay the costs, only that there is a transparent process for deciding whether or not they should pay. They should, however, ensure that the reason for providing a service is identified, rather than, for example, whether someone who might benefit from the service can be identified.

In developing new rules for splitting the costs of fisheries services between the Government and the industry, the Government was concerned that the rules would also be consistent with imminent Treasury guidelines for cost-recovery in the public sector. The Treasury guidelines for setting charges in the public sector were published in early 1999 (New Zealand Treasury 1999). Six objectives for cost-recovery were identified that addressed efficiency, equity and fiscal issues:

i. encouraging decisions on the volume and standard of services demanded, consistent with (a) the efficient allocation of resources generally and (b) the outcomes the government is seeking
ii. minimising the cost of supply over both the short and long-term when capital costs are significant
iii. keeping transaction costs low, and keeping evasion of user charges at acceptable levels
iv. reducing reliance on funding from general taxation (with its associated costs)
v. dealing equitably with the taxpayer, those who benefit from the output, and, or, those whose actions give rise to the output and
vi. looking for new ways to lower costs and find appropriate providers.

The guidelines noted that the assessment of cost-recovery options will often involve a trade-off between these objectives when they point in different directions. Thus, the guidelines could not "set out to be definitive; rather, they provide a check-list of issues on which to base a sound analysis" (New Zealand Treasury 1999).

When the Ministry of Fisheries looked at this "checklist" of objectives, it realised that the objectives relating to efficiency could not be applied, because the Government made all the decisions. The Government established a Joint Working Group, made up of officials from the Ministry of Fisheries and the Treasury and representatives of the fishing industry to develop a proposal for new rules for splitting the costs of fisheries services between the Government and the industry. The rules were to be consistent with the principles in the amended Fisheries Act, the Treasury guidelines and an earlier Government decision on the core roles of the Government and of rights-holders in fisheries management.

The core roles of the Government were considered to be:

i. to ensure sustainability

ii. to meet the Crown's Treaty of Waitangi and international obligations
iii. to enable efficient resource use in the fisheries sector including the better specification of the rights of recreational fishers and aquaculturalists so they are comparable to the rights of commercial and customary fishers and
iv. to ensure the integrity of fisheries management systems including criminal enforcement, setting standards and specifications, monitoring and auditing fisheries management plans and the delivery of services in consultation with rights-holders, environmentalists and the general public.

The core role of fisheries rights-holders was considered to be managing their harvesting rights within the sustainability and management frameworks determined by the Government.

4. REFORMING THE COST RECOVERY REGIME

The Joint Working Group quickly realised that looking at the cost-recovery rules in isolation would not address the fundamental problems of the cost-recovery regime. Two reforms were necessary: first, to the way in which decisions were made on the purchase of fisheries services, and second, how the costs of those services were to be split.

The Government should continue to purchase, or provide, the services that are its core responsibility, but rights-holders should be allowed to purchase other non-core services at their own expense so long as the services delivered satisfy the quality standards that allow the Government to deliver on its core mandate. While non-core services are purchased, or provided, by the Government, their costs should be recovered from rights-holders to enable them to make rational decisions as to whether they could more cost-effectively purchase the non-core services. They are in a better position to make this judgement than the Government is. The Government should fund the provision of core services, including policy advice, supporting international arrangements, criminal enforcement and prosecutions. It is the party that determines the precise nature and extent of those services. Recovering some of the costs of core services would leave the Government open to industry pressure to scale down the provision of those services which were cost recovered, and to pressure on Treasury to scale down those services not cost recovered.

The Joint Working Group's recommendations thus fell into two groups: (a) changes to the processes under which services are determined and delivered and (b) new cost-recovery rules reflecting the changes to the processes. The new cost-recovery rules and the amended levies calculated from them are now in the process of being put into effect. Reforming the way in which service purchase decisions are made could prove a bigger challenge. Bureaucracies that have developed around centralised decision-making can find it difficult to change their focus. But a start has been made; the Ministry of Fisheries has begun a review of its own systems and processes to see whether they need to be re-aligned to meet the demands of the new environment.

5. CONCLUSION

Cost-recovery is not, in itself, a solution to fisheries management problems. It may even make things worse. If one believes that governments are best at managing fisheries, cost-recovery has little to commend it. Resource-rentals may be a preferable approach in these circumstances. If a government receives revenue for the use of a resource and meets its own management costs from that revenue it will have some incentive to control the costs, as it gets to keep the remainder.

If one believes that rights-holders are better at managing fisheries, cost-recovery can be a useful tool, but only if the objective of the cost-recovery regime implemented is to enable rights-holders to manage their rights effectively. The worst of all worlds is the one in which the Government manages fisheries and recovers the costs from rights-holders. Government then has no incentive to manage efficiently. This is the world from which New Zealand has begun to emerge.

6. LITERATURE CITED

Kaufmann, B. and G. Geen 1997. Cost-recovery as a fisheries management tool, *Marine Resource Economics*, 12, 57-66.

New Zealand Officials Committee 1993. Report to Ad Hoc Ministerial Fisheries Review Committee.

New Zealand Treasury 1999. Guidelines for setting charges in the public sector.

FISHER OBLIGATIONS IN CO-MANAGED FISHERIES: THE CASE FOR ENFORCEMENT

J.P. McKinlay* and P.J. Millington**
* Fisheries Western Australia, WA Marine Research Laboratories
PO Box 20, North Beach, Perth W.A. 6020
<jmckinlay@fish.wa.gov.au>
** Fisheries Western Australia
Locked Bag No.39, Cloisters Square Post Office, Perth W.A. 6850
<pmillington@fish.wa.gov.au>

1. INTRODUCTION

Fisheries management is a difficult juggling act between protecting resource sustainability, ensuring equitable resource access between competing user groups and promoting economic efficiency in exploitation of the resource. How to best achieve these objectives continues to be a topic of vigorous debate (Hannesson 1996; Stephenson and Lane 1995, Caddy 1999), but most analysts agree that formal or informal property rights go a long way toward ameliorating the problems of open access to common resources, particularly with regard to resource over-exploitation and non-compliance with fishery rules. Much of this paper focuses on the relative merits of input controls (restrictions on effort) versus output controls (catch quotas)[1]. National fisheries resources in Australia, like many developed countries worldwide, are generally regulated by one or both of these approaches, namely (a) access rights – typically access is granted to particular fishing areas or to specific stocks, through licensing systems and (b) harvesting rights – quota systems allow particular fishers to remove specific amounts of fish from a stock, also through licensing systems.

Both approaches establish a continuing legal fishing right for those involved and such rights are thought advantageous since they may promote responsible resource-use among participants. Specifically, if fishers have a guaranteed stake in the future of a fishery resource there is an expectation that they will actively work to ensure the continued sustainability of the resource. This process is enhanced when fishers perceive management goals and fishing regulations to be fair, equitable, and necessary to maintain the biological integrity of the resource. To this end, resource stewardship is thought to be greatly enhanced when fishers are actively involved in the management process, both in terms of setting management priorities and designing fishery rules. In Australia, the principles of co-management and cooperative management are well established and fishers in many states provide substantial input into the management process (Exel and Kaufmann 1997, Fisheries WA 1998). Fishers also contribute financially to the management process, especially in Australia's most valuable fisheries where a large proportion of management expenses are cost-recovered from industry through licence fees (Sutinen 1994, Penn et al. 1997). As a result, the costs of management, research and compliance are closely monitored by industry and fisheries agencies are increasingly required to justify their expenditure. Under this financial incentive it is little wonder that fishers themselves are entering into the property rights debate. In this paper we propose that property rights, combined with co-management, create an atmosphere in which fishers can be encouraged to assist management agencies in the enforcement of fishery rules. Further, we believe that stewardship may be fostered under either access rights, or harvest rights approaches to management, and that fisher cooperation with compliance personnel is equally likely under either system. First, however, we make some brief comments about the merits of quota systems, since these systems are often (perhaps unjustifiably) promoted as the prime method for engendering resource stewardship.

Proponents of catch quotas suggest this management system optimises economic efficiency and fosters resource stewardship among fishers (Hannesson 1996). They claim that allocating a proportion of the total allowable catch[2] among a limited number of fishers protects the value of capital investment in the fishery. This most often takes the form of individual transferable quotas (ITQs) (Grafton 1996), although other systems involving individual transferable effort units have been used in Western Australia (WA) fisheries for more than three decades (Bowen 1994). Other approaches, such as the share-based fishing-rights system in the New South Wales fisheries of Australia (Young 1995), are emerging. Catch quota approaches are thought to be important because many poorly-managed input systems of management have resulted in over-capitalisation, over-fishing, and the collapse of fisheries world-wide, typically because proper attention was not given to latent effort and efficiency increases. Responsible resource-use is apparently better engendered under ITQs since fishers own a continuing right to harvest a specific amount of fish, and as such will fish responsibly to ensure the ecological health of their resource. For instance, Walters and Pearce (1996) suggest that ITQ's, particularly when rights are long-term, encourage fishers to participate in the research and enforcement processes since their input is reflected in the future value of the catch. Others criticise the use of catch quotas, arguing that the often deleterious social consequences that such systems have on fishing-dependent communities are unacceptable (Rennie 1998, Davis

---

[1] Other types of property rights, such as territorial use rights, privatisation of use rights, and community property rights are not considered in this paper; see Symes (1998) for a review of these regulatory systems.

[2] The total allowable catch (TAC) is determined from biological information, but is often greatly influenced by socio-political circumstances (Hutchings et al. 1997).

1996). Fishers themselves are often critical of quota management for this reason, claiming that market-driven allocation of property rights inevitably leads to rationalisation and concentration of ownership that progressively forces "traditional" fishers from the industry (McCay 1995, Charles 1992). This is important, since it would appear *prima facie* that responsible resource-use under ITQs is most likely to occur among owner-operators, and may not be as easily encouraged among contract skippers who hold no property rights themselves. Fishers also often lament that fisher competence becomes secondary under ITQs, although it is questionable whether the incentive to compete amongst one another is removed, possibly leading to increased incentives to under-report catch. Evidence for increased compliance with conservation measures under ITQs appears mixed, with practices such as high-grading, discarding, and under-reporting of catches common in some fisheries (McCay 1995, McCay *et al.* 1995, Grafton 1996, Rennie 1998). The misreporting of commercial catch may have serious consequences for determining biologically appropriate TACs (Walters and Pearce 1996).

Illegal fishing practices are critical to any debate about the relative merits of input versus output approaches to fisheries management. Under an assumption of optimum enforcement of fishery rules and near-perfect compliance by fishers, quota-management appears an attractive solution to the problem of simultaneously promoting resource sustainability and economic efficiency[3]. However, this is rarely the case in any fishery, making the issues of enforcement and compliance central to both sides of the ITQ debate. It is generally recognised that enforcement procedures and costs may be substantially different between the two approaches to management (O'Boyle and Zwanenburg 1997, McLaughlan 1994). While enforcement costs typically account for a large proportion of any management budget, experience has shown that effective enforcement under quota management may be far more expensive than under effective effort control schemes (Buck 1995; McLaughlan 1994). It is therefore surprising that discussion of enforcement under ITQs is not more prevalent in fisheries literature than it otherwise appears. There are perhaps three reasons for this. First, non-compliant behaviour is notoriously difficult to measure with any certainty and evidence of illegal activity is often difficult to gather, anecdotal, or gathered after the fact. Second, the shift in enforcement resources needed to accommodate quota monitoring may mean that other areas of potential non-compliance (such as at-sea fisher behaviour) may be neglected. Third, fishers and management agencies may be reluctant to advertise the fact that non-compliance with fishery rules may have increased under quota management, or that enforcement procedures and capabilities may be ill-equipped (or under-equipped) to deal with illegal activity. Proponents of quota management often suggest improved compliance is a primary reason for considering quotas in the first place; this argument is questionable if evidence suggests that compliance problems and enforcement costs increase under quota management.

The issue of quota management is of particular interest in Western Australia, where there has been much discussion within management and industry about the relative merits of introducing such a system to the Western Australian rock lobster fishery (Bowen 1994). While there is no clear consensus among professional fishers on the value of quota management, a recent industry survey has shown that a majority of fishers are opposed to any change from the current individual transferable effort (ITE) management arrangements (The Marketing Centre 1996, see also Davis 1996 for a discussion of the opposition to quota management in the Maine lobster fishery). This begs the question: would compliance with regulations be better under quota management? In fact, do catch-based property rights in the long term provide fishers with any increased incentives for implementing conservation measures than those existing under well-managed effort-based access rights? If the proponents of quota management are to be believed, the answer is yes. However, there is little information to substantiate claims that either effort or catch-based property rights should be favoured because of their ability to encourage responsible fisher behaviour. Theoretically, responsible behaviour can be engendered under either system of management, particularly when fishers are provided with substantial input to the management process (Hønneland 1999).

This paper therefore seeks to address the issue of responsible resource-use among fishers and fisher involvement in the enforcement process in rights-based fisheries. We do not presuppose that effort controls or quota management necessarily engenders more or less resource stewardship, but rather that under any management system small numbers of fishers will regularly break regulations regardless of real or perceived incentives for resource conservation. We examine the role of compliant fishers, who usually form the majority in any fishery, in assisting with the enforcement of regulations within their fishery. Fisher organisations frequently advocate increased management responsibility for their members, but seldom extend this request to the realm of enforcement. Increased management involvement for the fishing industry also brings responsibilities, one of which is a need for fishers, individually and collectively, to participate in the enforcement of rules within their fishery. We outline ways in which fisher-involvement in enforcement should be encouraged, around the management table and on the water. We draw many of our examples from our experience with the Western Australian rock lobster fishery, but the ideas we advocate could apply to many input or output controlled fisheries.

2. THE WESTERN AUSTRALIA ROCK LOBSTER FISHERY

2.1 Overview

The Western Australian rock lobster (*Panulirus cygnus*) fishery is Australia's most valuable single-species fishery, with annual commercial catches currently ranging between 9000 and 13000t for a ex-vessel value of $A200-

---

[3] This also presupposes that quotas are set on the basis of near-perfect biological information and that they may be adjusted in a timely fashion as updated biological information is obtained.

$A300 million. In 1997/98 this corresponded to around 39% of the total value of Western Australian fisheries production, and approximately 11% of the total value of national fisheries production (ABARE 1998). The fishery is estimated to have a total capitalisation of $A2 billion, with market values for individual fishing operations of $A2 and $A3 million. Live lobster exports to Japan, Taiwan, and China account for the majority of the value of the catch (Marec Pty Ltd 1997).

The fishery is managed as a limited-entry individual transferable effort fishery; *i.e.* individuals control rights to use a certain number of lobster pots, and these may be bought and sold among existing fishery participants. There are currently no direct output controls in the form of an annual TAC, the management instead controls the exploitation rate to ensure escapement of animals to maintain the breeding-stock. Additional management measures comprise a variety of restrictions, including closed seasons, spatial closures, limits on the total number of pots fished by an individual fisher, gear restrictions, size-limits on lobster, and breeding-stock protection. Many of these restrictions also apply to recreational fishers. In the 1998/99 season there were around 600 licensed commercial vessels fishing a total of 56 800 pots over 1600km of coastline. There is a substantial recreational sector of approximately 32 000 licensed participants, with annual catches of 5-10% of total commercial landings. Recreational fishers may dive, or use up to two pots to take rock lobster, with a maximum daily catch limit of 8 lobsters per licence-holder. Entry to the commercial fishery has been limited since 1963 (Hancock 1981), and there are currently no restrictions on entry to the recreational fishery.

The management measures for the rock lobster fishery have largely succeeded in limiting fishing pressure so that the latent effort has been removed and the fishery is biologically sustainable. Declining breeding-stock indices and predictions of environmentally driven low puerulus-stage settlement in the early 1990s prompted the introduction in 1993/94 of a management plan designed to boost breeding-stock levels. These changes were designed to increase egg production to the level it was in the late 1970s and early 1980s (*i.e.* about 25% of the estimated virgin levels), and indications are that this target has largely been achieved (Donohue 1998). The fishery is economically sustainable, with resource-rents maintained at around $A30 million annually (Lindner 1994). In 1995/96 the commercial rock lobster fishery began operating on the basis of partial cost-recovery from license fees, with full cost-recovery expected to be implemented by 2001/02. Enforcement costs are currently around $A2 million per year, accounting for approximately 50% of the total costs recovered from industry. Research accounts for the second largest expenditure at 20-30% of total costs (Fisheries WA 1999).

2.2 Co-management

In recent years fisher participation in the management process has become commonplace in many fisheries around the world (Wilson *et al.* 1994, Nielsen and Vedsmand 1999). Australian fisheries are no exception, with the establishment of many fisheries of management advisory committees comprised of representatives from major resource stake-holders, the management authority and relevant scientific bodies. In Western Australia stakeholder membership of these committees is drawn mainly from the commercial catching and processing sectors, with some representation from the recreational sector of the fishery. Advisory committees are usually charged with making recommendations about issues affecting the fishery to the government Minister responsible for fisheries. Although final responsibility rests with the Minister, to a large degree management advisory committees steer the future direction of many of the management, research and marketing activities in their fishery.

There is a long history of industry participation in management of the Western Australian rock lobster fishery. In 1966 the Fisheries Act was amended to establish the Crayfish Industry Advisory Committee, an expert body charged with providing advice and recommendations about matters affecting the rock lobster fishery[4] (Anon 1968). This committee comprised a Chair (appointed by the Minister), fisher representatives, scientific advisors, and one person to represent interested parties not engaged in the commercial fishing industry. Over the last 30 years this committee has undergone many changes to bring it to its present form, the Rock Lobster Industry Advisory Committee (RLIAC). The RLIAC was established under the *Fish Resources Management Act 1994*, and is an expertise-based statutory advisory committee with a membership of eight commercial fishers, two rock lobster processors, the Executive Director of Fisheries WA (plus one other staff member) and an expert in recreational fishing. Fisheries WA research and management personnel are granted observer status at RLIAC meetings to provide advice, as is a representative from the Western Australian Fishing Industry Council. An independent Chair is appointed by the Minister and a full-time Executive Officer runs the day-to-day business. The RLIAC itself establishes a number of sub-committees, each chaired by a RLIAC member, which include additional industry and Fisheries WA representation. These include the Marketing Research, Finance, Compliance and Research Subcommittees.

In addition to identifying issues that affect rock lobster fishing and providing advice and recommendations to the government Minister responsible for fisheries, the RLIAC itself has defined its role as:

> '*to primarily provide advice for the sustainability of the rock lobster resource, taking into account the social, economic and other implications of its advice on the benefits or otherwise derived from the fishery.*'

Information is disseminated to the fishing industry by a variety of mechanisms including popular magazines,

---

[4] Prior to 1966 a Fishermen's Advisory Committee, which included rock lobster industry representatives, provided advice on all Western Australian fisheries to the Minister responsible for fisheries.

discussion papers, management reports, newsletters, and coastal tours where fishers are provided with up to date fishery information from fisheries scientists and access to managers and RLIAC members. The success in implementing many of the current management measures in the rock lobster fishery attests to the overall success of the consultative and collaborative relationship that exists between Fisheries WA and the fishing community.

2.3 Enforcement and compliance

The fishery operates over a wide geographic area ranging between latitudes 21°44'S and 34°24'S. Most commercial and recreational fishing occurs within 60 and 2 nautical miles of the coast. Fisheries WA employs a total of 85 fisheries officers to service the State's fisheries and of these, 45 officers are actively involved in enforcing rock lobster regulations. Fisheries Officers use three large patrol vessels (greater than 20m), 8 small patrol vessels (up to 8 m), and make trips aboard commercial fishing vessels. At-sea inspections mainly check for gear and licence breaches while land-based inspections check catch composition. A Special Investigations unit of between five and ten officers investigates serious fisheries offences.

Compliance with fisheries regulations is generally high (Donohue 1998), although for some fishery rules the extent of non-compliance is difficult to measure with confidence. One reason for high compliance, at least with respect to catch regulations, is that more than 90% of commercial catches are processed through rock lobster processing factories. Since there are only a limited number of processing factories it is relatively easy for fisheries personnel to monitor a large proportion of the total catch. Another reason for high compliance is the "three strikes" regulations. These rules enable the licences of fishers to be suspended for serious fisheries offences, with the possibility of complete licence-cancellation for three serious offences ("black marks") in a 10 year period. The high value of rock lobster licences means this law provides a substantial deterrent against serious breaches of fishery regulations. Gray (1999) speculates that the generally high levels of prosperity in the fishery may also reduce the incentive to fish illegally. While we agree this is likely the case for many fishers, there are also a small but significant number who may be poorly capitalised, mediocre fishers, or contract fishers working for smaller profit margins than owner-operators. Incentives to break fishery rules are likely to remain for these operators. Recreational fishers or "shamateurs" (unlicenced illegal operators) undertaking illegal fishing activity, are not considered here except to note that commercial fishers have little hesitation in reporting such activity to enforcement personnel.

A research initiative to assess how compliance within the fishery may be improved through optimising enforcement activity is currently in progress. The objectives of the project are to: (a) examine spatial and temporal trends in non-compliant activity; (b) determine factors which may deter or encourage non-compliant behaviour; (c) develop performance indicators to allow the success of enforcement services to be monitored through time; and, (d) determine stakeholder perceptions about enforcement and compliance. The fishing community is generally supportive of this project, part of the funding for which is provided by industry.

3. FISHER INVOLVEMENT IN ENFORCEMENT
3.1 Main constraints

Traditional fisheries management operates on a "command and control" model of governance (Dubbink and van Vliet 1996); *i.e.* management authorities attempt to regulate fishing processes by means of legal and administrative means. This approach has long been considered inadequate, primarily because it does not foster an environment where fishers feel they have part-ownership of management decisions (Jentoft 1989, Nielsen and Vedsmand 1997). Jentoft (1989) suggests that the legitimacy of any regulatory scheme is subject to at least four constraints:

i. Content of the regulations: greater legitimacy occurs when fishers perceive regulations to coincide with their view of the issues
ii. Distributional effects: the more equitably regulations are imposed, the greater their legitimacy
iii. Formulation of the regulations: the more that fishers are involved in developing regulations, the more legitimate the regulatory process will be regarded and
iv. Implementation of regulations: regulations will be considered more legitimate when fishers are involved in their implementation and enforcement.

If management agencies can institute practices that encourage fisher-involvement in the formulation and implementation of regulations (iii and iv), then in some sense the content and distributional effects (i and ii) attend to themselves. We consider fisher involvement in the formulation and enforcement of fisheries regulations to be of the utmost importance. In this section we examine three ways in which the fishing community should be actively involved in the enforcement process.

3.2 Commercial fishers reporting illegal activity
Let's start with a story.

> At a compliance committee meeting between managers, fishers and compliance personnel a report was presented detailing an investigation that had uncovered substantial out-of-quota fishing in the valuable red herring[5] fishery. The reaction from fisher representatives on the committee, which included several well-respected fishers of considerable influence in the fishing community, was one of "Well done! Bravo! *But we could have told you about those guys years ago.*"

This true story, sadly could be about many of Australia's co-managed fisheries and probably many other fisheries around the world. The problem we are alluding to is the reluctance of many commercial fishers to report

---

[5] The red herring fishery is an imaginary fishery on a sea far-far away. On a more serious note, during research for this paper similar stories were reported to the authors by several Australian compliance managers.

other commercial fishers they know (or suspect) to be breaking fishing regulations. We do not suggest that fishers never report illegal activity; indeed, many compliance managers in Australia can cite notable instances where this has occurred, but in general these tend to be the exception rather than the rule[6]. This phenomenon may have detrimental implications for the management of a fishery, particularly in regard to the efficiency and effectiveness of enforcement programmes.

Working as a fisher may be arduous and sometimes dangerous. Few other professions (policing is an obvious exception) place participants in potentially life-threatening situations where they may rely on the help of others for their survival. The concept of loyalty is therefore important to many fishers. Traditional fishing communities are often close knit, with allegiances formed through common experience, shared ethnicity and family ties. Combine this with the fact that fishers and enforcement personnel have historically been pitted in adversarial roles and it is perhaps little wonder that many fishers are reluctant to assist in policing fishery rules. We argue, however, that co-management places an imperative on fishing communities to take responsibility for policing their fishery and this means co-operating with enforcement services. In our experience, fishers are often opposed to involvement in enforcement activity arguing that: "We are fishers, not enforcement officers. Through cost-recovery we contribute money toward enforcement, but that is where our role should end." There are problems with this line of argument, however.

First, people illegally fishing can, and do, go to great lengths to hide their activities (Anderson 1989). This may be helped by the fact that a fishery often ranges over vast geographic areas, but is serviced by a relatively small number of enforcement officers and vessels. Recent technological advancements assist those engaged in illegal fishing to avoid detection; powerful vessels allow fishers to range over greater distances, satellite telephones provide secure channels for communication and radar forewarns fishers of approaching vessels well before visual contact is made. Advances in fish finding technology in the Western Australian rock lobster fishery present particular problems for policing gear restrictions. Differential GPS allows fishers to place and retrieve gear with high precision. The consequence of this is that lobster pots are often placed in small, dispersed clusters of less than five pots, instead of the more traditional method "lines" of 20-30 pots. The implications for enforcement of gear restrictions, such as carrying out pot-counts, are obvious.

It is in this climate that enforcement officers must police fisheries regulations. It is not realistic to expect effective enforcement of regulations by means of random checks of fisher activity. While random checks are necessary for measuring overall compliance rates within a fishery, to a large extent enforcement effort must be targeted at known or suspected offenders (*i.e.* "intelligence" driven). These are usually fishers suspected of regularly breaking fisheries regulations, and as such are likely to have a greater deleterious effect (at least on an individual level) on the management goals for the fishery, compared with those who only occasionally or opportunistically break regulations. It is also likely that targeting, catching and prosecuting known offenders sends an important deterrent message to other fishers (Sutinen 1996). Targeted enforcement operations can only be initiated, however, with access to information about illegal activities and the fishing community is best placed to provide information about how and where illegal activity is occurring.

A second reason why fishers should provide enforcement personnel with information about illegal fishing is a practical one, namely cost-effectiveness. Enforcement, especially at-sea enforcement, is costly, and in Australia up to 90% of all enforcement expenses in cost recovered fisheries are met by industry (Sutinen 1994). This has led to increasing industry pressure for enforcement groups to justify their expenditure. An obvious way for industry to ensure that enforcement expenditure is used efficiently is to assist enforcement groups to effectively target their effort. It is perhaps ironic that the same fishers who claimed prior knowledge of "those guys" in the red herring fishery are also these charged with assessing the legitimacy of compliance budgets and expenditure.

How then can fishers be encouraged to report illegal activity undertaken by other commercial fishers? Experience in the rock lobster fishery has shown that commercial fishers are most likely to report illegal activity when they perceive other fishers to be "taking money directly from their pockets" (such as poaching lobster from other fishers' pots), or when they perceive illegal activities to be detrimental to the sustainability of the resource. Fortunately, almost all rules in the rock lobster fishery are designed to protect against these exact circumstances; *i.e.* to protect fair and equitable access to the resource and to protect the biological sustainability of the fishery[7]. If we accept this premise, then *a priori* it seems reasonable that, at least for rules which are perceived as legitimate, fishers can be convinced that it is in their interests to report on fishers they know break the rules. We suggest six practical ways in which this can be encouraged:

i. *Codes of conduct for ethical fisher behaviour*. Fishers who hold access rights to a fishery are responsible for the maintenance of a public resource and should be subject to the same ethical standards as the public officials who are seen to more directly manage it. Codes of conduct for ethical fisher-behaviour in a range of "scenario" situations should be developed

---

[6] Our experience has shown that commercial fishers will freely inform on recreational fishers they suspect of breaking regulations, and vice versa. Recreational fishers also seem willing to inform on other recreational fishers breaking regulations; in fact, most illegal activity reported to Fisheries WA personnel falls into this category.

[7] We do not suggest that all rules in all fisheries are implemented for these reasons, rather, that a majority of regulations in the rock lobster fishery have been enacted to ensure distributional justice and biological sustainability.

in conjunction with fisher representatives. Such a code would: (a) serve as a "plain english" interpretation of fishery rules in a variety of circumstances; and, (b) help establish the legitimacy of fishery rules by providing context to their interpretation. It is also desirable that similar codes of conduct be developed for all stake-holders in a resource (*e.g.* recreational fishers).

ii. *Education on the legitimacy of regulations.* Many fishers make subjective judgments about what constitutes "bad behaviour" among their peers – fishers deem some offences less important than others and may not consider particular offences worthy of reporting. Through education, fishers must be made aware that the cumulative effect of many small breaches of regulations may indeed endanger the sustainability of the resource and ultimately threaten their own livelihoods.

iii. *Illegal activity "hotlines".* Management agencies should establish mechanisms for fishers to easily report any illegal activity they witness. In Western Australia we have established the Fishwatch program, a 24 hour toll-free telephone service set up to receive public information about illegal fishing activity; such programmes have subsequently been adopted in most other Australian states. Internet reporting provides another method that is likely to become increasingly accessible for fishers. For example, an Australian fishing company (Austral Fisheries) and the Tasmanian Conservation Trust have established a website to monitor illegal fishing for Patagonian toothfish in the Southern Ocean.

iv. *Rewards for information.* In many law enforcement contexts, agencies offer rewards for the provision of information leading to the successful prosecution of offenders. Rewards are usually staggered to be commensurate with the severity of the offence and the magnitude of the penalty. In New South Wales, for instance, industry-funded rewards of up to $A500 may be paid for information leading to the conviction of persons undertaking illegal rock lobster fishing.

v. *Foster good relations between fishers and fisheries officers.* Good relations based on mutual trust and respect between fishers and fisheries officers is perhaps the most important mechanism for encouraging fishers to volunteer information about illegal activity. This is best engendered by one-to-one contact, and should not be restricted to those situations where fishers are inspected to check their compliance with fishery rules. A simple "Hello, how's it going?" on the wharf goes a long way. Such a strategy is enhanced by maximising the time spent by fisheries officers in the field and reducing their administrative burden.

vi. *Effective administration and legislation.* Central to the issue of commercial fishers reporting illegal activity is the effectiveness of the administrative and legal environment in which they do so. Fishers must feel confident that the information they provide will be acted on promptly, with concern for issues of confidentiality, and that the legal system is capable and willing to impose realistic sanctions. (See also Section 4 below: Management agency responsibilities).

3.3 Disciplinary committees

Legal action, although obviously necessary for serious fisheries offences, must be considered an undesirable outcome. Legal sanctions are often costly to impose, liable to fail because of minor technicalities, and variable in their effectiveness (Franzoni 1998). For example, judicial discretion exercised by different magistrates sometimes leads to quite different findings for similar offences. Courts often do not appreciate the seriousness of fisheries crimes, especially in high-value fisheries such as the WA rock lobster fishery[8]. Fishers themselves lament the fact that prosecutions often result in penalties that are not commensurate with the potential gains to be made from illegal activities. Another problem is that court proceedings are often lengthy, resulting in a loss of immediacy between crime and punishment. For many in the fishing community such delays are perceived as inaction on the part of the management agency and the potential deterrent value of prosecution is diminished as a result.

Peer review by disciplinary committees composed of management and fisher representatives may offer a viable alternative, or addition, to the judicial process when legal prosecution is deemed unwarranted or impractical. Indeed, it is the fishers and managers themselves who are best equipped to decide appropriate penalties for fisheries crimes. Peer review is likely to be more cost-effective than court proceedings, could take place in a more timely manner, and encompass a wide range of penalties designed to match the seriousness of the offence. Perhaps most importantly, peer review provides for greater legitimacy and consistency of outcomes, from the point of view of the fishing community and individual offenders. For example, fishers breaking regulations may do so under an assumption that they can foster doubt about their guilt in a judicial setting where magistrates have little knowledge of the fishing process. Such an approach is unlikely to succeed if judged by their peers. It is also possible that judgement by peers may create a deterrent effect in itself, since fishers are unlikely to relish having their dishonest behaviour paraded before fellow fishers. Such an approach parallels community policing initiatives where juveniles confront and possibly make atonement to their victims as an alternative to judicial approaches to disciplinary action.

In the case of Fisheries WA, the discretionary powers of the Executive Director allow some scope for a peer review process, although to date such proceedings have usually occurred after a judicial decision to decide additional administrative penalties (*e.g.* "black marks"). There has been strong opposition from fishers (and their representatives) against participating in such proceedings. The argument against involvement has centred around the fact that fishers feel peer review may create conflict within

---

[8] In Western Australia it has been suggested that one solution to this problem might be to establish Fisheries Tribunals with magistrates who have specialist knowledge of fisheries law.

fishing communities, and that undesirable pressure may be brought to bear on fishers serving on such committees. While these are legitimate concerns, in an environment of co-management, the onus falls upon fishers (the majority of whom are honest) to take collective responsibility to ensure such pressure is not brought against members of disciplinary committees. Nonetheless, to address these concerns we propose two mechanisms which may alleviate fisher anxiety about serving on disciplinary committees:

i. Retired fishers may be suitable candidates to participate in peer-review situations. They often maintain an active interest in their fishery, have the respect of existing fishers, but are independent of fisher organisations and management. Representatives should be chosen on the provision that they no longer hold a financial interest in the fishery.

ii. A number of fisher representatives could be chosen from different regions of the fishery. If a fisher caught conducting illegal fishing is brought before the disciplinary committee, then the representative from that area of the fishery could act as an observer only, or be excluded from the process altogether. The case would then be heard and determined by fishers from areas other than the region of the fishery where the offence took place.

There are a range of alternative punitive measures which may be suitable in lieu of legal prosecution, and whose application would depend on the statutory powers of individual management organisations. For example, the Executive Director of Fisheries WA has, under certain circumstances, discretionary power to cancel, suspend or refuse to renew fishing licenses. Other types of administrative sanctions may include probationary periods after an offence, community service, attendance at compulsory education programmes, or installation of compulsory electronic vessel-monitoring systems. There are particular legal implications for administering such penalties. Peer review processes do not have the traditional safeguards associated with the criminal justice system, such as the right to a jury trial (although this is rare for statutory offences or breaches of management plans) or the requirement that charges are proved beyond a reasonable doubt. It would be appropriate that any peer review process be subject to careful scrutiny to ensure that sanctions were issued with caution and discretion.

3.4 Involvement in compliance working groups

The National Fisheries Compliance Committee (1999) has a stated commitment to collaborate with fisheries stake-holders to develop and implement fisheries policies and laws. They also support co-management of fisheries through Management Advisory Committees, the membership of which can be held accountable for meeting duties and obligations as stake-holder representatives. In the Western Australian rock lobster fishery it is the RLIAC Compliance Subcommittee which operates as the compliance working group for the fishery. This subcommittee is responsible for reviewing enforcement budgets, assessing the compliance implications of changes to fishing rules and alerting the enforcement manager to trends in compliance. Involving fishers in this process is beneficial in a number of respects. High levels of voluntary compliance may be encouraged when fishers are involved in designing fisheries laws and compliance planning, since fishers are more likely to be responsive to self-developed regulations than rules imposed from an autonomous management agency (Jentoft 1989). Many fisheries have in place regulations which are difficult to enforce (Hemming and Pierce 1997), but equally fishers may have difficulty complying with some rules. This may occur when rules are developed without regard to the realities of the fishing process. At best, rules perceived as flawed may be pushed to the limit; at worst they will be openly ignored (Kesteven 1987). Involvement in formulation of rules therefore gives fishers the opportunity to contribute in developing rules with the practicalities of fishing foremost in mind.

Members of the RLIAC and its subcommittees are encouraged to report to industry, Fisheries WA and other interested parties on issues under its consideration. There is also a clear imperative for RLIAC to solicit industry input on alternate courses of action under consideration. The reasons and rationale behind committee recommendations are conveyed to representative groups by way of a Chair's Summary, most usually distributed to the fishing community as a newsletter after each RLIAC meeting. In these reports it is important to inform the fishing community about the discussions – the compromises and trade-offs – which lead to the final formulation of a decision or new rule. It is this process, perhaps more than any other, that makes or breaks the fishing community's perceived legitimacy of rules. It is also important that fisher representatives do not paint the management agency as the "bad guys" when unpopular decisions must be made. If fishers are part of the decision-making process, they should be equally accountable for difficult decisions that must be made in a fishery. If legitimate reasons exist for deciding on a particular course of action, then these reasons need to be put reasonably, but strongly, to fishers.

The attitudes and perceptions of fishers toward management, compliance and enforcement are vital to the effectiveness of any regulatory effort (Clay and McGoodwin 1995). It is important to know why fishers choose to break fisheries laws and it is not sufficient to rely solely on the information supplied by fisher representatives on management committees. Industry surveys to ascertain views on management measures and attitudes toward other stake-holder groups should be undertaken on a periodic basis to monitor trends over time. The effectiveness of education programmes can often only be measured in this way. By carrying out surveys of this kind management authorities can foster interest in group support – people like to have their opinions canvassed on issues they see as important to their livelihood. The fact alone that fishers are surveyed indicates a willingness on the part of management and enforcement to take notice of opinions. Finally, as well as the "why", surveys of this kind provide enforcement agencies with some of their most useful information regarding the who, how, where, and extent of illegal activity. In the Western Australian rock lobster fishery we have recently undertaken a compliance survey of 4000 recreational fishers, and are currently developing surveys for commercial skippers and crew, participants in the processing sector of the fishery, selected retailers and fisheries officers themselves.

4. MANAGEMENT AGENCY RESPONSIBILITIES

4.1 Responsiveness

Encouraging fishers to participate in enforcing fisheries law imparts certain responsibilities on management agencies. Governments need to ensure that supporting legislation and policy is provided so that fisher involvement in enforcement is encouraged and supported.

Fisheries enforcement services must make every effort to ensure they are responsive to information about illegal activities reported to them by the fishing community, both in terms of direct action and feedback to those providing the information. This should operate on a formal basis by reporting the results of investigations to Management Advisory Committees, but also on an informal basis between Fisheries Officers and the individual fishers who report the activity. Fishers in the rock lobster fishery have recently criticised Fisheries WA on this point, and we are currently implementing feedback mechanisms to ensure that all information received is acted on in a timely manner and that those involved are advised of the outcomes. There is also a responsibility to consult with fishers on priorities for patrolling activities and about the existence of problem areas within a fishery. This make good sense since it is usually fishers who are best informed about the nature and extent of illegal activities in a fishery. Priorities should be directed to those illegal activities commonly perceived as deleterious to the sustainability of the fishery.

4.2 Confidentiality

We believe that individual fishers have a responsibility to speak out when they hear of other fishers who are breaking regulations. In turn, fisheries agencies have a responsibility to ensure fishers may do so in a climate that is safe and free from recrimination. This is not always an easy task, but is greatly assisted by ensuring that information received from fishers is treated as strictly confidential. Most government agencies who rely on the receipt of information which would otherwise not be obtained except under circumstances of confidentiality have the power to suppress the source of such information. The rationale is that, if the flow of information was to cease, the effective operation of the agency may be prejudiced. This is true for investigations into fisheries crime, and agencies should ensure that appropriate legislation is in place to ensure confidentiality, both during inquiries and in any subsequent legal proceedings. Tasmania's illegal fishing telephone hotline service provides an interesting example. Tasmanian Fisheries offer monetary rewards for information leading to successful prosecutions of fisheries offences, however informants do not have to identify themselves in order to participate in the reward scheme; they are simply assigned a number and are able to collect any reward which may result from their information on this basis.

Enforcement agencies also have a "duty of care" to ensure they deal with confidential (and often anonymous) information in responsible ways. That is, information, and especially anonymous information, should be treated cautiously until such time as enforcement officers can independently determine the validity (or otherwise) of the intelligence received. Informants sometimes make mistakes about what constitutes illegal activity, either by misinterpreting events they have witnessed or by relying on circumstantial evidence. Malicious accusations, with no basis in fact, are also possible. Fisheries officers should ensure that when investigating suspected offenders they do so without prejudice.

4.3 Judicial process

Enforcement programmes should undertake all reasonable steps to inform stake-holders of their legal obligations and the consequences of not meeting them. But at the end of the day, they must be prepared to prosecute those who willfully break the law. Fishers must believe that if they cheat there is a reasonable chance they will be caught and, if caught, that prosecution will be successful. It is also important that deterrent penalties are larger than the gains made through illegal activity, taking into account the probability of detection and successful prosecution (Beddington *et al.* 1997). In the case of the rock lobster fishery, monetary sanctions are typically a nominal penalty, plus court costs, plus a fine (if applicable) approximately ten times the value of the illegal catch. Sanctions should also be dependent on the offenders' history, since this provides an additional deterrent not to repeatedly violate fishery laws.

Management agencies need to ensure that the judiciary is well educated with regard to the deleterious consequences of fisheries crime, which is itself a subset of "environmental crime". This requires a recognition on the part of magistrates that the judgments they make affect not simply the individual who has committed the crime, but also the wider fishing (and non-fishing) community and in turn the sustainability of the resource. To be effective, criminal sanctions must not only punish the individual, they should also deter others from engaging in similar activities. This point cannot be stressed enough; the legitimacy of fishery rules, and the willingness of honest fishers to report on those they know to be breaking regulations, hinges on fisher confidence in the legal system to adequately deal with fisheries crime. There must also be a willingness for fishers, fisheries officers and police to cooperate in ensuring that fishers who report fisheries crime are not unduly harassed or victimised in their communities.

Finally, it is important to inform fishers about the nature of successful prosecutions. Advertising successful prosecutions educates fishers about the types of penalties received for particular fisheries offences, and deters others from committing similar acts. It can also serve as a "shaming" penalty, as is the case for Western Australian fisheries where detailed outcomes (including names) of successful prosecutions are published in a quarterly magazine, *Western Fisheries*.

5. CONCLUSION

We have advocated increased management responsibility for fishers, particularly in the area of fisheries enforcement. In the Western Australian rock lobster fishery access is pursuant to a limited-entry management plan within which catch-shares are indirectly allocated through pot-holdings. There is the possibility of strengthening these access-rights, thereby ensuring for fishers continuity of access should management plans be amended or revoked. However, we feel that if fishers wish for greater security of access and increased devolution of management responsibility in general, it is important they demonstrate a responsible attitude toward compliance with, and enforcement of, fisheries rules. Fishers should be involved in the formulation, and assist in the implementation, of fisheries rules on both an individual and collective basis. In a real sense a lot of what we have discussed centres around explaining to the fishing community the legitimacy of fishery rules. This flow of information should operate both ways. Managers and scientists need to educate fisher representatives about the scientific and management processes; fishers, for their part, need to educate managers and scientists about the realities of the fishing process and how this affects compliance with fishery rules. It is perhaps in this exchange of information that the true value of co-management may be found. We have suggested three mechanisms for increasing fisher involvement in the enforcement process: (a) encouraging fishers to help enforcement staff by providing information about illegal activities; (b) participating in peer review of fishery offences; and (c) involvement in compliance working groups. These processes will only be effective, however, if management agencies can ensure appropriate administrative and legislative structures exist to encourage and support fisher involvement in enforcement.

6. ACKNOWLEDGMENTS

This research was funded by Fisheries Research and Development Corporation (Australia) (Project Number: 98/156) and the commercial rock lobster industry in Western Australia. The support and advice given by Nic Caputi, John Looby, and Bruce Webber is gratefully acknowledged. Thanks are also extended to those national compliance managers, fisheries officers and professional rock lobster fishers who provided us with their insights into the strengths and weaknesses of enforcement activities in Australia. A small part of this work is drawn from a presentation by the senior author to the 26th Australasian Fisheries Law Enforcement Conference in May 1999.

7. LITERATURE CITED

ABARE: Australian Bureau of Agricultural Resource Economics 1998. Australian Fisheries Statistics, Canberra, 51 pp.

Anderson, L.G. 1989. Enforcement issues in selecting fisheries management policy, *Marine Resource Economics, 6*, pp. 261-277.

Anon.. 1968. Fishery Advisory Committees, *Fishing Industry News Service Vol 1(2)*, pp.9.

Beddington, J.R., K. Lorenzen and I. Payne 1968. Limits to exploitation of capture fisheries. In: D.A. Hancock, D.C. Smith, A. Grant and J.P. Beumer (eds.) *World Fisheries Congress Proceedings No. 2*, pp. 529-536, CSIRO Publishing, Australia 1997.

Bowen, B.K. 1994. Long term management strategies for the Western rock lobster fishery: Evaluation of management options (Volume 1). *Fisheries Management Paper No. 67.*, 67 pp.

Buck, E.H. 1995. Individual transferable quotas in fisheries management. *Report for Congress by the Congressional Research Service*, The Committee for the National Institute for the Environment, Washington.

Caddy, J.F. 1999. Fisheries management in the twenty-first century: will new paradigms apply? *Reviews in Fish Biology and Fisheries 9*, pp. 1-43.

Clay, P.M. and J.R. McGoodwin 1995. Utilizing social sciences in fisheries management *Aquat. Living Resour. 8*, pp. 203-207.

Charles, A.T. 1992. Fisheries conflicts: a unified framework. *Marine Policy 16(5)*, pp.379-393.

Davis, A. 1996. Barbedwire and bandwagons: a comment on ITQ fisheries management. *Reviews in Fish Biology and Fisheries 6*, pp. .97-107.

Donohue, K. 1998. Western Rock Lobster management - options and issues. *Fisheries Management Paper No.113, Fisheries WA, Perth*. 68 pp.

Dubbink, W. and M. van Vliet 1996. Market regulation vs co-management: two perspective on regulating fisheries compared *Marine Policy 20(6)*, pp. 499-516.

Exel, M. and B. Kaufmann 1997. Allocation of fishing rights: implementation issues in Australia. In: Pikitch, E.K., D.D. Huppert, and M.P. Sissenwine (eds.) *Proceedings of the American Fisheries Society Symposium No.20. Seattle, 14-16 June 1994.*, 343 pp.

Fisheries Western Australia 1999. The Western Rock Lobster Fishery: cost recovery and managed fishery fees. *Information brochure,* Fisheries WA, Perth, 4 pp.

Fisheries Western Australia 1998. A guide for Management and Ministerial Advisory Committees (MACs) and the conduct of meetings issued by Monty House, Minister for Primary Industries; Fisheries. *Fisheries Management Guide No.1*, Fisheries WA, Perth.

Franzoni, L.A. 1998. Negotiated enforcement and credible deterrence, *Dept of Economics Working Series*, University of Bologna, Italy.

Franzoni, L.A. 1999. Negotiated enforcement and credible deterrence, *The Economic Journal*. 109(458) October 1999, pp. 509-535.

Grafton, R.Q. 1996. Individual transferable quota: theory and practice. *Reviews in Fish Biology and Fisheries 6*, pp.5-20.

Gray, H. 1999. "Skinnin' the pots" - a history of the Western Rock Lobster Fishery. *PhD thesis*, Murdoch University, Perth. 467 pp.

Hancock, D.A. 1981. Research for management of the rock lobster fishery of Western Australia. *Proc. Gulf Carribb. Fish. Inst. 33*, pp. 207-229.

Hannesson, R. 1996. On ITQ's: an essay for the Special Issue of Reviews in Fish Biology and Fisheries. *Reviews in Fish Biology and Fisheries 6*, pp. 91-96.

Hønneland, G. 1999. A model of compliance in fisheries: theoretical foundations and practical application. *Ocean and Coastal Management 42*, pp. 699-716.

Hemming, B. and B.E. Pierce 1997. Fisheries enforcement: our last fisheries management frontier *In:* D.A. Hancock, D.C. Smith, A. Grant and J.P. Beumer (eds.) *World Fisheries Congress Proceedings No. 2*, pp. 675-679, CSIRO Publishing, Australia.

Hutchings, J.A., C. Walters and R.L. Haedrich 1997. Is scientific inquiry incompatible with government information control. *Can.J.Fish.Aquat.Sci. 54*, pp. 1198-1210.

Jentoft, S. 1989. Fisheries co-management, *Marine Policy 13(2)*, pp. 137-154.

Kesteven, G.L. 1987. If I don't someone else will, *NAGA (October1987)*, pp. 13-14.

Lindner, B. 1994. Long term management strategies for the Western Rock Lobster industry: Volume 2 Economic efficiency of alternative input and output based management systems in the Western Rock Lobster Fishery. *Fisheries Management Paper No.68, Fisheries WA*, Perth, 36 pp. +appendices.

Marec Pty Ltd 1997. Optimising the worth of the lobster catch - options and issues. *Fisheries Management Paper No.101,* Fisheries WA, Perth, 83 pp.

McCay, B.J. 1995. Social and ecological implications of ITQ's: an overview, *Ocean and Coastal Management 28*, pp. 3-22.

McCay, B.J., C.F. Creed, R.A. Finlayson, R. Apostle and K. Mikalsen 1995. Individual transferable quotas (ITQs) in Canadian and US fisheries. *Ocean and Coastal Management 28*, pp. 85-115.

McLaughlan, N. 1994. Long term management strategies for the Western rock lobster fishery: Law enforcement considerations (Volume 4). *Fisheries Management Paper No. 70.*, Fisheries WA, Perth, 12 pp.

National Fisheries Compliance Committee 1999. *Strategic Direction for Australian Fisheries Compliance & Framework for Fisheries Agencies* Primary Industries and Resources S.A, 14 pp.

Nielsen, J.R. and T. Vedsmand 1997. Fishermen's organisations in fisheries management, *Marine Policy 21(2)*, pp. 277-288.

Nielsen, J.R. and T. Vedsmand 1999. User participation and institutional change in fisheries management: a viable alternative to the failure of "top down" driven control? *Ocean and Coastal Management 42*, pp. 19-37.

O'Boyle, R. and K.C.T. Zwanenburg 1997. A comparison of the benefits and costs of quota versus effort-based fisheries mannagement. *In:* D.A. Hancock, D.C. Smith, A. Grant and J.P. Beumer (eds.) *World Fisheries Congress Proceedings No. 2*, pp. 675-679, CSIRO Publishing, Australia.

O'Boyle, R. 1993. Fisheries management organizations: a study of uncertainty, In: S.J. Smith, J.J.Hunt and D. Rivard (eds.) Risk evaluation and biological reference points for fisheries management. *Can. Spec. Publ. Fish. Aquat. Sci. 120*, pp. 423-436

Penn, J.W., G.R. Morgan and P.J. Millington 1997. Franchising fisheries resources, an alternative model for defining access rights in Western Australian fisheries. *In:* D.A. Hancock, D.C. Smith, A. Grant and J.P. Beumer (eds.) *World Fisheries Congress Proceedings No. 2*, pp.675-679, CSIRO Publishing, Australia 1997.

Rennie, H.G. 1998. Geographical problems in implementing ITQ: New Zealands quota management system. Paper presented at "Crossing Boundaries", the 7th *Biennial Conference for the International Association for the Study of Common Property*, Vancouver, 10-14 June 1998.

Stephenson, R.L. and D.E. Lane 1995. Fisheries management science: a plea for conceptual change. *Can.J.Fish. Aquat.Sci. 52*, pp. 2051-2056.

Sutinen, J.G. 1994. Summary and conclusions of the workshop on enforcement measures, In: *Fisheries Enforcement Issues* OECD Workshop on Enforcement, Paris, 253pp.

Sutinen, J.G. 1996. Fisheries compliance and management: assessing performance, *A Report to the Australian Fisheries Management Authority*, 29 pp. August 1996.

Symes, D. 1998. Property rights, regulatory measures and the strategic response of fishermen, pp.3-16, In: D. Symes (ed.) *Property rights and regulatory systems in fisheries*, Blackwell Science, Oxford, 268 pp.

The Marketing Centre 1996. Fishermen's views on the future management of the rock lobster fishery, *Fisheries Management Paper No.89,* Fisheries WA., Perth, 47 pp. +appendices.

Walters, C. and P.H. Pearce 1996. Stock information requirements for quota management systems in commercial fisheries, *Reviews in Fish Biology and Fisheries 6*, pp. 21-42.

Wilson, J.A., J.M. Acheson, M. Metcalfe and P. Kleban 1994. Chaos, complexity and community management of fisheries. *Marine Policy 18(4)*, pp. 291-305.

Young, M.D. 1995. The design of fishing-right systems – the New South Wales experience. Ocean.Coast.Manage. 28(1-3) pp. 45-61.

ENFORCEMENT AND COMPLIANCE OF ITQs: NEW ZEALAND AND THE UNITED STATES OF AMERICA

W.J. Nielander and M.S. Sullivan
Fisheries Management Consultancy International, Ltd
116 Interlake Blvd, Lake Placid, Fla. 33852, USA
<wnielander@htn.net> and <mssa@ihug.co.nz>

1. INTRODUCTION

In the two decades since 1980 fisheries management on the international stage, has undergone an unprecedented transformation of a legal and practical nature. During this period fisheries have been principally administered under a régime of national exclusive economic zones provided for in the 1982 United Nations Convention on the Law of the Sea[1]. The advent of expanded fisheries jurisdiction, however, has not been the solution hoped for and the period has been noted for rising catches in low value species, soaring industry costs and over-capitalisation and growing management complexity[2].

It is in this context that both New Zealand and the United States introduced Individual Transferable Quota systems (ITQs). The United States has adopted a more cautious approach to ITQ based management and introduced ITQ fisheries on a limited scale[3]. New Zealand, on the other hand, has made ITQs, implemented in 1986[4], its main focus of fisheries management and the current system employed in New Zealand is widely recognised as one of the most innovative approaches to managing wild fisheries stocks in recent decades[5]. To understand the nature of enforcement issues that have been confronted under ITQs in New Zealand and the United States, it is first necessary to examine the background against which these systems were introduced.

2. BACKGROUND – NEW ZEALAND AND THE UNITED STATES

2.1 New Zealand

2.1.1 The New Zealand context

New Zealand's ITQ system, as introduced in 1986, was structured along a traditional model of defined "rules" set out in statute and subordinate regulations. When detected, non-compliance with those "rules" was to be met by financial/economic sanctions imposed by both the Courts and the empowering statute itself. Aspects of New Zealand's political and legal environment are extremely important in understanding how New Zealand was able to introduce a statutory and regulatory-based system of complex rules and requirements that has been readily amenable to "police" and "forensic" style enforcement methodologies.

2.1.2 Criminal law model of fisheries legislation

To enforce regulatory standards governing economic activities, New Zealand and other British common law countries have historically tended to use a criminal law legal model which sets out rules of acceptable and prohibited behaviour with criminal penalties for that prohibited behaviour. The criminal law has, however, been adapted to the demands of regulatory environments by way of a number of legal devices. These include alteration to traditional requirements relating to burdens and standards of proof. These devices are designed to simplify the task that agencies responsible for enforcing the "rules" might otherwise face in trying to establish matters that are within the unique knowledge of the offender or occur in circumstances that are difficult or costly to supervise.

Under New Zealand fisheries legislation offences are of strict liability in nature[6]. There are specified defences provided which the offender is required to prove on the balance of probabilities[7]. In addition, there is an array of various evidential presumptions and certification provisions, which greatly simplify the task of the Ministry in proving specific factual issues[8]. The 1986 ITQ legislation, when introduced, contained offences of a rather simplistic nature and relatively low maximum penalties (up to a maximum of $NZ10 000). The Act did, however,

---

[1] Part V, Articles 55-57, UN Doc A/CONF 62/122 21 ILM 1261.
[2] The State of World Fisheries and Aquaculture, FAO, 1995, pgs 8, 50-53. The picture remains largely unchanged in the latter part of the 1990's, though the expansion of fishing fleets is slowing. The State of World Fisheries and Aquaculture, FAO, 1998, pgs 7 & 14.
[3] Currently, three Federal ITQ programmes operate in the United States: surf clam and ocean quahog in Mid-Atlantic and New England waters; wreckfish along the South Atlantic coast; and halibut and sablefish off Alaska.
[4] By the Fisheries Amendment Act 1986, inserting various new provisions into the *Fisheries Act 1983*.
[5] Neher P., Arnson R. & Mollet N., 1989, Rights Based Fishing, NATO Advanced Science Institutes Series, Kliwer Academic Publishers, pg 1 and Boyd R. & Dewes C., 1992. Putting Theory Into Practice: Individual Transferable Quotas in New Zealand Fisheries, Society and Natural Resources, Vol 5, pg 179-198.

[6] See s105(1) *Fisheries Act 1983* and s240 Fisheries Act 1996.
[7] See s105(2) and s105A *Fisheries Act 1983* and s241 & s245 Fisheries Act 1996.
[8] *Eg* see s102 to s103A and s106 to s106A of the *Fisheries Act 1983* and s111, s193, s195 and s248, s249 of the Fisheries Act 1996.

retain the traditional emphasis in Commonwealth fisheries legislation on forfeiture of illegal fishing gear and vessels in addition to quota. Unlike other jurisdictions, however, forfeiture under the New Zealand legislation was an automatic consequence of conviction and not at the discretion of the Court. The Courts, however, retained a residual discretion to make "non-forfeiture" orders in the event that they found there to be "special reasons relating to the offence"[9]. After conviction, the legislation vested a discretion in the Minister of Fisheries to redeem forfeit property, on such terms and conditions as the Minister saw fit[10].

The *Fisheries Act 1983* has been the subject of a number of revisions relating to the structure of offences and penalties since its introduction[11]. In particular these have included:

i. the increase in maximum penalties to $NZ250 000
ii. the introduction of forfeiture as a minimum penalty
iii. the reclassification of forfeitable offences into 3 distinct categories and
iv. the introduction of more specific statutory guidelines in the exercise of the Minister's discretion to redeem forfeit property.

The new *Fisheries Act 1996* has, however, introduced a number of substantial changes to the offences and penalty structures for fisheries offences which were designed to address the compliance problems that plagued the *Fisheries Act 1983*. These provisions of the new Act (though not yet in force) do away with general offences arising under the Act, setting out specified offences throughout the Act. The radical change, however, between the new Act and its predecessor, has been in the penalty régime. For the first time since the Fisheries Act 1908, a category of fisheries offences now attract substantial levels of imprisonment. In addition while the forfeiture régime previously provided for under the *Fisheries Act 1983* has been substantially retained, it has been adjusted to remove some of the past inequities.

The forfeiture provisions continue to distinguish between property and quota. The level of forfeiture, similar to the *Fisheries Act 1983*, is directly linked to the nature of the offences and the maximum penalties imposed. The forfeiture of quota, however, is limited to the most serious category of offences[12]. The most significant change is the substitution of the discretion previously vested in the Minister to redeem forfeit property, with a provision that persons with registered interests in quota or legal or equitable interests in property that is forfeit to the Crown may apply to the Court within 35 days of the date on which the forfeiture occurred, for relief from forfeiture[13].

2.1.3 Impact of political and constitutional system

Critical to the viability of this criminal law model in ITQ management has been the fact that all New Zealand fisheries, unlike the United States (as well as Australia and Canada), are administered by a central government; this avoids the complications arising from state and federal jurisdictions. This has allowed for the adoption of relatively cohesive administrative procedures underpinning the ITQ system in New Zealand and a unified enforcement/monitoring strategy. In addition, the ITQ system has, until recently, been exclusively managed by the Ministry of Agriculture and Fisheries (MAF) and its successor, the Ministry of Fisheries (MOF) (hereinafter collectively referred to as "the Ministry").

Because of the uni-cameral nature of its Westminster Parliamentary style of government, combined with the absence of an entrenched written constitution and Bill of Rights[14], New Zealand has been able to more effectively limit the extent of the rights of participants in the fishing industry to be free from random inspection, search and entry to premises and equip fishery officers with a range of powers that in any other context would be considered excessive[15].

As a consequence of the above factors, Fishery Officers have some of the widest powers available to any

---

[9] s107B *Fisheries Act 1983* - "A special reason is one that is not found in the common run of cases. While not necessarily categorised as "exceptional" or "extraordinary", it is one that may properly be characterised as not ordinary or common or usual". *Basile v Atwill* [1995] 2 NZLR 537, 539 (CA).

[10] s107C *Fisheries Act 1983*. For a comprehensive examination of the legal nature and operation of forfeiture provisions under New Zealand and Fisheries legislation, see Sullivan M S, Forfeiture of Fishing Vessels in Australia and New Zealand, MLAANZ Journal, 14(1), 39-.

[11] Particularly by the Fisheries Amendment Act 1990 (1990 No 29).

[12] s255 Fisheries Act 1996.

[13] s256 Fisheries Act 1996. Section 256, however, sets out the factors that the Court must consider before it may order relief from forfeiture. The Court is required to determine whether the person making application has an interest in the property or quota and whether the interest was created solely or principally for the purposes of avoiding an application of the Act in respect of forfeiture. No order may be made under this provision unless the Court is satisfied that it is necessary to avoid manifest injustice.

[14] The rights and freedoms in Part II of the New Zealand Bill of Rights Act 1990 are not constitutionally entrenched and may be over-ridden by an ordinary enactment but, in interpreting an enactment, a consistent meaning is to be preferred to any other meaning. *Ministry of Transport v Noort, Police v Curran*, [1992] 3 NZLR 260. In recent times, however, the New Zealand Court of Appeal has been adopting a more restrictive view of the application of the Bill of Rights Act. See further, Schwartz H, The Short Happy Life and Tragic Death of the New Zealand Bill of Rights, New Zealand Law Review, Part II 1998, pgs 259-311.

[15] However, even in jurisdictions which have entrenched constitutional Bill of Rights, such as Canada and the United States, where the Act under which a search is exercised, or order to produce is made, is "regulatory" then lower standards are exacted. See *Thomson Newspapers v Canada* (1990) 67 DLR (4th) 161 (SCC), and *Lovgren v Byrne*, 787 F2d 857 (3rd Cir 1986).

enforcement officer in New Zealand, including powers of random entry, search and questioning[16]. In addition, unlike the police or other public officers, their powers of entry, search and seizure powers are largely conferred on the Fishery officer personally (by virtue of holding the appropriate warrant) rather than by a search warrant issued by a court. The only general restrictions placed on these powers are that:

i. all (except the power to seize) may only be exercised at reasonable times and
ii. none of these powers may be exercised in respect of a private dwelling house or Maori reservation without written authorisation of the Court.

These powers, with some additional refinements, are largely re-enacted under Part XI of the Fisheries Act 1996 (which is also not currently in force).

2.2 The United States
2.2.1 The US context

In the United States, enforcement of fishery management regulations is primarily processed under the *Magnuson-Fishery Conservation and Management Act 1976* (Magnuson – Stevens Act), that was considerably amended in 1996. Under the Magnuson–Stevens Act, the primary tool for imposition of fines is through an administrative process by the Department of Commerce and National Oceanic and Atmosperic Administration. However, the Magnuson–Stevens Act also provides for permit sanctions, forfeiture of vessels and quota, and jail time for specific fraudulent acts. In addition, there is federal criminal legislation, such as the Lacey Act and general criminal code that is also used in prosecutions relating to fisheries violations, and particulary, ITQ violations.

2.2.2 Magnuson-Stevens Fishery Conservation and Management Act 1996 - Civil Law System

The penalties under the Magnuson-Stevens Act for violation of various prohibitions can result in a civil penalty not exceeding $US100 000 for each violation. Each day of a continuing violation constitutes a separate offense. In addition, not only may an owner or operator be fined for violations, the fishing vessel used in the commission of the Act is liable *in rem* for any civil penalty assessed for such violation, and may be proceeded against in the United States District Court. The penalty constitutes a maritime lien on the vessel, which may be recovered in an action *in rem*, and the District Court of the United States having jurisdiction over the vessel.[17]

In addition to civil penalties and *in rem* liability, the Magnuson-Stevens Act also provides that in any case which a vessel has been used in the commission of a prohibited act, and the owner or operator of the vessel has been issued a permit, or if any civil penalty or criminal fine imposed on the vessel or the owner has not been paid or is overdue, the Secretary may:

i. revoke any permit with respect to such vessel or person, with or without prejudice to the issuance of subsequent permits
ii. suspend such permit for a period of time considered by the Secretary to be appropriate
iii. deny such permit or
iv. impose additional conditions and restrictions on any permit issued to or applied for by such vessel or person under the Act…[18].

The Magnuson-Stevens Act also provides a criminal offense for several prohibited acts under the Act. With respect to ITQ violations, the Act provides a criminal punishment for willfully submitting to the Council, Secretary or Governor false information.[19] The punishment for submitting false information to the Secretary of Commerce or Official Management Council or a State Governor regarding fishing information is $US100 000, or imprisonment for not more than six months, or both[20].

The current Administrative system used to enforce Fisheries Laws in the United States has the following key components:

i. Publication of present levels of penalties within specified bands for specified violations.
ii. Issuing of infringement notices directly from the fisheries officers for lower level violation.
iii. Proceedings are commenced by way of a notice of violation action (NOVA for unpaid infringement notices and more serious violations) similar in concept to a statement of claim. These NOVAs are issued by the agencies' attorneys directly.
iv. As the proceedings are civil in nature they may be compromised by way of settlement before hearing. Agreements may also be reached that relate to future compliance with rules or restriction of activities, which have the force of any civil settlement.
v. Additional sanctions such as permit revocation or suspension, banning, forfeiture of property used in the commission of the violation may also be sort, but are imposed only at the discretion of the Court.
vi. If the matter goes to hearing before an administrative law judge, it does so under civil procedural rules. Hearsay evidence is applicable (evidence being evaluated in terms of its weight as opposed to strict rules of admissibility).
vii. A defendant may appeal the decision of the administrative law judge to the United States District Court.

In the US this administrative system relieves considerable pressure on the limited resources of the Courts and the compliance agency. Several hundred, and perhaps thousands of actions, may be commenced each year under this administrative system. As a result, only 10-20% make

---

[16] s79 *Fisheries Act 1983*.
[17] 16 U.S.C. §1858(d).
[18] 16 U.S.C. §1858(g).
[19] 16 U.S.C. §1859(a).
[20] 16 U.S.C. §1859(b).

it past the administrative law judge phase and into United States District Court.

2.2.3 Nature and power of fisheries officers

The National Marine Fisheries Service has fisheries offices and special agents in each of its six regions and central headquarters in Silver Springs, Maryland, USA. An Enforcement Officer serves as a uniformed officer performing routine inspection, patrol and surveillance duties to detect illegal activity with respect to a variety of fishery and wildlife conservation laws. A Special Agent is responsible for initiating and conducting full-scale investigations of alleged criminal and civil violations under the various under the various fish and wildlife laws. This involves interrogating suspects and interviewing witnesses; conducting searches and seizures with and without warrants; securing and serving search warrants; making arrests; inspecting records and documents; developing evidence for the orderly presentation to United States Attorney and other legal officers; testifying in court; preparing detailed written reports such as witness briefs and trial books and carrying out undercover operations. Special Agents are usually responsible for ITQ investigations.

Whether a fishery officer or special agent, or other state or federal officer authorized by the agency or cooperating agreement, each such officer or agent has the powers as enumerated under the Magnuson Act. These powers enable him, with or without a warrant or other process, to:

i. arrest any person, (on reasonable cause) committing an act prohibited by the Act
ii. board, search or inspect, any fishing vessel which is subject to the provisions of the Act
iii. seize any fishing vessel used or employed in the violation of any provision of the Act and
iv. seize any fish taken or retained in violation of any provision of the Act[21].

2.2.4 Lacey Act

The Lacey Act[22] provides that it is unlawful for any person to import, export, transport, sell, receive, acquire, or purchase any fish or wildlife or plant taken or possess in violation of any law, treaty, or regulation of the United States or in violation of any Indian Tribal law, as well as in violation of any state or foreign law.

2.2.5 General criminal law

If a particular fisheries offence is of such serious nature, the agency (at its discretion) may bring the matter to a United States Attorney's office for prosecution under applicable criminal laws. A different set of procedures than those available in civil and administrative matters apply to criminal matters, such as criminal grand jury indictments and related search and seizure warrants.

Under the United States Criminal Code[23], it is unlawful for anyone to knowingly and willfully:

i. falsify, conceal, or cover up by any trick, scheme or device a material fact
ii. make any material false, fictitious, or faudulent statement or representation or
iii. make or use any false writing or document knowing the same to contain any materially false, fictitious, or fraudulent statement or entry[24].

A violation under 18 USC 1001 can carry imprisonment of not more than 5 years or a fine up to $US100 000 or both.

2.2.6 Forfeiture

In addition to civil or criminal penalties, a fishing vessel and its gear may be subject to forfeiture under the Magnuson–Stevens Act.[25] Any fishing vessel used, and any fish (or the fair market value thereof) taken or prohibited by the Act shall be subject to forfeiture to the United States. Such forfeiture must be undertaken in the United States District Court. Therefore, a civil enforcement action under the Magnuson–Stevens Act and a forfeiture action are separate actions undertaken in two distinct courts or tribunals.

Interestingly, the Magnuson–Stevens Act provides for two rebuttable presumptions relative to forfeitures. The first rebuttable presumption is that all fish found on board a fishing vessel which is seized in connection with a prohibited act is presumed to have been taken in violation of the Act. The second rebuttable presumption is that any vessel found shoreward of the outer boundary of the United States EEZ, or beyond the EEZ of any nation, that is capable of use for large scale drifnet fishing, is presumed to have been actually engaged in fishing in such area. The use of such presumptions shifts the burden from the prosecuting agency to the offender.

3. NEW ZEALAND EXPERIENCE: THE ITQ SYSTEM IN OPERATION

3.1 Introduction of ITQ

The ITQ system introduced in New Zealand in 1986 had six primary components:

i. the establishment of Quota Management Areas (QMAs) and setting of Total Allowable Catches (TACs) for those areas
ii. the allocation and issuing of ITQs and the maintenance of a registry relating to the subsequent holding and leasing of that quota
iii. a requirement that fish must be caught under the authority of quota[26] and information must be

---

[21] 16 USC 1861(b) 1996.
[22] 16 USC 3372(1998).
[23] 18 USC 1001 (1998).
[24] 18 USC 1001(a) (1998).
[25] 16 USC 1860 (1996).
[26] The requirement to have quota prior to fishing has now been abandoned under the new provisions of the Fisheries Act 1996 as recently provided by s27 to s29 of the *Fisheries Act 1996 Amendment Act 1999* (No 101).

furnished in returns to the New Zealand authorities in order to enable them to monitor how much fish has been caught against that quota

iv. except in limited circumstances, the sale of fish by commercial fishermen is limited to Licensed Fish Receivers (LFRs) who are subject to stringent regulatory and record-keeping controls

v. reliance on self-policing by industry participants with wide enforcement powers conferred on Government officers to conduct random checks and the provision of heavy penalties for taking or possession of fish otherwise than in accordance with the law where no defences are available and

vi. the introduction of a resource-rental based on the quantity of quota held[27].

In addition to the above, elements of the earlier management régime were retained in respect of both ITQ and non-ITQ species. In particular all fishers are required to fish under the authority of an appropriate fishing permit[28], all commercial fishing vessels must be registered with the Ministry[29] and the power to condition or regulate gear, methods, landings, size, seasons, areas and other forms of fishing restrictions is preserved[30].

3.2 The enforcement/monitoring debate

Throughout the mid 1980s, during which the policy governing the introduction of the ITQ system was developed, there was considerable debate in New Zealand over how best to monitor and enforce the requirement that catch taken matched the quota held[31]. In resolving this issue, policy makers noted that New Zealand fisheries and the fishing industry had a number of characteristics, which needed to be taken into account. These were:

i. strong industry support for the introduction of the ITQ
ii. an Exclusive Economic Zone (EEZ) that was relatively isolated and with few transboundary problems or fishstocks
iii. an export-oriented fishing industry
iv. a single fisheries management enforcement jurisdiction
v. a value-added tax (GST) on all goods and services including fish and fish product[32]

vi. a well-established commercial (catch/processing marketing) sector
vii. the significant number of vertically integrated fishing/processing/marketing companies in the commercial sector and
viii. the diverse number and remoteness of many landing points used by the catching sector.

The focus of the debate, however, primarily revolved around the practicality and cost of the real-time monitoring and enforcement of catch/quota on board and at the landing site (through comprehensive use of observers and dockside monitoring) versus the combination of retrospective documentary-based monitoring and random auditing[33]. Real-time physical monitoring, while considered to be effective, was ultimately rejected[34] as being too costly (both in terms of industry compliance and direct government enforcement costs) considering the nature, operation and structure of the New Zealand fishing industry at the time[35]. The revolution in hardware and software that has since taken place, particularly in the last 5 years, may well lead to some form of electronic real time monitoring being introduced in the future.

3.3 The principal enforcement/compliance components of the New Zealand ITQ

3.3.1 The role of records and returns

A unique aspect of the ITQ system introduced in New Zealand is the reliance placed on, and the interrelating nature of, the record-keeping and reporting requirements imposed on the participants in the industry. In addition, the New Zealand system adopts a partial "honesty box" approach, whereby the fishing industry was given the principal task of ensuring it complied with the legal requirements of the ITQ. To this end requirements were imposed on the various participants in the fishing industry to produce a series of cross-referencing returns furnished to the Ministry. The essential documents are the various types of Catch Landing Returns (CLRs)[36], Quota

---

[27] Resource-rentals have subsequently been abandoned and replaced by the cost-recovery régime under Part VII of the *Fisheries Act 1983* and Part XIV of the *Fisheries Act 1996*.
[28] s62 and s63 *Fisheries Act 1983*.
[29] s57 *Fisheries Act 1983*.
[30] s63(4) and s89 *Fisheries Act 1983*.
[31] Crothers G T, Manager Fisheries Compliance, MAF Fisheries Establishment Advisory Group, Briefing on Fisheries Law Enforcement, November 1993, pgs 10-11.
[32] The requirement for businesses to maintain basic accounting records for GST purposes meant that additional requirements to detail sales and purchases of fish would not constitute a significant additional burden on the fishing industry, thereby making a record intensive system more feasible.

[33] Briefing on Fisheries Law Enforcement, supra note 31.
[34] In recent times the debate concerning dockside monitoring has resurfaced in New Zealand and was taken up and supported by the Fisheries Task Force in its report to the Minister of Fisheries, April 1992, pg 68. As a result, s300 of the Fisheries Act 1996, which came into force on 1 October 1996, makes provision for regulations to be made for the purposes of dockside monitoring. To date no regulations have been promulgated under this section. Whether dockside monitoring becomes a significant aspect of fisheries management in New Zealand is doubtful, however. An independent report commissioned by the NZ Fishing Industry Board indicated that such a programme could cost upwards of $NZ70 to $NZ100 million per annum, depending on the number of authorised landing ports. Briefing on Fisheries Law Enforcement, supra note 31, pg 12.
[35] Briefing on Fisheries Law Enforcement, supra note 31, p. 11.
[36] Apart from a few limited exceptions, catch landing returns were not originally part of the documentary flow originally incorporated into the ITQ system. Catch landing data was retained in a log maintained by the master of the vessel and produced on demand by a fishery officer.

Management Reports (QMRs) and Licensed Fish Receiver Returns (LFRRs). The specific purpose of these documents was to enable the New Zealand authorities to monitor the flow of fish against quota holdings for the respective QMAs[37].

In addition to various reporting requirements LFRs and "dealers in fish"[38] are governed by strict record-keeping requirements[39]. In particular, LFRs are required to issue and keep various internal records detailing the unloading, receipt, internal processing and subsequent on-sale of the fish landed to them. These documents are available for inspection at any time by Fishery Officers enabling them to verify the accuracy of the information supplied in the monthly returns.

3.3.2 The crucial role of the licensed fish receiver

An essential component of the ITQ system introduced in 1986 was the provision relating to licensing of fish receivers and prohibition on commercial fishers selling fish to persons who are not licensed or deemed to be licensed under regulations made under the Act[40]. Other persons not authorised under the Act are also prohibited from acquiring or possessing fish for the purposes of sale unless the fish is obtained from a person authorised under the Act to be in possession of fish for the purposes of sale.

While the number of licences to receive fish is not subject to any limit, the statutory requirements relating to "fish receiving", in combination with the regulatory requirements relating to licensing[41] and record keeping[42], place an extremely effective choke point in the distribution of fish within New Zealand's ITQ. In theory at least, all fish sold within New Zealand should be able to be traced back to source and, in the event that it cannot, the person being investigated runs the risk of being prosecuted for mere possession.

3.4 Problems which were predicted to arise under the ITQ in New Zealand's fisheries and responses

3.4.1 Monitoring and data-fouling

At the time of the introduction of the ITQ in New Zealand it was recognised that specific enforcement/compliance problems could undermine the integrity of the system[43].

Monitoring harvests and prosecuting cheaters was seen as one of the essential requirements for maintaining the credibility of the ITQ. It was also recognised from the inception of the ITQs quota management system in New Zealand that a self-policing scheme made it more feasible to mis-record species and weights, particularly when there is collusion between the various parties responsible for completing the documentation. A similar aspect of concern was potential data-fouling and the falsification of data required for management purposes, resulting from fishermen fearing detection through cross-matching with administrative records.

3.4.2 High-grading and overfishing

High-grading or discarding lower quality fish was also seen as a particular problem in any system that constrained output rather than input. Of all the potential problems, however, it was the issue of discrepancies between the mix of quota holdings held by fishermen and the actual catch in the net that most concerned policy managers. The threat from direct overfishing of stocks due to seasonal variations and the vagaries of fishing itself was also recognised and was to be dealt with by the implementation of a device called "overs and unders"[44].

3.5 Post-1986 operation of ITQ

3.5.1 Initial enforcement teething problems

Notwithstanding the early confidence of the Ministry that the flexibility of the ITQ system would resolve most issues, soon after its introduction in 1986 problems began to surface. In particular, a number of enforcement/compliance problems developed which had not been fully foreseen.

Bycatch and fish without quota

One of the principal problems related to bycatch and fish taken in excess of quota holding. This problem had its origins in the fact that at the time of allocating quotas on the basis of catch histories the legislation allowed fishermen to include fish that had been bycatch of target species. Under the New Zealand ITQ system, however, all

---

[37] The QMR is furnished monthly by the actual quota holder. That document allows the Ministry to monitor which were fish taken against quota. The LFRR is furnished monthly by Licensed Fish Receiver (LFR), the fish processors, and allows the Ministry to monitor the actual product flowing through the system against the fish recorded in the QMRs and the CLRs. The CLRs are furnished at the end of a fishing trip or monthly (depending on the type of fishing operation) by the commercial fishermen who actually catch the fish. The principal function of the CLRs is to enable the Ministry to ensure that all fish taken are actually recorded, determine what species were targetted, what species are caught, areas the fish was taken from, the date the fish was taken, which quota the fish was caught against, and to which LFR the fish was landed to.

[38] A "Dealer in fish" is essentially a person engaged in the wholesaling or retailing of fish who is not a LFR or a commercial fishermen . Refer r2 of the Fisheries (Recordkeeping) Regulations 1990.

[39] Under the Fisheries (Licensed Fish Receivers) Regulations 1997.

[40] The legislation provides a limited exemption for traditional "wharf sales". Refer s67(2) *Fisheries Act 1983*.

[41] Under the Fisheries (Licensed Fish Receivers) Regulations 1997.

[42] Under the Fisheries (Recordkeeping) Regulations 1990.

[43] Undated anonymous MAF discussion document in possession of writer.

[44] In effect, in any fishing year a fisherman underfished his quota he could carry over 10% of his entitlement to the following year and fish that quota in addition to the quota he had available in the subsequent year. Conversely, in the event a fisher took up to 10% more fish under the authority of his quota in any year, that amount would be subtracted from the following year's entitlement.

quota, whether it was target or bycatch, could subsequently be targetted for creating effectively a "bycatch of bycatch" problem. In addition, so as to prevent wastage of fish and high grading of quota, fishers were prohibited under the 1986 legislation from dumping quota species at sea[45]. To provide a defence and prevent abuse of the system, fishers were able to obtain an immunity from prosecution for taking or possessing fish other than under the authority of quota if, as soon as practical after landing the fish, they surrendered the fish to the Ministry[46].

Inevitably the quantity of fish surrendered became too large for the Ministry to physically handle and an administrative system was adopted to allow fishers to retain the fish and pay an administrative "penalty". The level of this penalty was set at less than the landed value of the fish to ensure some incentive remained for fishers to actually bring the fish ashore[47].

Overs and unders - balancing fishing rights

Combined with the bycatch problem, the system of under and over fishing rights and how they attached to parcels of quota that were freely tradable under the New Zealand system became a logistical nightmare to administer. Disputes and differences of interpretation began to arise between the Ministry and fishers as to the exact nature and extent of available catching rights held by individuals. These issues made enforcement of restrictions on overfishing extremely difficult except in cases where the activity was particularly gross in nature.

Fishing on behalf

The legislation introduced in 1986 did not specifically deal with the issue of whether persons other than the quota holder could take fish under the authority of that person's quota. Following the introduction of the ITQ system in 1986, particularly as a consequence of the rationalisation of quota holdings into the hands of more efficient and larger fishing operators and companies, the practice of fishing on behalf of quota holders began to grow[48]. In effect, those individual fishers who either held insufficient amounts or no quota at all became contract fishers for those who held the quota. The nature of these arrangements was frequently informal and unwritten, giving rise to major difficulties in the New Zealand authorities determining who were legitimate fishers and who were not.

Area misreporting

New Zealand's ITQ was based around ten principal QMAs. While there were variations on these areas used in respect of different species, the basic principle remained that ITQs were allocated to fishers on the basis of their catch histories in that QMA and that the total of ITQ allocated equated to the commercially allowable and sustainable catch for that QMA.

With the introduction of the ITQ system in 1986, fishery officers began to detect examples of area misreporting in respect of certain QMA/species combinations. These combinations usually involved high value species (such as orange roughy, silver warehou or snapper) in which there were substantial differences between QMAs as to abundance and costs of catching quota. While most cases arose in the context of adjacent QMAs, the most notorious example of this type of offending arose in the context of "Operation Roundup" in 1990[49].

3.5.2 The legislative response

As a result of these and other technical problems the ITQ within a short time was subject to a major series of amendments by way of the *Fisheries Amendment Act 1990*. The principal amendments relevant to the enforcement of the ITQ were:

i. the introduction of a deemed value system[50]
ii. the introduction of a series of complex defences allowing for the retrospective counting of fish against quota or subsequent purchases of quota or payment or payment of a deemed value[51]
iii. a limited ability to create definitive quota balances[52] and
iv. the introduction of a requirement to furnish the New Zealand authorities with advance notice of the terms under which persons were authorised to take fish on behalf of other holders of quota[53].

In addition, as a result of a series of high profile cases involving quota busting, the 1990 legislation increased the maximum penalty for such offences from $NZ10 000 to $NZ250 000 per offence.

Subsequently, the Fisheries (Satellite Vessel Monitoring) Regulations 1993 were also passed to tackle the problem of area mis-reporting. As a result, after the 1 April 1994, an automatic location communicator had to be carried and operated on board foreign-licensed fishing

---

[45] s28ZB *Fisheries Act 1983*.

[46] s88(1)(c) *Fisheries Act 1983* as it then was (now substantially reproduced in s105A(1)). The potential problems with bycatch fishing and high-grading of fish had been identified as issues of major concern by industry and MAF personnel in the extensive interviews conducted by C.M. Dewes (1987), shortly after the introduction of the ITQ.

[47] For a more detailed description of the system that evolved, see Clark *et al* 1988, supra note 5, pgs 141-142.

[48] The adoption of a practice of authorising persons to fish quota on the owner or lessee's behalf led to a situation where fishery officers were unable to properly enforce the ITQ system and accordingly s28ZA(2) was inserted into the Fisheries Act 1983, requiring written notification of such arrangements to be furnished to MAF before fishing took place; Memorandum to Group Director, MAFisheries, 22 May 1992.

[49] Refer infra note 56.

[50] s28ZD of the *Fisheries Act 1983* made provision for the payment by fishers of an amount previously set by the Director-General in respect of fish landed otherwise than under the authority of or in excess of appropriate quota.

[51] s105 and s105A *Fisheries Act 1983*.

[52] s28ZCA *Fisheries Act 1983*.

[53] s28ZA *Fisheries Act 1983*.

vessels, foreign-chartered fishing vessels capable of engaging in trawling for fish and New Zealand fishing vessels exceeding 43 metres in overall length and capable of engaging in trawling for fish.

3.5.3 Refocusing of compliance group

Although the need to change the enforcement emphasis under a ITQ system from traditional (game warden/sea-borne) enforcement to that of monitoring quota/product flows had been foreseen early on, it was not until 1988, some two years after the introduction of the *1986 Amendment Act*, that MAF restructured its fisheries enforcement group in order to come to grips with the significant change in the enforcement environment under the ITQ.

In addition to restructuring the organisation of fisheries enforcement, significant numbers of the then-serving enforcement personnel were made redundant and replaced by experienced enforcement personnel, principally from other agencies such as Police and Customs. The change in skill-mix of enforcement personnel was also accompanied by a change in enforcement methodology. Under the ITQ system, enforcement effort has primarily become land-based, focusing on auditing paper trails, detecting illicit landings by means of random inspections/covert land-based observations and initiating targeted investigations based on intelligence gathering (including the use of informants). Most of the pre-ITQ sea based enforcement capability has been disposed of and only two small inshore patrol vessels have been retained. Most of the sea-borne surveillance in the EEZ is now done by the New Zealand Air Force (in terms of area restrictions) and on-board observers (in terms of transhipment supervision). MAF also introduced a centrally-based team of dedicated accountants. In addition to undertaking a scheduled number of random audits of fishing and LFR companies each year, this unit provided MAF with specialist forensic accountants who assisted in investigating, preparing and presenting evidence in complex quota-fraud cases.

To complement the introduction of new investigative skills fisheries enforcement officers were divided into two categories, surveillance and investigations. Surveillance officers, who constituted the visible uniformed presence in the field, were charged with the day to day enforcement of fisheries laws and other targeted field enforcement operations. While there was in practice some overlap and even joint operations with surveillance teams, fisheries investigators were primarily deployed in undertaking longer term/intensive investigations. Their model is similar to that found in most modern police forces incorporating uniformed and investigative, detective divisions.

MOF has largely persisted with the enforcement structures and methodologies it inherited from MAF in 1995. In addition to rationalising the number of compliance regions from three to two, however, the most significant change introduced by the Ministry of Fisheries has been the introduction of the Serious Offences Unit (SOU) based in Wellington. The SOU was a reaction to the demands placed on enforcement personnel in conducting and running large-scale complex quota-fraud cases and the difficulties in budgeting and resourcing such investigations. The SOU maintains a core compliment of three full time investigators and additional staff whose main focus is large-scale ITQ offences. But they also assist district offices in similar-type investigations.

3.5.4 Nature of offending confronted

New Zealand's experience with the ITQ system since its introduction has shown that output-constrained fisheries management systems, backed up by adequate record-keeping and reporting requirements, enable fisheries enforcement personnel to successfully pursue unlawfully-taken product past the catching stage. Under the previous input-management régimes in New Zealand, the majority of controls related to constraints on fishing effort rather than the product caught. Such provisions inevitably required direct surveillance and detection of the offence at the time the fish were taken in contravention of the law. What output constraints that existed prior to the introduction of the ITQ, were not adequately supported by documentary requirements under the relevant legislation at the time.

Contrary to observations made by several writers[54], the introduction of ITQs in New Zealand has clearly been accompanied by a significant increase in the level and scope of offending detected and prosecuted compared to pre-ITQ[55]. As was expected, the introduction of the ITQ in New Zealand created a range of economic incentives for some in the industry to cheat the system. What was not expected, perhaps, was the scale and extent of the "quota frauds" that has since been confronted.

The types of prosecutions undertaken since the advent of the ITQ in New Zealand have ranged from single small-scale fishers landing small quantities of fish into the black market, to large-scale complex conspiracies to misdeclare or fail to report hundreds of tonnes of high value species, which were exported overseas[56]. The counteract-

---

[54] See Boyd R. & Dewes C., 1992 supra note 5 at 185-186; Clarke I. *et al*, 1988 supra note 5 at p138.

[55] This observation is based on the writer's extensive personal experience in prosecuting and defending fisheries offences since 1984. Prior to the introduction of the ITQ in 1986, fisheries prosecutions were infrequent and almost exclusively minor in nature (excepting foreign fishing vessel cases). Following the introduction of the ITQs in 1986, there has been a major expansion in both the number and scale of fisheries prosecutions culminating in a series of gigantic prosecutions in the early 1990s. This expansion has also been reflected in a number of reported and unreported decisions originating from the District Court, High Court and Court of Appeal.

[56] The epitome of complex frauds encountered by MAF since the introduction of the ITQ in 1986 is the series of inter-related investigations and prosecutions dealt with between 1991 and 1994 known under the collective title "Operation Roundup". "Operation Roundup", which commenced in 1991 by way of

ing, benefit, of the ITQ system has been that in many cases, the nature or scope of the offending that has been brought to the attention of the Ministry by other quota-holders or participants in the Fishing Industry[57].

Against this background there has been little increase in actual enforcement costs associated with the introduction of the ITQ in New Zealand. In real terms, the cost of fisheries enforcement has declined as a percentage of the gross value of production of New Zealand fisheries[58]. ITQ compliance and poaching enforcement (which covers non-industry taking largely ITQ species) constitute the bulk of the enforcement budget, reflecting the principle focus of the compliance enforcement effort and the primacy of ITQ species in New Zealand.

Accompanying this relative stability in enforcement-related costs has been an overall decline in the number of dedicated front-line enforcement personnel since 1988. Small stations have been closed[59] accompanied by a redistribution of resources to the remaining offices and the overall impact was a reduction in front-line staff. By 1993 there were a total of 80 front-line staff employed as fishery officers responsible for the enforcement of Fisheries laws in all three regions[60]. These reductions in front-line officers has been partially offset by the addition of four accountants in Wellington who are responsible for conducting random audits of Licensed Fish Receivers and the expansion in the use of observers on industry vessels.

3.5.5 Continuing problem areas
Deemed values and bycatch fisheries

As expected, New Zealand authorities have encountered major difficulties in dealing with the problem of bycatch and target fish taken in excess of quota. The legislative solutions introduced in 1990, particularly the deemed value scheme, do not appear to have alleviated the problem. These systems appear unable to keep pace with the inevitable market fluctuations in values of various species or to be sufficiently precise to overcome what are significant regional and district variations in port prices or variations in the value of fish dependant on the state of processing.

In addition, the difficulty in proving what was legitimate bycatch or relabelled target fish has meant that, in the absence of an admission of targetting or extreme or gross offending, prosecutions undertaken for such offences have largely failed. This has led to recent amendments to the *Fisheries Act 1996* (not yet in force) which remove distinctions between target and bycatch fish and substitute a revised deemed value system in conjunction with end of year balancing[61].

Effectiveness of monitoring systems

The computerised monitoring system using data-input from the various cross-checking returns does not appear to have been particularly successful in operation. In a report to the OECD Committee in 1995, MOF claimed that while the ITQ system does provide incentives for fishers to overstate catches in non-ITQ species and understate ITQ species, "the system of cross-checks which operates through monitoring of LFR records and returns deters such practices"[62]. No evidence was offered in support of that statement whereas an independent review of the New Zealand ITQ commissioned by the Canadian Government concluded that the system appears to have been captured by the enormity of the data-entry required and the day-to-day operational requirements of the quota-registry system and in fact little actual monitoring of the quality and implications of the data received and processed is done[63].

The official Ministry line also appears to contradict the experience of fisheries enforcement personnel and the various large-scale prosecutions undertaken since the introduction of the ITQ[64]. In one case involving extensive

---

covert surveillance of Wellington and Napier wharf fronts, resulted in the detection of a series of large-scale on-going conspiracies to defraud the Quota Management System established in 1986. In the principal case involving landings by the fishing vessel *Perseverance,* the systems used by the fishing and LFR company to defeat the constraints imposed by the ITQ were multi-faceted and, for the period charged, involved a total of 574t of orange roughy misdeclared as cardinal fish or not declared at all. A more recent example of a large-scale ITQ prosecution was *Ministry of Fisheries v Abel Fisheries Ltd*, unreported, DC Wellington, CRN 7085005665 *et al.,* 23/02/98, involving 5 chartered Russian fishing vessels and several hundred of misdeclared and under-recorded quota.

[57] This was certainly the experience of the writer in the time he served as a Prosecutor with MAF Fisheries, and has been confirmed from practical experience since commencing private practice. Attempts to defeat the ITQ system, such as that at the centre of "Operation Roundup", strike at the property rights and investment that most legitimate fishers have under the system and there has been a very clear pattern in New Zealand of growing Industry rejection of those who set out to "cheat the system".

[58] Prior to the introduction of the ITQ system, MAF's annual direct enforcement budget was in the realm of $NZ12 to $NZ13 million per annum. MAF's 1993-94 operating budget showed direct enforcement costs remained at similar dollar levels. The total 1993-94 cost of fisheries enforcement being $NZ16.4 million (MAF $NZ10.9, Prosecution $NZ2.5 and Defence $NZ3.0).

[59] There has been a partial reversal of this trend with the opening of a single one-man station in New Plymouth in 1997.

[60] For example, Challenger District was reduced in 1988 from 16 fishery officers to 12 fishery officers (including a DCM and investigators). In 1999, this number has been further reduced to 7.

[61] Refer provisions of Fisheries Act 1996 Amendment Act 1999 No 101.

[62] OECD Committee Report, Ministry of Fisheries, 1995, p 45.

[63] Meltzer E, Report on the ITQ prepared on behalf of the Canadian Government: Enforcement under New Zealand Fisheries Quota Management System, Department of Fisheries and Oceans, 1991.

[64] For example, see the various "Operation Roundup" cases, *MAF v Basile* [1995] 1 NZLR 712, *MAF v William Rose*

and lopsided reporting of bycatch *vs* target catches, a District Court Judge noted the offending disclosed in that case had been:

> "... compounded by reason of the fact that the defendants continually reported what they were doing to MAF which accepted the information for monitoring purposes but did not initially analyse it. If the information had been analysed by MAF at an early stage the offending would have been detected at an early stage and the magnitude of it substantially diminished"[65].

Notwithstanding the enormous financial and physical resources that have been committed to the cross-checking/record based approach to the New Zealand QMS, both by the government and industry, there has never been a meaningful or detailed analysis undertaken as to the efficacy of the system. Nor has any assessment been made as to whether the fundamental premise on which the system was predicated, that it actually contributes to the effective enforcement and monitoring of ITQ, been validated. This is particularly the case in the face of the growing vertical integration of many companies and fishing operations and the use of centralised computer data bases which are used to generate the different reports furnished to the Ministry.

4. RECENT AND FUTURE DEVELOPMENTS IN NEW ZEALAND ITQ SYSTEM

4.1 The changing incentives: proportional quota and cost-recovery

In August 1991 the Minister of Fisheries began a comprehensive review of the QMS and fisheries management by appointing an independent Task Force to make recommendations on the future development of fisheries legislation and associated structures in New Zealand[66]. Although the Task Force delivered its report in April 1992, it was not until late 1994 that legislative and policy initiatives were forthcoming.

From 1 October 1994 an annual levy was introduced to recover the full costs of commercial fisheries management from the fishing industry[67]. Combined with the introduction of quota proportionality in 1990, cost-recovery has added a further dimension to the Fishing Industry's interest in ensuring compliance with fishery laws. Any change to the Total Allowable Commercial Catch (TACC), whether an increase or a decrease, now has a corresponding proportional affect on the underlying ITQ rights. The proportional system has transferred the financial risks associated with changes to the TACC from the Crown to the ITQ holders in the fishery. ITQ holders now face the direct costs of poor compliance through both an erosion of their ITQ rights and the costs directly attributable to enforcing the system.

The Fishing Industry in New Zealand, at least at the generic level, now has a strong and direct incentive to not only improve the compliance régime to ensure that it provides maximum protection to their rights, but also to ensure that funds spent on compliance are appropriately and cost-effectively spent. For the first time they also have a strong incentive to provide for the delivery of compliance and enforcement services (where benefits can be gained) outside of the traditional Government bureaucracy. In particular the fishing industry in New Zealand has a strong incentive to regulate its own activities under an ITQ system to preserve the value and integrity of their harvesting rights[68]. This new environment has been recognised by the Ministry of Fisheries in its Compliance Strategic Plan. One of the principle visions of the plan is that "those with the right to harvest fisheries gain responsibility to manage them within environmental limits and standards set by the Government"[69].

While many of these potential benefits have yet to be realised, the stage has been set at a policy level for a radical shift in at least the compliance aspects of the ITQ system in New Zealand, from the traditional government central/police model to much greater emphasis on the Industry itself implementing and enforcing the ITQ "rules" through self- regulation.

4.2 Legislative changes

In 1994 a new *Fisheries Act* was introduced into the New Zealand Parliament that significantly changes the operation of the ITQ system with much greater reliance on economic incentives and civil penalties. After the Bill was rewritten, the new *Fisheries Act 1996* Act took a more traditional approach to fisheries management, largely abandoning the civil penalty/retrospective balancing approach. The one particular innovation that did survive into the *Fisheries Act 1996* was the decision to separate the core quota-share in the fishery from the actual right to harvest a certain quantity of fish. The harvesting right, which is annually derived from the quota right will be a separately trade right known as Annual Catch Entitlement (ACE). As a result of growing concerns with the prescriptive nature of the 1996 Act and the complexity of implementing some of its provisions,

---

Trawling Ltd, 21/02/94, High Court, Napier, *Hill v MAF*, 17/012/91, Judgment No 3, Holland J, HC Christchurch, *MAF v Dong Won Fisheries Co Ltd*, 04/04/91, Heron J, High Court, Wellington, *Roach v Kidd*, 12/10/92, McGechan J., H.C. Wellington, *Aston v MAF*, 11CRNZ 478, *MAF v Lima*, unreported, HC Auckland, Ap 146/93, 103, 26/08/93, Fisher J., and *Ministry of Fisheries v Abel Fisheries Ltd*, unreported, DC Wellington, CRN 7085005665 *et al.*, 23/02/98.

[65] *MAF v Wellington Trawling Company Limited and Basile*, Unreported, DC Napier, 14/11/94, Hole DCJ.

[66] Sustainable Fisheries, Report of the Fisheries Task Force April 1992, pg iv.

[67] s107EB to s107ED *Fisheries Act 1983*.

[68] The often-cited example of this is the Challenger Scallop Enhancement Company and the Southern Scallop Fishery in Nelson, which operates its own compliance contracts and compliance manager.

[69] Ministry of Fisheries, Compliance Strategic Plan, 1997-2002.

recent amendments to the Act have re-introduced the civil penalty/annual balancing approach abandoned under the 1994 Bill.

Although the *Fisheries Act 1996* has been passed, the *Fisheries Act 1983* remains in force and has continued to be the primary Act. This state of affairs appears likely to continue for some time. The 1996 Act is being brought into force in incremental stages as supporting systems, procedures, forms, and regulations are developed to support it. While the 1996 Act contains some fundamentally new approaches to ITQ management in New Zealand, including some refinements to penalties and forfeitures imposed for offences under the Act and regulations, it largely persists with the Criminal Law/Government enforcement and compliance model of the ITQ system that was adopted in 1986.

5. THE UNITED STATES' EXPERIENCE
5.1 General enforcement issues

Enforcing the FMP regulations of any sort has proven to be a difficult challenge. A 1987 study by NOAA of the MFCMA raised the enforcement issue in the following terms:

> "In certain fisheries there is said to be widespread cheating on regulations. This is, to put it plainly, stealing valuable US property from law-abiding fishermen, the public owners, and from the taxpayer by increasing enforcement cost and diminishing revenues from unreported income. These unlawful practices seriously undermine the fundamental objectives of conservation and fair allocation. There are a number of contributing factors, including economic pressure, perception that risk of getting caught and punished is less than the rewards of violation, the respect of peers for large catches, inadequate funding, and a lack of understanding of the value of the management régime. Whatever factors contribute to these practices, it is plain that more aggressive and effective steps must be taken to discourage them."[70]

As a consequence, in the 1990 amendments, Congress made several changes to enhance the enforcement authority of the National Marine Fisheries Service and the Coast Guard. The Act now allows a maximum civil penalty of $US100 000 per violation, up from $US25 000 (16 USC sec 1858(a)). Most significantly, the Act now provides specific authority for the revocation or suspension of, or the imposition of conditions on, a fishing vessel permit, or the denial of a new permit[71].

Some fishing vessels are required to keep logbooks under regulations implementing the MFCMA similar to the New Zealand system. The logs are to contain, *inter alia*, information on catch and effort and are subject to scrutiny by "any authorised officer at any time"[72]. The Act makes it an offence to submit false information to the Secretary or to a Council[73].

One of the issues most fishery managers are concerned in regard to the implementation of ITQs and other output quota systems is discards. A report from Fisheries Information Systems in Juneau, Alaska, 20 July 1995 noted that, after the implementation of the IFQ, groundfish discards declined from 24% to less than 10% in the sablefish fishery. In addition, incidental catch declined, while discards of small sablefish declined from more than 3% to less than 2%. The presence of NMFS fishery observers on larger vessels in the Alaska IFQ programme undoubtedly restricts the opportunity to high-grade (smaller vessels are unable to carry observers). The initial flat prices offered by processors across different size classes of halibut during early 1996 suggest little incentive to high-grade. However, the increased landed-size of sablefish reported by the Canadian ITQ programme suggests that high-grading can be a concern.

ITQ shareholders have increased interest in fishery enforcement by the NMFS personnel who monitor ITQ landings since this enforcement effort protects the value (and possibly the size) of their future share in the fishery. Elements of the fishing industry advocate 100% observer coverage for all fishing vessels in ITQ programmes. Quota shareholders have an incentive to report on each other, since cheating directly harms individual quota holders. Additional incentives to report can be created by pooling quota shares revoked from cheaters and reallocating them to the remaining quota holders. The fear of losing ITQ shares, temporarily or permanently, may also provide an incentive that encourages compliance with regulations in ITQ fisheries. However, this is complicated by determinations of who is responsible for the illegal activity - those operating the vessel, the vessel owner, or the ITQ shareholder[74].

With an ITQ programme, a fisherman personally benefits from poaching, quota-busting and false catch-accounting (*ie*, under-reporting the quantity of fish landed); with open access management, only aggregate catches increase from false catch reports, and one fisherman filing a false report might not benefit. Thus, ITQs increase the incentive to operate illegally. ITQs may increase the incentive to cheat because unreported landings would supplement the short-term value of guaranteed quota shares. The increased dockside monitoring and enforcement staff across the North Pacific for halibut and

---

[70] Department of Commerce, NOAA Fishery Management Study 18 (30/06/87).
[71] 16 USC sec 1858(g). And see Sutinen, Rieser and Gauvin 1990.

[72] See 16 USC sec 1853(c); 50 CFR sec 611.9 (1978).
[73] 16 USC sec 1857(1)(I).
[74] Buck, Eugene H., Individual Transferable Quotas in Fishery Management, Report for Congress, September 1995.

sablefish, especially, makes enforcement expensive, while the sale of illegal halibut can be quite profitable.

5.2 Performance assessment: surf clam and ocean quahog fishery and the wreckfish fishery

In the surf clam and ocean quahog fishery, administration and enforcement costs have plummeted since the ITQ programme began. Before the ITQ programme, enforcement costs in this fishery were exceptionally high because unusually stringent management regulations were in effect - the Coast Guard closely monitored the number of trips and fishing hours of each individual vessel. Now extensive monitoring is no longer necessary; dockside monitoring alone is considered adequate. There have been no surf clam or ocean quahog enforcement cases in several years. This is also attributed to the fact that the resource is not overfished and there are a limited number of quota holders[75].

The South Atlantic Regional Council reports that wreckfish ITQ holders have also been co-operative, that compliance with ITQ programme regulations has been good and that administrative and enforcement costs are low.[76] However, this optimistic view is less persuasive given the small number of vessels and limited area fished in these fisheries. Simplified enforcement is more likely to be found in smaller fisheries arising from peer pressure and based on mutual interests of ITQ shareholders. It appears that effort has decreased dramatically because other fisheries are providing greater economic incentives than wreckfish.

5.3 Assessment of performance - Alaska Halibut and Sablefish IFQ Programme

There have been several cases prosecuted in the Alaska IFQ Halibut and Sablefish Programme in Alaska[77]. These cases, for the most part, have consisted of fishermen falsifying reporting by either under-reporting their catches or reporting catches from one area when the catches occurred in other areas. When IFQ violations occur in Alaska, the enforcement agency has several options available for prosecution. The violation may be brought under the civil administrative procedure of the Magnuson Act by issuance of a Notice of Violation and Assessment (NOVA) by the agency attorneys. In addition to a NOVA, the catch and vessel may be forfeited in a United States District Court. Forfeiture cases are filed on behalf of the Agency by the United States Attorney's office.

In Alaska, the Agency, through the US Attorney's office, has brought criminal actions under the Magnuson Act for false-reporting under the ITQ program. False-reporting under the Magnuson Act carries with it a possible jail time of six months or not more than $US100 000, or both. Criminal actions have also been brought under the Lacey Act and Title 18, General falsification statutes. Jail time under these Acts may be up to five years. Obviously, therefore, fishermen have argued that the criminal provisions of the Magnuson Act (only six months jail time) should be the exclusive criminal remedy for falsification of fisheries documents and not Title 18 of the US Code. However, the Courts have held that the Magnuson Act has not preempted Title 18 and the government may bring a prosecution under either Act.[78]

Fishermen have been jailed for violations of the Alaska IFQ programme for several reasons depending on the case. One case was brought criminally so that a grand jury indictment could be used as a means of obtaining otherwise difficult-to-reach information. The prosecution of IFQ fisheries cases based on fraud or falsification of documents has apparently been a strong deterrent since the number of cases in the past year has decreased.

6. SUMMARY

The Quota Management System introduced into New Zealand in 1986 has proved to be a qualified success. On the economic front, it has provided through the allocation of "property rights in the form of ITQs" a means of both allowing the free transfer of catching rights that has led to economic rationalisation of the industry and provided a "real" asset against which investments can be secured. The New Zealand experience has highlighted the validity of some the concerns that had been expressed concerning the introduction of ITQ based quota management systems. Quota-busting, data-fouling and bycatch/overfishing have proved difficult issues to deal with. Whether these problems prove to be inherent in the ITQ model or simply a consequence of how it was implemented may well be determined in the next decade as New Zealand continues to review the basic structure of its QMS.

On the enforcement side, the New Zealand experience has been that a system based on output-constraints supported by a paper-based trail of records and documents has significantly improved their ability to detect and prosecute illegal fishing activity by doing away with the need to be present at the commission of the offence. New Zealand has a number of advantages that have enable such a system to work effectively. New Zealand is isolated from other states were illegal fish might be landed. In addition it has no federal/state jurisdiction complications and has been able to impose strict record keeping/licensing requirements on the participants in the industry that may not be possible in other jurisdictions. There continue to remain real issues as to the overall cost-effectiveness of the current police/forensic model used to enforce the system and whether, and to what degree, this

---

[75] Telephone conversation with NOAA General Counsel Enforcement Attorney, North East Region, Charles Juliand, November 8, 1999.

[76] Telephone Conversation with Assistant Executive Director of the South Atlantic Fisheries Management Council, Greg Waugh, November 8, 1999. Also see paper.

[77] Telephone Conversation with Stephen Meyer, Special Agent in Charge, Alaska Region, November 8, 1999.

---

[78] United States v. Tomeny, 144 F.3d 749 (11th Cir. 1998).

model can be complemented or supplanted by a much greater emphasis on Industry itself implementing and enforcing the majority of ITQ "rules" through self-regulation.

In both the United States and the New Zealand some of the issues that continue to need addressing include:

Nature and levels of non-compliance

Developing better methods of evaluating whether levels of offending have decreased or increased, altered in nature or scope, or even whether the level and type of offending constitutes a threat to the overall efficacy of ITQs. A great deal of speculation is still involved in estimating the current levels of offending and their significance or impact. In addition, there is an on-going need to determine or measure the effectiveness of current enforcement/compliance strategies *vs* alternative approaches[79].

The role of records and returns

Notwithstanding the enormous financial and physical resources that have been committed to the cross-checking/record-based approach of the New Zealand QMS, both by government and stakeholders, there has never been a satisfactory analysis undertaken as to the efficacy of the system. Nor has any critical assessment been made of the fundamental premise on which the system was predicated, that it actually contributes to the effective enforcement and monitoring of the ITQ.

Nature of core enforcement/compliance skills required

Better strategic analysis of compliance methodologies that might be employed is needed. In New Zealand there is strong support among industry groups for a move away from the police-based model of law enforcement towards a more administrative and co-operative compliance methodology. Even within the Ministry of Fisheries in New Zealand there is an ongoing debate between those advocating the random compliance model (the Dockside monitoring/fishery officer on the wharf) and those supporting the targeted enforcement model (multi-disciplinary/paper trail investigative).

Administrative costs vs front line delivery

Development of effective operational auditing procedures to determine whether the current levels of expenditure on compliance are efficiently targeted at the priority/critical areas, the effectiveness of different delivery mechanisms employed by the compliance groups and the ratios of overheads/management relative to field expenditure are needed. Compliance expenditure is often dictated by historical budgeting practices and the prevailing skills of local staff and managers.

7. LITERATURE CITED

Boyd, R. and C.M. Dewees 1992. Putting Theory Into Practice: Individual Transferable Quotas in New Zealand Fisheries, Soc. Nat. Resour. 5(2) pp.179-198.

Clark, I., P. Major and N. Mollet 1988. The Development and Implementation of New Zealand's ITQ Management System. Rights Based Fishing. Eds. P.A. Neherand *et al.* p.117-147.

Dewees, C.M. 1989. Assessment of the Implementation of Individual Transferable Quotas in New Zealand's Inshore Fishery, *N. Am. J. Fish. Mgt*, 9(2). pp.131-139 (see pp135-136).

Sutinen J.G. 1996. Foundations of Compliance Strategy: A Framework for Policy, Report to the Compliance Business Unit, July 1996.

Sutinen, J., A. Rieser and J. Gauvin 1990. Measuring and Explaining Noncompliance in US Fisheries, 21(3), *Ocean Devt & Intl Law* pp335-372. Ministry of Fisheries, Compliance Strategic Plan, 1997-2002.

---

[79] In New Zealand the MOF has made substantial attempts to commence such a process (Sutinen 1996).

A MECHANISM TO ADDRESS SURPLUS GROWTH WITHIN QUASI-PROPERTY RIGHT SYSTEMS

C. Annand and F.G. Peacock
Department of Fisheries and Oceans, Scotia-Fundy Sector, Fisheries Management Halifax
PO Box 1035, Dartmouth, Nova Scotia, B2Y 4T3 Canada
<AnnandC@mar.dfo-mpo.gc.ca> and <PeacockG@mar.dfo-mpo.gc.ca>

1. INTRODUCTION

This paper provides a progress report on the introduction of *thresholds* in selected Canadian East Coast Fisheries and the degree to which the Department of Fisheries and Oceans (DFO) has been able to address multiple conflicting objectives of economic efficiency and socio-economic concerns. These objectives have been discussed in Angel *et al.* (1994 pp. 15-17). The economic performance objective has been to maintain an economically viable industry on an ongoing-basis where viability implies an ability to survive downturns with only a normal business failure rate and without government assistance. The employment objective has been to maximize employment subject to the constraint that those employed receive a reasonable income through earnings and fish-related transfer payments[1]. In consort with these objectives is the rising social issue related to the well-being of coastal communities. In this paper these goals will be viewed in light of resource-booms linked with possible later downturns and the possibility of sharing excess profits to meet demands on the resource from other than the original quota holders.

As is described in greater detail later in the paper thresholds are best described as mechanisms by which quasi-property rights (QPR) participants can define when, for how long and for how much resources will need to be compromised within the QPR format to ensure a devolution of activity to other participants. The case studies considered involve one groundfish stock (Georges Bank 5Z yellowtail flounder) and two invertebrate stocks (Scotian Shelf shrimp and Eastern Nova Scotia snow crab). The former is a case-study related to the possible implementation of the threshold concept for one species currently experiencing biomass growth, within the larger framework of an established long-term Individual Transferable Quota (ITQ) programme where most other species/stocks managed under the programme are only stable or in decline as a result of small spawning stock biomasses. A major objective in the decision to implement in 1991 ITQs for this fleet (vessels under 65' in length that use mobile gear, *i.e.* trawls) was to provide the fleet with an administrative means to resolve their excess capacity problems. The latter case describes the implementation of ITQs and thresholds simultaneously in new shellfish fisheries that have experienced considerable expansion since the collapse of many groundfish stocks in the early 1990s. These programmes were initiated and designed to match the resource abundance with the harvesting capacity, initially with the view to having a controlled-exit of vessels from the fishery if and when resources declined.

This paper describes the evolution of the threshold concept under a variety of different circumstances and provide a brief synopsis of the problems encountered. It examines the use of the Integrated Fisheries Management Plan (IFMP) process to provide a mechanism that is not only effective in reducing the political volatility associated with access and allocation issues within fisheries, but in doing so also garners support from industry as the decision-making process is devolved to them through these co-management approaches.

2. BACKGROUND

The use of quasi-property rights (QPR) in the form of individual quotas began in the Canadian East Coast fisheries in 1977 when Canada extended its Exclusive Economic Zone to 200 nm. The reduction of foreign fishing activity was heralded as going to provide a new frontier in prosperity for Canadian fish harvesting and processing. Even though several traditional stocks such as herring and groundfish were depressed due to high levels of exploitation, the "new era" anticipated growth in stocks of fish as well as all other venues of the fishing industry. Therefore at the announcement of the extended jurisdiction euphoria abounded in Canada as fishers, both traditional and new to the game, planned how to divide up the expected spoils of this good fortune.

The policy and decision-making processes of the day resulted in numerous and often conflicting principles of conservation, economic viability and more social objectives related to community support. By the late 1980s the real phenomenon of too many fishers and too few fish suggested that the vision of the government in its 1976 policy was wrong or that both industry and government were unable to make it work.

It was in this light, during the 1980s that government in concert with the industry began to seriously look at property right schemes as possible solutions for some of the problems. QPRs were seen as providing interesting solutions to some of the major problems in exploding fisheries in that they tend to make people accountable for their actions and can be successful, when used in a transferable format, at controlling capacity growth. In consort with stringent conservation controls for rebuilding resources QPRs can become formidable tools for adjustment in a way that industry supports.

Kirby (1982) recommended the allocation of non-transferable quotas to large fish companies which he termed Enterprise Allocations or EAs, as a means to encourage companies to live within their quotas. Since that

---

[1] In Canada, funds are transferred from the wealthier provinces to those with lower average incomes; they are referred to as 'transfer payments'.

time Individual Quota (IQs) and ITQs have been introduced for various fisheries from groundfish and herring to shrimp and crab, and within the Scotia Fundy region more than 50% of the landed value is now covered by these management approaches. In almost all cases those programmes introduced prior to 1996 have had as a major objective balancing capacity of the fleet with the resource, which coincidentally were in a state of decline at the point of introduction. Since their introduction, two salient points have become apparent among all QPRs, namely that fleet rationalization has occurred and stocks have stabilized or expanded after the time of introduction. The issue of rationalization is really a translation of economic self-sustainability or economic efficiency while conservation, leading to stock stabilization/growth, comes from the need by stakeholders to invest in the stocks for the future.

Despite the successes related to the introduction of limited access management, the question arises: does this type of management create any new problems? In the context of this discussion the real question might be the relationship of the technique to the socioeconomic concerns that confront managers on a day-to-day basis.

3. THE MOVE TO THRESHOLDS

Several factors can be identified as pertinent in developing a mechanism that could deal effectively with the boom and bust sequences seen in the East Coast Fisheries. These included ecological changes as well as economic and social considerations. An obvious question that arose related to the impact of resource fluctuations on management thinking. Two decades ago, the 200-mile limit promised a new dawn of prosperity largely to be based on groundfish. But the 1990s brought a codfish collapse and one of the largest employment losses and aid programmes in Canadian history. While more recently, unprecedented growth in invertebrate resources has resulted in shellfish becoming the largest source of revenue in the fishery.

Currently there is no established policy framework or regulatory mechanism to determine when and how wealth created in a given fishery might become subject to redistribution. Without a framework every improvement is subject to political lobbying by various interest groups. When no clear policy exists, Integrated Fishery Management Plans (IFMPs) do provide a process that allows industry to decide these issues with a minimum of political interference.

Thresholds are seen as one way of triggering a wider distribution of the resource in a way that is not as subject to political interference in the methods of reallocation in satisfying the needs or demands of others. Several issues have contributed to development of the threshold mechanism in QPR fisheries:

i. Much of Canada's commercial fishery is based in areas where there are few non-fishery employment opportunities and the commercial fishery has to reconcile the realities of a modern fishing industry in a global market environment with other public concerns including the support of coastal communities.

ii. The fisheries have achieved a high degree of efficiency, which manifests itself in several forms namely, (a) excess profits per individual and/or company and (b) concentration of activities among a few vessels and/or companies thereby affecting the coastal community network.

iii. In light of the groundfish collapse rapidly-growing resources have demanded a rethinking of old solutions as disputes among existing users and those wishing to again access to these lucrative resources intensified and both government and industry looked for ways to avoid the mistakes of the past and come up with innovative solutions that satisfy all concerned.

Central to this, and in the context of community infrastructure and support, are the small, inshore vessels which generally have not subscribed to such programmes in the past. This is changing as these fleets experience resource declines, reduced revenues and often community dislocations. The price paid by these groups is considered to be signifigant and something that political groups at all levels try to address. The issue of coastal community infrastructure is gaining popularity in many venues worldwide as the social values switch from those related solely to economic development to one encompassing both economic and social concerns with way and location of life being vitally important. The cry of "community death" and "save the community" are now common in Canada and in many nations worldwide. It is within this scenario that coastal communities networks, marine protected areas (MPAs), coastal infrastructure support *etc.*, have their followers and supporters which include many preeminent groups worldwide.

Within the threshold concept, questions will naturally arise including:

i. What is an excessive share of wealth under a QPR privilege?
ii. Who should decide when there is a resource surplus?
iii. What criteria should be used to trigger the use of thresholds in a particular fishery?
iv. What would be the future status of participants entering the fishery?
v. If abundance is increasing should the licence pool be extended and if so should the number of participants bear any relation to the long-term stability of the resource?
vi. Does the Department of Fisheries have a mandate to meet social objectives or is this better left to industry or other government departments to address this issue?

Rather than only being able to answer specific questions the case studies will likely consider far more as they assist in developing a clear policy framework. It may be that in the end 'a one-size-fits-all policy' will not be acceptable. However, the goal within a QPR system could be to attempt to allow market forces function while at the same time ensuring that at some point some level of protection, or constraining of market forces, at predefined thresholds. These thresholds, when defined, could include such things as super-profit limits and increases in resource abundance.

Thresholds can best be described as mechanisms by which QPR participants can define when, for how long and what resources will need to be compromised within the QPR format, to ensure a devolution of activity to new participants. This approach has the advantage of defining the long-term objectives of management plans as well as allowing the permanent stakeholders the opportunity to define its parameters. For coastal communities and fishers it provides a counter to the fear of consolidation/concentration and allows for income opportunities to a wider number of people who often live close to the resource under consideration.

4. THE PROCESS

While enjoying the benefits of QPRs, the fisheries in Canada remains common-property resources with no allocations having absolute durability although some have more than others and the terms have been spelled out more explicitly in the development of IQs and EAs. In most cases the percentage shares of the rights holders have not greatly changed in many years.

Even so, fleets hotly dispute the various shares. The management plan for Atlantic groundfish (DFO 1999), for example, says that "Allocation of fishery resources will be on the basis of equity taking into account adjacency to the resource, the relative dependence of coastal communities and the various fleet sectors upon a given resource, and economic efficiency and fleet mobility". But the plan gives no priority listing or weights to these factors. Industry members can argue that any of the options should prevail. Rival interests often lobby heatedly for higher quotas, not only through the advisory committee system, but directly to official's and the Minister's office. This in turn creates suspicions of undue influence and politicization of the fishery's management.

Besides the rivalries between existing users, others often demand a place in the fishery, particularly for newly-developing or rapidly-growing fisheries. When stock-abundance seems surplus to the needs of the existing fleet, several contentions arise: (a) whether other core fishers outside that sector should get a share; (b) whether new entrants from outside the fishery should be allowed in, (c) should communities be allocated shares or (d) should we allow fleet-efficiency to continue to increase at the possible expense of coastal communities' welfare.

The co-management approaches currently being adopted in Canada provide a process with which to deal with the multiple and conflicting rules that continue to exist, including access rules that are fragile, given that the power of allocating the resource that continues to reside with the Minister. Integrated Fisheries Management Plans (IFMPs) have been adopted in recent years to give industry a greater voice in the decision-making process and through a cooperative process involving science, management and Industry set the basic rules for management and allocation, sometimes on a multi-year basis.

Integrated Fisheries Management Plans (IFMPs) and Joint Project Agreements (JPAs) have encouraged industry to reduce the political volatility by bringing the decision-making process closer to the local level. Rather than accepting an external process that is subject to rigid rules, or political pressures, as stocks increase or decline, industry has defined its own mechanisms to avoid this. Through this process some groups have designed management responsibilities for administering their share of the total allowable catch (TAC), including biomass thresholds for management action, as well as other aspects of the fishery. These co-management arrangements provide industry with better security of access, clearer roles for government and industry, and more opportunity for industry to put its expertise to use in managing the fishery. This process, in fact, allowed the next logical step to occur *i.e.* the introduction of thresholds.

5. THRESHOLD CASE -STUDIES
5.1 Introduction

What exactly is meant in this article by thresholds, how are they determined and when are they used? To explain the current thinking several case studies will be used as examples of the rationale for their use and their intended goals. While it is evident that this approach can take several forms, themes come to the fore which suggest objectives for this approach.

5.2 The Scotian Shelf Shrimp Fishery
5.2.1 The setting

The fishery, long a relatively dormant operation because of bycatch problems, was revived in 1993 through the use of the Nordmor grate. Since that time the harvest from this resource has grown from a TAC of under 2000t established in 1995 to over 5000t in 1999. Projections through the Regional Advisory Process (RAP) are for further increases. It is prosecuted largely by groundfish draggers under 19.8m although a small (500t) shrimp trap fishery also exists. Between 1994 and 1996 both the Scotia Fundy and Gulf components of the fleet moved to ITQs. These draggers come from many localities within the Maritime Provinces of Canada with all vessels fishing the 6-9 month fishery within one or two locations (Canso and Petit de Gras) in Nova Scotia. Local fishers and processors do not participate in the fishery and landings are trucked to New Brunswick, Newfoundland or the USA.

Local fishers are mostly groundfish operators who have been hard hit by the 1993 moratorium on cod. The main consultative body for this fishery is the Scotian Shelf Advisory Committee in conjunction with the Atlantic Canadian Mobile Shrimp Association, which represents all 29 licence holders.

5.2.2 The problem

The fishery experienced its most profitable years in 1995 and 1996 due to high prices, increased stock abundance and gear selectivity in combination with the general downturn in the groundfish fishery and more recently a downturn in the inshore lobster fishery. With this situation demands for a "share of the resource" were made to the federal Minister by local and provincial politicians as well as fishers from the area, native groups and local merchants on the basis of adjacency to the resource and lack of other fishery related employment.

5.2.3 The solution

Faced with the prospect of arbitrated decisions by federal bureaucrats, the full-time licensed fishers set up a threshold scheme which would provide long term protection for vested licence holders and would provide for temporary sharing of stock growth on a predefined formula. One critical element was the determination of the average landed value and an open and reliable means of calculating it. Several objectives were put forward as part of the overall plan including: (a) protection and stability of the existing fleet; (b) incentives to the existing fleet to maintain stewardship of the resource; (c) and provision of economic relief in hard-pressed areas adjacent to the resource. The details of the programme are as follows:

i. Establish a quota minimum threshold equal to the 1997 quota base (3200t). As fishers think in terms of tonnes, a tonnage threshold was essential.
ii. Establish a minimum-revenue level which was equal to the 1997 tonnage and price, *i.e.* $Can5.29 million. Share revenue on amounts in excess to the 2 threshold levels. That is, if both thresholds are met, the sharing would be on the value on a predetermined basis. The value is established using a price per kilogram based on international accepted standards.
iii. Share the excess resource based on predetermined percentage levels of 50/50, 60/40 & 70/30 based on quota levels. New entrants, selected from applicants using a draw method would ultimately receive amounts equal to permanent licence holders over a 2-year period. Continued growth in value and tonnage could result in successive allocations to new groups of temporary licence holders over the duration of the 5-year plan.

The result of this approach has been a steady and controlled increase in licence holders under predefined conditions and in a manner that is consistent with the objectives of the long-term plan as developed by the industry of fulltime entrants.

5.3 The Eastern Nova Scotia Snow Crab Fishery
5.3.1 The setting

This fishery consists largely small vessels (under 19.8m length) based in four crab fishing areas in the eastern part of Nova Scotia. It operates from several local ports, which have a mix of fishers, with and without, crab licences. Through the 1970s the number of licence holders increased, with landings and licences keeping pace with one another. However, by 1982 quotas were not being reached and in 1984 quotas were removed as a management tool due to lack of scientific ability to predict stock biomass. Throughout the late 1980s increased biomass resulted in a resurgence of effort. Licences distributed remained stable until 1995 and biomass has continued to increase.

5.3.2 The problem

Perceived excess profits intra-port jealousies, the need for fishing alternatives given the groundfish downturn and the adjacency issue have all played a part in demands for more access, largely exerted through the political process. These factors were because of increased market prices due to the collapse in the Alaskan crab fishery and the growth in crab populations combined with the declines in both groundfish and lobster resource levels. The particular concerns in this fishery, apart from the knowledge of the status of the resource, are related to the efforts of many fishers to receive the benefits from ITQ fisheries where a level of great profitability is possible. Thus, there was a need to entrench a level of potential longer-term benefits while accepting the social need to have a process that delivers some benefits to local interests as surplus growth occurs. Research has indicated that with the stock increase certain areas within the crab zones were underfished by the traditional licence holders and it identified potential for new entrants.

5.3.3 The solution

Licence holders recognized that their fishery could tolerate additional effort on a temporary basis. Thresholds were then established in the inshore snow crab fishery in an attempt to protect a long-term viability position for existing IQ licence holders while at the same time allowing more fishermen to share resource surpluses. The long-term objectives for this fishery include (a) the continued biological and economic viability of these stocks and (b) the broader distribution of temporary access within the fishery to other core licence holders when both market and resource conditions are favourable, and in a manner that does not threaten the viability of the regular licence holders.

The threshold plan for each snow crab area is different but the principles are generally the same. These objectives are achieved through the following tactics:

i. Fulltime crab fishers identify location (traditional and non traditional areas) and threshold amounts (tonnage or tonnage and value).
ii. Licence holders identify a mechanism for sharing growth beyond a threshold. That is, sharing of access to temporary rights holders which may involve reciprocal zone-sharing, straight access-sharing or sharing of fishing zones.
iii. A complex tiered approach was developed, which if TACs continue to increase, will result in the temporary fleet receiving all the excess above the threshold and with equal access to the entire zone.
iv. Plans for crab are normally for a period of 5 years at which time issuance of thresholds based rights revert to a ground zero situation and negotiation recommence.

5.4 Georges Bank yellowtail flounder, a case-study in progress
5.4.1 The setting

In established multispecies demersalfisheries the concept of thresholds manifests itself in a different manner. Licence holders are being encouraged to develop a threshold procedure that protects the interests of the initial licence holders but also allows access to other ITQ fishers in a reasonable manner. Because of the complex nature of these demersal fisheries (they are Atlantic-wide in Canada and consist of many fleet sectors both competitive fishing using quotas ITQ) and the fact that resources are generally not increasing (as is the case for shellfish) the issue of thresholds has not surfaced. However in one area, notably Georges Bank, the yellowtail flounder resource is experiencing a major resurgence of abundance that is expected to continue at least in the near-term future. This involves a fishery where ITQs have been proposed for fewer than the total number of licence holders currently exploiting the other groundfish resources in the area.

Prior to 1993 the resource was not fished by any fleet however with the initial recovery of the resource, the Mobile Gear ITQ fleet of vessels under 65ft in length began a directed fishery. At that time approximately 176 licence holders with initial cod and haddock quotas had access to the resource as a bycatch although in reality fewer than half of the fishermen prosecuted the resources available on Georges Bank. In the short-term it was recognized that the resource would not support a directed fishery for all possible participants. The industry decided to restrict entry based on their criteria, leaving open the question of long-term benefits. This implied that there may be stock growth and the need for restructuring as it was known that the precautionary TACs established would in the long-term be scientifically based.

5.4.2 The problem

The resource has continued to increase, currently to 2000t with 93% available to the ITQ fleet whose members have now requested that a formal ITQ system be established to guarantee this level of harvest. With this request other ITQ licence holders have questioned the fairness of restricting involvment to the initial participants and have asked the government to intervene. Fearful that a bureaucratic solution would be imposed, the fleet owners were prepared to put forward a threshold mechanism that may be acceptable to all. As in the invertebrate fisheries the primary objective will be to protect the interests of the initial participants while offering some level of reasonable access to other licence holders and perhaps fleets, while resource levels remain high.

5.4.3 The solution

The situation in this case is more complex than the scenarios put forward under the single species invertebrate approach with the issue further complicated by the necessity of establishing an ITQ allocation formula and dealing with thresholds within a multispecies fishery. To date, the fleet, while agreeing to the general principle, has not recommended a consensual approach. Currently, members are considering a three tiered-formula dealing with the initial licence holders, other active IQ participants and the inactive vessels. A second tiered approach using yellowtail catch-history and initial cod and haddock allocations to provide the appropriate access opportunities is also being examined. Sharing with other fleets, *i.e.* EA fleets, has not been broached other than to ensure adequate bycatch is available. Market value will likely not be a part of the sharing formula.

6. CONCLUSIONS

The objective of thresholds over time in a true stock-growth situation would be to normalize the activities of all participants within the fishery, which would in so doing guarantee the long-term viability of the permanent IQ fleet. In situations of no stock-growth, shrinkage of temporary participants would occur such that to the degree possible to guarantee the viability of the permanent fleet (if stocks go below threshold amounts or value it will not be possible). The degree to which success of this objective can be measured is limited. Currently we are only experiencing the ascending part of the boom-and-bust cycle. However general indications suggest that the increased participation in decision-making delivered through the IFMP process aids both temporary and permanent licence holders in understanding the shared responsibility they have undertaken. This should enable them to deal with the stock fluctuations that will test this concept more fully in the future.

It appears thresholds offer solutions to specific problem, but in the context of ITQs other questions should be asked:

i. Are thresholds simply a way of addressing a social/political agenda?
ii. Should they be viewed as a global solution?
iii. Do they degrade ITQs *i.e.* economic efficiency?
iv. Should they only be used in IQ rather than ITQ fisheries? and
v. What is their security of access?

While we have moved forward in certain situations and have allowed the introduction of thresholds to alleviate specific problems, certain negative consequences may have to be taken into consideration if they are to obtain a broader application. The most obvious problem is that while they may be an effective means of addressing social and political concerns, thresholds can be seen as reducing the benefits that should accrue to property rights fisheries. In this sense, the adoption of this process, especially after establishing ITQ fisheries, could be seen as undermining what was seen to be a desirable move to management using property rights. Given the inherent boom-and-bust nature of the fishery, individual fisheries may react in differing degrees to the concept of redistribution or quota sharing. New fishers moving into ITQs in the aftermath of the groundfish collapse and coping with the social and economic pressures created with the growth in invertebrate fisheries, may be easily convinced by their fellow fishers and their communities that this is the appropriate course of action. Long standing fishers that have survived both fleet rationalisation and stock declines, and invested heavily to achieve the economic efficiency promised to those embracing QPRs, may be less amenable to the suggestion of thresholds.

In allowing fleets to choose their own path we could in effect create a tiered approach to QPRs. However, this whole topic will require further discussion before the threshold concept is adopted in any but special circumstances.

7. REFERENCES

Angel, J.R. 1994. Report of the workshop on Scotia Fundy Groundfish management from 1977 to 1993,

Canadian Technical Report of Fisheries and Aquatic Sciences, 1979 1-175.

DFO 1999. Integrated Atlantic Groundfish Management Plan, Guidelines pp 64, Communications Directorate, Fisheries and Oceans Canada, DFO/5536.

Kirby, M.J. 1982. Chairman. Navigating troubled waters – A new policy for Atlantic fisheries, Report of the Task Force on Atlantic fisheries, Highlights and Recommendations, 1-151.

THE MISSING T: PATH-DEPENDENCY WITHIN AN INDIVIDUAL VESSEL QUOTA SYSTEM – THE CASE OF NORWEGIAN COD FISHERIES

B. Hersoug, P. Holm and S.A. Rånes
The Norwegian College of Fishery Science, University of Tromsø
NFH, University of Tromsø, 9036 Tromsø, Norway
< bjoernh@nfh.uit.no>, <petterh@nfh.uit.no> and <steinr@nfh.uit.no>

*"Given the availability of such a marvellous system as the ITQs,
why have not all fishing nations implemented it?" (Hannesson 1992:93)*

1. INTRODUCTION

Individual Transferable Quotas (ITQs) or more generally individual (private) property rights, are now being introduced in fisheries management world-wide, not only as *a* solution but as *the solution* to the familiar management problems we find all-over: over-capacity, dramatically reduced stocks, declining profitability, rising management costs, unworkable MCS systems, lack of management legitimacy, *etc*. With a crude simplification we can group the solutions to the present problems in three camps, with economists generally arguing in favour of (a) individual property rights while social scientists, and in particular social anthropologists, are pointing to (b) co-management, community management or more generally to (c) user-group participation (Christy 1996, Neher *et al*. 1989, Hannesson 1998, Jentoft *et al*. 1998, Pinkerton 1989). In between we find the still reigning biologists, either defending the old models with more sophistication, or pointing to multi-species modelling, or to the even more complicated concept of "ecosystem management" (Degnbol 1998).

Even if rights-based fishing can be associated with all three views, the concept of ITQs is clearly connected to fisheries economists and to the market as the central agent in fisheries management. The basic idea is expressed by Hannesson (1992: 92): *"The advantages of ITQs are obvious....Dividing the TAC among all vessels participating in a fishery prevents a self-defeating race for the largest possible share of the total catch. And making the vessel quotas transferable makes it possible to minimise the cost of taking a given catch. In the short term, transferability ensures that the least efficient fishing vessels will not be used, as their quotas will be bought by the owners of the more efficient vessels at a price that benefits both buyer and seller. In the long run transferability means that the owners of fishing vessels can adjust their fishing capacity to the amount they may expect to be able to take, or vice versa."*

In this world *institutions, rights* and *efficiency* are simple and well defined concepts, given for everybody who are not blinded by "ideological or political reasons" (Hannesson 1992:93). Considering the theme of the conference: "Use of property rights in fisheries management", we feel as social scientists more like the devil's advocates; trying to show that there is more to institutional change than getting the incentives right, that property rights are constituted through complicated processes and that efficiency is highly dependent on the political setting. As a demonstrative case we shall use the (attempted) introduction of ITQs in the Norwegian fisheries in general and the introduction of individual vessel quotas (IVQs) in the cod fisheries in particular.

The interesting aspect of the Norwegian case is that the deliberate effort of introducing ITQs was flatly rejected, even though all actors acknowledged the need for a more flexible system. In the meantime, the introduction of a "temporary" IVQ-system, meant to be a crisis measure, in practice, has turned out to be something close to a permanent ITQ-system, where only the official recognition of the transferability is missing. Having introduced the "I" and the "Q" it will be argued that some form of "T" will follow, as a path-dependent process. This paradoxical situation, where planned intervention failed while the efforts of "muddling through" produced the unintended results, reminds us of the importance of politics and power, labelling and timing. By unfolding the story, we hope to shed some light on the construction, "selling", implementation and gradual acceptance of a quasi-ITQ system.

This paper will therefore concentrate on answering four basic questions:

i. Why was the deliberate attempt of introducing a Norwegian variant of the ITQ-system totally rejected, not only in the Norwegian Parliament but generally in the fishing industry?
ii. Why has the implicit attempt of introducing a temporary IVQ-system ended up as a "quasi ITQ-system"?
iii. What are the prospects of the existing system in terms of gaining transferability of the fishing rights (the path-dependency)?
iv. What are the lessons of this paradoxical case for institutional theory and, in more practical terms, for the introduction of new management systems?

In Section 2 we introduce the theoretical approach (institutional theory, embedded systems and nested systems). Section 3 describes the Norwegian fishing sector in general and the cod fishery in particular. In Section 4 we deal with the introduction of a Norwegian ITQ-system, while Section 5 describes the history and performance of the IVQ-system from its introduction in

1990 to the present. Section 6 deals with path-dependency, describing how the IVQ-system went from being provisional to permanent, from free entry to limited access, from a non-market situation to a "grey market" for fishing rights, and finally the "domino effect" whereby most other fisheries were closed as a result of the increasing pressure created by the closing of the cod fishery. A central point here is that the political prioritisation of a certain group of vessels, by necessity will require some form of recognised transferability. The last section deals with the analytical and practical lessons of the Norwegian experience in relation to the introduction of new management regimes.

2. INSTITUTIONAL CHANGE AS A POLITICAL PROCESS

In social science approaches to fisheries management it has become commonplace to state that management takes place through institutions, and even more important, that changes in management regimes (like introducing ITQs or rights-based fisheries more generally) happen through the establishment of new institutions or modification of old ones. Recognising that there are many competing forms of institutionalism, what are the minimum defining criteria – what are we talking about? According to Peters (1999:18) there are at least four defining characteristics of an institution:

i. It must be in some way a structural feature of society, formal or informal
ii. It must have some existence over time
iii. It must affect individual behaviour
iv. There should be some sense of shared values and meaning among its members.

With these defining characteristics in mind, there are a host of possible institutions in fisheries management, ranging from the formal Directorate of Fisheries to the informal network of co-operating fishermen fishing away from home. Institutions comprise complicated structures from *the scientific fisheries management institution* (including national as well as international research organisations, advisory boards, administrative as well as political entities) to simple management measures like *"the trawler ladder"*, the distribution key allocating quotas between trawlers and coastal vessels. It is, however, a long and sad tradition, also seen in the social sciences, to treat management only from an instrumental point of view, that is, institutions are seen as rules circumscribing the individual fishermen. By using the more comprehensive definition of Scott (1995:33) we also include the normative and the cognitive aspects: *"Institutions consist of cognitive, normative, and regulative structures and activities that provide stability and meaning to social behaviour. Institutions are transported by various carriers – cultures, structures and routines – and they operate at multiple levels of jurisdiction."*

In our case we are most concerned with institutional change and with the establishment of new institutions.

For the sake of simplicity we shall concentrate on two different institutional perspectives, one *instrumental*, which is quite common among fisheries economists, and a more sociological perception of institutions as *embedded* in a larger social structure.

Starting with the instrumental perspective, there are differences in approach, *e.g.* between economists and political scientists. Both will, however, be concerned with the manipulation of existing institutions and the design of new ones, like setting up ITQ regimes. Most often they will rely on some form of rational choice theory. Underlying this instrumental approach to institutional design and implementation is according to Peters (1998: 44): *"that utility maximization can and will remain the primary motivation of individuals, but those individuals may realize that their goals can be achieved most effectively through institutional action, and find that their behaviour is shaped by the institutions. Thus, in this view, individuals rationally choose to be to some extent constrained by their membership in institutions, whether that membership is voluntary or not".*

Institutions are consequently seen as sets of positive (inducements) and negative (rules) motivations for individuals, where the individual's utility maximisation is acting as the dynamic element in the institutional set-up. Although differing in detail, the different strains of rational-choice models are characterised by a common set of assumptions regarding the rational individual behaviour, and a common set of problems, relating to the classic challenge of how to make decisions relating to social welfare without having that decision imposed by a (central) authority. Most important, however, is the common assumption that institutions are formed from a *"tabula rasa"*. Past history is of little concern and new sets of incentives can produce the desired behaviour immediately, provided the right mix of inducements and constraints. The immediate focus will nevertheless vary.

A political scientist like Ostrom (1990) has been most concerned with extracting the necessary minimal requirements relating to the successful management of common property resources (CPRs). She is concerned with institutions as means of "prescribing, proscribing and permitting a certain type of behaviour". Economic historians, like North (1990) and economists, like Eggertsson (1990), have been more concerned with the particular institution of the market, and the rules prescribing property-rights regimes. All approaches within the rational choice paradigm have problems of explaining how the ultimate preference of maximising individual gain is made. While institutions can form most other preferences, the most important driving preference is somehow externally driven and (evidently) constant over time.

Against this instrumental perception of institutions we supply a more sociological concept, where management institutions are viewed as being embedded in a larger social structure. The key concept, "embedded" originated from Polanyi's (1944) famous study where he accounted for the social and cultural constraints on economic action in premarket societies. Thirty years later Granovetter (1985) resurrected the term in order to explain how rules, procedures and normative standards of conduct in various institutional realms, such as economic, cultural and social life, influence and shape each other (Apostle *et al.* 1998:236). While Polanyi was concerned with the disembedding features of modern market economies, Granovetter's intention was to show that these economies are indeed influenced by personal relations transmitted through networks. But as pointed out by Barber (1995), who traces the story of the embedded concept, Granovetter does not deal with the larger social systems in which all economies are located. This deficiency is the explicit starting point of Hollingsworth

to increase legitimacy, or more plainly, to increase support for a party or a position (Edelman 1985). Adherence *e.g.* to the Raw Fish Act is largely considered to be such a symbolic gesture, necessary for everyone seeking political support in the north. The symbolic use of institutions also allows for double book keeping, such as when fishermen flag one popular institutional solution with the public while participating in the development of another, contradicting institution in the more closed and concealed arenas of corporate management.

Having already indicated that there are a number of different types of *institutionalisms* we run the risk of distorting a complex issue by presenting two ideal types. Nevertheless, Table 1 points at some of the main differences between the two perspectives that we have briefly sketched.

Table 1
Properties of different institutional perspectives

Properties	Instrumental perspective	Embedded perspective
Definition	Rules/incentives	Cognitive, regulative and normative structures
Incentives	Rational self interest	Socially defined goals
Preferences	Exogenous	Endogenous
Actors	Organisational entrepreneurs	Social forces, mediated through institutional participants (with important time lags!)
Genesis	Deliberate construction	Normally a gradual social process through various stages
Change	Often, dramatic, based on bad performance	Gradual, incremental, conforming to social pressures

and Boyer (1997:3), who try to develop the argument that markets and other co-ordinating mechanisms are shaped by, and are shapers of, social systems of production. There is, in other words, an interconnectedness between social and economic institutions, working both ways. Dominant social values, rules and procedures may limit, or obstruct, what is planned in the economic sector, and vice versa, economic processes may, over time, influence social and economic beliefs. The important point is that institutional change is partly outside the realm of direct human intervention. Therefore, institutional reforms may give some quite unexpected results, or even no results in the short run, if they are totally out of context with what is considered socially and culturally acceptable. Whatever the case, institutional reform will most often require considerable time to show results, due to the inherent sluggishness in the system. (The more detailed logic, based on a nested systems perspective is spelt out in Holm, Rånes and Hersoug 1997).

Within the sociological perspective the use of symbols may play an important part in shaping a particular institution and not least in regulating the behaviour of the participants. In our case we should also include the symbolic use of institutions in politics, that is, the use of certain institutions as signal markers, in order

According to the embedded perspective, the introduction of a new management regime, like introducing ITQs, is more than designing the system, getting the incentives right and persuading the decision-makers. If institutions matter, politics matters even more! And politics is not an exogenous variable in fisheries management which can easily be eliminated. Politics is, whether we like it or not, the very essence of resource management, that is, allocating scarce resources (Easton 1953).

3. THE NORWEGIAN FISHING INDUSTRY

Norway is one of the larger fishing nations in the world, with catch volumes ranging from 2.5 to 3.5 million tonnes per year; this ranks Norway as number ten in the world according to FAO statistics. Norway's position arises from its areas along the coast belonging to a up-welling system (the Gulf Stream) and that these areas have been exclusively reserved for Norwegian fishermen. Most of the fish are caught within the Norwegian Exclusive Economic Zone (EEZ), an area encompassing more than 1.2 million square kilometres. In addition Norway is responsible for two fishing zones of approximately 1 million square kilometres around the islands of Spitzbergen and Jan Mayen. Nevertheless, 80%

of the total catches are based on shared stocks, with management responsibility shared with Russia, the European Union (EU), Iceland, the Faroe Islands and Greenland.

The fishing industry plays a relatively limited role in the overall Norwegian economy, being responsible for approximately 1.5% of GNP and near 2% of total employment. However, as an export industry fish and fish products are the second most important sector after oil (and before gas), being responsible for 9% of Norwegian exports. In 1998 the total export value was close to NOK 28 billion ($\cong$3.5 billion US$). This figure includes the sales of aquaculture salmon ($\cong$ NOK 10 billion), which has turned out to be an extremely important part of the Norwegian fishing industry, both in terms of production, employment and export. The exvessel value of the Norwegian catch is NOK 10 billion, distributed among 22 916 fishermen, of whom 6 257 are part-time and 16 659 have fishing as their main or sole occupation (Director of Fisheries 1998). The fishing industry plays a much more significant role, on the West Coast and in Northern Norway, where entire municipalities are based on fishing, processing, aquaculture and related activities.

The Norwegian fishing fleet comprises about 6658 active vessels of which about 90% are coastal vessels below 30 metres in length (Director of Fisheries 1998). While coastal vessels vary in size from 3m skiffs to 30m shelter-deck ships, the vessels in the 8 - 24.9m range account for 80% of the fleet's total landings. Seasonality is a characteristic feature of the coastal fishery since it exploits the different fish stocks as feeding and spawning migrations bring them close to shore.

As the mainstay of the traditional coastal economy, the fish resources have been regarded as the common property of the coastal people. In practical terms, this meant open access to the fisheries, but not for anyone. When capital intensive technologies - seine and trawl - were introduced in the groundfish fisheries towards the end of the last and the beginning of this century, the fishers resisted fiercely. They saw this as an attempt by merchants and industrial capitalists to take control of the fisheries. The fishers regarded them as outsiders with no legitimate right to harvest the resource. The Norwegian authorities have reluctantly accepted this viewpoint. Hence, seines were banned from groundfish fisheries with the adoption of the *1897 Lofoten Act* (Jentoft and Kristiansen 1989), while trawlers were banned with the adoption of the *1936 Trawler Act*, except for the few already established (Johansen 1972). Free access to the fishery only applied to people adhering to traditional fishing practices. This restriction was reconfirmed by the *1947 Ownership Act*, which reserved the right of owning fishing vessels to active fishermen (Mikalsen 1977).

During the post-war period the restrictions against capital intensive technologies, and hence against "outsiders", have gradually been weakened. As part of the attempt to rationalize the fisheries, both the Trawler and the Ownership Acts were made less restrictive (Sagdahl 1973; Mikalsen 1977). This meant the addition of an offshore trawler fleet in Norwegian fisheries, partly controlled by the processing industry. Instead of ending the traditional fishery, as the fishers had feared, the result has been a dual fleet-structure where the coastal and offshore sectors exist side by side. In addition to the differences in technology, operational patterns and ownership structure, the two fleet segments were subject to different regulatory regimes. While the coastal fishery remained under an open access fishery, the trawler fleet - as a direct consequence of its introduction in the face of massive resistance from established fishers - was subject to strict access controls right from the start. Not before the cod crisis of 1989/90 would the coastal fisheries be effectively closed, even if the principle was introduced by the mid 1980s.

4. THE INTRODUCTION OF AN ITQ-SYSTEM

In 1988/89 it was evident that a new cod crisis was looming. Following record catches in 1986-87 the researchers at the Institute of Marine Research openly admitted that the TACs had been fixed too high and that the stocks were rapidly declining. Consequently, the issue of over-capacity was put on the agenda.

The issue of ITQs (as distinct from non-transferable quotas) was introduced through the report of a working group on the structure of the harvesting sector. The group comprised representatives from the Ministry of Fisheries, the Directorate of Fisheries and the Norwegian Fishermen's Association (NFA). The original idea was to introduce enterprise allocations (*rederikvoter*) in the offshore fleet thereby making it possible for companies with two or more vessels to rationalise harvesting and by next the round, make it possible for two or more companies to co-operate in reducing fishing. This was considered by most fishermen and politicians alike to be more or less similar to ITQs. Although the NFA originally had endorsed the proposal, it was soon in a heated debate, with almost unified opposition from the coastal fishermen. On the outset everybody agreed that over-capacity was the main problem. According to the committee the costs of restructuring had to be borne by the fishermen themselves, as state subsidies had been dramatically reduced. This did not go well with the perception of fisheries policy as a regional development policy contributing to the coastal settlement pattern. Politically the new high flyer was shot down even before take-off.

Faced with overwhelming opposition the Minister backtracked and decided to initiate a white paper to Parliament, but dealing with a much larger range of management issues. Four officials from the Ministry of Fisheries were assigned to draft the first discussion paper, which (for the first time) was discussed not only with biologists from the Institute of Marine Research (the official adviser to the government) but with economists and social scientists from the Norwegian universities as well. Drawing heavily on the existing ITQ-schemes in

Australia, New Zealand, Iceland and Canada the group presented an overview of different forms of ITQs, ending up by recommending an ITQ-system with strong geographical limitations on transferability (Ministry of Fisheries 1991). The report discussed various forms of "transferability", including:

i. the traditional trading of quotas
ii. transfer of vessels with quotas
iii. enterprise allocations to be "traded" within the company
iv. the renting of quotas on an annual basis and
v. co-fishing where several owners may decide to use one boat to catch several quotas.

The Ministry's preferred version (pertaining to vessels larger than 8m) was based on TACs allocated to different groups (vessels and regions) based on their historical catch. Individual quotas given as shares, would be allocated for a limited period (5 years) and be subject to an annual resource fee, paid to the government. Quotas would be traded freely within groups and regions, while transfers across vessel groups and regions would require the permission of the Ministry.

By taking the demand for more flexibility and the need for regional stability into consideration, the Ministry thought the proposal would be accepted, both among fishermen and local and regional politicians; the opposite happened. *"The overwhelming majority of those consulted were strongly against ITQs, even in the modified version suggested in the draft"* (Apostle *et al.* 1998: 198). Looming large in the background was the fear that Norwegian harvesting rights would be bought up by European companies (Government of Norway 1991:126-7). Not only was the ITQ question connected to the coming debate of Norwegian accession to the EU, it was also immediately made an issue in local elections of 1991 where, especially, representatives of the north opposed "any privatisation of the commons". The issue threatened the political harmony of the Labour Party (now in government) and a task force within the party found that the question of ITQs was not on the political agenda, a position reinforced by the Prime Minister, evidently for expedient political reasons (Moldenæs 1993).

When the revised political version of the white paper appeared in Parliament, the question of transferability was considerably watered down. In the report from the Standing Committee on Fisheries the majority rejected outright an ITQ option while a minority would continue the work to introduce a programme for ITQs, a situation that was later reflected in the general debate in Parliament.

ITQs were, according to the winning coalition "a dead horse", thereby signifying a remarkable defeat for the former Minister, the top bureaucrats in the Ministry and the generally powerful employers organisation of Norway (NHO).The Ministry had evidently miscalculated not only the general political attitude, but misread the fishermen as well. Apostle *et al.* (1998) discuss the possible explanations, pointing to the short duration of the crisis (on recovery in 1992/93 overcapacity was no longer the pressing issue), the extraordinary process (where the industry organisations were not represented in the committee drafting the paper) and the lack of power on behalf of the government to make tough decisions in times of crisis. None of these explanations seen in isolation are satisfactory. Suffice to say that the horse was definitely not dead - it just took another route!

5. THE END OF OPEN ACCESS: THE IVQ SYSTEM

5.1 Beginning of the process

The open-access regime in the Norwegian coastal cod fishery came to the end with the collapse of the Northeast Arctic cod stock towards the end of the 1980s. Due to a sudden and unexpected decline in the size of the cod stock, the overall TAC was set to 340 000t in 1989, down from 630 000t the previous year. The coastal fleet's quota was reduced from 200 000t in 1988 to 116 000t in 1989. As it turned out, the combination of a small total quota and a competitive fishery produced unhappy results in the 1989 fishery. In contrast to most years during the latter half of the 1980s, the cod in that year proved easy to catch. This meant that the total quota for the whole year was finished and all fishing stopped as early as April 18, half-way through the traditional Lofoten fishery. This had severe distributional effects, with those who started early having good results, while the latecomers, often operating the smallest vessels, got little or nothing. Out of this emerged a strong resolve to avoid repetition of the 1989 situation: "Never again April 18!" On the basis of this experience, an individual vessel quota (IVQ) system was devised during the fall of 1989 and implemented during the 1990 season.

The political process by which the IVQ system was put together has been detailed by Holm and Raanes (1996). User groups, and particularly the fishers through the Norwegian Fishermen's Association, had a relatively strong position in this process and the key policy arena in the case of the negotiation of the IVQ regime was the Regulatory Council, established in 1983 as the meeting place between industry representatives and the fisheries authorities in resource management issues (Hoel, Jentoft and Mikalsen 1996). The Fishermen's Association formed the largest single group within the Council and appointed five of the nine industry representatives. While the Council formally only had an advisory role *vis-à-vis* the Fisheries Minister, the Council's decisions were usually very influential particularly if they were unanimous. In this case a government decision was made during the fall of 1989, based on the recommendation of the Regulatory Council.

The IVQ system was two-tiered: the most active vessels, as measured by the quantity of cod landed in the 1987-89 period were put under a vessel quota regime. These quotas were exclusive, so that the vessel owner (skipper) had full discretion to decide when and where to take it. The less active vessels were allowed to fish competitively under a group quota. There were no restrictions as to participation in this fishery; any registered fisher could join. However, the allocation to this group only amounted to about 20% of the total quota in the coastal cod fishery, and each vessel was subject to a small maximum quota (originally 2.5-3.5t, later to be increased).

In the 1990 season, the Individual Quota group (Group I - full rights) had 3534 vessels, while the Maximum Quota group (Group II - restricted rights) had 4172 vessels. Since then, the number of vessels in both groups have declined considerably, to 2766 and 3536 in 1999 for groups I and II respectively. Underneath this relative stability, substantial interchanges between the two groups have occurred. While transfers of vessel quotas by themselves are not allowed, such transfers happens when vessels change hands. A fisher can hence join Group I by buying a vessel with a right to fish in Group I. In addition, a certain movement in and out of this group is controlled by the fisheries authorities under the label of "recruitment". From 1994, continued participation in Group I was made subject to an activity requirement: To keep its quota, a vessel had to have fished at least 40% of its allocation the previous year, later to be adjusted to 10%. The vessel quotas that became available in this way, were set aside for the recruitment system within Group I.

Whether a particular vessel in 1990 qualified for Group I and got an individual quota, dependent on how much it had landed during the 1987-89 period. The principle of historical rights was also applied to decide the size of the vessel quota, although not in the same straightforward manner. First, it was not the catch record of each vessel that decided its quota, but the average catch records within the size group to which it belonged. Hence, every vessel within one size group (defined by length) would get the same quota, regardless of how much that particular vessel had caught. Second, the quota of each size group was not calculated, as one might have expected, as some constant share of historical records across size groups. Instead, the quotas were calculated on the basis of a scale decreased with increasing vessel size, so that the smallest vessels (under 8m) received 100% of their historical catches, while the largest (over 27.5m) only received 50%.

Two important points should be noted before we go to the question of how this system became permanent. First, the IVQ regime represented a departure from the traditional open-access regime in the coastal cod fishery. This observation is not only warranted by the fact that about 80% of the quota was allocated to Group I, the membership of which was strictly controlled. In addition, while Group II was open to almost anyone, the vessels within this group were severely restricted by the total group quota and the individual maximum quotas.

Second, the distributional consequences of the transition from an open-access to a rights-based regime were the key concern during the policy process. As is often the case, a principle of "historical rights" formed the basis of the quota-distribution mechanism. Such a principle is particularly suited to secure acceptance from the established parties in 'co-management' sectors, that is, sectors in which economic and political rights are linked. On top of this "grandfather clause", however, the smaller vessels were given preferential treatment, which is more unexpected. The main reason was that the smaller vessels were widely expected to be the main losers in a transition to a rights-based regime. Within this segment of the coastal fleet one would find a large concentration of vessels that only operated part of the year and fishers who would combine fishing with other occupations. The majority of those who would not qualify to Group I and therefore in practice would be barred from making a living in the fishery, were hence found here.

5.2 A temporary regime

An important reason why the IVQ regime could be adopted was that it was a temporary response to the resource crisis, and would be abandoned once the situation returned to normal. This was not explicitly stated neither in the regulations themselves nor in the Regulatory Council's recommendations to the Minister though all the major parties involved clearly expressed this view. The National Council of the Fishermen's Association in November 1989 reluctantly accepted the vessel-quota system, on condition that the excluded vessels be allowed to re-enter the cod fishery when the resource situation improved (NFA 1989b:7). One year later, the National Council still insisted that vessel-quota regime was temporary (NFA 1990a:21).

As the officially recognised representative of Norwegian fishers, such viewpoints from the Fishermen's Association would have carried weight even if the authorities had held a different opinion. In this case, however, they shared the NFA's standpoint at the outset. The Director of Fisheries, who was heavily involved in the design of the IVQ system as the chairman of the Regulatory Council, presented the IVQ system as a direct response to an exceptional situation in the coastal fishery (Director of Fisheries 1989a: II,18). Also the Ministry of Fisheries regarded the vessel-quota system as transitional.

While the fisheries authorities as well as the Fishermen's Association hence regarded the IVQ system as transitional, they were not in complete agreement as to which part of the system would have to change. The Fishermen's Association wanted to abandon access restrictions as well as individual quota rights and return to an open-access regime of the pre-1990 type (NFA 1989:7). The fisheries authorities, in contrast, wanted to get rid of the individual quota rights, but keep the strict entry controls (Director of Fisheries 1993a :II,8; Ministry of Fisheries 1992:136). In the debate over the quota system, this difference of opinions was not brought into the open, giving an appearance of complete agreement that the IVQ system was a temporary crisis measure.

5.3 The IVQ system becomes permanent

In spite of this agreement, the IVQ system became permanent: when the crisis passed, the established regime remained. We summarize this development in Table 2 which gives data on the coastal cod fishery for the 1990-1999 period. During the 1980s, the annual landings from the coastal fleet averaged 180 000t. Assuming this represented a "normal" situation in the fishery, then the crisis was over by 1993, which also was the perception within the industry. Despite this, the Ministry of Fisheries did not want to abandon the vessel quota system. With reference to over-capacity in the coastal fleet, it was argued that the quota was still insufficient to allow all vessels a normal level of operation (Ministry of Fisheries 1992). In spite of its earlier position, the Fishermen's Association supported this view (NFA 1992b), and the vessel quota system remained in place. The IVQ system remained also for 1994 which saw an innovation, however, in that 33% of the quota within Group I (full rights) was allocated on a competitive basis. Thus, the vessels got a maximum vessel quota of which only two thirds were guaranteed. For 1995, and the consecutive years, this arrangement was extended and the whole quota was allocated on a competitive basis.

The introduction of more competition within Group I did not mean a return to open access. The vessels that had been excluded from this group in 1990 were not allowed to re-enter as the Fishermen's Association originally had wanted. Two features of this system lead us to the conclusion that it should be interpreted as an *adjustment of the IVQ regime* rather than its abolition. First, although applied to allocate maximum rather than exclusive quotas, the arrangement for distributing quota allocations to individual vessels was not changed. Second, the shift from exclusive to maximum individual quotas was not permanent, but directly linked to the availability of cod on traditional fishing grounds. As operated at present, the system will produce exclusive vessel quotas when the availability is good, and maximum vessel quotas when it is not. Instead of a systemic change, the mechanism of "over-allocation" has injected a healthy dose of flexibility into the IVQ regime.

5.4 Political consequences of economic regulations

A main reason why the IVQ regime remained in place when the stock crisis passed away was a change in the position of the Norwegian Fishermen's Association. While the Association in 1989 saw the IVQ regime as a temporary derogation from open access, it actively supported the IVQ regime from 1994 onwards. How can the shift in the Norwegian Fishermen's Association's position towards the IVQ regime be explained? We argue that the answer lies in *the power of vested interests*. Once individual quota rights were established, the rights holders set out to protect their new-gained interest. To substantiate this interpretation, we must show that the vessel quotas and the fishing rights, represented important assets for the holders, and, second, that the interest in maintaining them became dominant in the Association's policy-making arenas.

Even if the majority of the participants in the cod fishery never have fished up to their technical ability, the exclusion of some 4000 vessels from full quota rights implied a dramatic improvement for the remaining 3500 rights holders in Group I. The price of their guaranteed

Table 2
Number of vessels and total quota for vessel quota group, maximum quota group and coastal fleet within the cod fishery 1990-1999 (Director of Fisheries 1990-1999d)

		1990	1991	1992	1993	1994	1995	1996	1997	1998	1999
Vessel quota group	No vessel	3 534	2 367	3 640	3 618	3 446	3 363	3 388	3 255	3 034	2 766
	Tot. quota (t)	61 750	70 375	101 800 [a]	133 420	184 425	195 460	192 780	237 330	196 025	143 490
Maximum quota group	No vessel	4 172	5 401	4 697	4 463	4 140	3 874	3 494	3 036	3 205	3 536
	Tot. quota (t)	12 000	17 000	11 000	17 000	21 000	21 000	21 000	25 000	20 000	15500
Total coastal fleet	No vessel	7 706	7 768	8 103	8 081	7 606	7 237	6 882	6 291	6 239	6 302
	Tot. quota [b] (t)	84 750	96 375	112 800	176 820	217 425	226 460	223 780	262 330	216 025	158 990

[a] After new advice from ICES Norway and Russia agreed the 17th July to increase the Norwegian TAC with 24 500t from 165 000t to 190 500t. Subsequently the total quota for the vessel quota group was increased to 118.800t.
[b] Total coastal fleet quota includes the bycatch quota and is hence larger than the sum of the total quotas for vessel quota and maximum quota groups.

quota (later fishing right) can be calculated by comparing similar vessels from Group I and Group II. Although officially there is no legal market for fishing rights, the fishing press has from the beginning shown advertisements for buying and selling vessels *with and without* fishing rights. As could be expected, the opening of a new, although "grey", market, initiated a dynamic process. During the 1990-1996 period, about 33% of Group I vessels changed hands (Director of Fisheries 1996). Today the figure has probably passed 50%. Having bought their rights, these fishermen are naturally unwilling to give them up, at least not without compensation. And the first generation of owners have long ago realised they stand to receive a windfall profit when selling, thus making this group less inclined to reverse the process and reverse to open access.

This attitude is also reflected in the question of rights transfers where we should have expected a restrictive stand on behalf of the NFA. However, when in 1991 a moratorium on sales outside the owner's home municipality was introduced (Director of Fisheries 1990c; 1991c), this led to massive protests from fishers who feared that the value of their vessels would fall. Within two weeks, the regional constraints were changed, to allow free transfers within counties. Still, the Norwegian Fishermen's Association was not satisfied and has ever since argued that a national market for vessels with quota rights is the only fair option (NFA 1992b).

This incident suggests that the interests created by the IVQ system had an immediate impact on policy-making within the Fishermen's Association. One factor that may explain this is the strong representational bias within the Association. While the proportion of rights holders relative to all fishermen is approximately 30%, the ratio of rights holders in the organisation's elite is 90% (Holm, Rånes and Hersoug 1996).

6. FROM IVQS TO ITQS?

In the opening paragraphs we claimed that the established IVQ system is gradually developing towards "something close to a permanent ITQ system". This statement requires qualification. The IVQ system is a property right system constituted by two types of rights. First, the IVQ system implies an *access right* to the coastal cod fishery. The legal authority for this right is given in accordance with the *Participation Act of 1972* (*Deltakerloven*). Second, the IVQ-system defines a *fishing-right* in the coastal cod fishery. This right is given in conformity with the *Marine Fisheries Act* (*Saltvannsfiskeloven*). Thus, legally the two types of rights within the IVQ regime are independent of each other. In practice, however, they are strongly linked, since a fisherman, or more precisely a vessel owner, holding an *access-right generally qualifies for a fishing-right*. The distinction is important because the two types of rights are different; the access-rights are more "complete" property rights than the fishing rights. Access-rights are relatively exclusive, freely transferable within each county (*fylke*), and seem to be rather secure. Even though they are only granted on an annual basis, they are now in the process of being renewed for the tenth consecutive year. The fishing-rights, on the other hand, are not exclusive. This is because the individual quotas within the IVQ system at present are distributed as competition quotas, and not as guaranteed quotas. Further, the fishing-rights are not divisible and consequently, not fully transferable. While it is possible to buy an access-right, and thus get a fishing right, it is not possible to buy or sell fractions of fishing rights or quotas. In addition, the security of the fishing-right is less than that of the access-right, since it has been subject to several modifications in the political process during the 1990s, and it is likely to be modified this year (Ministry of Fisheries 1998).

This brief outline of the characteristics of the IVQ-regime indicates that the present regime is not an ITQ-system in the Icelandic or New Zealand version of the system. Instead, it may be labelled as a system of *individual transferable access-rights*. While these access-rights give the owners fishing-rights (or quotas), it is not the fishing-rights or the quotas *per se* that are transferable. Thus, the regulation and allocation of individual fishing-rights, or quotas, is still within the political sphere, while access to the IVQ fishery is more or less managed through the marketplace, with important geographical limitations on transferability.

Many vessel owners, opposed to an ITQ-system like in Iceland or New Zealand, privately admit that the present system is too rigid and cumbersome. To fish more efficiently they must either circumvent (or bend) the regulations or engage in the time-consuming process of buying and selling vessels with rights. There are, however, indications of easier transfers in the future.

First, the road back to open access for the coastal fleet seems to be effectively closed. There will still be options but the choice of overshooting the recommended TACs as in the 1980s, will be more and more difficult as harvesting becomes constrained by a number of international treaties. An increasing awareness on behalf of the consumers regarding sustainable management, and numerous watchdogs on the national scene, will effectively block any irresponsible behaviour. Hence, the quota is fixed! Second, the "trawl ladder" (the allocation key for cod) and other allocation keys are up for revision in 2001. Although the ladder and the keys have been hotly debated in the previous years, they will, by 2001, have been in operation for twelve years and few believe there will be large changes in the system, although categories and percentages may change. Investments have been made over the last decade in the firm belief that the distribution is more or less fixed. Consequently, expansion through administrative reallocation will be difficult. Third, a system of licences and "unit quotas" (*enhetskvoter*) is under consideration for the larger coastal fleet. As soon as it is implemented it will be possible to merge existing rights. The same system has been proposed to be extended to all coastal vessels larger than 21m. And, there are alarming signs of dramatically reduced cod quotas for the next years. This means that a simpler system of merging existing rights has to be found.

One of the most serious challenges to the *status quo* will probably come from a new political alliance within the coastal cod fishery. This political alliance is constituted around the concept of the "robust coastal vessels". These are large (15-34m), new and modern coastal vessels, which have emerged within the coastal fisheries during the last years. The main reason for this particular development is a strong political support, symbolically as well as financially, for the realisation of such vessels into the coastal fisheries. The principal argument in favour of this policy has been that such vessels provide a steady supply of fish, thus contributing to permanent employment in the processing industry. In addition, such vessels will, according to their proponents, contribute to improved quality, better security and working environment and consequently, improved recruitment to the coastal fisheries (Ministry of Fisheries 1998).

The political basis for the robust coastal vessel has gained broad political support from the major parties within the Norwegian fisheries since the idea was launched in 1996. In spite of the favourable political support for such vessels, they are still few in number; only about 20 vessels have been built so far, while another 10 are under construction. The reason for this modest renewal is that the investments costs are formidable, each requires NOK million 25 to 45 per vessel, or ten times the costs of a traditional coastal vessel. These vessels will have problems of operating profitably in years with extremely generous quotas. Thus, it goes without saying that they will face considerable problems in future meagre years. There are two alternatives: either a political/administrative favouring of a particular group of vessels (which will be difficult under a system characterised by formal equal treatment), or a system which makes it possible for the new owners to buy or lease quotas in order to use their vessels' greater technical capacity, *i.e.* a system with greater transferability. Scrapping the whole political project of "robust coastal vessels" is highly unlikely, as it is widely endorsed, politically, administratively, and not least, among the processors, who stand to control this fleet in the future. Consequently, greater transferability seems to be inevitable.

7. LESSONS

Here the challenge is to understand why the deliberate effort of introducing an ITQ-system in Norway failed, while the temporary crisis measure of an IVQ-system survived and gradually developed towards an individual rights-based system.

Within the instrumental institutionalist perspective a standard explanation for the failure of introducing an ITQ-system would be that the incentives were not right. In addition there are, according to Hannesson (1992) two main reasons why "such a marvellous system as the ITQs" has not been implemented; either because anticipated side effects (such as regional inbalance and unemployment) have discouraged people from implementing the ITQ-system, or because "ideological and political reasons stand in the way". The first two explanations have considerable merit. Generally, the fishermen did not have much confidence in the incentives that were offered originally, but for different reasons. While the offshore fishermen felt that a 13-year advantage (before the quotas reverted back to the state) was not enough, the coastal fishermen did not believe in the strict separation of the different markets (created in order to keep some regional stability). Fear of regional inbalance and elimination of the small-scale fleet were also concerns of great importance in the political assault on the proposed ITQ-system.

The last factor, referring to ideological and political reasons, is definitely the weakest point, as it goes a long way in explaining away what is most important. Introducing an ITQ-system is a political and ideological act, whereby some actors obtain certain privileges while others lose them. Despite the possibility of compensating through side payments, the political reality is that a closing of a commons is not only an economic transaction, it is even more a transfer of political power. This is clearly demonstrated by the "unintentional" introduction of the IVQ system. According to the instrumentalist perspective this scheme was badly designed and even more haphazardly implemented. Nevertheless, this system stands a good chance of ending up as a transferable rights system. And again the explanation is found by applying the nested systems perspective, whereby the dynamics in the economic system gradually "force" a political (legal) solution.

In 1990 the adherents of ITQs in the catching sector were few and far between. Although they had support from parts of the processing industry and leading bureaucrats of the Ministry, the politicians balked out as soon as they understood the sentiment at the grassroot level. Privatisation of the marine commons was not on the agenda, at least not during an election campaign. In 1999 the situation has changed. Through the gradual development of an IVQ system in the coastal fleet, developed over a period of ten years, there has emerged a group of privileged rights holders. This group, which controls the most important positions not only in the Norwegian Fishermen's Association but also in the corporate co-management structure, has realised that they are best served with a continuation of a rights-based system. Consequently, most other fishing resources have also been allocated, according to size of vessel and their post-catch record participation. (The allocation keys may now be changed, as a result of political initiatives, but the principle of restricted access is not up for discussion).

On the other side, the adherents of free access, or more precisely of the coastal fishery as a commons-organised fishery, have gradually been marginalised. They have few commanding posts in the NFA and even fewer in the corporate structures. A limited number have organised in a splinter organisation (the Norwegian Association of Coastal Fishermen), but so far they have not succeeded in getting much official recognition. The majority of part-time fishermen remain unorganised. The last stronghold of opposition are the regional and local politicians who are strongly committed at the symbolic level against any "privatisation of the commons". So far they have been less interested in the technical details pertaining to the *de facto* selling of rights and accompanying quotas. And regarding the "new coastal fleet" local politicians have been just as uncritical as most of the fisheries establishment. They support local initiatives, even if these initiatives over time will contribute to the eradication of the small-scale fleet.

In the long run the present rights holders, and especially the group that are heavily represented in the NFA, will probably find that they stand to gain from easier transfers of rights and accompanying quotas. Then we will probably see a gradual transformation of the legal instruments regulating the economic activities. It is important to stress that Norway still does not have an ITQ-system of the Icelandic or New Zealand type. At most we have *a transferable access-rights system*. The present challenge is whether the accompanying quotas will be administratively allocated or distributed by the market. We have noted that the direction seems clear, bearing in mind the substantial political costs involved in removing allocated rights from a large number of small-scale fishermen and transfering them administratively to larger operators. It is then easier to let the larger buy out the smaller, but again, this depends on the degree to which the allocation system in the fisheries is considered consistent with society values. There is no doubt that fishing over the last ten years has moved closer to a position of being an ordinary industry, quite opposite to the former perception of being a way of life (Holm 1996). Nevertheless, fishing as a way of life, contributes to the maintenance of the settlement pattern and to the coastal culture, and is still of significant value in the political system. For this reason the NFA has to manoeuvre cautiously, balancing claims to be *an industry* (for example when fighting off the part-timers) and claims to represent *"a way of life"* (*e.g.* when NFA battles for the exclusive ownership of fishing vessels by active fishermen). According to the Danish author Storm P. "you need a strong morale to sell rubber bands by the yard!" That is probably what will be required by the Norwegian fishermen in the years to come.

In Norway the question of ITQs is either heaven or hell, the solution to most problems or the cause of even more problems to come. These positions have, to a large degree, hampered the understanding of what is going on through the silent process of creating a group of privileged rights holders. The paradox is evident; in 1990 the proponents of ITQs lost because they did not understand the political context. In 1999 the "hellfighters", primarily connected to the small-scale fleet and local/regional politics, stand to lose, due to lack of understanding of the underlying processes of the IVQ-system – a fact which underlines the usefulness of analysing the *political* process behind the new management schemes.

8. LITERATURE CITED

Apostle, R., G. Barret, P. Holm, S.Jentoft, L. Mazany, B. McCay, and K. Mikalsen 1998. Community, State, and Market on the North Atlantic Rim. Toronto, University of Toronto Press.

Barber, B. 1995. All economies are "embedded". The career of a concept, and beyond. Social Research, New York.

Christy, F. 1996. The Death Rattle of Open Access and the Advent of Property Rights Regimes in Fisheries. Marine Resource Economics. Vol. 11.

Degnbol, P. 1998. The end of short-term prognoses? Or what constitutes valid biological knowledge as a basis for management? Paper presented at ESSFIN Workshop on Multi-Disciplinary Research in Fisheries Management. Sophienberg Castle, Denmark 13-14 April.

Director of Fisheries (1977-1998a). Minutes of the meetings in the Regulatory Council. Bergen. Director of Fisheries.

Director of Fisheries (1989-1998c). Regulations in the cod fishery. Bergen. Director of Fisheries.

Director of Fisheries (1990-1999d). Statistics of quota allocation in the cod fishery. Bergen. Director of Fisheries.

Director of Fisheries (1996). Alterations in the quota register during the 1990-1996 period. Bergen. Director of Fisheries.

Director of Fisheries 1998. Statistics of the Norwegian Fisheries. Bergen. Director of Fisheries.

Easton, D. 1953. The Political system. An Inquiry into the State of Political Science. New York, Knopf.

Edelman, M.1985. The Symbolic Use of Politics. Urbana, University of Illinois Press.

Eggertsson, T. 1990. Economic Behaviour and Institutions. Cambridge, Cambrigde University Press.

Government of Norway 1991. Hoeringsnotat om strukturen i fiskeflåten. Oslo, Ministry of Fisheries.

Government of Norway 1991. St.Meld. Nr.58 Om Struktur og Reguleringspolitikken overfor Fiskeflåten. Oslo, Ministry of Fisheries.

Granovetter, M. 1985. Economic Action and Social Structure: The Problem of Embeddedness. American Journal of Sociology Vol 91 Number 3.

Hannesson, R. 1992. Trends in Fishery Management. World Bank Discussion Papers. Fisheries Series 217. Washington DC, The World Bank.

Hannesson, R. 1998. The role of economic tools in fisheries management, *in* Pitcher, T.J., P.B.Hart and D. Pauly. Reinventing Fisheries Management. London, Kluwer Academic Publishers.

Hoel, A.H., S. Jentoft and K. Mikalsen 1996. "User-Groups Participation in Norwegian Fisheries Management" *In* M. Chang Zhang, M.L. Windsor, B. McCay, L. Husak and R. Muth (Eds.) Fisheries Utilization and Policy. Themen 2. Procedings from the World Fisheries Congress. In press. New Dehli. Oxford and IBH Publishing Co.

Hollingsworth, J.R. and R. Boyer 1997. Contemporary capitalism. The Embeddedness of Institutions. Cambridge, Cambridge University Press.

Holm, P. 1996. "Fisheries Management and the Domestication of Nature" Sociologia Ruralis, 36(2).177-188.

Holm, P. and S.A. Rånes 1996. The Individual Vessel Quota system in the Norwegian Arctic Coastal Cod Fishery. Paper presented to "Voices from the Commons", The sixth Annual Conference of the IACP, Berkeley, California, 5-8 June 1996. Troms. Norwegian College of Fishery Science.

Holm, P., S.A. Rånes and B. Hersoug 1997. Political Attributes of Rights-based management Systems. The case of Individual Vessel Quotas in the Norwegian Coastal Cod Fishery. *In* Symes, D. (Ed.) Property Rights and Regulatory Systems in Fisheries.Oxford, Fishing News Books.

Jentoft, S. and T. I. Kristiansen 1989. "Fishermen's co-management: The Case of the Lofoten Fishery" Human Organization, 48(4). 355-365

Jentoft, S., B.J. McCay and D. Wilson 1998. Social theory and fisheries co-management. Marine Policy, Vol.22, No. 4-5, pp. 423-436.

Johansen, J. 1972. Traalfiske og traalerdebatt i 1930-årene. Hovedoppgave i historie. Trondheim. Universitetet i Trondheim.

Mikalsen, K. 1977. Lovgivning og interessekamp. Hovedoppgave, Institutt for Statsvitenskap, Universitetet i Oslo.

Ministry of Fisheries 1992. S.meld.nr 58 (1991-92) Om struktur og reguleringspolitikk overfor fiskeflaaten. Oslo, Ministry of Fisheries.

Ministry of Fisheries 1998. St meld nr 51 (1997-98) Perspektiver på utvikling av norsk fiskerinaering. Oslo, Ministry of Fisheries.

Moldenaes, T. 1993. Strukturmeldingen: interessekamp eller helhetsorientering og rasjonelle overveielser. Report. Troms. NORUT.

Neher, P.A., R. Arnason and N. Mollet (Eds.) 1989. Rights-based Fishing. NATOASI Series, Series E. Applied Sciences - Vol. 169. AH Dordrecht, Netherlands. Kluwer Academic Publishers.

NFA (1988-1994a). Minutes from the National Congress. Trondheim, Norwegian Fishermen's Association.

NFA (1989-1996b). Minutes from the National Committee. Trondheim, Norwegian Fishermen's Association.

North, D. 1990. Institutions, institutional change and economic performance. Cambridge, Cambridge University Press.

Ostrom, E. (1990). Governing the Commons: The Evolution of Institutions for Collective Action. Indiana University. Cambridge University Press.

Peters, B. G. 1999. Institutional theory in political science. The new institutionalism. London, Pinter.

Pinkerton, E. (Ed.) 1989. Co-operative management of local fisheries: New Directions for Improved Management & Community Development. Vancouver. University of British Columbia Press.

Polanyi, K. 1944. The Great Transformation. New York, Holt, Rinehart.

Sagdahl, B. 1973. Traalpolitikk og interessekamp. Hovedoppgave. Institutt for Statsvitenskap, Universitetet i Oslo.

Scott, W.R. 1995. Institutions and Organizations. Thousand Oaks. Sage publications.

THE SCALEFISH FISHERIES OF NORTHERN WESTERN AUSTRALIA – THE USE OF TRANSFERABLE EFFORT ALLOCATIONS IN THE MANAGEMENT OF MULTI-SPECIES SCALEFISH FISHERIES

L. Cooper* and L. Joll**
Fisheries Management Services Division
Fisheries Western Australian
*Northern Regional Office, P.O. Box 3064, Broome, Western Australia 6725
<lecooper@fish.wa.gov.au>
**Locked Bag 39, Cloisters Sq P.O., Western Australia 6854
<ljoll@fish.wa.gov.au>

1. INTRODUCTION

The demersal scalefish (Osteichthyes) stocks of the tropical waters of Northern Western Australia consist of a diverse range of species. The stocks are fished by a number of different managed fisheries - the Northern Demersal Scalefish Interim Managed Fishery (NDSF) in the waters around the Kimberley region (far north Western Australia), the Pilbara Fish Trawl Interim Managed Fishery (PFTF) and the Pilbara Trap Managed Fishery (PTF) in the Pilbara region (North West coast) (Figure 1). In addition, an open access line fishery also operates in the Pilbara.

The NDSF and PFTF are relatively new fisheries and the access and management arrangements provided in their developmental phases were inadequate for the level of management necessary for long term sustainability as they developed into fully developed fisheries. In seeking to develop new management and access arrangements both input and output options were considered. However, because of the issues associated with catch quotas in multi-species fisheries, such as dumping of catch of species of low value with overall quotas, over-quota dumping with species quotas and the remoteness of some of the localities in which vessels operate and land their

Figure 1
Western Australia

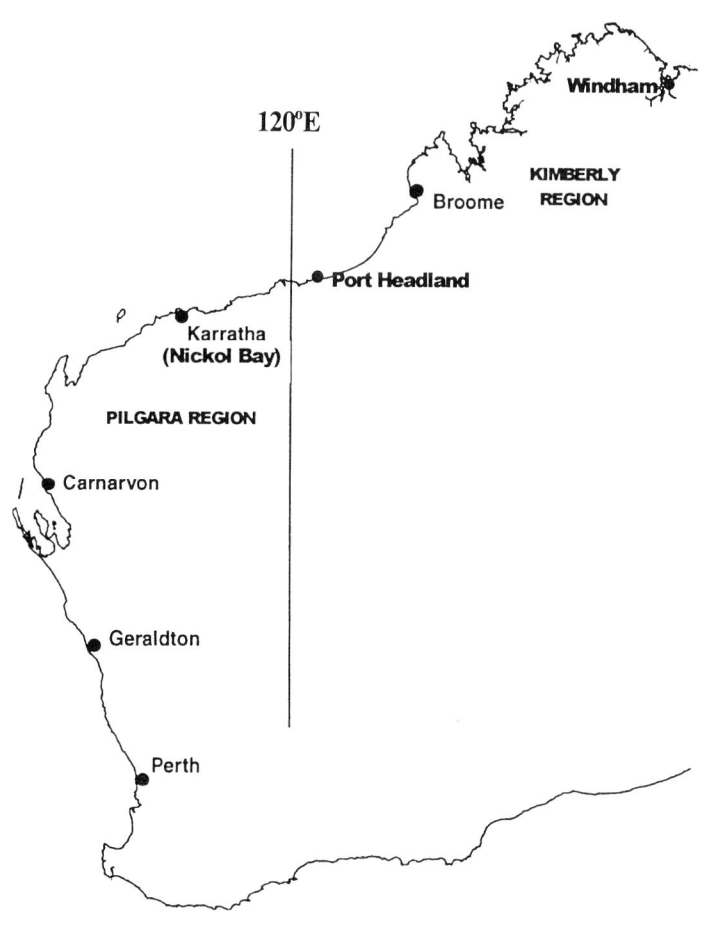

product, an output quota was not considered an appropriate management mechanism.

This paper describes the development of time-based input management arrangements for the NDSF and PFTF and the shift to new management arrangements for the PTF. Reference is made to how these arrangements have moved these fisheries from a relatively unmanaged status to one which more directly manages to ensure the sustainability of these fisheries. In the process, the form of the rights issued to licensees in these fisheries has created a more flexible access right and has developed a common interest in the long-term sustainability of the resource.

2. HISTORICAL BACKGROUND

2.1 Northern demersal scalefish interim managed fishery

The major fishes exploited in the NDSF include the snappers or sea perch (Lutjanidae), the emperors or nor-west snapper (Lethrinidae) and the cods or groupers (Serranidae). Presently the snappers or sea perch which includes red emperor (*Lutjanus sebae*) and jobfish (*Pristipomoides* species) dominate commercial catches by the trap and line fishermen in the NDSF (Fisheries Department of Western Australia, 1995).

Prior to 1987, the take of demersal scalefish by any means in Western Australian waters was restricted to people who held a Western Australian Fishing Boat Licence (WAFBL). A 'freeze' on the grant of this type of licence was implemented by the then Minister for Fisheries in 1983, prior to which a WAFBL was granted upon application. Today, in 1999 there are in excess of 1500 WAFBLs.

In June 1987 the first Constitutional Settlement Arrangements between the Commonwealth and Western Australia came into effect. In regard to trap and lining for demersal scalefish in the Kimberley region, this meant that management of trapping for all species (with the exception of rock lobster) within the waters extending to the 200 metre isobath came under State management. It also meant that line fishing in the waters from the baseline out to 12 nautical miles also came under State management, while line fishing outside of 12 nautical miles fell under Commonwealth management jurisdiction. Line fishing encompassed the take of all scalefish (with the exception of tuna) by means of hand lines, trolling and droplines. Previously, the state only had management jurisdiction over coastal waters *i.e.* from the baseline to 3 nautical miles. Consequently, there was overlapping management jurisdiction between the State and Commonwealth in the Kimberley region which resulted in different management arrangements regulating the exploitation of the same resources. In addition, within the State itself there were two commercial fisheries exploiting the same resource under different management regimes.

In 1988 in accordance with the provision of the *Fisheries Act 1905* a notice prohibiting the use of traps unless authorised to do so was implemented. Authorisation was granted in the form of a 'condition' on a WAFBL and was generally granted upon application. As a result approximately 20 holders of a WAFBL were authorised to take scalefish by means of trap. Lining for demersal scalefish in both Commonwealth and State waters at the same time was relatively unrestricted in comparison. Any person who held a fishing boat licence to fish in Commonwealth or State waters had a perceived right to take demersal scalefish by means of line. In the Western Australian fishing fleet alone, this meant in excess of 1500 fishermen had the ability to take demersal scalefish by line in the Kimberley region from the baseline out to 12 nautical miles.

In 1992, as a result of advice that the demersal scalefish resource in the Kimberley region could not sustain the fishing effort which might be exerted, both Commonwealth and State fisheries management authorities reviewed the management arrangements for the resource. In terms of the trap fishery, this meant the implementation of a catch history criteria that had to be satisfied prior to an authorisation being granted to continue fishing by trap. This resulted in the number of trap fishermen being restricted to nine. Further effort restrictions were placed on the number of traps which could be used by an authorised fishermen (20 traps/boat) as well as an area closure around the town site of Broome (Cape Bossut to Point Coulomb).

Also in 1992, the new *Commonwealth Fisheries Management Act 1991* came into effect. In accordance with that Act, an application for an authorisation to fish in relatively under-developed fisheries was generally approved providing that the applicant had the means to access the stocks. In respect to the Commonwealth line fishery outside of 12 nautical miles, which was considered at the time to be under-developed, authorisation to fish for demersal scalefish by means of line was granted to:

i. holders of Commonwealth Fishing Permits that authorised fishing in the Northern Shark Fishery or
ii. those who had an appropriate endorsement on their permit that authorised the use of dropline, handline or troll off the Kimberley coast east of 120°E.

There was no restriction on the number of lines, the length of the lines or the number of hooks per line. The only limitation was that permit holders had to specify the particular method of line-fishing which they wished to undertake; this was specified on the permit. In total approximately six Commonwealth Fishing Permits were granted which authorised the take of demersal scalefish in the Kimberley region outside 12 nautical miles by means of handline, dropline, trolling or longline. The State line-fishery inside of 12 nautical miles remained unregulated due to the limited interest and therefore limited fishing effort exerted on the resource in this area.

Further Offshore Constitutional Settlement Arrangements between the Commonwealth and Western

Australia in February 1995 saw the State obtain management authority over all waters off the Kimberley coast and extending to the Australian Fishing Zone for a number of species and methods of fishing, including the take of demersal scalefish by hook and line. Relinquishing of management jurisdiction to the State was prompted by the same demersal scalefish resource being targeted by two separate (and potentially 3) fisheries which it was considered could be better managed as one multi-species, multi-geared fishery.

In March 1995 the State Minister for Fisheries closed all fishing for demersal scalefish by means of line-fishing outside of 12 nautical miles. This was to enable the development of the *Kimberley Demersal Line Interim Managed Fishery Management Plan 1995* which was implemented in December 1995. This plan identified catch-history criteria which had to be satisfied prior to an authorisation to fish in the fishery being granted, as well as specifying the management arrangements for the fishery. These included restrictions on the type and amount of gear that could be used, restrictions on the use of automated hauling gear unless authorised and non-transferability of authorisation to fish in the fishery.

In mid-1995 the Minister for Fisheries appointed the Northern Demersal Scalefish Working Group (the 'working group') to review the management arrangements for the Kimberley Trap Fishery and the Kimberley Demersal Line Interim Managed Fishery, and to provide advice in respect to how these two fisheries could be managed under one management regime toprovide for long term sustainability of the resource. The working group consisted of an independent chairman, members from the trap and line fisheries, the recreational fishing sector, the indigenous community and Fisheries Western Australia (FWA).

The working group submitted their report to the Minister in late 1996. However, following consideration of the working group's report and as a result of a number of allegations that the working group process was inequitable to industry, the Minister sought further advice from FWA, the Kimberley trap fishermen and the Kimberley demersal line fishermen. In May 1997, the Minister finally approved the elements to form the basis of the *Northern Demersal Scalefish Interim Managed Fishery Management Plan 1997* which was subsequently drafted and implemented on 1 January 1998. The management arrangements with specific reference to the use of individually transferable effort unit allocations are discussed later in this document.

2.2 Pilbara fish trawl fishery interim managed fishery

The Pilbara Fish Trawl Interim Managed Fishery (PFTF) developed primarily out of the Nickol Bay Prawn Fishery (NBPF), when a number of licensees with surplus fishing capacity began to investigate the potential for demersal trawling for scalefish. Initial results were promising and, given the high variability in catches in the NBPF because of its variable recruitment of banana prawns (*Penaeus merguiensis*), a number of other operators were attracted to demersal fish trawling. Because the fish trawling activity was primarily developmental and an adjunct to another fishery, access was granted in a way that provided for a substantial level of effort to enter the fishery. A total of 84 months of access was granted, based on three pioneer boats having 12 months access and 8 later entrants having 6 months access. The thinking appeared to be that it was unlikely that fish trawling would prove to be economically viable in the long-term and that no harm would be done by providing the opportunity to examine the potential of fish trawling.

However, the fish trawling was indeed economically viable, and the catches in the fishery improved as prawn skippers became skilled at fish trawling (Figure 2). The relatively generous allocation of access into the fishery began to look as it might lead to recruitment overfishing of some of the longer-lived, late-maturing, low-fecundity species such as red emperor *(L. sebae)* and rankin cod (*Epinephelus multinotatus*). There were indications that some of the fast-growing species were still under-exploited. Scientific studies, which commenced in 1994, found this so (Stephenson and Dunk 1996) and consequently fishing effort on the key indicator species of red emperor and rankin cod was reduced to a sustainable level.

The reduction of fishing effort to sustainable levels required that a management plan be developed. Fishery access to that point had, with one exception, been derived from the vessel's access in the NBPF, but was attached to a WAFBL rather than the Managed Fishery Licence (MFL) for the prawn fishery. To develop an appropriate management regime it was necessary to define the form of access and to determine an appropriate set of management arrangements. Because the future of the fishery was still being explored, it was also decided to move through an interim managed fishery arrangement. This did not commit the access and management arrangements to the comparatively more permanent status of a managed fishery.

Discussions on new management arrangements commenced in 1996. However, because of a need for a substantial effort reduction to deal with the over-exploitation of red emperor and rankin cod, industry were initially loath to enter an arrangement. Discussions took place through 1996 and 1997, as the need for the reductions was examined and the proposed management regime developed. One of the issues in the development of the new arrangements was the relative inflexibility of the monthly unit of access. Boats with six monthly access units had to nominate which calender months they intended to operate in the fishery, but once a nomination had been made, the time was deducted from the vessel's access, whether it was fished or not. Time lost from bad weather or breakdown was simply forgone.

Figure 2
Annual landings in the pilbara fish trawl fishery 1989-1999 (1999 landings estimated)

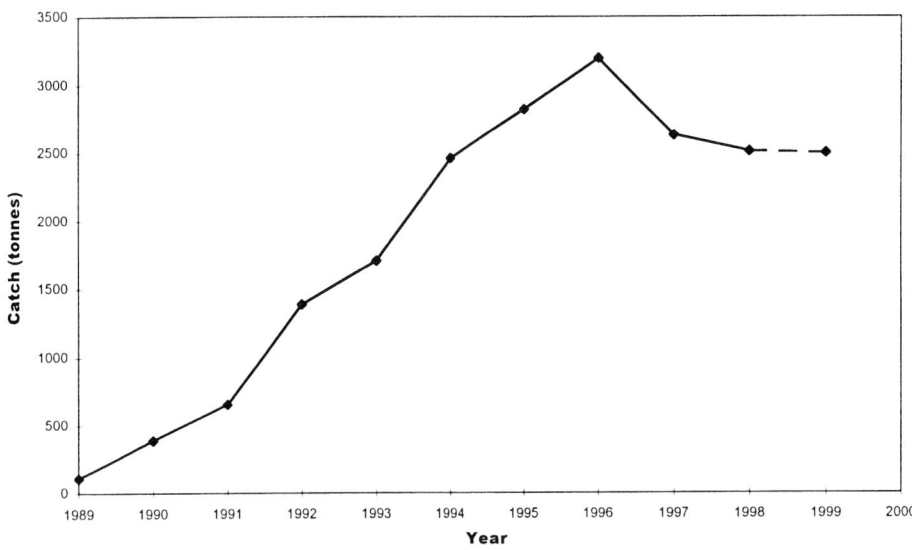

Industry's desire for a more flexible unit of access (such as weeks, days or hours) gave rise to consideration of the use of a vessel monitoring system (VMS), to monitor the use of vessels' time in the fishery. This desire for a VMS-management regime coincided with FWA's desire to sub-divide the fishery so that the spatial distribution of effort could be better controlled. The ability to control the spatial distribution of effort allowed implementation of a closed area to reduce effort in area which had been heavily exploited, focus of effort in an area of high abundance of fast-growing species and distribution some of the effort to more remote and under-fished areas of the fishery.

In addition to the development of a more flexible unit-of-time access, improved arrangements for transferring time-access between licences were developed. Transferability of access to a fishery is generally a normal component of management arrangements of most managed fisheries in Western Australia and the development of both permanent and temporary transfer arrangements provided the basis for a fairly simple mechanism for participants to adjust their holdings in the fishery in response to the proposed effort cuts. Smaller operators whose reduced access made the option of continuing to stay in the fishery unattractive were able to sell their access to larger operators seeking to regain access in order to maintain the profitability of their operation. Alternatively, the smaller operators were able to purchase access from other smaller operators to regain a more substantial level of access. Operators who wished to explore these avenues would be able to use the processes of temporary transfers to examine how their business ran at different levels of fishing activity, while permanent transfers could be used when operators finalized their decision to withdraw or to make a commitment to the fishery.

3. CURRENT MANAGEMENT REGIMES

3.1 The northern demersal scalefish interim managed fishery

The NDSF consists of an inshore and an offshore zone (Figure 3) which are managed by different management regimes. The inshore zone is regulated primarily by a limit on the number of permits authorising fishing in the zone (a total of 4) as well as a limitation on the type and quantity of handlines that may be used. Inshore permit holders may use up to five handlines with no more than six hooks per line. They cannot use automated hauling gear.

The offshore zone of the fishery is restricted to demersal trapping and lining and is regulated by means of individually transferable effort unit allocations as well as a number of other input controls. In respect to the individually transferable effort allocations, the capacity of the fishery in terms of the maximum number of trap fishing days and line fishing days is determined annually by the Executive Director of FWA. This is done after receiving advice from the Director of the Fisheries Research Division, the NDSF Management Advisory Committee (MAC) and the NDSF permit holders.

The capacity of the fishery is calculated in accordance with the following equation:

$$\text{Capacity of fishery} = \frac{\text{total allowable catch}}{\text{catch rate of trap or line}} \quad (1)$$
(maximum no. trap days or line days) kg/trap/day or kg/line/day

The estimated Total Allowable Catch (TAC) is a function of the estimated Total Sustainable Catch (TSC)

Figure 3
The northern demersal scalefish interim managed fishery

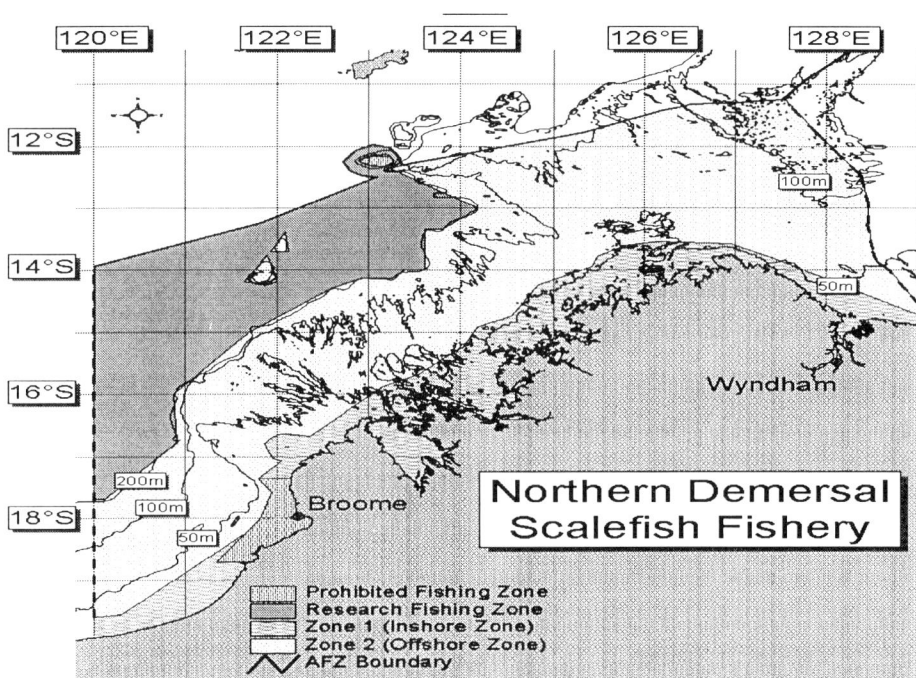

for the offshore zone of the fishery, which is estimated from available stock assessment data on the fishery. The catch rate of trap and line vessels (kg/boat/day) is determined by FWA's Research Division annually from the catch and effort statistics that the permit holders are required by statute to submit on a monthly basis. This is further refined using the catch rate for a trap or a line. Effectively, the capacity of the fishery varies annually according to the quantity of effort exerted by the offshore zone fishermen during previous years. The TAC may also vary on an annual basis.

Upon implementation of the *Northern Demersal Scalefish Interim Managed Fishery Management Plan 1997*, the successful applicants were granted a permit which conferred either trap-units or line-units, depending on their historical method of operation. The unit allocation was determined in accordance with equation (2) and is referred to here as the permanent unit holding.

$$\text{Permanent unit holding (trap-units or line-units)} = \frac{\text{capacity of fishery at commencement of plan}}{\text{number of permits granted at commencement of plan}} \quad (2)$$

The holder of a permit is currently authorised to use either a trap or line depending on the type of units conferred by a permit (*i.e.* trap- or line-units). Notwithstanding, the ability to transfer trap-units onto a permit which confers line units or vice versa is provided for within the current management arrangements. This provides flexibility in using traps and lines, although not simultaneously, according to prevailing weather conditions and behavioural aspects of the key target species.

A permit holder may temporarily or permanently transfer units from, or to, a permit. This enables permit holders to adjust their unit holding on a temporary or permanent basis according to the extent of their fishing operation hence increasing economic efficiency in the fishery without affecting the sustainability of the resource. The management system further provides a mechanism for the internal restructuring of the fishery by allowing less viable operators to sell their interests in the fishery and they maximise the economic viability of the fishery. This results in fewer permit-holders with larger unit holdings, which it is anticipated will result in remaining permit-holders having stronger commitment to the long term sustainability of the fishery.

The extent to which a permit-holder may fish the permanent unit holding is determined by the value of a trap-unit or a line-unit which is calculated on an annual basis using equation (3). Upon the commencement of the plan, 1 trap-unit was specified to have the value of 1 trap-day for the first licensing period. Similarly, 1 line-unit had a specified value of 1 line-day upon the commencement of the plan for the first licensing period.

$$\text{Annual unit value (trap-days or line-days)} = \frac{\text{capacity of fishery (max no. trap-days or line-days)}}{\text{total no. units in fishery (trap-units or line-units)}} \quad (3)$$

Therefore the extent to which a permit holder may fish in any one year can be determined with the following equation:

$$\text{Extent to which can fish (trap-days or line-days)} = \text{permanent unit holding} \times \text{Annual unit value} \quad (4)$$

Any change in the catch rate in the fishery automatically effects the unit value and therefore the extent to which the permanent unit holding can be fished in the following year. In 1999, each offshore zone permit holder could fish to the equivalent of 156 fishing days during the licensing period, assuming that 20 traps or 5 lines are used per day. This was an increase from 132 allowable fishing days (assuming 20 traps or 5 lines used per day) per permit-holder allocated in 1998.

The NDSF is monitored via a Vessel Monitoring System. That is, all vessels operating in the offshore zone of the fishery are statutorily required to have installed an approved Automatic Location Communicator (ALC). The ALC consists of two components (a) a transceiver that relays to the base monitoring system, the location, the speed and bearing of the vessel at any given point in time, and (b) a computer which enables Fisheries Western Australia (FWA) to communicate with the vessel and vice-versa. Prior to leaving port the master of the boat must inform FWA via the ALC of date of departure, the date fishing will commence and the number of traps or lines which will be used for that trip. In addition, the master of the boat is required to submit to FWA, via the ALC, a declaration that fishing has ceased and the estimated time of arrival in port. All details provided in the nominations and declarations are verified via the ALC or Fisheries Officers. As the management mechanism is self correcting a lack of integrity on the master's behalf will adversely affect the extent to which entitlement can be fished in future years.

The unit consumption is calculated on a trip by trip basis according to equation (5) and is directly correlated to the length of a fishing trip and the traps or lines used for each day of that trip:

Unit consumption number = length of trip (days) x amount of gear used/day   (5)

Although the individually transferable effort-unit allocation management regime offers a number of benefits there are a number of additional factors that need to be addressed. These include:

i. the complexity of the legislative framework which gives affect to the system
ii. the cost of providing management, compliance and research services to ensure the success of the system and
iii. that sufficient bycatch provisions for the key target species have not been introduced or are not enforced in other fisheries which incidentally take those species.

These issues are being addressed and it is anticipated that the management system will become cost-effective.

Other management arrangements regulating the offshore zone of the fishery include an area closure around the town site of Broome, restrictions on the maximum number of hooks per handline and droplines, restriction on the maximum internal volume of a trap and restriction on the size of mesh used in the trap. Provision has also been made in the management plan for the Executive Director of FWA to close the fishery if the TAC is exceeded.

3.2 Pilbara fish trawl interim managed fishery

The present fishery management regime is an Interim Managed Fishery Management Plan, and has been operating since 1 January 1998. The regime is effectively a time-quota, with spatial controls on the use of that time. The fishery is broken up into six areas, with fishing effectively limited to four of the areas (Areas 1,2, 4 and 5) (Figure 4). Area 3 is currently closed to allow the stocks to rebuild and is expected to remain closed for the foreseeable future. Area 6 is closed to commercial fishing activity but is available for research fishing designed to provide data on the deeper water (100 - 200m) fishstocks. The gear controls in the fishery remain the same as those established in the initial allocation of access (the primary controls are maximum headrope length and sweep lengths, mesh size and boat size and engine power).

The time allowed in the various areas open to commercial fishing has varied over the lifetime of the plan. These variations have been in response to the need to reduce fishing effort in Area 1 to reduce mortality levels of the indicator species, as well as to reflect more correctly the conversion of the time-access in Areas 4 and 5 from the previous monthly-access management system. The total levels of access into the fishery in 1998 and 1999 are shown in Table 1.

Table 1
Total hours of access the various areas of the pilbara fish trawl fishery in 1998 and 1999

Area	1	2	3*	4	5	6*
1998	17 135	3360	0	3360	5712	0
1999	11 481	3360	0	3058	5198	0

* Area 3 closed to trawling
** Area 6 only open to research trawling under an agreed plan

The major outcomes of the move to interim managed fishery status, a reduction in time-access and the issue of a transferable access right have been:

i. a reduction in the fishing mortality of the indicator species (red emperor) from unsustainable levels of $F=0.26$ in 1996 to the agreed limit level of $F=0.1$ by 1999
ii. a marked reduction in the number of active participants in the fishery as the larger operators have made arrangements to buy the time allocations of the smaller operators and
iii. strong interest in the long-term viability of the fishery.

4. NATURE OF RIGHTS

Section 136 of the *Fish Resources Management Act 1994* (FMRA) clearly specifies that "a person is not entitled to the grant of an authorization as of right". But, it is often perceived by the Western Australian fishing fleet, including the NDSF and PFTF, that previous fishing history in a particular fishery incurs some right to continue fishing in that fishery. Further, Section 71 of the FRMA states that prior fishing, or the use of a boat for fishing, in the fishery does not confer any right to the grant of an authorisation if a management plan is subsequently determined for the fishery. That is, Section 71 provides that no substantive rights are created. However, Section 71 also specifies that the Executive Director of FWA must take into account "the fact that a person held an authorisation when determining whether or not to grant the person another authorisation".

In accordance with Section 73 of the FRMA, to fish in a managed fishery or an interim managed fishery the appropriate authorisation must be held, *i.e.* an interim managed fishery permit (IMFP) or MFL. That is holding a WAFBL, a Commercial Fishing Licence which allows an individual to engage in fishing activities, or any other licence (other than the appropriate IMFP or MFL) granted subject to the FRMA or the *Fish Resource Management Regulations 1995* does not authorise the use of a boat for fishing or engage in a fishing activity in a managed fishery or an interim managed fishery.

It is also perceived that the grant of an IMFP "strengthens" the 'right' to fish more than an endorsement ('condition') on a WAFBL. Similarly, it is perceived that the grant of a MFL further strengthens the right to fish in the fishery. These perceptions often dictate the 'value' an individual places on the long term interest that they have in the fishery and its sustainability and therefore its profitability.

If a management plan for a managed fishery or an interim managed fishery is revoked or expires, "any authorisation in force in respect of the fishery ceases to have effect" (Section 70, FRMA). Further, subject to Section 55 of the FRMA, any instrument which revokes (*i.e.* the determining of a management plan for a fishery) a previous instrument of management for the fishery is subject to tabling and potential rejection by Parliament. In terms of the NDSF and the PFTF, the holder of an authorisation has the right in any one licensing period (12 months) to fish to the extent of the entitlement conferred by that permit for that 12 month period. There is no provision for carrying-over unused entitlement from one licensing period to the next. However, provision is made in the management plans for both the fisheries, for the Executive Director FWA following consultation with the permit-holders and the relevant Management Advisory Committee, to close either fishery or any part of either of those fisheries if the Executive Director considers that it is in the better interest of either fishery to do so.

The holder of a NDSF or PFTF authorisation has the right to apply to the Executive Director FWA for the renewal of that authorisation. The Executive Director may refuse to renew that authorisation on grounds that the applicant has been convicted of a fisheries offence under the FRMA or the relevant Commonwealth fisheries act. The Executive Director may also refuse to renew the authorisation if the applicant has contravened a 'condition' of a authorisation or the relevant management plan. Further, the renewal of the authorisation may also be refused on the basis that the holder has not used the authorisation in the previous 2 years or that the holder has failed to keep any record, or submit any return that is required to be kept or submitted under the FRMA. Failure to pay the relevant fees, charges or levies payable in respect of renewal or any other grounds that may be specified in the management plan for the fishery may also result in a similar refusal.

In addition, the holder of an NDSF or PFTF authorisation also has the right to apply to transfer the authorisation to another person or part of an entitlement under the authorisation to another authorisation. The Executive Director FWA may refuse to grant the application to transfer the authorisation or part of the entitlement if the grounds for transferability specified in the FRMA or in a management plan for the fishery have not been satisfied. The holder of an NDSF or PFTF authorisation has the right to apply to the Executive Director to transfer part of an entitlement under the authorisation for a limited period providing the management plan for that particular fishery authorises such a transfer.

5. FURTHER DEVELOPMENTS
5.1 Northern demersal scalefish interim managed fishery

The *Northern Demersal Scalefish Interim Managed Fishery Management Plan 1997* will expire on 31 December 1999. FWA has sought Ministerial approval to extend the management plan for a further 12 months to enable finalisation of the draft *Northern Demersal Scalefish Managed Fishery Management Plan*. This plan will then be forwarded to the Minister for his consideration and approval and is scheduled for implementation before the the end of 2000.

The draft *Northern Demersal Scalefish Managed Fishery Management Plan 1999* attempts to move away from the complexity of the previous *Northern Demersal Scalefish Interim Managed Fishery Management Plan 1997*. By doing this only one type of unit will exist under the plan - a fishing-day unit. This will remove the requirement to individually calculate the total trap fishing-days, line fishing-days, the value of a trap-unit and the value of a line-unit on annual basis. An average catch rate for the fishery (trap and line boats) will be calculated annually and used in determining the capacity, in terms of the total number of fishing days, for the offshore zone.

Moving to a singular unit-type fishery will enable all offshore zone permit holders to use traps or lines, although only one method of fishing is permitted to be undertaken for the duration of a single trip. FWA must be notified of the type of fishing to be undertaken on a particular fishing trip via the ALC prior to leaving port. The

Figure 4
Management areas of the pilbara fish trawl fishery

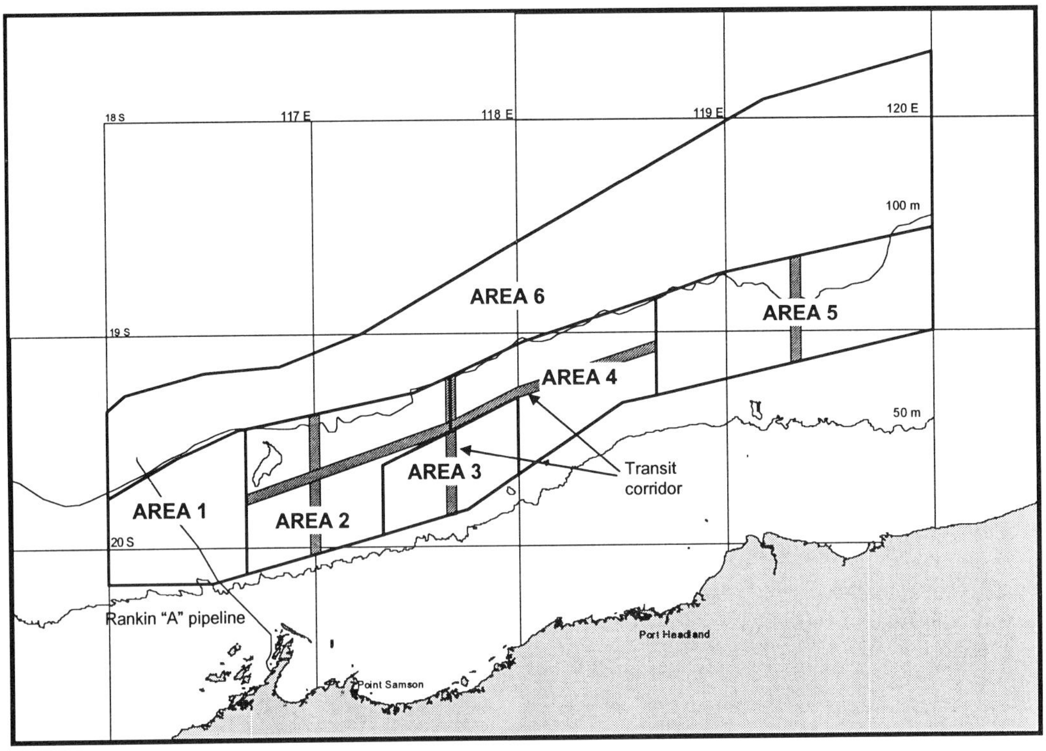

other nomination requirements specified in *the Northern Demersal Scalefish Interim Managed Fishery Management Plan 1997* have been maintained in the draft *Northern Demersal Scalefish Managed Fishery Management Plan 1999*.

This further-developed individually transferable effort-allocation management system will be less complex and therefore cheaper to administer. The movement to fully-managed fishery status will also strengthen the permit-holders' perceived access rights and therefore their long-term commitment to the fishery and hence the sustainability of the fishery. The improved management system will provide the permit-holders with additional flexibility which should improve the economic viability of the fishery.

5.2 Pilbara fish trawl interim managed fishery

The present Interim Managed Fishery Management Plan runs until 31 December 2000, when it is expected that the fishery will move to a fully-managed fishery status. Future management developments are likely to include the development of more selective fishing gear to enhance the catch of the under-exploited species while holding the catch of the more vulnerable species at agreed mortality levels. Possible changes in the management controls on boat size and power will facilitate a move to different gear types, including the possible unitisation of gear to allow for different gear allocations and the development of mid-water trawling gear, which is likely to be more selective for the under-exploited species.

5.3 Pilbara trap managed fishery

The Pilbara Managed Trap Fishery (PTF) is a small fishery of six licensees which may operate in the same general area as the PFTF and take many of the same species. However, the effort of the PTF is directed more at red emperor and rankin cod, the same large long-lived species which are used as key indicator species in the PFTF. The PTF has been barely profitable for many years, but in recent years a number of dedicated and skilled operators have begun to make the fishery viable. This has resulted in a number of previously under-utilised access entitlements becoming more fully used and has raised the likelihood of the fishery over-exploiting red emperor and rankin cod stocks. Given the large cutbacks in the PFTF to lower fishing mortality to sustainable levels, a sudden increase in fishing mortality on these species from the trap fishery is not desired.

Although the PTF is a fully managed fishery, the mechanics of its management plan were unwieldy in their ability to respond to the issue of the mobilisation of latent effort. The only tool available in the management plan was a reduction in the number of traps. This would immediately make a number of operators unprofitable and even make the economics of the more efficient operators fairly marginal. While this situation could have been resolved by transfering traps from departing operators to the remaining efficient operators, it was considered a rather heavy-handed response. Discussions with the licensees in the fishery were undertaken to explore other options to a direct cut in trap allocations which would allow them to continue to operate, albeit at a reduced total level of access, and to allow the adjustment of access to occur over a longer period.

After these discussions it was agreed to change the basis of access to the fishery from one based solely on traps to one based on a unit of time-gear access, using VMS to monitor access use. The use of such a system allows operators more flexibility in their response to the changes in access and provides for temporary and permanent transfers of units of time-access. In many ways the proposed changes to the Management Plan mirror the management arrangements for the NDSF and provide a common thread to the arrangements for both of these northern scalefish trap fisheries. The proposed changes are due to come into place on 1 January 2000.

5.4 Pilbara line fishery

The line fishery for demersal scalefish in the Pilbara currently remains largely unregulated, with approximately 1500 WAFBLs potentially able to operate in the fishery. In practice around 40 vessels have operated in the fishery over the last few years, some being active every year, while others move in and out of the fishery.

Within the context of the demersal scalefish resource and the high level of regulation of the activities of the PFTF and PTF, it is clearly undesirable to have a virtually unregulated sector. Proposals to limit access to the capture of scalefish by line have been publicised and benchmark dates have been published beyond which any history of line fishing will not be considered. The further work required to move the line fishing sector to a more managed state have not yet been finalised because of the urgent need to deal with the PFTF and the PTF, but it is expected that now the new management arrangements in the trap fishery have been determined, attention will be focussed on the line sector.

6. CONCLUSION

The development of transferable effort allocations in the multi-species scalefish fisheries of Northern Western Australia has resulted in management arrangements that manage for sustainability, but do not have some of the drawbacks of output controls such as catch-grading and dumping. However, the use of a notional global TAC (in reality a total expected catch) in the NDSF and PFTF as the means of determining the annual allocation of access has some deficiencies, as it is possible to over-exploit vulnerable species in the species mix while under-exploiting others. Consideration may need to be given in the future to mechanisms which more explicitly control the catch of the more vulnerable species. The use of fishing mortality reference points in the PFTF for particular indicator species (red emperor and rankin cod) does manage for the sustainability of these vulnerable species, but results in the more productive species being under-exploited. The use of VMS creates opportunities to control the spatial distribution of effort, which can allow the total effort levels to be targeted according to the distribution of species and so produce the optimum catch composition within the other constraints.

Given the nature of the localities where these fisheries operate, and the resources available to carry out compliance checks, the use of a VMS-monitored effort-based management system provides a practical sustainable management solution for these fisheries. In addition, the transferability of the access entitlements has created a market for the trading of access entitlements, which has allowed operators to efficiently adjust their levels of access according to their circumstances. Through the provision of this tradeable entitlement licensees develop a direct interest in the long-term sustainability of the fishery.

7. LITERATURE CITED

Fisheries Department of Western Australia 1995. Kimberley demersal line interim managed fishery, *Fisheries Management Plan*, 1 i-vi.

Stephenson, P.C. and I. Dunk 1996. Relating fishing mortality to fish trawl effort on the north west slope of Western Australia, *FRDC Final Report 93/25*.

THE FISHING-RIGHTS ON MARINE RESOURCES IN CHINA

Z. Wu

Bureau of Fisheries Administration and Fishing Port
Superintendence of South China Sea, Ministry of Agriculture
No 11 Nonghanguan Nan, Beijing 100026 China
<inter-coop@agri.gov.cn>

1. INTRODUCTION

China is a coastal country with a large area of marine waters and rich natural resources, especially in inshore areas. There are many commercial fisheries that have undergone rapid development, particularly in the last 20 years, with adoption of an open policy and development of appropriate fishing techniques and industries. By 1998, the total output was 39.06 million tonnes of which the production from the Exclusive Economic Zone was 14.05 million tonnes.

China has the largest fisheries labour-force and national population in the world. The full-time marine labour-force is 2.71 million people and there are in the fisheries over 5.37 million people. Twelve million people are part-time fishers engaged in fisheries-related side production.

With development in China, the fisheries have also been managed by legislation. More than 500 fisheries regulations have been formulated for the purposes of fisheries management, providing legal foundation and support for sustainable fisheries development. But China is also a developing country with a large population and fisheries products are essential food-stuffs. How to utilize the fisheries resources rationally is still a new concept, e.g. what should be rights of access to fisheries resources. Property-rights for fisheries resources is a crucial issue in relation to the survival of the population of several dozen million along the coast of China's maritime provinces. Under the current circumstances, how to allocate fisheries resources rationally is one of the most important issues in fisheries management. Therefore, it is very necessary to find a better and more effective way to set up a new mode of fisheries management – property-rights to fisheries resources - for sustainable fisheries development.

2. THE STATUS OF MARINE FISHING RIGHTS

Until now, there has been no universal definition of property-rights in fisheries in China. Property-rights are commonly understood as a fishing-right, *i.e.* the right to fish the marine fisheries resources.

Historically, the marine fisheries resources were exploited by labourers along the coast. Before the 1950s, when fishing skills and productivity were poor, while the fisheries resources were rich, there was no clear division between fishers and non-fishers because both could freely catch fish. In the early 1950s, with development of the economy, the division between fishers and non-fishers started becoming more obvious. The non-fishers gradually turned into farmers and had the user-rights to land, while fishers had fishing-rights. Since then, the clear division between the fishers and farmers has become more formal, that is, fishers make a living on sea and farmers on land. Meanwhile, the government control led prices of fish products in the markets and fishers had no right to market fish products. Under the planned economy, grain was allocated to fishers by the government at fixed prices.

After the adoption of an open-market policy, radical economic reforms started. As a result, the fisheries system under the planned economy was decentralized. The new system of fisheries companies and private fishing units was started everywhere. Even non-fishers were engaged in fisheries activities and fisheries investment. The previous pattern of fishing-rights had also to change accordingly.

3. THE CURRENT STATUS OF FISHING RIGHTS

The Chinese government put fisheries-management as a top priority. Fishing-rights in management has affected legislation and has played an important role in fisheries development.

The manner of assigning fishing-rights is "the fishing licence system". Under *The Fishery Law of the People's Republic of China*, anybody engaged in inland and inshore fishing operations must apply for a fishing licence from the fisheries authorities. Enforcement regulations have also been issued by the Minister of Agriculture to regulate fishing vessel construction, numbers of vessels, their tonnage, horse-power, gear, operation time, fishing grounds, species taken, as well as allowable catch levels. The fisheries authorities can control and identify the fishing-rights and sort of fish species to be taken, by ratifying, and issuing, fishing licences. The integrated procedures for ratifying and issuing fishing licences have been set out to ensure regulatory controls.

China has legal regulations on the price of access-rights to the fisheries resources. Anyone who is involved in fishing production has to pay tax for fisheries resources enhancement (resources tax) in addition to other taxes. According to "The Fee Collection for Fisheries Resources Enhancement and Protection", anyone and any organization that undertakes fisheries production in inland, mudflats, territories and other waters subject to Chinese sovereignty has to pay the Fisheries Resources Enhancement and Protection fee which is 1-3% of the total production value. If the value is especially high, then the tax will be 3-5% of the value. The taxes will be used for the conservation of fisheries resources.

After these changes in the legal sector, the fisheries in China have basically been developing in a healthy way. But some problems have appeared in the management of fishing-rights. One of them is that the qualifications of users' fishing-rights are not clear and lack of legal definition. Under the incentive for economic gain, more fishers start fishing and this causes excess fishing capacity to develop. The other problem is that the many other new businesses also involved in fishing compete with traditional fishers now reducing their incomes. The third problem is the abuse of rights in issuing fishing licences through "flexible" fishing-rights management.

4. DISCUSSION OF FISHING-RIGHTS
4.1 The quota system of fishing

The fisheries resources are the base of fish production. In the initial stage of exploitation, the fisheries resources may be so rich that no one doubted quantity-production in large-scale fisheries - but this is a mistake. But after uncontrolled development and these fisheries resources have been over-exploited, it has been realized that quantity-oriented production just led the degeneration of the resources. Therefore, some better and more reasonable management mechanisms have to be studied to formulate new policies and measures.

From the late 1970s, a fishing moratorium was initiated for the fisheries resources production in China. In the 1980s, the fishing licence system began to be implemented. Up to the 1990s, the fishing moratorium spread to the whole country. With implementation of these regulations, the fisheries resources have been well conserved and fishing intensity has been greatly controlled. The biomass of some fishes have recovered but these measures are not perfect and the most reasonable ones, cannot completely solve the problems of fishing capacity and allocation of fisheries resources. The production should depend on the sustainability of the fisheries resources, and fishing-capacity should be according to the total allowable catch level, that is, by a quota system.

4.2 To commercialize fishing-rights of fisheries resources

According to Chinese law, the marine fisheries resources are state-owned property and belong to the government. But under economic market-driven policies, the fisheries resources, to some extent, must meet market demands. In order to acquire the biggest share of the resource, some people have made every effort to expand and raise the efficiency of fisheries production. This will demand more fishing-rights accordingly. On the other hand, China has a large fisheries-related population and the demand for fishing-rights is gradually getting stronger. All of these aspects have formed considerable pressure on both the fisheries authorities and the marine fisheries resources. If management fails to solve these problems in time, they will trigger a social and political crisis. Therefore, a proper and effective management system must resolve the conflicts between fishers, fishing-capacity imbalance and fisheries resources productivity. When the total allowable catch level is certain and defined, it will be possible to specify a fleet fishing capacity. The most difficult issue is how to solve the conflicts among the fishers. The best way is to introduce the market mechanism to commercialize the fishing-rights which could then be openly sold among those who have qualification to own fishing-rights. That means that fishing-rights could be sold through auctions. And the licence of fishing-rights as a kind of a commodity could be transferred and circulated among the fishers.

4.3 The strict integrated management of user's rights of fisheries resources

Due to the migration and reproduction of the fisheries resources, the incentives for fishing activity and commercialized fishing-rights, it is necessary to conduct strict integrated management of the user's rights for access to fisheries resources. It has been proved that the previous decentralized-management mode is no longer suited to the current fisheries management problems. The original purpose of decentralized management was to allow the fisheries authorities at different levels to be involved actively in fisheries resources management. Meanwhile management of this kind has permitted local fisheries authorities to pursue their own local interests and has resulted in local protectionism, leading to an imbalance of interests in the whole region. This is harmful to sustainable fisheries resource-utilization and fisheries economic development. Therefore, fishing-rights management has to be the strictly integrated and subject to the fisheries authority of the central government, in cooperation with the necessary supervision and inspection measures.

THE IMPLEMENTATION OF FISHING-RIGHTS SYSTEMS IN SOUTHEAST ASIA: A CASE STUDY IN THAILAND

S. Anuchiracheeva
SEAFDEC Training Department
P.O. Box 97 Phrasamutchedi, Samutprakan 10290 Thailand
<supaporn@seafdec.org>

1. INTRODUCTION

The contribution of fisheries to food security, employment and income is recognized worldwide. Globalization is manifested in the fisheries sector through expanding trade, a greater reliance on market forces in policy-making and a rapid increase in the amount and international mobility of private investment capital. One result is that growth in the demand for fish products, no matter where it occurs, may affect fish production anywhere in the world through the mechanisms of foreign private investment and, or, trade.

It is estimated that between 15 and 20% of all animal protein comes from aquatic animals. Fish is highly nutritious and serves as a valuable supplement in diets providing essential vitamins and minerals. The world's oceans, lakes and rivers are harvested by artisanal fishers who provide vital nourishment for poor communities, not only in Africa and Asia, but also in many parts of Latin America, the Pacific islands and Indian ocean. Of the 30 countries most dependent on fish as a protein source, all but four are in the developing world.

Not only fish is a vital food, it is also a source of work and money for millions of people around the globe. In 1996, an estimated 30 million men and women derived an income from fisheries. An overwhelming majority of them - some 95% - were in developing countries.

In Southeast Asia, fisheries development is an integral part of the countries' economic and social development plans. The general policy objectives regarding the development of the fisheries sector are to increase fish production, to ensure food security and to increase employment, income and export earnings.

The marine and coastal waters of the Southeast Asian Region are some of the world's most productive regions. The region constitutes a rich area in which shallow water marine plants and animals reach a peak of species diversity. This diversity is associated with a high production of organic matter, which in turn is converted into high fishery yields. Coastal ecosystems such as upwelling areas are capable of producing over ten times as much organic matter per unit time as offshore waters. This high production of organic matter is transformed into a tremendous variety of economically valuable products that are used by the people in the region.

Marine fisheries of the Southeast Asian region are characterized by the use of multifarious fishing gear by a large number of small-scale fishermen, estimated at more than four million. However, since the 1960s, fishing pressure has increased particularly in coastal areas, which has led to a depletion of fishery resources and conflicts among the users of the resources. In addition, the 1982 United Nations Convention on the Law of the Sea (UNCLOS) has declared the new management regime of 200 nautical miles to be the Exclusive Economic Zones (EEZs) in which coastal states have national jurisdiction.

There have been efforts to devise management schemes to effectively develop the fisheries sector particularly in the coastal areas. However, the existing problems and constraints mentioned above have become even more critical. This may be due to the use of the unsuccessful management regime of open-access in developing the sector. It is hoped that using the fishing-rights system under community-based fisheries management, through the strengthening of fisher's groups, may offer a better solution to properly manage the fishery sector, particularly in coastal fisheries.

2. PROBLEMS AND CONSTRAINTS OF THAI FISHERY DEVELOPMENT AND HOW THEY REFLECT UPON THE SOUTHEAST ASIAN REGION

2.1 Impact of past fishery-development policies

There has been a focus by governments on increasing fishery production, but issues obstructing the development of fisheries toward sustainability are becoming critical. An analytical clarification of the major issues in fisheries, such as the depletion of marine fishery resources, unequal sharing of benefits, social conflicts, *etc.* may be required to further explore a better approach to manage the fisheries sector.

During the late 1960s and early 1970s, fishing was considered an attractive profitable industry. Fisheries development policies in the Southeast Asian Region focused upon the commercial fisheries sector and these fishery policies encouraged the use of modern fishing technologies and provided access to investment funds to support these innovations. The government policy centred on credit schemes and technological innovation, both of which favoured the commercial sector. Thailand, Indonesia, Malaysia and the Philippines received over US$ 590 million in fisheries aid, 88% of which was for capital investment, primarily mechanization and modernization of fishing vessels and technologies (Christy 1986; Lampe 1991 as cited by Pomeroy and Cruz-Trinidad 1996).

Artisanal fisheries management policy in the Thailand, including other countries in Southeast Asia, have been based mostly on advice derived from biological, resource-oriented studies. Thus, the fisheries policy-makers and managers have little knowledge about the economic, social and cultural aspects of the fishers and limited understanding of the linkage between fishery resources and fisher communities.

2.2 Depletion of marine fishery resources

The common-property nature of most renewable resources, like fisheries, implies that users have free and open access for their exploitation. Such exploitative conditions lead to problems of over-capitalization and incompatability between artisanal/small-scale fishermen[1] and commercial fishermen[2]. The main cause of the decline of marine resources in Thai waters is "overfishing", which is a direct result of rapid development of fishing industry. It has been estimated that overexploitation by trawlers in the Gulf of Thailand started in the late 1960s or early 1970s. The number of fishing boats and their efficient gear seriously depleted what had been abundant marine resources. Statistical analysis found that the sustainable-production capacity of demersal fish resources in the Gulf of Thailand, from the shore out to 50m isobath depth, is 750 000t, corresponding to about 8.6 million operation hours of trawlers (Muntana and Somsak 1982 in SEAFDEC 1987). However, trawlers have exceeded this catch of demersal fish since 1973.

In 1982, the total production of demersal fish was about 990 000t, which is 31% higher than the sustainable production capacity. In 1986, the catch was about 648 560t but the total trawler operation had reached 11.9 million hours. These figures show declining resource biomass and an even greater fishing effort being used to catch a declining yield.

Other obvious evidence indicating declining marine resources in Thai waters was the marine catch-rate of trawlers (CPUE) which in 1961 was 297.6kg/h, but fell to 54kg/h in 1985. In 1988 the CPUE declined even further to 38kg/h. The production consisted of 33.3% commercially-valuable species and 66.7% trash-fish. Of the trash-fish, 30.1% were small individuals of economically important species (Chantawong 1993). This is further evidence of the serious decline of the marine fishery.

Fisheries destruction is happening on a wider front. The mangrove forests, which provide the nursery habitat for fingerlings and juveniles of fish and other marine animals, decreased by an estimated 50% from 3679 km^2 to 1687km^2 during 1961-1993. There are two reasons for the decline of the mangrove forest. The first is the cutting of trees for charcoal and the second is tiger prawn aquaculture. By law, the owners of these businesses cannot trespass upon the conservation of mangrove forest. Unfortunately, because of the weakness of the law it fails to protect the mangrove forests. In addition, the wastewater draining from tiger prawn farms, which is mixed with pesticide and chemicals from prawn feed, has polluted the coastal areas, which are the fishing grounds of the artisanal fishermen and the nursery grounds for young fish.

2.3 Law and enforcement

The most important regulation is the *Fisheries Act, B.E. 2490 (1947)*, which relates to fishermen and is usually used to control fishing activities in Thai waters. When the laws were written the marine resources were abundant. The main purpose of enacting this law was to manage the fisheries and collect taxes from fishermen. However, many of its sections are out-dated and it appears necessary to make changes because the situation of the fishery industry has changed. Alternatively, new legislation may be required. However, to achieve sustainable use of the marine resources, regulations should be accompanied by the creation of awareness and education of both the artisanal and commercial fishermen to conserve the sea's resources. Weakness in law enforcement is not only the case in Thailand but it also occurs in all the countries in the Southeast Asia.

2.4 Unequal sharing of benefits and social conflicts

It may be seen that marine catch by commercial fishermen is far more than that caught by small-scale fishermen, who comprised (1990) 87% of the total fishermen in Thailand (Figure 1).

This unequal sharing between small and large-scale fishermen of the benefits from the marine resource leads to conflict among them. For instance, even though there is a regulation that trawlers and push netters are not allowed to operate in areas within 3 km of the shore, which is the fishing ground of small-scale fishermen, in practice control cannot be achieved. This is because the fine for illegal fishing is much lower than the benefit that the fishermen get from illegal fishing operations. The gear used also destroys the fish traps and small gill nets, which used in these inshore areas by the small-scale fishermen.

3. ATTITUDES OF THAI FISHERMEN TOWARD FISHERY MANAGEMENT AND FISHING-RIGHT SYSTEMS

3.1 Surveys of attitudes

Community-based fisheries management has become the focus of countries wishing to achieve better management of coastal fisheries and the fishing-right system as opposed to an open-access regime is an integral part of this. There have been efforts at testing the approach in many of Southeast Asian countries. In

---

[1] Small-scale fishermen can be classified using these criteria; *location of fishing grounds*: within 3 kilometers from shore and fishing expedition is a one day trip, *size and type of fishing boats*: under 15m in length and without a wheelhouse, *type of fishing gear*: traditional fishing gear including small push-nets, baby trawls, beach seines, gill nets, cast nets, scoop nets, hook and line and traps, and *labor force*: normally are only members of the family and not more than two persons.

[2] Commercial fishermen including the owners of the boats and crews. The income of the crews is paid in terms of a percentage of the total catch value of each trip.

Figure 1
Production volume of small-scale and large-scale fisheries
Source: Fishery Statistical Bulletin for the South China Sea Area, 1980, 1985, 1990, 1992

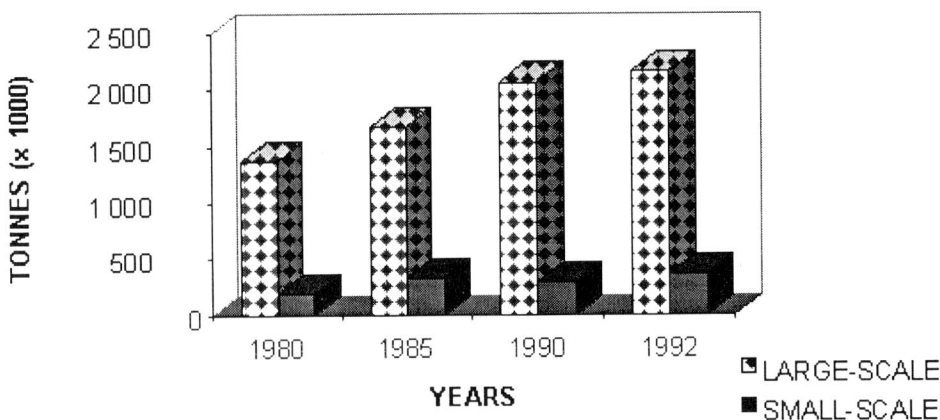

Thailand, there have been various research studies on this topic to determine the attitude of fishermen toward the problems and constraints of the present situation as well as to fishery management and the fishing-right system in their fishing activities.

A research study on the attitude of small-scale fishermen toward the fishing-right system has been conducted in Chantaburi Province in the Eastern Part of Thailand by Kasetsart University in collaboration with SEAFDEC in 1995. Data were collected from 300 randomly selected small-scale fishermen in 4 subdistricts in the Province. The personal-interview technique was used to collect the data and the definition of the fishing-right system was explained.

According to the definition of the Department of Fishery of Thailand, *"The fishing right is a kind of a property right, by which fishermen will have exclusive rights to use the sea areas and resources, which have been specified in each fishing right. In this system, a Territorial Use Right in Fishery will be granted to a fishermen's group based on a legal framework (law) established by the government. With the Fishing Right System, fishermen themselves may create their own fisheries management systems, which should result in the conservation of fishery resources as well as an improvement to their income and living conditions.*

The results of the interviews show that:

i. 79% of the interviewed fishermen agreed that the fishing right system, as defined by the Department of Fisheries, would ease conflicts between them and the commercial fishermen.
ii. More that 50% of the interviewed fishermen believed that the system would increase their catch and reduce fishing costs.
iii. 82% of interviewed fishermen agreed to take responsibilities for conserving and managing the fisheries resources.
iv. About 80% of the interviewed fishermen agreed that the government should continue law enforcement.
v. Almost 100% of the interviewed fishermen were uncertain whether the present fishery law is applicable to the system.
vi. The majority of the interviewed fishermen agreed that the government should urgently issue the laws and regulations that are required for the system.
vii. Most of the fishermen interviewed accepted the establishment of fishermen's management groups.
viii. 74% of the fishermen interviewed indicated an intention to participate in the activities of such groups.

In 1995 another research study on the problems and constraints associated with the participation of local fishermen in fishery management was conducted. Interviews with fishermen and community leaders were conducted by the author in six villages[1] in Trang Province, on the west coast of Thailand. The personal interview method was selected for data collection. The results of interviews were analyzed drawing the following responses:

3.2 National policy support

Sustainable management of coastal marine resources should be the most important issue in fishery development policy and should take precedence over the need to increase marine production. However, both issues should be planned together so as to move in the same

---
[3] There were six villages, Had Chao Mai, Ban Pra Muang, Ban Modtanoi and another three villages on Libong Island.

direction and to be mutually supportive. This means that the government should plan to increase marine production in the long-term by conserving and restoring marine resources to ensure that they will continue to be productive. The government should also have clear plans to support artisanal fishermen in this new role of taking care and protecting resources in the coastal areas. Any activities that work against the plans should be prohibited; including issuing of mangrove forest concessions for charcoal-making and tiger prawn farming, which are damaging the mangrove forests and the coastal areas. Trawling and push-netting operations, which damage the fishing grounds and destroy fish stock should also be banned.

3.3 The right to protect the coastal fisheries resources

In the fishery villages the artisanal fishermen must help to conserve the coastal resources in front of their villages by stopping the use of illegal fishing gear. However, because they have no authority to do so, they have often found that it is difficult to stop other fishermen who still use illegal gear. If they cannot stop them, the problem of declining of marine resources will not be solved. Local fishermen need the right to protect the sea. However, it should not be interpreted that they require authority to investigate and arrest fishermen who practice illegal fishing. They just need acceptance from other fishermen who are damaging the sea and that they have the authority to protect the resources used by their villages.

3.4 Support from extension agencies and organizations related to coastal marine resources management projects

There are many coastal marine resources management projects that involve the artisanal fishermen in certain villages in the southern part of Thailand; these projects include Mangrove Forest Planting, Sea Grass and Dugong Protection, Sea Turtle Conservation, Educational Projects for Village Members to Stop Illegal Fishing Gear and Artificial Reef Construction. These types of project need support from the government in terms of finance, scientific knowledge, advice, authorization, and the enforcement of legislation.

3.5 Raising awareness of artisanal fishermen toward the use of marine fishery resources

Many artisanal fishermen in the study area are still using illegal and destructive fishing gear, namely, dynamite, poison, push nets and baby trawls because such practices catch a lot of fish compared to the time and money invested. The fishermen do not know, or understand, that they are damaging the fishing grounds and fish stock and they are not aware of the importance to conserve the marine resources. Under these circumstances, they need to be made aware and understand the ecological web of this complex problem. If they can understand and accept the importance of marine conservation and the benefits they can get in the long-term, compliance will be assured.

3.6 Linkages with other local organizations

Due to the migratory nature of some marine resources, management of the coastal fishery resources cannot be done by an individual fisherman or a village within a particular area. The establishment of linkages between the local fishers' organizations between villages has become necessary because they can coordinate among themselves through such linkages in management systems, sharing experiences and learning from each other how to cope with the problems. In some cases they need mutual support to exert their power and influence on national fishery policy.

In addition, the implementation of fishing-rights in Thailand is currently being conducted as a pilot project. This is in the nature of a feasibility study; studying fishermen's attitudes toward the fishing-right system, studying the law and regulations; plus the implementation of a pilot project in Bangsapan district, Prajuab Kirikhan province. Here, open access to the sea and marine resources has been used for a long time. Fishermen are used to this system and afraid to use the new system. It takes time to educate fishermen to understand and accept a new system.

4. FISHING RIGHTS DEFINITIONS AND ESTABLISHMENT OF FISHING RIGHTS IN COASTAL FISHERY MANAGEMENT

There are many types of fishing rights are being used in various countries, which may be summarized as followed:

i. *A fishing right* is the right to fish, but also has a wider meaning. It can be described in terms of the licensing of fishing gear and boats, etc.

ii. *Property right/ territorial use right (TURF)* is the right to the resources within an area. This right can be granted to appropriate groups, giving the authority to use, manage and control harvesting. The definition is similar to the definition of the Department Fishery of Thailand mentioned above, which appears appropriate in the context of coastal fisheries management in Thailand.

iii. *User rights* are the rights to the resources within a particular area that may be granted to the original users of those resources. The users are the stakeholders in the exploitation of the resources.

4.1 Establishment of fishing rights in coastal fishery management

The result of the IPFC Symposium 1987 (Piumsombun 1994) concluded that *"Although there are various techniques for controlling excess capacity, it seems that the two techniques most relevant for Southeast Asia are the decentralization of management authority to local fishermen groups as for example, through territorial use rights in fisheries (TURFs), which are generally more suitable for small-scale fisheries, and the limitation of fishing units through a licensing system, generally more suitable for large scale fisheries".*

In fishery management in the past, fishermen were forced to follow the fishery laws and regulations; seasonal closures, mesh size limit, fishing gear restriction, *etc*. This method of management was not successful because there was no participation from the local fishermen who are considered as resource users in this management. The fishermen lack the sense of ownership of the marine resources. They do not only take as much as they can from the sea, but also nobody takes care and accepts responsibility for the resources. Fishing Rights (TURFs) can be organized under the umbrella of a fishery management arrangement. Granting Fishing Rights (TURFs) to the fishermen, will enable them to create their own management system, which can solve the problems of fishery management in the past. "Resource users are resource managers", is an approach that can allow fishery management to succeed. Involving local fishermen in the management can be achieved by encouraging and allowing them to participate in such activities. If they involve, or participate in these activities, awareness of the need to sustain marine resources will be created as Pretty (1995) states that: *"One views community participation as a means to increase efficiency, the central notion being that if people are involved, then they are more likely to agree with and support the new development or service. The other sees community participation as a right, in which the main aim is to initiate mobilization for collective action, empowerment and institution building".*

The limited number of personnel and patrol boats may not be the reason for the problem of management failure. It should be understood that without conservation-awareness toward the marine resources of the fishermen themselves, there will never be enough manpower and patrol boats to make law enforcement effective. In some cases social sanctions are far more effective than legal sanctions for achieving compliance. Peer pressure increases compliance and the strong participation of fishers in management planning and implementation of the coastal fishery management activities reduces management costs including those for monitoring, control and surveillance of fishing activities. Social sanctions can also reduce excessive competition among fishers (both the small-scale fishery sector and the commercial sectors) and thus enable management of resources for optimum economic benefits for all concerned.

In the Regional workshop on Coastal Fisheries Management based on Southeast Asian Experiences, organized by SEAFDEC in 1996; it was concluded that the participatory approach must be given a high priority in coastal fisheries management. It also pointed out that community-based fisheries management can be developed and successful only when fishers understand that they own the fishery resources. Granting fishing rights to fishers, and a limited-entry scheme provides the best opportunity for them to establish their own organizations.

5. FISHERY COOPERATIVES IN FISHERY MANAGEMENT

5.1 Requirements for success

To manage coastal fishery resources with the participation of the local fishermen, it is necessary *to encourage the local fishermen to build their own organization* to participate in organizing management activities, to coordinate with the government and other institutions, and to share benefits from the resources among themselves. A fishing right can be granted to such organizations, and the government can devolve its authority to manage the marine resources to them. Such local organizations should belong to the fishermen by regulation and practice. They should have control over their organization in terms of management and sharing of the benefits. Fishery cooperatives are one of the possible means of organization.

5.2 Advantages of using fishery cooperatives to manage coastal fishery resources

The members of a cooperative are real members by law and practice. The members have rights to be involved in the cooperative's activities and to monitor and select their own leaders. This can guarantee that the benefit of the cooperatives will be shared equally among the members. According to the principle of cooperatives, members will be encouraged to participate in and be responsible for the cooperative's activities.

If management of the coastal fishery resources is successful, the production of the fishermen will increase. Marketing and processing of the products will be the issues that the fisher organizations have to deal with in the near future. The fishery cooperative is a legal organization that has the authority to run businesses concerned with fishery production and marketing. Mostly, activities of the cooperatives are involved with marketing and processing of fish and fishery products, providing their members with fishing equipment and other necessities at cheaper prices compared to the market, and in providing loans at low interest rates to their members.

Cooperatives can make full use of the local knowledge and experience of local fishers in formulating management regulations that fit local conditions. Rules and regulations are agreed upon by fishermen in advance, and thus do not need to be enforced by an outside agency. In the Regional workshop on Coastal Fisheries Management based on Southeast Asian Experiences organized by SEAFDEC in 1996 it was concluded that the establishment of fishers' organizations or fishery cooperatives could lead to the success of cooperative-based fisheries management and fishery cooperatives can work if they are allowed to.

5.3 Factors to be considered in establishing and organizing cooperatives

Cooperatives should have clear benefits for the fishermen. Their activities should be concerned more with fishery management such as implementing community-base fishery management including the establishment of territorial use right fisheries (TURFs). Providing fishing rights will encourage the fishermen to become members of the cooperative.

The direction of cooperatives should focus on responding to the immediate problems or needs of the fishermen. They should concern themselves with both economic and fishery management activities. Providing support only for inputs for fishing operations may lead to the fishermen becoming more indebted if problems of fishery-resource decline are not yet solved. Involvment in coastal fishery management will allow cooperatives to have real power and be attractive to the fishermen.

A fishery cooperative law should be issued to facilitate the new direction and role of fishery cooperatives in fishery management. Most cooperatives concern themselves with economic activities to improve the living standards of their members. The objectives of fishery cooperatives should be both the short and long-term. Among short-term objectives should be to responses to the immediate problems of the fishermen, namely the development of fishing grounds. Long-term objectives should focus on the improvement of social conditions of fishermen and the sustainable development of fishery resources.

The current problems and constraints in the operation of fishery cooperatives, namely, a lack of understanding by the members of cooperative principles, lack the managerial skills of the committee members and cooperative workers, or low salaries for the cooperative workers should be minimized to ensure their success.

5.4 The role of fishery cooperatives in coastal fishery management

To represent the local fishermen in management of the coastal fishery resources, cooperatives should play an extra role beyond their normal activities. These extra roles could be as follows:

i. To be authorized and responsible for management of the fishing areas granted by the government.
ii. To delegate the fishing rights, *e.g.* TURFs, among the members of the cooperative.
iii. To use and manage the coastal fishery resources in fishing for the benefit of the members in the long-term.
iv. To set up coastal fishery management regulations for the fishing area with the involvement of the members.
v. To plan the coastal fishery development and management programmes.
vi. To create awareness of the members on sustainable fishery development.
vii. To coordinate between the government and the local fishermen.

6. CONCLUSION

There are many key conditions, which can make community-based fisheries management successful. The first and most important condition is the establishment of TURFs for fishing. It is important to establish the physical boundaries of such areas to be managed. This means the creation of a property right will grant the authority to the local fishermen to manage the area. The second condition is that the community-based fishery management concepts and regulations should be integrated into a framework of national fisheries policy and legislation. The third condition is support from the government, without which the management of the coastal marine resources with the participation of the artisanal fishermen may be unsuccessful. The support can come in terms of decentralization and delegation by the central government authority to local government and fishers' organizations as well as the issuance of laws and regulations, that support the fishers' organizations, such as fishery cooperatives. In addition, support may also include the provision of financial help, information and training to improve the skills and knowledge of the local fishermen in resource management and creating awareness of management needs among fishermen. The last important condition is the active participation of the fishermen in management planning and implementation in fishery management activities.

Apart from promoting fishing rights in coastal areas, control of the investments and licenses for commercial fisheries should also be done. And, development of the commercial fishery should be in line with coastal fishery management.

7. LITERATURE CONSULTED

Anuchiracheeva S. 1996. Dissertation for MSc course on "The Development of a Sustainable Fishery in Thailand Through Increased Participation of Artisanal Fishermen", UK, The University of Reading, AERDD, 78 pp.

Chantawong, T. 1993. Monitoring in Phang-Nga Bay. Technical Paper No. 17/1993 Marine Resources Survey Unit Andaman Sea, Fishery Department Center, Marine Division, Department of Fisheries, Thailand. pp16

FAO 1997. "The Role of Fishery Cooperatives in Promoting Sustainable Coastal Fisheries in Thailand", Report of the National Seminar, Thailand, 5-7 November 1997, 115 pp.

Fisheries Engineering Division 1995. "Fisheries Extension in Thailand". Thailand: Department of Fisheries.

Menasvata, D. and P.I. Andhi 1968. "Country Paper: Thailand". A paper submitted to the Seminar on Possibilities and Problems of Fisheries Development in Southeast Asia, Berlin, 10–30 September, 1968, FAO, 20 pp.

Piumsombun, S. 1994. Report on "the Socio-Economic Feasibility of introducing Fishing Right System in the Coastal Waters of Thailand". Submitted to the IPFC Symposium, Thailand, 23-26 November 1993, FAO, 15 pp.

Pomeroy, R.S. and C. T. Annabele 1996. "Socio-economic Aspect of Artisanal Fisheries in Asia", Perspective in Asia Fisheries, Philippines, ICLARM, 19 pp.

Pretty, J.N. 1995. Regenerating Agriculture: Policies and Practice for Sustainability and Self-Reliance. London, Earthscan Publications, pp320

SEAFDEC 1980. "Fishery Statistical Bulletin for the South China Sea Area", Thailand SEAFDEC, Training Department.

SEAFDEC 1985. "Fishery Statistical Bulletin for the South China Sea Area", Thailand SEAFDEC, Training Department.

SEAFDEC 1987. The Future of Thai Fisheries, Thailand, 4-6 June 1987. in Thai

SEAFDEC 1990. "Fishery Statistical Bulletin for the South China Sea Area", Thailand SEAFDEC, Training Department.

SEAFDEC 1992. "Fishery Statistical Bulletin for the South China Sea Area", Thailand SEAFDEC, Training Department.

SEAFDEC 1997. "Proceeding of The Regional Workshop on Coastal Fisheries Management based on Southeast Asian Experiences". Thailand, 19-22 November 1996, 348 pp.

Wongsanga, P., K. Jantarashote, S. Aujimangkul, M. Kaewnern and P. Suanrattanachai 1997. "Attitudes of Small-Scale Fishery Toward the Fishing Rights System, A case Study on Fishery Households in Chantaburi Province". Thailand, SEAFDEC, Training Department, 101 pp.

AUTHOR INDEX

Alden, D.	288
Andaloro, F.	206
Annand, C.	428
Anuchiracheeva, S.	456
Arbuckle, M.	370
Bailey, D.J.	352
Battaglene, T.	296
Bess, R	331, 390
Borg, J.	209, 313
Brayford, H.G	338
Campbell, D.	188, 296
Clement, G	254
Connor, R.	29, 267, 288
Cooper, L	445
Cox, A.	304
Craig, T.	17
Darcy, G.H.	96
Davidse, W.P	258
Drummond, K	327, 370
Edwards, R.	320
Fitzpatrick, D	53
Ford, W.	289
Gislason, G.S	118, 383
Gissurarson, H.H.	1
Goulstone, A.	78
Gwazani, R.	170
Hansen, J.	160
Harte, M.	331
Haward, M.	155
Heizer, S.	226
Hersoug, S.A.	173, 434
Holm, P.	173, 434
Hooper, M.	199
Jensen, C.	47
Joll, L.	445
Keithly, W.	219
Kemp, A.	304
Kenchington, E.	239
Kirk, P.	327
Leblanc, J.	215
Liew, D.S.K.	279
Lynch, T.	199
MacFarlane, A.	343
Matlock, G.C.	96
McCormack, S.	84
McFarlane, B.	39
McIlgorm, A	88, 148
McIntosh, L.	57
McKinlay, J.P.	405
McLoughlin, R.	84
McMurran, J.	184
Merla, A.	26
Millington, P.J.	405
Nelson, J.	239
Nelson, L.	327
Nielander, W.J	59, 415
Nordmann, C.	23
Paust, G.	338
Peacock, F.G.	160, 239, 428
Penn, J.P.	297
Rånes, S.A.	434
Rogers, P.P.	297
Sellers, R.	313
Smith, P.J.	348
Sproul, J.T.	348
Stevens, G.	239
Stokes, A.	107
Sullivan, M.S.	59, 415
Swanson, D.	127
Taylor-Moore, N	72
Truelove, K.	355
Tsamenyi, M.	88, 148
Tunesi, L.	206
Valatin, G.	137
Ward, J,	219
Watkins, D.	234
Wertheimer, A.C.	127
Wilson, M.	155
Wu, Z.	454
Wyatt, N.	402
Wylynko, B.	57
Zacharin, W.	320

SPECIES INDEX

Abalone	81,82,150,234
Alaska halibut	130,426
Atlantic Surf Clam	59,105,128,426
Bluefin Tuna	28
Capelin	5
Cod	1,2,5,140,258
Coral	359
Crab	150
Gemfish	359
Geoduck	226
Greenland Halibut	5
Haddock	5,140
Herring	1,2,5,140
Lobster	81,82,273,289
Northern zone, SA	321
Rock lobster, NZ	273
Western Australia	406
Mackerel	140
Mussels	140
Nephrops	1,3
North Pacific pollock	217
Ocean quahog	59,128,426
Orange roughy	116,254,273
Oreos	273
Pacific Halibut	61,106,349
Pacific Whiting	215
Pearls	338
Pilchards	S150
Plaice	140,258
Prawn	81,82,150
Western Australia	297,447
Redfish	5,9
Rock lobster	150
Sablefish	61,130,349,426
Saithe	5
Sauger	120
Scallops	1,21,239
Challenger Enhancement Co.	370
Port Phillip Bay	84-87
Shark	106
Shrimp/Prawns	3,297
Gulf of Mexico	219
Scotian Shelf	430
Snapper	150
Snow crab	431
Sole	140,258
Southern Bluefin Tuna	116,296
Tuna	150
Walleye	120
Whitefish	120
Whiting	140
Wreckfish	60,105,129
Yellowtail Flounder	431

SUBJECT INDEX

Access limitation 96
Access rules 23
Access rights
    AFMA 308
    Tasmanian rock lobster 291
Administration 9,322
    Alaska 350
    BC geoduck 229
    New Zealand 185-186,419
Alaska 348
Allocation 23,63,124,134,140,306
    AFMA 306
    Auctions 310
    Catch history 80
    Economic efficiency 309
    Formula 59
    Geoduck – British Columbia 228
    Lake Winnipeg 122,124
    Legal challenges 59
    Netherlands 140
    New South Wales - abalone 237
    New South Wales 80
    New Zealand 59,63,68,393
    Pacific halibut 385
    Recreational fishing 185
    Shark Bay prawn fishery 298
    South Africa 174,180
    Tenders 310
    Western Australia 298
    United Kingdom 140
    United States 59,66,134
Alaska IFQ 64
Annual catch entitlements 7-9
Appeals 395
    Alaska 64-65
    Atlantic surf clam 65-66
    New Zealand 65
    NSW abalone 237
    Ocean quahog 65
    Shark Bay prawn fishery 298
    South East Trawl Fishery 288
    United States 66
Aquaculture rights 231,327,331,340
Area licences 228,359
Argentina 27
Artisanal fisheries 459
Auctions 310
Australia 88-95,116,148,157,320,366
    Constitutional settlements 446
    Fishing licences 39,53
    Oceans policy 367
    Property rights 39,53
    Rights-based management 148
    Southeast Trawl Fishery 288
    State's legislation 43
    States rights 40
Biodiversity 26

Buyback programmes 99-100,139,148,211,261
Canada
    Community management 160
    Geoduck 226
    Pacific halibut 383
    Scallop fisheries 239
    Scotia-Fundy 160,279-,430
Catch validation 229
Challenger Group 21,370,374
Chatham Islands 204
China 454
Coasian conditions 198
Coastal area management 380
Coastal communities 328,341
Cod equivalents 10
Co-management 155,186,201,231,329,365,405,407
Commercial competitiveness 390
Commissions of Enquiry 5
Common Fisheries Policy 25,138,173, 258-260
Commons – privatization 6
Communities 85,132,429
Community Boards 163
Community Groups 164
Community management 155,160,178,405
Community property 355
Community rights 199
Compensation 89-92,95
Compliance 144,192,408,411,422
Confidentiality 412
Concentration of Ownership 51,274,280
    Measurement of 279
    New Zealand 267-278,394
    Netherlands 262
    Norway 440
    Scotia-Fundy demersals 279-
Conflicts 19
Conservation 247
Consultation
    EU 25
Cooperatives 99,179215,460-461
Costs 235,323,348
Cost Recovery 148,229,235,306,323,
    348-9,385,402,424
Court cases 30
    Iceland 5
    Australia 42-46,53-56
    Port Phillip 86
    New Zealand 59
    United States 59
Crew Earnings 248
Crew Participation 5
Croker Is Decision 188
Decommissioning 139,261
Definition of rights 18
Developing fisheries 314-315
Divisibility 339
    Definition 398,335

Duration	89-94,306,308,334
Economic benefits	255
Economic policy	391
Economic efficiency	261,306,309
Economic rights	190
Effort Quotas, Iceland	1
Employment	249
Enforcement	9,142,167,192,228,230,405, 408,415,419,420
New Zealand	419
United States	418,425
Western Australia	405
Enhancement	231,370
Enterprise allocations	241,399,428
Entitlements	166
Equity	361
Estuary fisheries	209
Exclusivity	308,336
Exploratory fishing	305
European Union	23,137
Access rules	23
Common Fisheries Policy	25
Consultation	25
ITQs	25
Management objectives	23
Multi-annual Guidance Plans	24
Policy	23
Regulation	23
TACs	23
Fishermen's attitudes	458
Fishery management systems	49
Fishing licences	39,50,53
Fishing rights	72
China	454
Coastal fisheries	459
Commercial fishing	184
Customary fishing	184,327
Definitions	459
New South Wales	90
Northern Territories	90
Queensland	89
Recreational fishing	185
South Australia	91
Tasmania	92
Victoria	93
Western Australia	94,406
Fleet Capacity	248
Canadian scallop fleet	245
Gulf of Mexico	220
Netherlands	259,261,265
New Zealand	267-278
New Zealand – trends	277
Shark Bay prawn fishery	301
Flexibility	334
Forfeiture	418
Freshwater fisheries	118,170
Ghana	27
Gini Index	268,281
Global Environment Facility	26
Governance	22,370
Herfindahl-Hirschman Index	268
High seas	8,107
Hatchery rights	339
Iceland	1
Ass. of Fishing Vessel Owners	11
Administration	9
Enforcement	9
High seas	8
Marine Research Institute	9
Problems	13
Transparency	13
Fairness	13
Property rights	10
TACs	8,11,15
ITQs	1-11
Legislation	1,3,4,10,14
Implementation	392
Input Controls	301
Indigenous land use	193-195
Indigenous fisheries	170,188,199,232
(see also Māori)	
Individual Fishing Quotas	127,228,384,438
Industry support	254,301
Industry responsibility	302
Infractions	422
Input controls	301
Institutional structures	148,190-193
Institutional change	435
International Pacific Halibut Commission	383
Italy	206
ITQs	
Allocation formula	59
Australia	148-150
Benefits	254-255
Characteristics	29,432
Customary fishing rights	203
Definition	29-37
EU	25
Fish prices	150
Iceland	1,2,4-6,10
Implementation	81,420
Lake Winnipeg	118
Netherlands	259
New Zealand	267,331-334,415,418
Norway	434
Orange roughy	254
Overfishing	359
Performance	6,11
Scotia-Fundy	279,432
Social goals	34
Southern bluefin tuna	296
Thresholds	428
United States	415
Judicial processes	412
Justice	13,33,361,457
Lake Kariba	170
Lake Winnipeg	118
Latent effort	213
Legal challenges	5,53
Allocation formulae	59
Legal definition – Natural resources	42
Legal rights	190
Legislation	252
Australia	53-56,88-95,235,305,367

China	454
Constitutional systems	416
Criminal law	415
Fishing rights	89
Iceland	1,3,4,10,14
Lacey Act	418
Magnuson-Stevens Act	97,104,135,348,350,417
National Standard Guidelines (US)	104
New South Wales	235
New Zealand	17,18,61,70,92,96,331-333, 392,395,416,424,421,424
Northern demersal scale fish	450
Pearling	338,341
Port Phillip	86
South Africa	353
Thailand	457
UNCLOS	108
United States	348,417
Western Australia	338,446
Mabo Decision	42,188

Management Plans
- Kimberley Demersal 447
- Pilbara fish trawl fishery 452
- Pilbara trap managed fishery 452

Management rules 362
Māori 185,202,268,327,199
Marine Farming 327-328,338
Marine protected areas 206
Multi-annual Guidance Plans 24
National Academy of Science 133
National policy 458
Netherlands 258
New South Wales 78,90
New Zealand 17,267-278, 327,390
- ITQs 334,418
- Challenger Group 19,370,374
- Chatham Is. 204
- Cost recovery 403
- Customary fishing rights 203
- Enforcement 415,420
- Indigenous fishing rights 201
- ITQs 21
- Legislation 17,18,332,395
- Māori 202,327
- Marine Farming 329,331
- Quota management system 17,19,21,372
- Recreational fishing 184
- Role of Government 19,22
- Role of Industry 19
- Seafood Industry Council 19
- Stakeholder input 20
- TACs 17

Netherlands 137-147,258
New fisheries 304
New South Wales 234
North Sea 260
Norway 434
Phillip Bay 84
Pilbara Trawl Fishery 447
Planning - annual
Precaution 3114,363
Prisoner's dilemma 190

Profit á Prendre 57
Property
- Concepts 43,47-52
- Legal definition 88,360

Property Rights 237
- Australia 39,53,88
- Benefits 255-257,293
- Characteristics 48,89,113,190-191
- Cost recovery 402
- Customary rights 184
- Definition 10,17,29,72-77,88,185,199,355
- Economic characteristics 30,190
- Effects 219
- Fisheries management 17
- High Seas 107
- ITQs 29
- Legal definition 29,53,190
- New South Wales 79,90
- Northern demersal scale fish 450
- Northern Territories 90
- Properties 48
- Problems 294
- Queensland 89
- Fisheries research 256
- Recreational fishing 184
- Responsibilities 256
- Versus access rights 292
- Scallop fisheries 239
- South Australia 91
- Strategic behaviour 114
- Sustainability 358
- Theory 111
- Value 142

Public company concept 321
Queensland 90
Quota Management 455
Quota markets 288
Recreational fishing 184,185
Redistribution 174,180
Rent 14,323
Right to Fish 57,74
Rights-based management 73
- Australia 148
- Co-management 155
- Objectives 212

Research 256
Resource sharing 212
Safety 231
Shark Bay, WA 297
Social conflict 457
Social costs 189
Social implications 353
Social justice 33
South Africa 173,353-354
- Legislation 353
- Policy 352

South Australia 91
South East Trawl Fishery 288
Southern Scallop fishery 21
Sovereignty 41,47-50
- Models 47

Stakeholder participation 20,134

Strategic behaviour	114
Strategy	165,214
Subsidies	343
Sustainability	26,73,357-360
Sustainable development	73,255,355-357
TACs	17
Tasmania	92
Rock lobster	289
Tenders	310
Thailand	456
Tragedy of the Commons	156
Transactions costs	191
Transferability	89-94,213,236,308, 316
China	455
Definition	308,335
Geoduck	228
Lake Winnipeg	123-124
New South Wales abalone	234
New Zealand	397
Netherlands	261-263
Pacific Halibut	385
Shark Bay prawn fishery	301
Trap fisheries	445
UNCLOS	108,360,415
UN Straddling Stocks Agreement	109
United Kingdom	137,147
United States	127-136,215-218
Access limitation	96-103
Alaska	348
Bluefin Tuna	128
Buyback programmes	99
Cooperatives	99
Enforcement	415
History of management	96
Individual Fishing Quotas	127
Legislation	417
National Academy of Science	133
User fees	385
User's rights	
China	455
Value of licences/rights	142-143,145,150,228,301, 263,294
Vessel Catch Quotas	2
Vessel Earnings	248
Victoria	93,84-87
Western Australia	94,209,313,338
Constitutional settlements	446
Estuarine fisheries	209
Legislation	313,338
Lobster	406
Northern demersal scalefish	448
Pearls	338
Pilbara Trawl Fishery	447
Scale fisheries	445
WTO	346